Natural Products as Source of Molecules with Therapeutic Potential

Valdir Cechinel Filho
Editor

Natural Products as Source of Molecules with Therapeutic Potential

Research & Development, Challenges and Perspectives

Springer

Editor
Valdir Cechinel Filho
University of Vale do Itajaí - UNIVALI
Itajaí, SC, Brazil

ISBN 978-3-030-00544-3 ISBN 978-3-030-00545-0 (eBook)
https://doi.org/10.1007/978-3-030-00545-0

Library of Congress Control Number: 2018962876

This Springer imprint is published by the registered company Springer Nature Switzerland AG
The registered company address is: Gewerbestrasse 11, 6330 Cham, Switzerland

This book is dedicated to:

- *My family (Valdir Cechinel*, Amélia Copetti Cechinel*, Emílio Cecconi*, and Bilmar Canarin Cecconi—* in memoriam)*
- *My wife Lenita Cecconi Cechinel and my daughters Camile C. Cechinel and Milene C. Cechinel*
- *My scientific fathers, Franco Delle Monache and Rosendo A. Yunes*

for all the support, trust, and understanding.

Acknowledgments

I thank all the authors and co-authors for the relevant participation as well as all the collaborators of the University of Vale do Itajaí (UNIVALI) and other partners. Special thanks to my daughter Camile for her excellent technical support.

Contents

About the Editor

Valdir Cechinel Filho holds a bachelor's degree in Organic Chemistry from the Federal University of Santa Catarina (1987), a master's degree from the Federal University of Santa Catarina (1991), and a PhD from the Federal University of Santa Catarina (1995). He was Pro-rector and Vice-Rector of Postgraduate, Research, Extension and Culture of the University of Vale do Itajaí (UNIVALI) (2002–2018) and is currently a professor/researcher of the Postgraduate Program (M/D) in Pharmaceutical Sciences of UNIVALI. He conducts research projects in collaboration with national and international researchers. He is also the Coordinator of the RIBIOFAR/CYTED/CNPq Network and RIBECANCER/CYTED/CNPq Network. He has experience in the field of Organic Chemistry, focusing on the following topics: bioactive natural products, synthesis of molecules of biological interest, and investigation of the relationship between chemical structure and biological activity. He has more than 360 published articles, 50 books and book chapters, and 10 patents in his name. He has supervised around 50 dissertations and theses. He serves as a Reviewer for more than 50 journals and as an Associate Editor of *Pharmaceutical Biology*. In 2012, he was awarded first place in the Caspar Stemmer Innovation Prize (FAPESC) for the Protagonist Innovation category. In 2013, he was a Finalist for the Santander Student Guide Award (partnership with the private sector) for his project developed in partnership with the pharmaceutical company Eurofarma. In 2015, he was a Finalist for the Santander 2015 Award, for the category Science and Innovation, with his project entitled "Search for new and effective anticancer agents from the Brazilian biodiversity." In 2017, he received the highest commendation of Academic Merit of the Council of Rectors of Brazilian Universities

(CRUB), supported by the CNPq, for his contribution to the Brazilian Higher Education. In March 2018, he became President of the Foundation UNIVALI and Rector of UNIVALI, with a mandate to head the Institution until 2022.

Medicinal Plants and Phytomedicines

Rivaldo Niero, Valdir Cechinel Filho, and Rosendo Augusto Yunes

Abbreviations

AIDS	Acquired immune deficiency syndrome
ANVISA	Agência Nacional de Vigilância Sanitária
BC	Before Christ
BfArM	Bundesinstitut für Arzneimittel und Medizinprodukte
BPH	Benign prostatic hyperplasia
CNS	Central nervous system
COX	Cyclooxygenase
CPQBA	Research Center for Chemical, Biological and Agricultural
CYP 450	Cytochrome P450
DNA	Deoxyribonucleic acid
ED_{50}	Effective dose that produces a therapeutic response in 50%
EMSA	Electrophoretic mobility assay
ESI-MS	Electrospray ionization mass spectrometry
FDA	Food and Drug Administration
GABA	Gamma-aminobutyric acid
HE	Hydroalcoholic extract
HET-CAM-test	Chorioallantoic membrane test
HPLC	High-performance liquid chromatography
HPTLC	High-performance thin-layer chromatography
IC_{50}	Inhibitory concentration that produces a therapeutic response in 50%
IL	Interleukin
iNOS	Nitric oxide synthase

R. Niero (✉) · V. Cechinel Filho
Núcleo de Investigações Químico-Farmacêuticas (NIQFAR) e Programa de Mestrado em Ciências Farmacêuticas, Universidade do Vale do Itajaí (UNIVALI), Itajaí, SC, Brazil
e-mail: niero@univali.br; cechinel@univali.br

R. A. Yunes
Departamento de Química, Universidade Federal de Santa Catarina (UFSC), Florianópolis, SC, Brazil

V. Cechinel Filho (ed.), *Natural Products as Source of Molecules with Therapeutic Potential*, https://doi.org/10.1007/978-3-030-00545-0_1

LC-MS	Liquid chromatography mass spectrometry
L-NOARG	L-NG-nitro arginine
mRNA	Messenger RNA
MRSA	Methicillin-resistant *Staphylococcus aureus*
NF-kB	Nuclear factor kappa B
NMR	Nuclear magnetic resonance
PGE_2	$Prostaglandin_2$
SFE	Supercritical fluid extraction
SGLT1	Sodium-glucose transport proteins
SRT	*Serenoa repens* extract
TLC	Thin-layer chromatography
TNFalpha	Tumor necrosis factor alpha
WHO	World Health Organization

1 Introduction

The use of plants for therapeutic purposes is one of the oldest practices of mankind. The Egyptians recorded the analgesic use of opium, as well as the use of fungi with antibiotic properties. Other civilizations, such as India and China, also left records on the use of medicinal plants, with a collection of 700 species, and still play an important role in traditional medicine (Barreiro and Bolzani 2009; Cechinel Filho and Yunes 2016). Today, according to the World Health Organization (WHO), approximately 80% of the world's population uses plants to treat basic illnesses, mostly in the form of extracts or their active ingredients (WHO 2018). The marketing of these phytomedicines has expanded considerably throughout the world, particularly in the European countries, such as Germany, France, Italy, the United Kingdom, and Spain, and, more recently, the United States. Brazil has an important role in this field, as it has the largest biodiversity in the world, with more than 35,000 catalogued species, of a total of between 350,000 and 550,000 plant species identified worldwide (Dutra et al. 2016). Thus, expertise in the areas related to the development of phytomedicines, including organic chemistry, preclinical pharmacology, clinical pharmacology, and pharmaceutical sciences, has increased in recent years (Newman and Cragg 2016; Leonti and Verpoorte 2017).

Phytomedicines are defined as the use of a crude drug (dried herb), an essential oil, an extract or fraction of it for medicinal properties, and quite often complex mixture of compounds that generally occur in low (variable) concentrations. The most commonly used phytomedicines are plant extracts obtained by the use of solvents and by maceration or percolation of the dried plants (Pferschy-Wenzig and Bauer 2015; Azwanida 2015). The extracts can be used as liquid preparations or in powdered form. The solvents most commonly used for extraction are water and alcohol. In some cases, fractions are used, which contain more concentrated levels of the active principles, and are generally obtained by partition with solvents of increasing polarity. However, due to the increasing popularity and expanding global market for phytotherapeutics, the safety of plant products has become a major public health concern. A lack of regulation and distribution channels (Internet sales) may result in poor quality products and, consequently, adverse reactions. The most common causes of adulteration are products with undeclared potent pharmaceutical substances, substitution or misidentification with toxic plant species, incorrect doses, and interactions with conventional medicines (Gurib-Fakim 2006).

The availability and quality of the raw materials are frequently problematic because the active principles are often unknown, and standardization and stability, though feasible, are not easy. Compared with modern medicine, herbal medicines cost less, are more often used to treat chronic diseases, and appear to have less frequent undesirable side effects. Thus, modern techniques have received attention in recent years, and the number of publications produced annually in this field has increased considerably. The most notable technique in this field is the hyphened analytical technique, which has enabled a reliable fingerprint to be obtained (Patel et al. 2010; Kulkarni et al. 2014). This has led to a growing class and a promising market, as it generates revenue of $ 21.7 billion a year.

2 Quality Control of Phytomedicines

2.1 Quality and Efficacy of Plant Material

The use of plants with medicinal purposes involves the action of multiple compounds generally in a very low concentration. Thus, its safety is known and accepted, but not its effectiveness. A disadvantage of phytotherapy is that in most cases, there is a lack of clearly defined, complete information on the composition of the extracts. Furthermore, phytomedicines require a thorough, in-depth assessment of their pharmacological qualities, which can now be done through the use of new biological technologies. Consequently, the development of fast and effective analytical methods for fingerprinting plant extracts is of high interest. In this context, several studies have shown that ESI-MS and LC-MS methods are particularly effective for characterizing plant extracts (Van Der Kooy et al. 2009).

Synergy between compounds is a basic principle of medicinal plants, which will be discussed in this chapter. However, efforts are especially dedicated to studying single molecules, rather than identifying synergies among different compounds. From a scientific perspective, the extracts of medicinal plants as a whole constitute the "active principles." Thus, the measurement of one or more components as markers is necessary. Also, in the case of extracts in which an active constituent has been determined, there is generally a group of substances that are active. It is therefore necessary to obtain a "fingerprint" of the extract in which all the possible constituents can be characterized and/or identified. This is now possible through various modern technological methods. Phytotherapy requires plant material with a standardized composition; however, natural material growing in the wild does not always have the same quality, due to different affecting factors such as climate, soil, genetic constitution, etc. Therefore, it is more efficient to cultivate the plants, in order to reduce variations in the constituents and to ensure controlled content of the pharmaceutically relevant constituents (Wang et al. 2004).

To improve the accuracy and consistency of control of phytomedicine preparations, regulatory authorities worldwide are requesting research into new analytical methods, for more rigorous standardization of phytomedicines. Significant differences have been observed in chamomile extracts by NMR-based metabolomics, which combine high-resolution ^{1}H-NMR spectroscopy with chemometric analysis, showing that the origin, purity, and preparation methods contributed to these differences (Mattoli et al. 2006; Bansal et al. 2014; Kumar 2016).

Over the last 10 years, numerous cases of intoxication, leading in most cases to end-stage renal failure, have been reported after consumption of weight loss diets containing Chinese herbal preparations. These intoxications were associated with species of the *Aristolochia* genus, such as *A. fangchi*, known to contain very nephrotoxic and carcinogenic metabolites called aristolochic acids (**1** and **2**, Fig. 1). Several dietary supplements, teas, and phytomedicines used in weight loss diets were analyzed. The presence of aristolochic acid I was confirmed by HPLC/UV-DAD/MS analysis (Ioset et al. 2003). These products were immediately recalled from the Swiss market.

Pesticides, which are mainly applied to crops to protect the plants, have been found in medicinal plants, as well as in infusions, decoctions, tinctures, and essential oils (Zuin and Vilegas 2000). This important aspect of medicinal plants is reviewed in a review article spanning more than 30 years. Other studies indicate the incidence of toxigenic fungi and their mycotoxins in 152 Argentinean medicinal and aromatic dried plants belonging to 56 species, which are used as raw material for phytomedicines (Rizzo et al. 2004). Fortunately, over the years, the specific pharmacognostical tools have changed dramatically, and most recently, DNA-based techniques have become another element of our spectrum of scientific methods (Heinrich and Anagnostou 2017).

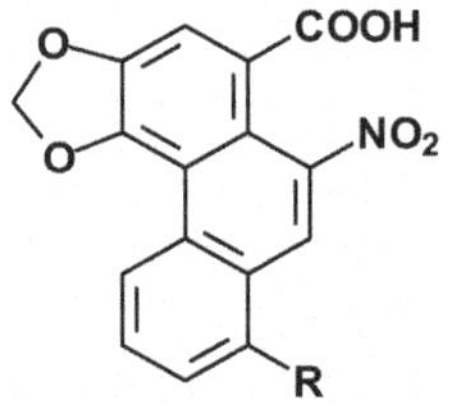

(1) R=H; aristolochic acid I
(2) R=OMe; aristolochic acid II

Fig. 1 Major carcinogenic and nephrotoxic compounds from *Aristolochia* genus

2.2 Production Methods

The nature of the solvents and the extraction and drying methods affects the composition of the extracts. For liquid phytomedicines, water and alcohol are the principal solvents. Polar compounds are soluble in water, and nonpolar lipophilic compounds are soluble in alcohol. However, when identical solvents are used, the extraction methods can yield extracts with different pharmacological actions. Studies have indicated, in the case of essential oils, that water steam distillation in acidic medium can be more advantageous than the traditional method if the volatile terpene derivatives present in the plants are in the form of glycosides or dimeric lactones (i.e., oregano, wormwood oils, siderites). Comparing the composition of essential oils obtained by water steam distillation and supercritical fluid extraction (SFE), it was found that SFE fractions are richer in ester constituents because the possibility of hydrolysis is reduced and the oils are more valuable than the classic oils. However, when the transformation processes are important (chamomile), the distillation should be the appropriate method (Lemberkovics et al. 1998). A Korean researcher has patented a method for the quantitative standardization of medicinal herbs by (i) preparing the sample of medicinal herbs and measuring the weight of the sample, (ii) roasting the sample by controlling the intensity of a fire, and (iii) classifying the sample into three types: a low, medium, and strong flame (Choudhary and Sekhon 2011). However some authors have indicated problems in the standardization of flavonoids in crude drugs and extracts from medicinal plants and in the application of HPLC, HPTLC-densitometry, and spectrophotometry in standardization. This fact demanded new and more efficient methods of analysis for controlling the quality of the extracts.

3 Medicinal Teas Today

Tea is the second most commonly consumed liquid on earth, after water. It has been drunk socially and regularly since 3000 BC. Its medicinal effects date back almost 5000 years. Tea is an infusion obtained from dried leaves of different plants or roots, herbs, spices, fruits, or flavors in hot water (Sharangi 2009). Scientific research around the world has provided clear evidence of the health benefits associated with drinking herbal tea. Today there are teas that prevent cholesterol, high blood pressure, fatigue, diabetes, excess weight, and detoxification, among others (Barnes 2010; Khan and Mukhtar 2013). A related compound found in tea is theophylline (**3**), a licensed medicine for the treatment of respiratory diseases such as asthma (Fig. 2). Tea infusions can be prepared from individual plants or from plant mixtures. A basic distinction is made between:

- Non-medicinal teas that are consumed for pleasure, such as black tea, flavored teas, etc.
- Medicinal teas

Fig. 2 Major bioactive compound from medicinal teas

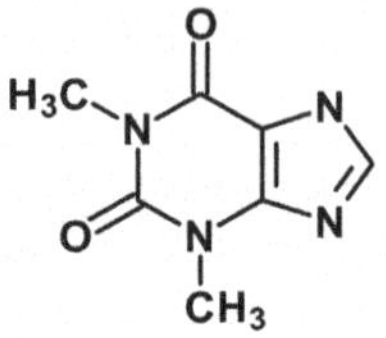

(3)

The indications for the use of medicinal teas are psychosomatic disorders, colds, urinary problems, constipation and diarrhea, and gastrointestinal disorders, among others (Schulz et al. 2001; Khan and Mukhtar 2013). There are no controlled clinical studies on the efficacy of medicinal teas, and their medicinal values are based largely on empirical evidence. In addition, the placebo effect must contribute to their efficacy. Several studies describe the findings of these ethnomedicine research efforts throughout the world over time (Khan and Mukhtar 2013; Boozari and Hosseinzadeh 2017; Naveed et al. 2018).

4 Modern Phytotherapy

Studies have indicated that in order to investigate an extract, which is a complex system, a reductionist method should be applied to determine each active compound separately, but according to previous discussion above, such research must be carried out with caution, as it will never quite explain the efficacy of the entire extract. The saying that "the whole is more than the sum of all the individual parts" is applicable to phytopharmaceuticals. After analyzing the progress introduced by the modern analytical method and in vitro and in vivo pharmacological assays with models of biological molecular test, the phytotherapy was significantly improved.

4.1 New Scientific Screening

There has been a significant increase in news and important methods for biological screening. Assays on Alzheimer's disease have focused on agents that counteract the loss of cholinergic activities. The Ellmann microplate assay and silica gel thin-layer chromatography were used to screen extracts from plants as possible new sources of AchE inhibitors (activity in the NF-kB and the HET-CAM-test) (Salles et al. 2003; Pohanka et al. 2011).

Peruvian medicinal plants were analyzed with respect to their antibacterial activity using a versatile microplate bioassay for rapid and sensitive determination of the organic compounds (Langfield et al. 2004). Grifolin (**4**) and grifolic acid (**5**), which are *S. aureus* and *S. epidermidis* growth inhibitors, were determined as the main

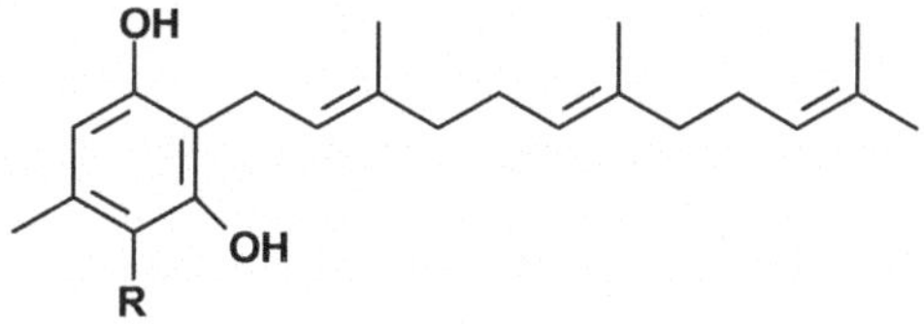

(4) R=H; Grifolin
(5) R=COOH; Grifolic acid

Fig. 3 Main active principles from Peruvian medicinal plants

active principles (Fig. 3). A recently developed method studied the correlation between the chemical composition and bioactivity of herbal medicine and identified the active components from the complex mixture. The advantage of this method compared with bioassay-guided isolation was demonstrated by its application on a typical herbal drug. Several modern methods of bioassay-guided fractionation have been promising alternatives for valuable information on the biological effects of complex materials with successful application in many fields (Weller 2012; Zhuo et al. 2016; Ren et al. 2017).

4.2 Mechanism of Action

Drug resistance has been a major obstacle in cancer chemotherapy. Active principles from plants used in traditional Chinese medicine may act by different molecular targets from those of clinically used antitumor drugs, making them attractive candidates for new therapeutics (Wu et al. 2011). Five hundred thirty-one natural products were tested for correlation with the microarray-based mRNA expression of six genes involved in nucleotide excision repair. The results showed no evidences associated with the expression of these genes, suggesting that mRNA expression is not related to resistance of the cell lines of these substances. In addition, other genes were identified, but none of these appear to be involved in DNA repair (Konkimalla et al. 2008).

Studies have been carried out using the electrophoretic mobility assay (EMSA) as a suitable technique for the identification of plant extracts that alter the binding between transcription factors and the specific DNA elements (Hellman and Fried 2007). These studies demonstrate that low concentrations of *H. indicus*, *P. longifolia*, *M. oleifera*, and *L. speciosa* inhibit the interactions between nuclear factors and target DNA elements, mimicking sequences recognized by the nuclear factor kappa B (NF-kB). Extracts of *P. foetida*, *C. sophera*, and *O. sanctum* were unable to inhibit NF-kB/DNA interactions. Extracts that inhibit both NF-kB binding activity and tumor cell growth might be a source for antitumor compounds, while those that inhibit NF-kB/DNA interactions with lower effects on cell grow could be of interest in inflammatory diseases. More recently, the progress in new methodologies and simulation of biomolecular processes to develop new drugs has been emphasized. The contribution of recent techniques, such as a network of specific metabolisms, has helped to understand the complexity of the tumor as well as the mechanisms of anticancer drugs (Olgen 2018). In addition, research findings shed light on the potential novel applications and formulation of TCMs via regulation of autophagy, which may be an important mechanism underlying the therapeutic effect of TCMs in treating diseases. Autophagy dysregulation is implicated in the pathogenesis of multiple diseases, such as aging, cancers, and diabetes, which may contribute to the discovery of potential drug targets (Cui and Yu 2018).

4.3 Synergy

The term synergism means that the effect of two or more substances causes better biological activity than the pure substances administered in a single dose (Wagner and Ulrich-Merzenich 2009). In this context, this effect occurs by different chemical and biological means. There are many examples of mono- and multi-extract combinations used presently, which exhibit synergistic efficiency based on multi-target mechanisms of action (Ma et al. 2009; Wagner 2011). Some phytotherapeutics available on the market in different forms (extract or fraction) generally exhibit synergism that has an important effect in improving the therapeutic potency. However, a contrary effect can sometimes be observed, leading to a biological activity decrease, called antagonism.

Although only a few clinical studies have confirmed the existence of synergism, preclinical studies have been extensively described, and large amounts of experimental evidence can be found in the literature, exhibiting health benefits to humans, improving consumer acceptance and economic value (Malongane et al. 2017).

To better illustrate the phenomenon of synergism, three examples are detailed:

1. *Tabebuia avellanedae* bark extract and β-lapachone (**6**) were combined to investigate the hematopoietic response of tumor-bearing mice (Fig. 4). Administration of extract (30–500 mg/kg) and β-lapachone (1–5 mg/kg) in distinct combinations caused a dose-dependent reversion of these effects. The best combination was that of 120 mg/kg extract and 1 mg/kg β-lapachone, which prolonged the life span of tumor-bearing mice, both producing the same rate of extension in the duration of survival (Queiroz et al. 2008). Toxic effects were evidenced by the higher doses of β-lapachone in normal and tumor-bearing mice. The studies by

Fig. 4 A bioactive compound that exhibits synergism with medicinal plants

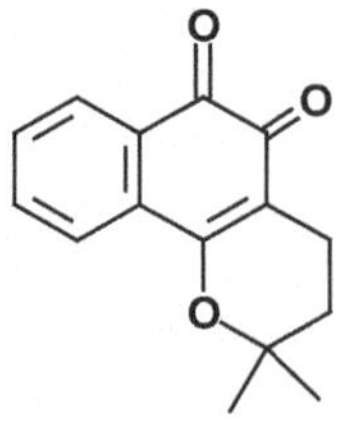

(6)

TLC and HPLC suggested that the antitumor action of extract and β-lapachone also act synergistically with other factors, such as specific cytokines. More recently, a study demonstrated that, in general, β-lapachone combined with beta lactams antimicrobials, fluoroquinolones and carbapenems acts synergistically inhibiting MRSA strains (Macedo et al. 2013).

2. Studies show that distinct plant extracts have caused synergistic effects against human pathogenic microorganisms (Amenu 2014). Although the traditional antibiotics have exhibited effective therapeutic action in the treatment of infectious diseases caused by fungi or bacteria, resistance to these drugs has re-emerged in old diseases. The use of combined drugs or plant extracts with drugs has been used as a strategy for decreasing this resistance, like the described use of β-lactams associated with β-lactamase inhibitors. Synergy has been confirmed between some components extracted from plants, such as flavonoids and essential oils, and synthetic antibiotics used to inhibit bacterial, fungal, and mycobacterial infections (Hemaiswarya et al. 2008; Cheesman et al. 2017). The potency and/or mechanisms of action of some types of combination are very different from that of drugs used pure, demonstrating the existence of synergism in these cases.
3. A paper describes the antiplasmodial potential of 13 plant species (and combination of plants to determine a possible synergism) used in traditional folk medicine in Kenya, for the treatment of malaria. Twenty-five percent of the tested plants were highly active, and 46% exhibited moderate activity against the malaria parasite (Gathirwa et al. 2011). Both synergism and antagonism were demonstrated for combination of some studied extracts. *Uvaria acuminata* and *Premna chrysoclada* presented the highest synergy, while the interaction between *Grewia plagiophylla* and *Combretum illairii* caused a pronounced antagonistic effect.

5 Some Selected Phytomedicines

Although the scientific results are not yet sufficient to establish the safety and efficacy of most herbal products, some of them provide scientific evidence suggesting that these are important medical therapies. However, it is still necessary to improve some aspects related to these points and to regulate and standardize herbal products. We have selected and shown some important aspects of herbal products among the most commonly studied plants (Table 1), such as *Ginkgo biloba* (ginkgo), *Serenoa repens* (saw palmetto), *Valeriana officinalis* (valerian), *Hypericum perforatum* (St. John's wort), *Allium sativum* (garlic), *Piper methysticum* (kava kava), and *Glycine max* (soy). Some selected medicinal plants currently under used and studied in Brazil are exemplified: *Matricaria chamomilla*, *Maytenus ilicifolia*, *Ilex paraguariensis*, *Phyllanthus niruri*, and *Cynara scolymus*.

Table 1 Biological activity and main classes of compounds from selected plants

Species	Biological activity	Active principles	Ref.
Ginkgo biloba	Anti-inflammatory, antitumor, antiaging, CNS disorders	Terpenoids (ginkgolides), polyphenols (biflavonoids), organic acids, carbohydrates	Chan et al. (2007), Van Beek and Montoro (2009) and Zuo et al. (2017)
Serenoa repens	Anti-infectious, anti-inflammatory, spasmolytic, antiandrogen, estrogenic effect	Fatty acids and phytosterols	Goldenberg et al. (2009), Plosker and Brogden (1996) and Saidi et al. (2017)
Valeriana officinalis	Analgesic, sedative, anticonvulsant	Alkaloids, sesquiterpenes, flavonoids, amides	Wichtl (2004a), Mennini et al. (1993) and Savage et al. (2018)
Hypericum perforatum	Antidepressant, antibacterial	Anthrones, phloroglucinol derivatives, flavonoids	Butterweck and Schmidt (2007), Marrelli et al. (2016) and Agapouda et al. (2017)
Allium sativum	Cardiovascular action, antimicrobial, antihypertensive, antilipidemic, anticancer	Sulfur-containing compounds	Wichtl, (2004b), Petrovska, Cekovska (2010) and Sharifi-Rad et al. (2016)
Piper methysticum	Sedative, anxiolytic, anti-stress, analgesic, local anesthetic, anticonvulsant, neuroprotective	Kavalactones	Gary (1997), Sarris et al. (2011) and Chua et al. (2016)
Glycine max	Antidiabetic, anticancer antioxidant	Flavonoids, saponins	Anupongsanugool et al. (2005) and Wang et al. (2013)
Matricaria chamomilla	Anti-inflammatory, antimutagenic, anticancer, antispasmodic, anxiolytic	Flavonoids, terpenoids	McKay and Blumberg (2006), Srivastava and Gupta (2015) and Miraj and Alesaeidi (2016)
Maytenus ilicifolia	Gastroprotective analgesic, contraceptive	Terpenoids, flavonoids	Carlini (1988), Mariot and Barbieri (2007), Niero et al. (2011) and Tabach et al. (2017a, b)
Ilex paraguariensis	Hypocholesterolemic, hepatoprotective, diuretic, CNS stimulant	Flavonoids, triterpenoids, alkaloids, caffeoyl derivatives	Bracesco et al. (2011), Luz et al. (2016) and Ronco et al. (2017)
Cynara scolymus	Antidiabetic, diuretic, hepatoprotective, lipid lowering, antispasmodic, hypocholesterolemic	Sesquiterpene lactones, phenolic acids, flavonoids	Wegener and Fintelmann (1999), Noldin et al. (2003), Ben Salem et al. (2015) and Elsebai et al. (2016)
Phyllanthus niruri	Analgesic, anti-hepatotoxic, anti-HIV, anti-hepatitis B, antihypertensive	Lignans, flavonoids, alkaloids, coumarins, saponins, tannins	Calixto et al. (1998), Bagalkotkar et al. (2006) and Kaur et al. (2017)

5.1 Ginkgo biloba *(Ginkgo)*

Ginkgo biloba is one of the world's most important plants with recognized therapeutic potential. The World Health Organization indicates the medicinal uses supported by clinical data, which have been used for symptomatic treatment of mild to moderate cerebrovascular insufficiency (dementia syndromes in primary degenerative dementia, vascular dementia, and mixed forms of both) with the following symptoms: memory deficit, concentration disturbances, depressive emotional condition, dizziness, tinnitus, and headaches (Kleijnen and Knipschild 1992; Diamond et al. 2000; Isah 2015). Ginkgo composition has several types of components, mainly flavonoids and terpenoids. Bilobalide (**7**) and ginkgolides A (**8**) and B (**9**) are the main bioactive substances (Fig. 5). The crude extracts from ginkgo and their constituents have been investigated as scavenging free radicals, lowering oxidative stress, reducing platelet aggregation, having anti-inflammatory properties, and others (Chan et al. 2007; Van Beek and Montoro 2009). It is estimated that *Ginkgo biloba* existed on earth for 200 million years, and it is considered as a "living fossil." Its natural habitats are in China, Japan, and Korea, and it was first introduced to Europe in the eighteenth century (Heather and Smith 2004). Because of its therapeutic and economic interest, this plant has been cultivated on a large scale in several countries, including France and the United States (Singh et al. 2008; Isah 2015). Today, its standardized extract is available throughout the world, and it is one of the most sold phytomedicines. A number of review articles concerning the different aspects of *G. biloba* have been published, including recent papers focusing on its chemical and biological aspects (Pedroso et al. 2011; Mohanta et al. 2014; Zuo et al. 2017).

5.2 Serenoa repens *(Saw Palmetto)*

Serenoa repens, commonly known as saw palmetto, is currently classified as part of the genus *Serenoa*. It is known by a number of synonyms, including *Sabal serrulatum*, under which name it still often appears in alternative

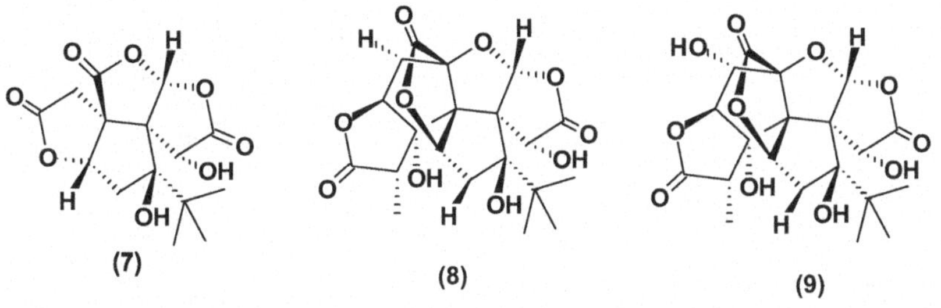

Fig. 5 Major bioactive compounds from *Ginkgo biloba*

medicine (Wilt et al. 1998). It is endemic to the southeastern United States, particularly along the Atlantic and Gulf Coastal plains, but also as far inland as southern Arkansas. This small palm has dark red berries that contain volatile oil consisting of 65% free fatty acids and 35% their ethyl esters. The berries also contain phytosterols (β-sitosterol and derivatives), which are the main components responsible for the activity in the treatment of benign prostatic hyperplasia (BPH) (Plosker and Brogden 1996). This disease is a nonmalignant enlargement of the prostate common in men and can lead to obstructive and irritative lower urinary tract symptoms (Gordon and Shaughnessy 2003; Tacklind et al. 2009). The synthetic drug finasteride (**10**), a 4-azasteroid (Fig. 6), was designed considering the phytosterols. These molecules inhibit the 5-α-reductase, an enzyme that converts testosterone into 5-α-dihydrotestosterone, which binds to androgen receptors in the prostate cells, stimulating their growth and division. Its mechanism of action is unclear, but the effects are generally attributed to a combination of the spasmolytic, antiandrogen, and anti-inflammatory activities of the extract (Goldenberg et al. 2009). Recently an investigation about the mechanism of 5α-reductase inhibition by extract through computational methods has been performed. The results showed that sterols and fatty acids can play a role in the inhibition of the enzyme, suggesting a competitive mechanism (Governa et al. 2016). More recently, a study evaluated the effects of *Serenoa repens* alcohol extract in patients with BPH symptoms. The study was performed on 70 men aged 40–79 divided into 2 groups. Forty patients were treated with *Serenoa repens* extract (SRT), and 30 were observed only. The treated group showed statistically significant differences between patient controls in all observed parameter. The mild improvements of the urine flow and prostate size indicate the efficiency of this phytotherapeutics agent in patients with BPH (Saidi et al. 2017).

Fig. 6 The structure of finasteride, a synthetic drug for the treatment of BHP

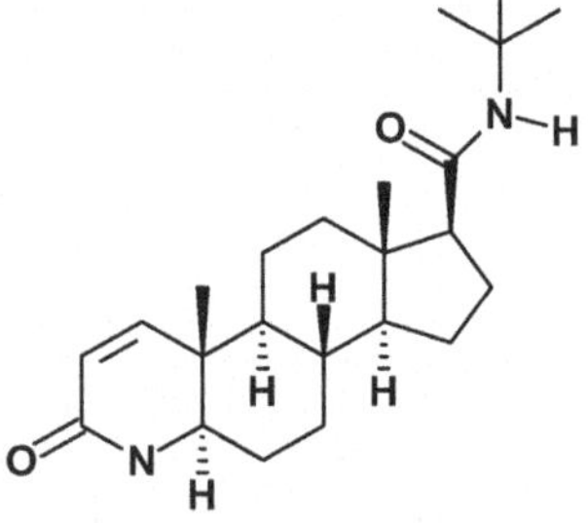

(10)

5.3 Valeriana officinalis *(Valerian)*

Valerian (*Valeriana officinalis*, Valerianaceae) is a hardy perennial flow*ering* plant, with heads of sweetly scented pink or white flowers. Valerian is a native plant of Europe and parts of Asia and has been introduced to North America (Plushner 2000; Hadley and Petry 2003). Several active principles were determined for valerian (Fig. 7), including actinidine (**11**), valerianine (**12**), isovaleramide (**13**), valeric acid (**14**), acevaltrate (**15**), isovaltrate (**16**), valtrate (**17**), acetoxyvalerenic acid (**18**), valerenic acid (**19**), hesperidin (**20**), 6-methylapigenin (**21**), and linarin (**22**) (Wichtl 2004a). The historical use of valerian to treat diseases related to the CNS prompted scientists to investigate the interaction of its components with the GABA neurotransmitter receptor system (Mennini et al. 1993; Savage et al. 2018). However, these studies remain inconclusive, and the action mechanism remains obscure. Valerian is used in herbal medicine as a sedative and is currently used as a remedy for insomnia. Studies have showed a significant reduction in anxious behavior when valerian extract or valerenic acid is exposed, suggesting as a potential alternative to the traditional anxiolytics (Murphy et al. 2010; Felgentreff et al. 2012; Leach and Page 2015). However, the literature has demonstrated some conflicts with respect to the medicinal properties in preclinical studies, indicating no significant differences between valerian and placebo, either in healthy individuals or in individuals with general sleep disturbances or insomnia (Taibi et al. 2007). A review evaluating the therapeutic potential of 1000 plants concluded that only 9 plants, including valerian, demonstrated considerable evidence of a therapeutic effect. On the other hand, some adverse effects related to the use of valerian have been reported. Large doses or chronic use may result in stomachache, apathy, and a feeling of mental dullness or mild depression (Cravotto et al. 2010; Leach and Page 2015).

5.4 Hypericum perforatum *(St. John's Wort)*

Hypericum perforatum is among the most commonly used and preferred herbal drugs and is still the only herbal alternative to classic synthetic antidepressants for the treatment of mild to moderate depression (Butterweck and Schmidt 2007). A great number of preclinical and clinical studies have been conducted to verify the effectiveness of the plant extract (Vacek et al. 2007). The literature data indicates that several organic compounds contribute to the antidepressant efficacy of the plant extract, particularly hypericin (**23**) and hyperforin (**24**), which are considered the main active principles of this plant (Agapouda et al. 2017). Some flavone derivatives, including the well-documented rutin (**25**), seem to contribute to the antidepressant efficacy (Fig. 8). On the other hand, despite the great number of studies focusing on the clinical efficacy of this plant, only a few have been published concerning its oral bioavailability and pharmacokinetic data. Besides the recognized antidepressant potential of *H. perforatum*, other important biological actions are

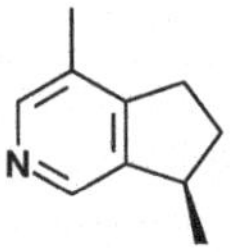

(11)

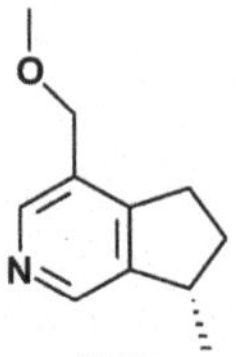

(12)

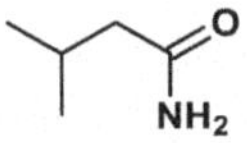

(13)

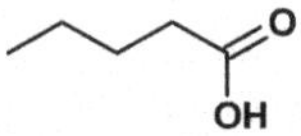

(14)

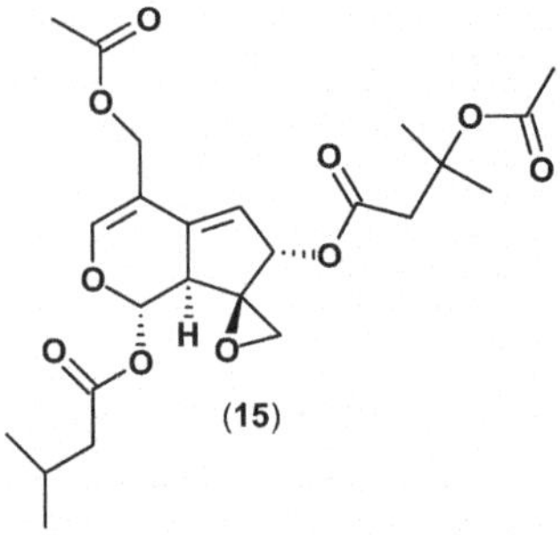

(15)

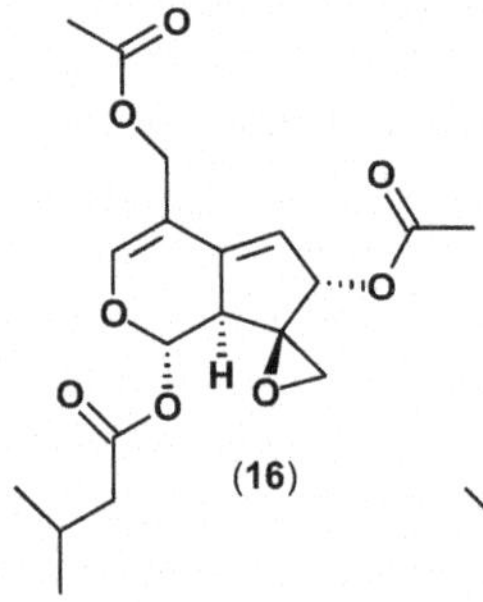

(16)

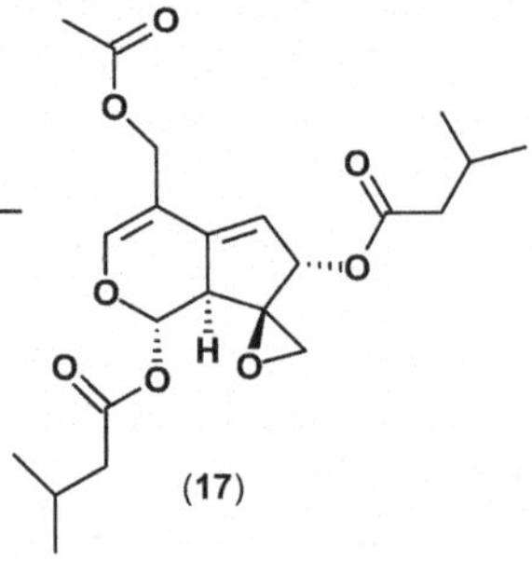

(17)

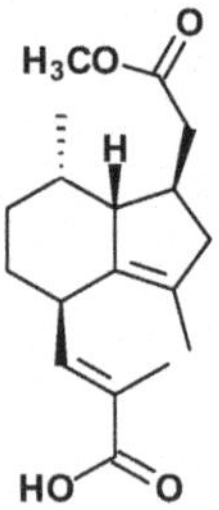

(18)

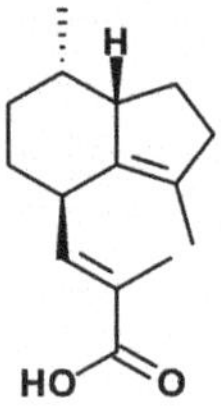

(19)

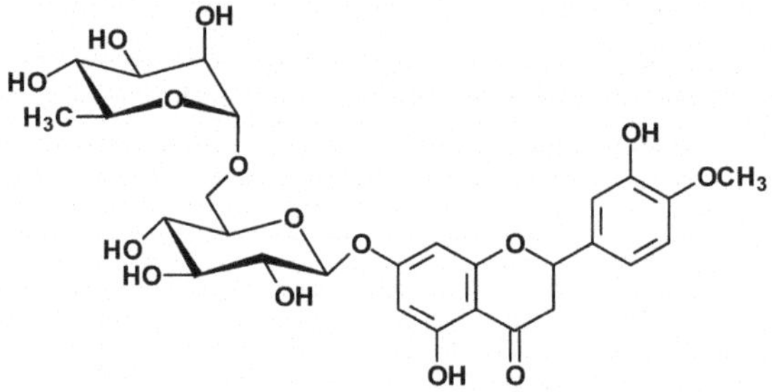

(20)

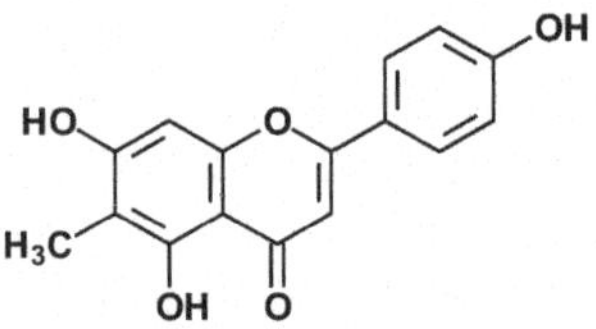

(21)

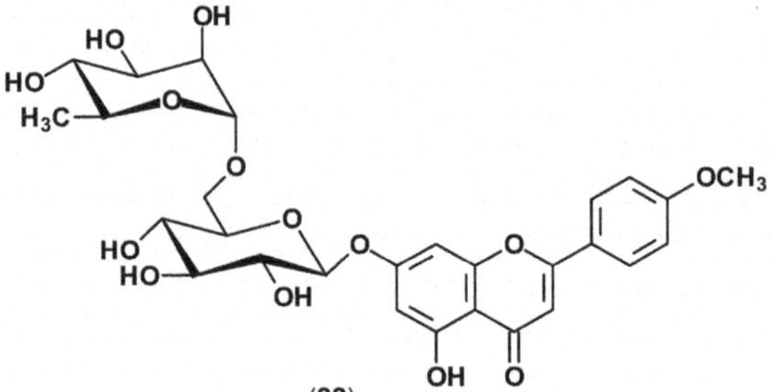

(22)

Fig. 7 Major active principles from *Valeriana officinalis*

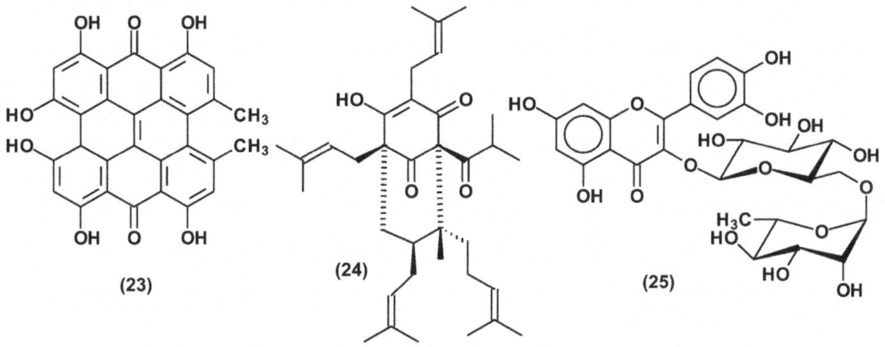

Fig. 8 Major active principles from *Hypericum perforatum*

Fig. 9 Main active principle from *Allium sativum*

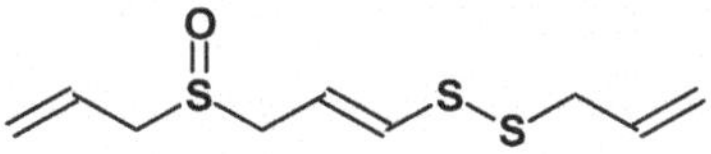

(26)

attributed to its extracts and/or fractions and constituents (Marrelli et al. 2016). The literature reports several review articles on the therapeutic properties and active principles of this plant, such as those recently published and indicated in bibliography (Sarris 2013; Wölfle et al. 2014; Agapouda et al. 2017).

5.5 Allium sativum *(Garlic)*

Allium sativum, popularly known as garlic, exhibits a great number of medicinal properties (Petrovska and Cekovska 2010; Sharifi-Rad et al. 2016). The literature reports that about 2000 active compounds are produced by this plant. Because of its widespread medicinal and nutritional use, many different forms are currently available in the worldwide market, generally with variation of sulfuric compounds, which are considered the main active components of this plant (Wichtl 2004b). Ajoene (**26**) was identified as one of the main active principle of this plant (Fig. 9), with several confirmed biological effects, including antithrombotic, antitumor, antifungal, and antiparasitic properties (Adaki et al. 2014; Bayan et al. 2014; Sharifi-Rad et al. 2016; Foroutan-Rad et al. 2017). Dietary patterns in the Mediterranean, characterized by high consumption of fruits and vegetables, especially garlic, are believed to be beneficial, according to regional patterns of atherosclerotic disease (El-Sabban and Abouazra 2008). Although preclinical studies have demonstrated promising results for the treatment of several ailments, the evidence based on rigorous clinical trials of garlic is not convincing. For hypercholesterolemia, the reported effects are small and may not, therefore, be of clinical relevance. Few studies are available on its use for reducing blood pressure, and the reported effects are too minor to be clinically meaningful (Pittler and Ernst 2007). For all other conditions, there is not enough available data for clinical recommendations. Studies suggest anticancerogenic effect of *Allium* and their associated organosulfur components against several cancer types. However, there are several limitations, including difficulty to establish a dose-risk relationship, which suggest caution in the interpretation (Li et al. 2013; Guercio et al. 2014).

5.6 Piper methysticum

Piper methysticum, well-known as kava kava, is a native plant of the Pacific Islands and has been used in ceremonies for thousands of years. Traditionally, a beverage is prepared, which is drunk before the evening meal (Gary 1997). Indigenous methods of mastication of the kava root have given way to grinding or pounding the plant substance, which is then mixed with water or coconut milk/water (Kava-kava 1998; Sarris et al. 2011). For the past few decades, kava has also gained popularity due to its anxiolytic and sedative properties (Bilia et al. 2002). However, in recent years, it has been involved in several cases of liver failure, which has led to its being banned in several countries and prompted widespread discussion of its benefits and risks as a social beverage and herbal remedy. This plant was well-tolerated and considered devoid of major side effects up until 1998, when the first report about its hepatotoxicity appeared (Teschke 2010). Causality of hepatotoxicity co-medicated drugs was evident after the use of predominantly ethanol and acetonic extract of kava in Germany, Switzerland, the United States, and Australia. Risk factors were found in most patients and include daily overdose, prolonged therapy, and co-medication with up to five other herbal remedies, dietary supplements, and synthetic drugs. On November 2001, the German health authority Bundesinstitut für Arzneimittel und Medizinprodukte (BfArM) published evidence that suggested an association between kava consumption and liver damage in 24 cases. However in 2002, BfArM withdrew the marketing authorizations of all medicinal products containing kava and kavain, including homeopathic products, which had to be recalled immediately (Mills and Steinhoff 2003). The active constituents consist of a group of lactones (about 18 compounds) of which kavain (**27**), yangonin (**28**), methysticin (**29**), dihydrokavain (**30**), demethoxyyangonin (**31**), and dihydromethysticin (**32**) account for about 95% of the lipid extract (Fig. 10) (Chua et al. 2016). It has been demonstrated that several kavalactones, the assumed active principles of the extracts, are potent inhibitors of several enzymes of the CYP 450 system (CYP1A2, 2C9, 2C19, 2D6, 3A4, and 4A9/11). This indicates that kava has high potential to cause pharmacokinetic drug interactions with other herbal products or drugs, which are metabolized by the CYP 450 enzymes (Anke and Ramzan 2004; Olsen et al. 2011; Chua et al. 2016).

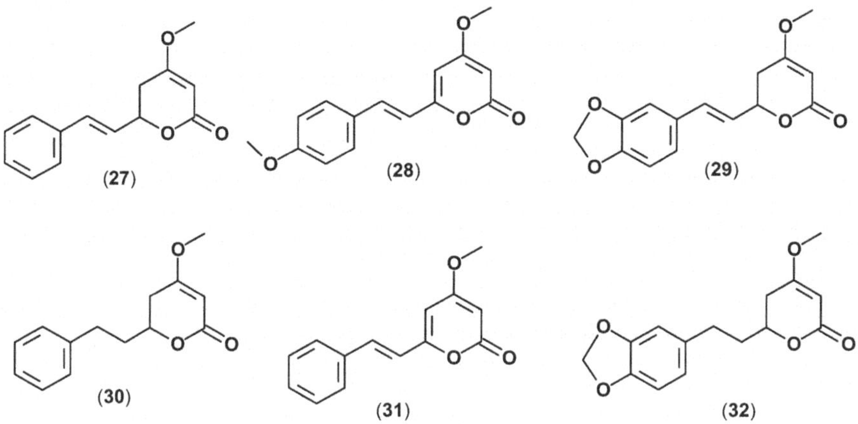

Fig. 10 Major active constituents from *Piper methysticum*

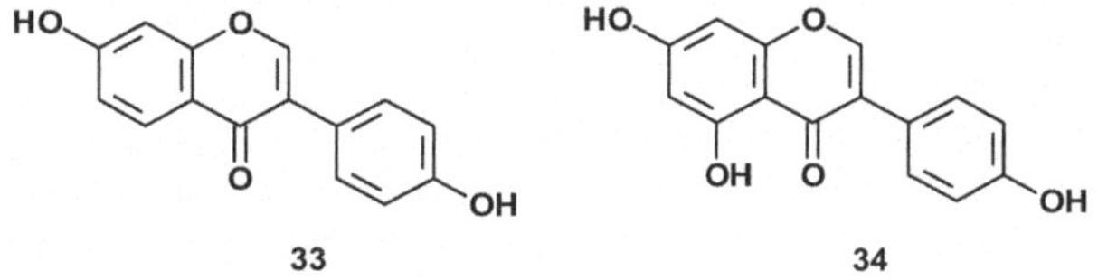

Fig. 11 Major active constituents from *Glycine max*

5.7 Glycine max *(Soy)*

For some 5000 years, the soybean was cultivated by the Chinese, but it was not until the twentieth century that it was commercially cultivated in the United States, achieving a rapid growth in production, with the development of the first cultivars (Shurtleff and Akiko 2015). In Brazil, it was brought over by the earliest Japanese immigrants in 1908, but its expansion occurred in the 1970s, through the interest of oil industry and the expanding international market. Soya is a well-known source of natural isoflavones, especially daidzein (**33**) and genistein (**34**) as the main active principles (Fig. 11) (Anupongsanugool et al. 2005). Such compounds are phytoestrogen mimetics and are considered by some nutritionists and physicians to be useful in cancer prevention while, in some cases, as carcinogenic and disruptive to the endocrine system (Fitzpatrick 2003). Several investigations of the potential of soy foods to reduce the risk of cancer, in particular breast cancer, have been reported (Messina and Wood 2008; Fritz et al. 2013). Most interest in this relationship is due to the fact that it is essentially a unique dietary source of isoflavones. This has led to increasing popularity and the commercial availability of supplements. Some disagreements concerning the therapeutic potential and toxic action have been evidenced, such as studies suggesting that isoflavones may stimulate the growth of existing estrogen-sensitive breast tumors (Messina and Wood 2008; Wang et al.

2013). In addition, a study conducted to evaluate and compare the influence of diets containing genistein (**34**) and soy extract on the growth of estrogen-independent human breast cancer cells demonstrated inhibition of tumor growth, presumably resulting from the synergistic effect of various bioactive substances. Isoflavones such as daidzein (**33**) and genistein (**34**) are found only in some plant families, because most plants do not have an enzyme, the isoflavone synthase, which converts a flavanone precursor into an isoflavone. In contradiction to well-known benefits of isoflavones, genistein (**34**) acts as an oxidant (stimulating nitrate synthesis) as well as blocking the formation of new blood vessels. Some studies have shown that genistein (**34**) acts as an inhibitor of the activity of substances in the body that regulate cell division and cell survival (growth factors).

5.8 Matricaria chamomilla

Chamomile (*Matricaria chamomilla*) is a widely used medicinal plant dating back to the times of ancient Egypt, Greece, and Rome, when it was used to treat fever, heat stroke, and many other conditions. It is one of the most commonly used herbal ingredients in teas or tisanes. Chamomile tea, obtained from the dried flower, has been traditionally used for medicinal purposes due to its calming, carminative, and spasmolytic properties (Srivastava and Gupta 2015; Miraj and Alesaeidi 2016). The main constituents of the flowers include phenolic compounds such as apigenin (**35**), quercetin (**36**), patuletin (**37**), luteolin (**38**), and their glycosides (Fig. 12). The principal components of the essential oil from the flowers are the terpenoids α-bisabolol (**39**) and their oxides (Singh et al. 2011). Evidence-based information regarding the bioactivity of this herb has been published (McKay and Blumberg 2006). Chamomile

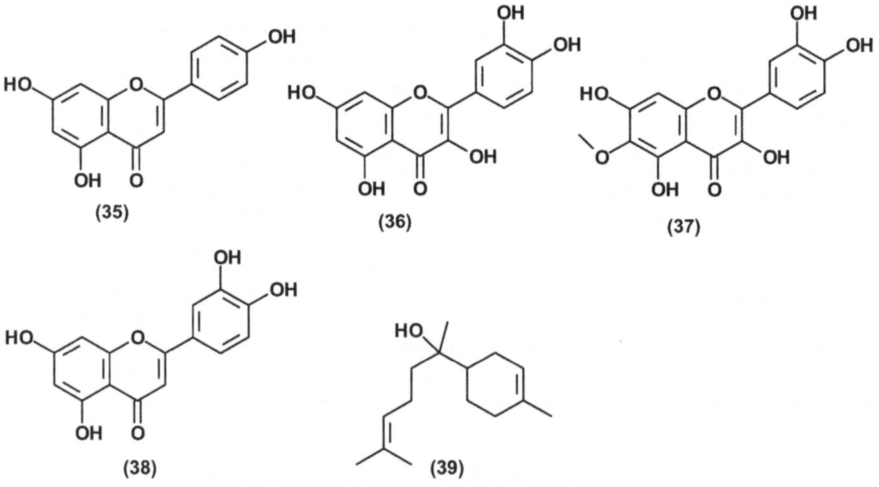

Fig. 12 Main active constituents from *Matricaria chamomilla*

has moderate antioxidant and antimicrobial activities and significant in vitro antiplatelet activity. Studies indicate potent anti-inflammatory action, some antimutagenic and cholesterol-lowering activities, and antispasmodic and anxiolytic effects. However, studies on humans are limited, and there have been clinical trials that examine the purported sedative properties of chamomile tea (McKay and Blumberg 2006). Adverse reactions to chamomile, consumed as a tisane or applied topically, have been reported among individuals with allergies to other plants. In addition, the anticancer properties of aqueous and methanol extracts were evaluated against various human cancer cell lines. Exposure of chamomile extracts caused minimal growth inhibitory responses in normal cells, whereas a significant decrease in cell viability was observed in various human cancer cell lines (Srivastava and Gupta 2007).

5.9 Maytenus ilicifolia

The family Celastraceae comprises about 60 genera and 200 species distributed in tropical and subtropical regions around the world (Carlini 1988; Mariot and Barbieri 2007). *Maytenus ilicifolia* Mart. ex. Reissek (Celastraceae), popularly known in Brazil and other South American countries, including Paraguay and Argentina, as "Espinheira Santa," "Espinheira divina," or "Cangorosa," is commonly used to treat stomach disorders and as a contraceptive (Niero et al. 2011). The therapeutic effects of *M. ilicifolia* are manifested not only in the stomach, for combating dyspepsia and gastrologer hyperchloric acidity, but also in the intestine, where it has analgesic action. *M. ilicifolia* is a phytotherapic compound, sold in various markets and drugstores (dry extract standardized), which contains principally tannins in its composition. Various extracts have shown gastroprotective properties in experimental ulcer models using rodents after oral and intraperitoneal administration (Tabach and Oliveira 2003; Leite et al. 2010). However, the compound responsible for its antiulcerogenic action remains unknown. Studies have shown that after several purification steps from tea, it consists of a polysaccharide arabinose, galactose, galacturonic acid, 4-O-methylglucuronic acid, rhamnose, and glucose mixture (Cipriani et al. 2006). This compound significantly inhibited ethanol-induced gastric lesions in rats with an ED_{50} of 9.3 mg/kg, suggesting that the arabinogalactan liberated from the infusion has a protective anti-ulcer effect. Several studies on its chemical constitution, and quality control, have been performed by hyphenated analysis as LC-MS and HPLC/ESI-MS and NMR. These techniques were useful for identifying the isomers present in complex flavonoid mixtures and for better phytochemical characterization (Cipriani et al. 2006; De Souza et al. 2008; Preto et al. 2013). The studies also demonstrated that the triterpenes friedelin (**40**) and friedelan-3-ol (**41**) are the major bioactive components of this plant (Fig. 13). More recently, a study evaluates their clinical and toxicological effects in order to establish its maximum safe dose. All selected volunteers completed the study without significant changes in the evaluated parameters (Tabach et al. 2017a, b).

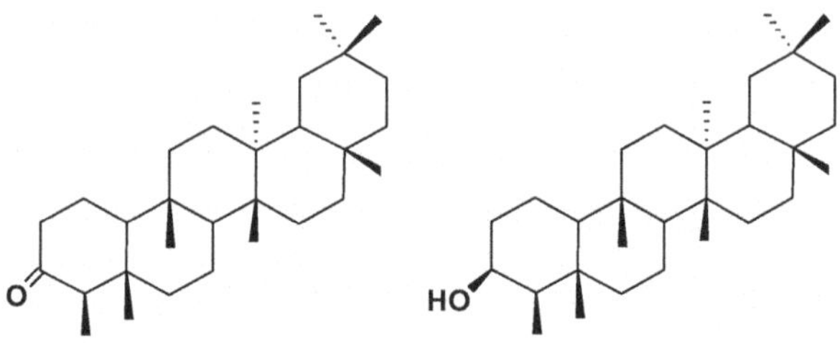

(40) (41)

Fig. 13 Major bioactive components from *Maytenus ilicifolia*

5.10 Ilex paraguariensis

Ilex paraguariensis St. Hilaire (Aquifoliaceae) is a plant that is widely cultivated in South America and has been used for centuries to prepare a tealike beverage with the aim of improving cognitive function (Bracesco et al. 2011). This function is attributed to the presence of caffeine (**42**), the main constituent of the leaves (Fig. 14). Caffeine was rapidly introduced to the market in the form of tea or as an ingredient in formulated foods or dietary supplements. This plant is rich in phenolic compounds, particularly flavonoids and caffeoyl derivatives. However, other classes, such as triterpenes, alkaloids, and fatty acids, have also been found (Saldaña et al. 1999; Filip et al. 2001; Bracesco et al. 2011; Mateos et al. 2018). Several studies have demonstrated hypocholesterolemic, hepatoprotective, central nervous system stimulant, and diuretic properties and benefits to the cardiovascular system (Schinella et al. 2005; Mosimann et al. 2006). In addition, several studies have suggested its action in protecting the DNA from oxidation, and in vitro low-density lipoprotein lipoperoxidation, as well as its high antioxidant capacity. It has also been reported that tea co-associated can prevent some types of cancers (Gonzalez et al. 2005; Heck and Mejia 2007). However some reviews have shown that consumption of tea made from this plant is associated with increased levels of oral, oropharyngeal, esophageal, and laryngeal cancers. Other evidence in the literature suggests that the tea is carcinogenic and that it plays a role in the development of cancers of the oral cavity, pharynx, larynx, and esophagus (Goldenberg et al. 2003; Goldenberg 2002; Loria et al. 2009; Bracesco et al. 2011). On the other hand, *Ilex paraguariensis* has gained public attention in South America, the United States, and Europe, and research into its biological activity and chemical composition has been increasing. A study demonstrated that the hydroalcoholic extract promote specific improvement of the short-term social memory, as well as facilitating the step-down inhibitory avoidance short-term memory, evaluated 1.5 h after training (Prediger et al. 2008). Other studies have evaluated the clastogenic and aneugenic potential of "mate" (tea made from dried leaves of yerba mate) in human lymphocyte culture. The results showed no statistical differences between infusion concentrations and, therefore, are deemed inactive due to the cytokinesis-block in the micronucleus

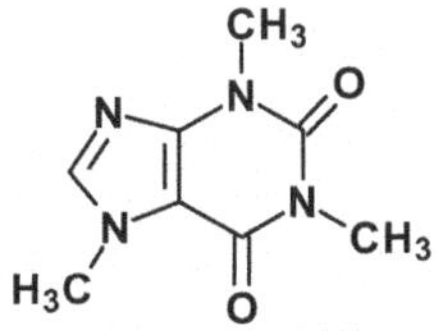
(42)

Fig. 14 Major bioactive component from *Ilex paraguariensis*

assay (Alves et al. 2007). In addition, the antidiabetic properties of the extract were evaluated in alloxan-induced diabetic Wistar rats. The results indicate that the bioactive compounds present in this plant might be capable of interfering in glucose absorption, by decreasing SGLT1 expression (Oliveira et al. 2008). More recently studies have demonstrated several other activities associated with the tealike beverage and extracts obtained from this species including the evidence of hot infusion intake in the risk of colorectal cancer (Santos et al. 2015; Piovezan-Borges et al. 2016; Luz et al. 2016; Ronco et al. 2017).

5.11 Cynara scolymus *L.*

Cynara scolymus, known as "artichoke," has acquired a high commercial value due to its uses as a food and in traditional folk medicine (Wegener and Fintelmann 1999; Ben Salem et al. 2015). In Brazil, known as "alcachofra," the leaves are frequently used to treat liver and gallbladder problems, diabetes, high cholesterol, hypertension, anemia, diarrhea, fevers, ulcers, inflammation, and pain, among other disorders (Noldin et al. 2003). This plant is also used to treat liver and gallbladder disorders in Europe and in several other countries and is sometimes available as a standardized herbal drug, indicated for high cholesterol and digestive and liver disorders (De Falco et al. 2015). Several preclinical and clinical investigations have demonstrated the efficacy and safety of artichoke extracts in the treatment of hepatobiliary dysfunction and digestive complaints, such as a sensation of fullness, loss of appetite, nausea, and abdominal pain. Moreover, earlier findings on its lipid-lowering and hepatoprotective effects have been confirmed (Wider et al. 2009; Nassar et al. 2013; Ben Salem et al. 2015). In general, flavonoids and caffeoylquinic acids are mainly responsible for the activities observed. Previous studies have reported the chemical composition and biological activities of artichoke cultivated in different regions of Brazil. These studies demonstrate that glycosyl flavonoids cynaroside (**43**) and scolymoside (**44**) are the major constituents along with cynaropicrin (**45**), a sesquiterpene lactone, and the triterpene lupeol (**46**) (Fig. 15). Cynarin (**47**), which is the main compound described for artichoke, was detected in very low concentrations, while the hexanic fraction exhibited considerable cytotoxicity and diuretic activities (Noldin et al. 2003). It was also demonstrated that the methanol

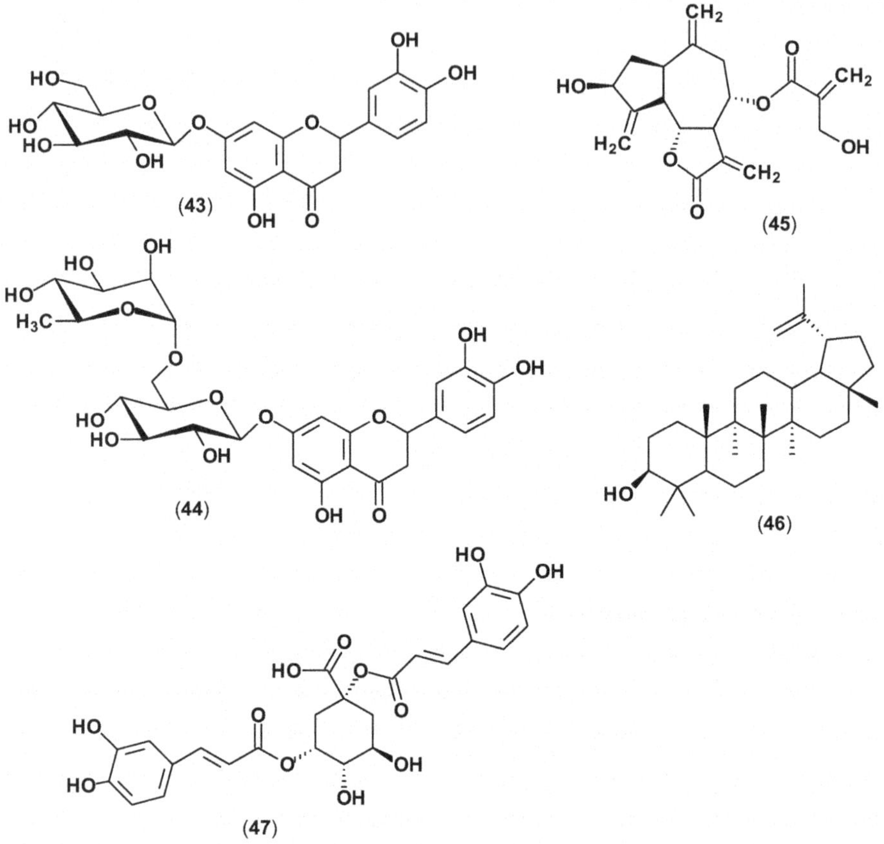

Fig. 15 Major bioactive components from *Cynara scolymus*

extract of *Cynara scolymus* exhibits probable antispasmodic activity when evaluated in isolated guinea pig ileum and rat duodenum. Furthermore, it was confirmed that some fractions and cynaropicrin (**45**), from *Cynara scolymus* cultivated in Brazil, cause antispasmodic effects against guinea pig ileum contractions induced by acetylcholine. The dichloromethane fraction showed the most promising biological effects, with cynaropicrin (**45**) being the main active substance present in this fraction (Emendörfer et al. 2005a). Cynaropicrin has also shown a wide range of other pharmacologic properties such as antihyperlipidemic, antitrypanosomal, antimalarial, antifeedant, antispasmodic, anti-photoaging, and antitumor action, as well as activation of bitter sensory receptors and anti-inflammatory properties (Emendörfer et al. 2005b; Elsebai et al. 2016).

5.12 Phyllanthus niruri

Phyllanthus niruri Linn. (Euphorbiaceae), known in Brazil as "quebra pedra," is a medicinal plant used in the folk medicine of several countries to treat a wide range of diseases, especially kidney and bladder disorders, intestinal infections, diabetes, and hepatitis B (Calixto et al. 1998; Bagalkotkar et al. 2006; Kaur et al. 2017). The main active principles, characterized as flavonoids, alkaloids, terpenoids, lignans, polyphenols, tannins, coumarins, and saponins, have been identified from various parts of this plant (Lee et al. 2016). However, the tannin corilagin (**48**, Fig. 16) seems to be one of the main bioactive compounds showing relevant antinociceptive and gastroprotective effects (Moreira et al. 2013; Klein-Júnior et al. 2017). Different extracts, fractions, and pure substances from the plant have been proven to have important therapeutic effects in many preclinical and clinical investigations. Some of its most commonly studied therapeutic properties include anti-hepatotoxic, antilithic, antinociceptive, antihypertensive, anti-HIV, and anti-hepatitis B effects (Mao et al. 2016). Previous studies focusing on the mechanisms underlying the analgesic effects of hydroalcoholic extract of *P. niruri* have been conducted against formalin-induced nociception in mice, as well as against capsaicin-mediated pain. The results confirm the antinociceptive properties of this plant against neurogenic pain, suggesting that the mechanisms of antinociceptive action do not involve activation of the opioid receptor, synthesis of prostaglandin endogenous release of glucocorticoids, or interaction with either the ctl or ct2 adrenoceptors or the serotoninergic system and do not appear to involve L-arginine-derived nitric oxide or the nitric oxide-related pathways (Santos et al. 1995; Moreira et al. 2013). Further studies demonstrated the chemistry, pharmacology, and therapeutic potential of the plants of the *Phyllanthus* genus, including *P. niruri*, and found evidence, through studies on the extracts and purified compounds of these plants, to support most of their reported uses in folk medicine as an antiviral agent, for the treatment of genitourinary disorders, and as an antinociceptive agent. Although well-controlled, double-blind clinical trials are still unavailable, the data indicated in the literature strongly

Fig. 16 Major bioactive components from *Phyllanthus niruri*

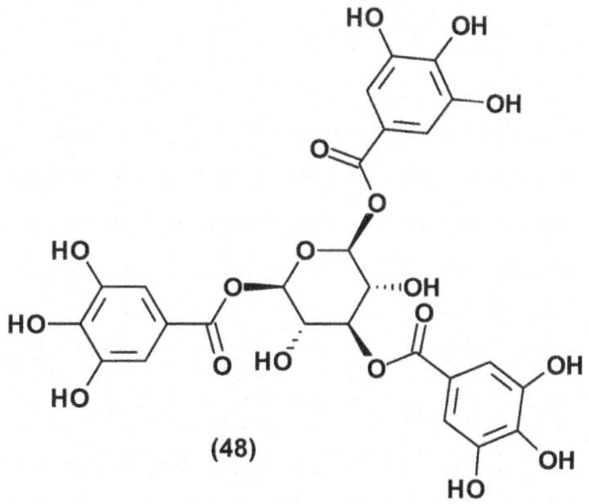

(48)

support the view that the plants belonging to the *Phyllanthus* genus have potential beneficial therapeutic action for the control of hepatitis B, nephrolithiasis, and pain (Bagalkotkar et al. 2006; Mao et al. 2016; Kaur et al. 2017).

6 Case Study: Development of the Phytomedicine from Biodiversity

Victor Siaulys, one of the founders of the Aché Company, noted that football players used the plant (*Cordia verbenacea*) in the form of an infusion, to treat edema and inflammation, as it enabled them to recover much more quickly from their injuries. This finding led to the creation of a division within the company in 1989, for research and development of a phytomedicine from the plant (Ereno 2005; Calixto 2005a). The project continued until 1998, when it was conducted with the effective participation of the university. In 2001, it was demonstrated that the active principle responsible for its anti-inflammatory action was not artemetin, but α-humulene, a sesquiterpene present in the essential oil (Calixto 2005b; Medeiros et al. 2007). The anti-inflammatory property of α-humulene was demonstrated in preclinical and clinical experiments in mice and humans, respectively. Since most of the essential oils are found in the leaves, a study on the cultivation and extraction of the active principle, including its agronomic, chemical, and phytochemical development, was carried out at the Research Center for Chemical, Biological and Agricultural (CPQBA). Parallel studies were carried out with chemical markers, to ensure the quality control of the extract. More than a hundred professionals were involved with the project between 1998 and 2004, including agronomists, biochemists, and pharmaceutical and medical professionals. Besides the preclinical studies involving pharmacological and toxicological tests on mice in the laboratory, clinical trials were conducted in three stages, involving nearly 700 patients. In phase I, the product was tested in about 290 healthy volunteers, in phase II it was tested in around 90 patients with chronic tendinitis and myofascial pain, and phase III involved approximately 280 patients with the same diseases. Having completed all the stages of the preclinical tests and clinical trials, the company decided to apply to ANVISA to register Acheflan® for commercial use, with indication for tendinitis and myofascial pain. Within 7 years, the Aché laboratory had invested more than $15 million on the research and development of Acheflan®. For the time being, the launch of Acheflan is restricted to the domestic market, which has a turnover of US $400 million per year on phytotherapy drugs alone, and the company's next goal is to gain a share of the overseas market.

Cordia verbenacea is a Brazilian specie and popularly known as "Erva-baleeira." In the traditional medicine, it is used as anti-inflammatory, anti-arthritic, analgesic, and anti-ulcer properties (Sertié et al. 2005). In addition, the leaf infusions are used against prostatitis, contusions, and anti-inflammatory and antimicrobial activity. The principal classes of constituents have been identified as flavonols, terpenoids, and steroids (Carvalho et al. 2004; Sciarrone et al. 2017). Artemetin (**49**), a

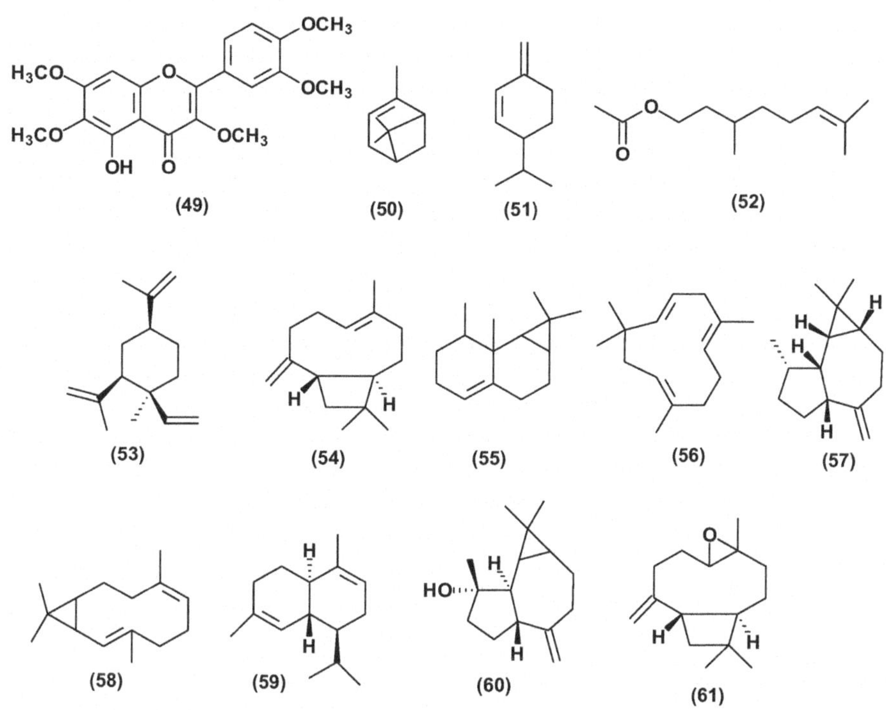

Fig. 17 Major bioactive components from *Cordia verbenacea*

5-hydroxy-3,6,7,3',4'-pentamethoxyflavone, has been identified as the compound responsible for the anti-inflammatory activity. In the volatile oil, α-pinene (**50**), β-phellandrene (**51**), citronellol acetate (**52**), β-elemene (**53**), *trans*-caryophyllene (**54**), β-gurjunene (**55**), α-humulene (**56**), *allo*-aromadendrene (**57**), bicyclogermacrene (**58**), δ-cadinene (**59**), spathulenol (**60**), and epoxy-caryophyllene (**61**) have been identified (Fig. 17). Acheflan® is the first phytomedicine totally developed in Brazil approved by ANVISA (a national authority similar to FDA) and launched in June 2005 by Aché Pharmaceutical Laboratories (Calixto 2005a; Dutra et al. 2016). It is indicated for topical use in the treatment of chronic tendonitis and myofascial pain. Moreover, the studies with the lyophilized crude leaf extract revealed inhibition of nystatin- and miconazole-induced edema, and the effects were similar to naproxen and dexamethasone. In addition, systemic treatment with the essential oil reduced carrageenan-induced rat paw edema, myeloperoxidase activity, and mouse edema induced by carrageenan, bradykinin, substance P, histamine, and platelet-activating factor (Medeiros et al. 2007). Other studies on the anti-inflammatory properties of two sesquiterpenes isolated from the essential oil, α-humulene (**56**) and (-)-*trans*-caryophyllene (**54**), revealed that oral treatment displayed marked inhibitory effects in different inflammatory experimental models in mice and rats (Fernandes et al. 2007). Both were effective in reducing platelet-activating factor, bradykinin- and ovoalbumin-induced mouse paw edema, while only α-humulene (**56**) was able to diminish the edema formation caused by histamine injection, as well as having important inhibitory effects on carrageenan-induced paw edema in mice and rats. Systemic treatment with α-humulene largely prevented both tumor necrosis factor alpha (TNFalpha) and interleukin-1beta (IL-1beta) generation in carrageenan-injected rats, while (-)-*trans*-caryophyllene (**54**) decreased only the release of TNFalpha. Furthermore, both compounds reduced the production of prostaglandin E_2 (PGE_2), as well as inducible nitric oxide synthase (iNOS) and cyclooxygenase (COX-2) expression, induced by intraplantar injection of carrageenan in rats. The anti-inflammatory effects of α-humulene (**56**) and (-)-*trans*-caryophyllene (**54**) were comparable to those observed in dexamethasone-treated animals, used as a positive control drug. A pharmacokinetic study was carried out to assess the plasma and tissue levels, tissue distribution, and skin absorption of α-humulene (**56**) after oral, intravenous, and topical administration in mice. α-Humulene (**56**) exhibited rapid and good absorption following oral and topical administration (Chaves et al. 2008). All these findings help to explain the topical and systemic anti-inflammatory and antinociceptive properties previously reported and provide support for the clinical studies conducted with the phytomedicine Acheflan®.

7 Conclusions

In the nineteenth century, medicinal plants were one of the principal sources of medicines. The synthesis of acetylsalicylic acid marked the pivotal point for the development of Western medicine based on pure molecules obtained specially for synthesis.

The discovery of some natural products, such as the antibiotic and antileukemia drugs, changed the assessment of natural products for some time.

In the 1980s, the introduction of powerful synthetic methods and screening led the pharmaceutical industries to produce synthetic drugs, and the research into natural products declined.

However, Chinese medicine and its recognition by the WHO once again led to widespread acceptance of medicinal plants among people of different countries.

Thus, there is a growing need to assess the quality, safety and efficacy of these products, given that many pharmaceutical industries, especially in Germany, have initiated marketing campaigns to encourage the use of herbal medicines.

The number of plant species in the world is nearly half a million. Approximately 5% to 10% of these may have medicinal properties that require further studies. In fact, the nature has led to the production of an array of bioactive molecules that may interact with the most pharmacologically important targets.

It is known that the traditional medicine of developing countries (such as South America, Africa, India, and Asia) have a huge number of plants (approximately 300,000) that still have not been chemically and pharmacologically studied. We hope to encourage studies of this type, which will be of great human importance for these countries and for the world.

Acknowledgments The authors are grateful to UNIVALI, CNPq, and FAPESC-SC (Brazil) for their financial support and to all the authors of our research groups who participated in the papers cited in this chapter.

References

Adaki S, Adaki R, Shah K, Karagir A (2014) Garlic: review of literature. Indian J Cancer 51(4):577–581

Agapouda A, Booker A, Kiss T et al (2017) Quality control of *Hypericum perforatum* L. analytical challenges and recent progress. J Pharm Pharmacol. https://doi.org/10.1111/jphp.12711

Alves RJ, Jotz GP, Amaral VS et al (2007) The evaluation of mate (*Ilex paraguariensis*) genetic toxicity in human lymphocytes by the cytokinesis-block in the micronucleus assay. Toxicol In Vitro 22:695–869

Anke J, Ramzan I (2004) Pharmacokinetic and pharmacodynamic drug interactions with Kava (*Piper methysticum* Forst). J Ethnopharmacol 93:153–160

Azwanida NN (2015) A review on the extraction methods use in medicinal plants, principle, strength and limitation. Med Aromat Plants 4:3. https://doi.org/10.4172/2167-0412.1000196

Amenu D (2014) Antimicrobial activity of medicinal plant extracts and their synergistic effect on some selected pathogens. Am J Ethnomed 1(1):18–29

Anupongsanugool E, Teekachunhatean S, Rojanasthien N, Pongsatha S, Sangdee C (2005) Pharmacokinetics of isoflavones, daidzein and genistein, after ingestion of soy beverage compared with soy extract capsules in postmenopausal Thai women. BMC Clin Pharmacol 5(2):1–10

Bagalkotkar G, Sagineedu SR, Saad MS et al (2006) Phytochemicals from *Phyllanthus niruri* Linn. and their pharmacological properties: a review. J Pharm Pharmacol 58:1559–1570

Bansal A, Chhabra V, Rawal RK et al (2014) Chemometrics: a new scenario in herbal drug standardization. J Pharm Anal 4(4):223–233

Barnes PJ (2010) Theophylline. Pharmaceuticals (Basel) 3(3):725–747

Barreiro EJ, Bolzani VS (2009) Biodiversidade: Fonte potencial para a descoberta de fármacos. Quim Nova 32(3):679–688

Bilia AR, Gallon S, Vincieri FF (2002) Kava-kava and anxiety: growing knowledge about the efficacy and safety. Life Sci 70:2581–2597

Boozari M, Hosseinzadeh H (2017) Natural medicines for acute renal failure: a review. Phytother Res 31(12):1824–1835

Bracesco N, Sanchez AG, Contreras V et al (2011) Recent advances on Ilex paraguariensis research: minireview. J Ethnopharmacol 136(3):378–384

Butterweck V, Schmidt M (2007) St. John's wort: role of active compounds for its mechanism of action and efficacy. Wien Med Wochenschr 157:356–361

Bayan L, Koulivand PH, Gorji A (2014) Garlic: a review of potential therapeutic effects. Avicenna J Phytomed 4(1):1–14

Ben Salem M, Affes H, Ksouda K, Dhouibi R, Sahnoun Z, Hammami S, Zeghal KM (2015) Pharmacological studies of artichoke leaf extract and their health benefits. Plant Foods Hum Nutr 70(4):441–453

Calixto JB (2005a) Cordia verbenácea. Arquivos Brasileiros de Fitomedicina Científica 2:5–8

Calixto JB (2005b) Twenty-five years of research on medicinal plants in Latin America: a personal view. J Ethnopharmacol 100:131–134

Calixto JB, Santos AR, Cechinel Filho V et al (1998) A review of the plants of the genus *Phyllanthus*: their chemistry, pharmacology, and therapeutic potential. Med Res Rev 18:225–258

Carlini ELA (1988) Estudo da ação antiúlcera gástrica de plantas brasileiras: *Maytenus ilicifolia* (Espinheira Santa). CEME/AFIP, Brasília

Carvalho PM, Rodrigues RFO, Sawaya ACHF, Marques MOM, Shimizu MT (2004) Chemical composition and antimicrobial activity of the essential oil of *Cordia verbenacea* D.C. J Ethnopharmacol 95:297–301

Cechinel Filho V, Yunes RA (2016) Novas perspectivas dos produtos naturais na química medicinal moderna. In: Yunes RA, Cechinel-Filho V (eds) Química de produtos naturais, novos fármacos e a moderna farmacognosia, 5th edn. da Univali, Itajaí, pp 13–39

Chan PC, Xia Q, Fu PP (2007) *Ginkgo biloba* leave extract: biological, medicinal, and toxicological effects. J Environ Sci Health C Environ Carcinog Ecotoxicol Rev 25:211–244

Chaves JS, Leal PC, Pianowisky L et al (2008) Pharmacokinetics and tissue distribution of the sesquiterpene alpha-humulene in mice. Planta Med 74(14):1678–1683

Cheesman MJ, Ilanko A, Blonk B et al (2017) Developing new antimicrobial therapies: are synergistic combinations of plant extracts/compounds with conventional antibiotics the solution? Pharmacogn Rev 11(22):57–72

Choudhary N, Sekhon BS (2011) An overview of advances in the standardization of herbal drugs. J Pharm Educ Res 2(2):55–70

Chua HC, Christensen ET, Hoestgaard-Jensen K et al (2016) Kavain, the major constituent of the anxiolytic kava extract, potentiates GABAA receptors: functional characteristics and molecular mechanism. PLoS One 11(6):e0157700

Cipriani TR, Mellinger CG, de Souza LM et al (2006) A polysaccharide from a tea (infusion) of *Maytenus ilicifolia* leaves with anti-ulcer protective effects. J Nat Prod 69(7):1018–1021

Cravotto G, Boffa L, Genzini L et al (2010) Phytotherapeutics: an evaluation of the potential of 1000 plants. J Clin Pharm Ther 35:11–48

Cui B, Yu JM (2018) Autophagy: a new pathway for traditional Chinese medicine. J Asian Nat Prod Res 20(1):14–26

De Souza LM, Cipriani TR, Serrato RV et al (2008) Analysis of flavonol glycoside isomers from leaves of *Maytenus ilicifolia* by offline and online high performance liquid chromatography-electrospray mass spectrometry. J Chromatogr A 17(1–2):101–109

Diamond BJ, Shiflett SC, Feiwel N et al (2000) *Ginkgo biloba* extract: mechanisms and clinical indications. Arch Phys Med Rehabil 81(5):668–678

Dutra RC, Campos MM, Santos AR et al (2016) Medicinal plants in Brazil: pharmacological studies, drug discovery, challenges and perspectives. Pharmacol Res 112:4–29

De Falco B, Incerti G, Amato M, Lanzotti V (2015) Artichoke: botanical, agronomical, phytochemical and pharmacological overview. Phytochem Rev 14(6):993–1018

El-Sabban F, Abouazra H (2008) Effect of garlic on atherosclerosis and its factors. East Mediterr Health J 14:195–205

Elsebai MF, Mocan A, Atanasov AG (2016) Cynaropicrin: a comprehensive research review and therapeutic potential as an anti-hepatitis C virus agent. Front Pharmacol 7:472

Emendörfer F, Emendörfer F, Bellato F et al (2005a) Antispasmodic activity of fractions and cynaropicrin from *Cynara scolymus* on guinea-pig ileum. Biol Pharm Bull 28:902–904

Emendörfer F, Emendörfer F, Bellato F et al (2005b) Evaluation of the relaxant action of some Brazilian medicinal plants in isolated guinea-pig ileum and rat duodenum. J Pharm Pharm Sci 8:63–68

Ereno D (2005) Da natureza para a farmácia. Rev Pesquisa Fapesp 110:1–3

Felgentreff F, Becker A, Meier B et al (2012) Valerian extract characterized by high valerenic acid and low acetoxy valerenic acid contents demonstrates anxiolytic activity. Phytomedicine 19(13):1216–1222

Fernandes ES, Passos GF, Medeiros R et al (2007) Anti-inflammatory effects of compounds alpha-humulene and (-)-trans caryophyllene isolated from the essential oil of *Cordia verbenacea*. Eur J Pharmacol 569(3):228–326

Filip R, López P, Giberti G et al (2001) Phenolic compounds in seven South American Ilex species. Fitoterapia 72(7):774–778

Fitzpatrick LA (2003) Soy isoflavones: hope or hype? Maturitas 44:S21–S29

Foroutan-Rad M, Tappeh KH, Khademvatan S (2017) Antileishmanial and immunomodulatory activity of *Allium sativum* (garlic): a review. J Evid Based Complement Altern Med 22(1):141–155

Fritz H, Seely D, Flower G et al (2013) Soy, red clover, and isoflavones and breast cancer: a systematic review. PLoS One 8(11):e81968

Gary PND (1997) Kava kava: gift of the islands. Altern Med Rev 2(5):355–364

Gathirwa JW, Rukunga GM, Mwitari PG et al (2011) Traditional herbal antimalarial therapy in Kilifi district, Kenya. J Ethnopharmacol 134(2):434–442

Goldenberg D (2002) Maté: a risk factor for oral and oropharyngeal cancer. Oral Oncol 38(7):646–649

Goldenberg D, Golz A, Joachins HZ (2003) The beverage maté: a risk factor for cancer of the head and neck. Head Neck 25(7):595–601

Goldenberg, L, So A, Fleshner N et al (2009) The role of 5-alpha reductase inhibitors in prostate pathophysiology: is there an additional advantage to inhibition of type 1 isoenzyme? Can Urol Assoc J 3(3 Suppl 2):S109–S114

Gonzalez ME, Song YS, Ramirez-Mares MV et al (2005) Effect of yerba mate (*Ilex paraguariensis*) tea on topoisomerase inhibition and oral carcinoma cell proliferation. J Agric Food Chem 53(6):1966–1973

Gordon AE, Shaughnessy AF (2003) Saw palmetto for prostate disorders. Am Fam Physician 67(6):1281–1283

Governa P, Giachetti D, Biagi M et al (2016) Hypothesis on *Serenoa repens* (Bartram) small extract inhibition of prostatic 5α-reductase through an in silico approach on 5β-reductase x-ray structure. Peer J 22(4):e2698

Guercio V, Galeone C, Turati F, La Vecchia C (2014) Gastric cancer and *allium* vegetable intake: a critical review of the experimental and epidemiologic evidence. Nutr Cancer 66(5):757–773

Gurib-Fakim A (2006) Medicinal plants: traditions of yesterday and drugs of tomorrow. Mol Asp Med 27:1–93

Hadley S, Petry JJ (2003) Valerian. Am Fam Physician 67(8):1755–1758

Heather B, Smith M (2004). *Ginkgo biloba*. In The complete natural medicine guide to the 50 most common medicinal herbs, 2a edn. Robert Rose Inc., Toronto, pp 140–150

Heck CI, De Mejia EG (2007) Yerba mate tea (*Ilex paraguariensis*): a comprehensive review on chemistry, health implications, and technological considerations. J. Food Sci 72:138–151

Heinrich M, Anagnostou S (2017) From pharmacognosia to DNA-based medicinal plant authentication – pharmacognosy through the centuries. Planta Med 83(14–15):1110–1116

Hellman LM, Fried MG (2007) Electrophoretic mobility shift assay (EMSA) for detecting protein-nucleic acid interactions. Nat Protoc 2(8):1849–1861

Hemaiswarya S, Kruthiventi AK, Doble M (2008) Synergism between natural products and antibiotics against infectious diseases. Phytomedicine 15:639–652

Ioset JR, Raoelison GE, Hostettmann K (2003) Detection of aristolochic acid in Chinese phytomedicines and dietary supplements used as slimming regimens. Food Chem Toxicol 41(1):29–36

Isah T (2015) Rethinking *Ginkgo biloba* L.: Medicinal uses and conservation. Pharmacogn Rev 9(18):140–148

Kaur N, Kaur B, Sirhindi G (2017) Phytochemistry and pharmacology of *Phyllanthus niruri* L.: a review. Phytother Res 31(7):980–1004

Kava Kava (1998) Monograph. Alternative Med Review 3:459–461

Khan N, Mukhtar H (2013) Tea and health: studies in humans. Curr Pharm Des 19(34):6141–6147

Klein-Júnior LC, da Silva LM, Boeing T et al (2017) The protective potential of *Phyllanthus niruri* and corilagin on gastric lesions induced in rodents by different harmful agents. Planta Med 83(1–02):30–39

Konkimalla VS, Wang G, Kaina B et al (2008) Microarray-based expression of DNA repair genes does not correlate with growth inhibition of cancer cells by natural products derived from traditional Chinese medicine. Cancer Gen Proteom B5(2):79–84

Kulkarni KM, Patil LS, Khanvilkar VV et al (2014) Fingerprinting techniques in herbal standardization. Indo Am J Pharm Res 4(02):1049–1072

Kumar D (2016) Nuclear magnetic resonance (NMR) spectroscopy for metabolic profiling of medicinal plants and their products. Crit Rev Anal Chem 46(5):400–412

Kleijnen J, Knipschild P (1992) *Ginkgo biloba*. The Lancet 340:1136–1138

Langfield RD, Scarano FJ, Heitzman ME et al (2004) Use of a modified microplate bioassay method to investigate antibacterial activity in the Peruvian medicinal plant *Peperomia galioides*. J Ethnopharmacol 94:279–827

Leach MJ, Page AT (2015) Herbal medicine for insomnia: a systematic review and meta-analysis. Sleep Med Rev 24:1–12

Lee NY, Khoo WK, Adnan MA, Mahalingam TP, Fernandez AR, Jeevaratnam K (2016) The pharmacological potential of *Phyllanthus niruri*. J Pharm Pharmacol 68(8):953–969

Lemberkovics E, Kéry A, Marczal G et al (1998) Phytochemical evaluation of essential oils, medicinal plants and their preparations. Acta Pharm Hung 68(3):141–149

Leonti M, Verpoorte R (2017) Traditional Mediterranean and European herbal medicines. J Ethnopharmacol 199:161–167

Li L, Sun T, Tian J et al (2013) Garlic in clinical practice: an evidence-based overview. Crit Rev Food Sci Nutr 53(7):670–681

Loria D, Barrios E, Zanetti R (2009) Cancer and yerba mate consumption: a review of possible associations. Rev Panam Salud Publica 25(6):530–953

Luz ABG, da Silva CHB, Nascimento MVPS et al (2016) The anti-inflammatory effect of Ilex paraguariensis A. St. Hil (mate) in a murine model of pleurisy. Int Immunopharmacol 36:165–172

Leite JP, Braga FC, Romussi G, Persoli RM, Tabach R, Carlini EA, Oliveira AB (2010) Constituents from *Maytenus ilicifolia* leaves and bioguided fractionation for gastroprotective activity. J Braz Chem Soc 21(2):248–254

Ma XH, Zheng CJ, Han LY et al (2009) Synergistic therapeutic actions of herbal ingredients and their mechanisms from molecular interaction and network perspectives. Drug Discov Today 14(11–12):579–588

Macedo L, Fernandes T, Silveira L, Mesquita A, Franchitti AA, Ximenes EA (2013) β-Lapachone activity in synergy with conventional antimicrobials against methicillin resistant *Staphylococcus aureus* strains. Phytomedicine 21(1):25–29

Mao X, Wu LF, Guo HL et al (2016) The genus phyllanthus: an ethnopharmacological, phytochemical, and pharmacological review. Evid Based Complement Alternat Med 2016:7584952

Mariot MP, Barbieri RL (2007) Metabólitos secundários e propriedades medicinais da espinheira-santa (*Maytenus ilicifolia* Mart. ex Reiss. e *M. aquifolium* Mart.). Rev Bras Pl Med 9:89–99

Marrelli M, Statti G, Conforti F et al (2016) New potential pharmaceutical applications of hypericum species. Mini Rev Med Chem 16(9):710–720

Mateos R, Baeza G, Sarriá B, Bravo L (2018) Improved LC-MSn characterization of hydroxycinnamic acid derivatives and flavonols in different commercial mate (Ilex paraguariensis) brands. Quantification of polyphenols, methylxanthines, and antioxidant activity. Food Chem 241:232–241

Mattoli L, Cangi F, Maidecchi A et al (2006) Metabolomic fingerprinting of plant extracts. J Mass Spectrom 41:1534–1545

McKay DL, Blumberg JB (2006) A review of the bioactivity and potential health benefits of chamomile tea (*Matricaria recutita* L.). Phytother Res 7:519–530

Medeiros R, Passos GF, Vitor CE et al (2007) Effect of two active compounds obtained from the essential oil of *Cordia verbenacea* on the acute inflammatory responses elicited by LPS in the rat paw. Brit J Pharmacol 151:618–627

Messina MJ, Wood CE (2008) Soy isoflavones, estrogen therapy, and breast cancer risk: analysis and commentary. Nutr J 7:17

Mills SY, Steinhoff B (2003) Kava-kava: a lesson for the phytomedicine community. Phytomedicine 10:261–262

Miraj S, Alesaeidi S (2016) A systematic review study of therapeutic effects of *Matricaria recuitta* chamomile (chamomile). Electron Physician 8(9):3024–3031

Mohanta TK, Tamboli Y, Zubaidha PK (2014) Phytochemical and medicinal importance of *Ginkgo biloba* L. Nat Prod Res 28(10):746–752

Mosimann AL, Wilhelm-Filho D, Da Silva EL (2006) Aqueous extract of *Ilex paraguariensis* attenuates the progression of atherosclerosis in cholesterol-fed rabbits. Biofactors 26(1):59–70

Murphy K, Kubin ZJ, Shepherd JN et al (2010) *Valeriana officinalis* root extracts have potent anxiolytic effects in laboratory rats. Phytomedicine 17(8–9):674–678

Malongane F, McGaw LJ, Mudau FN (2017) The synergistic potential of various teas, herbs and therapeutic drugs in health improvement: a review. J Sci Food Agric 97(14):4679–4689
Mennini T, Bernasconi P, Bombardelli E, Morazzoni P (1993) In vitro study on the interaction of extracts and pure compounds from *Valeriana officinalis* roots with GABA, benzodiazepine and barbiturate receptors in rat brain. Fitoterapia 64(4):291–300
Moreira J, Klein-Júnior LC, Cechinel Filho V, De Campos Buzzi F (2013) Antihyperalgesic activity of corilagin, a tannin isolated from *Phyllanthus niruri* L. (Euphorbiaceae). J Ethnopharmacol 146(1):318–323
Nassar MI, Mohamed TK, Elshamy AI et al (2013) Chemical constituents and anti-ulcerogenic potential of the scales of *Cynara scolymus* (artichoke) heads. J Sci Food Agric 93(10):2494–2501
Naveed M, BiBi J, Kamboh AA et al (2018) Pharmacological values and therapeutic properties of black tea (*Camellia sinensis*): a comprehensive overview. Biomed Pharmacother 100:521–531
Newman DJ, Cragg GM (2016) Natural products as sources of new drugs from 1981 to 2014. J Nat Prod 79:629–661
Niero R, Andrade SF, Cechinel FV (2011) A review of the ethnopharmacology, phytochemistry and pharmacology of plants of the Maytenus genus. Curr Pharm Des 17:1851–1871
Noldin VF, Cechinel Filho V, Delle Monache F et al (2003) Composição química e atividades biológicas das folhas de *Cynara scolymus* L. (alcachofra) cultivada no Brasil. Quim Nova 26:331–334
Olgen S (2018) Overview on anticancer drug design and development. Curr Med Chem 25(15):1704–1719
Oliveira DM, Freitas HS, Souza MF et al (2008) Yerba mate (*Ilex paraguariensis*) aqueous extract decreases intestinal SGLT1 gene expression but does not affect other biochemical parameters in alloxan-diabetic wistar rats. J Agric Food Chem 56(22):10527–10532
Olsen LR, Grillo MP, Skonberg C (2011) Constituents in kava extracts potentially involved in hepatotoxicity: a review. Chem Res Toxicol 24(7):992–1002
Patel KN, Patel JK, Patel MP et al (2010) Introduction to hyphenated techniques and their applications in pharmacy. Pharm Methods 1(1):2–13
Pedroso JL, Henriques Aquino CC, Escórcio Bezerra ML et al (2011) *Ginkgo biloba* and cerebral bleeding: a case report and critical review. Neurologist 2:89–90
Petrovska BB, Cekovska S (2010) Extracts from the history and medical properties of garlic. Pharmacogn Rev 4(7):106–110
Pferschy-Wenzig EM, Bauer R (2015) The relevance of pharmacognosy in pharmacological research on herbal medicinal products. Epilepsy Behav 52(part B):344–362
Piovezan-Borges AC, Valério-Júnior C, Gonçalves IL et al (2016) Antioxidant potential of yerba mate (*Ilex paraguariensis* St. Hil.) extracts in *Saccharomyces cerevisae* deficient in oxidant defense genes. Braz J Biol 76(2):539–544
Pittler MH, Ernst E (2007) Clinical effectiveness of garlic (*Allium sativum*). Mol Nutr Food Res 51:1382–1385
Plosker GL, Brogden RN (1996) *Serenoa repens* (Permixon). A review of its pharmacology and therapeutic efficacy in benign prostatic hyperplasia. Drugs Aging 9(5):379–395
Pohanka M, Hrabinova M, Kuca K et al (2011) Assessment of acetylcholinesterase activity using indoxylacetate and comparison with the standard Ellman's method. Int J Mol Sci 12(4):2631–2640
Prediger RD, Fernandes MS, Rial D et al (2008) Effects of acute administration of the hydroalcoholic extract of mate tea leaves (*Ilex paraguariensis*) in animal models of learning and memory. J Ethnopharmacol 120:465–473
Preto MS, Tavares MI, Sebastião PJ et al (2013) Determination of herb authenticity by low-field NMR. Food Chem 136(3–4):1272–1276
Plushner SL (2000) Valerian: *Valeriana officinalis*. Am J Health-System Pharm 57(4):328–333
Queiroz ML, Valadares MC, Torello CO et al (2008) Comparative studies of the effects of *Tabebuia avellanedae* bark extract and beta-lapachone on the hematopoietic response of tumour-bearing mice. J Ethnopharmacol 117:228–235

Ren H, Wang B, Zhao H (2017) Breaking the silence: new strategies for discovering novel natural products. Curr Opin Biotechnol 10(48):21–27

Rizzo I, Vedoya G, Maurutto S et al (2004) Assessment of toxigenic fungi on Argentinean medicinal herbs. Microbiol Res 159:113–120

Ronco AL, De Stefani E, Lasalvia-Galante E et al (2017) Hot infusions and risk of colorectal cancer in Uruguay: a case-control study. Eur J Clin Nutr 71:1429–1436

Saidi S, Stavridis S, Stankov O, Dohcev S, Panov S (2017) Effects of *Serenoa repens* alcohol extract on benign prostate hyperplasia. Pril Makedon Akad Nauk Umet Odd Med Nauki 38(2):123–129

Saldaña MD, Mohamed RS, Baer MG et al (1999) Extraction of purine alkaloids from maté (*Ilex paraguariensis*) using supercritical CO_2. J Agric Food Chem 47(9):3804–3808

Salles T, Maria T, Vanderlan V et al (2003) Seleção de plantas com atividade anticolinesterase para tratamento da doença de Alzheimer. Quim Nova 23:301–304

Santos AR, Filho VC, Yunes RA et al (1995) Analysis of the mechanisms underlying the antinociceptive effect of the extracts of plants from the genus *Phyllanthus*. Gen Pharmacol 26:1499–1506

Santos EC, Bicca MA, Blum-Silva CH et al (2015) Anxiolytic-like, stimulant and neuroprotective effects of *Ilex paraguariensis* extracts in mice. Neuroscience 292:13–21

Sarris J (2013) St. John's wort for the treatment of psychiatric disorders. Psychiatr Clin North Am 36(1):65–72

Sarris J, La Porte E, Schweitzer I (2011) Kava: a comprehensive review of efficacy, safety, and psychopharmacology. Aust N Z J Psychiatry 45(1):27–35

Savage K, Firth J, Stough C et al (2018) GABA-modulating phytomedicines for anxiety: a systematic review of preclinical and clinical evidence. Phytother Res 32(1):3–18

Schinella G, Fantinelli JC, Mosca SM (2005) Cardioprotective effects of *Ilex paraguariensis* extract: evidence for a nitric oxide-dependent mechanism. Clin Nut 24(3):360–366

Schulz V, Hänsel R, Tyler VE (2001) Rational phytotherapy: a physician's guide to herbal medicine, 4th edn. Springer, Berlin, pp 25–36

Sciarrone D, Giuffrida D, Rotondo A et al (2017) Quali-quantitative characterization of the volatile constituents in *Cordia verbenacea* D.C. essential oil exploiting advanced chromatographic approaches and nuclear magnetic resonance analysis. J Chromatogr A 1524:246–253

Sertié JAA, Woisky RG, Wiezel G et al (2005) Pharmacological assay of *Cordia verbenacea*. Oral and topical anti-inflammatory activity, analgesic effect and fetus toxicity of a crude leaf extract. Phytomedicine 12:338–344

Sharangi AB (2009) Medicinal and therapeutic potentialities of tea (*Camellia sinensis* L.) – a review. Food Res Int 42:529–535

Sharifi-Rad J, Mnayer D, Tabanelli G et al (2016) Plants of the genus *Allium* as antibacterial agents: from tradition to pharmacy. Cell Mol Biol 62(9):57–68

Singh B, Kaur P, Gopichand et al (2008) Biology and chemistry of *Ginkgo biloba*. Fitoterapia 79(6):401–418

Singh O, Khanam Z, Misra N et al (2011) Chamomile (*Matricaria chamomilla* L.): An overview. Pharmacogn Rev 5(9):82–95

Srivastava JK, Gupta S (2007) Antiproliferative and apoptotic effects of chamomile extract in various human cancer cells. J Agric Food Chem 55(23):9470–9478

Srivastava JK, Gupta S (2015) Chamomile: a herbal agent for treatment of diseases of the elderly. In: Ronald Ross Watson (ed) Foods and dietary supplements in the prevention and treatment of disease in older adults, Academic Press, Elsevier Inc., pp 171–183.

Shurtleff W, Akiko A (2015) The soybean, it's history, and it's opportunities. chapter one: U.S. Soy:International Buyers' Guide http://ussec.org/wp-content/uploads/2012/08/Chap1.pdf

Tabach R, Duarte-Almeida MJ, Carlini EA (2017a) Pharmacological and toxicological study of *Maytenus ilicifolia* leaf extract. part I – preclinical studies. Phytother Res 31(6):915–920

Tabach R, Duarte-Almeida MJ, Carlini EA (2017b) Pharmacological and toxicological study of *Maytenus ilicifolia* leaf extract part II – Clinical study (phase I). Phytother Res 31:921–926

Tacklind J, MacDonald R, Rutks I et al (2009) *Serenoa repens* for benign prostatic hyperplasia. Cochrane Database Syst Rev 2:CD001423

Taibi DM, Landis CA, Petry H et al (2007) Systematic review of valerian as a sleep aid: safe but not effective. Sleep Med Rev 11:209–230

Teschke R (2010) Kava hepatotoxicity – a clinical review. Ann Hepatol 9(3):251–265

Tabach R, Oliveira WP (2003) Evaluation of the anti-ulcerogenic activity of a dry extract of *Maytenus ilicifolia* Martius ex. Reiss produced by a jet spouted bed dryer. Pharmazie 58(8):573–576

Vacek J, Klejdus B, Kubán V (2007) Hypericin and hyperforin: bioactive components of St. John's Wort (*Hypericum perforatum*). Their isolation, analysis and study of physiological effect. Ceska Slov Farm 56:62–66

Van Beek TA, Montoro P (2009) Chemical analysis and quality control of *Ginkgo biloba* leaves, extracts, and phytopharmaceuticals. J Chromatogr A 1216(11):2002–2032

Van der Kooy F, Maltese F, Choi YH et al (2009) Quality control of herbal material and phytopharmaceuticals with MS and NMR based metabolic fingerprint. Planta Med 75:763–775

Wagner H (2011) Synergy research: approaching a new generation of phytopharmaceuticals. Fitoterapia 82(1):34–37

Wagner H, Ulrich-Merzenich G (2009) Synergy research: approaching a new generation of phytopharmaceuticals. Phytomedicine 16:97–110

Wang Y, Tang H, Nicholson JK et al (2004) Metabolomic strategy for the classification and quality control of phytomedicine: a case study of the chamomile flower (*Matricaria recutita* L.). Planta Med 70:250–255

Wang Q, Ge X, Tian X et al (2013) Soy isoflavone: the multipurpose phytochemical (review). Biomed Rep 1:697–701

Wegener T, Fintelmann V (1999) Pharmacological properties and therapeutic profile of artichoke (*Cynara scolymus* L.). Wien Med Wochenschr 149:241–247

WHO. WHO launches the first global strategy on traditional and alternative medicine. Available in http://www.who.int/globalchange/ecosystems/biodiversity/en/. Assessed 28 Mar 2018

Weller MG (2012) A unifying review of bioassay-guided fractionation, effect-directed analysis and related techniques. Sensors (Basel) 12(7):9181–9209

Wichtl M (2004a) Herbal drugs and phytopharmaceuticals: a handbook for practice on a scientific base, 3rd edn. Scientific Publishers/CRC Press, Stuttgart/Boca Raton, pp 630–634

Wichtl M (2004b) Herbal drugs and phytopharmaceuticals: a handbook for practice on a scientific base, 3rd edn. Scientific Publishers/CRC Press, Stuttgart/Boca Raton, pp 19–22

Wider B, Pittler MH, Thompson-Coon J et al (2009) Artichoke leaf extract for treating hypercholesterolaemia. Cochrane Database Syst Rev 4:CD003335

Wilt TJ, Ishani A, Stark G et al (1998) *Saw palmetto* extracts for treatment of benign prostatic hyperplasia. JAMA 280(18):1604–1609

Wölfle U, Seelinger G, Schempp CM (2014) Topical application of St. John's wort (Hypericum perforatum). Planta Med 80(2–3):109–120

Wu CP, Ohnuma S, Ambudkar SV (2011) Discovering natural product modulators to overcome multidrug resistance in cancer chemotherapy. Curr Pharm Biotechnol 12(4):609–620

Zhuo R, Liu H, Liu N et al (2016) Ligand fishing: a remarkable strategy for discovering bioactive compounds from complex mixture of natural products. Molecules 21(11):E1516

Zuin VG, Vilegas JH (2000) Pesticide residues in medicinal plants and phytomedicines. Phytother Res 14(2):73–88

Zuo W, Yan F, Zhang B et al (2017) Advances in the studies of *Ginkgo biloba* leaves extract on aging-related diseases. Aging Dis 8(6):812–826

Plant Products with Antifungal Activity: From Field to Biotechnology Strategies

Giovanna Simonetti, Alessio Valletta, Olga Kolesova, and Gabriella Pasqua

1 Introduction

Plant extracts have been traditionally used to treat a number of diseases due to the fact that they have various medicinal activities, such as anti-inflammatory, anti-androgenic, anti-proliferative, antifungal, antimicrobial, antioxidant, and others. Roots, stems, leafs, flowers, and fruits of medicinal plants are used in Ayurvedic medicine as well as European, Russian, and Asiatic folk medicine to treat different infections caused by bacteria, fungi, virus, parasite, as well as noninfectious metabolic disorders (Chuang et al. 2007). The compositions of plant extracts contain a plurality of pharmaceutical important biological active molecules, but only a small percentage of plants have been explored for their antifungal activity.

About 1.2 billion people worldwide are estimated to suffer from a fungal disease (Denning and Bromley 2015). There are an estimated three to six million fungal species. Of these, only very few (about 150–300) are known to cause disease in humans. Human fungal pathogens are a common underestimated cause of severe diseases associated with high morbidity and mortality. Four human fungal pathogens cause invasive infections as *Candida albicans*, *Cryptococcus neoformans*, *Aspergillus fumigatus*, and *Histoplasma capsulatum* (Kullberg and Arendrup 2015; Kim 2016).

Skin mycoses affect more than 20–25% of the world's population (Havlickova et al. 2008), and frequently they are associated with yeasts as *Candida* and *Malassezia* and dermatophytes such as *Trichophyton* and *Microsporum* (White et al. 2014). Although these skin-related infections are not generally life-threatening, they

G. Simonetti (✉) · O. Kolesova
Department of Public Health and Infectious Diseases, "Sapienza" University of Rome, Rome, Italy
e-mail: giovanna.simonetti@uniroma1.it

A. Valletta · G. Pasqua
Department of Environmental Biology, "Sapienza" University of Rome, Rome, Italy

V. Cechinel Filho (ed.), *Natural Products as Source of Molecules with Therapeutic Potential*, https://doi.org/10.1007/978-3-030-00545-0_2

Table 1 Fungal infections classified according to the site of infection

Site of infection	Fungi genus	Diseases
Superficial	*Malassezia*	Pityriasis versicolor
	Hortaea	Tinea nigra
	Trichosporon	White piedra
	Piedraia	Black piedra
Cutaneous	*Trichophyton, Epidermophyton, Microsporum*	Tinea corporis, tinea manuum, tinea pedis
	Candida	Cutaneous candidiasis
Subcutaneous	*Sporothrix*	Sporotrichosis
	Phialophora, Fonsecaea pedrosoi	Chromoblastomycosis
	Pseudallescheria, Madurella, and others	Mycetoma
	Exophiala, Bipolaris, Exserohilum, and other dematiaceous moulds	Phaeohyphomycosis
Systemic	*Candida*	Systemic candidiasis
	Histoplasma	Histoplasmosis
	Blastomyces	Blastomycosis
	Paracoccidioides	Paracoccidioidomycosis
	Cryptococcus	Cryptococcosis
	Aspergillus	Aspergillosis
	Zygomycetes	Zygomycosis
	Fusarium, Paecilomyces, Trichosporon	Hyalohyphomycosis
	Penicillium	Penicilliosis

represent a common global problem and can become chronic. Furthermore, the treatments often require long-term therapy and are not resolving in all (Pfaller 2012). Side effects and resistance are frequently due to the current antifungal agents (Table 1), such as the most widely used azole drugs (Pfaller 2012; Zavrel and White 2015).

Phytomedicine, which has been historically an important aspect of traditional medicine in non-industrialized countries, is becoming now an integral part of healthcare in these countries. Different strategies can be applied to improve the yields of bioactive metabolites in the plant and to obtain standardized extracts in chemical mode (Dias et al. 2016; Atanasov et al. 2015).

This chapter provides an overview of the published results on plant-derived natural products showing antifungal activity against human pathogens. Moreover, biotechnological approaches have been explored in order to increase the production of active extracts, in alternative to obtain extracts from plants directly collected from their natural habitat. The use of plants as source of bioactive compounds is related with the accessibility of the starting material. Cultivation of medicinal plants could be a sustainable alternative to wildcrafting, but until today, two thirds of the used plants are still collected in the wild. Moreover, some species are protected to conserve the biodiversity. The Nagoya Protocol on "access to genetic resources and the fair and equitable sharing of benefits arising from their utilization to the Convention on Biological Diversity" needs to be respected and should become a major tool for benefit sharing as well as the conservation and sustainable use of biological diversity (Buck and Hamilton 2011).

It should be emphasized that often the amount of the extract from plant is low, depending on the developmental degree of the biosynthetic organ, and season-dependent. Moreover, the chemical composition of the extract is not only dependent on the species but also on soil composition, processing, and storage (Atanasov et al. 2015). Cultivation of medicinal plants under controlled conditions could make possible to maintain the concentration of important compounds in the plant. The aim is also to increase potency, reduce toxin levels, and increase uniformity and predictability of extracts. The trend towards niche production of high-value species for non-food markets offers great opportunities (Lubbe and Verpoorte 2011). In recent years, different strategies have been developed to produce active compounds using plant cell and organ cultures. Moreover, in vitro propagation of plants, where possible, could solve the problems concerning the loss of genetic diversity and habitat destruction. The literature reports a significant progress in the use of plant tissue cultures, called "chemical factories" of secondary metabolites (Rao and Ravishankar 2002), also offering the opportunity to optimize yield and achieve a uniform, high-quality products.

Actually biosynthetic pathways of secondary metabolites are mostly poorly understood, and relatively few genes for key enzymatic or regulatory steps have been isolated (Canter et al. 2005). The production of plant secondary metabolites by means of large-scale culture of plant cells in bioreactors is technically feasible, but unfortunately, some of the most interesting products are only in very small amounts or not all produced in plant cell cultures. Screening, selection, and medium optimization may lead to 20- to 30-fold increase in case one has producing cultures. In case of phytoalexins, elicitation will lead to high production. Metabolic engineering offers new perspectives for improving the production of compounds of interest. A promising approach to improve the production of important metabolites is to upregulate the enzymes important for the synthesis of metabolites or to increase their precursors (Verpoorte et al. 2002).

2 Human Fungal Diseases

About 1.2 billion people worldwide are estimated to suffer from a fungal disease. An estimated 1.5 to two million people die of a fungal infection each year surpassing those killed by either malaria or tuberculosis (Denning and Bromley 2015). Four human fungal pathogens that cause invasive infections are *Candida albicans*, *Cryptococcus neoformans*, *Aspergillus fumigatus*, and *Histoplasma capsulatum* (Kim 2016). Fungal infections may be classified according to the site of infection, route of acquisition, and type of virulence. When classified according to the site of infection, fungal infections are designated as superficial, cutaneous, subcutaneous, and systemic (Table 1). Most are infections of the skin or mucosa. Superficial mycoses are limited to the stratum corneum and include common skin diseases as well as rare infections confined to specific geographical areas or groups of patients (Brooks et al. 2013). Cutaneous infections involve the integument and its appendages, including hair and nails. Cutaneous mycoses are most commonly caused by dermatophytes, which require keratin for survival, but they can be caused by non-dermatophyte moulds or *Candida* species. Dermatophytes can be classified as anthropophilic, zoophilic, or geophilic, depending on their primary habitat (humans, animals, and soil, respectively). The clinical picture of dermatomycoses is variable due to the degree of keratin destruction by the fungus and the inflammatory response of the host. For example, anthropophilic dermatophytes cause little inflammation but can cause recurrent or chronic infections, while zoophilic and geophilic dermatophytes tend to induce acute and highly inflammatory responses. Inflammatory symptoms such as pruritus, erythema, swelling, and burning can have a significant impact on the quality of life of the affected individual (Schaller et al. 2016). The deep mycoses are uncommon infections caused by fungi, and they are divided into subcutaneous and systemic mycoses. Subcutaneous mycoses include a range of different infections characterized by infection of the subcutaneous tissues usually at the point of traumatic inoculation. While skin manifestations always occur in subcutaneous mycoses, or mycoses of implantation, as they are also known, they are only occasionally seen in systemic mycoses (Carraco-Zuber et al. 2016). Systemic mycoses involve the lungs, abdominal viscera, bones, and/or central nervous system. The most common portals of entry are the respiratory tract, gastrointestinal tract, and blood vessels Brooks et al. 2013) (Table 1).

3 Antifungal Susceptibility Methods

In vitro antifungal tests are crucial in the screening process. Diffusion method is used in many clinical microbiology laboratories for routine antimicrobial susceptibility testing. In this procedure, antimicrobial agent diffuses into the agar and inhibits growth of the test microorganism, and then the diameters of inhibition growth zones are measured. The diffusion method is not appropriate to determine the minimum inhibitory concentration (MIC), as it is impossible to quantify the amount of the extracts and its constituents diffused into the agar medium. The disc diffusion method is a qualitative test, and its results should not be used for quantitative purposes.

Dilution methods are the most appropriate ones for the determination of MIC values, since they offer the possibility to estimate the concentration of the tested antimicrobial agent in the agar (agar dilution) or broth medium (macrodilution or microdilution). Dilution methods may be used to quantitatively measure the in vitro antimicrobial activity against fungi. MIC value recorded is defined as the lowest concentration of the assayed antimicrobial agent that inhibits the visible growth of the microorganism tested, and it is usually expressed in µg/mL or mg/L. The most recognized standards are provided by the CLSI and the European Committee on Antimicrobial Susceptibility Testing (EUCAST). The dilution assays give the most reliable results and are capable of determining the MIC and MBC of a particular sample (Balouiri et al. 2016).

*3.1 Anti-***Candida** *Plant Products*

Candida species are opportunistic fungal pathogens. These fungi belong to the normal microbiota of an individual's mucosal oral cavity, gastrointestinal tract, and vagina, but in particular, conditions such as immunocompromised patients and diabetes hemodialysed patients can cause cutaneous and systemic mycoses. *Candida* may spread in different parts of the body and cause a wide range of diseases such as meningitis, endocarditis, pyelonephritis, renal papillary necrosis, multiple parenchymal abscesses, endophthalmitis, septic arthritis and osteomyelitis, peritonitis, and pneumonia (Hope et al. 2012). Moreover, the use of devices including catheters and prolonged hospitalization increase the prevalence of invasive candidiasis. Catheter-related microbial biofilms are associated with 90% of *Candida* infections and considered as the major cause of morbidity and mortality among hospitalized patients (DiDone et al. 2011). The difficulty to eradicate *Candida* infections is owing to its unique switch between yeast and hyphae forms and more likely to biofilm formations that render resistance to antifungal therapy. Biofilms provide a safe haven for *Candida*, facilitate drug resistance, and act as sources for chronic infections (Donlan 2001). Systemic candidiasis is caused by different species of *Candida*. The most common causative agents of candidiasis are *C. albicans*, *C. glabrata*, *C. krusei*, *C. parapsilosis*, and *C. tropicalis*. In recent years, *C. dubliniensis*, *C. guilliermondii*, *C. kefyr*, *C. lusitaniae*, *C. pelliculosa*, and *C. zeylanoides* have been detected with increasing frequency (Kauffmann and Mandell 2010; El-Atawi et al. 2017). Azoles (fluconazole, itraconazole, voriconazole, and ketoconazole), echinocandins (micafungin, caspofungin), and amphotericin B are used to treat *Candida* human infections (Arendrup and Patterson 2017). Azole antifungal agents are most useful in the therapy for mucosal infections related to *Candida*. However, after a previous treatment with azole antifungal agent, the patient could show a microbiological resistance (Arendrup and Patterson 2017). Drugs currently on the market have been developed by studying the few known targets for fungal cells, such as the particular composition of the cell membrane and the related enzymes. Ergosterol, nucleic acids, and glucan are the most studied molecular targets to destroy *Candida* species, being considered the basis of the development of new antifungal drugs. Since the availability of these drugs is limited, resistances developed by the fungal cells represent an important problem to be managed. Although amphotericin B is the most well-known agent in terms of efficacy in serious *Candida* infections, it has the greatest potential toxicity. At the same time, fluconazole is less toxic but also has some side effects such as nausea, headache, and others (Table 2) (Grohskopf and Vincent 1996). The specific type of medication and length of treatment will depend

Table 2 Side effects of antifungal agents

Chemical group	Drugs	Side effects
Macrolides	Amphotericin B	Hypokalaemia, hypomagnesaemia, renal injury, nausea, vomiting, abdominal pain, rash, headache, hepatic necrosis (Laniado-Laborín and Cabrales-Vargas 2009)
Pyrimidine	Flucytosine	Hepatotoxicity, bone marrow depression, nausea, vomiting, and diarrhoea (Vermes et al. 2000)
Azole	Ketoconazole Fluconazole Itraconazole Voriconazole	Dizziness, drowsiness, insomnia, impaired consciousness, vision, hallucinations, paraesthesia, tremor, convulsions, flatulence, nausea, vomiting, diarrhoea, constipation, thrombophlebitis, chills, fever, aches, nausea/vomiting, hypotension, nephrotoxicity, hypokalaemia hypomagnesaemia, suppression erythro-thrombopoiesis (Verweij et al. 2009)
Echinocandins	Caspofungin Micafungin	Phlebitis and the histamine-like reaction marked by rash, urticaria, flushing, bronchospasm, hypotension and facial swelling, arrhythmias, and cardiac failure (Koch et al. 2015)

on many factors, including the age and health of the infected person, the location and severity of the infection, and the specific species of *Candida* causing the infection (Pappase et al. 2015).

Recent studies have shown that some plant extracts have anti-*Candida* activity like some antifungal synthetic drugs (Martins et al. 2015a; Soliman et al. 2017). Several of these showed promising minimum inhibitory concentration (MIC) such as peppermint (0.08 μg/mL), *Thymus villosus* (0.64 μg/mL), eucalyptus (0.05 μg/mL), lemongrass oil (0.06 μg/mL), *Cinnamomum zeylanicum* (0.01 μg/mL), ginger grass oil (0.08 μg/mL), and coriander (0.2 μg/mL) (Soliman et al. 2017).

Among them, the essential oils and extracts rich in different phenol compounds are being tested in the search of new antifungals with fewer side effects. Although the mechanisms of action are not yet completely clear, it has been speculated that plant products could represent a viable alternative to the antifungal drugs; however, up to date, none of these plant products is marketed for anti-*Candida* therapy (Güllüce et al. 2003; Raut and Karuppayil 2014).

Essential oils are complex volatile compounds whose composition varies depending on factors such as the climate, the age of the plant, and even the organ from which they are extracted. In the case of *Agastache rugosa,* the composition of the oil extracted from the flower is different from the one extracted from the leaf, and this means different antifungal activity (Shin and Kang 2003).

Curcumin is a particular polyphenol, which shows a marked activity against *Candida*. The study of curcumin against 14 strains of *Candida* including 4 ATCC strains and 10 clinical isolates showed that curcumin is a potent antifungal compound against *Candida* species with MIC values range from 250 to 2000 μg/mL (Neelofar et al. 2011). In another study, anti-*Candida* activity of curcumin was demonstrated against 38 different strains of *Candida* including some fluconazole-resistant strains and clinical isolates of *C. albicans*, *C. glabrata*, *C. krusei*, *C. tropicalis*, and

C. guilliermondii. The MIC_{90} values for sensitive and resistant strains were 250–650 and 250–500 μg/mL, respectively (Khan et al. 2012; Zorofchian Moghadamtousi et al. 2014).

The investigation of curcumin mediation for photodynamic therapy can reduce the biofilm biomass of *C. albicans*, *C. glabrata*, and *C. tropicalis*. The results demonstrated that association of four LED fluences for light excitation with 40 μM concentration of curcumin inhibited up to 85% metabolic activity of the tested *Candida* species. Photodynamic effect considerably decreased *C. albicans* viability in either planktonic or biofilm cultures probably through increasing the uptake of curcumin by cells (Dovigo et al. 2011).

Phenolic compounds are widely found in plant foods (fruits, cereal grains, legumes, and vegetables) and beverages (tea, coffee, fruits juices, and cocoa). The most common phenolic compounds are phenolic acids (cinnamic and benzoic acids), flavonoids, proanthocyanidins, coumarins, stilbenes, lignans, and lignins. The anti-*Candida* mechanisms of phenolic compounds reported in the literature include inactivation of enzyme production (Teodoro et al. 2015) and anti-biofilm activity (Evensen and Braun 2009).

Resistant *Candida* species to the current antifungal drugs have been observed; thus, alternative therapy based on plant extracts rich in phenolic compounds should be considered (Martins et al. 2015b).

Rangkadilok et al. (2012) demonstrated that *Dimocarpus longan* Lour. (longan) seed exhibited antifungal activity against *Candida* species. In contrast, longan pulp and whole fruit did not demonstrate any inhibitory effects. Ellagic acid showed the most potent antifungal activity followed by corilagin and gallic acid, respectively. Ellagic acid inhibited *C. parapsilosis* more effectively than *C. krusei* and also some *C. albicans* clinical strains. Baidam cultivar possessed higher antifungal activity (MIC = 500–4000 μg/ml) as it contained higher contents of ellagic acid and gallic acid (MIC = 1000–8000 μg/ml).

Mahmoudabadi et al (2007) studied the anti-*Candida* activity against 14 isolates of *C. albicans*, *C. parapsilosis*, *C. tropicalis*, and *C. glabrata* was of aqueous and alcoholic extracts of *Zataria multiflora*. Aqueous extract showed no remarkable activity against *Candida* species. Conversely, MIC of the methanolic and ethanolic extracts was 70.7 and 127 mg/L, respectively. The biological activity of *Zataria multiflora* was mainly associated with its main chemical components, including thymol, rosmarinic acid, flavonoids, and carvacrol.

Brighenti et al. (2017) screened 60 plant extracts from Brazilian Pantanal biome for *C. albicans* anti-biofilm activity. Effects on biofilm inhibition and disruption and cytotoxicity were also evaluated. The most active extract was chemically characterized. *Buchenavia tomentosa* ethanolic extract showed noticeable antifungal activity and was selected for biofilm experiments. Subinhibitory concentration of extract inhibited fungal adhesion. Maximum killing reached 90% of *C. albicans* cells in suspension and 65% of cells in biofilms. The active extract was noncytotoxic. Chemical characterization showed the presence of phenols. Ellagic and gallic acids showed activity on *C. albicans*.

Punica granatum is a plant with worldwide application in folk medicine. Polyphenols extracted from pomegranate fruit were active against phytopathogenic fungi. The extract of *P. granatum* showed good results as a topical antifungal agent for the treatment of candidosis associated with denture stomatitis (Bassiri-Jahromi et al. 2017; Vasconcelos et al. 2003).

The tannin punicalagin is the major component of pomegranate fruit peel. This substance was isolated not only from *Punica granatum* but also was described from *Terminalia mollis* and *Terminalia brachystemma*, as having antifungal activity against *C. albicans*, *C. krusei*, and *C. parapsilosis* (Liu et al. 2009).

Alcoholic and water hot extracts of the *Punica granatum (pomegranate*) peels as well as the dried powder were prepared. The antifungal activity of the extracts containing gallotannic acid was evaluated by means of the agar-well diffusion assay. The extract exhibited potent activity against *C. albicans* and *C. tropicalis* (Shaokat et al. 2017).

In vitro antifungal activity of acetonic extracts of *Punica granatum* L., *Quercus suber* L., and *Vicia faba* L. against seven pathogen fungi and the in vivo antifungal activity against *C. albicans* have been studied. The phytochemical screening was also carried out and showed that the extracts contained mainly proanthocyanidins. Other polyphenols were also present but in low quantity. The acetone extract of *V. faba* L. showed in vitro activity, and it was the most active for treating candidiasis in mice (Akroum 2017).

The antifungal activity of extracts from 10 different plants, commonly used in folk medicine, was evaluated against 19 *Candida* strains, including *C. albicans*, *C. glabrata*, *C. parapsilosis*, and *C. tropicalis* species. Although the majority of the extracts had no antimicrobial effect, *Juglans regia* extract was very effective, exerting an inhibitory effect against all the tested *Candida* strains, while *Eucalyptus globulus* was effective against 17 of them. *Pterospartum tridentatum* and *Rubus ulmifolius* presented similar antifungal effects, being effective against six *Candida* strains. The diameter of halo ranged, respectively, between 9–14 mm and 9–21 mm to the mentioned plant extracts. Both extracts showed similar MIC_{50} values for *C. albicans* strains, while *C. parapsilosis* and *C. glabrata* were more sensible to *E. globulus*. Otherwise, all the *C. tropicalis* strains were more sensible to *J. regia* (Martins et al. 2015b).

Recently, it has been demonstrated that extracts by *Vitis vinifera* seeds obtained from mature grapes, rich in polymeric flavan-3-ols, exhibit good antifungal activity against *Candida* species suggesting their use in mucocutaneous fungal infections (Pasqua and Simonetti 2016; Simonetti et al. 2014, 2017b). Moreover, it has been demonstrated a significant inhibition of *Candida albicans*, in an experimental murine model of vaginal candidiasis, using grape seed extract (GSE) with high content of polymeric flavan-3-ols (Simonetti et al. 2014). The antifungal activity of unripe grape extracts from agro-industrial wastes has been evaluated against several strains of *Candida* spp. All the extracts tested showed antifungal activity. The geometric mean MIC ranged from 53.58 to 214.31 μg/ml for *Candida* spp. (Simonetti et al. 2017a). It is important to highlight that grapeseed extracts (GSE), recognized safe by the Food and Drug Administration, are used as food additives and in cosmetics.

The high tolerability of plant products is another important aspect, unlike synthetic drugs that often have adverse effects on humans and animals, such as the nephrotoxicity of amphotericin B (Table 2).

3.2 Anti-dermatophytes Plant Products

Dermatophytes are a group of pathogenic fungi and the major cause of dermatophytosis infections of the human skin, hair, and nail. Cutaneous and subcutaneous mycoses caused by dermatophytes fungi affect keratinized structures of the body.

Dermatomycosis, depending on the primary localization of the damage (epidermis, nails, or hair), is divided into epidermomycoses, onychomycosis, and trichomycosis. The most frequently involved dermatophyte genera of humans and other animals are *Trichophyton*, *Epidermophyton*, and *Microsporum*. They have some common biological properties. The most common anthropophilic dermatophytes are *Trichophyton rubrum*, *Trichophyton tonsurans*, *Trichophyton violaceum*, and *Epidermophyton floccosum*; anthropozoophilic are *Trichophyton mentagrophytes*, *Microsporum canis*, and *Microsporum ferrugineum*; and geophysical are *Microsporum gypseum* and *Trichophyton ajelloi*. Dermatophytes even if they are not responsible for systemic mycoses are the etiologic agents of the most common human mycoses. Dermatophytes as pathogens have the ability to survive in the environment of the macroorganism, and they can introduce hyphae into intercellular spaces, facilitated by the production of enzymes that lyse keratin, collagen, and elastin. Moreover, it has been shown that the low immunogenicity of dermatophytes such as *T. rubrum* can be associated with the production of lipophilic toxin that inhibits cellular immunity and the proliferation of keratinocytes (Vorzheva and Chernyak 2004).

Nowadays, the fungal infections of the human skin, hair, and nail are treated by the oral or topical antifungal agents such as fluconazole, triazoles, and terbinafine which have a high spectrum of activity against dermatophytes. Despite the fact that many antifungal agents are available, their side effects (Table 2) and interactions with drugs, as well as the presence of resistant organisms, have created the need for safer and more effective treatment. In addition, treatment of dermatophytosis is usually expensive and should be applied for a long time.

Some extracts are active against dermatophyte fungi. Simonetti et al. (2017a, b) demonstrated that *Vitis vinifera* seed extracts obtained from different tables and wine cultivars have antidermatophytic activity against collection of strains of *T. mentagrophytes*, *M. gypseum*, and *M. canis*. Geometric minimal inhibitory concentration ranged from 20 to 97 μg/ml. The activity of the extracts was lower than terbinafine but comparable with that of fluconazole.

The antifungal activity of unripe grape extracts from agro-industrial wastes has been demonstrated against several strains of dermatophytes. All the extracts tested showed antifungal activity. The geometric mean MIC ranged from 43.54 to 133.02 μg/mL for dermatophytes. The highest negative significant correlation has been found between MICs and caffeoyl derivatives ($r = -0.962$, $p < 0.01$) (Simonetti et al. 2017a, b).

Endo et al. (2015) demonstrated the antifungal activity of *Rosmarinus officinalis* and *Tetradenia riparia* hydroalcoholic extracts against dermatophytes. According to the fluorescence microscopy and scanning electron microscopy results, *Rosmarinus officinalis* and *Tetradenia riparia* hydroalcoholic extracts cause inhibition of hyphal growth and irregular growth pattern. The MIC values range from 62.5 to 250 μg/mL (Morais et al. 2017). The dragon's blood from *Croton urucurana*, used by indigenous cultures of the Amazon River for the treatment of infected wounds, has shown to have antifungal activity. The results showed that dragon's blood MIC against *T. rubrum*, *T. mentagrophytes*, *M. canis*, and *E. floccosum* was 2.5 mg/ml and against *T. tonsurans* was 1.25 mg/ml (Gurgel et al. 2005).

Rodrigues et al. (2012) showed that the extract from aerial parts of *Pothomorphe umbellata*, a native Brazilian plant, is active against dermatophytes, in particular, against *T. rubrum*.

T. rubrum is one of the most common species of human dermatophytes which causes tinea pedis, nail infection, tinea cruris, and tinea corporis. MIC value of the ethanol extract of *Pothomorphe umbellata* was 156.25 μg/mL, while the methanol extract shows MIC value of 78.13 μg/mL. The antifungal activity of *Pothomorphe umbellata* could be due to the presence of ß-sitosterol, stigmasterol, and campesterol. In essential oil were identified several compounds as spathulenol, β-caryophyllene, caryophyllene oxide, germacrene D, bicyclogermacrene, b-elemene, b-pinene, a-cadinol, d-cadinene, a-copaene, and limonene.

*3.3 Anti-*Malassezia *Plant Products*

Malassezia spp. are normally present in the normal biota of a healthy human skin. Usually *Malassezia* spp. cause only chronic recurrent superficial mycoses without any life threat. In individuals with immunosuppression as well as endocrinopathies, chronic dermatoses and bacteria infections *Malassezia* could cause skin infections like dandruff, pityriasis versicolor, seborrhoeic dermatitis, and folliculitis. According to the recent studies, *Malassezia* spp. play a role in the pathogenesis of atopic dermatitis and psoriasis, especially in cases involving the scalp (Velegraki et al. 2015). The *Malassezia* genus includes 14 species: *M. furfur*, *M. pachydermatis*, *M. sympodialis*, *M. globosa*, *M. obtusa*, *M. restricta*, *M. slooffiae*, *M. equina*, *M. dermatis*, *M. japonica*, *M. nana*, *M. capre*, *M. yamatoensis*, and *M. cuniculi*; however, only 9 of them could infect humans (Cabanes et al. 2011). A number of different methods have already been proposed for the treatment of *Malassezia* skin mycoses, but they have many disadvantages, in particular, long-time course duration, insufficient therapeutic efficacy, and tolerance resistance to antifungal therapy. Nowadays, *Malassezia* systemic infections require prompt identification of the pathogenic agent and treatment with liposomal amphotericin B, itraconazole, or fluconazole. Fungi of the genus *Malassezia*, due to their cultural characteristics, have extraordinary resistance to environmental factors and natural and synthetic antimycotic agents of systemic and topical application, which is why it is possible to explain the existing problems in the treatment of patients (Velegraki et al. 2015; Bragutsa 2007).

Simonetti et al. (2017a, b) demonstrated anti-*Malassezia* activities of *Vitis vinifera* seed extracts obtained from different tables and wine cultivars. Geometric minimal inhibitory concentration ranged from 32 to 161 μg/mL for *M. furfur*. The MIC for *M. furfur* was inversely correlated with the amount of the polymeric fraction ($r = -0.7228$) and only weakly was correlated to the content of flavan-3-ol monomers.

Shams-Ghahfarokhi et al. (2006) studied the antifungal activity of aqueous extracts obtained from *Allium cepa* and *Allium sativum* against *M. furfur*. *Allium cepa* and *Allium sativum* are used in the folk medicine of many countries due to their antifungal, antiprotozoal, antihelminthic, antiviral, disinfectant, and antitumor properties as well as in the treatment of gastric and hepatic disorders, diabetes mellitus, hypertension, hypercholesterolaemia, and immunodeficiency syndromes. The results of Shams-Ghahfarokhi indicate that *Allium cepa* and *Allium sativum* extracts were active against *M. furfur* with MIC values ranging from 0.08 to 0.16 mg/ml.

Filip R. et al. (2010) have evaluated the effect of the aqueous extract of *Ilex paraguariensis* on the growth of *M. furfur*. *Ilex paraguariensis* is a plant that typically grows in north-eastern area of Argentina, Southern Brazil, and Eastern Paraguay. The results demonstrated that the aqueous extract of *Ilex paraguariensis* possesses inhibitory activity against *M. furfur* (MIC = 1000 μg/ml). Probably the presence of chlorogenic acid, caffeic acid, theobromine, and rutin in the extract provides observed anti-*Malassezia* activity.

Onlom et al. (2014) investigated antifungal activities of the extracts obtained from the roots of *Asparagus racemosus* Willd against *M. furfur* and *M. globosa*. *Asparagus racemosus* Willd or shatavari (Asparagaceae family) is an important medicinal plant in Ayurvedic medicine due to various activities including phytoestrogenic, antibacterial, anti-candidal, antidiarrhoeal, antioxidant, immunostimulant, anti-dyspepsic, and antitussive effects. It has been shown that defatted ethanolic extract from the roots of *Asparagus racemosus* has anti-*Malassezia* activity against both *M. furfur* and *M. globosa* with MIC value 25 mg/mL.

*3.4 Anti-***Aspergillus** *Plant Products*

Aspergillus spp. infections have grown in importance in the last years. The *Aspergillus* genus consists of about 40 pathogenic species. *Aspergillus fumigatus*, along with *Aspergillus flavus*, *Aspergillus niger*, and *Aspergillus terreus* are the main pathogens of aspergillosis (Kulko 2012). *A. fumigatus* is an opportunistic pathogen that usually affects cavities that have formed in the lungs from preexisting lung diseases. In the lungs, *A. fumigatus* forms tangled mass of fungus fibres and blood clots. The fungus mass gradually enlarges, destroying lung tissue in the process, but usually does not spread to other areas. *A. niger* is a causative agent causing invasive aspergillosis. Invasive aspergillosis in immunocompromised host is a major infectious disease leading to reduce the survival rate of world population. Until recent years, the only drugs available to treat aspergillosis were amphotericin B and itraconazole, the latter in oral and intravenous formulations. Recently, voriconazole, posaconazole, and caspofungin have also been approved for the treatment of aspergillosis. Infections caused by *A. terreus* resistant to amphotericin B, are treated with triazoles, voriconazole, and echinocandin. Although resistance to antifungal drugs is not as great a concern as resistance to antibacterial agents, there has been an increase in the number of reported cases of both primary and secondary resistance in human mycoses (Denning et al. 1997). Therefore, the resistance of the fungus to the drug or an inadequate concentration of the antifungal drug at the site of infection might contribute to the high mortality rate seen for these infections (Hedayati et al. 2007).

Bansod and Rai (2008) tested oils extracted from 15 medicinal plants that were screened for their activity against *A. fumigatus* and *A. niger*. The results showed that *Cymbopogon martini*, *Eucalyptus globulus*, and *Cinnamomum zylenicum* oils to control (miconazole nitrate) have antifungal activity. The oils of *Mentha spicata*, *Azadirachta indica*, *Eugenia caryophyllata*, *Withania somnifera*, and *Zingiber officinale* exhibited moderate activity. The *Cuminum cyminum*, *Allium sativum*, *Ocimum sanctum*, *Trachyspermum copticum*, *Foeniculum vulgare*, and *Elettaria cardamomum* oils demonstrated comparatively low activity against *A. niger* and *A. fumigatus* as compared to control. Mixed oils showed maximum activity as compared to standard. Arunkumar and Muthuselvam (2009) reported the activity of *Aloe vera* extracts against *A. flavus* and *A. niger*. The higher antifungal activity was observed with acetonic extract (15 ± 0.73 nm and 8 ± 0.37 nm).

*3.5 Anti-*Cryptococcus *Plant Products*

Cryptococcosis is an important systemic mycosis and the third most prevalent disease in human immunodeficiency virus HIV-positive individuals. About 8% of the patients with AIDS and HIV-infected have cryptococcosis. Fungi of the *Cryptococcus* genus are causative agents of cryptococcosis. Very often the disease affects people with depressed immune systems. Commonly this disease affects the central nervous system and, in some cases, less often, the lungs, mucous membranes, and skin. Cryptococcosis occurs by the inhalation of infectious cells and is considered a primary pulmonary infection, which may lead to a disseminated infection. The disseminated infection could affect the central nervous system, causing meningitis, encephalitis, or meningoencephalitis. Among all the fungi of the genus *Cryptococcus*, which includes a large number of species, only *C. neoformans* (in Europe and North America) and *C. gattii* (tropical and subtropical zones) are considered to be pathogenic in humans. *C. neoformans* has a spherical, round, or oval shape and an average cell size of 8–40 μm. The main antifungal agents for cryptococcosis treatment are amphotericin B, flucytosine, and oral triazole antifungal drugs, such as fluconazole and itraconazole (Gullo et al. 2013). However, the use of flucytosine in patients with AIDS has been controversial due to its toxicity.

Despite the existing methods of treatment, the main lines of modern clinical trials include a comparative analysis of the efficacy and safety of antimycotic drugs for various infectious agents, as well as the search for the most optimal systemic drugs (Piraccini and Gianni 2013).

Considerable interest of the scientific community has been attracted to plant-based fungicides, because plants have their own protection against fungal pathogens through the ability to produce antifungal compounds in order to be protected from biotic attack, which may be necessary for resistance to fungal infections (Gurgel et al. 2005).

Valente et al. (2013) showed that *Oenanthe crocata* L. essential oil have activity against *C. neoformans*. The oil was predominantly composed of monoterpene hydrocarbons (85.8%), being the main compounds trans-b-ocimene (31.3%), sabinene (29.0%), and cis-b-ocimene (12.3%). The oil was particularly active against dermatophytes and *C. neoformans*, with MIC values ranging from 0.08 to 0.16 μg/mL.

Rangkadilok and collaborators (2012) showed that longan (*Dimocarpus longan* Lour.) seed extract has antifungal activity against *C. neoformans*. This natural plant contains polyphenolic compounds which exhibit several pharmacological properties. Extract of longan fruit contained high levels of polyphenolic compounds such as corilagin, gallic acid, and ellagic acid. Longan seed extract exhibited antifungal activity against *C. neoformans* with MIC of 4000 μg/ml.

Ranganathan and Balajee (2000) demonstrated that extracts of *Cassia alata* and *Ocimum sanctum* have anti-*Cryptococcus* activity. The ethanolic extract of *O. sanctum* did not show any activity against all the strains up to a concentration of 1000 μg/mL. The MIC of ethanolic extract of *C. alata* ranged from 500 to 1000 μg/ml, and the extract showed fungicidal activity at 1000 μg/ml at acidic pH. Decreased activity at neutral pH and least activity at pH 8 were recorded for the extract. The combination of extract of *O. sanctum* and *C. alata* inhibited growth of the organism at a concentration ranging from 62.5 to 125 mg/ml. The combination of the extracts showed fungicidal activity at 125 μg/ml. It is known that the leaves of *C. alata* contain anthraquinones, flavonoids, quinones, and sterols which could be a reason of the effect of the extract combination on *C. neoformans* (Table 3).

4 Production of Antimicrobials Through Plant Tissue Cultures

As discussed above, plants are a vast and still largely unexplored source of antimicrobial secondary metabolites. Different methods can be adopted to obtain antimicrobial compounds of plant origin, and each one has its own advantages and disadvantages.

Since the ancient times, plant bioactive compounds have been obtained through direct extraction from wild plants. It should be noted that specialists are required for the harvesting of wild plants intended for human use, since many plants morphologically similar to the species of interest may have a different phytochemical profile corresponding to different biological activities and may contain toxic metabolites,

Table 3 Antifungal activity of extracts obtained from plants collected directly from their natural habitat

Plant family	Species name (plant common name)	Natural habitat	Main active constituents	Part of plant	Fungal species	Antifungal activity	Reference
Annonaceae	*Annona squamosa* (sugar-apples)	South and Central America, Africa, India, Southeast Asia, Australia	Alkaloids, Glycosides Flavonoids, Tannins Phenols Saponins	leaves		MIC (μg/ml)	Kalidindi et al. (2015)
					Alternaria alternata	800	
					Candida albicans	600	
					Fusarium solani	600	
					Microsporum canis	400	
					Aspergillus niger	400	
Apiaceae	*Oenanthe crocata* (water dropworts)	Mediterranean region	Terpenes	Aerial parts		MIC (μg/ml)	Valente et al. (2013)
					Candida albicans	0.64–1.25	
					Candida guilliermondii	0.64	
					Candida krusei	1.25	
					Candida parapsilosis	1.25	
					Candida tropicalis	1.25	
					Cryptococcus neoformans	0.16	
					Epidermophyton floccosum	0.08	
					Microsporum canis	0.08	
					Microsporum gypseum	0.08	
					Trichophyton mentagrophytes	0.16	
					Trichophyton mentagrophytes	0.16	
					Trichophyton rubrum	0.08	
					Trichophyton verrucosum	0.64–1.25	
					Aspergillus flavus	2.5	
					Aspergillus fumigatus	1.25	
					Aspergillus niger	1.25	

Apocynaceae	*Alstonia macrophylla* (hard alstonia)	Far East	Alkaloids	Leaves		MIC (μg/ml)	Chattopadhyay et al. (2001)
					Trichophyton mentagrophytes	64,000	
					Trichophyton rubrum	32,000	
					Microsporum gypseum	128,000	
Asclepiadaceae	*Cryptolepis buchanani* (kareballi)	India	Tannins Alkaloid Saponins Flavonoids	Leaves		Inhibition zone (mm)	Verweij et al. (2009)
					Chrysosporium keratinophilum	12	
					Trichophyton rubrum	14	
					Chrysosporium indicum	11	
Asteraceae	*Tridax procumbens* (coatbuttons)	Central America	Carboxylic acid Ethyl esters	Aerial parts		Inhibition zone (mm)	Policegoudra et al. (2014)
					Trichophyton mentagrophytes	7	
					Trichophyton rubrum	4	
					Trichosporon beigelii	2	
					Candida albicans	2	
	Baccharis trinervis (cambara-rebentao, casadinha preta, assapeixe-fino)	South America	Monoterpenes	Aerial parts		MIC (μg/ml)	Sobrinho et al. (2016)
					Trichophyton rubrum	150–310	
Ebenaceae	*Diospyros virginiana L.* (American persimmon)	Southeastern of the United States, China	Vitamins Carotenoids Carotenes	Fruits		MIC (μg/ml)	Ciric et al. (2014)
					Aspergillus fumigatus	5	
					Aspergillus versicolor	40	
					Aspergillus ochraceus	10	
					Aspergillus niger	40	
					Penicillium verrucosum	10	
					Penicillium ochrochloron	40	
					Penicillium funiculosum	10	
					Trichoderma viride	40	

(continued)

Table 3 (continued)

Plant family	Species name (plant common name)	Natural habitat	Main active constituents	Part of plant	Fungal species	Antifungal activity	Reference
Euphorbiaceae	*Croton urucurana Baill.* (red sap dragon's Dimocarpus blood)	South America	Flavans	Resin		MIC (µg/ml)	Gurgel et al. (2005)
					Trichophyton rubrum	2500	
					Trichophyton mentagrophytes	2500	
					Trichophyton tonsurans	1250	
					Microsporum canis	2500	
					Epidermophyton floccosum	2500	
Hypericaceae	*Hypericum perforatum* (St. John's wort)	Temperate regions	Xanthones	Roots		MIC (µg/ml)	Tocci et al. (2013b)
					Candia albicans	663.98	
					Candida parapsilosis	1024	
					Candida glabrata	1024	
					Candida tropicalis	406.37	
					Candida krusei	256	
					Cryptococcus neoformans	53.81	
					Trichophyton mentagrophytes	181	
					Microsporum gypseum	362.04	
Juglandaceae	*Argemone mexicana* (Mexican poppy)	India, Mexico, Nigeria	Terpenes	Stem and leaves		Inhibition zone (mm)	More and Kharat (2016)
					Mucor indicus	30	
					Aspergillus flavus	23	
					Aspergillus niger	21	
					Penicillium notatum	20	
Lamiaceae	*Zataria multiflora* (Zataria)	Middle East	Terpenes	Essential oils		MIC (ml/ml)	Mahboubi et al. (2017)
					Trichophyton rubrum	0.125	
					Trichophyton mentagrophytes	0.03	
					Microsporum canis	0.03	
					Microsporum gypseum	0.125	
					Trichophyton schoenleinii	0.06	

	Rosmarinus officinalis (Rosemary)	Africa, Iberian Peninsula	Flavonoids	Leaves		MIC (μg/ml)	Morais et al. (2017)
					Trichophyton rubrum	250	
					Trichophyton mentagrophytes	6.5	
					Microsporum gypseum	125	
	Tetradenia riparia (misty plume bush, ginger bush, iboza)	Southern Africa	Flavonoids	Leaves		MIC (μg/ml)	Morais et al. (2017)
					Trichophyton rubrum	62.5	
					Trichophyton mentagrophytes	125	
					Microsporum gypseum	62.5	
	Lavandula stoechas (lavender)	Iberian Peninsula, Italy	Monoterpenes	Essential oils from aerial parts		MIC (ml/ml)	Zuzarte et al. (2013)
					Candida albicans	2.5	
					Candida tropicalis	2.5	
					Candida krusei	2.5	
					Candida guilliermondii	1.25	
					Candida parapsilosis	2.5	
					Cryptococcus neoformans	0.64	
					Epidermophyton floccosum	0.32	
					Microsporum canis	0.64	
					Microsporum gypseum	0.64	
					Trichophyton mentagrophytes	0.64	
					Trichophyton mentagrophytes	0.64	
					Trichophyton rubrum	0.64	
					Trichophyton verrucosum	0.64	
					Aspergillus fumigatus	1.25	
					Aspergillus flavus	5	
					Aspergillus niger	2.5	

(continued)

Table 3 (continued)

Plant family	Species name (plant common name)	Natural habitat	Main active constituents	Part of plant	Fungal species	Antifungal activity	Reference
	Thymus herba-barona (Caraway thyme)	Mediterranean Region	Phenols Monoterpenoid phenols	Essential oils from aerial parts		MIC (ml/ml)	Zuzarte et al. (2013)
					Candida albicans	0.16	
					Candida tropicalis	0.16	
					Candida krusei	0.16	
					Candida guilliermondii	0.16	
					Candida parapsilosis	0.16	
					Cryptococcus neoformans	0.16	
					Epidermophyton floccosum	0.16	
					Microsporum canis	0.16	
					Microsporum gypseum	0.16	
					Trichophyton mentagrophytes	0.16	
					Trichophyton mentagrophytes	0.16	
					Trichophyton rubrum	0.16	
					Trichophyton verrucosum	0.16	
					Aspergillus fumigatus	0.16	
					Aspergillus flavus	0.32	
					Aspergillus niger	0.32	
Lamiaceae	*Zataria multiflora* (Avishan shirazi)	Southwestern Asia	Thymol Phenolic acids Monoterpenoid phenols	Aerial parts		MIC (μg/ml)	Mahmoudabadi et al. (2007)
					Candida albicans	125,000	
					Candida glabrata	126,000	
					Candida parapsilosis	125,000	
					Candida tropicalis	131,000	
					Candida albicans	76,000	
					Candida glabrata	66,000	
					Candida parapsilosis	64,000	
					Candida tropicalis	76,000	

	Salvia amplexicaulis (salvia)	Southeastern Europe	Phenolic acids Flavonoids (including flavones and flavonols) Polyphenol	Aerial parts, including stems Leaves, Inflorescences		MIC (μg/ml)	Alimpic et al. (2017)
					Candida krusei	64,000	
					Candida albicans	32,000	
					Candida parapsilosis	16,000	
					Aspergillus glaucus	8000	
					Trichophyton mentagrophytes	16,000	
Liliaceae	*Aloe vera*	South-west Arabian Peninsula	Tannins Saponins Flavonols	Leaves		Inhibition zone (mm)	Arunkumar and Muthuselvam (2009)
					Aspergillus flavus	11–15	
					Aspergillus niger	10–8	
	Allium ascalonicum (shallot, onion)	Worldwide	Flavonols Saponins	Bulbs		Inhibition zone (mm)	Mahmoudabadi and Nasery (2009)
					Microsporum gypseum	11	
					Trichophyton mentagrophytes	10.7	
					Trichophyton mentagrophytes	16	
					Epidermophyton floccosum	15.3	
					Epidermophyton floccosum	17.6	
Meliaceae	*Azadirachta indica* (neem)	Indian subcontinent	Flavonols Phytosterols	Leaves Seeds		MIC (μg/ml)	Mahmoud et al. (2011)
					Trichophyton rubrum	31	
					Trichophyton mentagrophytes	31	
					Microsporum nanum	31	
Moringaceae	*Moringa oleifera* (moringa, drumstick tree, horseradish tree)	Far East	Alkanes	Seed		MIC (μg/ml)	Chuang et al. (2007)
					Trichophyton rubrum	2500	
					Trichophyton mentagrophytes	2500	
					Epidermophyton floccosum	2500	
					Microsporum canis	2500	

(continued)

Table 3 (continued)

Plant family	Species name (plant common name)	Natural habitat	Main active constituents	Part of plant	Fungal species	Antifungal activity	Reference
Myrtaceae	*Eucalyptus camaldulensis* (river red gum)	Iran	Saponin Glycosides Tannins Phenols	Leaves		MIC (μg/ml)	Falahati et al. (2005)
					Microsporum canis	1600	
					Microsporum gypseum	1600	
					Trichophyton rubrum	1600	
					Trichophyton schoenleinii	400	
					Trichophyton mentagrophytes	400	
					Epidermophyton floccosum	400	
	Eucalyptus citriodora (lemon-scented gum, blue-spotted gum, lemon eucalyptus)	Australia	Terpenoids	Leaves		MIC (μg/ml)	Tolba et al. (2015)
					Microsporum canis	0.6	
					Microsporum gypseum	5	
					Trichophyton mentagrophytes	1.25	
					Trichophyton rubrum	0.6	
	Eucalyptus microcorys (tallowwood)	Australia	Phenolic acid Rutin	Leaves		MIC (μg/ml)	Bhuyan et al. (2017)
					Geotrichum candidum	4.88	
					Aspergillus brasiliensis	2.44	
					Candida albicans	1250	
Oxalidaceae	*Psidium guajava* (yellow lemon guava guava *L*)	Central America, South America	Carboxylic acid Terpenes Aromatic dicarboxylic acid Polyphenols	Leaves		MIC (μg/ml)	Morais-Braga et al. (2016)
					Trichophyton mentagrophytes	2670	
					Trichophyton rubrum	2670	
					Trichophyton tonsurans	16,000	

Piperaceae	*Piper regnellii* (piper)	Tropical and subtropical regions of the world	Neolignans conocarpan, eupomatenoid-3, eupomatenoid	Leaves		MIC (μg/ml)	Koroishi et al. (2008)
					Trichophyton mentagrophytes	250	
					Microsporum canis	250	
					Trichophyton rubrum	62.5	
					Microsporum gypseum	62.5	
	Dimocarpus longan (longan, lumyai)	Southern Asia	Carboxylic acids	Leaves		MIC (μg/ml)	Rangkadilok et al. (2012)
					Cryptococcus neoformans	500	
	Pothomorphe umbellatum (pariparoba)	Amazonian region	Phytosterols	Aerial parts		MIC (μg/ml)	Rodrigues et al. (2012)
					Trichophyton rubrum	1250	
Polypodiaceae	*Drynaria quercifolia* (Oakleaf fern)	India, Southeast Asia, Australia	Terpenes Phytosterols Flavonols	Leaves		Inhibition zone (mm)	Nejad and Deokule (2009)
					Trichophyton mentagrophytes	25	
Rutaceae	*Aegle marmelos* (Bengal quince, golden apple)	Indian subcontinent	Alkaloids Tannins Phlobotannins Flavans	Essential oil		MIC (μg/ml)	Balakumar et al. (2011)
					Trichophyton mentagrophytes	400	
					Trichophyton rubrum	400	
					Microsporum canis	400	
					Microsporum gypseum	400	
Sapindaceae	*Matayba guianensis* (Camboata de pombo)	South America	Phytosterols	Root bark		MIC (μg/ml)	Assis et al. (2014)
					Candida albicans	1.95	
					Candida parapsilosis	0.97	
					Trichophyton mentagrophytes	15.62	
					Trichophyton rubrum	31.25	
Vitaceae	*Vitis vinifera* (grapevine)	Mediterranean region	Flavan-3-ols	Seeds		MIC (μg/ml)	Simonetti et al. (2014), Simonetti et al. (2017a, b)
					Candida albicans	5.7–20.2	
					Candida species	1–32	
					dermatophytes	20–97	
					Malassezia furfur	32–161	

widely spread in the plant kingdom (Kinghorn 2010). Furthermore, since the natural resources are limited, they may be depleted at a rate faster than they regenerate. Due to the overexploitation, a number of plants producing bioactive metabolites have become endangered species (Chen et al. 2016).

The plant cultivation using conventional agricultural methods represents a possible alternative to the exploitation of wild plant resources. However, this approach is not always feasible or economically viable. Some species are distributed in certain areas and are difficult to cultivate outside of their local ecosystems (Mander e Liu 2010). Several woody plants are slow-growing, and their cultivation could be therefore economically disadvantageous. Finally, field-growing plants may be damaged by unpredictable adverse environmental conditions or pathogen attacks.

In both field-grown and wild-collected plants, the quality and quantity of metabolite production are often fluctuating and heterogeneous, depending on environmental conditions (Gerth et al. 2006). The synthesis of secondary metabolites often occurs in a particular stage of life cycle, or it is confined to specific organs, tissues, or cells.

The main alternatives to direct extraction from plants are the chemical synthesis and the biotechnological production.

Only in the last century, a restricted number of active ingredients were replaced by synthetic or semi-synthetic compounds, whose molecular structure has been often inspired by that of natural compounds. Well-known examples are salicylic acid and its derivative acetylsalicylic acid (Lichterman 2004). Several methods have recently been developed for the total synthesis of resveratrol (Nicolaou et al. 2010; Snyder et al. 2011; Klotter and Studer 2014), a stilbene compound whose antiviral activity has been demonstrated against several human and animal viruses, including influenza A virus (Palamara et al. 2005), Epstein-Barr virus (De Leo et al. 2012), herpes simplex virus (Docherty et al. 1999), respiratory syncytial virus (Zang et al. 2011), varicella zoster virus (Docherty et al. 2006), African swine fever virus (Galindo et al. 2011), and HIV-1 (Clouser et al. 2012). However, it must be said that the chemical synthesis, although of high scientific value, at present is rarely adopted for the large-scale production of bioactive plant metabolites, due to low yields that make it often economically unsuitable. One limiting factor for chemical synthesis of many secondary metabolites is their large size and the presence of multiple chiral centres (Wilson and Roberts 2012).

As regards the biotechnological production of natural antimicrobial compounds, in recent decades, great interest has been given to plant tissue cultures, a collection of techniques in which plant cells, tissues, organs, or whole plantlets are cultivated on synthetic media, in aseptic environment, under controlled physico-chemical conditions. In vitro plant cultures could be used as "bio-factories" for the production of high-value secondary metabolites (Wilson and Roberts 2012).

Different types of in vitro plant cultures can be distinguished based on the degree of differentiation. The lowest degree of differentiation is exhibited by cell cultures. These are initiated from surface sterilized explants (i.e. isolated plant tissues), which are inoculated on jellified media, containing growth regulators, and nutrients (Hall 2000). In appropriate growing conditions, some cells of the explant proliferate, forming disorganized masses of dedifferentiated cells called "calli", which can be grown indefinitely in a periodically renewed jellified culture media or transferred to liquid media to create suspension cultures (Wilson and Roberts 2012).

A higher degree of differentiation is observed in organ cultures. The regeneration of plant organs (organogenesis) can be induced by subjecting the explants or cultured cells to specific hormone combinations. The regeneration of roots (rhizogenesis) or shoots (caulogenesis) can proceed either directly or indirectly. The direct mode involves the development of organs directly from the differentiated tissues of explants in contrast to the indirect mode, where an intervening step of callus formation precedes regeneration (Pulianmackal et al. 2014).

An interesting biotechnological system for the production of high-value secondary metabolites is represented by the hairy root cultures, which are obtained through the infection of plant cells with *Agrobacterium rhizogenes*. Hairy root cultures are characterized by a high grow rate in hormone-free culture media and by a high degree of genetic and metabolic stability. These genetically transformed roots can produce levels of secondary metabolites comparable to that of intact plants (Srivastava and Srivastava 2007).

The main limitation of root cultures, both transformed and untransformed, is that they indefinitely maintain an anatomical primary structure. This represents a problem when the metabolites of interest are biosynthesized predominantly or exclusively in the root in secondary structure. An example is the biosynthesis of essential oils in *Angelica archangelica* L. (Apiaceae), which occurs specifically in secondary secretory ducts formed by vascular cambium activity and located in the secondary phloem (Pasqua et al. 2003). To the best of our knowledge, no strategies have been developed to induce the transition from the primary to the secondary structure in in vitro cultured roots. The highest degree of differentiation is exhibited by the in vitro plantlet cultures, which can be obtained either by rooting of cultured shoots or by in vitro seed germination.

4.1 Production of Antimicrobials Through Plant Cell Cultures

Callus cultures and suspension cultures of several species had been exploited for the biotechnological production of secondary metabolites with antimicrobial activity.

One of the most studied plant species is grapevine (*Vitis vinifera* L., Vitaceae), which produces a broad spectrum of polyphenols with proven antimicrobial activities, including stilbenes and flavan-3-ols (Mulinacci et al. 2008; Santamaria et al. 2011).

Stilbenes are a small family of polyphenols biosynthesized through the phenylpropanoid pathway, found in a number of unrelated plant species, including grapevine, sorghum (*Sorghum bicolor* L. Moench), peanut (*Arachis hypogaea* L.), bilberries (*Vaccinium myrtillus* L.), and several conifer species (*Pinus* spp. and *Picea* spp.) (Chong et al. 2009). It has been demonstrated that stilbenes play a role in plant chemical defence against microorganisms (Jeandet et al. 2002; Ahuja et al. 2012);

thus, it should not be surprising that some of them exhibit remarkable antimicrobial activity against phytopathogenic bacteria and fungi (Morales et al. 2000; Chong et al. 2009). Pinosylvin and its 3-O-methyl ether, which are naturally accumulated by conifers, showed a strong antifungal activity against *Coriolus versicolor* and *Gloeophyllum trabeum*, two wood-destroying fungi (Schultz et al. 1992). Resveratrol and its glucoside piceid exogenously applied to apples inhibited *Venturia inaequalis*, the causal agent of apple scab, reducing spore germination and inhibiting the penetration through cuticular membranes (Schulze et al. 2005). Resveratrol inhibited conidial germination of *Botrytis cinerea*, the grey mould agent on grapes (Adrian et al. 1997), and reduced the germination of sporangia of *Plasmopara viticola*, the downy mildew agent (Pezet et al. 2004). Methylated resveratrol derivatives, such as pterostilbene, showed a much higher antifungal activity than resveratrol (Pezet et al. 2004). In in vitro tests, also the resveratrol oligomers viniferins exhibited a significantly higher antifungal activity than resveratrol (Pezet et al. 2004). It has been hypothesized that the lower antifungal activity of resveratrol is related to its higher hydrophilicity that limits diffusion across biological membranes (Pezet and Pont 1995).

As discussed above, in addition to being active against phytopathogenic microorganisms, stilbenes exhibited interesting potential as active ingredients against animal and human viral pathogens (Abba et al. 2015). Stilbene antimicrobial activity has also been demonstrated against some human pathogenic bacteria (Chan 2002; Taylor et al. 2014), protozoa (Kedzierski et al. 2007), and fungi (Chan 2002). Most of the published studies are focused on resveratrol (Paulo et al. 2001); however, in recent years, resveratrol derivatives, both natural (Sakagami et al. 2007; Basri et al. 2014) and synthetic or semi-synthetic (Chalal et al. 2014), are being studied to evaluate their antimicrobial properties.

Several studies have demonstrated that grapevine cell cultures are able to biosynthesize stilbenes and that stilbene production can be greatly increased by the use of elicitors (Waffo-Teguo et al. 2001; Decendit et al. 2002; Larronde et al. 2005; Tassoni et al. 2005; Belhadj et al. 2008; Ferri et al. 2009; Santamaria et al. 2010, 2011, 2012; Belchí-Navarro et al. 2012). The term "elicitor" refers to physical or chemical factors capable of triggering an array of plant defence responses, including phytoalexin neosynthesis (Namdeo 2007; Naik and Al–Khayri 2016).

A number of chemical elicitors have been tested that enhance stilbene production in grapevine cell cultures. Among these, the most effective proved to be jasonates, signal molecules that mediate defence responses against herbivores and pathogens, in addition to alleviating abiotic stresses, including UV stress, salt stress, osmotic stress, heat stress, cold stress, heavy metal stress, and ozone stress (Dar et al. 2015). Both jasmonic acid and its methyl ester methyl jasmonate proved effective in enhancing stilbene biosynthesis in cell cultures of *V. vinifera* cvs. Red Globe and Michele Palieri (Santamaria et al. 2010) (max total stilbene production 647 and 1220 μmol kg^{-1} FW, respectively, with methyl jasmonate) and cv. *Italia* (Santamaria et al. 2011) (max total stilbenes 1.023 mg g^{-1} DW by with jasmonic acid).

Another elicitor frequently used to increase the productivity of cell cultures is chitosan, a polysaccharide derived from the partial deacetylation of chitin, the main structural component of the fungal cell wall and of the arthropod exoskeleton. On cell cultures of *V. vinifera* cv. *Italia*, chitosan showed a treasurable effect on stilbene biosynthesis (Santamaria et al. 2011); otherwise, it is shown to be an effective elicitor on cell cultures of *V. vinifera* cv. Barbera (Ferri et al. 2011). This is an example of how significantly stress responses can vary between different grapevine cultivars.

Few studies are available regarding the effect of physical elicitors on stilbene biosynthesis in grapevine cell cultures. In a recent paper, it has been compared the effect of methyl jasmonate (chemical elicitor) and low-energy ultrasounds (physical elicitor) on the production of viniferins in cell cultures of *V. vinifera* cv. Alphonse Lavallée (Santamaria et al. 2012). It has been observed that ultrasounds have an effect compared to methyl jasmonate and that the two elicitors, when used in combination, have a synergistic effect in enhancing δ-viniferin biosynthesis (1.43 mg g^{-1} DW). Another study compared the impact of methyl jasmonate, salicylic acid, and ultraviolet C radiation on stilbene production in cell cultures of *V. vinifera* L. cv. Cabernet Sauvignon (Xi et al. 2015). Once again, the best results were obtained by combining chemical elicitors (methyl jasmonate or salicylic acid) with physical elicitor (ultraviolet C radiation) (1.6–2 mg g^{-1} DW).

Another extensively studied species is *Hypericum perforatum* L. (St. John's wort, Hypericaceae), a medicinal plant used since ancient times for its numerous curative properties. Many studies are available regarding the antimicrobial activity of extracts obtained from *H. perforatum* plant (Reichling et al. 2001; Avato et al. 2004; Saddiqe et al. 2010; Naeem et al. 2010; Süntar et al. 2016). Only recently, the research has focused on extracts from *H. perforatum* cell cultures. These extracts are rich in xanthones, a class of non-flavonoid polyphenols whose antimicrobial activity has been reported in numerous studies (Suksamrarn et al. 2003; Pinheiro et al. 2003; Laphookhieo et al. 2006; Ahmad 2016).

Tocci and collaborators (2010) observed that cell suspensions of *H. perforatum* are able to produce different xanthones and that the total xanthone content increased in response to elicitation with chitosan, from 0.13 to about 0.56 mg g^{-1} DW. In control cells, only paxanthone was detected, while in treated cells, the emergence of 1,3,6,7- and 1,3,5,6-tetrahydroxyxanthone, cadensin G, and 1,7-dihydroxyxanthone was observed.

Conceição et al. (2006) investigated the impact of different elicitors on the production of xanthones and flavonoids in St. John's wort cultured cells. In the cells elicited with lyophilized powder of the phytopathogenic fungus *Colletotrichum gloeosporioides*, a significant increase in both the quantity and diversity of xanthones was observed, while flavonoids became undetectable. A further increase in xanthone concentration and diversity was registered in response to priming with methyl jasmonate prior to fungal elicitation. Conversely, the treatment with methyl jasmonate alone caused a decrease in xanthone content and induced the biosynthesis of a new class of flavonoids, the flavones. This finding points to the possibility of selectively increasing the production of different compounds by choosing the most appropriate elicitor/s.

Franklin et al. (2009) observed that after cocultivation of St. John's wort cell cultures with *Agrobacterium tumefaciens*, the flavonoid profile remained unaltered, while xanthone profile significantly changed with a 12-fold increase in total xanthone content and with the neosynthesis of several xanthones not detected in control cells (i.e. 1,3,6,7-tetrahydroxy-8-prenyl xanthone, 1,3,6,7-tetrahydroxy-2-prenylxanthone, 1,3,7-trihydroxy-6-methoxy-8-prenylxanthone, paxanthone). The massive presence of xanthones (over 4 mg g^{-1} DW) was related to the high antibacterial activity of the extracts obtained from cells cocultured with *A. tumefaciens*.

Other examples on the exploitation of plant cell cultures for the production of antimicrobials are reported in Table 4.

4.2 Production of Antimicrobials Through Plant Organ Cultures

Many secondary metabolites are biosynthesized and/or accumulated in specific organs, tissues, and cell types (Valletta et al. 2010); therefore, they should be produced at low levels or not produced at all in in vitro cultures of undifferentiated cells. By means of appropriate hormonal treatments and exploiting the totipotency of plant cells, it is possible to regenerate in vitro plant organs, thus obtaining root or shoot cultures (Murthy et al. 2014).

Among organ cultures, root cultures are the most investigated because of their potential in the biotechnological production of antimicrobials. The plant root resides in an environment where potentially pathogen microorganisms are massively present. On the other end, the root maintains important mutualistic relationships with edaphic microorganisms, such as mycorrhizal fungi and nitrogen-fixing bacteria. For these reasons, many secondary metabolites involved in plant-microorganism relationships, both antagonistic and mutualistic, exhibit root-specific biosynthesis and accumulation (Pasqua et al. 2005).

For example, xanthone accumulation and biosynthesis in *H. perforatum* are root-specific (Tocci et al. 2017). In the root of the plant, these polyphenols are accumulated at relatively low levels, not suitable for large-scale production (Valletta et al. 2016). Recently, it has been demonstrated that *H. perforatum* in vitro regenerated roots are able to constitutively produce xanthones at higher levels (about 4–5 mg g^{-1} DW) than the root of the plant (Tocci et al. 2011, 2012, 2013a, b; Valletta et al. 2016). Among different elicitors, chitosan resulted the most effective, capable of causing a fivefold increase in xanthone production. A similar increase was obtained by treating the root cultures with acetic acid at low concentration (Valletta et al. 2016).

Methanol extracts obtained from *H. perforatum* elicited roots were tested on several human pathogenic fungi, including *Candida* spp. and *C. neoformans*, and dermatophytes (Tocci et al. 2011, 2012, 2013a, b; Zubrická et al. 2015).

Simonetti et al. (2016), in collaboration with the German biotechnology company ROOTec, started experiments on preindustrial scale-up from the laboratory

Table 4 Examples of antimicrobial metabolites of plant origin produced by different types of in vitro cultures

Types of culture	Plant species (family)	Metabolite/s	Genera of fungi	Reference
Cell cultures	*Rauvolfia tetraphylla* L. (Apocynaceae) *Physalis minima* L. (Solanaceae)		*Aspergillus* *Aspergillus* *Fusarium* *Penicillium*	Shariff et al. (2006)
	Origanum acutidens (Hand.-Mazz.) Ietsw. (Lamiaceae)	Essential oils	*Candida* *Aspergillus* *Fusarium* *Microsporum* *Moniliania* *Mortieraula* *Penicillium Rhizopus* *Rhizoctonia* *Trichophyton*	Sökmen et al. (2004)
	Ephedra strobilacea (Ephedraceae) *Ephedra procera* C.A.Mey. (Ephedraceae) *Ephedra pachyclada* Boiss. (Ephedraceae)	Ephedrine Pseudoephedrine Norpseudoephedrine Other alkaloids	*Aspergillus* *Candida*	Parsaeimehr et al. (2010)
Untransformed root cultures	*Hypericum perforatum* L. (Hypericaceae)	Xanthones	*Candida* spp. *Cryptococcus neoformans* Dermatophytes	Tocci et al. (2011, 2012, 2013a, b)
			Malassezia furfur	Simonetti et al. (2016)
Hairy root cultures	*Ocimum basilicum* L. (Lamiaceae)	Rosmarinic acid	*Aspergillus niger*	Bais et al. (2002); Ahmad et al. (2016)
	Lithospermum erythrorhizon Siebold & Zucc. (Boraginaceae)	Shikonin derivatives	*Rhizoctonia, Nectria*	Brigham et al. (1999)
In vitro propagated plantlets	*Stevia rebaudiana* Bertoni (Asteraceae)	Not specified	*Sclerotonia* *Curvularia* *Alternaria* *Aspergillus* *Microsporum Rhizopus*	Debnath (2007)

scale to a larger scale of *H. perforatum* root cultures. In the mist bioreactor used in this study, roots are cultivated in modules included in a plastic bag. Each module is a net on which roots can grow through. At a certain frequency, culture medium is sprayed from the top of the system and collected from the bag to saturate the atmosphere. Thanks to a series of pumps, the culture is collected and sprayed again. The methanol extracts containing xanthones, obtained from the roots cultivated in bioreactor, showed an interesting activity against planktonic cells and biofilm of *M. furfur*. The minimal inhibitory concentration was 16 μg mL^{-1}, while the inhibition percentage of biofilm formation, at a concentration of 16 μg mL^{-1}, ranged from 14% to 39% (Simonetti et al. 2016).

Recently, hairy root cultures of *H. perforatum* had been obtained (Vinterhalter et al. 2006; Bertoli et al. 2008; Koperdáková et al. 2009; Tusevski et al. 2013a), and thorough analyses had been carried out to determine their polyphenolic profile (Tusevski et al. 2013a, b; Tusevski and Simic 2013); however, to the best of our knowledge, no data on the antimicrobial activity of *H. perforatum* hairy root extracts are available.

Rosmarinic acid is an ester of caffeic acid and 3,4-dihydroxyphenyllactic acid, commonly found in species belonging to families Boraginaceae and Lamiaceae (subfamily Nepetoideae) (Petersen and Simmonds 2003). Several studies indicated that rosmarinic acid and its derivatives possess antiviral and antibacterial activities (Petersen and Simmonds 2003; Swarup et al. 2007; Bulgakov et al. 2012; Abedini et al. 2013).

Basil (*Ocimum basilicum* L.) untransformed roots produce rosmarinic acid at low levels (< 0.1% g fresh weight basis) (Bais et al. 2002). Basil hairy root cultures showed three-fold increases in growth and rosmarinic acid production; in addition, in response to elicitation with cell wall extract of the fungus *Phytophthora cinnamomi*, the production was enhanced about 2.7-fold compared with the untreated control roots (Bais et al. 2002).

4.3 Production of Antimicrobials Through Plantlets Propagated In Vitro

At present, the in vitro plant propagation is a technique mainly exploited for the multiplication of species of agronomic interest and as ex situ conservation strategy. Few examples regarding the production of antimicrobials from in vitro cultured plantlets are available in the literature.

Stevia rebaudiana Bertoni (Asteraceae) is an endemic herbaceous plant indigenous to the mountains between Paraguay and Brazil. *S. rebaudiana* is currently used all over the world for the production of low-calorie sweeteners. Areal parts of *S. rebaudiana* contain diterpene glucosides, viz. stevioside and rebaudioside with a sweet taste, which are not metabolizable by the human body. The biggest part of the sweet glycosides consists of the stevioside molecule (Brandle et al. 1998). The sweetener stevioside (Nepovim and Vanek 1998) extracted from the plants is 300 times sweeter than sugar.

Debnath (2007) has developed a procedure for in vitro propagation *Stevia rebaudiana*, starting from nodal segments with axillary buds. Chloroform and methanol extract of leaves collected from in vitro propagated plants had been tested for their antimicrobial activity against several medically important bacteria (*Escherichia coli*, *Bacillus subtilis*, *Streptococcus mutans*, *Staphylococcus aureus*) and fungi (*Sclerotinia minor*, *Curvularia lunata*, *Alternaria alternata*, *A. niger*, *M. gypseum*, *Rhizopus* sp.). The methanolic extract was the most effective against all fungi and bacteria tested, followed by chloroform extract, while aqueous extract proved to be undefective. These results clearly indicate that the solvent, playing a crucial role in the solubilization of antimicrobial molecules, also affects the antimicrobial activity.

5 Conclusions

Currently, opportunistic fungal infections are considered a serious problem regarding public health. Due to the increasing incidence of drug-resistant fungi, the research of new antifungal agents is required. In the present review, the antifungal activity of natural extracts from different plants and plant matrices has been evaluated and compared. Several plant natural products have been tested against fungal human pathogens. Despite good antifungal activity of plant products, only few have been tested in vivo. Many of these extracts are "generally recognized as safe" (GRAS). Due to the complexity of the natural matrices, the use of advanced analytical techniques, such as mass spectrometry, nuclear magnetic resonance, and high-performance liquid chromatography, is necessary nowadays for the early detection and identification of new compounds in crude plant extracts. The search for new antifungal compounds of plant origin could significantly contribute to the development of emerging countries, which are particularly rich in natural resources. The production of phytochemicals with antifungal activity is also possible using biotechnology strategies. The use of this technology could bring a series of practical and ecological advantages for a sustainable production, avoiding the risk of extinction of some plant species.

It is desirable that in the near future, the research will intensify its efforts to discover new plant-based antifungals, focusing mainly on the intertropical flora, which today is the least investigated, although it has a very high biodiversity. It should also be stressed that the hot-humid climate promotes the development of fungi and therefore the plant's defences against fungal infections, including antifungal molecules. Great efforts must also be made to identify the main biologically active compounds contained in plant extracts, in order to obtain standardized products. An interesting field of study will concern the combined effects of different plant extracts, due to synergistic, additive, and antagonistic interactions between different bioactive molecules.

Finally, it is desirable that in the near future, the regulation on the use of bioactive plant extracts will be uniformed in order to promote their use and diffusion.

References

Abba Y, Hassim H, Hamzah H, Noordin MM (2015) Antiviral activity of resveratrol against human and animal viruses. Adv Virol 65(6):297–303

Abedini A, Roumy V, Mahieux S et al (2013) Rosmarinic acid and its methyl ester as antimicrobial components of the hydromethanolic extract of *Hyptis atrorubens* Poit. (Lamiaceae). J Evid Based Complement Alternat Med 2013:1–11

Adrian M, Jeandet P, Veneau J et al (1997) Biological activity of resveratrol, a stilbenic compound from grapevines, against *Botrytis cinerea*, the causal agent for gray mold. J Chem Ecol 23(7):1689–1702

Ahmad I (2016) Recent insight into the biological activities of synthetic xanthone derivatives. Eur J Med Chem 116:267–280

Ahmad K, Talha Khalil A, Somayya R (2016) Antifungal, phytotoxic and hemagglutination activity of methanolic extracts of *Ocimum basilicum*. J Tradit Chin Med 36(6):794–798

Ahuja I, Kissen R, Bones AM (2012) Phytoalexins in defense against pathogens. Trends Plant Sci 17:73–90

Akroum S (2017) Antifungal activity of acetone extracts from *Punica granatum* L., *Quercus suber* L. and *Vicia faba* L. J Mycol Med 27(1):83–89

Alimpić A, Knežević A, Milutinović M et al (2017) Biological activities and chemical composition of Salvia amplexicaulis Lam. extracts. Ind Crop Prod 105:1–9

Arendrup MC, Patterson TF (2017) Multidrug-resistant Candida: epidemiology, molecular mechanisms, and treatment. J Infect Dis 216:S445–S451

Arunkumar S, Muthuselvam M (2009) Analysis of phytochemical constituents and antimicrobial activities of *Aloe vera* L. against clinical pathogens. World J Agr Sci 5(5):572–576

Atanasov AG, Waltenberger B, Pferschy-Wenzig EM et al (2015) Discovery and resupply of pharmacologically active plant-derived natural products: a review. Biotechnol Adv 33(8):1582–1614

Avato P, Raffo F, Guglielmi G et al (2004) Extracts from St John's wort and their antimicrobial activity. Phytother Res 18(3):230–232

Bais HP, Walker TS, Schweizer HP et al (2002) Root specific elicitation and antimicrobial activity of rosmarinic acid in hairy root cultures of *Ocimum basilicum*. Plant Physiol Biochem 40(11):983–995

Balakumar S, Rajan S, Thirunalasundari T et al (2011) Antifungal activity of *Aegle marmelos* (L.) Correa (Rutaceae) leaf extract on dermatophytes. Asian Pac J Trop Biomed 1(4):309–312

Balouiri M, Sadiki M, Ibnsouda SK (2016) Methods for in vitro evaluating antimicrobial activity: a review. J Pharm Anal 6:71–79

Bansod S, Rai M (2008) Antifungal activity of essential oils from Indian medicinal plants against human pathogenic *Aspergillus fumigatus and A. niger*. World J Med Sci 3(2):81–88

Basri DF, Xian LW, Abdul Shukor NI et al (2014) Bacteriostatic antimicrobial combination: antagonistic interaction between epsilon-viniferin and vancomycin against methicillin-resistant *Staphylococcus aureus*. Biomed Res Int 2014:1–8

Bassiri-Jahromi S, Pourshafie MR, Ardakani EM et al (2017) *In vivo* comparative evaluation of the pomegranate (*Punica granatum)* peel extract as an alternative agent to nystatin against oral candidiasis. Iran J Med Sci 43(3):296–304

Belchí-Navarro S, Almagro L, Lijavetzky D et al (2012) Enhanced extracellular production of *trans*-resveratrol in *Vitis vinifera* suspension cultured cells by using cyclodextrins and methyljasmonate. Plant Cell Rep 31(1):81–89

Belhadj A, Telef N, Saigne C et al (2008) Effect of methyl jasmonate in combination with carbohydrates on gene expression of PR proteins, stilbene and anthocyanin accumulation in grapevine cell cultures. Plant Physiol Biochem 46(4):493–499

Bertoli A, Giovannini A, Ruffoni B et al (2008) Bioactive constituent production in St. John's wort *in vitro* hairy roots. Regenerated plant lines. J Agric Food Chem 56(13):5078–5082

Bhuyan DJ, Vuong QV, Chalmers AC et al (2017) Phytochemical, antibacterial and antifungal properties of an aqueous extract of *Eucalyptus microcorys* leaves. S Afr J Bot 112:180–185

Bragutsa EV (2007) Treatment of chronic resistant forms of fungal skin lesions (malasshesiosis and mycosis of the foot) using combined therapy with the Terbizil preparation. Ukr. J Dermatol, Venereal, Brach Metol 1:57–59

Brandle JE, Starratt AN, Gijzen M (1998) *Stevia rebaudiana*: its agricultural, biological, and chemical properties. Can J Plant Sci 78(4):527–536

Brigham LA, Michaels PJ, Flores HE (1999) Cell-specific production and antimicrobial activity of naphthoquinones in roots of *Lithospermum erythrorhizon*. Plant Physiol 119(2):417–428

Brighenti FL, Salvador MJ, Gontijo AVL et al (2017) Plant extracts: initial screening, identification of bioactive compounds and effect against *Candida albicans* biofilms. Future Microbiol 12:15–27

Brooks GF et al (2013) Jawetz Melnick & Adelbergs medical microbiology, 26th edn. McGraw-Hill, New York

Buck M, Hamilton C (2011) The Nagoya Protocol on access to genetic resources and the fair and equitable sharing of benefits arising from their utilization to the Convention on Biological Diversity. Rev Eur Community Int Environ Law 20(1):47–61

Bulgakov VP, Inyushkina YV, Fedoreyev SA (2012) Rosmarinic acid and its derivatives: biotechnology and applications. Crit Rev Biotechnol 32(3):203–217

Cabañes FJ, Vega S, Castellá G (2011) *Malassezia cuniculi* sp. nov., a novel yeast species isolated from rabbit skin. Mala Med Mycol 49(1):40–48

Canter PH, Thomas H, Ernst E (2005) Bringing medicinal plants into cultivation: opportunities and challenges for biotechnology. Trends Biotechnol 23(4):180–185

Carrasco-Zuber JE, Navarrete-Dechent C, Bonifaz A et al (2016) Cutaneous involvement in the deep mycoses: a literature review. Part I—subcutaneous mycoses. Actas Dermosifiliogr 107(10):806–815

Chalal M, Klinguer A, Echairi A et al (2014) Antimicrobial activity of resveratrol analogues. Molecules 19(6):7679–7688

Chan MMY (2002) Antimicrobial effect of resveratrol on dermatophytes and bacterial pathogens of the skin. Biochem Pharmacol 63(2):99–104

Chattopadhyay D, Maiti K, Kundu AP et al (2001) Antimicrobial activity of Alstonia macrophylla: a folklore of bay islands. J Ethnopharmacol 77(1):49–55

Chen SL, Yu H, Luo HM et al (2016) Conservation and sustainable use of medicinal plants: problems, progress, and prospects. Chin Med 11(1):37

Chong J, Poutaraud A, Hugueney P (2009) Metabolism and roles of stilbenes in plants. Plant Sci 177(3):143–155

Chuang PH, Lee CW, Chou JY et al (2007) Anti-fungal activity of crude extracts and essential oil of *Moringa oleifera* Lam. Bioresour Technol 98(1):232–236

Clouser CL, Chauhan J, Bess MA et al (2012) Anti-HIV-1 activity of resveratrol derivatives and synergistic inhibition of HIV-1 by the combination of resveratrol and decitabine. Bioorg Med Chem Lett 22(21):6642–6646

Conceição LFR, Ferreres Tavares RM, Dias ACP (2006) Induction of phenolic compounds in *Hypericum perforatum* L. cells by *Colletotrichum gloeosporioides* elicitation. Phytochemistry 67:149–155

Dar TA, Uddin M, Khan MMA et al (2015) Jasmonates counter plant stress: a review. Environ Exp Bot 115:49–57

De Assis PA, Theodoro PN, De Paula JE et al (2014) Antifungal ether diglycosides from *Matayba guianensis* Aublet. Bioorg Med Chem Lett 24(5):1414–1416

De Leo A, Arena G, Lacanna E et al (2012) Resveratrol inhibits epstein barr virus lytic cycle in Burkitt's lymphoma cells by affecting multiple molecular targets. Antivir Res 96(2):196–202

De Morais CB, Scopel M, Pedrazza GPR et al (2017) Anti-dermatophyte activity of Leguminosae plants from Southern Brazil with emphasis on *Mimosa pigra* (Leguminosae). J Mycol Med 27(4):530–538

Debnath M (2007) Clonal propagation and antimicrobial activity of an endemic medicinal plant *Stevia rebaudiana*. J Med Plants Res 2(2):045–051

Decendit A, Waffo-Teguo P, Richard T et al (2002) Galloylated catechins and stilbene diglucosides in *Vitis vinifera* cell suspension cultures. Phytochemistry 60(8):795–798

Denning DW, Bromley MJ (2015) How to bolster the antifungal pipeline. Science 347(6229): 1414–1416
Denning DW, Venkateswarlu K, Oakley KL et al (1997) Itraconazole resistance in *Aspergillus fumigatus*. Antimicrob Agents Chemother 41(6):1364–1368
Dias MI, Sousa MJ, Alves RC et al (2016) Exploring plant tissue culture to improve the production of phenolic compounds: A review. Ind Crop Prod 82:9–22
DiDone L, Oga D, Krysan DJ (2011) A novel assay of biofilm antifungal activity reveals that amphotericin B and caspofungin lyse *Candida albicans* cells in biofilms. Yeast 28(8):561–568
Docherty JJ, Fu MMH, Stiffler BS et al (1999) Resveratrol inhibition of herpes simplex virus replication. Antivir Res 43(3):145–155
Docherty JJ, Sweet TJ, Bailey E et al (2006) Resveratrol inhibition of varicella-zoster virus replication *in vitro*. Antivir Res 72(3):171–177
Donlan RM (2001) Biofilms and device-associated infections. Emerg Infect Dis 7(2):277
Dovigo LN, Pavarina AC, Carmello JC et al (2011) Susceptibility of clinical isolates of Candida to photodynamic effects of curcumin. Lasers Surg Med 43(9):927–934
El-Atawi K, Elhalik M, Kulkarni T et al (2017) Evolving invasive neonatal systemic candidiasis, a review. J Pediatr Neonatal Care 6(6):00271
Endo EH, Costa GM, Nakamura TU et al (2015) Antidermatophytic activity of hydroalcoholic extracts from Rosmarinus officinalis and Tetradenia riparia. J Mycol Med 25(4):274–279
Evensen NA, Braun PC (2009) The effects of tea polyphenols on *Candida albicans*: inhibition of biofilm formation and proteasome inactivation. Can J Microbiol 55(9):1033–1039
Falahati M, Tabrizib NO, Jahaniani F (2005) Anti dermatophyte activities of Eucalyptus camaldulensis in comparison with Griseofulvin. Iran J Pharmacol Ther 4(2):80–83
Ferri M, Tassoni A, Franceschetti M et al (2009) Chitosan treatment induces changes of protein expression profile and stilbene distribution in *Vitis vinifera* cell suspensions. Proteomics 9(3):610–624
Ferri M, Dipalo SC, Bagni N et al (2011) Chitosan elicits mono-glucosylated stilbene production and release in fed-batch bioreactor cultures of grape cells. Food Chem 124(4):1473–1479
Fernanda Lourenção Brighenti, Marcos José Salvador, Aline Vidal Lacerda Gontijo, Alberto Carlos Botazzo Delbem, Ádina Cléia Botazzo Delbem, Cristina Pacheco Soares, Maria Alcionéia Carvalho de Oliveira, Camila Miorelli Girondi, Cristiane Yumi Koga-Ito, (2017) Plant extracts: initial screening, identification of bioactive compounds and effect against biofilms. Future Microbiology 12(1):15–27
Filip R, Davicino R, Anesini C (2010) Antifungal activity of the aqueous extract of Ilex paraguariensis against Malassezia furfur. Phytother Res 24(5):715–719
Franklin G, Conceição LF, Kombrink E et al (2009) Xanthone biosynthesis in Hypericum perforatum cells provides antioxidant and antimicrobial protection upon biotic stress. Phytochemistry 70(1):60–68
Galindo I, Hernáez B, Berná J et al (2011) Comparative inhibitory activity of the stilbenes resveratrol and oxyresveratrol on African swine fever virus replication. Antivir Res 91(1):57–63
Gerth A, Schmidt D, Wilken D (2006) The production of plant secondary metabolites using bioreactors. In: XXVII international horticultural congress-IHC2006: international symposium on plant biotechnology: From Bench to 764, pp 95–104
Grohskopf LA, Vincent JL (1996) Systemic Candida infections. Yale J Biol Med 69(6):505–515
Gullo FP, Rossi SA, de CO Sardi J et al (2013) Cryptococcosis: epidemiology, fungal resistance, and new alternatives for treatment. Eur J Clin Microbiol Infect Diss 32(11):1377–1391
Güllüce M, Sökmen M, Daferera D et al (2003) *In vitro* antibacterial, antifungal, and antioxidant activities of the essential oil and methanol extracts of herbal parts and callus cultures of *Satureja hortensis* L. J Agric Food Chem 51(14):3958–3965
Gurgel LA, Sidrim JJC, Martins DT (2005) *In vitro* antifungal activity of dragon's blood from *Croton urucurana* against dermatophytes. J Ethnopharmacol 97(2):409–412
Hall RD (2000) Plant cell culture initiation. Mol Biotechnol 16(2):161–173
Havlickova B, Czaika VA, Friedrich M (2008) Epidemiological trends in skin mycoses worldwide. Mycoses 51(s4):2–15

Hedayati MT, Pasqualotto AC, Warn PA (2007) *Aspergillus flavus*: human pathogen, allergen and mycotoxin producer. Microbiology 153(6):1677–1692

Hope WW, Castagnola E, Groll AH et al (2012) ESCMID guideline for the diagnosis and management of *Candida* diseases 2012: prevention and management of invasive infections in neonates and children caused by *Candida* spp. Clin Microbiol Infect 18(s7):38–52

Jeandet P, Douillet-Breuil AC, Bessis R et al (2002) Phytoalexins from the Vitaceae: biosynthesis, phytoalexin gene expression in transgenic plants, antifungal activity, and metabolism. J Agric Food Chem 50(10):2731–2741

Kalidindi N, Thimmaiah NV, Jagadeesh NV et al (2015) Antifungal and antioxidant activities of organic and aqueous extracts of *Annona squamosa* Linn. leaves. J Food Drug Anal 23(4):795–802

Kauffmann CA, Mandell GL (2010) Atlas of fungal diseases. GEOTAR Media, Moscow, p 2010

Kedzierski L, Curtis JM, Kaminska M et al (2007) *In vitro* antileishmanial activity of resveratrol and its hydroxylated analogues against *Leishmania major* promastigotes and amastigotes. J Parasitol Res 102(1):91–97

Khan N, Shreaz S, Bhatia R et al (2012) Anticandidal activity of curcumin and methyl cinnamaldehyde. Fitoterapia 83(3):434–440

Kim JY (2016) Human fungal pathogens: Why should we learn? J Microbiol 54(2016):145–148

Kinghorn K (2010) Toxic plants. Columbia University Press, New York

Klotter F, Studer A (2014) Total synthesis of resveratrol-based natural products using a palladium-catalyzed decarboxylative arylation and an oxidative heck reaction. Angew Chem Int Ed Engl 53(9):2473–2476

Koch C, Uhle F, Wolff M et al (2015) Cardiac effects of echinocandins after central venous administration in adult rats. Antimicrob Agents Chemother 59(3):1612–1619

Koperdáková J, Komarovská H, Košuth J et al (2009) Characterization of hairy root-phenotype in transgenic *Hypericum perforatum* L. clones. Acta Physiol Plant 31(2):351–358

Koroishi AM, Foss SR, Cortez DA et al (2008) *In vitro* antifungal activity of extracts and neolignans from Piper regnellii against dermatophytes. J Ethnopharmacol 117(2):270–277

Kulko AB (2012) Pathogens spectrum of deep human mycosis. Oncohematology 7(3):55–61

Kullberg BJ, Arendrup MC (2015) Invasive *candidiasis*. N Engl J Med 373(15):1445–1456

Laniado-Laborín R, Cabrales-Vargas MN (2009) Amphotericin B: side effects and toxicity. Rev Iberoam Micol 26(4):223–227

Laphookhieo S, Syers JK, Kiattansakul R et al (2006) Cytotoxic and antimalarial prenylated xanthones from *Cratoxylum cochinchinense*. Chem Pharm Bull 54(5):745–747

Larronde F, Richard T, Delaunay JC et al (2005) New stilbenoid glucosides isolated from *Vitis vinifera* cell suspension cultures (cv. Cabernet Sauvignon). Planta Med 71(09):888–890

Lichterman BL (2004) Book: aspirin: the story of a wonder drug. Br Med J 329(7479):1408

Liu M, Katerere DR, Gray AI et al (2009) Phytochemical and antifungal studies on *Terminalia mollis* and *Terminalia brachystemma*. Fitoterapia 80(6):369–373

Lubbe A, Verpoorte R (2011) Cultivation of medicinal and aromatic plants for specialty industrial materials. Ind Crop Prod 34(1):785–801

Mahboubi M, HeidaryTabar R, Mahdizadeh E (2017) The anti-dermatophyte activity of *Zataria multiflora* essential oils. J Mycol Med 27(2):232–237

Mahmoud DA, Hassanein NM, Youssef KA et al (2011) Antifungal activity of different neem leaf extracts and the nimonol against some important human pathogens. Braz J Microbiol 42(3):1007–1016

Mahmoudabadi AZ, Nasery MG (2009) Anti fungal activity of shallot, Allium ascalonicum Linn. (Liliaceae) *in vitro*. J Med Plants Res 3(5):450–453

Mahmoudabadi AZ, Dabbagh MA, Fouladi Z (2007) *In vitro* anti-*Candida* activity of Zataria multiflora Boiss. J Evid Based Complement Altern Med 4(3):351–353

Mander L, Liu HW (2010) Comprehensive natural products II: chemistry and biology, vol 1. Elsevier

Martins N, Barros L, Henriques M et al (2015a) Activity of phenolic compounds from plant origin against *Candida* species. Ind Crop Prod 74:648–670

Martins N, Ferreira IC, Barros L et al (2015b) Plants used in folk medicine: The potential of their hydromethanolic extracts against *Candida* species. Ind Crop Prod 66:62–67

Morais-Braga MFB, Carneiro JNP, Machado AJT et al (2016) *Psidium guajava* L., from ethnobiology to scientific evaluation: Elucidating bioactivity against pathogenic microorganisms. J Ethnopharmacol 24(194):1140–1152

Morales M, Ros B, Pedreno MA (2000) Plant stilbenes: recent advances in their chemistry and biology. Advances in Plant Physiology 3:39–70

More NV, Kharat AS (2016) Antifungal and Anticancer Potential of *Argemone mexicana* L. Medicines 3(4):28

Mulinacci N, Santamaria AR, Giaccherini C et al (2008) Anthocyanins and flavan-3-ols from grapes and wines of *Vitis vinifera* cv. Cesanese d'Affile. Nat Prod Res 22(12):1033–1039

Murthy HN, Lee EJ, Paek KY (2014) Production of secondary metabolites from cell and organ cultures: strategies and approaches for biomass improvement and metabolite accumulation. Plant Cell Tissue Organ Cult 118(1):1–16

Naeem I, Saddiqe Z, Patel A et al (2010) Analysis of flavonoid and antimicrobial activity of extracts of *Hypericum perforatum*. Asian J Chem 22(5):3596

Naik PM, Al-Khayri JM (2016) Abiotic and biotic elicitors–role in secondary metabolites production through *in vitro* culture of medicinal plants. In: Shanker AK, Shanker C (eds) *Abiotic and biotic stress in plants—recent advances and future perspectives*. InTech, Rijeka, pp 247–277. http://doi.org/10.5772/61442.

Namdeo AG (2007) Plant cell elicitation for production of secondary metabolites: a review. Pharmacog Rev 1(1):69–79

Neelofar K, Shreaz S, Rimple B et al (2011) Curcumin as a promising anticandidal of clinical interest. Can J Microbiol 57(3):204–210

Nejad BS, Deokule SS (2009) Anti-dermatophytic activity of *Drynaria quercifolia* (L.) J. Smith. Jundishapur J Microbiol 2(1):25

Nepovím A, Vaněk T (1998) *In vitro* propagation of *Stevia rebaudina* plants using multiple shoot culture. Planta Med 64(08):775–776

Nicolaou KC, Kang Q, Wu T et al (2010) Total synthesis and biological evaluation of the resveratrol-derived polyphenol natural products hopeanol and hopeahainol. J Am Chem Soc 132(21):7540–7548

Onlom C, Khanthawong S, Waranuch N (2014) *In vitro* anti-*Malassezia* activity and potential use in anti-dandruff formulation of *Asparagus racemosus*. Int J Cosmet Sci 6(1):74–78

Palamara AT, Nencioni L, Aquilano K et al (2005) Inhibition of influenza A virus replication by resveratrol. J Infect Dis 191(10):1719–1729

Pappas PG, Kauffman CA, Andes DR (2015) Clinical practice guideline for the management of candidiasis: 2016 update by the Infectious Diseases Society of America. Clin Infect Dis 62(4):e1–e50

Parsaeimehr A, Sargsyan E, Javidnia K (2010) A comparative study of the antibacterial, antifungal and antioxidant activity and total content of phenolic compounds of cell cultures and wild plants of three endemic species of *Ephedra*. Molecules 15(3):1668–1678

Pasqua G, Monacelli B, Silvestrini A (2003) Accumulation of essential oils in relation to root differentiation in *Angelica archangelica* L. Eur J Histochem 47(1):87

Pasqua G, Monacelli B, Valletta A et al (2005) Synthesis and/or accumulation of bioactive molecules in the *in vivo* and *in vitro* root. Plant Biosyst 139(2):180–188

Paulo L, Oleastro M, Gallardo E et al (2001) Antimicrobial properties of resveratrol: a review. In: Science against microbial pathogens: communicating current research and technological advances 2:1225–1235

Petersen M, Simmonds MS (2003) Rosmarinic acid. Phytochemistry 62(2):121–125

Pezet R, Pont V (1995) Mode of toxic action of *Vitaceae* stilbenes on fungal cells. In: Daniel M, Purkayastha RE (eds) *Handbook of phytoalexin metabolism and action*. Dekker, New York, pp 317–331

Pezet R, Gindro K, Viret O et al (2004) Effects of resveratrol, viniferins and pterostilbene on *Plasmopara viticola* zoospore mobility and disease development. Vitis-Geilweilerhof 43(3): 145–148

Pfaller MA (2012) Antifungal drug resistance: mechanisms, epidemiology, and consequences for treatment. Am J Med 125(1):S3–S13

Pinheiro L, Nakamura CV, Dias Filho BP et al (2003) Antibacterial xanthones from *Kielmeyera variabilis* mart. (Clusiaceae). Mem Inst Oswaldo Cruz 98(4):549–552

Piraccini BM, Gianni C (2013) Update on the management of onychomycosis. G Ital Dermatol Venereol 148(6):633–638

Policegoudra RS, Chattopadhyay P, Aradhya SM et al (2014) Inhibitory effect of Tridax procumbens against human skin pathogens. J Herb Med 4(2):83–88

Pulianmackal AJ, Kareem AV, Durgaprasad K et al (2014) Competence and regulatory interactions during regeneration in plants. Front Plant Sci 5:142

Ranganathan S, Balajee SAM (2000) Anti-Cryptococcus activity of combination of extracts of Cassia alata and *Ocimum sanctum*. Mycoses 43(7–8):299–301

Rangkadilok N, Tongchusak S, Boonhok R et al (2012) *In vitro* antifungal activities of longan (Dimocarpus longan Lour.) seed extract. Fitoterapia 83(3):545–553

Rao SR, Ravishankar GA (2002) Plant cell cultures: chemical factories of secondary metabolites. Biotechnol Adv 20(2):101–153

Raut JS, Karuppayil SM (2014) A status review on the medicinal properties of essential oils. Ind Crop Prod 62:250–264

Reichling J, Weseler A, Saller R (2001) A current review of the antimicrobial activity of *Hypericum perforatum* L. Pharmacopsychiatry 34(1):116–118

Rodrigues ER, Nogueira NGP, Zocolo GJ et al (2012) *Pothomorphe umbellata*: antifungal activity against strains of Trichophyton rubrum. J Mycol Med 22(3):265–269

Saddiqe Z, Naeem I, Maimoona A (2010) A review of the antibacterial activity of *Hypericum perforatum* L. J Ethnopharmacol 131(3):511–521

Sakagami Y, Sawabe A, Komemushi S et al (2007) Antibacterial activity of stilbene oligomers against vancomycin-resistant Enterococci (VRE) and methicillin-resistant *Staphylococcus aureus* (MRSA) and their synergism with antibiotics. Biocontrol Sci 12(1):7–14

Santamaria AR, Antonacci D, Caruso G et al (2010) Stilbene production in cell cultures of *Vitis vinifera* L. cvs Red Globe and Michele Palieri elicited by methyl jasmonate. Natl Prod Res 24(15):1488–1498

Santamaria AR, Mulinacci N, Valletta A et al (2011) Effects of elicitors on the production of resveratrol and viniferins in cell cultures of *Vitis vinifera* L. cv Italia. J Agric Food Chem 59(17):9094–9101

Santamaria AR, Innocenti M, Mulinacci N et al (2012) Enhancement of viniferin production in *Vitis vinifera* L. cv. Alphonse Lavallée Cell suspensions by low-energy ultrasound alone and in combination with methyl jasmonate. J Agric Food Chem 60(44):11135–11142

Schaller M, Friedrich M, Papini M et al (2016) Topical antifungal-corticosteroid combination therapy for the treatment of superficial mycoses: conclusions of an expert panel meeting. Mycoses 59(6):365–373

Schultz TP, Boldin WD, Fisher TH et al (1992) Structure-fungicidal properties of some 3-and 4-hydroxylated stilbenes and bibenzyl analogues. Phytochemistry 31(11):3801–3806

Schulze K, Schreiber L, Szankowski I (2005) Inhibiting effects of resveratrol and its glucoside piceid against *Venturia inaequalis*, the causal agent of apple scab. J Agric Food Chem 53(2):356–362

Shams-Ghahfarokhi M, Shokoohamiri MR, Amirrajab N et al (2006) *In vitro* antifungal activities of *Allium cepa*, *Allium sativum* and ketoconazole against some pathogenic yeasts and dermatophytes. Fitoterapia 77(4):321–323

Shaokat SS, Hameed HA, Mohammad HJ (2017) Anti-fungal activity *of Punica granatum* I. peels powder and extracts from pathogenic samples. Indian J Pharm Sci 16(2):12–20

Shariff N, Sudarshana MS, Umesha S et al (2006) Antimicrobial activity of *Rauvolfia tetraphylla* and *Physalis minima* leaf and callus extracts. Afr J Biotechnol 5(10):946–950

Shin S, Kang CA (2003) Antifungal activity of the essential oil of *Agastache rugosa* Kuntze and its synergism with ketoconazole. Lett Appl Microbiol 36(2):111–115

Simonetti G, Santamaria AR, D'Auria FD et al (2014) Evaluation of anti-Candida activity of *Vitis vinifera* L. seed extracts obtained from wine and table cultivars. Biomed Res Int 2014:1–11

Simonetti G, Tocci N, Valletta A et al (2016) *In vitro* antifungal activity of extracts obtained from *Hypericum perforatum* adventitious roots cultured in a mist bioreactor against planktonic cells and biofilm of *Malassezia furfur*. Nat Prod Res 30(5):544–550. http://doi.org/10.1080/14786419.2017.1410811

Simonetti G, D'Auria FD, Mulinacci N et al (2017a) Phenolic content and in vitro antifungal activity of unripe grape extracts from agro-industrial wastes. Nat Prod Res:1–5. http://doi.org/10.1080/14786419.2017.1410811

Simonetti G, D'auria FD, Mulinacci N (2017b) Anti-dermatophyte and anti-*Malassezia* activity of extracts rich in polymeric flavan-3-ols obtained from *Vitis vinifera* seeds. Phytother Res 31(1):124–131

Snyder SA, Gollner A, Chiriac MI (2011) Regioselective reactions for programmable resveratrol oligomer synthesis. Nature 474(7352):461–466

Sobrinho ACN, de Souza EB, Rocha MFG et al (2016) Chemical composition, antioxidant, antifungal and hemolytic activities of essential oil from Baccharis trinervis (Lam.) Pers. (Asteraceae). Ind Crop Prod 84:108–115

Sökmen M, Serkedjieva J, Daferera D et al (2004) *In vitro* antioxidant, antimicrobial, and antiviral activities of the essential oil and various extracts from herbal parts and callus cultures of *Origanum acutidens*. J Agric Food Chem 52(11):3309–3312

Soliman S, Alnajdy D, El-Keblawy AA et al (2017) Plants' natural products as alternative promising anti-Candida drugs. Pharmacogn Rev 11(22):104

Srivastava S, Srivastava AK (2007) Hairy root culture for mass-production of high-value secondary metabolites. Crit Rev Biotechnol 27(1):29–43

Suksamrarn S, Suwannapoch N, Phakhodee W et al (2003) Antimycobacterial activity of prenylated xanthones from the fruits of *Garcinia mangostana*. Chem Pharm Bull 51(7):857–859

Süntar I, Oyardı O, Akkol EK (2016) Antimicrobial effect of the extracts from *Hypericum perforatum* against oral bacteria and biofilm formation. Pharm Biol 54(6):1065–1070

Swarup V, Ghosh J, Ghosh S et al (2007) Antiviral and anti-inflammatory effects of rosmarinic acid in an experimental murine model of Japanese encephalitis. Antimicrob Agents Chemother 51(9):3367–3370

Tassoni A, Fornalè S, Franceschetti M et al (2005) Jasmonates and Na-orthovanadate promote resveratrol production in *Vitis vinifera* cv. Barbera cell cultures. New Phytol 166(3):895–905

Taylor EJ, Yu Y, Champer J et al (2014) Resveratrol demonstrates antimicrobial effects against *Propionibacterium acnes in vitro*. Dermatol Ther 4(2):249–257

Teodoro GR, Ellepola K, Seneviratne CJ et al (2015) Potential use of phenolic acids as anti-*Candida* agents: a review. Front Microbiol 6(1420):2015

Tocci N, Ferrari F, Santamaria AR et al (2010) Chitosan enhances xanthone production in *Hypericum perforatum* subsp. *angustifolium* cell cultures. Nat Prod Res 24(3):286–293

Tocci N, Simonetti G, D'Auria FD et al (2011) Root cultures of *Hypericum perforatum* subsp. *angustifolium* elicited with chitosan and production of xanthone-rich extracts with antifungal activity. Appl Microbiol Biotechnol 91(4):977–987

Tocci N, D'Auria FD, Simonetti G et al (2012) A three-step culture system to increase the xanthone production and antifungal activity of *Hypericum perforatum* subsp. *angustifolium in vitro* roots. Plant Physiol Biochem 57:54–58

Tocci N, D'Auria FD, Simonetti G et al (2013a) Bioassay-guided fractionation of extracts from *Hypericum perforatum in vitro* roots treated with carboxymethylchitosans and determination of antifungal activity against human fungal pathogens. Plant Physiol Biochem 70:342–347

Tocci N, Simonetti G, D'Auria FD et al (2013b) Chemical composition and antifungal activity of *Hypericum perforatum* subsp. *angustifolium* roots from wild plants and plants grown under controlled conditions. Plant Biosyst 147(3):557–562

Tocci N, Gaid M, Kaftan F et al (2017) Exodermis and endodermis are the sites of xanthone biosynthesis in *Hypericum perforatum* roots. New Phytol 217(3):1099–1112

Tolba H, Moghrani H, Benelmouffok A et al (2015) Essential oil of Algerian Eucalyptus citriodora: Chemical composition, antifungal activity. J Mycol Med 25(4):e128–e133

Tusevski O, Simic SG (2013) Phenolic acids and flavonoids in *Hypericum perforatum* L. hairy roots. Int J Pharm Bio Sci 4(3):737–748

Tusevski O, Stanoeva J, Stefova M et al (2013a) Hairy roots of *Hypericum perforatum* L.: a promising system for xanthone production. Open Life Sci 8(10):1010–1022

Tusevski O, Petreska Stanoeva J, Stefova M et al (2013b) Phenolic profile of dark-grown and photoperiod-exposed *Hypericum perforatum* L. hairy root cultures. Sci World J 2013:1

Valente J, Zuzarte M, Gonçalves MJ et al (2013) Antifungal, antioxidant and anti-inflammatory activities of *Oenanthe crocata* L. essential oil. Food Chem Toxicol 62:349–354

Valletta A, Trainotti L, Santamaria AR et al (2010) Cell-specific expression of tryptophan decarboxylase and 10-hydroxygeraniol oxidoreductase, key genes involved in camptothecin biosynthesis in *Camptotheca acuminata* Decne (Nyssaceae). BMC Plant Biol 10(1):69

Valletta A, De Angelis G, Badiali C et al (2016) Acetic acid acts as an elicitor exerting a chitosan-like effect on xanthone biosynthesis in *Hypericum perforatum* L. root cultures. Plant Cell Rep 35(5):1009–1020

Vasconcelos LCS, Sampaio MCC, Sampaio FC et al (2003) Use of *Punica granatum* as an antifungal agent against candidosis associated with denture stomatitis. Mycoses 46:192–196

Velegraki A, Cafarchia C, Gaitanis G (2015) Malassezia infections in humans and animals: pathophysiology, detection, and treatment. PLoS Pathog 11(1):e1004523

Vermes A, Guchelaar HJ, Dankert J (2000) Flucytosine: a review of its pharmacology, clinical indications, pharmacokinetics, toxicity and drug interactions. J Antimicrob Chemother 46(2):171–179

Verpoorte R, Contin A, Memelink J (2002) Biotechnology for the production of plant secondary metabolites. Phytochem Rev 1(1):13–25

Verweij PE, Snelders E, Kema GH et al (2009) Azole resistance in Aspergillus fumigatus: a side-effect of environmental fungicide use? Lancet Infect Dis 9(12):789–795

Vinterhalter B, Ninković S, Cingel A et al (2006) Shoot and root culture of *Hypericum perforatum* L. transformed with *Agrobacterium rhizogenes* A4M70GUS. Biol Plant 50(4):767–770

Vorzheva II, Chernyak AB (2004) Allergy to dermatophyte fungies. Allergology 4:42–47

Waffo-Teguo P, Lee D, Cuendet M (2001) Two new stilbene dimer glucosides from grape (*Vitis vinifera*) cell cultures. J Nat Prod 64(1):136–138

White TC, Findley K, Dawson TL et al (2014) Fungi on the skin: dermatophytes and *malassezia*. Cold Spring Harb Perspect Med 4:a019802

Wilson AS, Roberts SC (2012) Recent advances towards development and commercialization of plant cell culture processes for the synthesis of biomolecules. Plant Biotechnol J 10(3):249–268

Xi HF, Ma L, Wang LN et al (2015) Differential response of the biosynthesis of resveratrols and flavonoids to UV-C irradiation in grape leaves. N Z J Crop Hortic Sci 43(3):163–172

Zang N, Deng XX et al (2011) Resveratrol-mediated gamma interferon reduction prevents airway inflammation and airway hyperresponsiveness in respiratory syncytial virus-infected immunocompromised mice. J Virol 85:13061–13068

Zavrel M, White TC (2015) Medically important fungi respond to azole drugs: an update. Future Microbiol 10:1355–1373

Zorofchian Moghadamtousi S, Abdul Kadir H, Hassandarvish P et al (2014) A review on antibacterial, antiviral, and antifungal activity of curcumin. Biomed Res Int 2014:1–12

Zubrická D, Mišianiková A, Henzelyová J et al (2015) Xanthones from roots, hairy roots and cell suspension cultures of selected *Hypericum* species and their antifungal activity against *Candida albicans*. Plant Cell Rep 34(11):1953–1962

Zuzarte M, Gonçalves MJ, Cavaleiro C et al (2013) Antifungal and anti-inflammatory potential of Lavandula stoechas and Thymus herba-barona essential oils. Ind Crop Prod 44:97–103

Liquid Chromatography for Plant Metabolite Profiling in the Field of Drug Discovery

Luiz Carlos Klein-Júnior, Johan Viaene, Amorn Slosse, and Yvan Vander Heyden

1 Introduction

"A new golden age of natural products discovery" (Shen 2015), "The re-emergence of natural products for drug discovery in the genomics era" (Harvey et al. 2015), and "The impending renaissance in discovery and development of natural products" (Pawar et al. 2017) are some review titles, used in recent publications, which highlight the importance of natural products in drug discovery nowadays.

The historic significance of natural products is very well described in the literature (Atanasov et al. 2015; Li and Weng 2017). Probably, one of the most important marks is the *Ebers Papyrus*, a scroll dating back to 1500 BC. It contains more than 800 medicinal preparations (mostly plant based) used in the ancient Egyptian medicine, stored since 1873 at the University of Leipzig (Atanasov et al. 2015; Universitätsbibliothek Leipzig 2016; Li and Weng 2017).

The use of complex mixtures for the treatment of pathological conditions lasted until the beginning of the nineteenth century. In 1817, the pharmacist Friedrich Wilhelm Adam Sertürner reported the isolation and evaluation of the *principium somniferum* from the opium poppy (*Papaver somniferum* L.), nowadays known as morphine (Atanasov et al. 2015). This boosted the isolation of many other important natural products, such as quinine, caffeine, and atropine, creating the foundations of the Western medicine.

Although history serves as a proof of concept for the importance of natural products in drug discovery, the Big Pharma companies abandoned most of their research

L. C. Klein-Júnior (✉)
Chemical-Pharmaceutical Investigation Nucleus, University do Vale do Itajaí – UNIVALI, Itajaí, SC, Brazil
e-mail: lcklein@univali.br

J. Viaene · A. Slosse · Y. Vander Heyden
Department of Analytical Chemistry, Applied Chemometrics and Molecular Modelling, Vrije Universiteit Brussel – VUB, Brussels, Belgium

V. Cechinel Filho (ed.), *Natural Products as Source of Molecules with Therapeutic Potential*, https://doi.org/10.1007/978-3-030-00545-0_3

programs in this field during the last decades (David et al. 2015; Harvey et al. 2015; Shen 2015). Constant taxonomic modifications, low yield, ecological and legal aspects, accessibility issues, and variation in quality of plant material are some of the reasons for the decline in natural product interest (Atanasov et al. 2015; Harvey et al. 2015).

Nevertheless, one of the most important challenges for pharmaceutical companies in natural product researching is also the main reason for the renewed interest in the area. In the late twentieth century, Big Pharma centered their studies in combinatorial synthesis and high-throughput screening (HTS). Unfortunately, plant extracts are not compatible with HTS campaigns, mainly due to their high viscosity, aggregation and/or precipitation, non-specific binding, and presence of fluorescent and/or quenching compounds, mostly requiring adaptation, purification, and additional steps for proper evaluation, hampering this methodology for drug discovery (Atanasov et al. 2015; Shen 2015). However, the number of new drugs approved from the use of combinatorial synthesis and HTS is considerably low, demonstrating that this approach was quite frustrating (Atanasov et al. 2015; Harvey et al. 2015).

To the poor results obtained in the last decades, the recent awarding of the 2015 Nobel Prize in Physiology or Medicine to Youyou Tu was added. Prof. Tu discovered the plant natural product artemisinin, used for the treatment of malaria. Together with recent analytical and chem-bioinformatical technological advances, the renewal of interest in natural products seems to be undeniable (Atanasov et al. 2015; Shen 2015).

Traditionally, bioactivity-guided fractionation is used in the study of natural products. In this approach, pharmacological assays are performed in order to drive the isolation of active compounds, theoretically avoiding the isolation of non-active compounds (Hubert et al. 2017). Usually, this workflow is very time- and money-consuming. In addition, it is not uncommon to isolate already known compounds with already known pharmacological properties (Atanasov et al. 2015; Hubert et al. 2017).

In this sense, new technologies have been introduced in order to avoid this very laborious methodology. An emerging approach is the metabolic profiling (Atanasov et al. 2015). Using chemometric tools, it is possible to correlate a chemical profile and a bioactivity of plant extracts, giving valuable information regarding the most active compounds. In addition, using hyphenated techniques, the early dereplication is also possible, avoiding the isolation of already known bioactive compounds (Allard et al. 2017; Begou et al. 2017).

Several analytical techniques have been used for metabolic profiling, such as direct nuclear magnetic resonance (NMR) and mass spectrometry (MS) analysis (Gemperline et al. 2016; Kumar 2016). However, because of the complexity of extracts, a preliminary separation, such as by liquid chromatography (LC), is often applied (Allard et al. 2017; La Barbera et al. 2017).

In this chapter, focus is given to the workflow in LC plant metabolic profiling for drug discovery, which includes sample preparation, development of the metabolite profile and pharmacological-assay considerations, as well as the overall correlation of metabolite profiles and bioactivity using chemometric tools.

2 Sample Preparation

Sample preparation is the first major step to be considered in the analysis of secondary metabolites, in the context of drug discovery, because it has a great impact on the metabolite contents and consequently the obtained results (Rates 2001; Huie 2002; Kim and Verpoorte 2010; Vuckovic 2012; Wu et al. 2013; La Barbera et al. 2017). However, this step is often still being done manually causing high costs. Therefore, to handle a large number of samples simultaneously (high throughput) and to minimize possible degradation of the metabolites, sample preparation should be simple and fast. Results of the chemical analysis generally depend on sampling and extraction within sample preparation. Usually sampling is correctly being performed with appropriate quenching (methods that prevent or minimize the enzymatic or biochemical processes of the plants because these could result in metabolic profile modification) to assure experimental reproducibility. However, much attention has to be paid to the improvement of the extraction step, as researchers use own experience-based protocols omitting adequate extraction evaluation (Kim and Verpoorte 2010; Mushtaq et al. 2014; Wen et al. 2014; Klein-Júnior et al. 2016a). Extraction, removal of possible interfering compounds, cleanup, enrichment of the metabolites of interest, and, if necessary, the transformation of analytes in a more suitable form that is compatible with, for instance, LC-MS analysis should be done in such a way that the composition of the components remains more or less constant.

Extraction is the main step in sample preparation; thus these methods have to be considered in regard to their suitability to meet the aims of a study (Kim and Verpoorte 2010; Sasidharan et al. 2011; Gupta et al. 2012). This important procedure, executed before chromatographic analysis, differs depending on the choice of the extraction solvent, which, among other things, is determined by the used analytical method and the chemical characteristics of the considered compounds. Moreover, different aspects have to be taken into account when extraction is carried out, e.g., solvent properties, the solvent-sample ratio, and the extraction time, pressure, and temperature, as these determine its success (Kim and Verpoorte 2010; Choi and Verpoorte 2014; Azmir et al. 2013). There is however no single solvent which dissolves all compounds in a sample, so multiple solvent extractions are needed to get a total view of the metabolome (Azmir et al. 2013; Martin et al. 2014; Mushtaq et al. 2014; La Barbera et al. 2017). Hence comprehensive methods, using a gradient of different solvents with increasing polarity, allow to efficiently extract a wide range of metabolites in one run. Moreover, these methods provide the potential to extract metabolites without degradation of any kind (De Monte et al. 2014; Yuliana et al. 2011; Mushtaq et al. 2014; Hill and Roessner 2015).

Extraction can be split in an actual extraction and the cleanup/enrichment of metabolites (fractionation of crude extracts of metabolite groups of interest) as a consequence of their low concentration in complex matrices. The conventional or classical extraction methods, such as maceration, percolation, and Soxhlet extraction, are the methods at first applied, making use of suitable solvents. The main drawbacks of these methods are a possibly protracted extraction time, consumption of large amounts of unhealthy and polluting solvents, high costs, low selectivity, and the potential degradation of metabolites (Sasidharan et al. 2011; Gupta et al. 2012; Azmir et al. 2013; Mushtaq et al. 2014; Brusotti et al. 2014; Azwanida 2015; Raks et al. 2018). Consequently alternative techniques, which give minimal sample degradation, are less hazardous solvent consuming, are less energy demanding, are time saving, have more environment-friendly properties, and provide better extraction efficiency and selectivity, were and are still being developed (Sasidharan et al. 2011; Gupta et al. 2012). Frequently used techniques are, for instance, microwave-assisted extraction (MAE), ultrasound-assisted extraction (UAE), supercritical fluid extraction (SFE), and pressurized solvent extraction (PSE). Such variety of techniques, however, requires careful evaluation as each technique may considerably influence the composition and hence the biological activity of the extract (Azwanida 2015; Atanasov et al. 2015).

UAE is the most commonly used technique; is very simple, fast, and cheap without solvent-choice limitation; and is environment-friendly. It is based on an increased surface contact between solvents and samples by using ultrasound waves, facilitating solvent transport in the plant matrix. High quantities of metabolites are obtained using small amounts of solvents. The application of ultrasound energy may have a possible negative impact on the active metabolites (Kim and Verpoorte 2010; De Monte et al. 2014; Azwanida 2015).

MAE, consisting of closed- or open-vessel MAE, makes use of microwave energy to facilitate the distribution of analytes from the plant matrix into the solvent. Advantages of this technique are lower solvent consumption, its speed and efficiency compared to Soxhlet and UAE, and its extraction quality and yield against traditional methods. MAE is less time-consuming but more expensive than UAE and cheaper and less environment-friendly than SFE. However, its technique cannot be used for temperature-labile components (De Monte et al. 2014; Raks et al. 2018).

SFE as extraction technique mainly uses CO_2 as supercritical extraction solvent. Thanks to the low critical temperature of CO_2 (31.1 °C), thermolabile constituents may be extracted without degradation or denaturation. CO_2 (greenhouse gas) is cheap, safe, and easily available. It is also less toxic, nonflammable, and easily removable when ending the extraction. It provides high-quality and reliable extracts. One of the main drawbacks is the high cost of the equipment.

PSE is based on the principle that at increased pressure the solvent remains in the liquid phase after heating, which allows fast and little solvent-consuming extractions. Simultaneously higher yields are obtained. H_2O is often used as a green solvent. This technique can be hyphenated with UAE to gain efficiency (De Monte et al. 2014; Gupta et al. 2012; Mushtaq et al. 2014; Klein-Júnior et al. 2016c; Raks et al. 2018).

For the cleanup and concentration step, solid-phase extraction (SPE) is extensively used because of its possible automation and the availability of a wide range of sorbents, usually applied in cartridges (Tu et al. 2010; Wen et al. 2014). It is based on the adsorption or partition of the extracted compounds on a solid phase, which initially retains the interested group of analytes. After washing of the unwanted components, the metabolites of interest are desorbed (eluted) with an appropriate solvent. Basically SPE is a fractionation rather than an extraction technique. Liquid-liquid extraction (LLE) was and still is a frequently used cleanup technique, but because it is too time- and solvent consuming, it is more and more replaced by SPE. A related SPE technique that was developed is the solid-phase micro-extraction. It is similar to SPE but miniaturized, faster, and greener (Mushtaq et al. 2014; Klein-Júnior et al. 2016c).

The actual extraction methods do not result in collecting the entire plant metabolome; thus compromises are to be made (Vuckovic 2012; Brusotti et al. 2014; Wen et al. 2014; Klein-Júnior et al. 2016a). Hence, the enormous number of plant metabolites (± 30,000) with considerable differences in polarity, stability, and chemical diversity, as well as the extraction reproducibility, makes the design of efficient protocols mandatory (Kaiser et al. 2009; t'Kindt et al. 2009; Choi and Verpoorte 2014; Mushtaq et al. 2014). Metabolite profiling studies must be designed in such a way that a maximum of metabolites are detected, in other words, i.e., that different metabolite groups are extracted (Sasidharan et al. 2011; Gupta et al. 2012).

An arising issue is the use of chemometrics and experimental-design approaches in optimizing the extraction and, occasionally, the fractionation processes (Klein-Júnior et al. 2016a). Several studies have been dealing with this optimization, where different factors, such as solvent concentrations, temperature, extraction time, sample-solvent ratio, particle size, and pH, are studied in order to obtain maximal responses, for, for instance, yield, amount of a specific class of metabolites, and biological activities (Souza et al. 2007; Das et al. 2013; Zhu and Liu 2013; Martin et al. 2014; Wang et al. 2014; Hammi et al. 2015; Wu et al. 2015; Izadiyan and Hemmateenejad 2016; Ćujić et al. 2016; de O Silva et al. 2017; Dary et al. 2017).

As a rule, when performing metabolome analysis, the number of extracted compounds reflects the best metabolomics conspectus. However, in case of targeted metabolite profiling, in order to obtain an accurate overview of the chemical diversity of the metabolite group of interest, other responses/approaches are being elaborated.

In this regard, Klein-Júnior et al. (2016a) extensively studied the optimization of indole alkaloid extraction and fractionation based on UPLC-DAD metabolite profiling with the aim to develop a less time- and solvent-consuming method, as well as to represent maximally the entire chemical composition of the plant being extracted. In other words, how can an efficient extraction method be developed to obtain a maximal number of metabolites? In a first step in the optimization of an UAE method, a fractional factorial screening design was executed to determine the significant effects of the selected factors. The evaluation of this design was done in two steps. Firstly, the response yield (obtained with LLE fractions), number of peaks, and sum of peak areas showed to have little meaning in relation to chemical diversity. In order to obtain a comprehensive picture of the metabolic profile, a new approach was presented to overcome the inadequate information of the abovementioned responses. Euclidean distance measurements between the metabolic profiles of the alkaloid fractions and the blank injection were calculated. This distance is an indication of the extract chemical diversity. In other words, the higher the distance, the higher the chemical diversity of the extract, i.e., the higher the metabolite content compared to the blank signal.

In a second step, the entire metabolic profiles were used as responses to determine detailed information of the factor effects on these profiles. Effect plots or effect fingerprints were calculated for each factor and graphically evaluated. From these plots thermolabile compounds could be indicated, and peaks (compounds) as well as the important factors were selected for further optimization. This was done performing a central composite design with temperature and extraction time as optimization factors. Here, the heights of the selected peaks were determined as responses. After response modeling (and avoiding degradation), the optimal combination of time and temperature was determined. For the optimization of the alkaloid fractionation, SPE was applied as alternative technique of LLE. Applying a Box-Behnken design, three factors were studied, and as response the sum of the peak areas of six metabolites in the profiles was used. Sample concentration, percentage of acetonitrile, and eluting volume were set to obtain the best fractionation conditions. It can be concluded that the Euclidean distance approach and the entire metabolic profiles are useful as responses for extraction optimization of specific component groups. This study allowed developing a time- and solvent-saving method as well as a reliable extraction and fractionation method of indole alkaloids, without component degradation.

3 Development of the Metabolite Profile

Herbal samples have a complex composition. This is related to the multitude of metabolic pathways involved in transforming the nutrients and compounds taken up by the plant through its root system and aerial parts, into compounds required by the plant at a given stage in its life cycle. Additionally, these processes highly depend on light exposure, rainfall, soil type, and numerous other external factors, resulting in a high variability in chemical composition between different samples of a same plant species (Wagner and Ulrich-Merzenich 2009; Lu et al. 2005; Li et al. 2010; Liu 2011). Further, when the sample extraction is set up to isolate compounds of a specific group (for instance, alkaloids), the compounds in the mixture may show a high degree of similarity for several physicochemical properties. Because in drug discovery, the aim is to find and fully characterize potential new drug candidates, it is essential that the chemical analysis allows separating the various compounds in the (cleaned-up) extracts. A wide range of chemical separation techniques is available to obtain this goal (Liu 2011).

Spectral techniques are also very popular tools in chemical analysis. They can be applied to characterize a compound in terms of its ultraviolet, visible or infrared radiation absorbance, fluorescence, mass spectrum, nuclear magnetic resonance spectrum, etc. (Gauglitz and Vo-Dinh 2003; Gunzler and Williams 2001). Although this spectral information has its value in various contexts, it is important to know that not all of these techniques do separate the information of the different compounds. As a result, the UV spectrum, for instance, of a mixture of compounds will contain information from all UV-absorbing compounds present. However, information from individual compounds is required in the context of drug discovery because of the need to find active compounds, which may develop as potential new lead compounds (Liu 2011). To assure that the information provided by spectroscopic techniques is specific for given compounds in a mixture, these compounds need to be separated prior to the spectral detection. Various separation techniques, including chromatography (Tistaert et al. 2011) and electrophoresis (Gunzler and Williams 2001), are available.

In this chapter, the discussion is restricted to chromatographic separation techniques. These separate compounds in a mixture based on their different interaction behavior in a two-phase system, called the stationary and the mobile phase, which are (relatively to one another) moving in opposite directions. Various forms exist, which will be briefly overviewed in this section, focusing on their use in the context of drug discovery.

3.1 Thin-Layer Chromatography

The simplest technique is thin-layer chromatography (TLC) (Fig. 1) (Sherma and Fried 2003; Tang et al. 2014). In TLC, a layer of the stationary phase, often bare silica or chemically modified silica (see further), is attached to a plate. Small volumes of sample solution are spotted on one side of the plate in little spots or bands. The analysis is started by placing the plate in a recipient, the development chamber, containing the mobile phase, which is usually a mixture of organic solvents, as shown in Fig. 1. The mobile phase then starts moving through the stationary phase by capillary forces, dragging also the sample compounds with it. Compounds with a relatively high affinity for the stationary phase will migrate slowly, while those with a higher affinity for the mobile phase will migrate faster. This results in a separation of the compounds according to their affinity differences for stationary and mobile phases. When the solvent has moved a given distance on the plate – when the separation is maximal at the conditions applied – the plate is removed from the

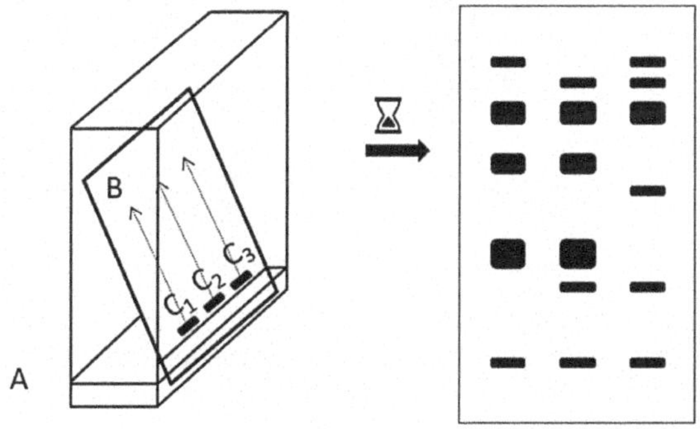

Fig. 1 Thin-layer chromatography. Left side: A, development chamber; B, TLC plate; C1–3, spots of samples or standard solutions. Right side: TLC plate after separation and revelation: pattern of spots is visualized

development chamber and dried and the compounds are revealed through the application of spray agents or by UV light radiation which causes quenching of fluorescence on pretreated plates or by densitometric measurements (Tang et al. 2014), which reveal a pattern of spots for each sample. TLC is often used in traditional drug discovery because it allows the simultaneous analysis of several samples, it is relatively cheap, and a wide range of stationary and mobile phases can be used. Since the plates are single use, applying aggressive solvents is not even an issue. A major drawback of the technique, however, is low efficiency and resolution, which means that compounds with similar properties are likely to show overlapping spots. Therefore, TLC is rather applied to separate groups of compounds, after which their spots can be individually scraped off the plate, redissolved in an appropriate solvent, and subjected to column chromatography (see next section), since on modern chromatography columns, compounds with similar properties can be better separated (Sherma and Fried 2003; Tang et al. 2014). However, given the latter, very often the TLC step may be skipped, and after sample preparation, immediately column chromatography is applied.

3.2 *Column Chromatography*

In column chromatography (Fig. 2), a small amount of sample, containing compounds that need to be separated, is injected in a constantly flowing mobile phase, which can be a liquid, a gas, or a supercritical fluid, called, respectively, liquid (LC), gas (GC), and supercritical fluid chromatography (SFC). Case studies applying GC and SFC can be found in refs (Liu et al. 2016; Li et al. 2013). However, in this chapter, we will further focus on LC. Although in the early days of LC, gravitation force

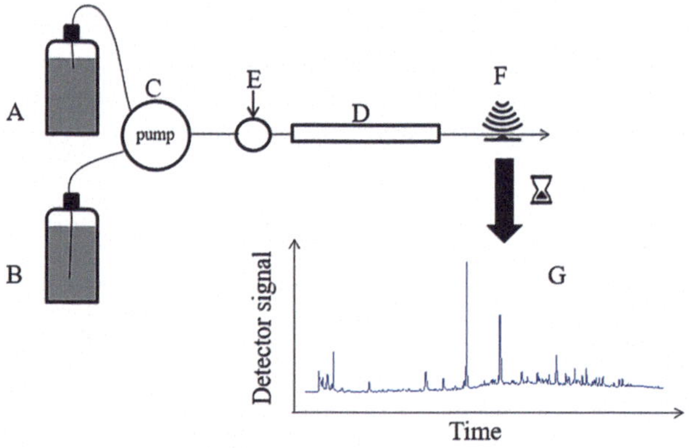

Fig. 2 Schematic representation of HPLC. A and B: solvents used as mobile phase components, C: pump moving the solvents at a constant flow rate, E: injector where the sample is introduced in the mobile phase flow, D: analytical column, F: detector, G: graphical output of the detector: chromatogram

was used as a motor for the separation process, nowadays, the mobile phase is pumped through the column at a tunable flow rate. This process is called high-pressure/high-performance liquid chromatography (HPLC) (Lough and Wainer 1996; Dong 2006; Waksmundzka-Hajnos and Sherma 2011). HPLC nowadays has become a benchmark technique in analytical chemistry and is a common tool in drug discovery. The mobile phase is thus pumped through a column containing the stationary phase, which is usually composed of particles that are mechanically immobilized in the column. Classically the particles are silica based, which has polar properties. Apolar and intermediately polar stationary phases have been created by chemically binding apolar or intermediately polar functional groups (for instance, C18 or C_3-CN chains) to the silanol groups on the surface of the silica particles. As a result, continuous developments have resulted in the availability of a large choice of stationary phases (Lough and Wainer 1996; Dong 2006; Waksmundzka-Hajnos and Sherma 2011).

A stationary phase should be selected in such a way that it has chemical properties that differ in some way (for instance, polarity) from the mobile phase. Mobile and stationary phases with various properties can be selected. In liquid chromatography, two modes of chromatography are classically defined, normal and reversed phase. In normal-phase chromatography, the mobile phase has an apolar nature, for instance, hexane-based, and the stationary phase a polar, for instance, silica. In reversed-phase chromatography, the opposite is true with polar mobile phases, i.e., water-based, to which methanol or acetonitrile (or mixtures) may be added, and apolar stationary phases, e.g., C18. A number of variations on these modes exist, including ion-pair and micellar chromatography (Lough and Wainer 1996; Dong 2006; Waksmundzka-Hajnos and Sherma 2011).

In modern drug discovery, reversed-phase LC is extensively used because of a more limited use of organic solvents and its compatibility with mass spectrometry (Watson and Sparkman 2007) (see further), which is progressively more frequently used as a detector in comprehensive metabolomics profiling.

The principle behind the chromatographic separation is that different compounds interact differently with the mobile and stationary phases. Compounds with a relatively higher affinity for the mobile phase will go rather fast through the column, while compounds with more affinity for the stationary phase will require more time to travel the same distance. As a result, the compounds are separated and elute from the column at different times. After the separation, a detector (see further) then continuously measures specific information, thus detecting what elutes from the column. Elution of only mobile phase results in a baseline signal. A deviation (peak) from the baseline signal indicates the presence of a compound (see Fig. 2). The graphic representation of the detector signal, which is called a chromatogram, shows a peak for each detected compound. Each peak has a maximum, which occurs at a time that is called the retention time of the corresponding compound.

In herbal analysis the composition of the mobile phase is often changed as a function of time, which is called gradient elution. As a result, each compound detaches from the stationary phase when the mobile phase reaches a composition necessary to overpower the compound's interaction with the stationary phase

(Lough and Wainer 1996; Dong 2006; Waksmundzka-Hajnos and Sherma 2011). The result is that mixtures of polar and apolar compounds can be separated and determined in one run. Optimizing chromatographic methods essentially means finding a set of conditions (mobile phase composition, stationary phase, gradient conditions, etc.) allowing an acceptable separation and a reasonable analysis time (Dejaegher et al. 2010; Alaerts et al. 2007). In drug discovery, in our opinion, priority usually is given to the quality of the separation, to allow a maximal separation of the compounds and their specific characterization.

Compounds may be separated by HPLC, when their interaction with mobile and stationary phases is sufficiently different. When the compounds are known, and when the sample is not too complex, the separation can be optimized by screening different stationary phase-mobile phase combinations to find the combination resulting in the best separation. However, when the compounds in a sample are not a priori known or very numerous (as is the case in drug discovery), even after separation optimization, it is often not possible to find mobile and stationary phase conditions where all compounds are completely baseline separated. Then the peaks of given compounds show (partial) overlap. The quality of separation between two consecutively eluting compounds can be quantified as the resolution (Lough and Wainer 1996). This parameter quantifies the ratio of the difference between the retention times and the sum of the peak widths of two consecutively eluting compounds. For a baseline separation of two peaks of similar height, a role of thumb specifies that resolution should be above 1.5.

The mobile and stationary phases; the mobile phase flow rate; the instrument itself; the column chemistry, dimensions, and brand; the size and shape of the stationary phase particles; and the column temperature all may influence the retention times of the compounds in a given mixture. The peak widths are also influenced by multiple factors, including column length, flow rate of the mobile phase, stationary and mobile phase chemistry, dimensions of the tubings, analysis temperature, and size of the particles. However, on a given system with a fixed set of conditions, the column properties play a key role in the observed peak width and thus the quality of the separation that can be achieved. A column's separation power is quantified as the efficiency and is expressed as the number of theoretical plates. The higher this number, the better the column is expected to perform, i.e., the more complex mixtures can be separated. The number of theoretical plates on a column increases when the theoretically defined height equivalent to a theoretical plate (HETP) decreases, and with increasing column length (Lough and Wainer 1996). Because of the complexity of herbal samples, the efficiency of a column is decisive for the quality of the fingerprint that is developed. Strategies to increase column efficiency are thus based on two principles: increasing the column length on the one hand and decreasing HETP on the other (Lough and Wainer 1996; Dong 2006; Waksmundzka-Hajnos and Sherma 2011).

The first option, however, is limited by another phenomenon, the back pressure that is generated by the column. Classical HPLC instruments and columns are built to withstand back pressures of up to 400 bar. Since increasing the column length fastly increases back pressure, this option is limited. However, a solution was provided with the development of monolithic columns. These columns have a rather low efficiency, but they have larger pores and thus a low back pressure. Consequently, they can be serially coupled, which in the end results in more efficient separations (Dejaegher et al. 2010; Alaerts et al. 2007).

A second option is modifying the column properties to decrease HETP. The Van Deemter equation (HETP = A + B/u + Cu) is very helpful to rationalize this process (Dong 2006). It expresses HETP as the sum of three processes causing dispersion or band broadening: Eddy diffusion, due to different paths followed by molecules through the porous particles and represented by the A-term; longitudinal diffusion, leading to diffusion along the axis of the column and expressed by the B-term; and a C-term, reflecting the mass transfer of the analyte between the mobile and stationary phases. The Eddy diffusion is independent from the linear velocity, u, while the longitudinal is inversely correlated and the mass transfer term directly. These coefficients can be affected either by reducing the permeable zone in the particles (which has led to the development of superficially porous or core-shell particles) (Hayes et al. 2014; Guiochon and Gritti 2011) or by decreasing the size of the particles. Especially the latter option causes again an increase in the back pressure, which resulted in the development of ultrahigh-pressure liquid chromatography (UHPLC) instruments and columns, which can be used at pressures till 1000 bar (Waksmundzka-Hajnos and Sherma 2011). These latter techniques show improved efficiencies and have the advantage that in the same time more compounds can be separated, either a similar separation can be obtained in a shorter analysis time. Therefore, UHPLC has made its way into herbal fingerprint analysis, including drug discovery.

3.3 Detectors

3.3.1 Spectroscopic Detection

Both classical HPLC and its core-shell, monolithic, and ultrahigh-pressure variants are applied in drug discovery. The amount of information that is collected depends very much on the detector used (Lough and Wainer 1996). UV detectors are very popular. The simplest can measure only the absorbance at one tunable wavelength. The output generated is a chromatogram, showing a two-dimensional peak pattern for each sample along a time axis and an absorbance axis.

The chromatographic fingerprints are of interest in drug discovery and were applied, for instance, by Ben Ahmed et al. (2017) to indicate the potentially antioxidant compounds in *Pistacia atlantica* leaf extracts. Nevertheless, the information obtained is too limited to allow identification of the compounds. Additionally, when not all compounds are fully separated, it is difficult to estimate the contribution of each individual compound. Technological developments have resulted in the development of a more advanced type of detectors that can register a spectrum at each time point. These detectors are called diode-array detectors (DAD), and they generate a three-dimensional output, with a time, a wavelength, and an absorbance axis (Lough and Wainer 1996).

Even though DAD offers a UV spectrum for each peak, which can be used for identification purposes, UV spectra do not always allow the unambiguous identification of compounds. It often allows identifying the group of compounds, but since UV spectra of closely related compounds tend to be very similar, in essence, they lack the required specificity to truly identify the compounds (Lough and Wainer 1996). Techniques that provide highly specific identification are those that can distinguish between compounds with small structural differences. Mass spectrometry (Watson and Sparkman 2007) and nuclear magnetic resonance (Garrido and Beckmann 2013; Qin et al. 2009) are nowadays progressively used in drug discovery since they provide characteristic structural information. Fluorescence (He et al. 2013), evaporative light scattering (Alaerts et al. 2007), and electrochemical (He et al. 2013) detectors are also occasionally used when developing fingerprint profiles.

3.3.2 Mass Spectrometric Detection

Mass spectrometry allows measuring chemically charged species in the gas phase. A number of processes are necessary to obtain and analyze the charged species: ionization, mass analysis (including fragmentation possibilities), and detection.

Ionization

Since the output of the HPLC system is a liquid, an interface is required to bring the compounds from this liquid to a charged gas phase. In mass spectrometry various ionizers exist to make this transfer. In modern instruments, a popular choice is to use electrospray ionization. Its principle is to lead the LC eluent through a capillary with a nanoscale diameter, which can be electrically charged. Additionally, a temperature increase and a gas flow are applied. This leads to the formation of progressively smaller and smaller droplets of liquid that finally explode due to an overload in charges of the same polarity. As a result, the compounds in the sample get a positive or negative charge and are brought in the gas phase. Once these charged species are formed, they can be manipulated by changing the electric charge (or magnetic fields) in various parts of the mass spectrometer. In order to avoid interferences in these processes and to result in the highest signal, these manipulations need to be conducted in a vacuum environment, making a vacuum pump an essential part of any mass spectrometer.

Mass Analysis

Another crucial process in mass spectrometry is separating the charged species according to their mass/charge ratio. This is done in the mass analyzer. Again, various mass analyzers are available. Most can either be applied in a mode allowing to scan charged compounds in a given m/z range or in a mode focusing on compounds with a specific m/z value. Quadrupole mass analyzers are extensively used, mainly to assay a number of known (targeted) compounds. These quadrupole analyzers contain four cylindrical rods on which charges are applied that are changed in polarity at a very high frequency. Charged species are attracted to the rods, followed by a repulsion when the polarity changes. For each m/z value, a given set of charges on the rods results in the charged compound passing through to the next zone in the mass spectrometer, which can be a collision cell, a second analyzer, or a detector. Other compounds with other m/z values either are attracted or repulsed by the charges and do not make it to the end of the analyzer. When such a quadrupole analyzer is applied in scanning mode, the charges on the rods are changed as a function of time to consecutively fulfill the requirements for all m/z values in a given m/z range. Some analyzers (so-called trap analyzers) are built in such a way that they even allow to immobilize selected compounds and eject them in a very precise way according to their m/z range. In principle, they also manipulate charged species through application of changes in electric fields in the mass analyzer, which is similar to the simple quadrupole analyzer.

An analyzer that does not allow the selection of ions with a specific m/z is the time-of-flight mass analyzer. In essence, it is a tube with a fixed distance. Since the m/z of an ion determines the time required to complete this distance, the m/z of the ion is determined from its travel time.

Fragmentation

When a compound is ionized through electrospray ionization, it gets either a positive or a negative charge, but it does not fall apart into fragments. From the obtained information, the molecular mass of small molecules can be determined. However, when fragmenting the ionized compound, a number of fragment ions are created, which are highly characteristic for a given compound. These fragment ions can then be used to elucidate the chemical structure of the compound. To create such fragments, energy is required. This energy is usually provided through a gas stream, which is kept at high energy in a collision cell. Through the invention of tandem mass spectrometry, nowadays, it is possible to select in a first analyzer a specific charged compound (called a mother or precursor ion), which can lead to the collision cell, where it is fragmented, followed by a separation of the fragment ions (also called product or daughter ions) in a second mass analyzer. The first mass analyzer is often a quadrupole; for the second several choices are available: another quadrupole analyzer or, for instance, a time-of-flight analyzer. Alternatively, in trap analyzers, the selection of the precursor ion and fragmentation and selection of the fragment ions can consecutively happen in one single space.

From MS Spectra to Compound Identification

Mass analyzers are usually grouped in low-resolution and high-resolution instruments. The difference is very important in drug discovery, since it determines how appropriate the instrument is for identification of unknown compounds. Low-resolution mass analyzers, for instance, quadrupole analyzers, can only determine the m/z value to unit value. For identification, this is often too limited, due to the occurrence of compounds that have the same mass at unit resolution but differ in the numbers behind the comma. Therefore, quadrupoles alone are usually not sufficient when the aim is to elucidate a compound's structure. This requires a high-resolution mass analyzer (for instance, time-of-flight or Orbitrap analyzer), which can determine a compound's mass, called the accurate mass, with a very good accuracy. Identification is based on the comparison of this measured accurate mass with the accurate masses of a range of compounds with some restrictions, for instance, on the number of carbon, hydrogen, and oxygen atoms. The differences of the measured accurate mass with the theoretical masses (also called the mass differences) are then determined and used as an identification criterion. Another aspect that is usually considered in compound identification is the correspondence of its isotopic pattern with that of candidate molecules. The most likely molecular formula is then obtained as the one with the smallest mass difference and the best correspondence in isotopic pattern (Watson and Sparkman 2007).

Commercial MS processing software packages (for instance, Masslynx and Excalibur) provided by instrument suppliers can be used in drug discovery. LC-MS data are processed in several steps. In a first step, the precursor ion data are processed in order to select characteristic marker compounds. In commercial software, the precursor ion LC-MS data of a sample is often presented as a list of intensities with given retention times and m/z values. Exploratory multivariate analysis techniques (for instance, principal component analysis) are used to visualize similarities and differences in the intensities at the observed retention time-m/z pairs in different samples. In a next step, multivariate discrimination models can be built, linking the intensity information of the samples to a given property. This property can be the plant species, plant part, growing region, harvesting season, or any other property that might relate to the chemical composition. The model's information is then used to determine m/z retention time pairs that are characteristic for each property, resulting in a list of characteristic marker compounds, for instance, to distinguish between two considered herbal species. The compounds are then tentatively identified based on the m/z values of their precursor ion.

Confirmation of the tentative identification of the compounds is done using the information obtained after fragmentation. This is done by comparison of the observed fragmentation pattern with the fragmentation pattern of a limited number of compounds with a good fit (corresponding isotopic pattern and low mass defect) for the precursor ion. This is done for each marker compound and often results in the identification of already known compounds. However, sometimes, potentially new compounds are detected, which can then be further characterized to confirm their chemical structure by complementary structural elucidation techniques, like NMR (Roessner and Dias 2013).

4 Pharmacological Evaluation

To use the metabolic profiling approach, the chemical analysis of a sample is linked to a biological activity. Although in vivo assays using rodents have already been used for this purpose (Cardoso-Taketa et al. 2008), it is not recommended. In vivo assays usually are less reproducible (Reardon 2016), a problem which will have impact on the mathematical model established to indicate the bioactive compounds. In fact, lack of reproducibility is a recurring concern in scientific publications (Plant et al. 2014; Munafò et al. 2017). In addition, in vivo assays go against the spirit of 3R (reduction, refinement, and replacement of animal experimental testing) that should be used in laboratories (Doke and Dhawale 2015). In this sense, in vitro assays can properly replace in vivo approaches in the screening of natural products.

Usually, in vitro assays are based on simple chemical reactions, such as antioxidant activity determination using 2,2-diphenyl-1-picrylhydrazyl (DPPH) (Ben Ahmed et al. 2017) or on a purified protein and molecule interaction, such as the inhibition of enzymes, e.g., of acetylcholinesterase (Klein-Júnior et al. 2016a). In addition, cell cultures can also be used as an alternative, such as in antiproliferative assays (Li et al. 2017).

As already highlighted in this chapter (see Sample Preparation), an important step in the metabolic profiling approach is the extraction procedure. This will also have impact on the pharmacological assays. Traditionally, in vitro assays are performed using organic-solvent extracts, such as ethanol or methanol, and/or organic-solvent mixture extracts, such as dichloromethane/methanol (1:1), and/or aqueous extracts (McCloud 2010). However, because of the extract chemical diversity, some problems are often faced, mostly related to pan-assay interference compounds (PAINS). These compounds are promiscuous artifacts, such as catechols, quinones, phenolic Mannich bases, and hydroxyphenylhydrazones (Baell 2016), that interact in a non-drug-like manner with the target (Baell and Walters 2014). In natural product research, the most usual problems are (i) tannins interaction with proteins, such as enzymes, leading to false-positive results; (ii) antagonism due to compounds with opposing pharmacological activities; (iii) active compound dilution in non-active compounds, not reaching a high enough concentration to elicit their activity in the extract; (iv) presence of naturally fluorescent and quenching products in the extract, which might affect fluorescence-based in vitro assays; (v) presence of colored compounds in the extract, which might affect colorimetric-based in vitro assays; (vi) micelle formation, due to amphipathic compounds, such as saponins, which might lead to cell death; and (vii) chelation of metals essential for biochemical assays, such as those estimating the activity of metalloproteases (Henrich and Beutler 2013).

To overcome these difficulties, some strategies can be applied. The most frequent is prefractionation, when complex extracts are semi-purified usually based on their polarity to obtain a simpler sample (Henrich and Beutler 2013). This can be achieved using liquid-liquid (LLE) or solid-phase extraction (SPE), as well as chromatographic methods. For a metabolic profiling approach, prefractionation by chromatographic methods is less used since the extract composition could be oversimplified, losing some valuable information, especially related to natural product synergy. However, using LLE or SPE, it is possible to retain most of the extract complexity, while removing interfering compounds. Klein-Júnior et al. (2016a, b) used SPE cartridges to obtain an alkaloid fraction from *Psychotria nemorosa* which was able to inhibit both butyrylcholinesterase and monoamine oxidase B.

However, some compounds that initially can be considered as interfering, such as phenols, may also be relevant in some studies. Ben Ahmed et al. (2016) showed that phenolic compounds, such as galloylquinic acid and gallic acid, were the ones responsible for the antioxidant activity of extracts of *Pistacia atlantica*. Thus, prefractionation may then also not be a good approach. To avoid PAINS, some modifications to the in vitro methodology can be proposed, such as multiple concentration testing (also essential for IC_{50} estimation), choice of the method based on the extracts' fluorescent or colorimetric characteristics, and the use of additional agents to reduce aggregation and other non-specific bindings (Henrich and Beutler 2013). Butler et al. (2014) suggested the use of generic inactive extracts, obtained by the degradation of randomly selected specimen extracts, as negative controls. To predict the matrix effect, this inactive extract could be spiked with the positive control, giving additional information regarding PAINS or PAINS-like compounds in the extract.

In vitro assays have also their own limitations, which include the lack of bioavailability and metabolism information. In this context, in vivo experiments using alternative organisms could additionally be performed; *Danio rerio* (zebra fish), *Drosophila melanogaster* (fruit fly), and *Caenorhabditis elegans* (a nematode) are some alternative species that can be used in medium- to high-throughput approaches. However these organisms are not that frequently used for the metabolic profiling approach (Atanasov et al. 2015; Doke and Dhawale 2015).

5 Multivariate Data Handling of Chromatographic Fingerprints

After chemical analysis (fingerprint development) and pharmacological evaluation of the samples, both datasets must be linked in order to indicate the bioactive constituents in the extract. In this sense, the chromatographic fingerprints are subjected to a multivariate data handling procedure. However, raw fingerprint data usually need to be *pretreated* appropriately. *Peak alignment* is often necessary. It is imperative that peaks corresponding to the same molecule occur at the same retention time in different samples. However, in practice, retention time shifts are observed between analyses. These are due to several minor variations in experimental conditions, such as temperature, mobile phase composition, and flow rate (Alaerts et al. 2010a; Korifi et al. 2014).

Typically, *warping* is the most appropriate alignment methodology for chromatograms obtained from the same species. In warping, peaks are shifted, stretched, and/or compressed along the x-axis (time), without changing their sequence, aiming to enhance similarity between profiles (Bloemberg et al. 2013). In general, the profile to be aligned (P) is fitted to a target chromatogram (T) (Bloemberg et al. 2013). Different techniques can be used for warping, such as dynamic time warping (DTW), parametric time warping (PTW), and fuzzy warping (FW) (Alaerts et al. 2010a; Bloemberg et al. 2013).

One of the most used methods is correlation optimized warping (COW). COW was created by Vest Nielsen et al. (1998) and can be applied to either single or multi-trace profiles. By this approach, both P and T are equally segmented, allowing each segment in P to be compressed or stretched to better resemble T. Two parameters must be optimized: the segment length m and the number of points *per* segment that a segment might be compressed or elongated, also called slack t (Bloemberg et al. 2013; Korifi et al. 2014). Klein-Júnior et al. (2016a), for instance, used this approach to warp UPLC-DAD data. In Fig. 3a, untreated data are observed, where shifts are quite visual in the chromatograms. This is reinforced by the correlation graphic (bottom). After warping (Fig. 3b), it is evident that the correlation was improved, highlighted by both the chromatograms and the plot.

Other preprocessing methods, with other goals, might also be used. In fact, *autoscaling* is often applied. It aims to make all metabolites equally important using standard variation as scaling factor. It can be calculated as follows:

$$\tilde{x} = \frac{x_{ij} - \overline{x}_i}{s_i}$$

where x_{ij} is the measured variable, $\overline{x}_i$ is the column mean, and s_i is the column standard deviation (Van den Berg et al. 2006).

Column centering is another pretreatment frequently used, since it removes the vertical offset from the data. It removes from each column the column mean, as follows:

$$\tilde{x}_{ij} = x_{ij} - \overline{x}_i$$

where x_{ij} is the variable and $\overline{x}_i$ is column mean (van den Berg et al. 2006).

Some other preprocessing techniques can also be applied, such as other scaling approaches, normalizations, and standard normal variate (SNV) (Van den Berg et al. 2006; Zeaiter and Rutledge 2009; Alaerts et al. 2010a). However, there is no recipe which method is best used, since each data has its own particularities. Moreover, a combination of preprocessing steps can also be applied, and its outcome should be evaluated to be able to select the best combination (Bloemberg et al. 2013; Gerretzen et al. 2015).

As a second step in the data treatment, *unsupervised data analysis* is performed to explore the data structure. It only takes into account matrix **X**, containing p fin-

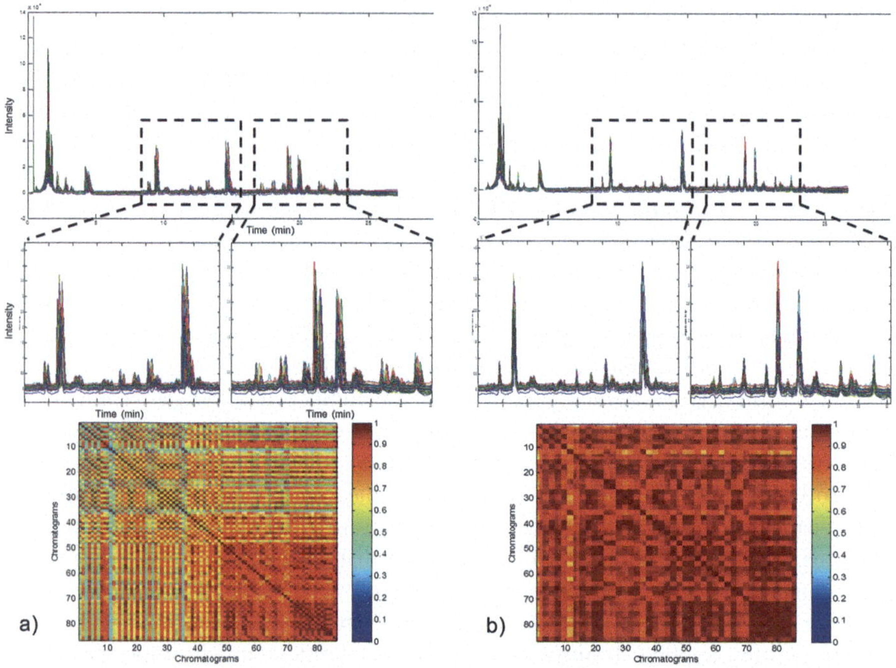

Fig. 3 Chromatographic fingerprints and correlation graphics (at the bottom) before (**a**) and after (**b**) warping, data obtained from the UPLC-DAD analysis of alkaloid fractions of *Psychotria nemorosa* (Klein-Júnior et al. 2016a)

gerprints, consisting of *n* measurement points. It aims to detect cluster formation tendency, giving a general idea of the data structure (Alaerts et al. 2010a; Goodarzi et al. 2013; Ren et al. 2015). For that purpose, two main techniques are used: *principal component analysis* (PCA) and *hierarchical cluster analysis* (HCA).

PCA is a technique that allows variable reduction, making it easier to visualize the data. It involves computation of new variables, known as principal components (PCs), orthogonal to each other, that retain most of the remaining variation information in the original or pretreated matrix **X**. The first PC explains most of the data variation, and each following PC contains less information regarding the variance in the dataset. The objects are then projected on (the first) two or three PCs and visualized in a score plot, used to observe cluster tendency in the samples (Fig. 4a), while the contribution of each original variable to the PC score can be visualized in a loading plot (Fig. 4b), used to evaluate the influence of each variable on the clustering (Esbensen and Geladi 2009; Goodarzi et al. 2013; Ren et al. 2015).

Another technique is HCA. It creates a dendrogram, which represents the cluster formation in the matrix **X**, organized as a hierarchical tree. Usually, the most similar objects are merged firstly. In an iterative process, objects are merged by similarity until all objects form only one cluster containing the entire dataset. This similarity is measured either as a distance, such as Euclidean distance and Mahalanobis dis-

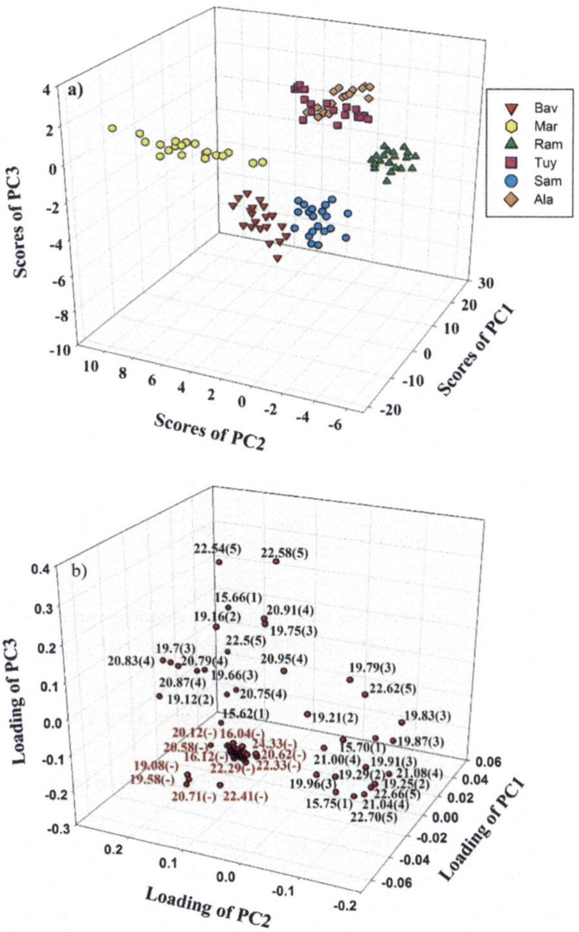

Fig. 4 Principal component analysis (PCA) score plot (**a**) and loading plot (**b**) for the fatty acid fingerprints of walnut samples from six regions in Iran (Esteki et al. 2017)

tance, or as a correlation, such as Pearson correlation coefficients. Another item that influences the dendrogram is the linkage function, which gives the similarity metric for pairs of groups (Fig. 5). Usually applied linkage functions include, among others, single linkage, unweighted average linkage, and centroid (Lee and Yang 2009; Goodarzi et al. 2013; Ren et al. 2015). HCA and PCA are commonly performed together to highlight cluster tendency in the dataset.

As a final step in the data handling for metabolic profile approach, *supervised data analysis techniques* are used. Specifically, *multivariate calibration* approaches are often applied. Then, both matrix **X** information (fingerprints) and the response vector **y** (biological activity, usually given as continuous values – e.g., IC_{50}) are used. Their main purpose is to build a regression model between the response (**y**) and the predictors (**X**). The regression coefficients of the model enable to indicate peaks in the fingerprint associated to the considered activity (Goodarzi et al. 2013; Ren et al. 2015). The model is described as:

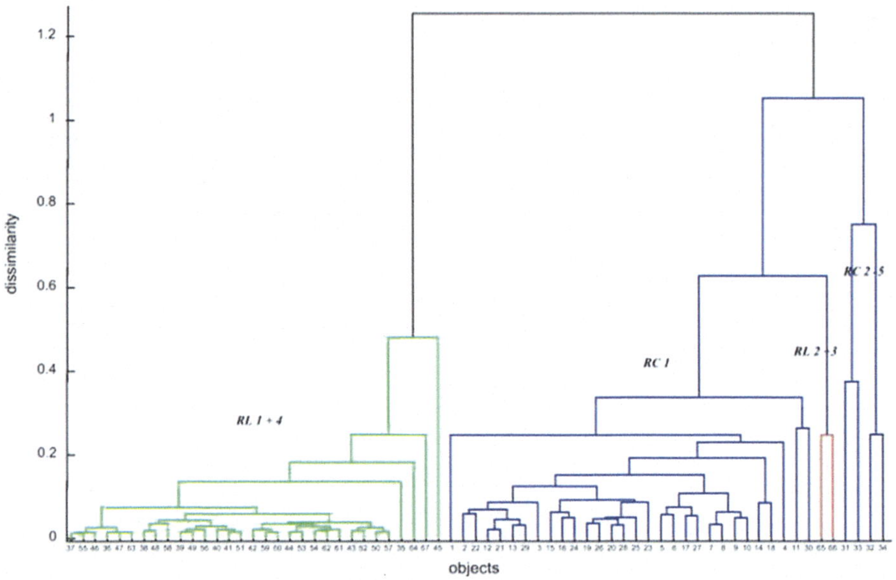

Fig. 5 Hierarchical cluster analysis (HCA) dendrogram for the fingerprints of two Chinese herbs: rhizome *Chuanxiong* (RC) and rhizome *Ligustici* (RL) (Alaerts et al. 2010b)

$$\boldsymbol{y} = \boldsymbol{X}\boldsymbol{b} + \boldsymbol{e}$$

where $\boldsymbol{b}$ is a $p \times 1$ vector of regression coefficients and $\boldsymbol{e}$ an $n \times 1$ residual vector (Tistaert et al. 2009).

Usually, a training set and a test set are applied, when enough samples are available. The training set is used to build a model and the test set to evaluate the predictive properties of the model. However, sometimes, not enough samples are available to establish a calibration and test set, which are both sufficiently representative. Therefore, in such case, almost the whole matrix **X** is used to build the model, and its predictive properties are evaluated using cross-validation (CV), in which the root-mean-squared error of cross-validation (RMSECV) is determined, often from one sample at the time that is left out (leave-one-out cross-validation). The model with the lowest RMSECV, found for different models constructed, usually is selected as the best (Alaerts et al. 2010a; Ren et al. 2015). RMSECV is calculated as follows:

$$RMSECV = \sqrt{\sum_{i=1}^{N} \frac{\left(\hat{y}_{CV,i} - y_i\right)2}{N}}$$

where N is the number of calibration samples, y_i the experimentally obtained response of the ith sample, and $\hat{y}_{CV,i}$ the response predicted by the model for the ith sample.

Another feature that is used to evaluate the model is the simplicity, which is directly correlated to the lower number of components. Finally, interpretability, related to the regression coefficients, is also important, since it must be easy to determine the contribution of the variables to the model (Alaerts et al. 2010a). Therefore, the number of (latent) variables in the model is a compromise between a model complexity and a low RMSECV.

Different techniques can be used to build the calibration models, such as stepwise multiple linear regression (stepMLR), principal component regression (PCR), partial least squares (PLS), and orthogonal projection to latent structures (OPLS) (Alaerts et al. 2010a). One of the most used methods is PLS. A PLS model can be written as:

$$\boldsymbol{X} = \boldsymbol{T}\boldsymbol{P}^T + \boldsymbol{E}$$

$$\boldsymbol{y} = \boldsymbol{T}\boldsymbol{P}^T\boldsymbol{b} + \boldsymbol{f} = \boldsymbol{T}\boldsymbol{q} + \boldsymbol{f}$$

$$\boldsymbol{b} = \boldsymbol{P}\boldsymbol{q}$$

where $\boldsymbol{T}$ gives the score matrix of **X** and **y**, $\boldsymbol{P}$ the loading matrix of **X** on $\boldsymbol{T}$ and $\boldsymbol{P}^T$ its transposal, $\boldsymbol{E}$ the residual matrix, $\boldsymbol{b}$ the regression coefficients, $\boldsymbol{q}$ the loading vector of **y** on $\boldsymbol{T}$, and $\boldsymbol{f}$ the residual vector of **y** (Nguyen Hoai et al. 2009).

The overlay plot of the regression coefficients with the original chromatograms is important for peak indication. In this sense, if using IC_{50} as response, negative regression coefficient values indicate active peaks, since lower IC_{50} indicates higher activity. Based on this prediction, peaks that match with "negative peaks" on the regression coefficient plot indicate active compounds (Figs. 9 and 10). In this sense, hyphenated techniques are important for the early dereplication of the active extract. LC-MS and LC-NMR techniques avoid the tedious and costly isolation of already known bioactive compounds. However, if the compound had never been assayed for a given activity, its isolation is mandatory to confirm the model prediction. Usually, modern techniques, such as medium-pressure liquid chromatography (MPLC) and/or preparative HPLC, are used for this purpose. Finally, the chemical structure can be determined and its activity determined.

6 Applications of Metabolite Profile

6.1 Linked to In Vitro Reaction Assays

Kvalheim et al. (2011) applied PLS regression to model the ferric reducing antioxidant power (FRAP) assay value (as a response) as a function of the obtained fingerprints (gradient HPLC with UV detection at one selected wavelength) for 60 mixtures, prepared according to an experimental design approach (no extraction involved). The mixtures were composed of 12 compounds with antioxidant properties, varying in compound-dependent concentration and composition. Forty mixtures were used as training set to build the model and the remaining 20 as test samples to validate the model. Except for two outlying mixtures in the training set, the prediction results obtained by PLS showed very good correspondence with the measured FRAP measurements, both for training and test set (not shown here; for the figure the reader is referred to Kvalheim et al. 2011).

The FRAP measurements of the pure compounds indicated that they could be sorted according to their decreasing antioxidant capacity, the most active compounds having a high FRAP value and vice versa. A typical fingerprint is shown in Fig. 6a. Three approaches were compared to retrieve the same sequence of the antioxidant potential. In a first approach, the regression coefficients of the PLS model were evaluated. Based on the regression coefficient plot, the most important peaks were indicated. The regression coefficient plot is shown in Fig. 6b. The size of the regression coefficients depended largely on the size of the peaks, making it impossible to rank the compounds according to their antioxidant capacity in the same sequence as obtained through the FRAP assay of the pure compounds. In the second approach (not shown here; for the figure the reader is referred to Kvalheim et al. 2011), target projection loadings (Kvalheim et al. 2011) were calculated and visualized in a profile. The size of the peaks resulted in less concentration-dependent conclusions; however it did not yet fully correspond to the sequence of the antioxidant capacity determined by FRAP. The third alternative multiplied the target projection loadings with the selectivity ratio (Kvalheim et al. 2011) and is shown in Fig. 6c. The size of the peaks in this profile showed the same trend as the FRAP results of the pure compounds. The third method showed the best correspondence with the results of the FRAP assay performed on pure standards.

As a result, the latter method is presented as a good basic procedure when the compounds are unknown, as is the case in the drug discovery context. Since in this study the compounds were all a priori known, UV spectra were sufficient for unambiguous identification. However, when the compounds would not be known in advance, techniques providing structural information (MS, NMR) would be necessary to identify the compounds after isolation or separation of the compounds, for instance, by means of liquid chromatography.

Xu et al. (2015) developed a data-driven method to determine the bioactive components in HPLC profiles of a set of Radix *Puerariae lobatae* samples and in a synthetic sample set. Like Kvalheim et al. (2011), they also focused on the antioxidant activity, measured by the FRAP assay. Fingerprint profiles were registered by gradient HPLC, and one detection wavelength was selected per sample set. Pretreatment of the HPLC fingerprint data was done using asymmetric least squares (background correction) and COW (retention time shift correction). They applied a variant on classic PLS modeling, called sure independence screening interval PLS (SIS-iPLS). This algorithm eliminates the variables that are only weakly correlated to the modeled response. The rationale for this method is that uninformative variables included in the PLS model, as well as the absence of important variables, will decrease the model's predictive properties (while the best model is more complex). The algorithm initially selects a number of m inter-

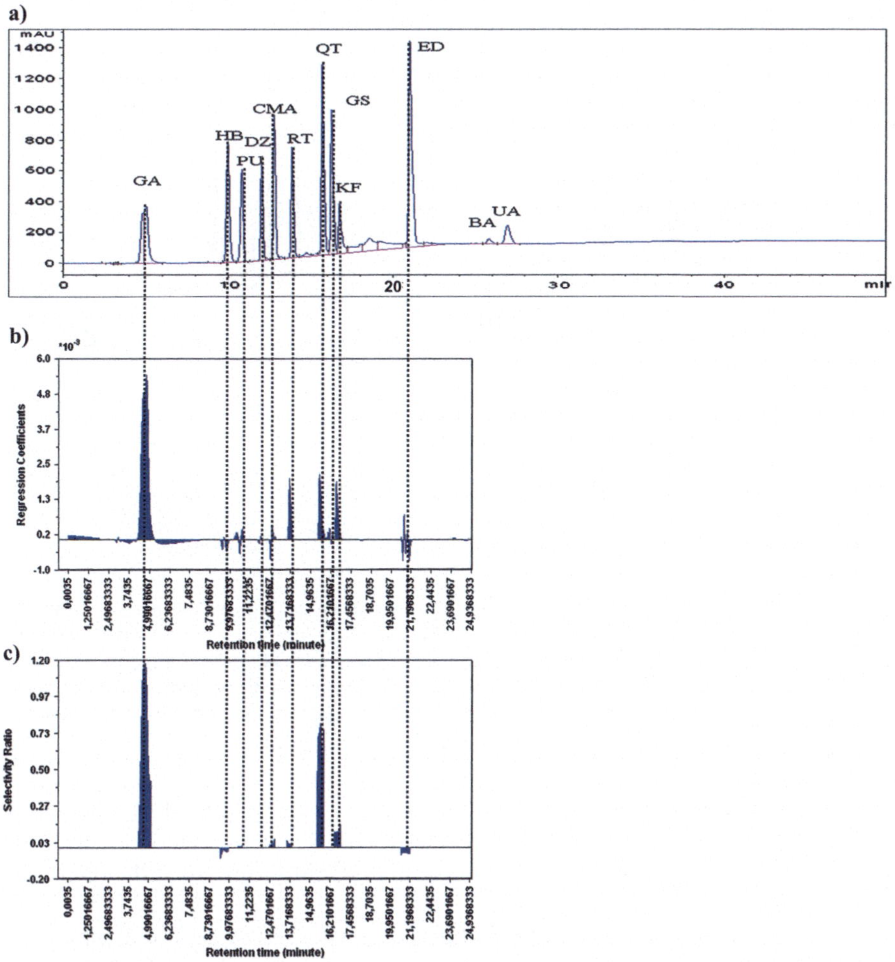

Fig. 6 (**a**) Typical fingerprint of the 12-compound mixture, (**b–c**) indication of important time zones in the PLS model by evaluation of the regression coefficients (**b**) and the multiplication of the selectivity ratio and the target projection loadings (**c**). Fine gray vertical lines highlight retention times of the 12 compounds. (Adapted from Kvalheim et al. 2011)

vals (coinciding with peaks in the chromatograms) which have the highest correlation with the FRAP value (see Fig. 7a).

The initially selected intervals are iteratively eliminated one by one (according to decreasing correlation with y), and their predictive value is assessed through the calculation of the RMSEP (root-mean-squared error of prediction) and plotted as a function of the eliminated intervals. Which variables are finally retained (Fig. 7c) is decided evaluating the minimum in this graph (Fig. 7b).

Next, PLS models were made leaving out one of the included components (intervals) and were compared to the performance of the PLS model with all components,

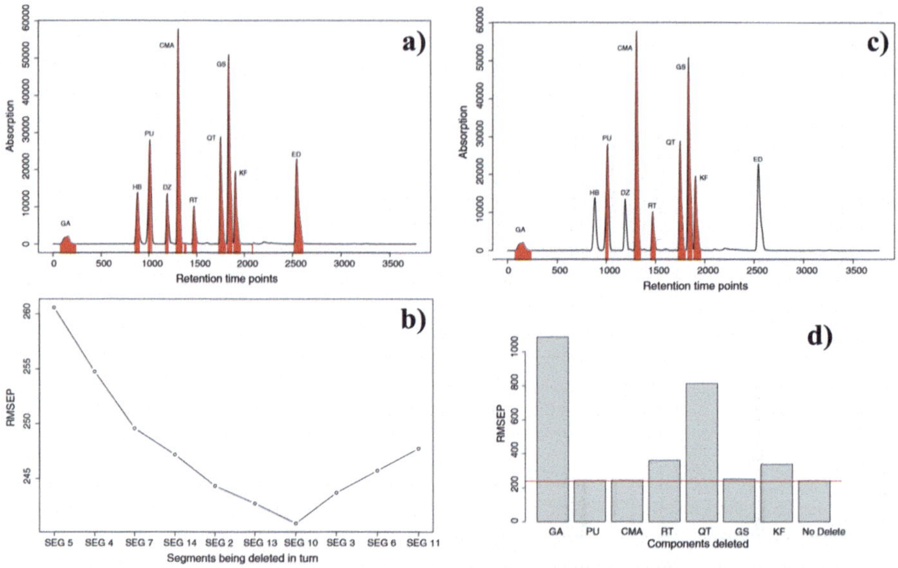

Fig. 7 SIS-iPLS procedure to link fingerprint information to the antioxidant activity. Initially selected m intervals (**a**): red peaks, RMSEP as a function of the eliminated intervals (**b**), the intervals maintained after iterative backward variable selection (**c**): red peaks, RMSEP increase for PLS models created after deletion of the specific intervals: *GA* gallic acid, *PU* puerarin, *CMA* coumaric acid, *RT* rutin, *QT* quercetin, *GS* genistein, *KF* kaempferol, No delete: model with all seven intervals (=reference level, indicated with the red line), (**d**) (Xu et al. 2015)

providing the reference RMSEP. The importance of each individual interval is assessed by the influence of eliminating it from the PLS regression model. The most important variables lead to the highest increase in RMSEP relative to the reference level (Fig. 7d).

For the synthetic sample set, the findings were checked against and confirmed the wet chemistry measurements. For the real herbal samples, the method's validity was studied differently. The intervals were compared to those obtained by other variable selection algorithms. The quality of all these models was compared by studying the distributions of the selected variables for the samples with the 30% lowest and 30% highest similarity values with the average profile and overlaying these with the FRAP distribution of the samples. Consequently, Xu et al. (2015) concluded that their method included more relevant variables than the other models. For the real plant samples, the study did not include identification of the compounds. This could be done by LC-MS or NMR, which provide more detailed structural information.

Ben Ahmed et al. (2016) studied the antioxidant compounds in *Pistacia atlantica* leaf extracts. The antioxidant activity of 28 samples was determined by two in vitro methods. The first assay used 2,2-diphenyl-1-picrylhydrazyl (DPPH), the other potassium ferricyanide (PFC). Both assays are complimentary, since they measure different aspects of antioxidant properties. Additionally, gradient HPLC profiles, using two serially coupled monolithic columns, were developed for the samples. The raw HPLC profiles showed retention time differences, which were corrected with correlation optimized warping. This resulted in well-aligned HPLC profiles (see Fig. 8a). As a result, a matrix **X**, containing the (aligned) HPLC profiles of the 28 samples, and two column vectors, **y**, containing the antioxidant activities of the samples determined by DPPH and PFC, respectively, were obtained. Multivariate calibration models (PLS and OPLS) were built based on column-centered, normalized plus column-centered, and standard-normal-variate transformed plus column-centered fingerprint data and used to predict the antioxidant activities of the samples from their fingerprint data. The predictive properties of the models were assessed based on the correspondence of the predicted and the measured antioxidant activities for the samples and expressed as the root-mean-squared error of cross-validation (RMSECV). The best models are those with a low RMSECV value.

The models were mainly used to evaluate the contribution of sample compounds to the measured activities. The antioxidant activity of a sample can be calculated as the summed products of the measured detector signal at a given time point in a sample's fingerprint (depending on the sample and the time point) and the regression coefficient characteristic for that time point. The regression coefficients of the different time points also constitute a profile (see Fig. 8b–d for column-centered, normalized plus column-centered, and standard-normal-variate transformed plus column-centered fingerprints, respectively). The peaks in this profile will be high (in absolute value) for compounds that have a large influence on the antioxidant activity and are small or absent for those that are not important for the antioxidant activity. Depending on the direction (positive or negative peaks), the compound's increase relates to an increase or decrease of the antioxidant activity. Comparing the regression coefficient profiles of the two assays with the fingerprints allowed identifying compounds influencing mainly the PFC measurement, others mainly influencing the DPPH measurement, and a third group influencing both. The figures of the regression coefficients for the DPPH and PFC models, overlaid with the fingerprints, are shown in Fig. 8a, b, respectively.

Identification of the indicated compounds was done by LC-ESI-QToF-MS in negative ionization mode, registering MS data in MS^E mode. In this mode, low-energy precursor ion information is registered, next to fragment ion information. Combination of both precursor and fragment ion information allowed the authors to tentatively identify 12 of the 13 compounds, yielding one unidentified compound. Future research should focus on the isolation and structural elucidation of this compound using structural elucidation approaches.

6.2 *Linked to In Vitro Enzymatic Assays*

To indicate the peaks responsible for the inhibition of butyrylcholinesterase and monoamine oxidase-A in an alkaloid fraction of the leaves of *Psychotria nemorosa*, Klein-Júnior et al. (2016a) used OPLS models. These enzymes are related to the

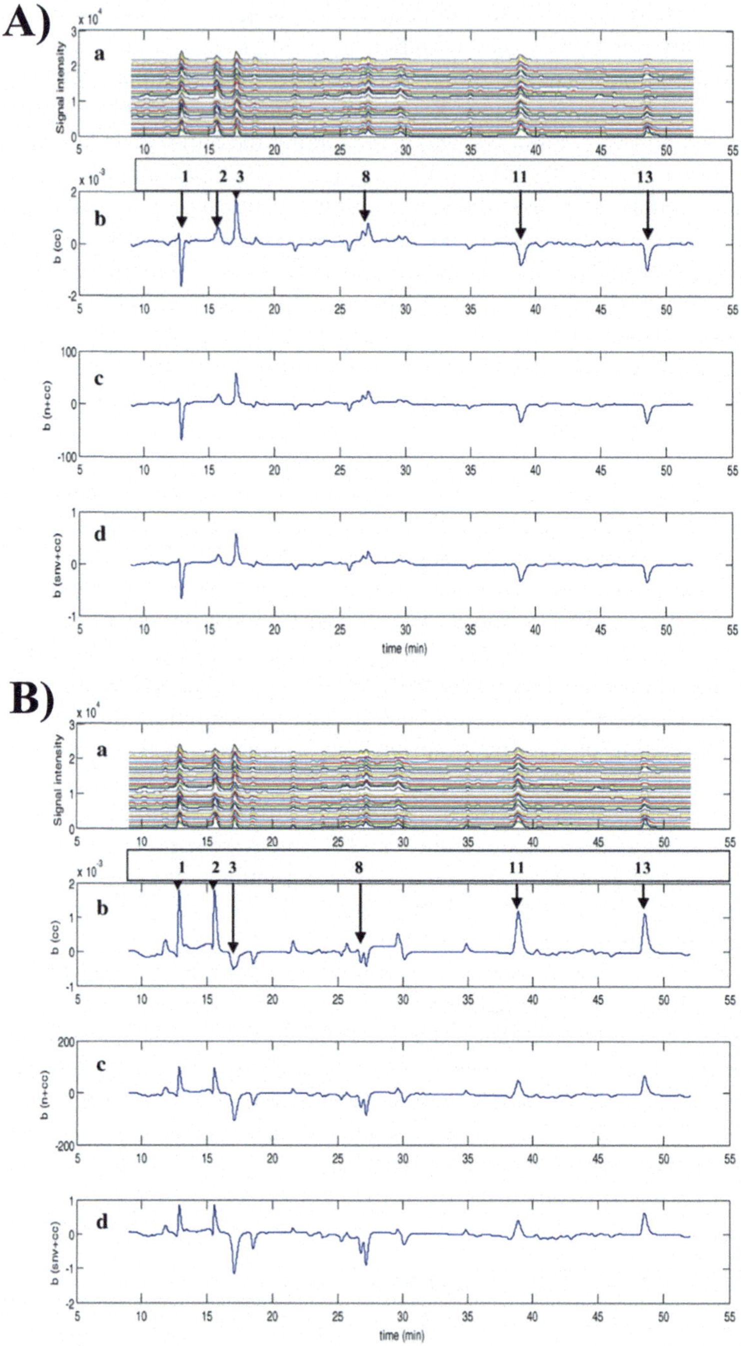

Fig. 8 OPLS regression modeling of the antioxidant activity determined using the DPPH (**a**) and PFC (**b**) methods. Preprocessed *Pistacia atlantica* fingerprints (a), OPLS regression coefficients obtained with column-centered (b), normalized plus column-centered (c), and standard-normal-variate transformed plus column-centered (d) fingerprint data. Numbers 1, 2, 3, 8, 11, and 13 refer to the peaks identified with LC-MS (Ben Ahmed et al. 2016)

remediation of the symptoms of neurodegenerative diseases, such as Alzheimer's and Parkinson's disease. Forty-three samples were collected from five different locations and were extracted using methanol-assisted micro-extraction by an ultrasonic bath. These extracts were fractionated by solid-phase extraction to elute alkaloid fractions. These fractions were evaluated for their modulation of enzymatic activity, and their chemical profiles were determined by ULPC-DAD, detected at 280 nm. The matrix **X**, consisting of 43 rows (samples) and 36,001 columns (time points), was submitted to COW alignment, followed by SNV and column centering. To model the activity as a function of the fingerprints, Klein-Júnior et al. evaluated different techniques: PLS and OPLS (Fig. 9). For both enzymes, the best results were obtained by OPLS, since it showed less noisy regression coefficients than PLS. OPLS removes the variation in matrix **X** that is not correlated to the response **y**. Four compounds were indicated as multifunctional, meaning able to inhibit both butyrylcholinesterase and monoamine oxidase-A (Fig. 10). Although these compounds were not isolated yet, their indole nature was confirmed by LC-SPE-NMR, and a fraction, enriched in these compounds (as well as others), was able to significantly inhibit the enzymatic activity.

Kang et al. (2013) applied the metabolic profiling approach to study the skin whitening effect of *Morus alba*. Since tyrosinase plays a pivotal role in the production of melanin pigment, its inhibition was measured to evaluate the in vitro effect of the *M. alba* at 10 μg/mL. The extracts were obtained from six different samples of the root bark using pressurized liquid extraction. Methanol, methanol/water (8:2, 5:5, 2:8,

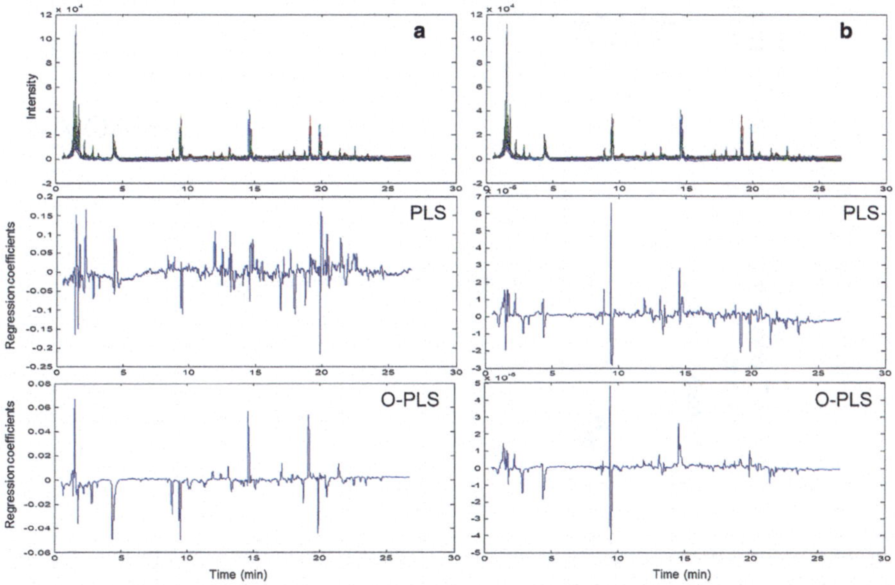

Fig. 9 Chromatographic fingerprints (top figure) and the regression coefficients from PLS and O-PLS models for butyrylcholinesterase inhibitory activity (**a**) and for monoamine oxidase-A inhibitory activity (**b**) (Klein-Júnior et al. 2016a)

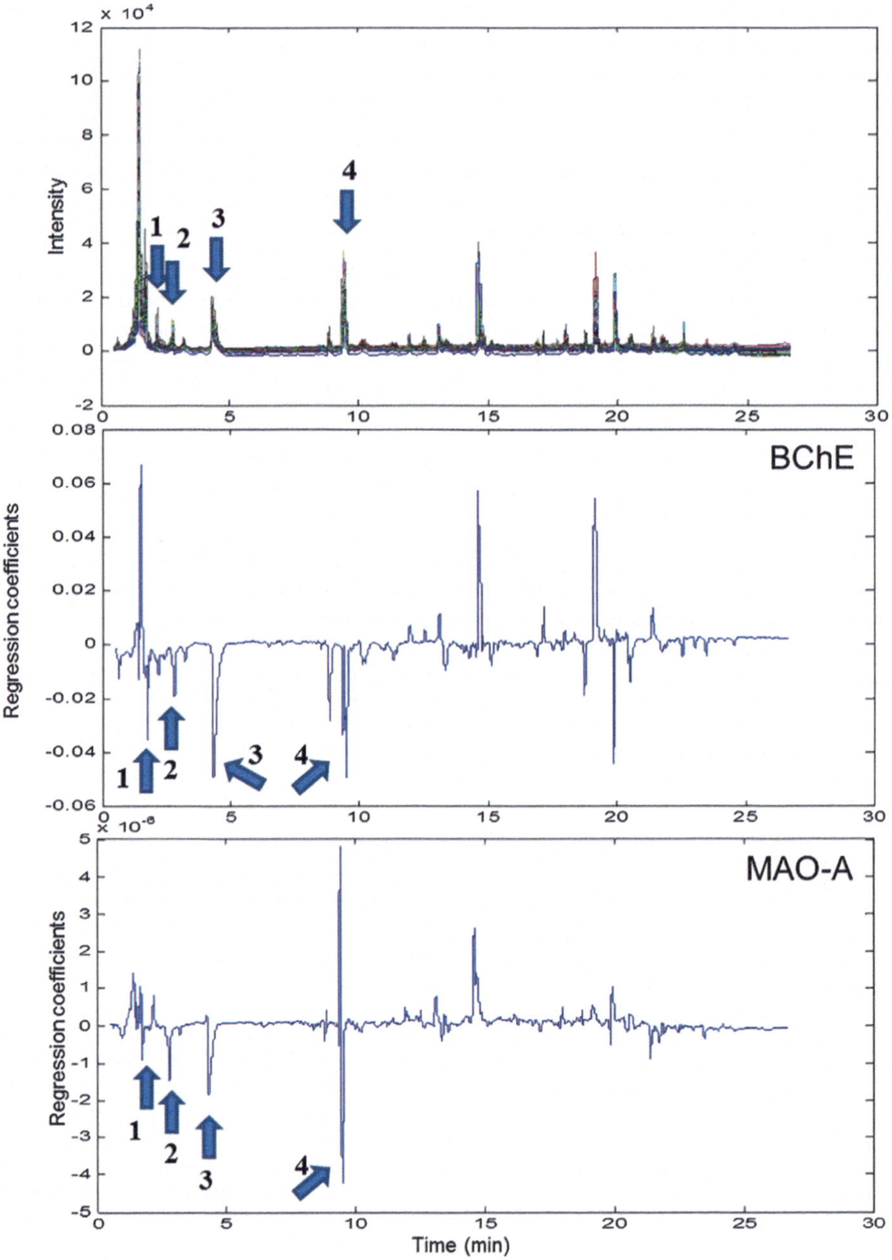

Fig. 10 Chromatographic fingerprints and the regression coefficients from O-PLS models for butyrylcholinesterase (BChE) inhibitory activity and for monoamine oxidase-A (MAO-A) inhibitory activity. The arrows indicate potentially multi-target compounds (Klein-Júnior et al. 2016a)

v/v), methanol/ethyl acetate (9:1, 7:3, v/v), and methanol/ethyl acetate/water (7:1,5:1,5, v/v) were used as solvents, totalizing 42 samples. The chemical analysis was performed by HPLC-DAD, giving chromatograms with 14,400 time points at 254 nm. Before linking fingerprint and bioactivity, the chromatograms were aligned using COW. Finally, using PLS modeling, it was possible to correlate the tyrosinase inhibitory effect of each extract to the chromatographic profile. To validate the model, two approaches were used: one where the samples were divided according to their origin (six groups) and the other one where the samples were divided according to the extraction solvent (seven groups). Then, a PLS model was created from a calibration set in which one of the groups was removed. The removed group, called the prediction set, was used to verify the prediction ability of the model. It was observed that, regardless of the source of difference between the samples, PLS was able to satisfactorily predict the activity of the extracts. Therefore, comparing the regression coefficient plot of the model with the chromatogram, it was possible to indicate that oxyresveratrol and its mono- and diglycosides, mulberrofuran G, kuwanon G, kuwanon H, and morusin exhibited positive regression coefficients, in different intensities, which are in line with previous studies, according to the authors.

Ben Ahmed et al. (2018) studied the leaves of *Pistacia atlantica*, an Algerian medicinal plant (previously studied for its antioxidant activity; Ben Ahmed et al. 2016), used for its antidiabetic and antihypertensive effects. The group evaluated the α-amylase, α-glucosidase, and angiotensin I-converting enzyme inhibitory effects of phenolic compound-enriched fractions. The plant material, collected in different periods and regions, was extracted with acetone/water (7:3) by maceration. After removal of acetone, the aqueous fraction was defatted, and ethyl acetate was used to obtain the evaluated fractions. The chromatographic fingerprints were obtained by HPLC-DAD, with a detection wavelength set at 254 nm. Prior to multivariate calibration, fingerprints were aligned by COW, and the matrix **X** was further pretreated by SNV followed by column centering. Twenty-eight samples were used to build a PLS model, using the inhibitory activity, expressed as IC_{50}, as response **y**. Through evaluation of the regression coefficient plots and using LC-MS-based identification, glucogallin, quinic acid, and galloylquinic acid were indicated as α-amylase inhibitors; methyl gallate and tetragalloylglucoside as α-glucosidase inhibitors; gallic acid, gentisic acid, and digalloylquinic acid as α-amylase and α-glucosidase inhibitors; and glucogallin, gallic acid, galloylshikimic acid, methyl gallate, digalloylquinic acid, digallic acid, trigalloylglucose, and tetragalloylquinic acid as angiotensin I-converting enzyme inhibitors.

6.3 Linked to In Vivo Assays

Li et al. (2017) studied how in various fruit parts of *G. xanthochymus* differences in fingerprints obtained by UHPLC-QToF-MS could correlate with bioactivities measured with zebra fish models, in order to determine biflavonoids with anti-angiogenic activity. To prepare crude extracts, the pericarp, aril (= sheath that encloses the seed), and seed of the fruits were separately extracted in a 70% methanol/H_2O solvent and then with a subsequent ultrasonic extraction and finally a combination of these extracted parts were put together. After redissolution in methanol, nine samples (three different collections) were analyzed using UHPLC. Zebra fish embryos were subdivided in six groups of twenty embryos that were treated with water, either containing extracts or not. Growth of the subintestinal vessels (SIVs) of the embryos were then microscopically investigated. Pericarp and aril extracts at different low concentrations show considerable length decreases in vessel growth. However, seed extracts have no significant influence. These results indicated that in pericarp and aril extracts, comparable bioactive components are expected to occur, different from these in seed. From the raw chromatographic data, retention time, exact mass, and ion intensity of the compounds in the fingerprint were used as variables for the chemometric study. Prior to this analysis, fingerprints were preprocessed, making use of peak identification, peak integration, and peak alignment. The obtained data was then further normalized.

The principal component analysis (PCA) score plot showed clustering of aril and pericarp samples, separated from the seed extracts. Calculation of goodness of fit and predictability of the PCA values stated that there were similar metabolic components in aril and pericarp. The PCA score plot clearly indicated that the chemical composition of the seeds is considerably different. This trend was also observed in the total ion chromatograms and the PCA loading plot, where differences could be seen in the variables. In the latter plot, marker ions from seed can be discriminated from those seen in the aril-pericarp. To compare the inactive seed group and the active aril-pericarp group, orthogonal partial least square-discriminant analysis (OPLS-DA) with scatterplot (S-plot) was applied.

Important marker ions that differentiate the two groups were determined. This way 13 markers, at the top of the S-plot, were indicated as candidates that potentially have anti-angiogenic activity. From these 13 components, two biflavonoids, xanthochymol and amentoflavone, were further studied for their potential bioactivity in zebra fish embryos. No effect of xanthochymol on the growth of SIV was seen. However, amentoflavone importantly decreased the vessel growth at certain concentrations. Moreover, both compounds showed downregulation of the expressions of certain genes. The study results show that amentoflavone has anti-angiogenic effects. Further study of seven other biflavonoids for their bioactivity-structure relation in the in vivo method displayed that only fukugetin inhibited the growth of SIVs, in other words had anti-angiogenic activity.

Wen et al. (2018) studied the targeted isolation and identification of compounds with potential antihyperlipidemic activity in crabapple using UHPLC-DAD-MS-SPE/NMR fingerprints, mice experiments, and PLS-DA. Twelve crude extracts from three crabapple species, four extracts of each variety, were prepared, making use of different solvents and solvent concentrations. The in vivo mice model was set up, consisting of eighteen groups (eight animals per group), three control groups and fifteen test groups each containing randomly chosen mice, in which obesity was induced. Twelve of the 15 groups were treated with (12 groups) or without (three groups) the 12 different extracts in order to investigate their possible cholesterol-reducing effect. Only six of the 12 extracts showed antihyperlipidemic activity, so the extracts were classified in an active and an inactive group. In order to determine which compounds in the extracts correlate with the results of the mice experiments, the 12 samples were analyzed by LC-MS of which the raw data were preprocessed through alignment and autoscaling before chemometric analysis.

PCA, PLS-DA, and independent sample t-test of the UHPLC data were carried out. In order to generate appropriate classification models, the fingerprints were measured with + and − electrospray ionization. PCA positive and negative score plots clearly showed the two distinct groups. No outliers were observed; therefore all data were used for PLS-DA. Performing the latter analysis, significant values for goodness of fit and predictability were calculated, indicating a well-defined model to distinguish between active and inactive samples. After evaluation of the variables through Variable Importance in Projection, the Pearson correlation coefficient, and the p-value in the independent samples t-test, 22 differentiating variables were selected.

After automated MS-guided SPE trapping, ten enriched target compounds with potential cholesterol-reducing activity were obtained, eluted with deuterated methanol and introduced in an NMR system. By means of the obtained NMR and MS/MS spectra, seven compounds were identified by comparing with existing data from a local database, while the remaining three constituents went through the usual chemistry research channels for identification. Out of these ten constituents, six were already studied earlier and reported to have cholesterol-reducing capacities.

7 Conclusions

The study of plant extracts, aiming the discovery of new chemical entities with therapeutic potential, involves a complex process, merging areas, such as chemistry, pharmacy, biology, and pharmacology. To rationalize this intricate procedure, given methodologies have been defined, avoiding the trial-and-error approach. One methodology drawing the attention is the metabolic profiling approach, where the chemical fingerprint is correlated chemometrically with a specific activity, affording valuable information regarding the active compounds in a given extract.

As highlighted in this book chapter, although this methodology aims guiding (and theoretically, simplifying) the identification of the active compounds, several steps are important and highly influenced by different factors, demanding a well-designed experimental procedure. From sample preparation till data analysis, each step has a given number of possible approaches, making it challenging to choose the most appropriate combination.

The first important step is the plant-material extraction. It will directly impact the final result. Too complex extracts may contain several artifacts that may trouble the LC analysis, as well as impair the in vitro evaluation, mainly because of the presence of PAINS. In addition, greener and faster extraction techniques, such as supercritical fluid extraction, are preferred over the classical methods, e.g., maceration.

Therefore, in order to prevent future problems, the extraction procedure and, eventually, the cleanup methodology should be optimized, avoiding meaningless information to be obtained.

In the chromatographic analysis, usually reversed-phase LC is used. Even so, a huge number of possibilities can be screened, such as different stationary and mobile phases, flow rates, and detectors. This will be limited, in the end, mainly by the resources of the laboratory. However, detectors that can give structural information, e.g., mass spectrometry, may be recommended to indicate potentially active compounds, thus avoiding the re-isolation of already known (bioactive) secondary metabolites.

In parallel, the pharmacological evaluation methodology also plays an important role in the output. Classical in vivo experiments may not be a good choice, since they are less reproducible, weakening the mathematical model. Last but not least, ethical aspects must also be taken into account here. Therefore, in vitro assays are preferred, giving responses with lower error and providing a high-throughput possibility. However, because of its own limitations, e.g., lack of pharmacokinetic information, alternative organisms for in vivo assays have been proposed, such as zebra fish.

As a final step, chemometric tools are employed to link chemical and pharmacological information. In this sense, multivariate calibration techniques are used. However, most of the time, data pretreatment is a must. Peak alignment procedures, such as COW, are often applied. Then, unsupervised data analysis techniques, e.g., PCA and HCA, are used to detect outliers and to observe trends in the (pretreated) matrix $\mathbf{X}$. Finally, in order to correlate the (pretreated) matrix $\mathbf{X}$ with the response $\mathbf{y}$, calibration methodologies are applied. Usually PLS is used, often giving good prediction models. However, in some cases, variations of this method, such as OPLS, are used. Regardless of the method, the regression coefficient plot comparison to the LC fingerprints can be used to indicate the potentially active compounds. These might be identified by hyphenated techniques, e.g., LC-MS and LC-NMR. However, the isolation of the bioactive metabolites, if not already described in the literature, is mandatory to validate the prediction, as well as for identification purposes in case unknown compounds are involved.

In the end, the metabolic profiling approach works as a sieve. It retains (and eliminates) unimportant information (inactive compounds) and lets the important data (active compounds) pass over the modeled mesh. However, one must keep in mind that the mesh selectivity is based on earlier knowledge. Therefore, a poor matrix $\mathbf{X}$ will cause deformations in the sieve, retaining either more or less information than it should and allowing to pass either more or less important data. Ultimately, inactive compounds thus can be indicated, and valuable compounds could be missed. When it happens, and when it is recognized what is not evident, there is no other way than to "sift" everything again.

References

Alaerts G, Matthijs N, Smeyers-Verbeke J et al (2007) Chromatographic fingerprint development for herbal extracts: a screening and optimization methodology on monolithic columns. J Chromatogr A 1172:1–8
Alaerts G, Dejaegher B, Smeyers-Verbeke J et al (2010a) Recent developments in chromatographic fingerprints from herbal products: set-up and data analysis. Comb Chem High Throughput Screen 13:900–922
Alaerts G, Merino-Arévalo M, Dumarey M et al (2010b) Exploratory analysis of chromatographic fingerprints to distinguish rhizoma chuanxiong and rhizoma ligustici. J Chromatogr A 1217:7706–7716
Allard PM, Genta-Jouve G, Wolfender JL (2017) Deep metabolome annotation in natural products research: towards a virtuous cycle in metabolite identification. Curr Opinion Chem Biol 36:40–49
Atanasov AG, Waltenberger B, Pferschy-Wenzing EM et al (2015) Discovery and resupply of pharmacologically active plant-derived natural products: a review. Biotechnol Adv 33:1582–1614
Azmir J, Zaidul ISM, Rahman MM et al (2013) Techniques for extraction of bioactive compounds from plant materials: a review. J Food Eng 117:426–436
Azwanida N (2015) A review on the extraction methods used in medicinal plants, principle, strength and limitation. Med Aromat Plants 4(2–6):196
Baell JB (2016) Feeling nature's PAINS: natural products, natural product drugs, and pan assay interference compounds (PAINS). J Nat Prod 79:616–628
Baell JB, Walters MA (2014) Chemical con artists foil drug discovery. Nature 513:481–483
Begou O, Gika HG, Wilson ID et al (2017) Hyphenated MS-based targeted approaches in metabolomics. Analyst 142:3079–3100
Ben Ahmed Z, Yousfi M, Viaene J et al (2016) Antioxidant activities of *Pistacia atlantica* extracts modeled as a function of chromatographic fingerprints in order to identify antioxidant markers. Microchem J 128:208–217
Ben Ahmed Z, Yousfi M, Viaene J et al (2017) Seasonal, gender and regional variations in total phenolic, flavonoid, and condensed tannins contents and in antioxidant properties from *Pistacia atlantica* ssp. leaves. Pharm Biol 55:1185–1194
Ben Ahmed Z, Yousfi M, Viaene J et al (2018) Potentially antidiabetic and antihypertensive compounds identified from *Pistacia atlantica* leaf extracts by LC fingerprinting. J Pharm Biomed Anal 149:547–556
Bloemberg TG, Gerretzen J, Lunshof A et al (2013) Warping methods for spectroscopic and chromatographic signal alignment: a tutorial. Anal Chim Acta 781:14–32
Brusotti G, Cesari I, Dentamaro A (2014) Isolation and characterization of bioactive compounds from plant resources: the role of analysis in the ethnopharmacological approach. J Pharm Biomed Anal 87:218–228
Butler MS, Fontaine F, Cooper MA (2014) Natural product libraries: assembly, maintenance, and screening. Planta Med 80:1161–1170
Cardoso-Taketa AT, Pereda-Miranda R, Choi YH et al (2008) Metabolic profiling of the Mexican anxiolytic and sedative plant *Galphimia glauca* using nuclear magnetic resonance spectroscopy and multivariate data analysis. Planta Med 74:1295–1301
Choi YH, Verpoorte R (2014) Metabolomics: what you see is what you extract. Phytochem Anal 25:289–290
Ćujić N, Šavikin K, Janković T et al (2016) Optimization of polyphenols extraction from dried chokeberry using maceration as traditional technique. Food Chem 194:135–142
de O Silva E, Borges LL, da Conceição EC et al (2017) Box–Behnken experimental design for extraction of artemisinin from *Artemisia annua* and validation of the assay method. Rev Bras Farmacogn 27:519–524

Dary C, Baghdikian B, Kim S et al (2017) Optimization of ultrasound-assisted extraction of bioactive alkaloids from *Stephania cambodica* using response surface methodology. C R Chim 20:996–1005

Das AK, Mandal V, Mandal SC (2013) A brief understanding of process optimisation in microwave-assisted extraction of botanical materials: options and opportunities with chemometric tools. Phytochem Anal 25:1–12

David B, Wolfender JL, Dias DA (2015) The pharmaceutical industry and natural products: historical status and new trends. Phytochem Rev 14:299–315

De Monte C, Carradori S, Granese A et al (2014) Modern extraction techniques and their impact on the pharmacological profile of *Serenoa repens* extracts for the treatment of lower urinary tract symptoms. BMC Urol 14(63):1–11

Dejaegher B, Alaerts G, Matthijs N (2010) Methodology to develop liquid chromatographic fingerprints for the quality control of herbal medicines. Acta Chromatogr 22:237–258

Doke SK, Dhawale SC (2015) Alternatives to animal testing: a review. Saudi Pharm J 23:223–229

Dong MW (2006) Modern HPLC for practicing scientists. Wiley, Hoboken, 304pp

Esbensen KH, Geladi P (2009) Principal component analysis: concept, geometrical interpretation, mathematical background, algorithms, history, practice (Chapter 2.13). In: Brown SD, Tauler R, Walczak B (eds) Comprehensive chemometrics, vol 2. Elsevier, Amsterdam, pp 211–226

Esteki M, Farajmand B, Amanifar S et al (2017) Classification and authentication of Iranian walnuts according to their geographical origin based on gas chromatographic fatty acid fingerprint analysis using pattern recognition methods. Chem Intell Lab Syst 1217:7706–7716

Garrido L, Beckmann N (eds) (2013) New applications of NMR in drug discovery and development. RSC Publishing, Cambridge, UK, 530pp

Gauglitz G, Vo-Dinh T (eds) (2003) Handbook of spectroscopy. Volume 1. Wiley-VCH, Weinheim, 538pp

Gemperline E, Keller C, Li L (2016) Mass spectrometry in plant-omics. Anal Chem 88:3422–3434

Gerretzen J, Szymańska E, Jansen JJ et al (2015) Simple and effective way for data preprocessing selection based on design of experiments. Anal Chem 87:12096–12103

Goodarzi M, Russell PJ, Vander Heyden Y (2013) Similarity analyses of chromatographic herbal fingerprints: a review. Anal Chimica Acta 804:16–28

Guiochon G, Gritti F (2011) Shell particles, trials, tribulations and triumphs. J Chromatogr A 1218:1915–1938

Gunzler H, Williams A (eds) (2001) Handbook of analytical techniques. Wiley-VCH, Weinheim, 1182pp

Gupta A, Naraniwal M, Kothari V (2012) Modern extraction methods for preparation of bioactive plant extracts. Int J Appl Nat Sci 1:8–26

Hammi KM, Jdey A, Abdelly C et al (2015) Optimization of ultrasound-assisted extraction of antioxidant compounds from Tunisian *Zizyphus lotus* fruits using response surface methodology. Food Chem 184:80–89

Harvey L, Edrada-Ebel R, Quinn RJ (2015) The re-emergence of natural products for drug discovery in the genomics era. Nat Rev Drug Discov 14:111–129

Hayes R, Ahmed A, Edge T et al (2014) Core-shell particles: preparation, fundamentals and applications in high performance liquid chromatography. J Chromatogr A 1357:36–52

He Y, Wu Q, Hansen SH et al (2013) Differentiation of tannin-containing herbal drugs by HPLC fingerprints. Pharmazie 68:155–159

Henrich CJ, Beutler JA (2013) Matching the power of high throughput screening to the chemical diversity of natural products. Nat Prod Rep 30:1284–1298

Hill CB, Roessner U (2015) Advances in high-throughput LC-MS analysis for plant metabolomics. In: de Vooght-Johnson R (ed) Advanced LC-MS applications for metabolomics. Future Science Ltd, London, pp 58–71

Hubert J, Nuzillard JM, Renault JH (2017) Dereplication strategies in natural product research: how many tools and methodologies behind the same concept? Phytochem Rev 16:55–95

Huie CW (2002) A review of modern sample-preparation techniques for the extraction and analysis of medicinal plants. Anal Bioanal Chem 373:23–30

Izadiyan P, Hemmateenejad B (2016) Multi-response optimization of factors affecting ultrasonic assisted extraction from Iranian basil using central composite design. Food Chem 190:864–870

Kaiser KA, Barding GA Jr, Larive CK (2009) A comparison of metabolite extraction strategies for ^{1}H-NMR-based metabolic profiling using mature leaf tissue from the model plant *Arabidopsis thaliana*. Magn Reson Chem 47:S147–S156

Kang KB, Lee DY, Kim TB et al (2013) Prediction of tyrosinase inhibitory activities of *Morus alba* root bark extracts from HPLC fingerprints. Microchem J 110:731–738

Kim HK, Verpoorte R (2010) Sample preparation for plant metabolomics. Phytochem Anal 21:4–13

Klein-Júnior LC, Viaene J, Salton J et al (2016a) The use of chemometrics to study multifunctional indole alkaloids from *Psychotria nemorosa* (*Palicourea comb. nov.*). Part I: extraction and fractionation optimization based on metabolic profiling. J Chromatogr A 1463:60–70

Klein-Júnior LC, Viaene J, Tuenter E et al (2016b) The use of chemometrics to study multifunctional indole alkaloids from *Psychotria nemorosa* (*Palicourea comb. nov.*). Part II: indication of peaks related to the inhibition of butyrylcholinesterase and monoamine oxidase-A. J Chromatogr A 1463:71–80

Klein-Júnior LC, Vander Heyden Y, Henriques A (2016c) Enlarging the bottleneck in the analysis of alkaloids: a review on sample preparation in herbal matrices. Trends Anal Chem 80:66–82

Korifi R, Le Dréau Y, Dupuy N (2014) Comparative study of the alignment method on experimental and simulated chromatographic data. J Sep Sci 37:3276–3291

Kumar D (2016) Nuclear magnetic resonance (NMR) spectroscopy for metabolic profiling of medicinal plants and their products. Crit Rev Anal Chem 46:400–412

Kvalheim OM, Chan HY, Benzie IFF et al (2011) Chromatographic profiling and multivariate analysis for screening and quantifying the contributions from individual components to the bioactive signature in natural products. Chemom Intell Lab Syst 107:98–105

La Barbera G, Capriotti AL, Cavaliere C et al (2017) Liquid chromatography – high resolution mass spectrometry for the analysis of phytochemicals in vegetal-derived food and beverages. Food Res Int 100:28–52

Lee I, Yang J (2009) Common clustering algorithms (Chapter 2.27). In: Brown SD, Tauler R, Walczak B (eds) Comprehensive chemometrics, vol 2. Elsevier, Amsterdam, pp 557–618

Li FS, Weng JK (2017) Demystifying traditional herbal medicine with modern approaches. Nat Plants 3(17109):1–7

Li Y, Wu T, Zhu J et al (2010) Combinative method using HPLC fingerprint and quantitative analyses for quality consistency evaluation of a herbal medicinal preparation produced by different manufacturers. J Pharm Biomed Anal 52:597–602

Li JR, Li M, Xia B et al (2013) Efficient optimization of ultrahigh-performance supercritical fluid chromatographic separation of *Rosa sericea* by response surface methodology. J Sep Sci 36:2114–2120

Li P, Yue GGL, Kwok HF et al (2017) Using ultra-performance liquid chromatography quadrupole time of flight mass spectrometry-based chemometrics for the identification of anti-angiogenic biflavonoids from edible *Garcinia* species. J Agric Food Chem 65:8348–8355

Liu WJH (ed) (2011) Traditional herbal medicine research methods. Identification, analysis, bioassay, and pharmaceutical and clinical studies. Wiley, Hoboken, 466pp

Liu CT, Zhang M, Yan P et al (2016) Qualitative and quantitative analysis of volatile components of Zhengtian pills using gas chromatography mass spectrometry and ultra-high performance liquid chromatography. J Anal Methods Chem 2016:1–8

Lough WJ, Wainer IW (1996) High performance liquid chromatography: fundamental principles and practice, 1st edn. Blackie Academic and Professional, Glasgow, 288pp

Lu GH, Chan K, Liang YZ et al (2005) Development of high-performance liquid chromatographic fingerprints for distinguishing Chinese Angelica from related umbelliferae herbs. J Chromatogr A 1073:383–392

Martin AC, Pawlus AD, Jewett EM et al (2014) Evaluating solvent extraction systems using metabolomics approaches. RSD Adv 4:26325–26334

McCloud TG (2010) High throughput extraction of plant, marine and fungal specimens for preservation of biologically active molecules. Molecules 15:4526–4563

Munafò MR, Nosek BA, Bishop DVM et al (2017) A manifesto for reproducible science. Nat Human Behav 1:0021

Mushtaq MY, Choi YH, Verpoorte R et al (2014) Extraction for metabolomics: access to the metabolome. Phytochem Anal 25:291–306

Nguyen Hoai N, Dejaegher B, Tistaert C et al (2009) Development of HPLC fingerprints for *Mallotus* species extracts and evaluation of the peaks responsible for their antioxidant activity. J Pharm Biomed Anal 50:753–763

Pawar SV, Ho JC, Yadav GD et al (2017) The impending renaissance in discovery & development of natural products. Curr Top Med Chem 17:251–267

Plant AL, Locascio LE, May WE et al (2014) Improved reproducibility by assuring confidence in measurements in biomedical research. Nat Method 11:895–898

Qin HL, Deng AJ, Du GH et al (2009) Fingerprinting analysis of *Rhizoma chuanxiong* of commercial types using 1H nuclear magnetic resonance spectroscopy and high performance liquid chromatography method. J Integr Plant Biol 51:537–544

Raks V, Al-Suod H, Buszewski B (2018) Isolation, separation, and preconcentration of biologically active compounds from plant matrices by extraction techniques. Chromatographia 81:189–202

Rates SMK (2001) Plants as source of drugs. Toxicon 39:603–613

Reardon S (2016) A mouse's house may ruin studies. Nature 530:264

Ren S, Hinzman AA, Kang EL et al (2015) Computational and statistical analysis of metabolomics data. Metabolomics 11:1492–1513

Roessner U, Dias DA (eds) (2013) Metabolomics tools for natural product discovery: methods and protocols. Humana Press, New York, 311pp

Sasidharan S, Chen Y, Saravanan D et al (2011) Extraction, isolation and characterization of bioactive compounds from plants' extracts. Afr J Tradit Complement Altern Med 8:1–10

Shen B (2015) A new golden age of natural products discovery. Cell 163:1297–1300

Sherma J, Fried B (eds) (2003) Handbook of thin-layer chromatography, 3rd edn. Marcel Dekker, New York, 1331pp

Souza CRF, Bott RF, Oliviera WP (2007) Optimization of the extraction of flavonoid compounds from herbal material using experimental design and multi-response analysis. Lat Am J Pharm 26:682–690

t'Kindt R, Morreel K, Deforce D et al (2009) Joint GC-MS and LC-MS platforms for comprehensive plant metabolomics: repeatability and sample pre-treatment. J Chrom B 877:3572–3580

Tang TX, Guo WY, Xu Y et al (2014) Thin-layer chromatographic identification of Chinese propolis using chemometric fingerprinting. Phytochem Anal 2:266–272

Tistaert C, Dejaegher B, Nguyen Hoai N et al (2009) Potential antioxidant compounds in *Mallotus* species fingerprints. Part I: Indication, using linear multivariate calibration techniques. Anal Chim Acta 652:189–197

Tistaert C, Dejaegher B, Vander Heyden Y (2011) Chromatographic separation techniques and data handling methods for herbal fingerprints: a review. Anal Chim Acta 690:148–161

Tu Y, Jeffries C, Ruan H et al (2010) Automated high-throughput system to fractionate plant natural products for drug discovery. J Nat Prod 73:751–754

Universitätsbibliothek Leipzig (2016) Available at https://papyrusebers.de. Accessed Dec 2017

Van den Berg RA, Hoefsloot HCJ, Westerhuis JA et al (2006) Centering, scaling, and transformations: improving the biological information content of metabolomics data. BMC Genomics 7:142

Vest Nielsen N-P, Carstensen JM, Smedsgaard J (1998) Aligning of single and multiple wavelength chromatographic profiles for chemometric data analysis using correlation optimised warping. J Chromatogr A 805:17–35

Vuckovic D (2012) Current trends and challenges in sample preparation for global metabolomics using liquid chromatography-mass spectrometry. Anal Bioanal Chem 403:1523–1548

Wagner H, Ulrich-Merzenich G (2009) Synergy research: approaching a new generation of phytopharmaceuticals. Phytomedicine 16:97–110

Waksmundzka-Hajnos M, Sherma J (eds) (2011) High performance liquid chromatography in phytochemical analysis. CRC Press, Boca Raton, 996pp

Wang D, Wang S, Du Q et al (2014) Optimization of extraction and enrichment of steroidal alkaloids from bulbs of cultivated *Fritillaria cirrhosa*. Biomed Res Int 2014:1–15

Watson JT, Sparkman OD (2007) Introduction to mass spectrometry: instrumentation, applications, and strategies for data interpretation, 4th edn. Wiley, Chichester, 864pp

Wen Y, Chen L, Li J et al (2014) Recent advances in solid-phase sorbents for sample preparation prior to chromatographic analysis. Trends Anal Chem 59:26–41

Wen C, Wang D, Li X et al (2018) Targeted isolation and identification of bioactive compounds lowering cholesterol in the crude extracts of crabapples using UPLC-DAD-MS-SPE/NMR based on pharmacology-guided PLS-DA. J Pharm and Biomed Anal 150:144–151

Wu H, Guo J, Chen S et al (2013) Recent developments in qualitative and quantitative analysis of phytochemical constituents and their metabolites using liquid chromatography-mass spectrometry. J Pharm Biomed Anal 72:267–291

Wu J, Yu D, Sun H et al (2015) Optimizing the extraction of anti-tumor alkaloids from the stem of *Berberis amurensis* by response surface methodology. Ind Crop Prod 69:68–75

Xu J, Xu QS, Chan CO et al (2015) Identifying bioactive components in natural products through chromatographic fingerprint. Anal Chim Acta 870:45–55

Yuliana ND, Khatib A, Verpoorte R et al (2011) Comprehensive extraction method integrated with NMR metabolomics: a new bioactivity screening method for plants, adenosine a1 receptor binding compounds in *Orthosiphon stamineus* Benth. Anal Chem 83:6902–6906

Zeaiter M, Rutledge D (2009) Preprocessing methods (Chapter 3.04). In: Brown SD, Tauler R, Walczak B (eds) Comprehensive chemometrics, vol 3. Elsevier, Amsterdam, pp 121–231

Zhu C, Liu X (2013) Optimization of extraction process of crude polysaccharides from pomegranate peel by response surface methodology. Carbohydr Polym 92:1197–1202

Functional Foods as Source of Bioactive Principles: Some Marked Examples

Adriana Campos, Elisa Brasili, Camile Cecconi Cechinel-Zanchett, and Valdir Cechinel Filho

Abbreviations

ABG	Aged black garlic
AChE	Acetylcholinesterase
AGE	Aged garlic extracts
AK	Autophosphorylation-activated protein kinase
AP-1	Protein-1
ATP	Adenosine triphosphate
BDMCur	Bisdemethoxycurcumin
CAT	Catalase
CCur	Cyclocurcumin
CDPK	Ca^{2+} -dependent protein kinase
COX-2	Cyclooxygenase-2
CREB	CAMP-response element-binding protein
CVD	Cardiovascular disease
DADS	Diallyl disulfide
DHA	Docosahexaenoic acid
DMCur	Demethoxycurcumin
DPA	Docosapentaenoic acid
DPPH	2,2-Diphenyl-1-picrylhydrazyl
DAS	Diallyl sulfide
DATS	Diallyl trisulfide

A. Campos
Departamento de Nutrição, Faculdade de Saúde Pública, Universidade de São Paulo-USP, São Paulo, Brazil

E. Brasili
Department of Environmental Biology, Sapienza University of Rome, Rome, Italy

C. C. Cechinel-Zanchett · V. Cechinel Filho (✉)
Programa de Pós-Graduação em Ciências Farmacêuticas and Núcleo de Investigações Químico-Farmacêuticas (NIQFAR), Universidade do Vale do Itajaí – UNIVALI, Itajaí, Santa Catarina, Brazil
e-mail: cechinel@univali.br

V. Cechinel Filho (ed.), *Natural Products as Source of Molecules with Therapeutic Potential*, https://doi.org/10.1007/978-3-030-00545-0_4

DM	Diabetes mellitus
EC	Epicatechin
EGC	Epigallocatechin
EGCG	Epigallocatechin gallate
EGFR	Epidermal growth factor receptor
EPA	Eicosapentaenoic acid
EVOO	Extra virgin olive oil
FDA	Food and Drug Administration
FFDCA	Federal Food, Drug, and Cosmetic Act
GRAS	Generally recognized as safe
GR	Glutathione reductase
GPx	Glutathione peroxidase
GST	Glutathione- S -transferase
GSH	Reduced glutathione
HDL	High-density lipoprotein
HO-1	Heme-oxygenase 1
IFG	Impaired fasting glucose
IFN-γ	Interferon-γ
IGF	Insulin-like growth factor
IkB	Inhibitory factor I-kappa B kinase
iNOS	Inducible nitric oxide synthase
IRAK	IL-1 receptor-associated kinase
JAK	Janus kinases (JAK) through inhibition of
LDL	Low-density lipoprotein
LPa	Lipoprotein A
MAP	Mitogen-activated protein
MAPKs	Mitogen-activated protein kinases
MCP	Monocyte chemoattractant protein
MIC	Minimum inhibitory concentration
MDA-MB-231	Metastatic process in human breast cancer cells
MMPs	Matrix metalloproteinases
NF-kB	Nuclear factor-kappa B
NF-jB	Nuclear factor-jB
Nrf2	Nuclear factor erythroid-derived 2
NQO1	NAD(P)H:quinone oxido-reductase 1
PhK	Phosphorylase kinase
PUFAs	Polyunsaturated fatty acids
ROS	Reactive oxygen species
SAC	S-allylcysteine
SAMC	S-allylmercaptocysteine
SCFA	Short-chain fatty acids
SIRT	Silent information regulator
SOD	Superoxide dismutase
TAG	Triacylglycerols
T2DM	Type2 diabetes mellitus

TNF-α	Tumor necrosis factor
USA	United States of America
US	United States
USDA	US Department of Agriculture
USDHHS	US Department of Health and Human Services
VLDL	Very-low-density lipoprotein
γ-GCL	γ-glutamyl-cysteine ligase

1 Introduction

Functional foods are those that have a beneficial effect on health beyond the basic function of nutrition, helping to promote better health conditions and to reduce the risk of various diseases (Granato et al. 2017).

This concept was created in Japan in the 1980s and is widely used around the world. The regulation of these products is made by different authorities and has not been defined yet in many countries. For example, in Europe new products have appeared, and the interest to define standards and guidelines to the development and promotion of these products are growing, but the European Union still regulates compounds, ingredients, and plants in national levels (Van Buul and Brouns 2015). In the United States, the Federal Food, Drug, and Cosmetic Act (FFDCA) does not provide a legal definition of functional foods; thus, the Food and Drug Administration (FDA) has no authority to establish a formal regulatory category for such foods (Ross 2000).

Some functional foods are more predominant in markets, such as green tea, soybean, flaxseed, broccoli, grapes, cabbage, tomatoes, watermelon, psyllium, oats, onions, garlic, and ginger, among others. Such foods/plants have a rich phytochemistry, with bioactive compounds that are responsible for their health-promoting potential (Butt and Sultan 2011).

The present chapter deals with different aspects, including the use and therapeutic potential of some selected main functional foods and their active principles, such as curcuma, garlic, olive oil, grape, broccoli, and probiotic, among others.

2 Functional Foods

2.1 Curcuma

Turmeric (*Curcuma longa* or *Curcuma domestica*) is a rhizomatous herbaceous perennial plant of Zingiberaceae family. It is widely cultivated in India, China, and Indonesia and received much interest from both the scientific and culinary world (Stanić 2017). For example, in India, turmeric has been used in curry; in China, it is used as a colorant; in Thailand, it is used in cosmetics; in Korea, it is served in drinks; and in the United States, it is used in mustard sauce, cheese, butter, and chips as a colorant in addition to capsules and powders forms (Gupta et al. 2013).

Turmeric plant consists of several constituents isolated from the rhizome, including curcumin (1,7-bis(4-hydroxy-3-methoxyphenyl)-1,6-heptadiene-3,5-dione), the most active polyphenolic compound also called diferuloylmethane. The curcuminoids also include demethoxycurcumin (DMCur), bisdemethoxycurcumin (BDMCur), and the recently discovered cyclocurcumin (CCur) (Stanić 2017) (Fig. 1).

Curcuminoids have been approved by the US FDA as "generally recognized as safe" (GRAS) (Gupta et al. 2013), and their good tolerability and safety profiles have been shown by clinical trials, even at doses between 4000 and 8000 mg/day (Basnet and Skalko-Basnet 2011). It is commonly proven that all constituents of turmeric plant play a synergistic action in promoting health; nevertheless, curcumin as the most abundant of the curcuminoids group is also the most investigated of the phenolic compounds.

The medicinal properties of curcumin have widely been recognized through its long history and mainly associated with multiple health benefits. Curcumin has been shown to benefit inflammatory conditions and to help in the management of

Fig. 1 Chemical structures of curcuminoids: curcumin (**a**), demethoxycurcumin (**b**), bisdemethoxycurcumin (**c**), and cyclocurcumin (**d**)

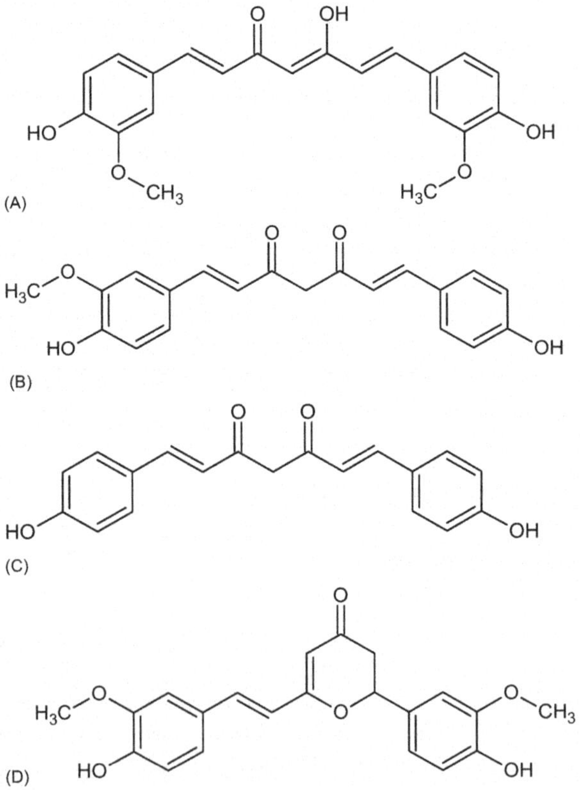

various malignant diseases, i.e., diabetes, allergies, arthritis, metabolic syndrome, Alzheimer's disease, cancer, and other chronic illnesses (Perrone et al. 2015; Pulido-Moran et al. 2016; Hewlings and Kalman 2017).

The biochemical properties and biological effects of curcumin are associated to its chemical structure. Different types of functional groups in the curcumin molecule, including β-diketone group, carbon-carbon double bonds, and phenyl rings that contain varying numbers of hydroxyl and methoxy substituents, allow this compound to exhibit different mechanisms of action. In this scenario, the ability of curcumin to affect many targets in networked signaling pathways results in a pleiotropic behavior and accounts for its definition of multitarget agent (Ghosh et al. 2015).

Most of its benefits are due to the antioxidant and anti-inflammatory properties that explain the majority of the effects of this diferuloylmethane. The most important mechanisms by which curcumin is able to promote the majority of its activities as antioxidant involve the inhibition of superoxide radicals, hydrogen peroxide, and nitric oxide radical as well as the regulation of different antioxidant proteins (Ak and Gülçin 2008; Barzegar and Moosavi-Movahedi 2011) and the preservation of mitochondrial function, by maintaining mitochondrial redox potential (Correa et al. 2013). In particular, curcumin is able to upregulate the expression of cytoprotective proteins, including superoxide dismutase (SOD), catalase (CAT) (Pan et al. 2012), glutathione reductase (GR), glutathione peroxidase (GPx) (Yarru et al. 2009), heme-oxygenase 1 (HO-1) (Jeong et al. 2006), glutathione-S-transferase (GST), reduced glutathione (GSH) (Lavoie et al. 2009), NAD(P)H:quinone oxido-reductase 1 (NQO1), and γ-glutamyl-cysteine ligase (γ-GCL) (Rushworth et al. 2006). In vitro studies have shown that curcumin induces the expression of these proteins through activation of the nuclear factor erythroid-derived 2 (Nrf2) that is the master regulator of the antioxidant response against oxidative stress and is involved in the inducible expression of cytoprotective genes (Cuadrado et al. 2009; Trujillo et al. 2013; Rezaee et al. 2017).

There are evidences showing that curcumin can modulate a variety of enzymes that are closely associated with inflammation and cancer, including cyclooxygenase-2 (COX-2), lipoxygenase, and inducible nitric oxide synthase (iNOS) enzymes, as well as mitogen-activated and Janus kinases (JAK) (Zhou et al. 2011) through inhibition of nuclear factor kappa B (NF-kB) activation (Surh et al. 2001), a eukaryotic transcription factor which regulates inflammation, cellular proliferation, transformation, and tumorigenesis. In this context, it has been proposed that curcumin inhibits NF-kB activation and proinflammatory genes expression via suppressing the phosphorylation of inhibitory factor I-kappa B kinase (IkB). Consequently, the enzymes involved in inflammatory processes are downregulated.

The effects of curcumin are also mediated by the inhibition of protein kinases such as autophosphorylation-activated protein kinase (AK), phosphorylase kinase (PhK) (Reddy and Aggarwal 1994), Ca^{2+}-dependent protein kinase (CDPK) (Hasmeda and Polya 1996), IL-1 receptor-associated kinase (IRAK) (Jurrmann et al. 2005), and mitogen-activated protein kinases (MAPKs) (Rafiee et al. 2009). These proteins are involved in the regulation of various aspects of cell functions, including cell growth, differentiation, metabolism, and apoptosis (Battaini and Mochly-Rosen 2007). Curcumin has also been found to inhibit the expression of several inflammatory cytokines including tumor necrosis factor (TNF-α), the major mediator of inflammation in most diseases, interferon-γ (IFN-γ), interleukins (IL)-1β, 2-, 5-, 6-, 8-, and 12-monocyte chemoattractant protein (MCP), as well as migration inhibitory proteins that play an major role in the local and systemic inflammation and have been implicated in a variety of inflammatory diseases (such as rheumatoid arthritis, Crohn's disease, multiple sclerosis and cancer) (Karuppagounder et al. 2017). It was demonstrated that the suppression of inflammatory cytokine production is mediated through regulation of NF-kB, protein-1 (AP-1) and by downregulating intercellular signaling proteins such as protein kinase C (Julie and Jurenka 2009).

Recently it has been shown that curcumin inhibits tumor growth alongside by inhibiting cell-cycle progression or by inducing apoptosis and also by inhibiting angiogenesis, the expression of anti-apoptotic proteins, multiple cell survival signaling pathways, and their cross-communication and by modulating immune responses (Sa and Das 2008; Kumar et al. 2016; Rivera et al. 2017). Curcumin induces the initiation of both p53-dependent and p53-independent G2/M phase cell-cycle arrest, thereby restricting cell proliferation and tumor progression in a number of human cancer cell lines, including melanoma cancer cells (Kocyigit and Guler 2017; Liao et al. 2017), Hela cancer cells (Ahmadi et al. 2017), prostate cancer (Chen et al. 2018), and thyroid cancer cells (Allegri et al. 2018).

It is well known that some phytochemicals "switch on" or "turn off" specific signaling molecules, depending on the nature of targeted signaling cascade. As reviewed by Brasili and Cechinel-Filho (2017), curcumin shows different hormetic dose responses in relation to carcinogenesis.

Anticancer activity of curcumin is also mediated by a modulation in expression of the silent information regulator 2 (SIRT) family of proteins, known as sirtuins, that modulate several diverse functions such as cellular reprogramming, a key factor in cancerous transformation (Kumar et al. 2016). Recently, it has been observed that the alteration in cancer progression is often associated with the alteration in oncogenic and anti-oncogenic miRNAs, represented by single-stranded, noncoding, endogenous molecules of 18–25 nucleotides that are involved in regulation of post-transcriptional gene expression (Melo and Esteller 2011). Many evidences have established a direct link between miRNAs and cancer and the ability of curcumin to downregulate the expression of pro-oncogenic miRNAs, such as miR-27a, miR-21, miR-20a, miR-17-5p, and miR-34a (Gandhy et al. 2012; Mudduluru et al. 2011) and upregulate the expression of anti-oncogenic miRNAs, including miR-15a/miR16-1, miR22, miR-145, miR-203, and let-7a in bladder cancer (Saini et al. 2011), in leukemic cells (Gao et al. 2012), and in human retinoblastoma cells (Sreenivasan et al. 2012).

The capacity of curcumin to modulate several pathways and target multiple genes, transcription factors, inflammatory cytokines, enzymes, growth factors, receptors, adhesion molecules, anti-apoptotic proteins, and cell-cycle proteins, leading to apoptosis and inhibition of cell proliferation and migration, makes this bioactive compound a potential therapeutic candidate to be used in targeting various types of cancers such as multiple myelomas; pancreatic, lung, breast, oral, prostate, and colorectal cancers; and head and neck squamous cell carcinoma (Kumar et al. 2016; Jalili-Nik et al. 2017; Panda et al. 2017). As a result, there is significant interest in developing adjuvant chemotherapies to augment currently available treatment protocols, which may allow decreased side effects and toxicity without compromising therapeutic efficacy.

To date, most of curcumin studies in humans have been in populations with existing health problems. Perhaps this is because studies on healthy people can be challenging in that benefits may not be as immediate and measurable if biomarkers are normal at baseline. Therefore, following subjects over time may provide the best insight into any potential health benefits in healthy people, although such studies can be time-consuming and costly. One study on healthy subjects has demonstrated that curcumin increases high-density lipoprotein (HDL) and reduces LDL cholesterol. In particular, daily oral administration of curcumin (20 mg) for 30 days increased the HDL cholesterol by 72% and apolipoprotein-A levels by 24% while decreasing the LDL cholesterol by 38.4% and apolipoprotein-B from an average 109.25 mg/dL to 90.75 mg/dL ($P < 0.05$) levels (Ramírez-Boscá et al. 2000). It is well known that increased levels of apolipoprotein-B are related to CVD, whereas apolipoprotein-A is cardioprotective. The efficiency of curcumin to modulate cholesterol metabolism was also demonstrated in patients with an alterated lipid metabolism. Chuengsamarn et al. (2014) showed that curcumin lowered the body weight, body mass index, total cholesterol, LDL cholesterol, and fasting plasma glucose with a simultaneous increase in HDL cholesterol in patients with type 2 diabetes. Since abnormal lipid metabolism and increased glucose production principally contributes to the pathogenesis of atherosclerosis in diabetes patients, these observations suggest the potentially protective role of curcumin in diabetes-induced cardiovascular diseases.

Despite its reported benefits, one of the major problems with ingesting curcumin by itself is its poor bioavailability, which appears to be primarily due to poor absorption, rapid metabolism, and rapid elimination. To increase the bioavailability of curcumin, including among others longer circulation and resistance to metabolic processes as well as an enhanced absorption by improved permeability, several strategies were adopted such as the encapsulation in starch nanoparticles, liposomes, micelles, and phospholipid complexes (Prasad et al. 2014). These formulations retain the great antioxidant ability of curcumin and display significantly higher water solubility and stability in comparison with raw curcumin. Moreover, curcumin encapsulated with polymers such as poly n-butyl cyanoacrylate, poly (lactide-co-glycolide), chitosan, albumin, or acrylamide polymers showed to enhance bioavailability and peak serum levels of curcumin, compared with curcumin alone (Maradana et al. 2013).

2.2 Garlic

Allium sativum (garlic), a member of the Liliaceae family, is widely used as a valuable spice for flavoring and a popular remedy for various ailments and disorders. Ancient cultures frequently ingested garlic as a folk medicine to treat flatulence, intestinal disorders, respiratory infections, skin diseases, and many other disorders. According to scientific-based evidence, *A. sativum* has a wide array of therapeutic effects associated to antibacterial (Harris et al. 2001), antiviral, antidiabetic, antihypertensive (Hosseini and Hosseinzadeh 2015), cardioprotective, hepatoprotective, hypolipidemic, and antioxidant properties, as well as immune enhancement including antiplatelet and procirculatory activities (Ryu and Kang 2017). *In vitro* and *in vivo* studies have also shown that garlic can suppress the incidence of many cancers, such as breast, blood, bladder, gastric, colorectal, skin, uterus, esophagus, and lung cancers (Li et al. 2017; Ma et al. 2017; Yin et al. 2018).

Garlic contains more than 2000 biologically active substances including volatile, water-soluble, and oil-soluble organosulfur compounds along with essential oils, dietary fiber, sugars 32% (included inulin), flavonoids, and pectin that work synergistically to provide various health benefits (Amagase 2006). The medicinal properties of garlic are mainly attributed to a high concentration of organosulfur compounds that give to garlic a characteristic flavor. The major sulfur-containing compounds in intact garlic are γ-glutamyl-S-allyl-L-cysteines and S-allyl-L-cysteine sulfoxides (alliin) in addition to several novel cyclic sulfoxides called garlicnins recently isolated from acetone extracts of garlic (Nohara et al. 2013).

When the garlic bulb is crushed, minced, or otherwise processed, alliin is released from compartments and interacts with the enzyme alliinase. Hydrolysis and immediate condensation of the reactive intermediate (allylsulfenic acid) form allicin (diallyl thiosulfinate or diallyl disulfide), a substance that is a stronger antibiotic then penicillin or tetracycline (Fig. 2). Allicin itself is an unstable product and is further metabolized to vinyldithiines, depending on environmental and processing conditions.

Despite the numerous health benefits of garlic, some people are reluctant to eat raw garlic because of its pungent taste and smell, and the raw garlic can cause gastrointestinal discomfort. For this reason, aged garlic preparations have been developed. Aged black garlic (ABG) is known as functional food and is produced by application of high temperature and humidity (Ryu and Kang 2017). During the aging process, the odorous, harsh, and irritating compounds

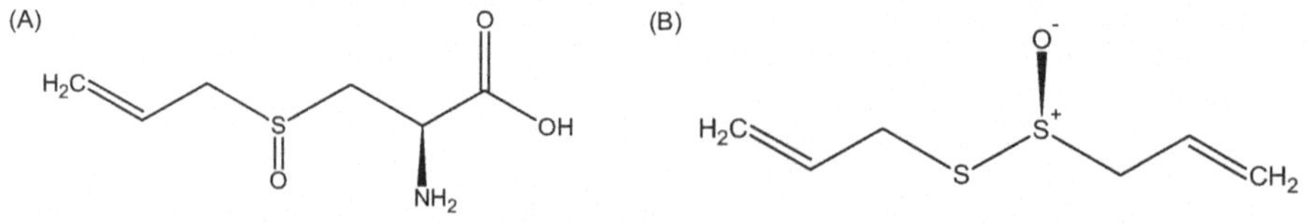

Fig. 2 Alliin (**a**) and allicin (**b**)

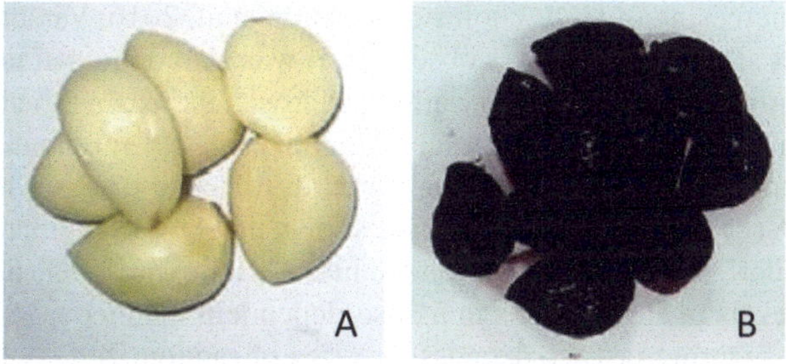

Fig. 3 Fresh raw garlic (**a**) and aged black garlic (**b**). (Adapted from Ryu and Kang 2017)

in fresh raw garlic are converted naturally into stable and safe compounds. As a result, ABG has a sweet and sour taste and jelly-like texture. The heating process leads to a Maillard reaction, creating the typical dark brown color, and produces antioxidant compounds (Fig. 3).

Several bioactive compounds were detected in ABG preparations including phenols, flavonoids, pyruvate, thiosulfate, S-allylcysteine (SAC), and S-allylmercaptocysteine (SAMC) (Ryu and Kang 2017). Aged garlic extracts (AGE) obtained from ABG contain water-soluble and lipid-soluble organosulfur compounds, but not allicin. SAC and SAMC are the major unique water-soluble organosulfur compounds in AGE, while diallyl sulfide (DAS), diallyl disulfide (DADS), diallyl trisulfide (DATS), and diallyl tetrasulfide are lipid-soluble compounds in AGE.

Many favorable experimental and clinical studies on the consumption of garlic, especially of aged garlic preparations, demonstrate a wide spectrum of biological activities. Firstly, garlic has been shown to have antibacterial, antifungal, antiviral, and antiprotozoal activities. In vitro studies showed garlic activity against many types of Gram-negative and Gram-positive bacteria including species such as *Escherichia*, *Salmonella*, *Staphylococcus*, *Streptococcus*, *Klebsiella*, *Proteus*, *Bacillus*, *Clostridium*, *Mycobacterium tuberculosis*, and *Helicobacter pylori* (Harris et al. 2001; Yadav et al. 2015; Sheppard et al. 2018). The minimum inhibitory concentration (MIC) values for Gram-positive strains were between 5 and 10 μg/ml of allicin, while for Gram-negative bacteria, they were between 15 and 30 μg/ml.

An antifungal activity of garlic was demonstrated against *Candida albicans* (Khodavandi et al. 2011), while an antiparasitic activity was observed against some major human intestinal protozoan parasites including *Entamoeba histolytica*, *Giardia lamblia*, and *Leishmania* (Ankri and Mirelman 1999; Metwally et al. 2016). Allicin is considered the most potent compound of garlic with significant antibacterial and fungicidal properties which have been confirmed through in vitro test with allicin in pure form (Reiter et al. 2017), while *in vivo* activity has not been well confirmed with preclinical and clinical studies so far (Marchese et al. 2016).

Several reports have revealed a significant effect of garlic consumption on the cardiovascular system. It includes a reduction of the risks of cardiovascular disease by inhibiting platelet aggregation and lowering cholesterol and blood pressure, based on the fact that allicin (diallyl thiosulfinate) is degraded into diallyl polysulfides by H2S preventing myocardial injury and dysfunction. In addition, an improving of lipid metabolism through a lowering of blood LDL cholesterol and an increasing of HDL cholesterol (Atkin et al. 2016; Varshney and Budoff 2016; Wang et al. 2017b) was also observed. It was also demonstrated that AGE can reduce the progression of coronary atherosclerosis, increasing the brown adipose and decreasing the white adipose tissue (Ahmadi et al. 2013; Bahadoran et al. 2017). A common and independent risk factor for cardiovascular diseases is represented by lipoprotein A (LPa). Ma et al. (2017) have shown that DADS regulates lipid metabolism through ERK1/ERK2 signaling and inhibits LPa expression, both the mRNA and protein levels in HepG2 cells in a dose-dependent manner.

During last the decades, a series of randomized controlled trials of high quality were designed to investigate garlic efficacy in the management of type 2 diabetes mellitus. In particular, garlic has been shown to have antihyperglycemic and lipid-lowering properties as well as an efficacy in lowering of C-reactive protein and serum adenosine deaminase levels suggesting that garlic can be a valuable agent in providing good glycemic control and the prevention of long-term complications (Kumar et al. 2013; Wang et al. 2017a, b). Moreover, it was demonstrated that garlic supplementation in combination with antidiabetic medication such as metformin provides a better control of diabetes type 2 (Ashraf et al. 2011).

Among the natural organosulfur compounds from garlic, the diallyl trisulfide (DATS), it has been shown to exhibit effective antitumor properties. Li et al. (2017) have demonstrated that DATS inhibits the viability of breast cancer stem cells by reducing tumorspheres formation and decreasing the expression of breast cancer stem cells markers (CD44, ALDH1A1, Nanog, and Oct4), as well as inhibiting proliferation and inducing apoptosis. Furthermore, they showed that DATS suppressed the activity of the canonical Wnt/β-catenin signal pathway that is critical for maintaining cancer stem cells characteristics. On the other hand, organosulfur compounds have been shown to have multi-targeted antitumor activities in a variety of other cancer cells. Esophageal-gastric junction adenocarcinoma is an aggressive tumor with high incidence and dismal prognosis worldwide. Yin et al. (2018) demonstrated that DADS inhibited the metastasis of type II esophageal-gastric junction adenocarcinoma cells via NF-kB and PI3K/AKT signaling in vitro. Some evidences have suggested that allicin induced cell death in human hepatoma cells through either autophagy or apoptosis and might be a potential novel complementary gene therapeutic agent for the treatment of apoptosis-resistant cancer cells. Chu et al. (2013) demonstrated that allicin induced apoptotic cell death through caspase-dependent and caspase-independent pathways by reactive oxygen species (ROS) overproduction in human HCC Hep 3B (p53(mutation)) cells.

2.3 Omega-3

The term omega-3 (ω-3 or n-3) is a structural descriptor for a family of polyunsaturated fatty acids (PUFAs). n-3 refers to the position of the double bond that is closest to the methyl terminus of the acyl chain of the fatty acid. All n-3 fatty acids have this double bond on carbon 3, counting the methyl carbon as carbon one. The simplest n-3 fatty acid is α-linolenic acid (18:3n-3) that is synthesized from the n-6 fatty acid linoleic acid (18:2n-6) by desaturation and catalyzed by delta-15 desaturase. α-linolenic acid can be converted to stearidonic acid (18:4n-3), and then stearidonic acid can be elongated to eicosatetraenoic acid (20:4n-3), which can be further desaturated to yield eicosapentaenoic acid (20:5n-3, known as EPA). From EPA, for addition of carbons, desaturation, and elongation reactions, it is possible to obtain docosapentaenoic acid (22:5n-3, known as DPA) and docosahexaenoic acid (DHA, known as 22:6 n-3) (Calder 2012).

EPA, DPA, and DHA were found in significant quantities in dietary plant and marine organisms. Plant-derived n-3 fatty acids are present in flaxseed oil, canola (rapeseed) oil, nuts, soybean oil, flaxseeds, and soybeans, while marine-derived n-3 PUFAs are the key components of flesh of both lean and oily fish and other seafood (Meyer et al. 2003). A single lean fish meal (e.g., one serving of cod) could provide about 0.2–0.3 g of marine n-3 fatty acids, while a single oily fish meal (e.g., one serving of salmon or mackerel) could provide 1.5–3.0 g of these fatty acids. Fish oil is prepared from the flesh of oily fish (e.g., tuna) or from the livers of lean fish (e.g., cod liver). In a typical fish oil supplement, EPA and DHA together comprise about 30% of the fatty acids present, so that a 1 g fish oil capsule will provide about 0.3 g of EPA + DHA (Calder 2012).

There are considerable evidences from clinical, experimental, and epidemiological studies that omega-3 fatty acids, particularly EPA and DHA, have a beneficial and protective role in a variety of human diseases, including diabetes, atherosclerosis, asthma, and arthritis (Mori and Beilin 2001; Mozaffarian and Wu 2011).

The major benefits of EPA and DHA are related to their capacity to reduce the cardiometabolic risk factors by modulating numerous physiological processes such as blood pressure and cardiac function (Geleijnse et al. 2002; Mori and Woodman 2006; Miller et al. 2014), arterial compliance (Nestel et al. 2002; Pase et al. 2011), lipid metabolism (Okada et al. 2017), anti-inflammatory (Calder 2012), and antioxidative actions (Mas et al. 2010). These effects explain why fish oil, omega-3, EPA, and DHA have together become the most commonly consumed natural dietary supplement in the United States.

As molecules capable of reducing cardiovascular risk, omega-3 fatty acids are able to improve endothelial function and arterial compliance, decrease the heart rate, and exert an antihypertensive effect through a plethora of mechanisms including the suppression of vasoconstrictor prostanoids, the enhanced production and/or release of nitric oxide, the reduction of plasma noradrenaline, changes in calcium flux, an increased membrane fluidity, antioxidative actions of n-3 fatty acids, or an increase in HDL cholesterol (Mori 2017). In addition, n-3 fatty acids could affect heart rate through their incorporation into myocardial cells and altering electrophysiological function in a manner that reduces the vulnerability to ventricular fibrillation (Leaf et al. 2003). In this context, the anti-arrhythmic effects of n-3 fatty acids are due to their ability to inhibit the fast, voltage-dependent sodium current and the L-type calcium currents and to modulate potassium channels (Leaf et al. 2003).

Omega-3 fatty acids are also able to modulate lipid metabolism, reducing plasma triacylglycerols (TAG) (Chauhan et al. 2017; Karalis 2017). The fall in plasma triglycerides is due to a reduction in hepatic VLDL cholesterol synthesis. The mechanisms include reduced fatty acid availability for triglyceride synthesis as a result of decreased de novo lipogenesis, a reduction in the delivery of non-esterified fatty acids to the liver, increased fatty acid β-oxidation, altered enzymatic activity for triglyceride assembly in the liver, and increased hepatic synthesis of phospholipids instead of triglycerides (Harris and Bulchandani 2006). Omega-3 fatty acids have no effects on total blood cholesterol; they affect LDL-C, LDL particle size, and HDL-C; these effects are variable and depend on dose and population studied. When tested alone, DHA supplements increased HDL-C and LDL particle size, whereas EPA decreased HDL_3 (Cottin et al. 2011; Jacobson et al. 2012).

Observational and interventional studies have confirmed that intake of n-3 PUFAs has an emerging clinical utility for the treatment of several inflammatory diseases (Mozaffarian and Wu 2011; Harris et al. 2013; Jones and Roper 2017). The anti-inflammatory and immunomodulatory effects of n-3 fatty acids are most probably related to their attenuating actions on inflammatory eicosanoids including altered leucotriene formation, cytokines, oxidative stress, endothelial and cell-cell activation, and immune cell function (Mori 2017). n-3 Fatty acids were shown to reduce ex vivo production of proinflammatory cytokines including TNFα, IL-1, and IL-6 following lipopolysaccharide stimulation of monocytes/lymphocytes (Calder 2012). De Caterina et al. (2000) showed that DHA was more potent than EPA in inhibiting the expression of vascular cell adhesion molecule-1, intercellular adhesion molecule-1, and E-selectin, after stimulation. The EPA and DHA-induced reduction in adhesion molecule expression was accompanied by decreased binding of human lymphocytes and monocytes to cytokine-stimulated endothelial cells. The anti-inflammatory effects of n-3 fatty acids are in part mediated by a novel family of local lipid mediators generated during self-limited resolution of inflammation. Serhan (2017) described the E-series resolvins derived from EPA and D-series resolvins, protectins, neuroprotectins, and maresins derived from DHA as mediators acting via G-coupled protein receptors. Two series of resolvins and protectins have been identified: those derived via lipoxygenase metabolism of EPA and DHA and a second series derived from aspirin-triggered cyclooxygenase-2 or cytochrome P450 metabolism of EPA and DHA. These mediators have potent anti-inflammatory and pro-resolving actions and increase with time during the inflammatory process. Mas et al. (2012) have shown that the resolvin and protectin pathway precursors 18R/S-HEPE and 17R/S-HDHA, as well as the resolvins 17S-RvD2, 17S-RvD1, and 17R-RvD1, were present in human plasma following n-3 fatty acid supplementation, at concentrations that are known to have potent anti-inflammatory effects.

The US Department of Health and Human Services and Agriculture (USDHHS/USDA) and the American Heart Association (AHA) have issued guidelines recommending regular intake of omega-3 as part of a healthy dietary pattern aimed at reducing the population burden of cardiovascular diseases (Lloyd-Jones et al. 2010). Current guidelines suggest that individuals should consume approximately 500 mg/day of EPA and DHA (Gebauer et al. 2006), which is achievable with at least two

100 g serves of fish per week, preferably oil fish species such as fresh tuna, salmon, mackerel, herring, and sardines (Kris-Etherton et al. 2002) and preferably broiled or baked but not fried (Mozaffarian et al. 2005). The most practical recommendation for increasing dietary n-3 fatty acids is to incorporate fish as part of a healthy diet that includes increased fruits and vegetables and moderation of salt intake.

2.4 *Olive Oil*

The olive tree (*Olea europaea* L.) is the only representative of the Oleaceae family producing edible fruits. Olive tree fruits represent valuable sources of nutrients and non-nutrients, responsible for the nutritional and sensory properties, as well as for the biological activities and health benefits attributed to edible olives and olive oils (Rotondi et al. 2010). Olive oil, mainly virgin olive oil, is among the most important components of the Mediterranean diet. According to chemical composition and degree of acidity, the International Olive Council (2017) has defined that the best brand corresponding to extra virgin olive oil (EVOO) must contain less than 0.8 g of free acids per 100.0 g, expressed as oleic acid, besides presenting no noticeable organoleptic defects.

The chemical composition of olive oils concerning the bioactive compounds and quality parameters results from the complex interaction between several factors, including genetic factors (cultivar), geographical origin, agroclimatic conditions (abiotic and biotic stress, such as cultivation practices and pathogens outbreaks), and processing methods. In addition, the maturation stage of olives constitutes a relevant issue to determine the quality of olive oil. The relevance of this factor has been attributed to the changes occurring at different maturation stages including physiological, biochemical, metabolic, and enzymatic features. Generally, advanced maturation results in an increase of phenolic content to a maximum level that decrease according to the progress of the season.

The sensory characteristics of olive oil are due to a complex mixture of volatile compounds, including aldehydes, alcohols, ketones, esters, and hydrocarbons, which have been lately identified and quantified (Gómez-Romero et al. 2012).

Virgin olive oil is mainly constituted by triacylglycerols, in addition to other minor compounds which include almost 230 different chemicals (Servili et al. 2009). Glyceride fraction that represents from 90.0% to 99.0% of total oil weight is mainly composed by the phospholipids and mono-/di- and triacylglycerols. The wide diversity of fatty acids is represented by a number of mono- and polyunsaturated fatty acids including myristic (C14:0), palmitic (16:0), palmitoleic (C16:1), heptadecanoic (C17:0), heptadecenoic (C17:1), stearic (C18:0), oleic (C18:1), linoleic (C18:2), linolenic (C18:3), arachidic (C20:0), eicosenoic (C20:1), behenic (C22:0), and lignoceric (C24:0) acids. From these, oleic acid (C18:1, *n*-9) and linoleic acid (C18:2, *n*-9, *n*-12) are the main components, representing from 55% to 83% and from 5% to 15% of the total fatty acids, respectively. On the other hand, the olive oil composition presents only low quantities of saturated fatty acids (8.0–

20.0%). Concerning the non-glyceride fraction (0.5–1.5%), the chemical analysis has revealed the presence of hydrocarbons, aliphatic alcohols, sterols, pigments, and several volatile and phenolic compounds (Waterman and Lockwood 2007; Bakhouche et al. 2013; Franco et al. 2014; Gouvinhas et al. 2017).

Among the hydrocarbons, it is important to highlight the presence of squalene (or 2,6,10,15,19,23-hexamethyl- 2,6,10,14,18,22-tetracosahexaene), a natural occurring terpenoid in olive, that represents the main component of the hydrocarbon fraction (more than 90%). To date, besides its role as valuable dietary component constituting a precursor in cholesterol biosynthesis, squalene has been recognized as oxidation inhibitor, exerting a valuable contribution to the stability of olive oil (Moreda et al. 2001). The most important aliphatic alcohols are fatty alcohols, diterpene alcohols, and benzyl esters of hexacosanoic and octacosanoic acids that show interesting biological activities in several chronic diseases. An important class of lipids that is related to the quality of the oil and used to check its genuineness is constituted by sterols with amounts ranging from 855 to 2185 mg kg^{-1}. The main sterols present in olive oil are β-sitosterol, campesterol, stigmasterol, clerosterol, sitostanol, and δ-5-avenasterol (Alves et al. 2005). The importance of these components is linked to their chemical structure being similar to cholesterol (excluding the addition of an extra methyl or ethyl group), which reduces the cholesterol absorption and thus reducing the circulation levels of cholesterol (Servili et al. 2009). The characteristic color of olive oil is due to two types of pigments, chlorophylls and carotenoids (mainly lutein and β-carotene), contributing for the greenness and yellowness, respectively.

Interesting, olive oil is characterized by a valuable source of bioactive phytochemicals, which are responsible for the biological activity attributed to this food matrix, being mainly represented by polar and lipophilic phenolic compounds. The most important polar phenolic compounds are cinnamic and benzoic acids, phenolic alcohols, secoiridoids, lignans, hydroxy-isochromans, and flavonoids, while tocopherols and tocotrienols are the main lipophilic phenolic constituents (Gouvinhas et al. 2017). The group of benzoic and cinnamic acids includes mainly gallic, vanillic, and syringic acids as well as caffeic, ferulic, and sinapic acids. The three most important phenolic compounds in olive oil are represented by oleuropein, the most abundant secoiridoid compound, and the phenolic alcohols hydroxytyrosol and tyrosol that result from hydrolysis of secoiridoid compounds (Brenes et al. 2000). Oleuropein, hydroxytyrosol (3,4-(dihydroxyphenyl)-ethanol), and tyrosol ((*p*-hydroxyphenyl)-ethanol) represent up to 90% of the total phenolic content of olive oil and are related structurally. Some flavonoids are also present in olive oils, mainly luteolin, apigenin, quercetin, and taxifolin. The lipophilic or nonpolar phenols are mainly represented by tocopherols and tocotrienols. The α-tocopherol is the most abundant (90.0%), although β- and γ-tocopherols are also present (Gouvinhas et al. 2017). It is generally accepted that exists a complex and concrete relationship between the molecular structure of olive oil bioactive compounds and their possible health effects. Olive oil bioactive phytochemicals contribute to the maintenance of the normal physiological status but also to the prevention of distinct pathological conditions related to oxidative stress, such as cancer, cardiovascular disease, metabolic disorders, and inflammation (Gómez-Romero et al. 2012).

The most relevant function of olive oil phytochemicals was associated to the antioxidant power. The most important antioxidant activity of olive oil due to phenolic compounds has been related to a lower incidence of coronary heart disease as well as to a lower risk of some type of cancers, since they can reduce DNA damage, lipid peroxidation, and also the amount of ROS generated, diminishing the inflammation and the inhibition of platelet-activating factor (Omar 2010). The free radical-scavenging ability of phenolic compounds is exerted by stopping the propagation chain during the oxidation process through the donation of a radical hydrogen to alkylperoxyl radicals (produced by lipid oxidation) and the formation of stable derivatives during this reaction. Also secoiridoids such as oleuropein aglycon and hydroxytyrosol as well as lignans containing an *ortho*-diphenolic structure display high antioxidant activity, due to the improved radical stability through the formation of intramolecular hydrogen bonds formed during the reaction with free radicals compounds (Visioli and Galli 1998). These activities allow to prevent the degenerative diseases linked to oxidative stress in separate tissues and organic systems.

The biological activity of olive oil compounds is related to their bioavailability that is referred to the degree in which it is liberated from food and absorbed in the intestinal tract. The highest plasma concentration of derivatives from olive oil is recorded from 1 to 3 h after ingestion, indicating that the major absorption of these compounds takes place in the small intestine. Several studies have shown that some of the most important bioactive compounds present in olive oil can reach noticeable concentrations in plasma, after intake, pointing to the feasibility of the direct benefits of olive oil consumption. For example, the bioavailability of polyphenols, which are dose-dependently absorbed and extensively metabolized mainly as glucuronides, varies among the different classes, and the plasma concentration of polyphenols has been reported to be lower than 10 μM (Scoditti et al. 2012).Concerning the absorption of hydroxytyrosol and tyrosol in human subjects, some authors have been reported that was 30–60% and 20–22%, respectively, of total intake as determined in 24 h urine samples (Quiles et al. 2002). Therefore, up to 49 μM hydroxytyrosol and 38.6 μM tyrosol from virgin olive oil might be absorbed, while according to other authors' claim that regarding these compounds, the physiological concentration after oral ingestion of olive oil is in the range of 10–100 μM (Quiles et al. 2002). Oleuropein, that is rapidly absorbed from the intestine, reaches a maximum (peak) of plasma concentration of 370 μM after 2 h of oral administration of 20 mg $kg^{-}1$ in animal models (Al-Azzawie and Alhamdani 2006).

Another compound with special relevance present in olive oil is oleocanthal. The daily consumption of 50 g of EVOO containing up to 200 μg per mL oleocanthal corresponding to an intake of up to 9 mg/day, with absorption values of 60–90%.

It is important to remark that doses assayed in vitro and in experimental animals are frequently higher that those applied in humans, and in addition, it should be also considered that the metabolism of bioactive compounds may differ in these different models.

Several epidemiological studies have indicated the potential of dietary olive oil to reduce the incidence, severity, and progression of cancer process in humans. The anticancer activity of dietary olive oil in humans has been mainly attributed to the capacity of its nutritional and non-nutritional components to inhibit the onset of the cancer process and its progression at different stages by protecting against oxidative DNA damage; modulation of biosynthesis of the colon cancer promoters bile acids; decreasing estrogen synthesis in adipose tissue; antiestrogen effect by structural competition; decrease of free estradiol; changes in cell membrane fluidity, structure, and degree of peroxidation; modulation of genes involved in cell proliferation; and anti-inflammatory and immunomodulatory effects, which were extensively reviewed by Escrich et al. (2007). The antitumor properties of bioactive compounds present in olive oil have been demonstrated in vitro against diverse human malignant cell lines, namely, HT-29 (colon adenocarcinoma), MCF-7 (breast adenocarcinoma), urinary bladder carcinoma, and hepatocellular carcinoma (HepG2) cell lines (Goulas et al. 2009). However, the efficiency of olive oil bioactive compounds against cancer differs depending on the target cell type. For instance, hydroxytyrosol and tyrosol induce an increase of the radical oxygen species level in breast epithelial cells. These results point out that hydroxytyrosol and tyrosol attenuate oxidative stress in normal cells, preventing the malignization by protecting against DNA damage (Scotece et al. 2012), although they are inefficient when the malignancy has occurred. Oleocanthal displayed the capacity to inhibit carcinogenesis in concentrations of around 3 μM (Khanal et al. 2011), also presenting the ability to induce apoptotic cell death in human breast and colon carcinomas, besides the inhibition of the metastatic process in human breast cancer cells (MDA-MB-231), at the concentration of 15 μM (Elnagar et al. 2011).

In addition to the cytotoxic effect demonstrated in vitro against cancer cells, oleuropein and hydroxytyrosol display antiangiogenic activity inhibiting the creation of the surrounding interacting network that provides the microenvironment required for the malignant cells growth, migration, and invasion (Owen et al. 2000). Oleuropein has been defined as the main responsible for the anti-tumor activity of olive oil against skin and breast cancer as demonstrated in vivo studies with experimental animals (Selvaggini et al. 2006; Kimura and Sumiyoshi 2009). This activity may be due to the inhibition of the expression of vascular endothelial growth factor and metalloproteinase 2, 9, and 13 through the reduction of COX-2 level (Scoditti et al. 2012), while regarding breast cancer, oleuropein was efficient in minimizing the tumor size, being able to completely remove 9–12 days tumors by disturbing the actin cytoskeleton of tumor cells in vivo (Escrich et al. 2007). On the other hand, oleocanthal has been identified as a potent COX (cyclooxygenase) inhibitor, which turns it into a valuable candidate to be tested regarding anticancer activity. Among the biological functions demonstrated in vitro for oleocanthal against cancer cells, it has been stressed the attenuation of monocyte chemoattractant protein 1, which is a critical instigator of malignant lesions (Scotece et al. 2012). Moreover, oleocanthal displays an antiproliferative effect via the inhibition of extracellular signal-regulated kinases 1/2 and p90RSK phosphorylation, promotes cell apoptosis by activating caspase-3 and PARP, and induces DNA fragmentation in human malignant cells (Khanal et al. 2011). Additionally, in vivo, oleocanthal displayed an inhibitory effect on migratory and invasive actions characteristic of cancer cells, responsible for the metastatic process, possibly as a result of its ability to inhibit c-Met phosphorylation (Elnagar et al. 2011). Some epidemiological studies have evaluated the capacity of dietary olive oil to modify the plasma fatty acids profile. Mayneris-Perxachs et al. (2014) evaluated the effects of 1-year intervention with a Mediterranean diet (supplemented with virgin olive oil) on plasma fatty acid composition and metabolic syndrome in a population at high cardiovascular risk. The results obtained evidenced that the dietary supplementation with olive oil increases the plasma concentrations of palmitic and oleic acids while lowering the level of margaric, stearic, and linoleic acids.

Oleuropein has been also stressed on their ability to inhibit the atherosclerosis process, by the downregulation of the expression of TNF-α and the consequent inhibition of the expression of the monocyte chemotactic protein-1 and vascular cell adhesion molecule (Bogani et al. 2007).

2.5 Tea

Tea is one of the most commonly consumed beverages in the world (Graham 1992). Originated in China and Southeast Asia, tea has been cultivated and consumed for more than 2000 years. In traditional Chinese medicine and Ayurvedic practices, tea has been used extensively as a stimulant, diuretic, and astringent. Other traditional uses of green tea include promoting digestion, improving mental health, and regulating blood sugar as well as body temperature (Cooper et al. 2005). Tea is prepared from the processed leaf of *Camellia sinensis*, an herbal plant belonging to the Theaceae family. Two varieties are mainly used to prepare tea beverage: *C. sinensis* var. *sinensis*, a small-leaved, bush-like plant originated from China, grown in several countries of Southeast Asia experiencing a cold climate, and *C. sinensis* var. *assamica*, a large-leaved tree discovered in the Assam region of India and introduced in several countries enjoying a semitropical climate (De Mejia et al. 2009). The *sinensis* variety accounts for most green tea production, while the *assamica* variety is particularly used for black tea. The differences among teas arise from production methods, growth conditions, and geographic origin. Tea types can be classified based on processing or harvested leaf development in green (non-fermented), black (fermented), oolong (semi-fermented), and white (harvested leave buds with white trichomes, non-fermented, or semi-fermented). The chemical composition of tea is characterized by the presence of large amounts of flavonoids including catechin, epicatechin (EC), epigallocatechin (EGC), and epigallocatechin gallate (EGCG) (Fig. 4).

Green tea, which represent 20% of world consumption, and oolong tea have high level of EGCG (50–80% of total catechins). EGCG can be found abundantly in green tea leaves (7.1 g per 100 g green tea leaves), oolong tea (3.4 g per 100 g oolong tea), and black tea leaves (1.1 g per 100 g black tea leaves)

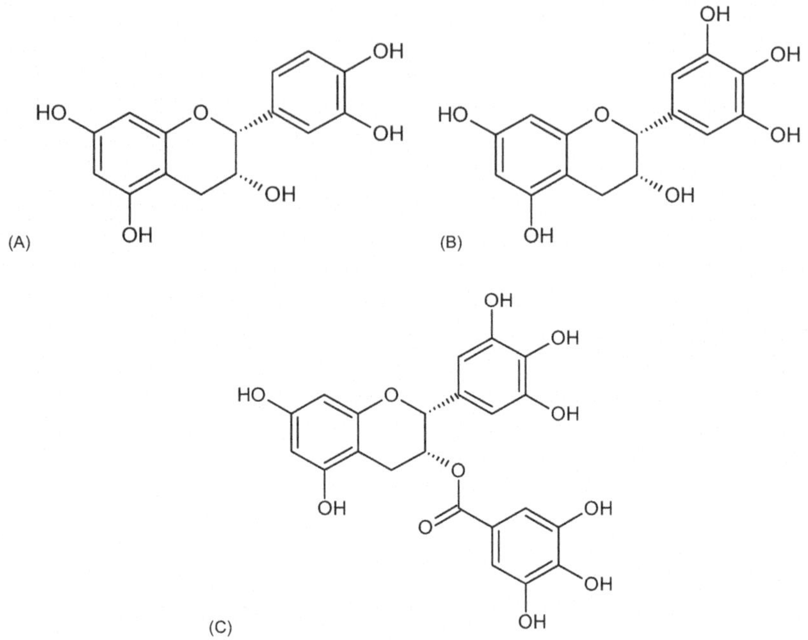

Fig. 4 Epicatechin (**a**), epigallocatechin (**b**), and epigallocatechin gallate (**c**)

(Chacko et al. 2010). Green tea contains approximately 70% catechins (monomeric flavonoids), 10% minor flavanols (mainly quercetin, kaempferol, myricetin, and their glycosides), and 20% polymeric flavonoids. Tea contains also several amino acids, but L-theanine (c-glutamylethylamide), specific to the tea plant, is the most abundant, accounting for 50% of the total amino acids. Volatile fractions of tea leave contain more than 600 different molecules. In addition, tea contains carbohydrates, caffeine, adenine, gallic acids, tannins, gallotannins, quercetin glycosides, carotenoids, tocopherols, vitamins (A, K, B, C), small amounts of aminophylline, and a yellow volatile oil that is solid at 25°C and has a strong aromatic odor and taste (Jayabalan et al. 2008).

During fermentation of fresh tea leaves, some catechins are oxidized or condensed to larger polyphenolic molecules (dimers or polymers), such as theaflavins (theaflavin, theaflavin-3-gallate, theaflavin-3′-gallate, and theaflavin-3-3′-digallate) (3–6%) and thearubigins (12–18%). These polymers are responsible for black tea's bitter taste and dark color. Black tea contains mainly thearubigins (70%), theaflavins (12%), flavonols (10%), and catechins (8%). The total polyphenol content of green and black teas is similar but with different types of flavonoids present, due to the degree of oxidation during processing (Stangl et al. 2006).

Extensive *in vivo* and *in vitro* studies have demonstrated that tea polyphenols, especially EGCG, have preventive effects against chronic diseases including heart disease, diabetes, neurodegenerative disease, and cancer (Higdon and Frei 2003; Khan and Mukhtar 2013).

A series of population-based cohort studies have showed that tea consumption was associated with reduced risk of metabolic syndrome and diabetes mellitus (DM) (Hamer et al. 2008; Grosso et al. 2015; Siddiqui et al. 2015). Interesting, various kinds of tea showed different effects on DM. For instance, consumption of unfermented green tea or semi-fermented oolong tea was considered to protect against the development of T2DM in Chinese men and women. Tea or tea extract has been demonstrated to reduce insulin resistance and improve glycemic control (Yu et al. 2017). Green tea consumption was associated with a lower risk of impaired fasting glucose (IFG), probably as a result of its high level EGCG, which had insulin mimetic effects. In this scenario, EGCG can inhibit the hepatic glucose production and promote tyrosine phosphorylation of the insulin receptor and insulin receptor substrate-1 (IRS-1). In addition, EGCG controls gluconeogenesis by suppressing the expression of genes phosphoenolpyruvate carboxykinase (PEPCK) and glucose-6-phosphatase (G6Pase), ameliorates cytokine-induced β-cell damage, and improves insulin sensitivity (Han 2003). Finally, EGCG regulates the expression of genes involved in the insulin signal transduction pathways and glucose uptake (Cao et al. 2007).

Green tea has been also proposed as an antioxidant which plays a role in reducing the number of free radicals involved in numerous diseases states including cardiovascular disease (CVD) (Lobo et al. 2010). This makes scientists believe that EGCG could be a potential therapeutic agent against CVD, which are mainly caused by oxidative stress. As reviewed by Eng et al. (2018), EGCG was found to exhibit a wide range of therapeutic properties including anti-atherosclerosis, anticardiac hypertrophy, anti-myocardial infarction, antidiabetes, anti-inflammatory, and antioxidant. These therapeutic effects are mainly associated with the inhibition of LDL cholesterol (anti-atherosclerosis), inhibition of NF-κB (anticardiac hypertrophy), inhibition of MPO activity (anti-myocardial infarction), reduction in plasma glucose and glycated hemoglobin level (antidiabetes), reduction of inflammatory markers (anti-inflammatory), and the inhibition of ROS generation (antioxidant).

Both observational and intervention studies have provided evidence in support of a protective role of green tea intake in the development of oral-digestive tract cancer and gastric, liver, lung, and colorectal cancers (Yuan 2013). Green tea consumption was inversely associated with rate of recurrence, especially in early stages of breast cancer (Inoue et al. 2001; Zhang et al. 2007). Bettuzzi et al. (2006) have shown that the consumption of green tea catechin capsules after 1 year inhibited the conversion of high-grade prostate intraepithelial neoplasia to cancer, in comparison with a placebo.

Numerous mechanisms have been proposed to account for the cancer-preventive effects of green tea in epidemiological as well as laboratory studies. These mechanisms include the inhibition of growth factor signaling, inhibition of key cellular enzymes, inhibition of gene transcription, and induction of tumor suppressor genes (Khan et al. 2006; Tachibana 2009; Yang et al. 2009). Khan and Mukhtar (2008) highlighted the pathways related to cancer chemoprevention by tea polyphenols, specially EGCG as causing inhibition of mitogen-activated protein (MAP) kinases and activator protein-1 (AP-1), nuclear factor-jB (NF-jB) signaling pathway, epidermal growth factor receptor (EGFR)-mediated pathways, insulin-like growth factor (IGF)-1-mediated signal transduction pathway, proteasome activities, matrix metalloproteinases (MMPs), urokinase-plasminogen activator, and induction of apoptosis and cell-cycle arrest.

The effects of tea polyphenols on inflammation have also relevance to cancer prevention (Beltz et al. 2006). Tea polyphenols, especially EGCG, appear to modulate at different targets the anti-inflammatory activities related to arachidonic acid-dependent pathways such as COX inhibition. Within the arachidonic acid-independent pathways, NOS and NF-kB are targets of polyphenols (Miles et al. 2005). Tea phytochemicals inhibit COX-2 and iNOS expression by blocking NF-kB activation. Particularly EGCG suppresses activation of NF-kB by repression of degradation of the inhibitory unit IkB, which hampers subsequent nuclear translocation of the functionally active subunit of NF-kB (Kundu and Surh 2008).

Several studies presented evidence that green tea influences psychopathological symptoms (e.g., reduction of anxiety), cognition (e.g., benefits in memory and attention), and brain function (e.g., activation of working memory seen in functional MRI). In these studies, it was demonstrated that the effects of green tea cannot be attributed to a single constituent of the beverage. Conversely, beneficial green tea effects on cognition were observed under the combined influence of both caffeine and L-theanine, whereas separate administration of either substance was found to have a lesser impact (Mancini et al. 2017). The neuroprotective activities of green tea polyphenols were demonstrated in a wide array of cellular and animal models. Mandel et al. (2008) observed that tea consumption was inversely associated with the incidence of age-related dementia and Alzheimer's and Parkinson's diseases showing that tea polyphenols, particularly EGCG, are bioavailable to the brain and can act at multiple targets, via antioxidant, iron-chelation, signal transduction modulation, and other mechanisms, to produce neuroprotective and/or neurorescue action. These authors suggested that the neuroprotective/neurorestorative action of EGCG goes beyond antioxidant/radical-scavenging capacity, including activation of protein kinase C signaling pathways preventing apoptosis and mitochondrial membrane collapse. Agents such as flavonoids in green tea are capable of inducing pathways leading to activation of the transcription factor cAMP-response element-binding protein (CREB) with the potential to enhance both short-term and long-term memory. In addition, black tea ingestion seemed to produce a rapid increase in alertness and self-reported improvements in mood (Hindmarch et al. 1998).

2.6 Ginger

Zingiber officinale Roscoe belongs to the family Zingiberaceae and is commonly known as ginger. It originated in Southeast Asia, most common in India and southern China, and the rhizomes of the plant are extensively used in many countries as a spice and condiment to add flavor to food and in traditional herbal medicine (Park and Pizzuto 2002; Mashhadi et al. 2013; Semwal et al. 2015).

Phytochemical analysis verified the presence of more than 50 compounds in fresh ginger, grouped into two broader categories, volatile and nonvolatile compounds. Volatiles include sesquiterpene and monoterpenoid hydrocarbons, providing the distinct aroma and taste of ginger. Nonvolatile pungent compounds include gingerols, shogaols, paradols, and zingerone (Jolad et al. 2004; Butt and Sultan 2011).

Gingerols and shogaols, the latter are formed from gingerols when ginger is dried or cooked, are two phenolic substances extracted from ginger and give its characteristic odor and flavor. 6-gingerol (Fig. 5a) is the most abundant compound in fresh ginger. The content of 6-shogaol (Fig. 5b) is very low in fresh ginger but much higher in the processed ginger (Wang et al. 2015).

Besides its use in culinary uses, ginger and its major components, gingerols and shogaols, are known to have beneficial pharmacological properties, helping to treat and prevent diseases such as diabetes, obesity, diarrhea, ulcer, allergies, pain, fever, rheumatoid arthritis, inflammation, cardiovascular disorders, and various types of cancer and possess antioxidant and immunomodulatory activity (Butt and Sultan 2011; Mashhadi et al. 2013; Semwal et al. 2015).

The antioxidant capacity of ginger and its components have been explored in various *in vitro* and *in vivo* tests. Ginger exerts significant protective effects against oxidative stress in rats by increasing antioxidant defense mechanisms and lowering lipid peroxidation (Ahmed et al. 2008). In vitro antioxidant activity assay demonstrated that 6-gingerol, 8-gingerol, 10-gingerol, and 6-shogaol possess scavenging activities with IC_{50} values of 26.3, 19.47, 10.47, and 8.05 μM against DPPH radical; IC_{50} values of 4.05, 2.5, 1.68, and 0.85 μM against superoxide radical; and IC_{50} values of 4.62, 1.97, 1.35, and 0.72 μM against hydroxyl radical, respectively. These results showed that 6-shogaol exhibited the most potent antioxidant and anti-inflammatory and justifies the use of dry ginger in traditional systems of medicine (Dugasani et al. 2010). The ethyl acetate fraction (with high phenolic content) of

Fig. 5 Molecular structures of 6-gingerol (**a**) and 6-shogaol (**b**)

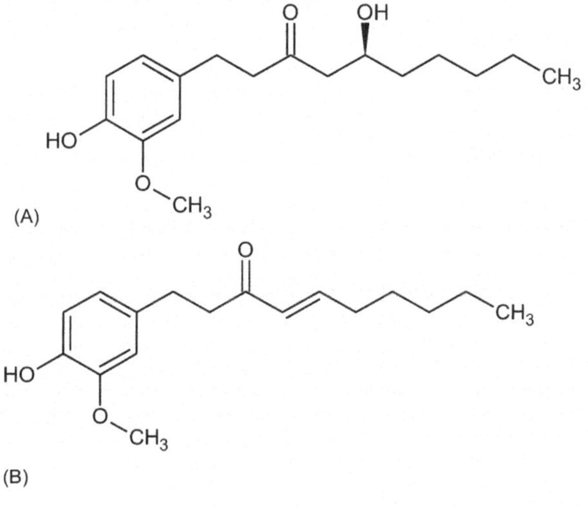

ethanolic extract of *Z. officinale* roots also showed strong antioxidant activity, besides presenting acetylcholinesterase enzyme (AChE) inhibitory activities (Tung et al. 2017).

The cytotoxic effect of ginger and its metabolites against a variety of cancer cell lines (lung, colon, skin, pancreas, prostate, liver, ovary, colon, breast, kidney) have been proved by many studies (Semwal et al. 2015).

Ginger supplementation plays an important role in improving the antioxidant system and adiponectin in obese women diagnosed with breast cancer (Karimi and Roshan 2013). Its compounds, 6-gingerol, 6-shogaol, and 6-paradol, exhibit antioxidant, anti-inflammatory, anticarcinogenic, antiproliferative, and antitumorigenic activities (Surh 1999; Hung et al. 2009; Jeong et al. 2009).

The anticancer potential of ginger is well documented, and its molecular aspects embrace inhibition of angiogenesis, metastasis, and cell-cycle progression, induction of apoptosis, and decreasing tumor initiation, promotion, and progression. It can also be used as an antiemetic agent against chemotherapy-induced nausea and improves cardiovascular disorders, diabetes mellitus, and gastrointestinal health (Baliga et al. 2011; Butt and Sultan 2011).

The consumption of raw and cooked ginger extracts can reduce blood glucose in normal and high-fat diet-induced diabetic rats (Oludoyin and Adegoke 2014). Arzati et al. (2017) showed that daily consumption of 2000 mg of ginger reduced fasting blood sugar levels and LDL/HDL ratio in 10 weeks.

Recent study indicated that ginger hydroalcoholic extract has the ability to improve liver functional biomarkers and hepatic marker enzymes, decrease iron deposition, and reduce renal functional disorders and renal histological damages. These effects could be due to its antioxidant potential by scavenging free radicals and chelating iron (Gholampour et al. 2017).

Ginger can be considered as a potential functional food for being source of bioactive compounds with various biological activities. The supplementation with ginger may be considered as a novel nutritional approach to prevent and treat, in a safe way, chronic diseases, which will benefit general population, researchers, health professionals, students, and industrialists (Trinidad et al. 2012; Mohd Yusof 2016).

2.7 Broccoli

Brassica oleracea (Brassicaceae) is a plant species that comprises several common foods, such as broccoli, which has the capacity to prevent various diseases, such as cancer and cardiovascular diseases (Finley 2003; Mahn and Reyes 2012).

There are several bioactive compounds evidenced in broccoli that may be responsible for its biological activity. Among these, the most studied are glucosinolates, sulforaphane, phenolic compounds, vitamin C, and some minerals, such as calcium, magnesium, selenium, and zinc (Moreno et al. 2006; Mahn and Reyes 2012).

Some factors influence the bioavailability of these compounds, and broccoli is mostly consumed as a processed food (steam blanching or cooking), and then its functional properties can be severely affected (Mahn and Reyes 2012).

Several studies have already proven the antioxidant activity of broccoli, mainly due to the large presence of flavonoids (Mageney et al. 2017). Bachiega et al. (2016) confirmed its antioxidant potential and demonstrated its antiproliferative activity. Furthermore, the study demonstrated that the fortification with selenium in this vegetable can reduce the mineral deficiency and increase the antioxidant and antiproliferative activity and the number of phenolic compounds.

The consumption of broccoli is inversely correlated with colorectal cancer risk, and this is believed to be at least partially attributable to the presence of isothiocyanates in this vegetable, such as sulforaphane. These compounds have been shown to have anti-inflammatory, antioxidant, and cytoprotective effects through the induction of Nrf2 (Dacosta and Bao 2017).

Sulforaphane (Fig. 6), derived from glucoraphanin, the principal glucosinolate of broccoli, has demonstrated several pharmacological actions, mainly anticancer and anti-inflammatory activity. This compound can modulate the NF-kB pathway, reducing the production of inflammatory cytokines and preventing cancer (Li and Zhang 2013). In addition, some studies verified that sulforaphane possesses the capacity to inhibit tumor progression, by the activation of apoptosis and cell-cycle arrest induction and capacity to be selectively toxic to malignant cells. It can be considered an anticancer agent alone and in combination with other therapeutic and management strategies (Tortorella et al. 2015).

Others broccoli components, indole-3-carbinol and 3,30-diindolylmethane, presented anticancer activity, mainly in prostate cancer (Wang et al. 2012).

Animal and human experiments have identified that sulforaphane possesses low level of toxicity and can protect a range of CVD, including hypertension, atherosclerosis, ischemia/reperfusion injury, diabetes, and its complications (Bai et al. 2015). A recent study has shown that this compound also presents potential effects for the treatment of obesity (Martins et al. 2018).

Several studies (*in vitro* and *in vivo*) and a smaller number of clinical trials with humans have suggested that the consumption of food rich in sulforaphane is safe and can be considered an effective strategy to reduce the risk of atherosclerosis, cancer, diabetes, and gastric, heart, neurodegenerative, ocular, and respiratory diseases. These health benefits may be attributable to Nrf-2-mediated induction of phase 2 detoxification enzymes, protecting cells against oxidative damage and to the inhibition of CYP enzymes involved in carcinogen activation, induction of apoptosis and cell-cycle arrest, and anti-inflammatory activity (Elbarbry and Elrody 2011).

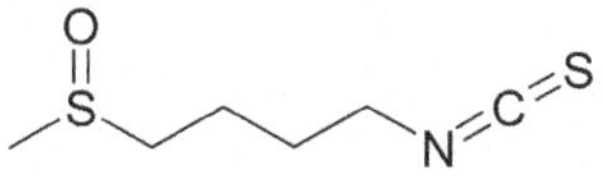

Fig. 6 Molecular structure of sulforaphane

The inclusion of broccoli as part of the daily diet can improve human nutrition and prevent chronic inflammatory pathologies, for presenting antioxidant and anti-inflammatory activities (Hwang and Lim 2014).

2.8 Grape

Grapes (*Vitis vinifera*) is one of the most popular and cultivated fruits in the world and have been used for health for their medicinal and nutritional values. They are considered natural antioxidants, due to the large presence of polyphenols, including flavonoids, presented in the seeds and skin of grapes and in wines, especially in red wines (Bertelli and Das 2009; Yang and Xiao 2013).

Several epidemiological studies have associated the consumption of grapes with lowered risk of chronic diseases (Yang and Xiao 2013).

Some biological activities have been proven in grape seeds. They can protect the organism against oxidative damage and possess antidiabetic, anticholesterol, anticancer, anti-inflammatory, and antiplatelet properties (Ma and Zhang 2017). Wang et al. (2017a, b) verified that the treatment with grape-seed polyphenols could completely inhibit the development of abdominal aortic aneurysm.

Neto et al. (2017) verified that red grape juice promotes a reduction in blood pressure at rest and is also capable of improving post-exercise hypotension in individuals with hypertension.

Resveratrol (Fig. 7) is a natural polyphenol found in grapes and red wine. Recent review concluded that this compound presents several health-promoting benefits and potential to cure and prevent multiple human diseases such as cancer, diabetes, obesity, cardiovascular and neurological diseases, and asthma. It can reduce the incidence of arterial hypertension, heart failure, and ischemic cardiac disease and can improve insulin sensitivity. This compound is considered an antioxidant, because it promotes nitric oxide production, suppresses platelet aggregation, and enhances HDL cholesterol (Rauf et al. 2017).

2.9 Black Pepper

Piper nigrum Linn, popularly known as "black pepper," belongs to the family Piperaceae that comprises over 1000 species with tropical and subtropical distribution. Is originated in the coastal areas of India but nowadays can be found in other

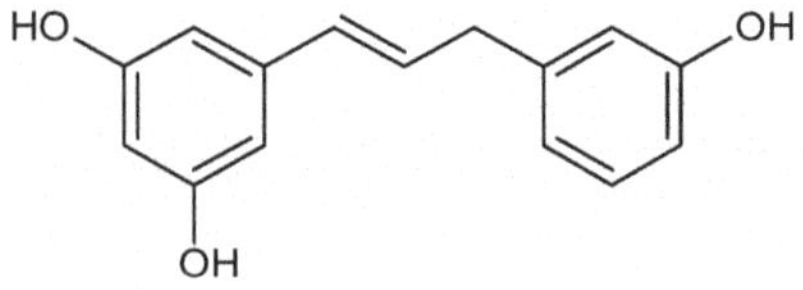

Fig. 7 Molecular structure of resveratrol

countries such as Brazil, Vietnam, Malaysia, and Indonesia (Butt et al. 2013; Rehman et al. 2015).

Black pepper, the dried unripe fruit, is commonly used as a condiment in various cuisines worldwide and is important for its medicinal value (Ahmad et al. 2012; Rehman et al. 2015).

Phytochemical analysis on black pepper have determined the presence of various active compounds: alkaloids (piperine, piperidine, piperolein, capsaicin, 2-dihydrocaspaicin), sterols (β-sitosterol, terpenoids, sesquiterpenes), fatty acids (linoleic acid), volatile oils (camphenes, pinenes), organic acids (hexadecanoic acid, octadecanoic acid), phenolic compounds (benzamides, gallic acid, kaempferol, coumarins, quercetin), polysaccharides, vitamins, and minerals (Rehman et al. 2015). Its main active principle is piperine (Fig. 8), which is known to possess many interesting pharmacological activities (Damanhouri and Ahmad 2014).

In conventional medicines, black pepper is used as a functional food to improve digestive system through various modules: better appetite and absorption and controlling dyspepsia and weight gain (Butt et al. 2013).

Several *in vitro* and *in vivo* studies prove the effectiveness of black pepper and its compounds in various biological activities.

Vijayakumar et al. (2004) demonstrated in rats that the supplementation with black pepper or its the active principle, piperine, lowered thiobarbituric acid reactive substances and conjugated dienes levels and maintained SOD, CAT, GPx, GST, and GSH levels to near those of control rats, reducing oxidative stress induced by high-fat diet. According to the results of another study, water and ethanol extracts of black pepper have high antioxidant activity and radical scavenging activity against various antioxidant systems in vitro (Gülçin 2005).

Recent study proved that piperine, hexane, and ethanol extracts of *P. nigrum* L. fruits possess potent analgesic and anti-inflammatory activities and excellent safety profile, demonstrating 0% quantal incidence of mortality in rats (Tasleem et al. 2014).

The anti-inflammatory activity of piperine is well known (Mujumdar et al. 1990); however, other alkaloids from *P. nigrum* have also shown the same potential. The alkaloid chabamide exert anti-inflammatory effects via the activation of the Nrf2/heme-oxygenase-1 pathway and can be a promising candidate for the treatment of inflammatory diseases (Ngo et al. 2017).

A recent review article reported that piperine possesses effective anticancer activity. This compound presents a potent antioxidant system, enhancing the activity of detoxifying enzymes and suppressing stem cell self-renewal, inhibits invasion,

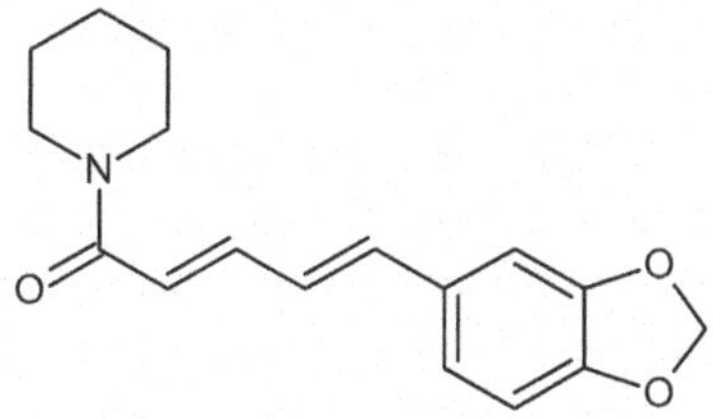

Fig. 8 Molecular structure of piperine

metastasis, angiogenesis, proliferation, and survival of various cancer cell lines and. Besides, piperine possesses selective cytotoxic activity against cancerous cells in comparison with normal cells. Clinical trials should be conducted to prove that piperine can be considered a promising anticancer (Manayi et al. 2017).

P. nigrum also has anticancer action, being a novel therapeutic spice for the treatment of colorectal carcinoma (Prashant et al. 2017). Its ethanolic extract can be cytotoxic and presents antiproliferative effect on MCF-7 cells and was able to damage DNA through intercalation and oxidative cleavage. It also demonstrated anticancer effect *in vivo* by cell-cycle arrest and apoptosis induction (Grinevicius et al. 2016).

Black pepper extracts possess good antibacterial and antimicrobial activity (Khan and Siddiqui 2007; Karsha and Lakshmi 2010). Chloroform extract inhibits *Escherichia coli* and *Staphylococcus aureus* growth by assessing cell morphology, respiration, pyruvic acid content, and ATP level (Zou et al. 2015).

The essential oils found in black pepper essential oil presents antioxidant, anti-inflammatory, and antinociceptive potential (Jeena et al. 2014).

Other activities have already been evidenced in studies with black pepper: immunomodulatory, stimulant, hepatoprotective, antifertility, anti-ulcer, antifungal, antihyperlipidemic, antihypertensive, and antiasthmatic. Some clinical trials have demonstrated the safety and efficacy of pepper in human subjects (Meghwal and Goswami 2013; Prashant et al. 2017).

3 Probiotics

Probiotics can be described as live microorganisms isolated from human and animal intestinal tracts and then administered orally as live cultures. They produce benefits to the host when given in sufficient amounts from food or supplements (Wallace 2009). Probiotics act by protecting the host due to their capability to compete with pathogens and by adhering to intestinal epithelial cells. Their diversified health and prophylactic properties led to their application as functional foods (Sarkar 2013).

Many cultures worldwide hold the tradition of fermented beverages consumption, and it has been empirically developed in ancient times as a process of raw food preservation (Baschali et al. 2017). Probiotic is a word derived from Latin that means "for life." A long time ago, fermented products as beer, cheese, bread, wine, and kefir had been commonly used for nutritional and therapeutic function. It is curious that these products were possibly found by chance, and probiotics goes alongside human race history, being traced back to ancient times about 10.000 years ago (Ozen 2015).

In 1965 the term probiotic was first used, by Lilly and Stillwell, to describe substances secreted by one organism which stimulate the growth of another. The introduction of bacterial species with beneficial effects in the gastrointestinal tract is an interesting option to reestablish the microbiota homeostasis and to prevent several diseases (Gupta and Garg 2009).

Recently the interest and research related to probiotics has increased. Multiple reports have suggested their beneficial effects for a variety of complex acute and chronic diseases (Wallace 2009). These microorganisms have an important role in the maintenance of immunologic equilibrium through interaction with immune cells. Also, the probiotic effectiveness can be specific to the goal needed (Wilkins and Sequoia 2017).

The dense microorganism population in the intestinal tract is called gut microbiota (Ríos-Covián et al. 2016) and contains beneficial bacteria that occur naturally. These microorganisms are also called probiotics and can be taken within foods or as supplementation. These bacteria have been widely studied in a variety of diseases, specially *Lactobacillus, Bifidobacterium*, and *Saccharomyces* species (Wilkins and Sequoia 2017). Probiotics can also be found as dietary supplements in different forms, such as capsules, tablets, granules, or liquids and in different dosage, with a number of colony-forming units ranging from 0.1 to 10 billion. In general, these supplements contain *Lactobacilli* and *Bifidobacteria* strains and are readily available in pharmacies and food stores (Wallace 2009).

The most commonly used bacterial genera in probiotic preparations are *Lactobacillus, Bifidobacterium, Escherichia, Enterococcus, Bacillus*, and *Streptococcus* (Gupta and Garg 2009). It is important to emphasize that probiotics mode of action of probiotics cannot be generalized to all strains and will lean on other elements such as microbiota balance, prebiotic use, and health status (Seidel et al. 2017).

Probiotics exert biological benefits by through different mechanisms, for example, due to their ability to attach to enterocytes, they inhibit the binding of enteric pathogens by a process of competitive exclusion. Also, they can influence on commensal microorganisms by the production of lactic acid and bacteriocins, which inhibit growth of pathogens and alter the ecological balance of enteric commensals (Kailasapathy and Chin 2000).

Also, they ferment nonabsorbed carbohydrates and proteins and produce a vast range of metabolites, such as short-chain fatty acids (SCFA). SCFA from microbiota fermentation may act by modulating the energy metabolism and the intestinal barrier operation. Their production depends on food intake and diet-mediated changes in the gut microbiota. SCFA have distinct physiological effects, including the contribution to shaping the gut environment and influence in the physiology of the colon. They can be used as energy sources by host cells and the intestinal microbiota. They also participate in different host-signaling mechanisms (Ríos-Covián et al. 2016; Hodzic et al. 2017).

In addition, the production of butyric acid by some probiotic bacteria affects the turnover of enterocytes and neutralizes the activity of dietary carcinogens, such as nitrosamines, that are generated by the metabolic activity of commensal bacteria in subjects consuming a high-protein diet. Therefore, inclusion of probiotic bacteria in fermented dairy products enhances their value as better therapeutic functional foods (Kailasapathy and Chin 2000). In the United States, probiotics can be found within different forms, such as yogurts, milk, cheese, cereal, nutrition bars, smoothies, and juice (Wallace 2009).

These beneficial bacteria are progressively combined into food products intended to grant health benefits in the human gut and body in general. It is important to say that the food format can affect probiotic survival, physiology, and efficacy. It is important to investigate bioactive components present in the foods and how it may interact with probiotics and, consequently, in human health (Sanders and Marco 2010).

Probiotics have been shown to be effective in several clinical conditions as infantile diarrhea, necrotizing enterocolitis, antibiotic-associated diarrhea, *Helicobacter pylori* infections, female urogenital infection, surgical infections, and obesity, among others (Gupta and Garg 2009). Evidences show that they are effective to treat different types of diarrhea, hepatic diseases, ulcerative colitis, irritable bowel syndrome, intestinal mucositis, and other diseases (Wilkins and Sequoia 2017). Other benefits from probiotic use are normalization of intestinal flora, anticarcinogenesis, hypocholesterolemic effect, and alleviation of lactose malabsorption and allergy (Sarkar 2013). There are a variety of mechanisms in which bacteria can signal to the brain and influence several processes related to neurotransmission, neurogenesis, and behavior (Cerdó et al. 2017).

An interesting research field that deserves to be mentioned is the association of probiotics strains and plants and its beneficial effects in health, for example, blueberry (*Vaccinium corymbosum* L) husks and three probiotic strains (*Bifidobacterium infantis* DSM 15159, *Lactobacillus gasseri* DSM 16737, and *Lactobacillus plantarum* DSM 15313) to attenuate colorectal inflammation, oncogenesis, and liver injuries (Håkansson et al. 2012), blueberry juice and probiotics effects reducing alcoholic fatty liver in mice (Zhu et al. 2016), synbiotic effect of yogurts with added probiotics combined with fruits in intestinal, endocrinological and cardiovascular health (Fernandez and Marette 2017), and antioxidant and antibacterial effects by probiotic-mediated blueberry fruit fermentation (Oh et al. 2017), among others. Furthermore, phytochemicals prevenient from diet can affect microbiota composition through synergistic effects, being able to enhance their bioactivity (Seidel et al. 2017).

Regarding its safety, Wilkins and Sequoia (2017) affirmed that probiotics are safe for infants, children, adults, and older patients, but caution is advised in immunologically vulnerable patients. But Wallace (2009) stands out that the secure use of probiotic in children is controversial particularly duo to the risk of sepsis from bacterial translocation.

3.1 Yogurt

Dairy products with probiotic cultures may be recommended for consumption as functional foods. These microorganisms' health effects led to a fast-growing interest in probiotics as functional foods in the current era of self-care and complementary medicine (Sarkar 2013).

Yogurt is an ancient food that has been a part of the human diet for thousands of years and has been promoted as a healthy food for providing highly bioavailable protein as well as probiotics. It goes by many names throughout the world. It is rich in calcium and potassium, which is especially important for Asian, African-American, and American Indian populations in which lactose intolerance dominates, and is a deterrent to consumption of dairy foods (Fisberg and Machado 2015).

The most common food product containing live bacterial strains is yogurt, being *Lactobacillus bulgaricus* and *Streptococcus thermophiles* are the main strains used. But only yogurts containing 10^8 live lactic acid organisms per gram can use the declaration "live active cultures" (Wallace 2009). While yogurt consumption varies greatly around the world, consumption is generally low. In the United States and Brazil, for example, only 6% of the population consume yogurt daily (Fisberg and Machado 2015). Its consumption is related to several benefits as hypocholesterolemic effects and reduction of breast and pancreatic cancer risk. Also, the lactose content in yogurt suffers reduction remaining only about 0.1%, meaning that yogurts containing higher levels of viable population could be tolerated by lactose maldigestors (Sarkar 2013).

Another good aspect of yogurt is that its lactose is more efficiently digested compared to others dairy products, especially because the bacteria in yogurt assist with its digestion. When the product exhibits a sufficient numbers of *S. thermophilus* and *L. bulgaricus*, it is very well tolerated even by lactose maldigesters, because it is effectively analogous to taking an enzyme supplement with a dairy food (Savaiano 2014).

Nowadays, yogurt is milk that has been fermented and acidified with viable bacteria, resulting in a thickened and frequently flavored product containing nutrients and is often a vehicle for fortification (added fibers, vitamins, and minerals). It is also easily altered by addiction of sweeteners, fruits, and flavors to affect consistency, aroma, and taste. It can also be produced from rice, soy, or nuts (Fisberg and Machado 2015).

When compared to natural yogurts, the flavored ones may exhibit somewhat reduced lactase activity but are still well tolerated (Savaiano 2014).

Recently studies have been done to incorporate extracts and compounds in yogurt. This product has been used as a suitable choice to be "functionalized," with addiction of other substances, as echium oil, phytosterols, and sinapic acid (Comunian et al. 2017); barley bran (Hasani et al. 2017); purple grape juice, grape skin flour, and oligofructose (Karnopp et al. 2017); and *Plukenetia volubilis* seeds (SIS) and β-glucans from *Ganoderma lucidum* (Vanegas-Azuero and Gutiérrez 2018).

Fig. 9 Kefir grains from milk. (Source: The author)

3.2 Kefir

A natural complex fermented milk product called kefir (Fig. 9) has recently gained prominence. It is a probiotic beverage that contains more than 50 species of probiotic bacteria and yeast and has been demonstrated to have several biological benefits as anti-obesity, anti-hepatic steatosis, antioxidative, antiallergenic, antitumor, anti-inflammatory, cholesterol-lowering, constipation-alleviating, and antimicrobial potential (Kim et al. 2018). Kefir is a term used for a fermented dairy beverage produced by the actions of the microorganisms encased in the "kefir grain" on the carbohydrates found in milk. It contains many bacteria strains known for their probiotic properties. It is a mixture of yeast and bacteria, living in a synbiotic association (Nielsen et al. 2014; Sharifi et al. 2017). But kefir can also be made by brown sugar and water solution incubation (Laureys and De Vuyst 2017).

Their composition differs in relation to the presence of some bacteria, for example, *Lactococcus, Leuconostoc, Lactobacillus* (L.), and *Oenococcus. Leuconostoc mesenteroides, Lactobacillus kefiri*, and *Lactobacillus kefiranofaciens* were isolated only from milk grains, whereas *Lactobacillus perolens, Lactobacillus parafarraginis, Lactobacillus diolivorans*, and *Oenococcus oeni* were isolated exclusively from sugar water grains (Zanirati et al. 2015).

The commercial use of kefir is now increasing, and a variety of kefir based-products can be found in the market, as kefir is an effective probiotic delivery (Prado et al. 2015). In Eastern Europe it is commonly used in hospitals patients. In the United States, it is gaining fame as a healthy probiotic beverage or as artisanal beverage. It can be home fermented from shared grains, but it can also be found as a commercial product (Nielsen et al. 2014).

The kefir beverage is a source of probiotics but also other molecules with healthy effects in its composition, as kefiran (Fig. 10), an exopolysaccharide found in kefir

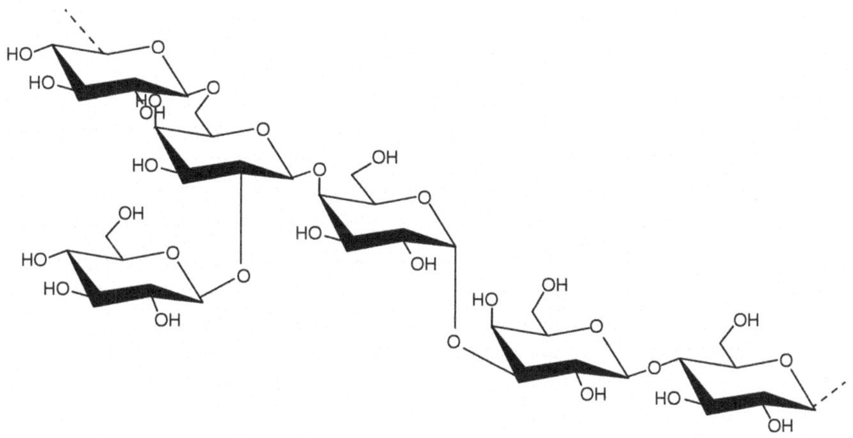

Fig. 10 Chemical structure of kefiran found in kefir (Prado et al. 2015)

that has a long list of biological activities such as antioxidant, antitumor agent, antimicrobial agent, and immunomodulatory. Kefiran can also be found separately in food and pharmaceutical products (Prado et al. 2015).

In Brazil, kefir is a homemade fermented beverage that can be obtained by kefir grains incubation in milk or brown sugar solution. Both forms of preparation demonstrated several lactic acid bacteria that could be used in combination with yeasts as starter cultures (Zanirati et al. 2015). In addition, kefir grains have a complex composition of microbial species such as the predominance of lactic acid bacteria, acetic bacteria, yeasts, and fungi (Prado et al. 2015). It is important to emphasize that it has a great resentence through gastrointestinal tract and is able to colonize it, showing a significant modulatory effect on host gut by both microbiota and mycobiota population (Kim et al. 2018).

Tung et al. (2018) showed that kefir peptides treatment had beneficial effects in high-fat diet-induced obese rats, improving obesity by lipogenesis inhibition, oxidative damage modulation, and lipid oxidation stimulation, suggesting that it may act as an anti-obesity agent.

Rafie et al. (2015) conducted a review of kefir and observed that in vitro studies on different cell lines and experimental studies consistently demonstrated the beneficial effects of kefir regarding cancer prevention and treatment. Its protection may be related to kefir bioactive components as peptides, polysaccharides, and sphingolipids.

It is suggested that some of the bioactive compounds found in kefir such as polysaccharides and peptides have great potential to inhibit the proliferation and to induce apoptosis in tumor cells (Sharifi et al. 2017).

3.3 Kombucha

Kombucha is a slightly sweet, slightly acidic tea beverage consumed globally but specially in China, Russia, and Germany. It is prepared by the fermentation of sweetened black tea infusions with a synbiotic culture of yeasts and bacteria (Greenwalt et al. 2000). It is fermented by a synbiosis of bacteria and yeast embedded within a cellulosic pellicle, which forms a floating mat in the tea and generates a new layer in each fermentation (Marsh et al. 2014). The microbial community of Kombucha tea consists of bacteria and yeast in two mutually non-exclusive compartments: the beverage and the biofilm floating on it (Chakravorty et al. 2016).

It is a nonalcoholic beverage prepared with water, sugar, tea, and a Kombucha culture that is also called "tea fungus," maintained in open vessels at room temperature for 1–3 weeks, resulting in a sharp acidity and specific flavor beverage (De Roos and De Vuyst 2018). Duo to the fermentation, ethanol, and carbon dioxide, a high amount of acids (gluconic, acetic, and lactic) as well as many other metabolites can be found in Kombucha (Marsh et al. 2014).

Although it is mainly produced with black and green tea derived from *C. sinensis* infusion, it can also be made with others plant species, as ginger (Salafzoon et al. 2017), Chinese herbal extracts (Fu et al. 2015), and *Litsea glaucescens* or *Eucalyptus camaldulensis* (Gamboa-Gómez et al. 2016). Kombucha tea benefits have been related to be mainly due to the presence of fermentation products such as flavonoids and other polyphenols with inhibition of hydrolytic and oxidative enzymes and anti-inflammatory effects (Pakravan et al. 2017).

Different studies have demonstrated different compositions for Kombucha. Marsh et al. (2014) evaluated the bacterial and fungal populations of five distinct pellicles and its fermented Kombucha. It was established that the major bacterial genus present was *Gluconacetobacter* (>85%), with only trace populations of *Acetobacter* detected (<2%). A predominant *Lactobacillus* population was also identified (up to 30%), with many subdominant genera, not previously associated with Kombucha. The yeast populations were found to be dominated by *Zygosaccharomyces* (>95%) in the fermented beverage, with a greater fungal diversity present in the cellulosic pellicle.

According to Chakravorty et al. (2016), the yeast community of the biofilm was dominated by *Candida sp*. And the beverage showed a significant shift in dominance from *Candida sp.* to *Lachancea sp.* on the 7th day of fermentation. The authors highlight that the biochemical properties of Kombucha changed with the fermentation progression, meaning that the beneficial properties of the beverage, such as the radical scavenging ability, increased significantly, with a maximum increase at day 7 (Chakravorty et al. 2016). The most prevalent bacteria found in Kombucha by De Roos and De Vuyst (2018) were acetic acid, including *Komagataeibacter* and *Gluconobacter* species.

Kombucha can be homemade or can also be found commercially with different flavors in natural products stores. It has demonstrated interesting biological properties, as antidiabetic (Bhattacharya et al. 2013) and hepatoprotective agent (Wang et al. 2014), in prevention against broad-spectrum metabolic and infective disorders (Vina et al. 2014) and to manage foot and mouth disease virus (Fu et al. 2015).

Also, its curative effects on hypercholesterolemic, particularly in terms of liver-kidney functions in rats (Bellassoued et al. 2015), antifungal properties and treatment of infections caused by *Malassezia* (Mahmoudi et al. 2016), and antihypertensive (Gamboa-Gómez et al. 2016) and antibacterial activity against enteropathogenic bacterial infections (Bhattacharya et al. 2016, 2018), have been described. The potential health effects have created an increased interest in Kombucha (Greenwalt et al. 2000).

Another use of Kombucha is in beauty industry, as demonstrated by Pakravan et al. (2017) by the improvement of aging-related skin abnormalities and regeneration of aged skin.

Although Kombucha had become a very popular fermented beverage, in a systematic review by Ernst (2003), no clinical studies were found relating to its efficacy, but several case reports and case series raise doubts about its safety, including suspected liver damage, metabolic acidosis, and cutaneous anthrax infections. Also, one fatality was on record. Gedela et al. (2016) highlights the caution related to possible hepatotoxicity. A recent report by Holbourn and Hurdman (2017) also stands out the concern about Kombucha use, describing the case of a woman with history of breathlessness, in which investigations revealed severe metabolic lactic acidosis that was linked to her use of Kombucha tea.

4 Prebiotics

Prebiotic is defined as a non-digestible food ingredient that confers benefits on the host by selectively stimulating one bacterium or a group of bacteria in the colon with probiotic properties (Gupta and Garg 2009). Most human trials present that prebiotics significantly increased levels of *Bifidobacteria* in the gut and consequently presented systemic health benefits (Roberfroid et al. 2010).

Literature shows that some food ingredients, specially milk components and prebiotics, may improve probiotic survival during the shelf-life of food products, which is important because it can intensify probiotic effects through increased dose effects (Sanders and Marco 2010). Prebiotics may be used as an additional support to probiotics, or even as an alternative, as prebiotics have long-term stability during the shelf-life of food and drinks. Besides resistance to processing, physical and chemical properties that display positive effects on flavor and consistence of products may promote prebiotics as a competition to probiotics (Markowiak and Slizewska 2017).

The food area designated to produce products to improve gastrointestinal health such as probiotics, prebiotics, and synbiotic has been the most important segment within functional foods. Most of these products are dairy-based, so the development of nondairy gut improvement products has been of great interest for the food industry, resulting in the rise of cereal-based probiotic and synbiotic products (Salmerón 2017). Also, the incorporation of prebiotics in a wide range of products that food industry offers on shelf is an innovative way to replace fat and sugars (Singla and Chakkaravarthi, 2017).

Functional products are being increasingly produced, such as prebiotic white chocolate with goji berry (Morais Ferreira et al. 2017) and fruit-yogurt gathering their probiotic and prebiotic properties (Fernandez and Marette 2017).

The industry of infant formula is also betting in prebiotics, because the gastrointestinal microbiota of breast-fed babies differs from classic standard formula-fed infants. Human breast milk naturally has in its composition a large number of prebiotic oligosaccharides and small amounts of probiotics, while standard infant formula does not. Duo to this, the industry adds different prebiotic oligosaccharides to infant formula, such as galacto-oligosaccharides, fructo-oligosaccharide, polydextrose, and mixtures of these, to mimic breast milk. There is evidence that the addition of prebiotics in infant formula alters the gastrointestinal microbiota, leading to a lower stool pH, a better stool consistency, frequency, and a higher concentration of *Bifidobacteria.* It also reduces the risk of gastroenteritis and infections, improves general well-being, and reduces the incidence of allergic symptoms such as atopic eczema (Roberfroid et al. 2010; Vandenplas et al. 2014).

The prebiotic beneficial effects are a well-established scientific fact, and the more data are accumulating, the more it will be recognized that such changes in the microbiota's composition, especially increase in *Bifidobacteria*, can be regarded as a marker of intestinal health (Roberfroid et al. 2010). Their benefits to human health when taken as supplement are related to an immune regulation and bacterial metabolite production, as short-chain fatty acid. This supplementation leads to increased growth of specific gut microbiota, especially *Bifidobacteria* (Wilson and Whelan 2017).

It can be found naturally in food or can be added to improve their nutritional and health value. Some examples are inulin, fructooligosaccharides, lactulose, soy oligosaccharides, and derivatives of galactose and *b*-glucans. Those substances may serve as a medium for probiotics. They stimulate their growth and contain no microorganisms (Markowiak and Slizewska 2017). Among prebiotics, the inulin-type fructans and galacto-oligosaccharide are the most investigated (Wilson and Whelan 2017). Different types of prebiotics can stimulate the growth of specific gut bacteria. Prebiotics have a huge capacity to transform the microbiota, although it happens at the level of individual strains and species and is not easily predicted a priori (Markowiak and Slizewska 2017). Many reports can be found in literature demonstrating its beneficial effects in health.

Experimental studies have also shown beneficial effects of prebiotics in tumor and cancer incidence reduction, increased calcium absorption, bone calcium accretion, and bone mineral density. Besides prebiotic properties on energy homeostasis, satiety regulation and body weight gain have also been proven, and together with data in obese animals and patients, these studies support the hypothesis that gut microbiota composition may contribute to modulate metabolic disorders (Roberfroid et al. 2010).

Dietary fiber as prebiotics is the main source of energy for microbiota population. *In vitro, in vivo*, and clinical trials have shown that it has positive effects in models of colitis, with improvement of inflammatory markers (Rasmussen and Hamaker 2017). Low doses of prebiotic supplementation may generate modulation of the gut bacteria and reduction of symptoms in irritable bowel syndrome, but larger doses may have neutral or negative impact on symptoms. Regarding Crohn's disease, prebiotics have not shown benefit to bacterial modulation or inflammatory response (Wilson and Whelan 2017).

5 Synbiotics

Probiotics and prebiotics are together called synbiotic. Some fungal strains belonging to *Saccharomyces* have also been used (Gupta and Garg 2009). Synbiotic is defined as nutritional supplement that contain probiotics and prebiotics that work in a form of synergism. This product is suggested to manage metabolic profiles of patients suffering from diseases, such as metabolic syndrome (Tabrizi et al. 2017).

When composing a synbiotic formula, it is important to select appropriate probiotic and prebiotic, which will exert a positive effect on the host's health when used separately. A prebiotic should selectively stimulate the growth of microorganisms, having a beneficial effect on health, with simultaneous absent or limited stimulation of other microorganisms (Markowiak and Slizewska 2017).

Bifidobacterium or *Lactobacillus* genus bacteria with fructooligosaccharides combination in synbiotic products seem to be the most popular used combination (Markowiak and Slizewska 2017). Furthermore, synbiotic products may have a greater effectiveness than either probiotics or prebiotics alone (Sanders and Marco 2010).

Tabrizi et al. (2017) published a systematic review and meta-analysis of randomized controlled trials evaluating the effects of synbiotic supplementation on glucose and lipid profile in diabetic patients. They conclude that it may improve insulin, HOMA-IR, HOMA-B, triglycerides, total cholesterol, and VLDL cholesterol levels, but did not affect LDL cholesterol and HDL cholesterol levels in patients with diabetes.

According to Markowiak and Slizewska (2017), the development of formulas containing both appropriate microbial strains and synergistic prebiotics may lead to the enhancement of the probiotic effect in the small intestine and the colon.

6 Conclusions

The concept of functional foods was defined since about 40 years, but the regulation of these products still is made by different authorities and has not been consolidated in many countries. It englobes foods that have a beneficial effect on health beyond the basic function of nutrition helping reduce the risk of diseases. The examples selected in this chapter, including curcuma, garlic, ômega-3, olive oil, tea, ginger, broccoli, grape, and black pepper, are relevant because of the promising therapeutic potential confirmed in preclinical and clinical trials. On the other hand, it is also important to mention the biological effects of probiotics, which include yogurt, kefir, and kombucha products, in addition to prebiotics and synbiotics. To summarize, these foods should be introduced in the daily consumption of the population, not only with nutritive purposes, but to prevent several diseases, increasing the life quality.

References

Ahmad N, Fazal H, Abbasi BH et al (2012) Biological role of *Piper nigrum* L. (black pepper): a review. Asian Pac J Trop Biomed 2(3):1–10

Ahmadi N, Nabavi V, Hajsadeghi F et al (2013) Aged garlic extract with supplement is associated with increase in brown adipose, decrease in white adipose tissue and predict lack of progression in coronary atherosclerosis. Int J Cardiol 168(3):2310–2314

Ahmadi F, Ghasemi-Kasman M, Ghasemi S et al (2017) Induction of apoptosis in HeLa cancer cells by an ultrasonic-mediated synthesis of curcumin-loaded chitosan-alginate-STPP nanoparticles. Int J Nanomedicine 12:8545–8556

Ahmed RS, Suke SG, Seth V et al (2008) Protective effects of dietary ginger *(Zingiber officinales* Rosc.) on lindane-induced oxidative stress in rats. Phytother Res 22:902–906

Ak T, Gülçin I (2008) Antioxidant and radical scavenging properties of curcumin. Chem Biol Interact 174(1):27–37

Al-Azzawie HF, Alhamdani MSS (2006) Hypoglycemic and antioxidant effect of oleuropein in alloxan-diabetic rabbits. Life Sci 78(12):1371–1377

Allegri L, Rosignolo F, Mio C et al (2018) Effects of nutraceuticals on anaplastic thyroid cancer cells. J Cancer Res Clin Oncol 144(2):285–294

Alves MR, Cunha SC, Amaral JS et al (2005) Classification of PDO olive oils on the basis of their sterol composition by multivariate analysis. Anal Chim Acta 549:166–178

Amagase H (2006) Clarifying the real bioactive constituents of garlic. J Nutr 136(3):716S–725S

Ankri S, Mirelman D (1999) Antimicrobial properties of allicin from garlic. Microbes Infect 1(2):125–129

Arzati MM, Honarvar NM, Saedisomeolia A et al (2017) The effects of ginger on fasting blood sugar, hemoglobin A1c, and lipid profiles in patients with type 2 diabetes. Int J Endocrinol Metab 15(4):1–7

Ashraf R, Khan RA, Ashraf I (2011) Garlic (*Allium sativum*) supplementation with standard antidiabetic agent provides better diabetic control in type 2 diabetes patients. Pak J Pharm Sci 24:565–570

Atkin M, Laight D, Cummings MH (2016) The effects of garlic extract upon endothelial function, vascular inflammation, oxidative stress and insulin resistance in adults with type 2 diabetes at high cardiovascular risk. A pilot double blind randomized placebo controlled trial. J Diabetes Complicat 30(4):723–727

Bachiega P, Salgado JM, De Carvalho JE et al (2016) Antioxidant and antiproliferative activities in different maturation stages of broccoli (*Brassica oleracea* Italica) biofortified with selenium. Food Chem 190(1):771–776

Bahadoran Z, Mirmiran P, Momenan AA et al (2017) Allium vegetable intakes and the incidence of cardiovascular disease, hypertension, chronic kidney disease, and type 2 diabetes in adults: a longitudinal follow-up study. J Hypertens 35(9):1909–1916

Bai Y, Wang X, Zhao S et al (2015) Sulforaphane protects against cardiovascular disease via Nrf2 activation. Oxidative Med Cell Longev 2015:1–13

Bakhouche A, Sánchez J, Beltrán-Debón R et al (2013) Phenolic characterization and geographical classification of commercial Arbequina extra-virgin olive oils produced in southern Catalonia. Food Res Int 50(1):401–408

Baliga MS, Haniadka R, Pereira MM et al (2011) Update on the chemopreventive effects of ginger and its phytochemicals. Crit Rev Food Sci Nutr 51(6):499–523

Barzegar A, Moosavi-Movahedi AA (2011) Intracellular ROS protection efficiency and free radical-scavenging activity of curcumin. PLoS One 6(10):1–7

Baschali A, Tsakalidou E, Kyriacou A et al (2017) Traditional low-alcoholic and non-alcoholic fermented beverages consumed in European countries: a neglected food group. Nutr Res Rev 30(1):1–24

Basnet P, Skalko-Basnet N (2011) Curcumin: an anti-inflammatory molecule from a curry spice on the path to cancer treatment. Molecules 16(6):4567–4598

Battaini F, Mochly-Rosen D (2007) Happy birthday protein kinase C: past, present and future of a superfamily. Pharmacol Res 55(6):461–466

Bellassoued K, Ghrab F, Makni-Ayadi F et al (2015) Protective effect of kombucha on rats fed a hypercholesterolemic diet is mediated by its antioxidant activity. Pharm Biol 53(11):1699–1709

Beltz LA, Bayer DK, Moss AL et al (2006) Mechanisms of cancer prevention by green and black tea polyphenols. Anticancer Agents Med Chem 6(5):389–406

Bertelli AA, Das DK (2009) Grapes, wines, resveratrol, and heart health. J Cardiovasc Pharmacol 54(6):468–476

Bettuzzi S, Brausi M, Rizzi F et al (2006) Chemoprevention of human prostate cancer by oral administration of green tea catechins in volunteers with high-grade prostate intraepithelial neoplasia: a preliminary report from a one-year proof-of-principle study. Cancer Res 66(2):1234–1240

Bhattacharya S, Gachhui R, Sil PC (2013) Effect of kombucha, a fermented black tea in attenuating oxidative stress mediated tissue damage in alloxan induced diabetic rats. Food Chem Toxicol 60:328–340

Bhattacharya D, Bhattacharya S, Patra MM et al (2016) Antibacterial activity of polyphenolic fraction of kombucha against enteric bacterial pathogens. Curr Microbiol 73(6):885–896

Bhattacharya D, Ghosh D, Bhattacharya S et al (2018) Antibacterial activity of polyphenolic fraction of kombucha against *Vibrio cholerae*: targeting cell membrane. Lett Appl Microbiol 66(2):145–152

Bogani P, Galli C, Villa M et al (2007) Postprandial anti-inflammatory and antioxidant effects of extra virgin olive oil. Atherosclerosis 190(1):181–186

Brasili E, Cechinel-Filho V (2017) Metabolomics of cancer cell cultures to assess the effects of dietary phytochemicals. Crit Rev Food Sci Nutr 57(7):1328–1339

Brenes M, Hidalgo FJ, García A et al (2000) Pinoresinol and 1-acetoxypinoresinol, two new phenolic compounds identified in olive oil. J Am Oil Chem Soc 77(7):715–720

Butt MS, Sultan MT (2011) Ginger and its health claims: molecular aspects. Crit Rev Food Sci Nutr 51(5):383–393

Butt MS, Pasha I, Sultan MT et al (2013) Black pepper and health claims: a comprehensive treatise. Crit Rev Food Sci Nutr 53(9):875–886

Calder PC (2012) The role of marine omega-3 (n-3) fatty acids in inflammatory processes, atherosclerosis and plaque stability. Mol Nutr Food Res 56(7):1073–1080

Cao H, Hininger-Favier I, Kelly MA et al (2007) Green tea polyphenol extract regulates the expression of genes involved in glucose uptake and insulin signaling in rats fed a high fructose diet. J Agric Food Chem 55(25):6372–6378

Cerdó T, Ruíz A, Suárez A et al (2017) Probiotic, prebiotic, and brain development. Nutrients 9(11):1–19

Chacko SM, Thambi PT, Kuttan R et al (2010) Beneficial effects of green tea: a literature review. Chin Med 5(23):1–9

Chakravorty S, Bhattacharya S, Chatzinotas A et al (2016) Kombucha tea fermentation: microbial and biochemical dynamics. Int J Food Microbiol 220:63–72

Chauhan S, Kodali H, Noor J et al (2017) Role of omega-3 fatty acids on lipid profile in diabetic dyslipidaemia: single blind, randomised clinical trial. J Clin Diagn Res 11(3):OC13–OC16
Chen S, Nimick M, Cridge AG et al (2018) Anticancer potential of novel curcumin analogs towards castrate-resistant prostate cancer. Int J Oncol 52(2):579–588
Chu YL, Ho CT, Chung JG et al (2013) Allicin induces anti-human liver cancer cells through the p53 gene modulating apoptosis and autophagy. J Agric Food Chem 61(41):9839–9848
Chuengsamarn S, Rattanamongkolgul S, Phonrat B et al (2014) Reduction of atherogenic risk in patients with type 2 diabetes by curcuminoid extract: a randomized controlled trial. J Nutr Biochem 25(2):144–150
Comunian TA, Chaves IE, Thomazini M et al (2017) Development of functional yogurt containing free and encapsulated echium oil, phytosterol and sinapic acid. Food Chem 237:948–956
Cooper R, Morré DJ, Morré DM (2005) Medicinal benefits of green tea: part I. Review of noncancer health benefits. J Altern Complement Med 11(3):521–528
Correa F, Buelna-Chontal M, Hernandez-Resendiz S et al (2013) Curcumin maintains cardiac and mitochondrial function in chronic kidney disease. Free Radic Biol Med 61:119–129
Cottin SC, Sanders TA, Hall WL (2011) The differential effects of EPA and DHA on cardiovascular risk factors. Proc Nutr Soc 70:215–231
Cuadrado A, Moreno-Murciano P, Pedraza-Chaverri J (2009) The transcription factor Nrf2 as a new therapeutic target in Parkinson's disease. Expert Opin Ther Targets 13:319–329
Dacosta C, Bao Y (2017) The role of MicroRNAs in the chemopreventive activity of sulforaphane from cruciferous vegetables. Nutrients 9(8):1–19
Damanhouri ZA, Ahmad A (2014) A review on therapeutic potential of *Piper nigrum* L. (Black Pepper): the king of spices. Med Aromat Plants 3(3):1–6
De Caterina R, Liao JK, Libby P (2000) Fatty acid modulation of endothelial activation. Am J Clin Nutr 71:213s–223s
De Mejia EG, Ramirez-Mares MV, Puangpraphant S (2009) Bioactive components of tea: cancer, inflammation and behavior. Brain Behav Immun 23(6):721–731
De Roos J, De Vuyst L (2018) Acetic acid bacteria in fermented foods and beverages. Curr Opin Biotechnol 49:115–119
Dugasani S, Pichika MR, Nadarajah VD et al (2010) Comparative antioxidant and anti-inflammatory effects of [6]-gingerol, [8]-gingerol, [10]-gingerol and [6]-shogaol. J Ethnopharmacol 127(2):515–520
Elbarbry F, Elrody N (2011) Potential health benefits of sulforaphane: a review of the experimental, clinical and epidemiological evidences and underlying mechanisms. J Med Plants Res 5(4):473–484
Elnagar AY, Sylvester PW, El Sayed KA (2011) (-)-Oleocanthal as a c-Met inhibitor for the control of metastatic breast and prostate cancers. Planta Med 77(10):1013–1019
Eng QY, Thanikachalam PV, Ramamurthy S (2018) Molecular understanding of Epigallocatechin gallate (EGCG) in cardiovascular and metabolic diseases. J Ethnopharmacol 210:296–310
Ernst E (2003) Kombucha: a systematic review of the clinical evidence. Forsch Komplementarmed Klass Naturheilkd 10(2):85–87
Escrich E, Moral R, Grau L et al (2007) Molecular mechanisms of the effects of olive oil and other dietary lipids on cancer. Mol Nutr Food Res 51:1279–1292
Fernandez MA, Marette A (2017) Potential health benefits of combining yogurt and fruits based on their probiotic and prebiotic properties. Adv Nutr 8(1):155S–164S
Finley JW (2003) Reduction of cancer risk by consumption of selenium-enriched plants: enrichment of broccoli with selenium increases the anticarcinogenic properties of broccoli. J Med Food 6(1):19–26
Fisberg M, Machado R (2015) History of yogurt and current patterns of consumption. Nutr Rev 73(1):4–7
Franco MN, Galeano-Días T, López Ó et al (2014) Phenolic compounds and antioxidant capacity of virgin olive oil. Food Chem 163:289–298

Fu N, Wu J, Lv L et al (2015) Anti-foot-and-mouth disease virus effects of Chinese herbal kombucha *in vivo*. Braz J Microbiol 46(4):1245–1255

Gamboa-Gómez CI, González-Laredo RF, Gallegos-Infante JA et al (2016) Antioxidant and angiotensin-converting enzyme inhibitory activity of *Eucalyptus camaldulensis* and *Litsea glaucescens* infusions fermented with kombucha consortium. Food Technol Biotechnol 54(3):367–374

Gandhy SU, Kim K, Larsen L et al (2012) Curcumin and synthetic analogs induce reactive oxygen species and decreases specificity protein (Sp) transcription factors by targeting microRNAs. BMC Cancer 12:1–12

Gao SM, Yang JJ, Chen CQ et al (2012) Pure curcumin decreases the expression of WT1 by upregulation of miR-15a and miR-16-1 in leukemic cells. J Exp Clin Cancer Res 31:27

Gebauer SK, Psota TL, Harris WS et al (2006) n-3 fatty acid dietary recommendations and food sources to achieve essentiality and cardiovascular benefits. Am J Clin Nutr 83:1526s–1535s

Gedela M, Potu KC, Gali VL et al (2016) A case of hepatotoxicity related to kombucha tea consumption. S D Med 69(1):26–28

Geleijnse JM, Giltay EJ, Grobbee DE et al (2002) Blood pressure response to fish oil supplementation: metaregression analysis of randomized trials. J Hypertens 20:1493–1499

Gholampour F, Ghiasabadi FB, Owji SM et al (2017) The protective effect of hydroalcoholic extract of Ginger (*Zingiber officinale* Rosc.) against iron-induced functional and histological damages in rat liver and kidney. Avicenna J Phytomed 7(6):542–553

Ghosh S, Banerjee S, Sil PC (2015) The beneficial role of curcumin on inflammation, diabetes and neurodegenerative disease: a recent update. Food Chem Toxicol 83:111–124

Gómez-Romero M, García-Villalba R, Carrasco-Pancorbo A et al (2012) Metabolism and bioavailability of olive oil polyphenols. In: Dimitrios B (ed) Olive oil – constituents, quality, health properties and bioconversions. InTech, Rijeka, pp 333–356

Goulas V, Exarchou V, Troganis NA et al (2009) Phytochemicals in olive-leaf extracts and their antiproliferative activity against cancer and endothelial cells. Mol Nutr Food Res 53(5):600–608

Gouvinhas I, Machado N, Sobreira C et al (2017) Critical review on the significance of olive phytochemicals in plant physiology and human health. Molecules 22(11):1–35

Graham HN (1992) Green tea composition, consumption, and polyphenol chemistry. Prev Med 21(3):334–350

Granato D, Nunes DS, Barba FJ (2017) An integrated strategy between food chemistry, biology, nutrition, pharmacology, and statistics in the development of functional foods: a proposal. Trends Food Sci Technol 62:13–22

Greenwalt CJ, Steinkraus KH, Ledford RA (2000) Kombucha, the fermented tea: microbiology, composition, and claimed health effects. J Food Prot 63(7):976–981

Grinevicius VMS, Kviecinski MR, Santos Mota NS et al (2016) *Piper nigrum* ethanolic extract rich in piperamides causes ROS overproduction, oxidative damage in DNA leading to cell cycle arrest and apoptosis in cancer cells. J Ethnopharmacol 189:139–147

Grosso G, Stepaniak U, Micek A et al (2015) Association of daily coffee and tea consumption and metabolic syndrome: results from the Polish arm of the HAPIEE study. Eur J Nutr 54(7):1129–1137

Gülçin I (2005) The antioxidant and radical scavenging activities of black pepper (*Piper nigrum*) seeds. Int J Food Sci Nutr 56(7):491–499

Gupta V, Garg R (2009) Probiotics. Indian J Med Microbiol 27(3):202–209

Gupta SC, Patchva S, Aggarwal BB (2013) Therapeutic roles of curcumin: lessons learned from clinical trials. AAPS J 15(1):195–218

Håkansson A, Bränning C, Molin G et al (2012) Blueberry husks and probiotics attenuate colorectal inflammation and oncogenesis, and liver injuries in rats exposed to cycling DSS-treatment. PLoS One 7(3):1–14

Hamer M, Witte DR, Mosdøl A et al (2008) Prospective study of coffee and tea consumption in relation to risk of type 2 diabetes mellitus among men and women: The Whitehall II study. Br J Nutr 100(5):1046–1053

Han MK (2003) Epigallocatechin gallate, a constituent of green tea, suppresses cytokine-induced pancreatic β-cell damage. Exp Mol Med 35(2):136–139
Harris WS, Bulchandani D (2006) Why do omega-3 fatty acids lower serum triglycerides? Curr Opin Lipidol 17(4):387–393
Harris JC, Cottrell SL, Plummer S et al (2001) Antimicrobial properties of *Allium sativum* (garlic). Appl Microbiol Biotechnol 57(3):282–286
Harris WS, Dayspring TD, Moran TJ (2013) Omega-3 fatty acids and cardiovascular disease: new developments and applications. Postgrad Med 125(6):100–113
Hasani S, Sari AA, Heshmati A et al (2017) Physicochemical and sensory attributes assessment of functional low-fat yogurt produced by incorporation of barley bran and *Lactobacillus acidophilus*. Food Sci Nutr 5(4):875–880
Hasmeda M, Polya GM (1996) Inhibition of cyclic AMP-dependent protein kinase by curcumin. Phytochemistry 42(3):599–605
Hewlings SJ, Kalman DS (2017) Curcumin: a review of its' effects on human health. Foods 6(10):1–11
Higdon JV, Frei B (2003) Tea catechins and polyphenols: health effects, metabolism, and antioxidant functions. Crit Rev Food Sci Nutr 43:89–143
Hindmarch I, Quinlan PT, Moore KL et al (1998) The effects of black tea and other beverages on aspects of cognition and psychomotor performance. Psychopharmacology 139(3):230–238
Hodzic Z, Bolock AM, Good M (2017) The role of mucosal immunity in the pathogenesis of necrotizing enterocolitis. Front Pediatr 5(40):1–17
Holbourn A, Hurdman J (2017) Kombucha: is a cup of tea good for you? BMJ Case Rep. https://doi.org/10.1136/bcr-2017-221702
Hosseini A, Hosseinzadeh H (2015) A review on the effects of *Allium sativum* (garlic) in metabolic syndrome. J Endocrinol Investig 38:1147–1157
Hung JY, Hsu YL, Li CT et al (2009) [6]-Shogaol, an active constituent of dietary ginger, induces autophagy by inhibiting the AKT/mTOR pathway in human non-small cell lung cancer A549 cells. J Agric Food Chem 57:9809–9816
Hwang JH, Lim SB (2014) Antioxidant and anti-inflammatory activities of broccoli florets in LPS-stimulated RAW 264.7 cells. Prev Nutr Food Sci 19(2):89–97
Inoue M, Tajima K, Mizutani M et al (2001) Regular consumption of green tea and the risk of breast cancer recurrence: follow-up study from the Hospital-based Epidemiologic Research Program at Aichi Cancer Center (HERPACC), Japan. Cancer Lett 167:175–182
International Olive Council. Designations and definitions of olive oils. Available online: http://www.internationaloliveoil.org/. Accessed on Aug 2017
Jacobson TA, Glickstein SB, Rowe JD et al (2012) Effects of eicosapentaenoic acid and docosahexaenoic acid on low-density lipoprotein cholesterol and other lipids: a review. J Clin Lipidol 6(1):5–18
Jalili-Nik M, Soltani A, Moussavi S et al (2017) Current status and future prospective of curcumin as a potential therapeutic agent in the treatment of colorectal cancer. J Cell Physiol. https://doi.org/10.1002/jcp.26368
Jayabalan R, Subathradevi P, Marimuthu S et al (2008) Changes in free-radical scavenging ability of kombucha tea during fermentation. Food Chem 109(1):227–234
Jeena K, Liju VB, Umadevi NP et al (2014) Antioxidant, anti-inflammatory and antinociceptive properties of black pepper essential oil (*Piper nigrum* Linn). JEOP 17:1–12
Jeong GS, Oh GS, Pae HO et al (2006) Comparative effects of curcuminoids on endothelial heme oxygenase-1 expression: ortho-methoxy groups are essential to enhance heme oxygenase activity and protection. Exp Mol Med 38:393–400
Jeong CH, Bode AM, Pugliese A et al (2009) [6]gingerol suppresses colon cancer growth by targeting leukotriene a4 hydrolase. Cancer Res 69:5584–5591
Jolad SD, Lantz RC, Solyom AM et al (2004) Fresh organically grown ginger (*Zingiber officinale*): Composition and effects on LPS-induced PGE2 production. Phytochemistry 65:1937–1954

Jones GJB, Roper RL (2017) The effects of diets enriched in omega-3 polyunsaturated fatty acids on systemic vaccinia virus infection. Sci Rep 7:15999

Julie S, Jurenka M (2009) Anti-inflammatory properties of curcumin, a major constituent. Altern Med Rev 14:141–153

Jurrmann N, Brigelius-Flohe R, Bol GF (2005) Curcumin blocks interleukin-1 (IL-1) signaling by inhibiting the recruitment of the IL-1 receptor-associated kinase IRAK in murine thymoma EL-4 cells. J Nutr 135:1859–1864

Kailasapathy K, Chin J (2000) Survival and therapeutic potential of probiotic organisms with reference to *Lactobacillus acidophilus* and *Bifidobacterium* spp. Immunol Cell Biol 78:80–88

Karalis DG (2017) A review of clinical practice guidelines for the management of hypertriglyceridemia: a focus on high dose omega-3 fatty acids. Adv Ther 34(2):300–323

Karimi N, Roshan VD (2013) Change in adiponectin and oxidative stress after modifiable lifestyle interventions in breast cancer cases. Asian Pac J Cancer Prev 14(5):2845–2850

Karnopp AR, Oliveira KG, De Andrade EF et al (2017) Optimization of an organic yogurt on sensorial, nutritional, and functional perspectives. Food Chem 233:401–411

Karsha PV, Lakshmi OB (2010) Antibacterial activity of black pepper *Piper nigrum* Linn. with special reference to its mode of action on bacteria. Indian J Nat Prod Resour 1(2):213–215

Karuppagounder V, Arumugam S, Giridharan VV et al (2017) Tiny molecule, big power: multitarget approach for curcumin in diabetic cardiomyopathy. Nutrition 34:47–54

Khan N, Mukhtar H (2008) Multitargeted therapy of cancer by green tea polyphenols. Cancer Lett 269:269–280

Khan N, Mukhtar H (2013) Tea and health: studies in humans. Curr Pharm Des 19:6141–6147

Khan M, Siddiqui M (2007) Antimicrobial activity of piper fruits. Nat Prod Radiance 6:111–113

Khan N, Afaq F, Saleem M et al (2006) Targeting multiple signaling pathways by green tea polyphenol (–)-epigallocatechin-3-gallate. Cancer Res 66(5):2500–2505

Khanal P, Oh WK, Yun HJ et al (2011) p-HPEA-EDA, a phenolic compound of virgin olive oil, activates AMP-activated protein kinase to inhibit carcinogenesis. Carcinogenesis 32(4):545–553

Khodavandi A, Harmal NS, Alizadeh F et al (2011) Comparison between allicin and fluconazole in *Candida albicans* biofilm inhibition and in suppression of HWP1 gene expression. Phytomedicine 19(1):56–63

Kim DH, Jeong D, Kim H et al (2018) Modern perspectives on the health benefits of kefir in next generation sequencing era: improvement of the host gut microbiota. Crit Rev Food Sci Nutr 1–12. https://doi.org/10.1080/10408398.2018.1428168

Kimura Y, Sumiyoshi M (2009) Olive leaf extract and its main component oleuropein prevent chronic ultraviolet B radiation-induced skin damage and carcinogenesis in hairless mice. J Nutr 139(11):2079–2086

Kocyigit A, Guler EM (2017) Curcumin induce DNA damage and apoptosis through generation of reactive oxygen species and reducing mitochondrial membrane potential in melanoma cancer cells. Cell Mol Biol 63(11):97–105

Kris-Etherton PM, Harris WS, Appel LJ (2002) Fish consumption, fish oil, omega-3 fatty acids, and cardiovascular disease. Circulation 106:2747–2757

Kumar R, Chhatwal S, Arora S et al (2013) Antihyperglycemic, antihyperlipidemic, anti-inflammatory and adenosine deaminase- lowering effects of garlic in patients with type 2 diabetes mellitus with obesity. Diabetes Metab Syndr Obes 6:49–56

Kumar G, Mittal S, Sak K et al (2016) Molecular mechanisms underlying chemopreventive potential of curcumin: current challenges and future perspectives. Life Sci 148:313–328

Kundu JK, Surh J (2008) Inflammation: gearing the journey to cancer. Mutat Res 659:15–30

Laureys D, De Vuyst L (2017) The water kefir grain inoculum determines the characteristics of the resulting water kefir fermentation process. J Appl Microbiol 122(3):719–732

Lavoie S, Chen Y, Dalton TP et al (2009) Curcumin, quercetin, and tBHQ modulate glutathione levels in astrocytes and neurons: importance of the glutamate cysteine ligase modifier subunit. J Neurochem 108:1410–1422

Leaf A, Kang JX, Xiao YF et al (2003) Clinical prevention of sudden cardiac death by n-3 polyunsaturated fatty acids and mechanism of prevention of arrhythmias by n-3 fish oils. Circulation 107:2646–2652

Li Y, Zhang T (2013) Targeting cancer stem cells with sulforaphane, a dietary component from broccoli and broccoli sprouts. Future Oncol 9(8):1097–1103

Li X, Meng Y, Xie C et al (2017) Diallyl trisulfide inhibits breast cancer stem cells via suppression of Wnt/β-catenin pathway. J Cell Biochem. https://doi.org/10.1002/jcb.26613

Liao W, Xiang W, Wang FF et al (2017) Curcumin inhibited growth of human melanoma A375 cells via inciting oxidative stress. Biomed Pharmacother 95:1177–1186

Lloyd-Jones DM, Hong Y, Labarthe D et al (2010) Defining and setting national goals for cardiovascular health promotion and disease reduction: the American Heart Association's Strategic Impact Goal through 2020 and beyond. Circulation 121:586–613

Lobo V, Patil A, Phatak A et al (2010) Free radicals, antioxidants and functional foods: impact on human health. Pharmacogn Rev 4:118–126

Ma ZF, Zhang H (2017) Phytochemical constituents, health benefits, and industrial applications of grape seeds: a mini-review. Antioxidants (Basel) 6(3):1–11

Ma X, Liu Y, Tan Y et al (2017) Diallyl disulphide inhibits apolipoprotein(a) expression in HepG2 cells through the MEK1-ERK1/2-ELK-1 pathway. Lipids Health Dis 16(1):223

Mageney V, Neugart S, Albach DC (2017) A guide to the variability of flavonoids in *Brassica oleracea*. Molecules 22(2):E252

Mahmoudi E, Saeidi M, Marashi MA et al (2016) *In vitro* activity of kombucha tea ethyl acetate fraction against *Malassezia* species isolated from seborrhoeic dermatitis. Curr Med Mycol 2(4):30–36

Mahn A, Reyes A (2012) An overview of health-promoting compounds of broccoli (*Brassica oleracea* var. italica) and the effect of processing. Food Sci Technol Int 18(6):503–514

Manayi A, Nabavi SM, Setzer WN et al (2017) Piperine as a potential anti-cancer agent: a review on preclinical studies. Curr Med Chem. https://doi.org/10.2174/0929867324666170523120656

Mancini E, Beglinger C, Drewe J et al (2017) Green tea effects on cognition, mood and human brain function: a systematic review. Phytomedicine 34:26–37

Mandel SA, Amit T, Weinreb O et al (2008) Simultaneous manipulation of multiple brain targets by green tea catechins: a potential neuroprotective strategy for Alzheimer and Parkinson diseases. CNS Neurosci Ther 14:352–365

Maradana MR, Thomas R, O'Sullivan BJ (2013) Targeted delivery of curcumin for treating type 2 diabetes. Mol Nutr Food Res 57(9):1550–1556

Marchese A, Barbieri R, Sanches-Silva A et al (2016) Antifungal and antibacterial activities of allicin: a review. Trends Food Sci Technol 52:49–56

Markowiak P, Slizewska K (2017) Effects of probiotics, prebiotics, and synbiotics on human health. Nutrients 9(1021):1–30

Marsh AJ, O'Sullivan O, Hill C et al (2014) Sequence-based analysis of the bacterial and fungal compositions of multiple kombucha (tea fungus) samples. Food Microbiol 38:171–178

Martins T, Colaço B, Venâncio C et al (2018) Potential effects of sulforaphane to fight obesity. J Sci Food Agric. https://doi.org/10.1002/jsfa.8898

Mas E, Woodman RJ, Burke V et al (2010) The omega-3 fatty acids EPA and DHA decrease plasma F(2)-isoprostanes: results from two placebo-controlled interventions. Free Radic Res 44(9):983–990

Mas E, Croft KD, Zahra P et al (2012) Resolvins D1, D2, and other mediators of self-limited resolution of inflammation in human blood following n-3 fatty acid supplementation. Clin Chem 58:1476–1484

Mashhadi NS, Ghiasvand R, Askari G et al (2013) Anti-oxidative and anti-inflammatory effects of ginger in health and physical activity: review of current evidence. Int J Prev Med 4(1):36–42

Mayneris-Perxachs J, Sala-Vila A, Chisaguano M et al (2014) Effects of 1-year intervention with a mediterranean diet on plasma fatty acid composition and metabolic syndrome in a population at high cardiovascular risk. PLoS One 9(3):1–11

Meghwal M, Goswami TK (2013) *Piper nigrum* and piperine: an update. Phytother Res 27(8):1121–1130

Melo SA, Esteller M (2011) Dysregulation of microRNAs in cancer: playing with fire. FEBS Lett 585:2087–2099

Metwally DM, Al-Olayan EM, El-Khadragy MF et al (2016) Anti-leishmanial activity (*in vitro and in vivo*) of allicin and allicin cream using *Leishmania major* (sub-strain Zymowme LON4) and Balb/c Mice. PLoS One 11(8):1–11

Meyer BJ, Mann NJ, Lewis JL et al (2003) Dietary intakes and food sources of omega-6 and omega-3 polyunsaturated fatty acids. Lipids 38:391–398

Miles EA, Zoubouli P, Calder PC (2005) Effects of polyphenols on human Th1 and Th2 cytokine production. Clin Nutr 24:780–784

Miller PE, Van Elswyk M, Alexander DD (2014) Long-chain omega-3 fatty acids eicosapentaenoic acid and docosahexaenoic acid and blood pressure: a meta-analysis of randomized controlled trials. Am J Hypertens 27:885–896

Mohd Yusof YA (2016) Gingerol and its role in chronic diseases. Adv Exp Med Biol 929:177–207

Morais Ferreira JM, Azevedo BM, Luccas V et al (2017) Sensory profile and consumer acceptability of prebiotic white chocolate with sucrose substitutes and the addition of goji berry (*Lycium barbarum*). J Food Sci 82(3):818–824

Moreda W, Pérez-Camino MC, Cert A (2001) Gas and liquid chromatography of hydrocarbons in edible vegetable oils. J Chromatogr A 936(1–2):159–171

Moreno DA, Carvajal M, López-Berenguer C et al (2006) Chemical and biological characterisation of nutraceutical compounds of broccoli. J Pharm Biomed Anal 41(5):1508–1522

Mori TA (2017) Marine omega-3 fatty acids in the prevention of cardiovascular disease. Fitoterapia 123:51–58

Mori TA, Beilin LJ (2001) Long-chain omega 3 fatty acids, blood lipids and cardiovascular risk reduction. Curr Opin Lipidol 12:11–17

Mori TA, Woodman RJ (2006) The independent effects of eicosapentaenoic acid and docosahexaenoic acid on cardiovascular risk factors in humans. Curr Opin Clin Nutr Metab Care 9(2):95–104

Mozaffarian D, Wu JH (2011) Omega-3 fatty acids and cardiovascular disease: effects on risk factors, molecular pathways, and clinical events. J Am Coll Cardiol 58:2047–2067

Mozaffarian D, Bryson CL, Lemaitre RN et al (2005) Fish intake and risk of incident heart failure. J Am Coll Cardiol 45(12):2015–2021

Mudduluru G, George-William JN, Muppala S et al (2011) Curcumin regulates miR-21 expression and inhibits invasion and metastasis in colorectal cancer. Biosci Rep 31(3):185–197

Mujumdar AM, Dhuley JN, Deshmukh VK et al (1990) Anti-inflammatory activity of piperine. Jpn J Med Sci Biol 43(3):95–100

Nestel P, Shige H, Pomeroy S et al (2002) The n-3 fatty acids eicosapentaenoic acid and docosahexaenoic acid increase systemic arterial compliance in humans. Am J Clin Nutr 76(2):326–330

Neto MM, da Silva TF, de Lima FF et al (2017) Whole red grape juice reduces blood pressure at rest and increases post-exercise hypotension. J Am Coll Nutr 36(7):533–540

Ngo QM, Tran PT, Tran MH et al (2017) Alkaloids from *Piper nigrum* exhibit anti-inflammatory activity via activating the Nrf2/HO-1 pathway. Phytother Res 31(4):663–670

Nielsen B, Gürakan GC, Unlü G (2014) Kefir: a multifaceted fermented dairy product. Probiotics Antimicrob Proteins 6(3–4):123–135

Nohara T, Fujiwara Y, Ikeda T et al (2013) Cyclic sulfoxides garlicnins B2, B3, B4, C2, and C3 from *Allium sativum*. Chem Pharm Bull (Tokyo) 61(7):695–699

Oh BT, Jeong SY, Velmurugan P et al (2017) Probiotic-mediated blueberry (*Vaccinium corymbosum* L.) fruit fermentation to yield functionalized products for augmented antibacterial and antioxidant activity. J Biosci Bioeng 124(5):542–550

Okada LSDRR, Oliveira CP, Stefano JT et al (2017) Omega-3 PUFA modulate lipogenesis, ER stress, and mitochondrial dysfunction markers in NASH – proteomic and lipidomic insight. Clin Nutr. https://doi.org/10.1016/j.clnu.2017.08.031
Oludoyin AP, Adegoke SR (2014) Efficacy of ginger (*Zingiber officinale* Roscoe) extracts in lowering blood glucose in normal and high fat diet-induced diabetic rats. Am J Food Nutr 2(4):55–58
Omar SH (2010) Cardioprotective and neuroprotective roles of oleuropein in olive. Saudi Pharm J 18:111–121
Owen RW, Giacosa A, Hull WE et al (2000) Olive-oil consumption and health: the possible role of antioxidants. Lancet Oncol 1:107–112
Ozen M (2015) The history of probiotics: the untold story. Benef Microbes 6(2):159–165
Pakravan N, Mahmoudi E, Hashemi SA et al (2017) Cosmeceutical effect of ethyl acetate fraction of Kombucha tea by intradermal administration in the skin of aged mice. J Cosmet Dermatol. https://doi.org/10.1111/jocd.12453
Pan Y, Wang Y, Cai L et al (2012) Inhibition of high glucose-induced inflammatory response and macrophage infiltration by a novel curcumin derivative prevents renal injury in diabetic rats. Br J Pharmacol 166:1169–1182
Panda AK, Chakraborty D, Sarkar I et al (2017) New insights into therapeutic activity and anticancer properties of curcumin. J Exp Pharmacol 9:31–45
Park EJ, Pizzuto JM (2002) Botanicals in cancer chemoprevention. Cancer Metast Rev 21:231–255
Pase MP, Grima NA, Sarris J (2011) Do long-chain n-3 fatty acids reduce arterial stiffness? A meta-analysis of randomised controlled trials. Br J Nutr 106(7):974–980
Perrone D, Ardito F, Giannatempo G et al (2015) Biological and therapeutic activities, and anticancer properties of curcumin. Exp Ther Med 10(5):1615–1162
Prado MR, Blandón LM, Vandenberghe LPS et al (2015) Milk kefir: composition, microbial cultures, biological activities, and related products. Front Microbiol 6:1177
Prasad S, Tyagi AK, Aggarwal BB (2014) Recent developments in delivery, bioavailability, absorption and metabolism of curcumin: the golden pigment from golden spice. Cancer Res Treat 46:2–28
Prashant A, Rangaswamy C, Yadav AK et al (2017) In vitro anticancer activity of ethanolic extracts of *Piper nigrum* against colorectal carcinoma cell lines. Int J Appl Basic Med Res 7(1):67–72
Pulido-Moran M, Moreno-Fernandez J, Ramirez-Tortosa C et al (2016) Curcumin and health. Molecules 21(3):264
Quiles JL, Farquharson AJ, Simpson DK et al (2002) Olive oil phenolics: effects on DNA oxidation and redox enzyme mRNA in prostate cells. Br J Nutr 88:225–234
Rafie N, Golpour Hamedani S, Ghiasvand R et al (2015) Kefir and cancer: a systematic review of literatures. Arch Iran Med 18(12):852–857
Rafiee P, Nelson VM, Manley S et al (2009) Effect of curcumin on acidic pH-induced expression of IL-6 and IL-8 in human esophageal epithelial cells (HET-1A): role of PKC, MAPKs, and NF-kappaB. Am J Physiol Gastrointest Liver Physiol 296(2):388–398
Ramírez-Boscá A, Soler A, Carrion MA et al (2000) An hydroalcoholic extract of *Curcuma longa* lowers the apo B/apo A ratio. Implications for atherogenesis prevention. Mech Ageing Dev 119(1–2):41–77
Rasmussen HE, Hamaker BR (2017) Prebiotics and inflammatory bowel disease. Gastroenterol Clin N Am 46(4):783–795
Rauf A, Imran M, Suleria HAR et al (2017) A comprehensive review of the health perspectives of resveratrol. Food Funct 8(12):4284–4305
Reddy S, Aggarwal BB (1994) Curcumin is a non-competitive and selective inhibitor of phosphorylase kinase. FEBS Lett 341(1):19–22
Rehman A, Mehmood MH, Haneef M et al (2015) Potential of black pepper as a functional food for treatment of airways disorders. J Func Foods 19:126–140
Reiter J, Levina N, Van Der Linden M et al (2017) Diallylthiosulfinate (allicin), a volatile antimicrobial from garlic (*Allium sativum*), kills human lung pathogenic bacteria, including MDR strains, as a vapor. Molecules 22(10):1–14

Rezaee R, Momtazi AA, Monemi A et al (2017) Curcumin: a potentially powerful tool to reverse cisplatin-induced toxicity. Pharmacol Res 117:218–227

Ríos-Covián D, Ruas-Madiedo P, Margolles A et al (2016) Intestinal short chain fatty acids and their link with diet and human health. Front Microbiol 7(185):1–9

Rivera M, Ramos Y, Rodríguez-Valentín M et al (2017) Targeting multiple pro-apoptotic signaling pathways with curcumin in prostate cancer cells. PLoS One 12(6):1–25

Roberfroid M, Gibson GR, Hoyles L et al (2010) Prebiotic effects: metabolic and health benefits. Br J Nutr 104(2):S1–S63

Ross S (2000) Functional foods: the Food and Drug Administration perspective. Am J Clin Nutr 71(6):1735S–1738S

Rotondi A, Alfei B, Magli M et al (2010) Influence of genetic matrix and crop year on chemical and sensory profiles of Italian monovarietal extra-virgin olive oils. J Sci Food Agric 90(15):2641–2648

Rushworth SA, Ogborne RM, Charalambos CA et al (2006) Role of protein kinase C delta in curcumin-induced antioxidant response element-mediated gene expression in human monocytes. Biochem Biophys Res Commun 341(4):1007–1016

Ryu JH, Kang D (2017) Physicochemical properties, biological activity, health benefits, and general limitations of aged black garlic: a review. Molecules 22(6):1–14

Sa G, Das T (2008) Anticancer effects of curcumin: cycle of life and death. Cell Div 3(1):14

Saini S, Arora S, Majid S et al (2011) Curcumin modulates micrRNA-203 mediated regulation of the Src-Akt axis in bladder cancer. Cancer Prev Res 4:1698–1709

Salafzoon S, Mahmoodzadeh Hosseini H et al (2017) Evaluation of the antioxidant impact of ginger-based kombucha on the murine breast cancer model. J Complement Integr Med 15(1):1–8

Salmerón I (2017) Fermented cereal beverages: from probiotic, prebiotic and synbiotic towards Nanoscience designed healthy drinks. Lett Appl Microbiol 65(2):114–124

Sanders ME, Marco ML (2010) Food formats for effective delivery of probiotics. Annu Rev Food Sci Technol 1:65–85

Sarkar S (2013) Probiotics as functional foods: documented health benefits. NFS 43(2):107–115

Savaiano DA (2014) Lactose digestion from yogurt: mechanism and relevance. Am J Clin Nutr 99(5):1251S–1255S

Scoditti E, Calabriso N, Massaro M et al (2012) Mediterranean diet polyphenols reduce inflammatory angiogenesis through MMP-9 and COX-2 inhibition in human vascular endothelial cells: a potentially protective mechanism in atherosclerotic vascular disease and cancer. Arch Biochem Biophys 527(2):81–89

Scotece M, Gómez R, Conde J et al (2012) Further evidence for the anti-inflammatory activity of oleocanthal: inhibition of MIP-1 and IL-6 in J774 macrophages and in ATDC5 chondrocytes. Life Sci 91(23–24):1229–1235

Seidel DV, Azcárate-Peril MA, Chapkin RS et al (2017) Shaping functional gut microbiota using dietary bioactives to reduce colon cancer risk. Semin Cancer Biol 46:191–204

Selvaggini R, Servili M, Urbani S et al (2006) Evaluation of phenolic compounds in virgin olive oil by direct injection in high-performance liquid chromatography with fluorometric detection. J Agric Food Chem 54:2832–2838

Semwal RB, Semwal DK, Combrinck S et al (2015) Gingerols and shogaols: Important nutraceutical principles from ginger. Phytochemistry 117:554–568

Serhan CN (2017) Treating inflammation and infection in the 21st century: new hints from decoding resolution mediators and mechanisms. FASEB J 31(4):1273–1288

Servili M, Esposto S, Fabiani R et al (2009) Phenolic compounds in olive oil: antioxidant, health and organoleptic activities according to their chemical structure. Inflammopharmacology 17(2):76–84

Sharifi M, Moridnia A, Mortazavi D et al (2017) Kefir: a powerful probiotics with anticancer properties. Med Oncol 34(11):183

Sheppard JG, McAleer JP, Saralkar P et al (2018) Allicin-inspired pyridyl disulfides as antimicrobial agents for multidrug-resistant *Staphylococcus aureus*. Eur J Med Chem 143:1185–1195

Siddiqui FJ, Avanc BI, Mahmudd S et al (2015) Uncontrolled diabetes mellitus: prevalence and risk factors among people with type 2 diabetes mellitus in an Urban District of Karachi, Pakistan. Diabetes Res Clin Pract 107(1):148–156

Singla V, Chakkaravarthi S (2017) Applications of prebiotics in food industry: a review. Food Sci Technol Int 23(8):649–667

Sreenivasan S, Thirumalai K, Danda R et al (2012) Effect of curcumin on miRNA expression in human Y79 retinoblastoma cells. Curr Eye Res 37:421–428

Stangl V, Lorenz M, Stangl K (2006) The role of tea and tea flavonoids in cardiovascular health. Mol Nutr Food Res 50:218–228

Stanić Z (2017) Curcumin, a compound from natural sources, a true scientific challenge – a review. Plant Foods Hum Nutr 72(1):1–12

Surh YJ (1999) Molecular mechanisms of chemopreventive effects of selected dietary and medicinal phenolic substances. Mutat Res 428(1-2):305–327

Surh YJ, Chun KS, Cha HH et al (2001) Molecular mechanisms underlying chemopreventive activities of anti-inflammatory phytochemicals: down-regulation of COX-2 and iNOS through suppression of NF-kappa B activation. Mutat Res 480–481:243–268

Tabrizi R, Moosazadeh M, Lankarani KB et al (2017) The effects of synbiotic supplementation on glucose metabolism and lipid profiles in patients with diabetes: a systematic review and meta-analysis of randomized controlled trials. Probiotics Antimicrob Proteins. https://doi.org/10.1007/s12602-017-9299-1

Tachibana H (2009) Molecular basis for cancer chemoprevention by green tea polyphenol EGCG. Forum Nutr 61:156–169

Tasleem F, Azhar I, Ali SN et al (2014) Analgesic and anti-inflammatory activities of *Piper nigrum* L. Asian Pac J Trop Med 7(14):60275–60273. https://doi.org/10.1016/S1995-7645

Tortorella SM, Royce SG, Licciardi PV et al (2015) Dietary sulforaphane in cancer chemoprevention: the role of epigenetic regulation and HDAC inhibition. Antioxid Redox Signal 22(16):1382–13424

Trinidad PT, Rosario SS, Marco PDL et al (2012) *Zingiber officinale* and *Curcuma longa* as potential functional foods/ingredients. Food Public Health 2(2):1–4

Trujillo J, Chirino YI, Molina-Jijón E et al (2013) Renoprotective effect of the antioxidant curcumin: recent findings. Redox Biol 1:448–456

Tung BT, Thu DK, Thu NTK et al (2017) Antioxidant and acetylcholinesterase inhibitory activities of ginger root *(Zingiber officinale* Roscoe) extract. J Complement Integr Med 14(4). https://doi.org/10.1515/jcim-2016-0116

Tung YT, Chen HL, Wu HS et al (2018) Kefir peptides prevent hyperlipidemia and obesity in high-fat-diet-induced obese rats via lipid metabolism modulation. Mol Nutr Food Res 62(3). https://doi.org/10.1002/mnfr.201700505

Van Buul VJ, Brouns FJ (2015) Nutrition and health claims as marketing tools. Crit Rev Food Sci Nutr 55(11):1552–1560

Vandenplas Y, De Greef E, Veereman G (2014) Prebiotics in infant formula. Gut Microbes 5(6):681–687

Vanegas-Azuero AM, Gutiérrez LF (2018) Physicochemical and sensory properties of yogurts containing sacha inchi (*Plukenetia volubilis* L.) seeds and β-glucans from *Ganoderma lucidum*. J Dairy Sci 101(2):1020–1033

Varshney R, Budoff MJ (2016) Garlic and heart disease. J Nutr 146:416S–421S

Vijayakumar RS, Surya D, Nalini N (2004) Antioxidant efficacy of black pepper (*Piper nigrum* L.) and piperine in rats with high fat diet induced oxidative stress. Redox Rep 9(2):105–110

Vina I, Semjonovs P, Linde R et al (2014) Current evidence on physiological activity and expected health effects of kombucha fermented beverage. J Med Food 17(2):179–188

Visioli F, Galli C (1998) Olive oil polyphenols and their potential effects on human health. J Agric Food Chem 46:4292–4296

Wallace B (2009) Clinical use of probiotics in the pediatric population. Nutr Clin Pract 24(1):50–59
Wang TT, Schoene NW, Milner JA et al (2012) Broccoli-derived phytochemicals indole-3-carbinol and 3,3'-diindolylmethane exerts concentration-dependent pleiotropic effects on prostate cancer cells: comparison with other cancer preventive phytochemicals. Mol Carcinog 51(3):244–256
Wang Y, Ji B, Wu W et al (2014) Hepatoprotective effects of kombucha tea: identification of functional strains and quantification of functional components. J Sci Food Agric 94(2):265–272
Wang CZ, Qi LW, Yuan CS (2015) Cancer chemoprevention effects of ginger and its active constituents: potential for new drug discovery. Am J Chin Med 43(7):1351–1363
Wang C, Wang Y, Yu M et al (2017a) Grape-seed polyphenols play a protective role in elastase-induced abdominal aortic aneurysm in mice. Sci Rep 7(1):9402
Wang J, Zhang X, Lan H et al (2017b) Effect of garlic supplement in the management of type 2 diabetes mellitus (T2DM): a meta-analysis of randomized controlled trials. Food Nutr Res 61:1–9
Waterman E, Lockwood B (2007) Active components and clinical applications of olive oil. Altern Med Rev 12(4):331–342
Wilkins T, Sequoia J (2017) Probiotics for gastrointestinal conditions: a summary of the evidence. Am Fam Physician 96(3):170–178
Wilson B, Whelan K (2017) Prebiotic inulin-type fructans and galacto-oligosaccharides: definition, specificity, function, and application in gastrointestinal disorders. J Gastroenterol Hepatol 32(1):64–68
Yadav S, Trivedi NA, Bhatt JD (2015) Antimicrobial activity of fresh garlic juice: an *in vitro* study. Ayu 36(2):203–207
Yang J, Xiao YY (2013) Grape phytochemicals and associated health benefits. Crit Rev Food Sci Nutr 53(11):1202–1225
Yang CS, Wang X, Lu G et al (2009) Cancer prevention by tea: animal studies, molecular mechanisms and human relevance. Nat Rev Cancer 9(6):429–439
Yarru L, Settivari R, Gowda N et al (2009) Effects of turmeric (*Curcuma longa*) on the expression of hepatic genes associated with biotransformation, antioxidant, and immune systems in broiler chicks fed aflatoxin. Poult Sci 88(12):2620–2627
Yin X, Feng C, Han L et al (2018) Diallyl disulfide inhibits the metastasis of type II esophageal gastric junction adenocarcinoma cells via NF-κB and PI3K/AKT signaling pathways in vitro. Oncol Rep 39(2):784–794
Yu J, Song P, Perry R et al (2017) The effectiveness of green tea or green tea extract on insulin resistance and glycemic control in type 2 diabetes mellitus: a meta-analysis. Diabetes Metab J 41(4):251–262
Yuan JM (2013) Cancer prevention by green tea: evidence from epidemiologic studies. Am J Clin Nutr 98(6):1676S–1681S
Zanirati DF, Abatemarco M Jr, Sandes SH et al (2015) Selection of lactic acid bacteria from Brazilian kefir grains for potential use as starter or probiotic cultures. Anaerobe 32:70–76
Zhang M, Holman CD, Huang JP et al (2007) Green tea and the prevention of breast cancer: a case-control study in Southeast China. Carcinogenesis 28:1074–1078
Zhou H, Beevers CS, Huang S (2011) The targets of curcumin. Curr Drug Targets 12(3):332–347
Zhu J, Ren T, Zhou M et al (2016) The combination of blueberry juice and probiotics reduces apoptosis of alcoholic fatty liver of mice by affecting SIRT1 pathway. Drug Des Devel Ther 10:1649–1661
Zou L, Hu YY, Chen WX (2015) Antibacterial mechanism and activities of black pepper chloroform extract. J Food Sci Technol 52(12):8196–8203

The Role of Flavonoids as Modulators of Inflammation and on Cell Signaling Pathways

Liliana V. Muschietti, Jerónimo L. Ulloa, and Flavia DC. Redko

Abbreviations

4′-HW	4′-hydroxywogonin
5B	(E)-3-(3,4-dimethoxyphenyl)-1-(5-hydroxy-2,2-dimethyl-2H-chromen-6-yl) prop-2-en-1-one
67LR	67-kDa laminin receptor
AA	Arachidonic acid
Afla	Amentoflavone
Akt	Protein kinase B
Alp	Alpinetin
Amp	Ampelopsin
AMPK	Adenosine monophosphate-activated protein kinase
AP-1	Activator protein-1
Api	Apigenin
Ast	Astragalin
BBB	Blood-brain barrier
BMDM	Bone marrow-derived macrophages
C3G	Cyanidin-3-O-glucoside
CAMs	Cell surface adhesion molecules
CAT	Catalase
Cat	Catechin
Chr	Chrysin
CNS	Central nervous system
COMT	Catechol-O-methyltransferase
COX	Cyclooxygenase

L. V. Muschietti (✉) · J. L. Ulloa · F. DC. Redko
Universidad de Buenos Aires, Facultad de Farmacia y Bioquímica: Departamento de Farmacología/ Cátedra de Farmacognosia, Buenos Aires, Argentina

Universidad de Buenos Aires, Consejo Nacional de Investigaciones Científicas y Técnicas, Instituto de Química y Metabolismo del Fármaco (IQUIMEFA), Facultad de Farmacia y Bioquímica, Buenos Aires, Argentina
e-mail: lmusch@ffyb.uba.ar

V. Cechinel Filho (ed.), *Natural Products as Source of Molecules with Therapeutic Potential*, https://doi.org/10.1007/978-3-030-00545-0_5

Dai	Daidzein
DMH	1,2-dimethyl hydrazine
DNA	Deoxyribonucleic acid
EGCG	Epigallocatechin-3-gallate
EGF	Epidermal growth factor
eNOS	Endothelial nitric oxide synthase
EpRE	Electrophile-responsive element
ERK	Extracellular signal-regulated kinases
Eup	Eupatilin
Fis	Fisetin
FlkA	Flavokawain A
Gen	Genisteína
GEN-27	5-hydroxy-7-[2-hydroxy-3-(piperidin-1-yl) propoxy]-3-{4-[2-hydroxy-3-(piperidin-1-yl) propoxy] phenyl}-4H-chromen-4-one
GPx	Glutathione peroxidase
HaCaT cells	Human keratinocytes
hAs	Human astrocytes
hBMEC	Injured human brain microvascular endothelial cell
HCT116	Human colon tumour
HGF	Human gingival fibroblasts
HIF-1α	Hypoxia-inducible factor 1-α
HMGB	High-mobility group box
HMGB1	High-mobility group box 1 protein
HO-1	Heme oxygenase-1
hPBMCs	Human peripheral blood mononuclear cells
HUVEC	Human umbilical vein endothelial cell
Ibc	Isobavachalcone
Ica	Icariin
ICAM	Intercellular adhesion molecule
ICT	3,5,7-trihydroxy-4′-methoxy-8-(3-hydroxy-3-methylbutyl)-flavone
IFN	Interferon
Ig	Immunoglobulin
IKK	IκB kinase
IL	Interleukin
iNOS	Inducible nitric oxide synthase
IRAK	IL-1 receptor-associated kinase
IκB	Inhibitor of kappa-B
JAK	Janus kinase
JNK	c-Jun N-terminal kinases
L2H17	1-(3,4-Dihydroxyphenyl)-3-(2-methoxyphenyl)prop-2-en-1-one
LicoC	Licochalcone C
LOX	Lypooxygenase
LPH	Lactase phlorizin hydrolase
LPS	Lipopolysaccharide
LT	Leukotriene

Lut	Luteolin
Mal	Malvidin
Mal3OG	Malvidin-3-O-glucoside
MALP-2	Macrophage-activating lipopeptide 2-kDa
MAPK	Mitogen-activated protein kinase
MCAO	Middle cerebral artery occlusion
MCP	Monocyte chemoattractant protein
MIP	Macrophage inflammatory protein
mMEC	Mouse mammary epithelial cell
MMP	Matrix metalloproteinase
MPO	Myeloperoxidase
mRNA	Messenger ribonucleic acid
Nag	Naringin
Nar	Naringenin
NF-κB	Nuclear factor kappa B
nNOS	Neuronal NOS
NO	Nitric oxide
NOS	Nitric oxide synthase
Nrf2	Nuclear factor-erythroid-related factor 2
NSAIDs	Non-steroidal anti-inflammatory drugs
Ono	Ononin
OroA	Oroxylin A
OVA	Ovalbumin
PAI-1	Plasminogen activator inhibitor 1
PCB	Polychlorinated biphenyl
PDGF	Platelet-derived growth factor
Pel	Pelargonidin
Peo	Peonidin
PG	Prostaglandin
Phl	Phloretin
PI3K	Phosphatidylinositol-3 kinase
Pin	Pinocembrin
PKC	Protein kinase C
poly[I:C]	Polyriboinosinic polyribocytidylic acid
PPAR	Peroxisome proliferator-activated receptor
Pru	Prunetin
Pue	Puerarin
Quer	Quercetin
RAGE	Receptor for advanced glycation end products
RANTES	Regulated upon activation normal T-cell expressed and secreted
ROS	Reactive oxygen species
Rut	Rutin
SG	Sophoraflavanone
SIRT	Sirtuin
SOCS	Suppressors of cytokine signaling

SOD	Superoxide dismutase
STATs	Signal transducer and activator of transcription
SULTs	Sulfotransferases
TACR-1	Tachykinin receptor 1
Tax	Taxifolin
TBARS	Thiobarbituric acid reactive substances
TGF	Tumour growth factor
TLR	Toll-like receptor
TNF-α	Tumour necrosis factor-α
Tollip	Toll-interacting protein
Tri	Tricin
TX	Thromboxane
UgoM	Ugonin M
UGTs	Uridine 5′-diphospho-glucuronosyltransferases
UV	Ultraviolet
VCAM	Vascular cell adhesion molecule
VEGF	Vascular endothelial growth factor
Vel	Velutin
Vix	Vitexin
Won	Wogonin

1 Introduction

Flavonoids are naturally occurring polyphenolic compounds widely distributed in the plant kingdom and found in all vascular plants. Not only are they present in plant organs such as flowers, fruits, barks, roots and seeds but also in different products including tea and wine (Middleton and Kandaswami 1994).

These compounds give the flowers the yellow, orange, blue and red colours. Flavonoids play a role in the plant growth; they act as visual attractors for pollination and protect plants against stressor factors such as ultraviolet radiation and the attack of insects and microorganisms (Hassan and Mathesius 2012). Flavonoids are low molecular weight compounds having a benzo-γ-pyrone moiety in their structure and are synthetised through the phenylpropanoid pathway. Their function has been demonstrated to be highly structure-dependent (Bakhtiari et al. 2017).

The chemical structure of flavonoids is based on a 15-carbon skeleton constituted by two benzene rings (A and B) which are linked via a heterocyclic pyrane ring (C). Flavonoids are mainly found either as aglycones (their basic structure), as glycosides or as methylated derivatives. Based on the different substitution and the oxidation pattern of ring C, flavonoids are classified into different subclasses: flavones, flavonols, flavanols, flavanones, isoflavones, anthocyanidins, chalcones and flavanonols (Fig. 1). The hydroxyl group substitution often occurs at C-3, C-5, C-7, C-3′, C-4′ and C-5′. When glycosides are formed, the glycosidic linkage is normally located at positions 3 or 7, and the carbohydrate can be rhamnose, glucose,

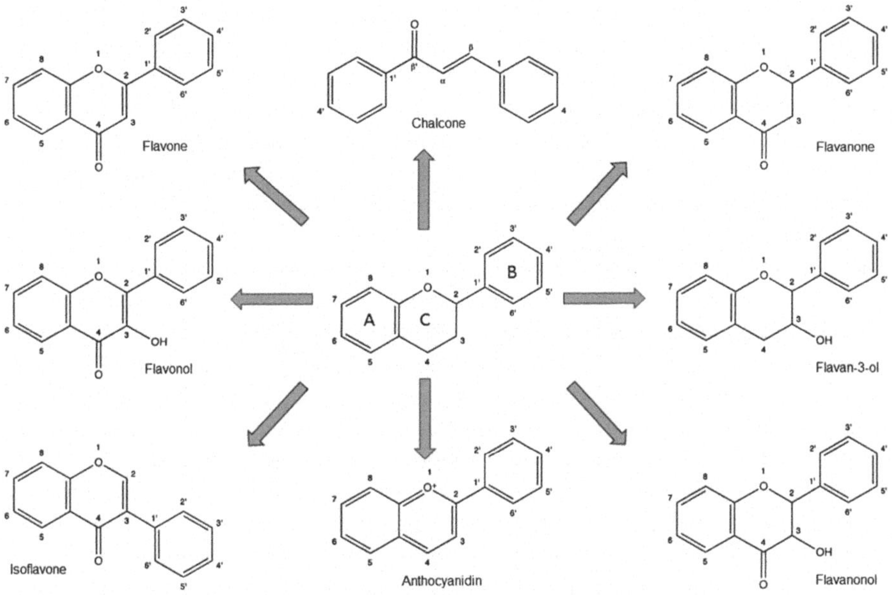

Fig. 1 Chemical structures of the main classes of flavonoids

glucorhamnose, galactose or arabinose (Xiao 2017). Although they are not regarded as nutrients, flavonoids are important constituents of the human diet, being flavonols, flavones, anthocyanidins, catechins, flavanones and isoflavones the major subgroups. Flavonols are mainly present in leafy vegetables, apples, onions, broccoli and berries. Flavones and anthocyanidins are found in relatively small quantities in grains, leafy vegetables and herbs. Catechins are abundant in tea, apples, grapes, chocolate and red wine. Flavanones are found in citrus fruit, and isoflavones are mainly found in soybeans (Wang et al. 2009). It is estimated that the total number of flavonoids known is around 8000 (Bode and Dong 2013), and this number is increasing considering their great structural diversity.

The interest in the biological activities of these compounds arose around 1930 when a mixture of the flavonoids eriodictyol and hesperidin called citrin (isolated from *Citrus* spp. juice) was found to have vitamin-like activity and was designated as ‘vitamin P’. This term was coined to indicate that this mixture decreased capillary permeability, prolonged the life span of Guinea pigs and reduced the signs of hypovitaminosis C in scorbutic experimental animals. Later on, the term vitamin P was abandoned because these compounds did not meet the requirements to be considered a vitamin (Middleton and Kandaswami 1994). When flavonoids were determined as the compounds responsible for these biological activities, research studies were undertaken in order to isolate such compounds and to study their mechanism of action.

In the late 1980s, the research on flavonoids received an additional impulse with the discovery of a phenomenon known as the 'French paradox'. Epidemiological studies indicate that French people have a relatively low incidence of cardiovascular disease and increased longevity while having a diet rich in saturated fats. This finding correlated with a diet replete in flavonoid-rich foods in association with red wine consumption. It has been suggested that the flavonoid intake is inversely correlated with mortality due to coronary heart disease (Formica and Regelson 1995; Knekt et al. 1996).

Flavonoids have long been recognized to possess a broad spectrum of biological activities such as antioxidant, anti-inflammatory, hepatoprotector, antibacterial, antiviral, antidiabetic, antiproliferative and anticarcinogenic (Chen et al. 2017). Epidemiological studies have indicated that a high dietary intake of flavonoids is associated with a decreased risk of a wide range of diseases including cardiovascular disease (Kuriyama et al. 2006). In this sense, flavonoids may influence lipid metabolism by inhibiting low-density lipoprotein oxidation, thus reducing atherosclerotic lesion formation. They are also known to inhibit platelet aggregation, to decrease vascular cell adhesion molecule expression, to improve endothelial function and to reduce blood pressure (Vauzour et al. 2010). Their consumption has also been associated with a reduced risk of lung cancer, breast cancer, renal cancer, non-Hodgkin's lymphoma and colorectal cancer (Fink et al. 2007; Frankenfeld et al. 2008; Gerd et al. 2008; Tang et al. 2009), better cognitive outcomes and with a reduced risk of dementia (Letenneur et al. 2007; Commenges et al. 2000). According to Williamson (2017), in the 1990s, the antioxidant activity of polyphenols was considered a panacea. However, over the last two decades, the attention has been focused on the concept of flavonoids as potential modulators of intracellular signaling cascades that are vital for cell functioning.

2 Absorption and Metabolism of Flavonoids

It is estimated that the daily intake of flavonoids contributed by the diet ranges from 50 to 800 mg/day (Pietta 1998, 2000), though some authors state that it can be up to 1 g (Middleton and Kandaswami 1994). However, the amount of polyphenols that should be consumed to derive maximum benefit is difficult to estimate. A cup of green tea or a glass of red wine can provide up to 200 mg of total flavonoids: one onion, 40 mg/100 g; a green salad, 1 mg/100 g; one apple, 6–10 mg; a peach, 1–2 mg; and an orange, 10 mg (Pietta 1998). In the UK, the mean intake of flavonols per day (mainly present in tea, cocoa, apples and broad beans) is 590 mg/day, and the intake of flavanones (citrus fruit) and flavonols (tea, apples or onions) is 25 and 61 mg/day, respectively. However, the intakes are dependent on individual diets and are highly variable (Williamson 2017).

The absorption of dietary flavonoids may depend upon the structure of the flavonoid (i.e. glycosides or aglycones), molecular size, molecular configuration, lipophilicity, solubility and pKa (Kumar and Pandey 2013). The absorption of these compounds may take place either in the small intestine, which is an efficient route that leads to high plasma levels or in the colon. Aglycones can be absorbed by the small intestine, while glycosides are considered too hydrophilic to be absorbed by passive diffusion in this site. Some flavonoid glycosides are enzymatically hydrolysed by either lactase-phlorizin hydrolase (LPH) or by β-glucosidase, and then the aglycones enter epithelial cells by passive diffusion.

However, those glycosides which are not substrates for these enzymes (e.g. flavonoids linked to a rhamnose moiety) are transported to the colon where the intestinal microflora degrade them to simple phenolic acids, which may be absorbed and further metabolized in the liver. The enzymatic deglycosylation driven by LPH and β-glucosidase is recognized as the first and determinant step in the absorption of flavonoids (known as phase I deglycosylation). In this sense, pharmacokinetic data suggest that quercetin (*Quer*) glucoside is absorbed in the small intestine, whereas quercetin rutinoside is absorbed in the colon after deglycosylation, showing that the presence of the sugar moiety determines the site of absorption (Day et al. 2000; Marín et al. 2015). However, barely 5–10% of total flavonoids may be absorbed in the small intestine, while unabsorbed flavonoids reach the colon to be excreted in the faeces (Gleichenhagen and Schieber 2016).

Taking into account that the absorption capacity of the colon is far less efficient than that of the small intestine, only a minimum absorption of these glycosides is to be expected. According to Hollmann (2004), two compartments are to be considered in the metabolism of flavonoids: the first one comprising the small intestine, the liver and the kidneys, and the other one, the colon. Flavonoids that are unabsorbable in the small intestine and flavonoids that have been absorbed and then secreted with bile will ultimately reach the colon.

Once absorbed (in either the small intestine or the colon), the metabolism of flavonoids is dominated by phase II enzymes, such as catechol-O-methyltransferase (COMT), sulfotransferases (SULTs) and uridine 5′-diphospho-glucuronosyltransferases (UGTs). UGTs are the major contributors, followed by SULTs and COMT (Chen et al. 2014). Flavonoid metabolites enter the bloodstream by the portal vein and are transported to the liver, where they may undergo further phase II transformations, and then are transported back to the bloodstream to be secreted in urine (Kumar and Pandey 2013; Marín et al. 2015) (Fig. 2). The complexity of the flavonoid metabolism implies that after their consumption, a wide variety of metabolites can be generated. Thus, the bioactive forms of flavonoids are not those found in plants, such as the glycosides or aglycones. Instead, circulating glucuronides, sulphates and O-methylated derivatives (formed only with flavonoids bearing a catechol B-ring) are believed to be those most likely to exert the biological effects and express beneficial effects in humans and animals (Spencer et al. 2001, 2003, 2004).

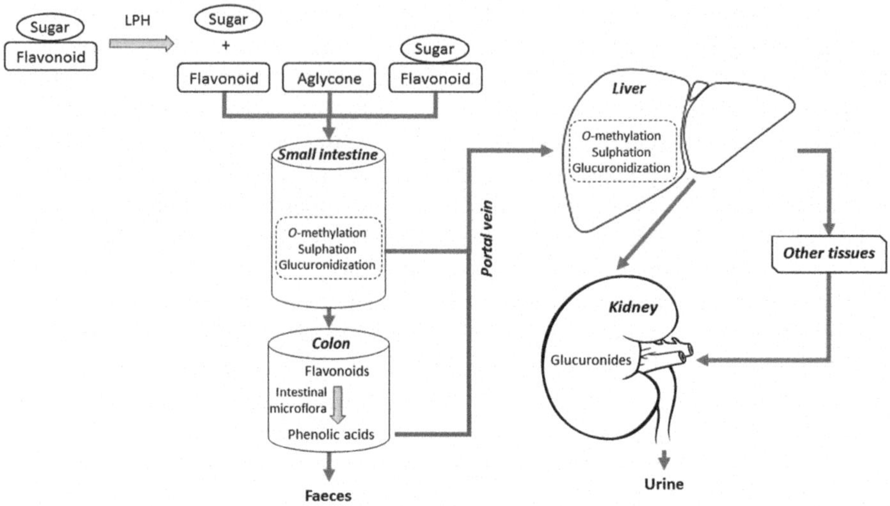

Fig. 2 Schematic diagram of the absorption and metabolism of flavonoids in humans. Aglycones can be absorbed in the small intestine. Flavonoids glycosides may be deglycosilated by LPH, β-glucosidase or intestinal microflora to aglycones and simple phenolic acids, respectively. The aglycones and phenolic acids enter the portal vein and are further metabolized in the liver

In this sense, Jaeger et al. (2017) have stated that it is important to use flavonoid metabolites when the mechanisms of action are studied in vitro. They also stated that the limited concentration of dietary flavonoid metabolites present in the circulation following ingestion and the key role played by the gut microbiota in the biotransformation of flavonoids in humans should also be taken into consideration.

3 Inflammation

Inflammation is a complex host response of body tissues to harmful stimuli, such as pathogens (bacteria, fungi and viruses), trauma or toxic compounds. It is a protective response involving host's cells, blood vessels, proteins and other molecular mediators (Kumar et al. 2013). The function of inflammation is to eliminate the initial cause of cell injury, clear out necrotic cells and damaged tissues and initiate tissue repair. The inflammatory status involves endothelial and epithelial cells, neutrophils, monocytes, macrophages and lymphocytes. The local and recruited cells are stimulated to release numerous mediators that amplify the inflammatory response and recruit additional cells (Firestein 2012). There are two types of inflammation, i.e. acute and chronic. Acute inflammation is the initial response of the body to harmful stimuli. It has a rapid onset and is short-lived (few hours or a few days). It is characterized by the release of numerous chemical mediators, fluid and plasma protein exudation and the migration of leukocytes (Kumar et al. 2013). When the stimulus persists, chronic inflammation may develop, which may be more insidious and long-lasting (weeks to months).

Chronic inflammation is characterized by simultaneous tissue destruction, mainly induced by the products secreted by inflammatory cells, and tissue repair involving vessel proliferation. A wide range of progressive diseases, including rheumatoid arthritis, asthma, atherosclerosis, neurological diseases and cancer, are related to chronic inflammation (Ribeiro et al. 2015).

The inflammatory response is characterized by the coordinated activation of various signaling pathways that regulate the expression of both pro- and anti-inflammatory mediators in resident tissue cells and leukocytes recruited from the blood. During the inflammatory process, mediators such as histamine, serotonin, prostaglandins (PGs), leukotrienes (LTs), platelet-derived growth factor (PDGF), reactive oxygen species (ROS), nitric oxide (NO), cytokines and chemokines may either be produced locally by cells (tissue macrophages, mast cells, endothelial cells or leucocytes) at the site of inflammation or may be derived from circulating inactive precursors that become activated in situ (complement proteins and kinins) (Kumar et al. 2013; Agati et al. 2012).

During inflammation, macrophages are activated by interferon gamma (IFN-γ), complement, immune complexes, lipopolysaccharide (LPS) and cytokines, such as interleukin (IL)-1β, tumour necrosis factor alpha (TNF-α) and IL-6. LPS initiates a signaling cascade through its interaction with Toll-like receptor 4 (TLR4) (Lu et al. 2008). Activated macrophages also produce inflammatory cytokines such as IL-1β, TNF-α, IL-6 and IL-12 and chemokines, such as IL-8, monocyte chemoattractant proteins (MCP-1 and MCP-2), complement cascade proteins, PGE_2, thromboxane (TX) A_2 and leukotrienes (LTB4) that contribute to the propagation of inflammation (Ribeiro et al. 2015). The activation of the signaling pathway leads to the release of the nuclear factor kappa B (NF-κB), which activates genes associated with the transcription of proteins related to the inflammatory process, such as inducible nitric oxide synthase (iNOS), which is the enzyme involved in NO synthesis, cyclooxygenases (COXs) and cytokines such as TNF-α, IL-6 and IL-1β.

The activator protein 1 (AP-1) is another transcription factor that responds to a wide variety of stimuli, such as bacterial and virus infection, stress and growth factors. This factor is important during the inflammatory response since it regulates gene expression of pro-inflammatory mediators, including cytokines (Shaulian and Karin 2001). Thus, the suppression of the expression of these pro-inflammatory mediators allows the amelioration and serves as a key mechanism to prevent and control inflammation (Agati et al. 2012; Fan et al. 2017).

3.1 *Inflammatory Mediators*

3.1.1 Nitric Oxide

NO is a highly reactive free radical produced by many cell types which is involved in the regulation of the inflammatory cascade. Such regulation includes not only its own production by immunocompetent cells but also the recruitment of leukocytes. NO is synthesized from L-arginine by nitric oxide synthase (NOS), which exists in three different isoforms: endothelial NOS (eNOS), neuronal NOS (nNOS) and iNOS. While a small amount of NO, synthesized by nNOS and eNOS, is essential for the normal functioning of the organism, when NO is synthesized in considerable amounts by iNOS, it participates in inflammatory processes acting synergistically with other inflammatory mediators (Nathan 1992; Tuñón et al. 2009). The activity of iNOS is induced by IL-1β, TNF-α, IFN-α, viral antigens, bacteria, protozoa and fungi, as well as by a low oxygen tension and a low environmental pH.

3.1.2 Arachidonic Acid Metabolites

Eicosanoids derive from the metabolism of arachidonic acid (AA) and comprise PGs, LTs, TXs and lipoxins. They play a vital role in physiologic and pathologic processes in immunity and inflammation (Zurier 2013). AA metabolites can mediate every step of inflammation, and agents that inhibit their synthesis diminish inflammation. The arachidonate metabolism is mediated by COX isoenzymes and by lipoxygenases (LOXs). Products of the COX pathway include PGs and TXA and are produced by COX-1 and COX-2. The former is produced in response to inflammatory stimuli and is expressed in many tissues (endothelium, monocytes, platelets, renal collecting tubules and seminal vesicles) and participates in the synthesis of PGs, which regulate physiological processes in response to hormones and other stimuli (Smith and Langenbach 2001). COX-2 is expressed primarily in cells involved in inflammation (macrophages, fibroblasts and endothelial cells), and its expression is induced by various stimuli, including PDGF and epidermal growth factor (EGF) and pro-inflammatory cytokines (IL-1β and TNF-α) (Ribeiro et al. 2015). Nonsteroidal anti-inflammatory drugs (NSAIDs) such as aspirin and ibuprofen inhibit COX activity, thereby blocking the PGs synthesis.

The synthesis of LTs involves multiple steps and is produced by 5-LOX, the major AA-metabolizing enzyme in neutrophils (Kumar et al. 2013). LOXs are responsible for the generation of hydroxyl acids and LTs from AA. There are three distinct LOX isozymes, namely, 5-LOX, 12-LOX and 15-LOX. LTB_4 is produced by neutrophils and is a potent chemotactic agent for neutrophils. LTC_4, LTD_4 and LTE_4 are produced mainly in mast cells. These mediators cause bronchoconstriction and increase vascular permeability (Kumar et al. 2013).

3.1.3 Cytokines

Cytokines are proteins that are mainly produced by activated lymphocytes and macrophages. They are the major mediators of local and intercellular communications that are required for an integrated response to a variety of stimuli (Tuñón et al. 2009). The production and secretion of cytokines are transcriptionally regulated. Their major role is the regulation of the intensity and duration of the inflammatory response. The expression of cytokines may be triggered by different stimuli such as trauma, stress, ischemia, ultraviolet light, microbes, local complement activation, ROS and nitrogen species and cytokines themselves working in autocrine loops (Ribeiro et al. 2015). The main cytokines involved in the inflammatory response are TNF-α and IL-1. These cytokines induce the expression of adhesion molecules in endothelial tissue and participate in the synthesis of other cytokines, such as IL-6, chemokines (IL-8 and MCP-1), growth factors, eicosanoids and NO (Kumar et al. 2013).

Cytokines related to acute inflammation are IL-1, TNF-α, IL-6, IL-11, IL-8, IL-16 and IL-17, among others. These cytokines usually act locally, and they mediate multiple effects, mainly leukocyte recruitment and migration. Cytokines involved in chronic inflammation are those mediating humoral responses, like IL-4, IL-5, IL-6, IL-7 and IL-13, whereas cellular responses are usually governed by IL-1, IL-2, IL-3, IL-4, IL-7, IL-9, IL-10, IL-12, IFN-γ, transforming growth factor (TGF)-β and TNF-α (Feghali and Wright 1997). Among these mediators, IL-1, TNF-α and IL-6 are the most studied cytokines involved in chronic inflammation-related diseases. There are different cytokine structurally related receptors that mediate cytokine communication, i.e. type I and type II cytokine receptors, TNF receptor, chemokine receptors, TGF-*β* receptor and a Toll/IL-1 receptor. After binding to the receptors, cytokines mediate their effects through the activation of several intracellular signaling pathways, such as the Janus kinases (JAK) and their downstream transcriptional factors, including the signal transducers and activators of transcription (STATs), phosphatidylinositol-3 kinase (PI3K) and mitogen-activated protein kinases (MAPKs) signaling cascades and the NF-κB pathway. Once the signaling cascade is initiated, several transcription factors such as NF-κB, AP-1 and nuclear factor of activated T cells can be recruited to the cytokine promoter region (Ribeiro et al. 2015). There are evidences suggesting that inflammatory cytokines have potential as therapeutic targets to treat inflammatory diseases. In this sense, several drugs such as etanercept and infliximab and anakinra have been developed as inhibitors of TNF-α and IL-1β, respectively (Agati et al. 2012).

3.1.4 Chemokines

Chemokines are a family of small (8–10 kDa) proteins that act primarily as leukocyte chemoattractants. The major roles of chemokines are to recruit leukocytes to the site of the inflammation and to control the normal anatomic organization of cells in different tissues. They exert their biological effects by binding to specific G protein-coupled receptors on target cells. Chemokines are divided into four groups, being the CXC and CC chemokines the two major ones. The former act primarily on neutrophils. IL-8 is the main representative of this group, and it is produced mainly in response to microbial products and other cytokines, such as IL-1 and TNF-α. CC chemokines include MCP-1, macrophage inflammatory protein (MIP)-1α and MIP-1β among others (Kumar et al. 2013). Some chemokines and their receptors are up-regulated in both acute and chronic inflammatory diseases. This finding provided the pharmaceutical industry with new targets for therapeutic intervention against different diseases. There are several approaches that are being developed to block the effects of chemokines, including small-molecule antagonists of chemokine receptors, modified chemokines and antibodies directed against chemokine receptors (Wells et al. 2006).

3.1.5 Cell Adhesion Molecules

Cell surface adhesion molecules (CAMs) are proteins involved in cell-cell and cell-extracellular matrix contact in a process named cell adhesion.

CAMs play vital roles in numerous physiological and pathological processes (Cines et al. 1998) including cell growth, differentiation, embryogenesis, immune cell transmigration and response and metastasis. Adhesion molecules are also capable of transmitting information from the extracellular matrix into the cell. The endothelial dysfunction is closely related to inflammatory processes, in which the adhesion of circulating monocytes to vascular endothelial cells is a critical step in both inflammation and atherosclerosis (Tuñon et al. 2009). Endothelial cells respond to pro-inflammatory stimuli such as TNF-α, LPS and IL-1β and recruit leucocytes by selectively expressing adhesion molecules on the surface (Iiyama et al. 1999). CAMs are grouped into four families: immunoglobulin (Ig) superfamily, integrins, cadherins and selectins. Adhesion molecules include members of the Ig superfamily such as the intercellular adhesion molecules (ICAMs), the vascular-cell adhesion molecule (VCAM-1) and endothelial cell selectin (E-selectin), among others (Tuñón et al. 2009).

3.2 Inflammation-Associated Intracellular Signaling Pathways

The set of processes by which a cell converts a signal or external stimulus into another specific signal or response is known as the biochemical pathway of signal transduction. LPS is an inflammatory stimulator of macrophages that triggers the production of pro-inflammatory mediators. The stimulation of TLR4 receptors with LPS leads to the activation of various intracellular signaling pathways such as those involving the inhibitor of κB (IκB) kinase (IKK), PI3K, protein kinase B (Akt) and MAPKs. These molecules eventually lead to the activation of transcription factors such as NF-κB, AP-1 or signal transducers and STATs, whose deoxyribonucleic acid (DNA)-binding capacity is modified by the various protein kinases involved in signal transduction, including MAPKs (Kim et al. 2004; Komatsu et al. 2017).

Inhibitory or stimulatory effects on these biochemical pathways profoundly affect cellular functions, altering the state of phosphorylation of target molecules and modulating gene expression (Williams et al. 2004). Inflammatory cells also produce soluble mediators, such as metabolites of the arachidonic acid, cytokines and chemokines, which act by further recruiting inflammatory cells to the site of damage producing more reactive species.

3.2.1 Nuclear Transcription Factor Kappa-B (NF-kB) Pathway

The NF-κB pathway is the main pathway when inflammatory responses develop. This factor plays a central role in the expression of more than 150 genes involved in immune and inflammatory responses. NF-κB is found in almost all animal cell types and is involved in cellular responses to stimuli such as stress, cytokines, free radicals, heavy metals, ultraviolet irradiation, oxidized LDL and bacterial or viral antigens. NF-κB can occur as either a homo- or a heterodimer consisting of five different transcription factor proteins: (RelA), c-Rel, Rel-B, p50 and p52 (Fan et al. 2017), the most common association is that between p50 and p65.

In an inactivated state, NF-κB is located in the cytosol complexed with the inhibitory protein IκB. Five IκB-like proteins have already been identified: IκBα, IκBβ, IκBγ, IκBε and Bcl-3. The binding of inflammatory mediators to their respective receptors triggers a signaling cascade that leads to the phosphorylation and activation of the IKK complex (IKK α,β,γ). IKK, in turn, phosphorylates the IκB-α protein, which results in ubiquitination, dissociation of IκB-α from NF-κB and eventual degradation of IκB-α by the proteosome (Rabinovich et al. 2011). The activated NF-κB then translocates into the nucleus where it binds to specific sequences of DNA and induces the expression of pro-inflammatory mediators. NF-κB has been reported as one of the most remarkable pro-inflammatory gene expression regulators which mediates the synthesis of several cytokines, such as TNF-α, IL-1β, IL-6 and IL-8, as well as COX-2 (Lawrence 2009; Bertics et al. 2014) (Fig. 3).

3.2.2 Signal Transducer and Activator of Transcription (STAT) Protein Family

The STATs proteins are intracellular transcription factors that mediate many aspects of cellular immunity, proliferation, apoptosis and differentiation, taking part in the regulation of cellular responses to cytokines, chemoattractants and growth factors. In unstimulated cells, STAT proteins are inactive in the cytosol. After their association with activated receptors, STAT proteins are phosphorylated by members of the JAK family of non-receptor protein-tyrosine kinases, which are associated with cytokine receptors. The tyrosine phosphorylation promotes the dimerisation of STAT proteins, which then translocate to the nucleus, where they stimulate the transcription of their target genes. Further studies have shown that STAT proteins are also activated downstream of receptor protein-tyrosine kinases, where their

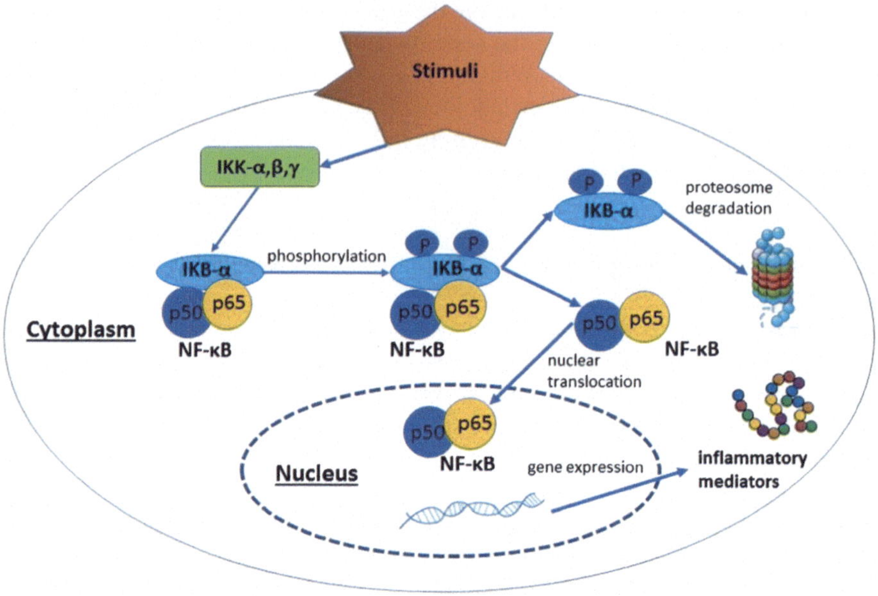

Fig. 3 Nuclear transcription factor kappa-B (NF-κB) pathway

phosphorylation may be catalyzed either by the receptors themselves or by associated non-receptor kinases. The STAT transcription factors thus serve as direct links between both cytokine and growth factor receptors on the cell surface and regulation of gene expression in the nucleus (Cooper 2000).

It has been demonstrated that the activation of the STAT3/5 pathways leads to subsequent COX-2 expression, while the activation of STAT1 correlates with the expression of iNOS and adhesion molecules (Kretzmann et al. 2008).

3.2.3 Activator Protein 1 (AP-1) Pathway

One of the most important signaling targets in the activation of T cells is the transcription factor AP-1. It is constituted by a set of structurally related dimers and formed by proteins of the Fos, Jun and ATF subfamilies (Rabinovich et al. 2011), which all have to dimerise before binding to their DNA target sites. AP-1 regulates many aspects of cell physiology in response to environmental changes, such as stress and radiation or to growth factor signals thereby acting like an environmental biosensor (Wagner 2001). In addition to the common regulation and activation of c-Jun by MAPKs, there are several other signaling pathways and interactions leading to c-Jun protein expression and thus AP-1 activation (Kappelmann et al. 2014).

3.2.4 Mitogen-Activated Protein Kinases (MAPKs) Pathway

Several studies have shown that the activation of NF-κB is triggered by MAPKs. There are three main subgroups of MAPKs: extracellular signal-regulated kinase (ERK), c-Jun N-terminal kinase (JNK) and p38. These kinases play a key role in the regulation of numerous cellular functions, including gene expression, mitosis, differentiation, apoptosis and cellular responses to inflammation (Cargnello and Roux 2011). It has also been demonstrated that they are involved in the signal transduction pathways that lead to the induction of pro-inflammatory mediators (Owuor and Kong 2002; Kaminska 2005; Kim et al. 2008; Xu et al. 2010). Several studies have shown that MAPKs play critical roles for the activation of NF-κB. MAPKs are important signaling components in the conversion of extracellular signals into intracellular responses through serial phosphorylation cascades. Upon stimulation, MAPKs are phosphorylated and activate the downstream protein kinases and transcription factors leading to the expression of pro-inflammatory mediators such as TNF-α, IL-6 and iNOS (Komatsu et al. 2017). Among the MAPK family members, the ERK route is frequently activated by mitogens and growth factors, while inflammation is a main trigger for JNK and p38 (Santangelo et al. 2007). Hence, the inhibition of MAPKs blocks inflammation through the modulation of the levels of pro- and anti-inflammatory mediators (Chen et al. 2017).

4 Flavonoids in the Inflammatory Response

In recent years, there has been an increasing progress in the elucidation of the mechanisms by which flavonoids exert their biological activities. A high intake of flavonoids has been associated with a reduced risk of cardiovascular disease, cancer and neurodegenerative disorders. In addition to their already known free radical scavenger effect, it has been demonstrated that flavonoids exert these beneficial effects through the interaction with cellular signaling pathways that mediate cell function under both normal and pathological conditions (Vauzour et al. 2010). It has been demonstrated that flavonoids are able to inhibit the expression of NOS, COX and LOX, which are responsible for the production of NO, PGs and LTs, respectively (Tuñón et al. 2009). Thus, the inhibition of these enzymes by flavonoids may be one of the most important mechanisms governing their anti-inflammatory activity (García-Lafuente et al. 2009) (Fig. 4). In a study carried out by Hämäläinen et al. (2007), the authors investigated the effects of 36 natural phenolic compounds on NO production in macrophages exposed to an inflammatory stimulus and evaluated their mechanisms of action. The most effective compounds were daidzein, genistein, isorhamnetin, kaempferol, quercetin, naringenin and pelargonidin, which

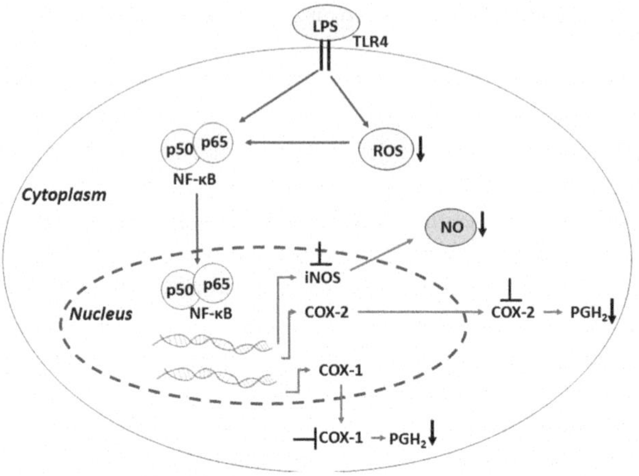

Fig. 4 Inhibitory effect of flavonoids on ROS, NO and PG. LPS binds to TLR4 and triggers the generation of ROS that activate the nuclear translocation of NF-κB. The NF-κB activation mediates iNOS and COX expression. These enzymes synthetise NO and PG, respectively. Black arrows represent a suppressive effect of flavonoids, and the T-shaped symbol represents the inhibitory activity. (Adapted from Leyva-López et al. 2016)

inhibited iNOS expression and NO production in a dose-dependent manner. The structural requirements for the inhibition of NO production were found to be the presence of a C-2,3 double bond, whereas the presence of sugar substituents either decreased or abolished the inhibitory effect. Hydroxyl groups in positions 7 and 4′ were found in all active compounds; such substitutions were not essential for the activity of the compound.

Flavonoids have been reported to act on the protein kinase and lipid kinase signaling cascades such as PI3K, Akt/PKB, tyrosine kinases, protein kinase C (PKC) and MAPKs (Spencer 2010; Park et al. 2011), inhibiting the transcription of factors as AP-1 or NF-κB. The inhibitory activity exerted on kinases is due to the competition with ATP for the binding to the catalytic sites on these enzymes, thus blocking signal transduction and cell activation processes in cells of the immune system (Ribeiro et al. 2015). Either the inhibitory or the stimulatory effects exerted on these pathways are likely to affect cell functioning by altering the phosphorylation state of target molecules and by modulating gene expression (Williams et al. 2004).

As anti-inflammatory agents, flavonoids have a similar mechanism of action to NSAIDs, since they inhibit the COXs responsible for the synthesis of PGs, which are also involved in physiological processes. The in vitro activity of flavonoids in the inflammatory response also involves other inflammatory mediators such as cytokines, adhesion molecules and chemokines (Agati et al. 2012; Leyva-López et al. 2016). Various flavonoids have been described as good modulators of cytokine production. The structural requirements for a flavonoid to exert a good inhibition of LPS-stimulated TNF-α release are the presence of a double bond at position C2-C3, with an 'oxo' function at position C4 and the presence of OH groups at positions 3′

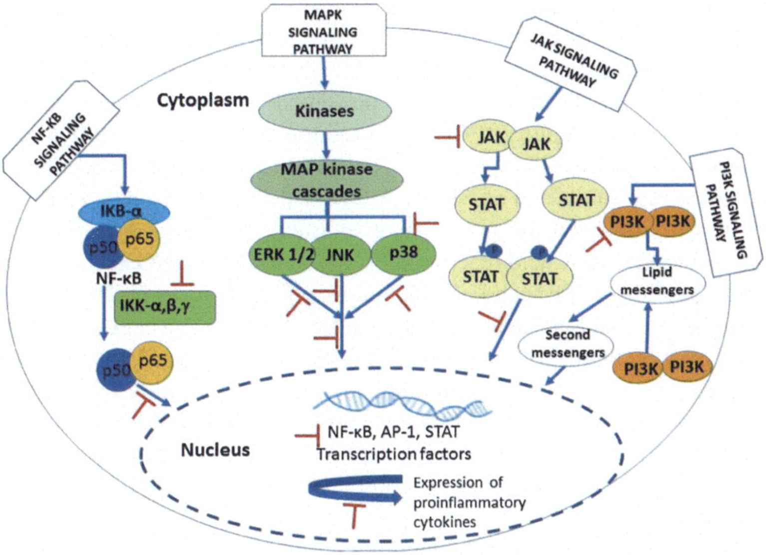

Fig. 5 Mechanism of action by which flavonoids block inflammation through inhibition of the function of NF-κB, MAPK, JAK and PI3K signaling pathways. The red T-shaped symbol indicates inhibition

and 4′ (Ribeiro et al. 2015). Molecular activities of flavonoids include the inhibition of transcription factors such as NF-κB and AP-1, as well as the activation of the nuclear factor erythroid 2-related factor 2 (Nrf2) (Tuñón et al. 2009; Serafini et al. 2010; Chen et al. 2017) (Fig. 5).

This chapter focuses on the results of recent studies assessing the role of the different subclasses of flavonoids as modulators of inflammatory mediators and on cell signaling pathways.

4.1 Flavones

In a mouse model of middle cerebral artery occlusion (MCAO), the pretreatment with chrysin (5,7-dihydroxyflavone, *Chr*) successfully decreased neurological deficit scores and infarct volumes, as compared with the control group. In this context, the up-regulation of NF-κB, COX-2 and iNOS caused by MCAO was inhibited by *Chr*. The increases in glial cell numbers and pro-inflammatory cytokine (IL-1β, IL-6, IL-12, IL-1α, IL-17A, IFN-γ and TNF-α) secretion usually caused by ischemia/reperfusion were significantly ameliorated by the pretreatment with *Chr* (Yao et al. 2014). Additionally, *Chr* prevents the increase in the number of inflammatory cells, IL-4 and IL-12 in an experimental model of asthma, which is a chronic airway inflammatory disorder. The decreased levels of IFN-γ were up-regulated, and the phosphorylation of Akt and ERK was decreased by *Chr*. Therefore, the authors hypothetised that *Chr* might have beneficial effects on chronic asthma (Yao et al. 2016). *Chr* significantly ameliorated the cardiac dysfunction in an induced myocardial injury model in diabetic rats that presented an up-regulated peroxisome proliferator-activated receptor (PPAR)-γ expression and a downregulation of receptor for advanced glycation end products (RAGE). In this model, inflammation was reduced through the inhibition of NF-κB p65/IKK-β and reduction of TNF-α levels. In addition, *Chr* inhibited the nitro-oxidative stress, as assessed by the levels of glutathione, thiobarbituric acid reactive substances (TBARS), NO and expression of superoxide dismutase (SOD) and eNOS, among others (Rani et al. 2016).

To evaluate the effect of flavones on diabetes mellitus, Wang et al. (2017) studied the effects of vitexin (8-D-glucosyl-4′,5,7-trihydroxyflavone, *Vix*) on pancreatic β-cell function in a model of LPS-stimulated rat islet tissue and in INS-1 cells. The authors demonstrated that both cell damage and apoptosis were decreased in cells treated with *Vix*. The pretreatment of cells with *Vix* reduced the production of TNF-α and attenuated the production of high-mobility group box (HMGB) in response to LPS stimulation.

It has been demonstrated that the treatment of ulcerative colitis with amentoflavone (3′,8′-biapigenin; *Afla*) decreases the levels of the inflammatory cytokines TNF-α, IL-1β and IL-6 together with the expression of iNOS and COX-2. It has also been observed that this flavone was able to inhibit the activation and nuclear translocation of NF-κB (p65/p50). These results allow postulating *Afla* as a potential protective compound in acetic acid-induced ulcerative colitis (Sakthivel and Guruvayoorappan 2013).

The neuroprotective effect of wogonin (5,7-dihydroxy-8-methoxyflavone, *Won*), a potent anti-inflammatory flavonoid, has been demonstrated through the reduction of the inflammatory response mediated by TLR4/NF-κB signaling pathway in mice with traumatic brain injury. A marked reduction in leukocyte infiltration, microglial activation, expression of TLR4, translocation of NF-κB to the nucleus and its DNA-binding activity, matrix metalloproteinase (MMP)-9 activity and expression of IL-1β, IL-6, inflammatory protein of macrophages-2 and COX-2 was observed after treatment with *Won* (Chen et al. 2012). The anti-inflammatory activity of 4′-hydroxywogonin (4′,5,7-trihydroxy-8-methoxyflavone, *4′-HW*) has also been demonstrated in vivo (Fan et al. 2017). In LPS-stimulated RAW 264.7 macrophages, *4′-HW* blocked the expression of COX-2 and iNOS, thus decreasing the levels of their products PGE_2 and NO, respectively. Moreover, in the same model, *4′-HW* suppressed the activation of TAK1 and TAB1, suggesting that TAK1/IKK/NF-κB signaling pathways were inhibited and downregulated the phosphorylation of MAPKs and PI3/Akt. This methoxyflavone also decreased the production of intracellular ROS. Furthermore, *4′-HW* also proved to have anti-inflammatory effects in a model of LPS-induced inflammation in an acute lung injury mice model (Fan et al. 2017).

Luteolin (3′,4′,5,7-tetrahydroxyflavone, *Lut*) has been demonstrated to inhibit the ROS increase, lipid peroxidation and glutathione depletion induced by short-term exposure of human bronchial epithelial cells (BEAS-2B) to Cr(VI). In these cells, the treatment with *Lut* decreased the Cr (VI)-induced promoter activity of

AP-1, hypoxia-inducible factor 1-α (HIF-1α), COX-2 and iNOS. An inhibition of the production of pro-inflammatory cytokines (IL-1β, IL-6, IL-8, TNF-α) and vascular endothelial growth factor (VEGF) was also observed. *Lut* inhibited multiple gene products including those related to inflammation: MAPK, NF-κB, COX-2, STAT-3, iNOS and TNF-α. *Lut* has been postulated as a potential chemopreventive agent against Cr (VI)-induced carcinogenesis (Pratheeshkumar et al. 2014). After TNF-α stimulation, *Lut* inhibited the adhesion of monocytes to endothelial cells and suppressed the expression of MCP-1, ICAM-1 and VCAM-1, which enhances the endothelial cell-monocyte interaction. In endothelial cells, inflammation is apparently prevented by suppression of the NF-κB pathway, since *Lut* decreased the NF-κB transcriptional activity, IκBα degradation, expression of IκB kinase β and subsequent NF-κB p65 nuclear translocation. *Lut* also proved to have anti-inflammatory effects in vivo, as assessed by histologic studies and chemokine levels (Jia et al. 2015). Besides, *Lut* has been evaluated as a potential therapeutic agent in the prevention and/or treatment of Alzheimer's disease in a human blood-brain barrier (BBB) model. In this model, the p38 MAPK-mediated NF-κB signaling pathway was examined by coculturing human brain microvascular endothelial cells (hBMECs) and human astrocytes (hAs) under fAβ1-40-damaged conditions (Zhang et al. 2017). *Lut* suppressed the production of inflammatory mediators and cytokines, such as COX-2, TNF-α, IL-1β, IL-6 and IL-8. However *Lut* did not display any scavenging effect on intracellular ROS in hBMECs and hAs.

Palmieri et al. (2012) have determined the effects of apigenin (4′,5,7-trihydroxyflavone, *Api*) on the TNF-α-induced endothelial dysfunction by evaluating the expression of eNOS and MMP-9. In this case, *Api* blocked the TNF-α-induced expression of eNOS and MMP-9 and the TNF-α-triggered activation of Akt, p38 MAPK and JNK signaling on endothelial. The use of specific Akt inhibitors, which presented Api-like effects on eNOS and MMP-9 expression, allowed demonstrating that the induction of eNOS and MMP-9 caused by TNF-α depends on Akt activation. The main mechanism of inhibition of Akt signaling involved 'classical' and 'nonclassical' ERs. A recent study has demonstrated that *Api* up-regulates the gene expression of inflammatory IL-17 cytokine family and LTA and the expression of the interferon beta 1 gene in BxPC-3 human pancreatic cancer cells (Johnson and De Mejia 2013). The effect of *Api* in a rodent model of diabetic nephropathy has also been evaluated (Malik et al. 2017). The administration of *Api* to streptozotocin-induced diabetic rats reduced ROS generation and restored the antioxidant status. Moreover, an anti-apoptotic effect was also demonstrated, since *Api* inhibited the MAPK/NF-κB/TNF-α and TGF-β1/MAPK/fibronectin pathways.

Another interesting methoxyflavone is velutin (3′,5-dihydroxy-4′,7′-dimethoxyflavone, *Vel*), isolated from the pulp of açaí fruit (*Euterpe oleracea* Mart.). This compound caused a significant reduction in the production of TNF-α and IL-6 in RAW 264.7 macrophages and in mouse peritoneal macrophages. *Vel* effectively inhibited the expression of pro-inflammatory cytokines through a significant reduction in the TNF-α and IL-6 messenger ribonucleic acid (mRNA) levels by inactivating NF-κB and by inhibiting p38 and JNK phosphorylation in the two macrophage models. In these cells, *Vel* displayed an inhibitory capacity on NF-kB activation that was higher than that of *Lut* and *Api* (Xie et al. 2012).

Tricin (4′,5,7-trihydroxy-3′,5′-dimethoxyflavone, *Tri*) isolated from Njavara rice (*Oryza sativa* L.) has been demonstrated to cause a significant downregulation of pro-inflammatory markers in human peripheral blood mononuclear cells (hPBMCs) stimulated with LPS. In that study, *Tri* reduced the NO production and iNOS expression; it attenuated LPS-induced COX-2 activity and PGE_2 production and blocked LPS-induced TNF-α and IL-6 production. Furthermore, *Tri* reduced the LPS-induced production of MMPs by hPBMCs and suppressed the LPS-induced activation of NF-κB and nuclear translocation of p65 (Shalini et al. 2012).

A flavone isolated from *Artemisia asiatica* Nakai (Asteraceae), eupatilin (5,7-dihydroxy-3′,4′,6′-trimethoxyflavone, *Eup*), has proved to have an anti-inflammatory effect in human bronchial epithelial cells affecting cell functionality and inflammatory cell adhesion in response to stimulation with TNF-α. In the study conducted by Jung et al. (2012), the authors demonstrated that *Eup* suppressed the expression of ICAM-1 and VCAM-1 mRNA in bronchial BEAS-2B epithelial cells stimulated with TNF-α. This effect was achieved by blocking the Akt-NF-κB signaling pathway, since a blockage of the IKK activity was detected. However, in BEAS-2B cells, the signaling of AP-1 was not affected, since no variations were detected in the levels of c-fos. These results established that, in bronchial epithelial cells, *Eup* caused a decrease in the adhesion of both monocytes and eosinophils to these cells due to the inhibition of Akt, thus suggesting that this flavone could modulate the pathogenesis of asthma as regards the generation of the inflammatory infiltrate.

Oroxylin A (5,7-dihydroxy-8-methoxyflavone, *OroA*) is the major flavonoid isolated from the roots of *Scutellaria baicalensis* Georgi. This compound is known as a potential anti-inflammatory agent. Song et al. (2012) have determined the action of *OroA* on LPS-induced angiogenesis in vitro and in ovo models. *OroA* affected negatively the expression of the LPS acceptor TLR4 and the activation of MAPKs, as well as the phosphorylation of JNK, p38 and ERK. Besides, the translocation of NF-κB dimers to the nucleus was limited after treatment with *OroA*. Kim et al. (2012) have evaluated the modulatory capacity of 5,6,7-trimethoxy- and 5,6,7-trihydroxyflavone derivatives on NO and PGE_2 production in LPS-stimulated RAW 264.7 cells. Thus, in this experimental model, 4′-bromo-5,6,7-trimethoxyflavone suppressed the expression of iNOS and COX-2. Furthermore, this compound downregulated the release of TNF-α, IL-6 and IL-1β as well as the expression of NO, PGE_2, TNF-α, IL-6 and IL-1β. These results suggested that the modulation exerted on the NF-κB signaling pathway would generate an anti-inflammatory response through the decrease in the degradation and phosphorylation rates of IκB-α.

The 3,5,7-trihydroxy-4′-methoxy-8-(3-hydroxy-3-methylbutyl)-flavone (*ICT*), a new derivative of the flavonol icariin, suppressed the LPS-induced TNF-α production in the human monocytic cell line THP-1, PBMCs and human monocytes in a dose-dependent manner. The pretreatment with *ICT* produced the downregulation of CD14/TLR4 by blocking the NF-κB and MAPK signaling pathways (Wu et al. 2012).

The modulation of the intestinal inflammatory response by flavones (*Chr*, *3′,4′-Dihydroxyflavone*, *Api*, *Lut* and *Quer*) and their unmethylated analogues have been evaluated by During and Larondelle (2013). The production of soluble pro-inflammatory mediators, such as IL-8, IL-6, MCP-1 and COX-2-derived PGE_2, and the activation of NF-κB in 3d-confluent and 21d-differentiated Caco-2 cells stimulated with IL-1β were evaluated after treatments with these flavones. The Caco-2 cell model allowed demonstrating that the O-methylation of *Chr* enhances its anti-inflammatory properties. Of all flavones, the demethylated form of *Chr* has displayed the highest anti-inflammatory activity. The effect of this derivative was achieved by a reduction of IL-8, IL-6, MCP-1 and COX-2-derived PGE_2 levels. The presence of hydroxyl groups on ring A (positions 5 and 7), the absence of methoxylation of the 3′-hydroxyl group on ring B and the methoxylation of the 3-hydroxyl group on ring C seemed to be responsible for the intestinal anti-inflammatory activity.

4.2 Flavonols

Lee et al. (2013) have studied the possible barrier protective effects of rutin (quercetin-3-O-rutinoside, *Rut*) on the secretion of pro-inflammatory mediators as well as the signaling pathways activated in human umbilical vein endothelial cells (HUVEC) stimulated with LPS. *Rut* blocked the disruption of the vascular barrier induced by LPS, the expression of CAM as well as the adhesion/transendothelial migration of monocytes to human endothelial cells. In addition, in the same model, *Rut* abrogated the permeability increase induced by acetic acid and the leukocyte migration induced by carboxymethyl cellulose. In addition, *Rut* reduced the expression of TNF-α and the activation of NF-κB induced by LPS. These findings allowed postulating *Rut* as a protective agent against inflammatory vascular diseases. In another study, Yoo et al. (2013) have observed that the treatment with *Rut* inhibited the up-regulation of VCAM-1, ICAM-1 and E-selectin caused by high-mobility group box 1 protein (HMGB1), and apparently this effect is mediated through attenuation of the HMGB1 signaling pathway. According to this study, *Rut* resulted in the reduction of HMGB1-induced mortality. *Rut* was also found to suppress the production of TNF-α and IL-6 and the activation of NF-κB and ERK1/2 by HMGB1.

Fisetin (3,3′,4′,7-tetrahydroxyflavone, *Fis*) has been demonstrated to be active in a mouse model of ultraviolet (UV) B-induced inflammation. In mice exposed to UV B radiation and then treated with *Fis* applied topically, a reduction of the hyperplasia and the infiltration of inflammatory cells as well as the levels of inflammatory mediators, such as TNF-α, IL-1β, IL-6 and PGE_2, and its receptors, and decreased COX-2 and myeloperoxidase (MPO) activities were observed. *Fis* inhibited UV B-induced expression of PI3K and Akt phosphorylation. The activation of the NF-κB signaling pathway was also inhibited in *Fis*-treated mice. *Fis* reduced the UV B-induced expression of IKKα/β and IκBα protein phosphorylation, thus restoring the IκBα protein levels. *Fis* also inhibited the activation of the p65 transcription factor and its nuclear translocation in UV B-exposed skin (Pal et al. 2015). The biological activity of *Fis* has also been evaluated in a murine model of acute pancreatitis where both pre- and post-treatment with this flavonol reduced the severity of acute pancreatitis and pancreatitis-associated lung injury. The pretreatment with *Fis* caused a decrease in pancreatic levels of TNF-α, IL-1β and IL-6. In vivo, *Fis* suppressed IκBα degradation and NF-κB activation, as well as activation of JNK, with similar in vitro effects on acinar pancreatic cells. In contrast, *Fis* did not affect the activation of ERK 1/2 and p38. Accordingly, the pretreatment with *Fis* inhibited the activation of JNK and the degradation of IκBα on pancreatic acinar cells. As observed in vivo, the treatment with *Fis* inhibited the production of TNF-α, IL-1β and IL-6 (Jo et al. 2014). In human gingival fibroblasts (HGFs) treated with *Porphyromonas gingivalis* LPS, *Fis* caused a significant reduction in the synthesis of PGE_2 and the expression of COX-2 without affecting cell viability. In this model, the treatment with the flavonoid inhibited the activation of ERK, JNK and p38 of the MAPK pathway, which is induced upon LPS treatment (Gutiérrez-Venegas et al. 2014). In a murine model of early brain injury after subarachnoid haemorrhage, high doses (50 mg/kg) of *Fis* improved neurological function parameters and reduced brain edema. TLR4 expression and NF-κB translocation to the nucleus were significantly reduced, as was the production of inflammatory cytokines such as TNF-α and IL-1β (Zhou et al. 2015a). In a study evaluating the antiseptic effects of *Fis* on HMGB1-mediated inflammation, this flavonoid proved to modulate pro-inflammatory responses. In HUVECs, HMGB1 augmented the phosphorylation of NF-κB, ERK1/2 and Akt, in addition to increasing TNF-α and IL-β production. These effects were significantly reduced by *Fis,* as was NF-κB p65 translocation to the nucleus (Yoo et al. 2014). In her 2015 review, Maher summarizes the effect of *Fis* on the central nervous system (CNS) functions. As regards inflammation, *Fis* proved to reduce LPS-induced microglial activation and neurotoxicity. Accordingly, the levels of TNF-α, PGE_2, iNOS and COX-2 were reduced after treatment with the flavonoid, and these effects seemed to be mediated by the inhibition of activation of NF-κB. *Fis* also suppressed other pro-inflammatory signaling pathways, such as JNK and p38 MAPK, in microglia in the temporary middle cerebral artery occlusion stroke model in mice. Results indicated that *Fis* has both in vitro and in vivo anti-inflammatory activity on the CNS immune system (Maher 2015). Other studies on microglial activation have shown that *Fis* inhibits cell migration and ROS production. Moreover, the expression of iNOS along with NO production was also reduced in cells stimulated with LPS plus IFN-γ and with peptidoglycan. The LPS/IFN-γ- or peptidoglycan-enhanced production of IL-1β was inhibited by *Fis*. This flavonol generated an endogenous increase in the anti-oxidative heme oxygenase-1 (HO-1) expression through the PI-3 kinase/Akt and the p38 signaling pathways, but not through ERK and JNK in microglia. *Fis* also significantly attenuated inflammation-related microglial activation and coordination deficit in mice in vivo (Chuang et al. 2014).

Icariin (4′-O-methyl-8-γ,γ-dimethylallyl kaempferol-3-rhamnoside-7-glucoside, *Ica*), a prenyl flavonoid glycoside, is the major active compound of *Herba epimedii*, which is a centuries-old traditional medicine herb. Formulations prepared with this herb are the most frequently prescribed ones (Zhang et al. 2014; Kong et al. 2015). The anti-inflammatory activity of *Ica* has been evaluated in a TNF-α/IFN-γ-induced inflammatory response in human keratinocytes (HaCaT cells). In HaCaT cells, the TNF-α/IFN-γ-induced production of IL-6, IL-8, IL-1β and MCP-1 and gene expression of IL-8, IL-1β, ICAM-1 and tachykinin receptor 1 (TACR1) were inhibited by *Ica*. The treatment with *Ica* produced a reduction in the phosphorylation of p38 MAPK and ERK that was augmented upon stimulation with TNF-α/IFN-γ. The abnormal expression of TNF-α-R1 and IFN-γ-R1 found in HaCaT cells after TNF-α/IFN-γ stimulation was modified by *Ica*, which downregulated the levels of the former and up-regulated the levels of the latter. These effects were mediated, at least partially, via the inhibition of the p38-MAPK signaling pathway, as well as by the regulation of the TNF-α-R1 and IFN-γ-R1-related signals (Kong et al. 2015). In an unpredictable chronic mild stress model of depression in rats, the chronic treatment with *Ica*, which can freely cross the BBB, reverted the increased levels of oxidative-nitrosative stress markers and inflammatory mediators like TNF-α and IL-1β. The activation of the NF-κB signaling pathway and increased iNOS mRNA expression in the hippocampus was also reverted by *Ica* (Liu et al. 2015). *Ica* modulates the activity of the histone deacetylase sirtuin (SIRT)6, with a maximum activating effect at 10 M. After treatment with *Ica*, the up-regulation of SIRT6 protein expression was observed, while the expression of NF-κB (p65) was downregulated in heart tissue and in aortic endothelial cells. An inhibitory effect of *Ica* on NF-κB inflammatory signaling pathways, as evidenced by decreased mRNA TNF-α, ICAM-1, IL-2, and IL-6 levels, was observed (Chen et al. 2015).

Astragalin (kaempferol-3-glucoside, *Ast*) is found in several plants, such as *Podophyllum peltatum*, *Paeonia lactiflora*, *Phytolacca americana*, *Cicer arietinum*, *Onobrychis arenorie*, *Phaseolus vulgaris*, *Rosa agrestis* and *Glycyrrhiza macedonica* (Li et al. 2014a; You et al. 2017). In primary-cultured mouse mammary epithelial cells (mMECs), *Ast* inhibited the production of TNF-α, IL-6 and NO, as well as expression of iNOS and COX-2 after LPS stimulation. The treatment of mMECs with *Ast* decreased the LPS-induced TLR4 expression, NF-kB activation, IκBα degradation and the phosphorylation of p38 and ERK (Li et al. 2014a) (Fig. 6).

4.3 Flavanones

The anti-inflammatory effect of alpinetin (7-hydroxy-5-methoxyflavanone, *Alp*), which is the main flavonoid of *Alpinia katsumadai* Hayata, has been investigated to find that *Alp* blocks the inflammatory process both in vitro, in LPS-stimulated RAW 264.7 cells, and in vivo in a LPS-induced acute lung injury model (Huo et al. 2012). The pretreatment with *Alp* induced a strong blockage of the production of TNF-α, IL-6 and IL-1β induced by LPS. In addition, in the in vitro model, *Alp* inhibited IκBα, p65, p38 and ERK phosphorylation. Besides, in the in vivo model, histopathologic studies demonstrated that the changes in the mouse lungs were minimal. Several findings suggest that *Alp* would act through the NF-κB and MAPK

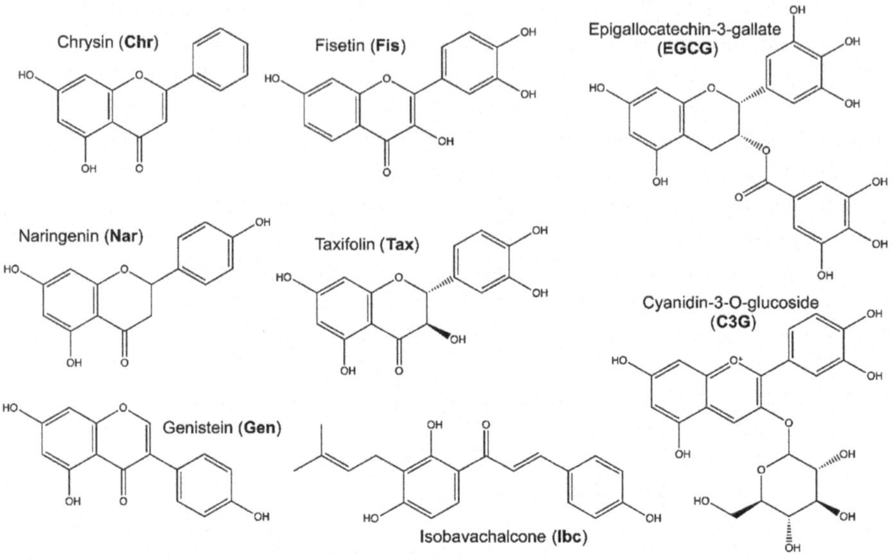

Fig. 6 Chemical structures of some flavonoids that act as modulators of inflammatory mediators and on cell signaling pathways

signaling pathways and that *Alp* would act as a potential protective agent in the acute lung injury model. Furthermore, the effect and mechanism of action of *Alp* was evaluated in a LPS-induced mouse mastitis model. In this in vivo study, *Alp* prevented the infiltration with neutrophils and the activation of myeloperoxidase and downregulated the expression of TNF-α, IL-1β and IL-6. Likewise, the phosphorylation of IκB-α and NF-κB p65 and the expression of TLR4, induced by LPS, were inhibited by the flavonoid. Additionally, in the in vitro model, *Alp* inhibited the expression of TLR4 and the production of TNF-α, IL-1β and IL-6 in LPS-stimulated primary mouse mammary epithelial cells. These results indicate that *Alp* could be considered a potential therapeutic agent for the treatment of mastitis, since it modulates the activation of the NF-κB signaling pathway mediated by the activation of TLR4 (Chen et al. 2013). Furthermore, Hu et al. (2013) have evaluated the signaling pathways involved in the anti-inflammatory activity of *Alp* in human THP-1 macrophages stimulated by LPS. In this case, *Alp* prevented the synthesis of TNF-α, IL-6 and IL-1β. *Alp* inhibited the activation of NF-κB, the degradation of IκBα and the phosphorylation of ERK, JNK and p38. Moreover, it was observed that the activation of PPAR-γ caused by *Alp* led to the decrease in the expression of TLR4 and the consequent inhibition of TLR4-dependent activation of NF-κB and MAPK. In turn, these events led to an inhibition of the release of pro-inflammatory cytokines.

Naringenin (4′,5,7-trihydroxyflavanone, *Nar*), a flavonoid derived from grapefruit and related citrus species, proved to have a protective effect in a model of LPS-induced human bronchial epithelium injury by suppressing the secretion of TNF-α, IL-6, SOD, NOS, MPO and NO. The LPS-induced up-regulation of NF-κB p65 mRNA expression was also reduced by *Nar*, and this flavonoid effectively suppressed NF-κB activation by inhibiting the degradation of IκB-α and the translocation of p65. The reduction in the secretion of TNF-α and IL-6 is possibly mediated by a blockage in the activation of the NF-κB and MAPK signaling pathways, since *Nar* inhibited the phosphorylation of ERK1/2, JNK and p38 MAPK (Yu et al. 2014). Furthermore, the suppressors of cytokine signaling (SOCS)-3 expression and the anti-inflammatory effects of *Nar* in microglial cells are regulated by adenosine monophosphate-activated protein kinase (AMPK) α and PKC δ. *Nar* downregulates the expression of iNOS and COX-2 and inhibits the release of NO. *Nar* has also displayed significant protective effects on microglial activation and improved the motor coordination function in a murine model (Wu et al. 2016).

Naringin (4′,5,7-trihydroxyflavanone 7-rhamnoglucoside, *Nag*), a flavanone-7-O-glycoside formed between naringenin and the disaccharide neohesperidose, is found as the major flavonoid glycoside in grapefruit. This flavone gives grapefruit juice its bitter taste (Pubchem 2017). In HaCaT cells, the pretreatment with *Nag* prevented UV B-induced apoptosis and the production of ROS and decreased the levels of inflammatory cytokines, such as IL-1β, IL-6, IL-8 and COX-2, as compared to UV B-exposed and non-treated cells. *Nag* inhibited the activation of p38 and JNK upon exposure of these cells to UV B. In a mouse model, the topical treatment prevented epidermal thickening, IL-6 production, apoptosis and the over expression of COX-2 caused by UV B irradiation. *Nag* also blocked the UV B-induced activation of p38. *Nag* would confer protection against UV B both in vitro and in vivo through inhibition of MAPK/p38 activation (Ren et al. 2016). Cisplatin, an effective chemotherapeutic agent, is known to cause a decline in the concentrations of reduced glutathione and ascorbic acid, a decrease in membrane-bound ATPases and glutathione peroxidase (GPx) activities and an increase in the activity of catalase (CAT) and SOD in striatum tissue of aged rats. The deterioration of striatum tissue was prevented by the treatment with *Nag*; the change in antioxidant enzymes was revoked, and the increase in malondialdehyde, protein carbonyls, NO and TNF-α levels was suppressed. Accordingly, *Nag* inhibited p53-, NF-κB- and TNF-α-mediated inflammation. Thus, *Nag* proved to have neuroprotective effects in this model (Chtourou et al. 2015). In an experimental diabetes mellitus rat model, the treatment with *Nag* improved the condition of the animals. In the cerebral cortex and hippocampus, the glucoside reduced the levels of oxidative stress markers and pro-inflammatory factors, such as TNF-α and IL-6. *Nag* also activated the expression of PPARγ, which inhibits the inflammatory response. The cognitive deficit in diabetic rats was also ameliorated by *Nag* through a decrease of oxidative stress marker levels and pro-inflammatory factors and activation of the PPARγ signaling pathway (Qi et al. 2015). The pretreatment with *Nag* of murine splenocytes exposed to ionizing radiation prevented intracellular ROS generation, thus preventing lipid peroxidation and nitrite production. A reduction in nuclear DNA damage and a recovery of cell viability were also observed after treatment with the flavonoid. *Nag* blocked the p38 phosphorylation and the downstream cascade of events involving inhibition of the NF-κB pathway (Manna et al. 2015).

In injured hBMECs, pinocembrin (5,7-dihydroxyflavanone, *Pin*), a flavonoid abundant in propolis, *Pinus* heartwood and *Eucalyptus*, reverts the cytotoxicity of β-amyloid peptides, which are known to be involved in Alzheimer's disease pathogenesis. In this model, the flavonoid increases cell viability and attenuates nuclear damage, and lower levels of LDH are released. *Pin* inhibits the inflammatory response through various mechanisms, including inhibition of MAPK activation, downregulation of IKK, a decrease in IκBα degradation, inhibition of NF-κB p65 nuclear translocation and the consequent reduction in the release of pro-inflammatory cytokines (TNF-α, IL-1 and IL-6). The anti-inflammatory effects of *Pin* in hBMECs are probably related to the inhibition of the MAPK and the NF-κB signaling pathways (Liu et al. 2014b). In LPS-stimulated BV2 microglial cells, *Pin* inhibited the production of TNF-α, IL-1β, NO and PGE_2 and the expression of iNOS and COX-2. PI3K and Akt phosphorylation and NF-κB activation were inhibited by *Pin*. Induction of nuclear translocation of Nrf2 and expression of HO-1 have also been observed after treatment with this flavonoid (Zhou et al. 2015b).

Sophoraflavanone G (5,7,D,2′,4′-tetrahydroxy-8-lavandulylflavanone, *SG*), isolated from *Sophora flavescens*, has been evaluated as a potential anti-inflammatory agent in LPS-stimulated RAW 264.7 macrophages. In these cells, *SG* blocked the expression of iNOS and COX-2, with the consequent decrease of NO and PGE_2. *SG* also reduced the production of pro-inflammatory cytokines, such as IL-1β, IL-6 and TNF-α. *SG* inhibited the phosphorylation of the p65 subunit of NF-κB, thus preventing its translocation to the nucleus. Although *SG* stimulated the synthesis of HO-1, it was observed that the activation of MAPK was downregulated, since the phosphorylation of ERK1/2, JNK and p38 did not occur. When cells were cocultured with *SG* and MAPK, lower activation levels of iNOS and COX-2 were observed. These results confirm the anti-inflammatory effect of *SG* evidenced by the negative modulation of NF-kB and MAPK signaling pathways (Wun et al. 2013).

Ugonin M (5,4′-dihydroxy-4″,4′-dimethyl-5′-methyl-5″H-dihydrofuran [2″,3″:6,7] flavanone, *UgoM*) has been isolated from *Helminthostachys zeylanica* (L.) Hook, which is a traditional Chinese medicine plant popularly used for the treatment of inflammation, among other applications. This flavanone suppresses the production of pro-inflammatory mediators such as NO, TNF-α, IL-1β and IL-6 and decreases cell counts and the protein content in the bronchoalveolar lavage fluid in LPS-induced acute lung injury in mice. In this context, *UgoM* attenuated pulmonary edema. Likewise, *UgoM* prevented the activation of iNOS and COX-2 in LPS-induced inflammation. *UgoM* blocked the translocation of NF-κB and the activation of MAPK through the degradation of NF-κB and IκB-α, as well as through the promotion of phosphorylation of ERK and p38 MAPK. In addition, in this model, the expression of TLR4 was blocked. On the other hand, in the same model, it was demonstrated that *UgoM* inhibited the expression of MPO and stimulated the expression of HO-1 and the antioxidant enzymes SOD, GPx and CAT (Wu et al. 2017).

4.4 Flavanonols

Taxifolin (2R,3R)-3,3′,4′,5,7-pentahydroxyflavanone, *Tax*) reverted the increase in mast cell infiltration caused by 1,2-dimethyl hydrazine (DMH) in a mouse colon cancer model. *Tax* also favoured the activation of antioxidant pathways through the increase in the levels of Nrf2, which activates the expression of cytoprotective genes in response to ROS. *Tax* downregulated the NF-κB and Wnt signaling pathways. The expression of NF-κB, TNF-α and COX-2 were reduced when compared to the group treated only with DMH. *Tax* would exert chemopreventive effects by modulating inflammatory, Wnt and antioxidant response pathways (Manigandan et al. 2015). The flavanonols 2′-hydroxy yokovanol and 2′-hydroxy neophellamuretin, isolated from the leaves and stems of *Desmodium caudatum*, along with other flavonoids, inhibited the production of IL-6, IL-12 and TNF-α in LPS-stimulated bone marrow-derived dendritic cells (Li et al. 2014c).

Ampelopsis grossedentata (Hand-Mazz) W.T. Wang, known as rattan tea, is popularly used in China for its anti-inflammatory and other pharmacological properties. It has been demonstrated that one of its main compounds is ampelopsin (3,5,7,3′,4′,5′-hexahydroxyflavanone, *Amp*). To understand the molecular mechanisms involved in the anti-inflammatory effects exerted by this flavonoid, the production of NO by RAW264.7 macrophages stimulated with LPS was evaluated. The pretreatment with *Amp* blocked the production of NO and the pro-inflammatory cytokines IL-1β, IL-6 and TNF-α. *Amp* blocked the activation of iNOS with the consequent inhibition in the translocation of NF-κB due to the inhibition of KKα/β and IκBα phosphorylation and nuclear translocation NF-κB p65. In addition, *Amp* inhibited the release of Akt without affecting MAPK phosphorylation. *Amp* also interfered with ROS-mediated PI3K/Akt phosphorylation. Thus, the anti-inflammatory effects of *Amp* are related to the inhibition of the Akt, IKK and NF-κB signaling pathways (Qi et al. 2012).

4.5 Flavan-3-ols

Epigallocatechin-3-gallate [(−)-cis-3,3′,4′,5,5′,7-hexahydroxy-flavane-3-gallate, *EGCG*)], the most abundant catechin in green tea infusions and one of the most active molecules known for its antioxidant properties, is known to downregulate the TLR4 signal transduction in LPS-stimulated endothelial cells. This downregulation is mediated by the 67-kDa laminin receptor (67LR) and by an up-regulation of the Toll-interacting protein (Tollip), which is a negative regulator of TLR signaling (Byun et al. 2014; Legeay et al. 2015). *EGCG* also modulates inflammatory responses in adipocytes through the 67LR, leading to a reduction of inflammatory mediator and cytokine levels (IKKβ, p-NF-κB, TNF-α and IL-6) after LPS stimulation. These data suggest that *EGCG* suppresses TLR4 signaling in LPS-stimulated adipocytes via 67LR (Bao et al. 2015). In an in vivo model of crescentic glomerulonephritis, the treatment with *EGCG* reduced mortality and markedly improved renal function and histology, when compared with vehicle-treated mice. More importantly, *EGCG* caused a decrease in p-Akt, p-JNK, p-ERK1/2 and p-P38 as well as restoration of PPARγ and SIRT1 levels. The Nrf2 signaling, which was impaired in vehicle-treated mice, was restored by *EGCG* (Ye et al. 2015). After stimulation of human hepatocytes with LPS, an increase in the production of TNF-α, regulated upon activation normal T-cell expressed and secreted (RANTES), MCP-1, ICAM-1, NO, VEGF and MMP-2 was observed. This effect was reduced by the pretreatment of cells with *EGCG*. The effects observed were related to the inhibition of NF-κB and MAPK signaling pathways through a downregulation of p-IκBα, p65, p-p65, p-p38, p-ERK1/2 and p-Akt, thus indicating that *EGCG* suppresses LPS-induced inflammatory response and oxidant stress and exerts hepatocyte-protective activity (Liu et al. 2014a). Besides, the exposure of human endothelial cells to environmental pollutants such as polychlorinated biphenyls (PCBs) increases the expression of vascular inflammatory mediators, including IL-6, CRP, ICAM-1, VCAM-1 and IL-1α/β. The pretreatment with *EGCG* prevents such increase together with an inhibition of nuclear import of p65, a decreased p65 NF-κB subunit and histone acetyltransferase p300 chromatin binding, as well as an increased chromatin binding of histone deacetylase HDAC1/2 and hypoacetylation of histone H3. Therefore, *EGCG* decreases PCB-induced vascular toxicity through epigenetic modifications (Liu et al. 2016). It has been postulated that *EGCG* might have renoprotective effects in a unilateral ureteral obstruction mice model. In the obstructed kidney, the induced oxidative stress and inflammatory response, as represented by inflammatory cytokines such as TNF-α, IL-6 and IL-1β, was prevented by *EGCG*, which was able to inhibit NF-κB, to enhance Nrf2 nuclear translocation and to promote HO-1 production (Wang et al. 2015b). In human HUVEC cells, *EGCG* suppressed the expression of IL-6, ICAM-1, TNF-α, and MCP-1 and the generation of ROS induced by uric acid. This suppression was achieved through the inhibition of Notch-1 signaling pathways (Xie et al. 2015). In a non-alcoholic fatty liver disease murine model, the treatment with *EGCG* caused downregulation in the expression of key pathological oxidative (e.g. nitrotyrosine formation) and pro-inflammatory markers (e.g. iNOS, COX-2 and TNF-α). *EGCG* inhibited the activity of TGF/SMAD, PI3K/Akt/FoxO1 and NF-κB pathways, thus reducing the severity of liver injury (Xiao et al. 2014).

(+)-Catechin [(2R,3S)-2-(3,4-dihydroxyphenyl)-3,4-dihydro-1(2H)-benzopyran-3,5,7-triol, *Cat*] reduced the levels of iNOS and COX-2 and the production of NO and ROS after stimulation of BV-2 (a mouse microglial cell line) with LPS. Even though the production of TNF-α and IL-6 was suppressed, IL-4 levels were increased. *Cat* inhibited IκB-α phosphorylation, thus inhibiting the nuclear translocation of NF-κB p65. On the other hand, the activation of Akt was inhibited and so was the phosphorylation of ERK1/2 and p38 MAPK. *Cat* also suppressed AMPK activity. It has been postulated that the anti-inflammatory activity exerted on this cell type was related to the suppression of pro-inflammatory mediators and inhibition of NF-κB activity through Akt, ERK, p38 MAPK and AMPK pathways (Hussein et al. 2015).

4.6 Anthocyanidins

Cyanidin-3-O-glucoside (3,3′,4,5,7-pentahydroxyflavylium-3-O-glucoside, *C3G*) is an anthocyanin commonly present in food and vegetables in the human diet. *C3G* has been demonstrated to have inhibitory capacity on the production of TNF-α, IL-6 and IL-1β both in vitro on HUVECs and in vivo in an acute respiratory distress syndrome model. The pretreatment with *C3G* improved histopathologic and clinical parameters in vivo. In the lung tissue, *C3G* has proved to suppress the LPS-induced NF-kB and MAPK signaling pathways activation by blocking the phosphorylation of IκB-α, NF-κB/P65, ERK, p38 and JNK (Ma et al. 2015). When HUVECs were exposed to palmitic acid, a significant increase in the levels of free radicals and oxidative stress markers occurred; however, this status was reverted upon treatment with *C3G*. The activation of NF-κB pro-inflammatory pathway and the expression of adhesion molecules induced by palmitic acid were inhibited by *C3G* possibly through the activation of the Nrf2/electrophile-responsive element (EpRE) pathway, since *C3G* induced Nrf2 nuclear localisation and activation of cellular antioxidant and cytoprotective genes (Fratantonio et al. 2015). Recent evidences have shown how, in the presence of *C3G*, TNF-α-stimulated intestinal cells can modify the physiological functioning of endothelial cells. The protective effects exerted by the anthocyanidin have also been demonstrated. In this in vitro non-contact coculture system with TNF-α-activated Caco-2 intestinal cells, E-selectin and VCAM-1 mRNA levels were increased as were leukocyte adhesion and NF-κB levels, which were inhibited by *C3G*. It has been observed that TNF-α stimulates the nuclear translocation of NF-κB and the expression of the genes encoding TNF-α and IL-8, whereas the pretreatment with *C3G* significantly reduces these effects by preventing the p38 translocation. In addition, *C3G* blocked the activation of TNF-α-stimulated HUVECs, in which the expression of E-selectin and VCAM-1 mRNA and increased levels of NF-κB were observed. This study has demonstrated that the main protective mechanism against chronic intestinal inflammatory diseases is related to the selective inhibition of the NF-κB pathway, making anthocyanidins important therapeutic agents to treat this disease (Ferrari et al. 2017). He et al. (2017) have evaluated the protective effects of *C3G* from sunlight UV radiation. In that study, *C3G* prevented apoptosis, the morphological changes and increased the viability of HaCaT cells exposed to UV B irradiation. In the same model, *C3G* also displayed a great ROS scavenging capacity. The expression of COX-2 in irradiated cells was also blocked by *C3G*. This flavonoid was also found to decrease the activation of EGF receptor in HaCaT, and this effect was mediated through the inhibition of Akt phosphorylation. It has been suggested that the photoprotective effects exerted by the flavonoid in UV B-irradiated keratinocytes were due to the interaction of the MAPK and Akt signaling pathways, since the nuclear translocation of p38, ERK and JNK were abrogated.

Taking into account that the evolution of atherosclerosis is related to the activation of the NF-κB pathway that leads to endothelial dysfunction and vascular inflammation and that anthocyanins are natural compounds with an important antioxidant activity, Paixão et al. (2012) have evaluated the effect of malvidin-3-O-glucoside (3,5,7-trihydroxy-2-(4-hydroxy-3,5-dimethoxyphenyl) chromeno-3O-glucoside, *Mal3OG*) on the biosynthesis of NO and on the activation of NF-κB induced by peroxynitrite in bovine arterial endothelial cells. The treatment with *Mal3OG* increased the release of NO by endothelial cells. In addition, *Mal3OG* facilitated both the phosphorylation of Akt and eNOS and decreased peroxynitrite-induced iNOS expression. Upon evaluating the activity of NF-κB in treated cells, a decrease in the nitration of IkB was observed. Moreover, a decrease in the peroxynitrite-induced expression of COX-2 and IL-6 production was also observed. These results allow postulating anthocyanidins as potential protective agents against cardiovascular diseases; and therefore, they are considered useful in the development of functional and nutraceutical foods.

The anti-inflammatory activity of malvidin (3,4′,5,7-tetrahydroxy-3′,5′-dimethoxyflavylium, *Mal*), the main constituent of wine, has been evaluated in LPS-stimulated RAW 264.7 macrophages. In these cells, the treatment with *Mal* blocked the activation of NF-κB induced by LPS and the ROS production. Besides *Mal* downregulated the activation of MAPK, stimulated the expression of MKP-1 and activated the PI-3-kinase-Akt pathway. Moreover, *Mal* maintained the mitochondrial membrane potential after LPS-induced depolarization in RAW 264.7 macrophages and reduced the nuclear translocation and the binding of NF-κB to DNA (Bognar et al. 2013). Furthermore, neither peonidin (3,4′,5,7-tetrahydroxy-3′-methoxyflavylium, *Peo*) nor *Mal* decreased the expression of inflammatory genes when added alone; however, the treatment of adipocytes with a combination of *Mal* and *Peo* (1:1) followed by LPS decreased the mRNA levels of IL-6, IL-1β, IL-8, MCP-1, TLR2, TNF-α, COX-2 and INF-γ-induced protein-10 (Mackert and McIntosh 2016).

Pelargonidin (3,4′,5,7-tetrahydroxyflavylium, *Pel*) and its glucoside form pelargonidin-3-glucoside (*P3G*), which are found in blue, purple and red fruits and vegetables, have antioxidant and antidiabetic activities in vivo. *Pel* inhibits the LPS-mediated secretion of HMGB1 by endothelial cells. HMGB1, a nucleosomal protein, mediates the production of TNF-α, IL-1α, IL-1β and IL-6 and activates NF-κB and ERK1/2 in HUVECs. In these cells, these effects are prevented by *Pel* (Min et al. 2016).

Byun et al. (2013) have evaluated the anti-inflammatory potential of procyanidin trimer *C1* in LPS-stimulated primary bone marrow-derived macrophages (BMDM) as an alternative to the tumorigenic RAW 264.7 cell line. The pretreatment with *C1* prevented the production of iNOS-derived NO and the pro-inflammatory cytokines IL-6 and TNF-α. Concurrently, in BMDM, it was observed that *C1* inhibited the release of PGE_2 and COX-2 and the expression of cell surface molecules (CD80, CD86 and MHC class II). It is believed that the downregulation of TLR4 would be responsible for the inhibition of MAPK and NF-κB signaling induced by LPS.

4.7 Isoflavonoids

Genistein (4′,5,7-trihydroxyisoflavone, *Gen*), an isoflavone derivative found in soy, has proved to reduce the secretion of IL-1β, IL-6 and IL-8 from TNF-α-stimulated MH7A cells (human synoviocytes). Upon TNF-α stimulation, NF-κB translocation to the nucleus and IκB kinase-α/β and IκBα phosphorylation were suppressed by *Gen*, and AMPK activity was inhibited. The inhibitory effect of *Gen* on TNF-α-induced pro-inflammatory cytokine production is dependent on AMPK activation. Data suggest that *Gen* would suppress TNF-α-induced inflammation through the inhibition of the ROS/Akt/NF-κB pathway and the promotion of AMPK activation in these cells (Li et al. 2014b). *Gen* also has anti-inflammatory effects on BV-2 microglia cells stimulated with the β-amyloid peptide 25–35 (Ab25–35). *Gen* has been demonstrated to revert the up-regulation of the mRNA and protein expression of IL-1β and iNOS and the downregulation of the expression of IL-10 caused by Ab25–35. This flavonoid also reverted the upregulation of TLR4 and NF-κB (p65 and p50) and inhibited the DNA binding and transcriptional activities of NF-kB (Zhou et al. 2014).

GEN-27 [5-hydroxy-7-[2-hydroxy-3-(piperidin-1-yl) propoxy]-3-{4-[2-hydroxy-3-(piperidin-1-yl) propoxy] phenyl}-4H-chromen-4-one] is a newly synthesized *Gen* derivative which reduces the secretion of pro-inflammatory cytokines IL-6 and IL-1 in THP-1 (human monocytes) and inhibits the nuclear translocation of NF-κB and phosphorylation of IκB and IKKα/β in both HCT116 (human colon tumour) and THP-1 cells. *GEN-27* modulates the NF-κB signaling pathway involved in inflammation-induced cancer cell proliferation (Wang et al. 2016).

Puerarin 8-(β-D-glucopyranosyl-7-hydroxy-3-(4-hydroxyphenyl)-4H-1-benzopyran-4-one, *Pue*) is an isoflavonoid isolated from the roots of *Pueraria lobate*, a plant used in the traditional Chinese medicine. *Pue* has been demonstrated to improve the histologic parameters of ovalbumin (OVA)-induced allergic inflammation in a murine asthma model. The increase in eosinophil counts and IL-4, IL-5 and IL-13 caused by OVA were prevented by the administration of *Pue*. On the other hand, IFN-γ levels, which were reduced after OVA induction, were restored by the flavonoid. *Pue* substantially inhibited eotaxin-3 levels, as compared with controls (Wang et al. 2015a).

Daidzein 4′,7-dihydroxyisoflavone (*Dai*) is an isoflavone found in soy. It has been demonstrated that *Dai* has effects on the adipocyte-macrophage crosstalk. When 3 T3-L1 adipocytes were cocultured with RAW 264.7 macrophages and treated with *Dai*, the increased mRNA levels of MCP1 and IL-6 were reduced. This phenomenon was also observed in RAW 264.7 macrophages cultured alone with *Dai*. *Dai* induced a significant inhibition of the palmitate-induced phosphorylation of JNK; however, no effects were observed on NF-κB activation after treatment with the flavonoid. *Dai* probably regulates pro-inflammatory gene expression by activating PPAR-α and PPAR-γ and by inhibiting the JNK pathway in adipocyte-macrophage cocultures (Sakamoto et al. 2016).

Dong et al. (2017) have determined the anti-inflammatory effects and molecular mechanisms of ononin (formononetin-7-glucoside, *Ono*) in LPS-stimulated RAW 264.7 macrophages. *Ono* has been isolated from the roots of *Astragalus membranaceus* (Fisch.) Bunge. This flavonoid did not alter cell viability. *Ono* downregulated mRNA expression of COX-2 and iNOS and inhibited the synthesis of PGE_2 and NO and the production of pro-inflammatory cytokines, such as TNF-α, IL-1β and IL-6. In addition, in LPS-treated cells, the phosphorylation of IκB-α, ERK, JNK and MAPKs proteins was significantly increased by *Ono*. This finding suggests that the anti-inflammatory activity is exerted through the modulation of the translocation of NF-κB and MAPK pathway-related proteins. Yang et al. (2013) have evaluated the anti-inflammatory activity of Prunetin (4′,5-dihydroxy-7-methoxyisoflavone, *Pru*) and elucidated its molecular mechanism of action. *Pru* effects were evaluated in LPS-stimulated murine macrophages. In vitro assays have demonstrated that *Pru* inhibits LPS-induced NO and PGE_2 production through the suppression of iNOS and COX-2 at the transcriptional level. Besides *Pru* avoided the activation of NF-κB and the subsequent downstream induction of pro-inflammatory mediators such as TNF-α, IL-6 and IL-1β by the negative modulation of phosphorylation of IKK-IκBα-NF-κB signaling. The treatment of RAW 264.7 macrophages with *Pru* decreased the expression of iNOS and COX-2 and pro-inflammatory mediators (NO and PGE_2). As a consequence, MAPK and NF-κB signaling pathways were affected by *Pru*.

4.8 Chalcones

1-(3,4-Dihydroxyphenyl)-3-(2-methoxyphenyl)prop-2-en-1-one (*L2H17*), a synthetic chalcone derivative, inhibits the expression of pro-inflammatory cytokines (TNF-α and IL-6), cell adhesion molecules (VCAM-1 and ICAM-1), chemokines and macrophage adhesion via modulation of the MAPK/NF-κB pathway in peritoneal macrophages in a hyperglycemia-induced inflammation murine model. Similar effects were observed in vivo, which contributed to a reduction of key markers for renal and cardiac dysfunction. In fact, in diabetic mice treated with *L2H17*, less fibrosis and pathological changes in both renal and cardiac tissues were observed (Fang et al. 2015b). In obesity-related glomerulopathy, it has been observed that *L2H17* protects against renal injury also by modulating the MAPK/NF-κB pathways and decreasing the expression of pro-inflammatory cytokines and cell adhesion molecules (Fang et al. 2015a).

Structure-activity relationship studies have shown that α-X-substituted 2′,3,4,4′-tetramethoxychalcones enhance the transcriptional activity of Nrf2 while inhibiting NF-κB. Inflammatory signaling pathways are known to be modulated by compounds that alkylate cysteinyl thiols. A positive correlation has been found between the anti-inflammatory and the thiol alkylating activity, that is, stronger electrophiles ($X = CF_3$, Br and Cl) are more potent. Nonetheless, the strongest electrophiles ($X = CN$ and NO_2) have been found to be ineffective (Rücker et al. 2015).

In brain endothelial cells, isobavachalcone (2′,4,4′-trihydroxy-3′-(3-methyl-2-butenyl)-chalcone, *Ibc*), which is a flavonoid present in *Psoralea corylifolia*, downregulates ICAM-1 expression and arrests NF-κB activity upon LPS stimulation, as well as after macrophage-activating lipopeptide 2-kDa (MALP-2) or polyriboinosinic polyribocytidylic acid (poly[I:C]) exposure. *Ibc* also downregulates LPS or poly[I:C]-induced expression of IFN-β, indicating that it can modulate both MyD88-dependent and TRIF-dependent signaling of TLR4 (Lee et al. 2015). *Ibc*, isolated from *Angelica keiskei*, has been demonstrated to modulate the inflammatory response. The modulation of iNOS expression by *Ibc* in murine macrophages stimulated with TLR agonists has been evaluated. *Ibc* suppressed the iNOS expression induced by MALP-2 (TLR2 and TLR6), poly [I:C] (TLR3) and LPS (TLR4). As *Ibc* was able to regulate the TLR signaling pathways, and considering that these receptors are known to be directly related to the induction of the innate immune response, *Ibc* could be considered a potential anti-inflammatory drug (Shin et al. 2013).

Chalcone glycosides are 4′-glycosidised-3′-oxychalcones and have been reported in *Brassica rapa* L. 'hidabeni', a popular Japanese turnip mainly cultivated and consumed as a traditional vegetable. The activities of various synthetic 'hidabeni' chalcones have been studied. Two compounds (3′,3,4,5-tetramethoxy-4′-hydroxychalcone and 3′,3,4,5-tetramethoxychalcone) have proved to inhibit NO production. The suppression of the LPS-induced iNOS expression caused by these compounds was due to the inhibition of STAT1, but not NF-κB, JNK or p38, pathways. 3′,3,4,5-tetramethoxychalcone also inhibited the activation of the MEK/ERK pathway (Hara et al. 2014).

Phloretin (2′,4′,6′-trihydroxy-3-(4-hydroxyphenyl)propiophenone, *Phl*) is a dihydrochalcone isolated from the apple tree and the pear tree. This flavonoid inhibits the release of PGE_2, the expression of COX-2 and the production of IL-8, MCP-1 and IL-6 in IL-1β-stimulated human lung epithelial A549 cells. ICAM-1 gene and protein expression along with monocyte adhesion to inflammatory A549 cells were suppressed by the flavonoid. *Phl* modified different signaling cascades causing inhibition of phosphorylation of Akt and MAPK and a reduction in nuclear translocation of NF-κB p65. *Phl* might exert an anti-inflammatory effect by inhibiting the synthesis of pro-inflammatory cytokines and COX-2 and ICAM-1 expression through the blockage of NF-κB and MAPK signaling pathways (Huang et al. 2015).

The chalcone (E)-3-(3,4-dimethoxyphenyl)-1-(5-hydroxy-2,2-dimethyl-2H-chromen-6-yl) prop-2-en-1-one (*5B*) has been demonstrated to reduce carrageenan-induced mouse foot edema and adjuvant-induced arthritis. In addition, the antiarthritic effects of *5B* have been evaluated in a collagen-induced arthritis in vivo model, while to investigate molecular mechanisms involved in the anti-inflammatory effects, the RAW 264.7 cell line was used. The pretreatment with *5B* prevented the advance of arthritis together with the blockade of the recruitment of $CD68^+$ cells in the knee joint. Moreover, a decrease in the secretion of TNF-α, IL-1β and IL-6 was observed. In LPS-stimulated macrophages, *5B* suppressed the expression of iNOS, COX-2, TNF-α, IL-6, IL-1β, NO and PGE_2. Besides, *5B* suppressed the activation of NF-κB induced by LPS; the latter effect was achieved by a modulation of IκB phosphorylation, the degradation of IκB and the nuclear translocation of p65 and p50. Likewise, *5B* suppressed the expression of TLR4 induced by LPS, the degradation of IL-1 receptor-associated kinase (IRAK) and the phosphorylation of JNK and ERK, but it had little positive effect on the activation of p38 kinase. Thus, *5B* could be a potential agent against rheumatoid arthritis, since its anti-inflammatory effect was found to be mediated by the TLR4, NF-κB and ERK/JNK signaling pathways in monocytes (Li et al. 2013).

Flavokawain A (2′-hydroxy-4,4′,6′-trimethoxychalcone, *FlkA*) is a chalcone derivative isolated from kava (*Piper methysticum*) extracts, which have been used as popular beverage in the Pacific islands. The suppressive effect of *FlkA* on the expression of pro-inflammatory mediators in LPS-stimulated macrophages and the molecular mechanisms responsible for these activities have been evaluated. *FlkA* inhibited the expression of iNOS and COX-2, together with the production of NO and PGE_2 in LPS-stimulated RAW 264.7 cells. The activation of the NF-κB and AP-1 signaling pathways were negatively affected when the cells were treated with *FlkA*. In the same experimental model, this flavonoid also attenuated the activation of JNK and p38 MAPK, which are responsible for the expression of iNOS and COX-2. In addition, *FlkA* blocked the expression of pro-inflammatory cytokines, such as TNF-α, IL-1β and IL-6. These findings allowed concluding that *FlkA* modulates the expression of pro-inflammatory mediators through NF-kB, AP-1 and JNK/p38 MAPK signaling pathways (Kwon et al. 2013). Another natural chalcone, *licochalcone C* ((2E)-3- (4-hydroxy-2-methoxy-3- (3-methyl-2-butenyl) phenyl)-1-(4-hydroxyphenyl) -2-propen-1-one, *LicoC*), has been found to inhibit NF-κB translocation and the generation of pro-inflammatory mediators, such as iNOS, ICAM-1 and VCAM-1. Furthermore, *LicoC* stimulated the phosphorylation of PI3K/Akt/eNOS with the consequent activation of the signaling pathway. As the protective effect of *LicoC* could be blocked with a specific PI3K inhibitor, the presence of this compound would be essential in the sepsis-induced inflammation (Franceschelli et al. 2017). The effects of flavonoids on intracellular signaling pathways and mediators associated with inflammation are summarized in Table 1.

5 Studies Performed in Humans

Studies assessing the evaluation of the effects of flavonoids in inflammation performed in either healthy human volunteers or in patients are scarce, as compared to in vitro and in vivo assays. Most of the studies have consisted in the administration of foods such as tea, fruit juices, grape extracts and red wine containing a mixture of flavonoids. Other studies evaluate the activity of pure polyphenolic compounds. Ribeiro et al. (2015) have reviewed the studies published before 2014. More recent research works include a systematic review carried out by Rangel-Huerta et al. (2015). In that review, authors examine the efficacy of phenolic compounds in cardiovascular diseases. Seventy-two articles were selected in which randomized

Table 1 Effects of flavonoids on intracellular signaling pathways and mediators associated with inflammation

Compound	Mechanism of action: Signaling pathway	Mechanism of action: Mediators	Reference
Flavones			
Chrisin	↓ NF-κB	↓ COX-2; ↓ iNOS; ↓ IL-1β; ↓ IL-6; ↓ IL-12; ↓ IL-1α; ↓ IL-17A; ↓ IFN-γ; ↓ TNF-α	Yao et al. (2014)
	↓ phosphorylation of Akt and ERK	↓ IL-4; ↓ IL-12; ↑ IFN-γ	Yao et al. (2016)
	↓ NF-κB	↓ TNF-α; ↓ SOD; ↓ eNOS; ↓ NO	Rani et al. (2016)
	↓ NF-κB	↓ IL-8; ↓ IL-6; ↓ MCP-1; ↓ PGE_2	During and Larondelle (2013)
Luteolin	↓ AP-1; ↓ NF-κB; ↓ MAPK; ↓ STAT-3	↓ HIF-1α; ↓ COX-2; ↓ iNOS; ↓ IL-1β; ↓ IL-6; ↓ IL-8; ↓ TNF-α; ↓ VEGF	Pratheeshkumar et al. (2014)
	↓ NF-κB	↓ MCP-1; ↓ VCAM-1; ↓ ICAM-1	Jia et al. (2015)
	↓ NF-κB; ↓ MAPK	↓ COX-2; ↓ TNF-α; ↓ IL-1β; ↓ IL-6; ↓ IL-8	Zhang et al. (2017)
Vitexin	↓HMGB1; ↓(p38) MAPK	↓ TNF-α	Wang et al. (2017)
Amentoflavone	↓ NF-κB (p65–50)	↓TNF-α; ↓ IL-1β; ↓ IL-6; ↓ COX-2; ↓ iNOS	Sakthivel and Guruvayoorappan (2013)
Wogonin	↓TLR4, ↓ NF-κB	↓MMP-9; ↓IL-1β; ↓IL-6; ↓IPM-2; ↓COX-2	Chen et al. (2012)
Velutin	↓ NF-κB; ↓ (p38) MAPK; ↓ JNK	↓TNF-α; ↓ IL-6	Xie et al. (2012)
Apigenin	↓Akt; ↓p38 MAPK	↓ TNF-α; ↓ MMP-9; ↓ eNOS	Palmieri et al. (2012)
	–	↓ IL17; ↓ LTA	Johnson and De Mejia (2013)
	↓MAPK; ↓NF-κB	↓ TNF-α; ↓ TGF-β1	Malik et al. (2017)
Tricin	–	↓TNF-α; ↓IL-6; ↓iNOS; ↓COX-2	Shalini et al. (2012)
Eupatilin	↓Akt; ↓NF-κB	↓ TNF-α; ↓ VCAM-1; ↓ ICAM-1	Jung et al. (2012)

(continued)

Table 1 (continued)

Compound	Mechanism of action		Reference
	Signaling pathway	Mediators	
3,5,7-trihydroxy-4′-methoxy-8-(3-hydroxy-3-methylbutyl)-flavone	↓ CD14 / TLR4	–	Wu et al. (2012)
4'-Hydroxywogonin	↓ TAK1 / IKK / NF-κB; ↓ PI3K / Akt	↓ iNOS; ↓ PGE2; ↓ TAK1; ↓ TAB1	Fan et al. (2017)
4′-bromo-5,6,7-trimethoxyflavone	↓NF-κB	↓ iNOS; ↓ COX-2; ↓ TNF-α, ↓ IL-6, ↓ IL-1β; ↓ NO, ↓ PGE_2	Kim et al. (2012)
Oroxylin A	↓TLR4; ↓ MAPKs; ↓ NF-κB	↓ JNK; ↓ p38 NF-κB; ↓ ERK	Song et al. (2012)
Flavonols			
Fisetin	↓ PI3K; ↓ phosphorylation of Akt; ↓ NF-κB	↓ COX-2; ↓ PGE_2; ↓ MPO; ↓ IL-1β; ↓ IL-6; ↓ TNF-α	Pal et al. (2015)
	↓ NF-κB; ↓ JNK	↓ IL-1β; ↓ IL-6; ↓ TNF-α	Jo et al. (2014)
	↓ ERK; ↓ JNK; ↓ MAPK (p38)	↓ COX-2; ↓ PGE_2	Gutiérrez-Venegas et al. (2014)
	↓ NF-κB; ↓ TLR-4	↓ IL-1β; ↓ TNF-α	Zhou et al. (2015a)
	↓ NF-κB; ↓ ERK; ↓ Akt	↓ IL-1β; ↓ TNF-α	Yoo et al. (2014)
	↓ NF-κB; ↓ JNK; ↓ MAPK (p38)	↓ COX-2; ↓ iNOS; ↓ PGE_2; ↓ TNF-α	Maher (2015)
	↑ PI3K/Akt; ↑ MAPK	↓ iNOS; ↓ NO; ↓ IL-1β; ↑ HO-1	Chuang et al. (2014)
Icariin	↓ ERK; ↓ MAPK (p38)	↓ IL-1β; ↓ IL-6; ↓ IL-8; ↓ MCP-1; ↓ ICAM-1	Kong et al. (2015)
	↓ NF-κB	↓ IL-1β; ↓ TNF-α; ↓ iNOS	Liu et al. (2015)
	↓ NF-κB	↓ IL-2; ↓ IL-6; ↓ TNF-α; ↓ ICAM-1	Chen et al. (2015)

Astragalin	↓ NF-κB; ↓ ERK; ↓ TLR-4	↓ IL-6; ↓ TNF-α; ↓ NO; ↓ COX-2; ↓ iNOS	Li et al. (2014a)
Rutin	↓ NF-κB	↓ TNF-α	Lee et al. (2013)
	↓ NF-κB	↓ TNF-α; ↓ IL-6	Yoo et al. (2013)
Flavanones			
Alpinetin	↓ NF-κB; ↓ MAPK	↓ TNF-α; ↓ IL-6; ↓ IL-1β; ↓ IκB-α; ↓ NF-κB p65; ↓ p38	Huo et al. (2012)
	↓ NF-κB	↓ TNF-α; ↓ IL-6; ↓ IL-1β; ↓ IκB-α; ↓ NF-κB p65; ↓ TLR4	Chen et al. (2013)
	↓ NF-κB; ↓ MAPK	↓ TNF-α; ↓ IL-6; ↓ IL-1β; ↓ IκBα; ↓ ERK; ↓ JNK; ↓ p38; ↓ p65	Hu et al. (2013)
Sophoraflavanone G	↓ NF-κB; ↓ MAPK;	↓ IL-1β, ↓ IL-6; ↓ TNF-α; ↓iNOS; ↓COX-2; ↓ ERK; ↓ JNK; ↓ p38	Wun et al. (2013)
Ugonin M	↓ NF-κB; ↓ MAPK (p38)	↓NO, ↓TNF-α, ↓ IL-1β; ↓ IL-6; ↓ iNOS; ↓ COX-2; ↓ IκB-α; ↓ ERK; ↓ TLR	Wu et al. (2017)
Naringenin	↓ NF-κB; ↓ ERK; ↓ JNK; ↓ MAPK (p38)	↓ IL-6; ↓ TNF-α; ↓ NO; ↓ SOD; ↓ NOS; ↓ MPO	Yu et al. (2014)
	↑ AMPK; ↑ PKCδ	↓ COX-2; ↓ iNOS; ↓ NO; ↑ SOCS-3	Wu et al. (2016)

(continued)

Table 1 (continued)

Compound	Mechanism of action		Reference
	Signaling pathway	Mediators	
Naringin	↓ JNK; ↓ MAPK (p38)	↓ IL-1β; ↓ IL-6; ↓ IL-8; ↓ COX-2	Ren et al. (2016)
	↓ p53; ↓ NF-κB	↓ SOD; ↓ CAT; ↓ NO; ↓ TNF-α	Chtourou et al. (2015)
	↑ PPARγ	↓ IL-6; ↓ TNF-α	Qi et al. (2015)
	↓ MAPK (p38); ↓ NF-κB	–	Manna et al. (2015)
Pinocembrin	↓ MAPK; ↓ NF-κB	↓ IL-1β; ↓ IL-6; ↓ TNF-α	Liu et al. (2014b)
	↓ PI3K; ↓ phosphorylation of Akt; ↓ NF-κB; ↑ Nrf2	↓ IL-1β; ↓ TNF-α; ↓ NO; ↓ PGE_2; ↓ COX-2; ↓ iNOS; ↑ HO-1	Zhou et al. (2015b)
Flavanonols			
Taxifolin	↑ Nrf2; ↓ NF-κB; ↓ Wnt	↓ TNF-α; ↓ COX-2	Manigandan et al. (2015)
2′-hydroxy yokovanol / 2′-hydroxy neophellamuretin	–	↓ IL-12; ↓ IL-6; ↓ TNF-α	Li et al. (2014c)
Ampelopsin	↓ NF-κB; ↓ MAPK; ↓ROS; ↓Akt; ↓IKK	↓NO, ↓TNF -α, ↓ IL-1β; ↓ IL-6; ↓ iNOS; ↓ NF-κB p65; ↓KKα / β; ↓IκBα	Qi et al. (2012)

Flavan-3-ols			
Epigallocatechin-3-gallate	↓ TLR-4; ↑ Tollip	–	Byun et al. (2014a)
	↓ NF-κB	↓ IL-6; ↓ TNF-α	Bao et al. (2015)
	↓ phosphorylation of Akt; ↓ ERK; ↓ JNK; ↓ MAPK (p38); ↑ PPARγ; ↑ Nrf2	–	Ye et al. (2015)
	↓ MAPK; ↓ NF-κB	↓ TNF-α; ↓ RANTES; ↓ MCP-1; ↓ ICAM-1; ↓ NO; ↓ VEGF; ↓ MMP-2	Liu et al. (2014a)
	↓ NF-κB	↓ IL-6; ↓ CRP; ↓ ICAM-1; ↓ VCAM-1; ↓ IL-1 α/β	Liu et al. (2016)
	↑ Nrf2; ↓ NF-κB	↓ IL-1β; ↓ IL-6; ↓ TNF-α; ↑ HO-1	Wang et al. (2015b)
	Notch-1	↓ IL-6; ↓ TNF-α; ↓ ICAM-1; ↓ MCP-1	Xie et al. (2015)
	↓ NF-κB; ↓ TGF/SMAD; ↓ PI3K/Akt	↓ TNF-α; ↓ COX-2; ↓ iNOS	Xiao et al. (2014)
Catechin	↓ Akt; ↓ ERK; ↓ NF-κB; ↓ MAPK (p38); ↓ AMPK	↓ COX-2; ↓ iNOS; ↓ ROS; ↓ NO; ↓ IL-6; ↓ TNF-α; ↑ IL-4	Hussein et al. (2015)
Anthocyanidins			
Cyanidin-3-glucoside	↓ NF-κB; ↓ ERK; ↓ JNK; ↓ MAPK (p38)	↓ IL-1β; ↓ IL-6; ↓ TNF-α	Ma et al. (2015)
	↓ NF-κB; ↑ Nrf2	–	Fratantonio et al. (2015)
	↓ NF-κB	↓ TNF-α	Ferrari et al. (2017)
	↓ MAPK (p38); ↓ Akt	↓ p38, ↓ ERK; ↓ JNK	He et al. (2017)

(continued)

Table 1 (continued)

Compound	Mechanism of action		Reference
	Signaling pathway	Mediators	
Peonidin Malvidin	↓ TLR-2	↓ IL-1β; ↓ IL-6; ↓ IL-8; ↓ TNF-α; ↓ MCP-1; ↓ COX-2	Mackert and McIntosh (2016)
Pelargonidin	↓ NF-κB; ↓ ERK	↓ IL-1α; ↓ IL-1β; ↓ IL-6; ↓ TNF-α	Min et al. (2016)
Malvidin	↓ NF-κB; ↓ MAPK; ↓ROS	↓ MKP-1; ↓ Akt	Bognar et al. (2013)
Malvidin 3O-Glucoside	↓ NF-κB; ↓Akt	↓ iNOS; ↓ COX-2; ↓ IL-6	Paixão et al. (2012)
Procyanidin trimer *C1*	↓ NF-κB; ↓ MAPK	↓ IL-6; ↓ TNF-α; ↓ PGE_2; ↓ COX-2; ↓ TLR4	Byun et al. (2013)
Isoflavonoids			
Genistein	↓ NF-κB; ↑ AMPK	↓ IL-1β; ↓ IL-6; ↓ IL-8; ↓ TNF-α	Li et al. (2014b)
	↓ NF-κB; ↓ TLR4	↓ IL-1β; ↓ iNOS; ↑ IL-10	Zhou et al. (2014)
GEN-27	↓ NF-κB	↓ IL-1; ↓ IL-6	Wang et al. (2016)
Puerarin	–	↓ IL-4; ↓ IL-5; ↓ IL-13; ↑ IFN-γ	Wang et al. (2015a)
Daidzein	↓ JNK; ↑ PPAR-α/−γ	↓ IL-6; ↓ MCP-1	Sakamoto et al. (2016)
Ononin	↓ NF-κB; ↓ MAPK	↓ COX-2; ↓ iNOS; ↓ IL-6; ↓ TNF-α; ↓ IL-1β	Dong et al. (2017)
Prunetin	↓ NF-κB; ↓ MAPK	↓ COX-2; ↓ iNOS; ↓ IL-6; ↓ TNF-α; ↓ IL-1β	Yang et al. (2013)
Chalcones			
L2H17	↓ MAPK; ↓ NF-κB	↓ IL-6; ↓ TNF-α; ↓ ICAM-1; ↓ VCAM-1	Fang et al. (2015a)
	↓ MAPK; ↓ NF-κB	↓ IL-6; ↓ TNF-α; ↓ ICAM-1; ↓ VCAM-1; ↓ IL-1β; ↓ IL-2; ↓ IL-12; ↓ IFN-γ	Fang et al. (2015b)

Isobavachalcone	↓ NF-κB	↓ ICAM-1; ↓ IFN-β	Lee et al. (2015)
	↓TLR4	↓ iNOS; ↓ MALP-2 (TLR2 and TLR6); ↓ poly [I: C]	Shin et al. (2013)
3′,3,4,5-tetramethoxy-4′-hydroxychalcone	↓ STAT-1	↓ NO	Hara et al. (2014)
3′,3,4,5-tetramethoxychalcone	↓ STAT-1; ↓ ERK	↓ NO	Hara et al. (2014)
Phloretin	↓ MAPK; ↓ NF-κB; ↓ phosphorylation of Akt	↓ IL-6; ↓ IL-8; ↓ IL-1β; ↓ MCP-1; ↓ COX-2; ↓ ICAM-1; ↓ PGE_2	Huang et al. (2015)
(E) -3- (3,4-Dimethoxyphenyl) -1-(5-hydroxy-2,2-dimethyl-2H-chromen-6-yl) prop-2-en-1-one	↓ NF-κB; ↓ MAPK	↓ IL-6; ↓ TNF-α; ↓ IL-1β; ↓ PGE_2; ↓ COX-2; ↓ iNOS; ↓ ERK; ↓ JNK	Li et al. (2013)
Flavokawain A	↓ NF-κB; ↓ AP-1; ↓JNK; ↓ p38 MAPK	↓ COX-2; ↓ iNOS; ↓ IL-6; ↓ TNF-α; ↓ IL-1β	Kwon et al. (2013)
Licochalcone C	↓ NF-κB	↓ iNOS, ↓ ICAM-1, ↓ VCAM-1	Franceschelli et al. (2017)

↓ reduces, inhibits or downregulates, ↑ increases, stimulates or up-regulates controlled trials with prospective, parallel or crossover designs in humans were included. Evidences indicate that flavonols are helpful in decreasing risk factors of cardiovascular disease, although further rigorous works are necessary to support that hypothesis. Cassidy et al. (2015) have conducted a study in a population of adults in the United States to assess if higher dietary flavonoid (anthocyanins, flavonols, flavanones, flavan-3-ols, polymers and flavones) intakes are associated with anti-inflammatory effects. The authors used an inflammation score that integrated 12 individual inflammatory biomarkers, which included CRP, TNF-α, IL-6, MCP-1 and MPO, among others. The authors concluded that there are evidences suggesting that the anti-inflammatory effect may be the central component underlying the reduction of risk of certain chronic diseases associated with higher intakes of anthocyanins and flavonols. The effects of (-)-epicatechin and quercetin-3-glucoside on some biomarkers of endothelial dysfunction and inflammation have been evaluated in a randomized double-blind, placebo-controlled, crossover trial in (pre)hypertensive adults. Results have shown that diet supplementation with pure epicatechin (100 mg/d) for a period of 4 weeks decreased soluble E-selectin levels, which is a marker of endothelial dysfunction. Supplementation with quercetin-3-glucoside (160 mg/d), during the same period, significantly decreased the levels of soluble E-selectin and IL-1β and the z score for inflammation (Dower et al. 2015). Recently, Javadi et al. (2017) have assessed the effects of *Quer* supplementation (500 mg/day, 8 weeks) on inflammatory factors and clinical symptoms. The study was a randomized, double-blind, placebo-controlled clinical trial of women with rheumatoid arthritis. The authors concluded that symptoms, including pain, early morning stiffness, disease activity and health assessment questionnaire score, were improved following *Quer* supplementation and demonstrated that *Quer* decreased TNF-α levels, possibly through suppression of cytokine gene expression. Kokkou et al. (2016) have carried out a study to evaluate the impact of grape juice supplementation on smoking-induced inflammatory processes and fibrinolytic impairment. The study has had a randomized, placebo-controlled, double-blind, cross-over design in which 26 healthy smokers received a 2-week oral treatment. Serum levels of ICAM-1 and plasminogen activator inhibitor 1 (PAI-1) were measured as markers of inflammatory and fibrinolytic status, respectively. The treatment with grape juice improved inflammatory and fibrinolytic status in healthy smokers and attenuated the acute smoking-induced increase of ICAM-1 and PAI-1 levels.

6 Conclusions

A high dietary flavonoid intake has been associated with a reduced risk and prevalence of cardiovascular and other inflammation-related diseases. Thus, over the past 10–15 years, research on flavonoids has received much attention in order to investigate their potential as new therapeutic drugs to treat these inflammatory disorders. Flavonoids have many advantages, as compared to synthetic drugs. These include fewer side effects and the fact that they are widely distributed in foods. Besides, they are readily absorbed in the intestine. As shown herein, the anti-inflammatory activity of flavonoids involves modulation of pro-inflammatory mediators through different intracellular pathways displaying a multitarget anti-inflammatory action. Research on this type of natural compounds has been carried out with the different classes of flavonoids; however, most of the studies are in vitro assays or animal models. According to the presented data, flavonoids could be considered candidates to proceed to the next phase in the drug development process. To date, human studies are scarce, but they provide some evidence of the efficacy of flavonoids as potential anti-inflammatory agents. Therefore, further well-designed in vivo experiments, along with good quality clinical studies, are needed to obtain conclusive results to determine if the findings obtained in vitro can be extrapolated to in vivo systems.

References

Agati G, Azzarello E, Pollastri S et al (2012) Flavonoids as antioxidants in plants: location and functional significance. Plant Sci 196:67–76

Bakhtiari M, Panahi Y, Ameli J et al (2017) Protective effects of flavonoids against Alzheimer's disease-related neural dysfunctions. Biomed Pharmacother 93:218–229

Bao S, Cao Y, Zhou H et al (2015) Epigallocatechin gallate (EGCG) suppresses lipopolysaccharide-induced toll-like receptor 4 (TLR4) activity via 67 kDa laminin receptor (67LR) in 3T3-L1 adipocytes. J Agric Food Chem 63:2811–2819

Bertics PJ, Koziol-White CJ, Gavala MI et al (2014) Signal transduction. In: Adkinson NF Jr, Bochner BS, Burks AW, Busse WW, Holgate ST, Lemanske RF, O'Hehir RE (eds) Middleton's allergy: principles and practice, 8th edn. Elsevier Saunders, Philadelphia, pp 184–202

Bode AM, Dong Z (2013) Signal transduction and molecular targets of selected flavonoids. Antioxid Redox Signal 19(2):163–180

Bognar E, Sarszegi Z, Szabo A et al (2013) Antioxidant and anti-inflammatory effects in RAW 264.7 macrophages of malvidin, a major red wine polyphenol. PLoS ONE 8(6):e65355

Byun EB, Sung NY, Byun EH et al (2013) The procyanidin trimer C1 inhibits LPS-induced MAPK and NF-κB signaling through TLR4 in macrophages. Int Immunopharmacol 15(2):450–456

Byun EB, SoYang M, Kim JH et al (2014) Epigallocatechin-3-gallate-mediated Tollip induction through the 67-kDa laminin receptor negatively regulating TLR4 signaling in endothelial cells. Immunobiology 219:866–872. https://doi.org/10.1016/j.imbio.2014.07.010

Cargnello M, Roux PP (2011) Activation and function of the MAPKs and their substrates, the MAPK-activated protein kinases. Microbiol Mol Biol Rev 75(1):50–83

Cassidy A, Rogers G, Peterson JJ et al (2015) Higher dietary anthocyanin and flavonol intakes are associated with anti-inflammatory effects in a population of US adults. Am J Clin Nutr 102(1):172–181

Chen CC, Hung TH, Wang YH (2012) Wogonin improves histological and functional outcomes, and reduces activation of TLR4/NF-κB signaling after experimental traumatic brain injury. PLoS One 7(1):e30294

Chen H, Mo X, Yu J et al (2013) Alpinetin attenuates inflammatory responses by interfering toll-like receptor 4/nuclear factor kappa B signaling pathway in lipopolysaccharide-induced mastitis in mice. Int Immunopharmacol 17(1):26–32

Chen Z, Zheng S, Li L et al (2014) Metabolism of flavonoids in human: a comprehensive review. Curr Drug Metab 15:48–61

Chen Y, Sun T, Wu J et al (2015) Icariin intervenes in cardiac inflammaging through upregulation of sirt6 enzyme activity and inhibition of the NF-kappa B pathway. Biomed Res Int 2015:1–12

Chen L, Teng H, Jia Z et al (2017) Intracellular signaling pathways of inflammation modulated by dietary flavonoids: the most recent evidence. Crit Rev Food Sci Nutr 6:1–17
Chtourou Y, Aouey B, Kebieche M et al (2015) Protective role of naringin against cisplatin induced oxidative stress, inflammatory response and apoptosis in rat striatum via suppressing ROS-mediated NF-κB and P53 signaling pathways. Chem Biol Interact 239:76–86
Chuang JY, Chang PC, Shen YC et al (2014) Regulatory effects of fisetin on microglial activation. Molecules 19:8820–8839
Cines DB, Pollak ES, Buck CA et al (1998) Endothelial cells in physiology and in the pathophysiology of vascular disorders. Blood 91(10):3527–3561
Commenges D, Scotet V, Renaud S et al (2000) Intake of flavonoids and risk of dementia. Eur J Epidemiol 16(4):357–363
Cooper GM (2000) The cell: a molecular approach, 2nd edn. Sinauer Associates, Sunderland
Day AJ, Canada FJ, Diaz JC et al (2000) Dietary flavonoid and isoflavone glycosides are hydrolysed by the lactase site of lactase phlorizin hydrolase. FEBS Lett 468(2–3):166–170
Dong L, Yin L, Zhang Y et al (2017) Anti-inflammatory effects of ononin on lipopolysaccharide-stimulated RAW 264.7 cells. Mol Immunol 83:46–51
Dower JI, Geleijnse JM, Gijsbers L et al (2015) Supplementation of the pure flavonoids epicatechin and quercetin affects some biomarkers of endothelial dysfunction and inflammation in (pre)hypertensive adults: a randomized double-blind, placebo-controlled, crossover trial. J Nutr 145(7):1459–1463
During A, Larondelle Y (2013) The O-methylation of chrysin markedly improves its intestinal anti-inflammatory properties: structure-activity relationships of flavones. Biochem Pharmacol 86(12):1739–1746
Fan C, Wu LH, Zhang GF et al (2017) 4′-Hydroxywogonin suppresses lipopolysaccharide-induced inflammatory responses in RAW 264.7 macrophages and acute lung injury mice. PLoS One 12(8):e0181191. https://doi.org/10.1371/journal.pone.0181191
Fang Q, Deng L, Wang L et al (2015a) Inhibition of mitogen-activated protein kinases/nuclear factor κB-dependent inflammation by a novel chalcone protects the kidney from high fat diet-induced injuries in mice. J Pharmacol Exp Ther 355:235–246
Fang Q, Wang J, Wang L et al (2015b) Attenuation of inflammatory response by a novel chalcone protects kidney and heart from hyperglycemia-induced injuries in type 1 diabetic mice. Toxicol Appl Pharmacol 288:179–191
Feghali CA, Wright TM (1997) Cytokines in acute and chronic inflammation. Front Biosci 2:12–26
Ferrari D, Cimino F, Fratantonio D et al (2017) Cyanidin-3-O-glucoside modulates the in vitro inflammatory crosstalk between intestinal epithelial and endothelial cells. Mediat Inflamm 2017:3454023
Fink BN, Steck SE, Wolff MS et al (2007) Dietary flavonoid intake and breast cancer risk among women on Long Island. Am J Epidemiol 165(5):514–523
Firestein GS (2012) Mechanisms of inflammation and tissue repair. In: Goldman L, Schafer A (eds) Goldman's Cecil medicine, vol 24. Elsevier Saunders, Philadelphia, pp 230–235
Formica JV, Regelson W (1995) Review of the biology of quercetin and related bioflavonoids. Food Chem Toxicol 33:1061–1080
Franceschelli S, Pesce M, Ferrone A et al (2017). Biological effect of licochalcone C on the regulation of PI3K/Akt/eNOS and NF-κB/iNOS/NO signaling pathways in H9c2 cells in response to LPS stimulation. Int J Mol Sci 18(4):pii:E690.
Frankenfeld CL, Cerhan JR, Cozen W et al (2008) Dietary flavonoid intake and non-Hodgkin lymphoma risk. Am J Clin Nutr 87(5):1439–1445
Fratantonio D, Speciale A, Ferrari D et al (2015) Palmitate-induced endothelial dysfunction is attenuated by cyanidin-3-O-glucoside through modulation of Nrf2/Bach1 and NF-κB pathways. Toxicol Lett 239:152–160
García-Lafuente A, Guillamón E, Villares A et al (2009) Flavonoids as anti-inflammatory agents: implications in cancer and cardiovascular disease. Inflamm Res 58(9):537–552

Gerd B, Leah BS, Paul SA et al (2008) Dietary flavonoids and colorectal adenoma recurrence in the polyp prevention trial. Cancer Epidemiol Biomark Prev 17(6):1344–1353

Gleichenhagen M, Schieber A (2016) Current challenges in polyphenol analytical chemistry. Curr Opin Food Sci 7:43–49

Gutiérrez-Venegas G, Contreras-Sánchez A, Ventura-Arroyo JA (2014) Anti-inflammatory activity of fisetin in human gingival fibroblasts treated with lipopolysaccharide. J Asian Nat Prod Res 16:1009–1017

Hämäläinen M, Nieminen R, Vuorela P et al (2007) Anti-inflammatory effects of flavonoids: genistein, kaempferol, quercetin, and daidzein inhibit STAT-1 and NF-kappa B activations, whereas flavone, isorhamnetin, naringenin, and pelargonidin inhibit only NF-kappa B activation along with their inhibitory effect on iNOS expression and NO production in activated macrophages. Mediat Inflamm 2007:45673

Hara H, Ikeda R, Ninomiya M et al (2014) Newly synthesized “Hidabeni” chalcone derivatives potently suppress LPS-induced NO production via inhibition of STAT1, but not NF-κB, JNK, and p38, pathways in microglia. Biol Pharm Bull 37:1042–1049

Hassan S, Mathesius U (2012) The role of flavonoids in root-rhizosphere signaling: opportunities and challenges for improving plant-microbe interactions. J Exp Bot 63:3429–3444

He Y, Hu Y, Jiang X et al (2017) Cyanidin-3-O-glucoside inhibits the UVB-induced ROS/COX-2 pathway in HaCaT cells. J Photochem Photobiol B 177:24–31

Hollman PCH (2004) Absorption, bioavailability, and metabolism of flavonoids. Pharm Biol 42:74–83

Hu K, Yang Y, Tu Q et al (2013) Alpinetin inhibits LPS-induced inflammatory mediator response by activating PPAR-γ in THP-1-derived macrophages. Eur J Pharmacol 721(1–3):96–102

Huang WC, Wu SJ, Tu RS et al (2015) Phloretin inhibits interleukin-1β-induced COX-2 and ICAM-1 expression through inhibition of MAPK, Akt, and NF-κB signaling in human lung epithelial cells. Food Funct 6:1960–1967

Huo M, Chen N, Chi G (2012) Traditional medicine alpinetin inhibits the inflammatory response in Raw 264.7 cells and mouse models. Int Immunopharmacol 2(1):241–248

Hussein SSS, Kamarudin MNA, Kadir HA (2015) (+)-Catechin attenuates NF-κB activation through regulation of Akt, MAPK, and AMPK signaling pathways in LPS-induced BV-2 microglial cells. Am J Chin Med 43:927–952

Iiyama K, Hajra L, Iiyama M et al (1999) Patterns of vascular cell adhesion molecule-1 and intercellular adhesion molecule-1 expression in rabbit and mouse atherosclerotic lesions and at sites predisposed to lesion formation. Circ Res 85(2):199–207

Jaeger BN, Parylak SL, Gage FH (2017) Mechanisms of dietary flavonoid action in neuronal function and neuroinflammation. Mol Asp Med S0098-2997(17):30111–30115

Javadi F, Ahmadzadeh A, Eghtesadi S et al (2017) The effect of quercetin on inflammatory factors and clinical symptoms in women with rheumatoid arthritis: a double-blind, randomized controlled trial. J Am Coll Nutr 36(1):9–15

Jia Z, Nallasamy P, Liu D et al (2015) Luteolin protects against vascular inflammation in mice and TNF-alpha-induced monocyte adhesion to endothelial cells via suppressing IKBα/NF-κB signaling pathway. J Nutr Biochem 26:293–302

Jo IJ, Bae GS, Choi SB et al (2014) Fisetin attenuates cerulein-induced acute pancreatitis through down regulation of JNK and NF-κB signaling pathways. Eur J Pharmacol 737:149–158

Johnson JL, de Mejia EG (2013) Flavonoid apigenin modified gene expression associated with inflammation and cancer and induced apoptosis in human pancreatic cancer cells through inhibition of GSK-3β/NF-κB signaling cascade. Mol Nutr Food Res 57:2112–2127

Jung J, Ko SH, Yoo DY et al (2012) 5,7-Dihydroxy-3,4,6-trimethoxyflavone inhibits intercellular adhesion molecule 1 and vascular cell adhesion molecule 1 via the Akt and nuclear factor-κB-dependent pathway, leading to suppression of adhesion of monocytes and eosinophils to bronchial epithelial cells. Immunology 137(1):98–113

Kaminska B (2005) MAPK signalling pathways as molecular targets for anti-inflammatory therapy—from molecular mechanisms to therapeutic benefits. Biochim Biophys Acta 1754:253–262

Kappelmann M, Bosserhoff A, Kuphal S (2014) AP-1/c-Jun transcription factors: regulation and function in malignant melanoma. Eur J Cell Biol 93(1–2):76–81

Kim HP, Son KH, Chang HW et al (2004) Anti-inflammatory plant flavonoids and cellular action mechanisms. J Pharm Sci 96(3):229–245

Kim JH, Na HJ, Kim CK et al (2008) The non-provitamin A carotenoid, lutein, inhibits NF-κB-dependent gene expression through redox-based regulation of the phosphatidylinositol 3-kinase/PTEN/Akt and NF-κB-inducing kinase pathway: role of H_2O_2 in NF-κB activation. Free Radic Biol Med 45(6):885–896

Kim DH, Yun CH, Kim MH et al (2012) 4′-Bromo-5,6,7-trimethoxyflavone represses lipopolysaccharide-induced iNOS and COX-2 expressions by suppressing the NF-kB signaling pathway in RAW 264.7 macrophages. Bioorg Med Chem Lett 22(1):70070–70075

Knekt P, Jarvinen R, Reunanen A et al (1996) Flavonoid intake and coronary mortality in Finland: a cohort study. BMJ 312:478–481

Kokkou E, Siasos G, Georgiopoulos G et al (2016) The impact of dietary flavonoid supplementation on smoking-induced inflammatory process and fibrinolytic impairment. Atherosclerosis 251:266–272

Komatsu W, Itoh K, Akutsu S et al (2017) Nasunin inhibits the lipopolysaccharide-induced pro-inflammatory mediator production in RAW264 mouse macrophages by suppressing ROS-mediated activation of PI3 K/Akt/NF-κB and p38 signaling pathways. Biosci Biotechnol Biochem 81:1956–1966

Kong L, Liu J, Wang J et al (2015) Icariin inhibits TNF-α/IFN-γ induced inflammatory response via inhibition of the substance P and p38-MAPK signaling pathway in human keratinocytes. Int Immunopharmacol 29:401–407

Kretzmann NA, Fillmann H, Mauriz JL et al (2008) Effects of glutamine on pro-inflammatory gene expression and activation of nuclear factor kappa B and signal transducers and activators of transcription in TNBS-induced colitis. Inflamm Bowel Dis 14(11):1504–1513

Kumar S, Pandey AK (2013) Chemistry and biological activities of flavonoids: an overview. Sci World J 2013:162750

Kumar V, Abbas AK, Aster JC (2013) Inflammation and repair. In: Robbins basic pathology, 9th edn. Elsevier Saunders, Philadelphia, pp 29–73

Kuriyama S, Shimazu T, Ohmori K et al (2006) Green tea consumption and mortality due to cardiovascular disease, cancer, and all causes in Japan: the Ohsaki study. JAMA 296(10):1255–1265

Kwon DJ, Ju SM, Youn GS (2013) Suppression of iNOS and COX-2 expression by flavokawain A via blockade of NF-κB and AP-1 activation in RAW 264.7 macrophages. Food Chem Toxicol 58:479–486

Lawrence T (2009) The nuclear factor NF-kappaB pathway in inflammation. Cold Spring Harb Perspect Biol 1(6):a001651

Lee W, Ku SK, Bae JS (2013) Barrier protective effects of rutin in LPS-induced inflammation in vitro and in vivo. Food Chem Toxicol 50(9):3048–3055

Lee KM, Kim JM, Baik EJ et al (2015) Isobavachalcone attenuates lipopolysaccharide-induced ICAM-1 expression in brain endothelial cells through blockade of toll-like receptor 4 signaling pathways. Eur J Pharmacol 754:11–18

Legeay S, Rodier M, Fillon L et al (2015) Epigallocatechin gallate: a review of its beneficial properties to prevent metabolic syndrome. Nutrients 7:5443–5468

Letenneur L, Proust-Lima C, Le Gouge A et al (2007) Flavonoid intake and cognitive decline over a 10-year period. Am J Epidemiol 165(12):1364–1371

Leyva-López N, Gutierrez-Grijalva EP, Ambriz-Perez DL et al (2016) Flavonoids as cytokine modulators: a possible therapy for inflammation-related diseases. Int J Mol Sci 17(6):921

Li X, Peng F, Xie C et al (2013) (E)-3-(3,4-Dimethoxyphenyl)-1-(5-hydroxy-2,2-dimethyl-2H-chromen-6-yl)prop-2-en-1-one ameliorates the collagen-arthritis via blocking ERK/JNK and NF-κB signaling pathway. Int Immunopharmacol 17(4):1125–1133

Li F, Wang W, Cao Y et al (2014a) Inhibitory effects of astragalin on lipopolysaccharide-induced inflammatory response in mouse mammary epithelial cells. J Surg Res 192:573–581
Li J, Li J, Yue Y et al (2014b) Genistein suppresses tumor necrosis factor α-induced inflammation via modulating reactive oxygen species/Akt/nuclear factor κB and adenosine monophosphate-activated protein kinase signal pathways in human synoviocyte MH7A cells. Drug Des Devel Ther 8:315–323
Li W, Sun YN, Yan XT et al (2014c) Anti-inflammatory and antioxidant activities of phenolic compounds from *Desmodium caudatum* leaves and stems. Arch Pharm Res 37:721–727
Liu Q, Qian Y, Chen F et al (2014a) EGCG attenuates pro-inflammatory cytokines and chemokines production in LPS-stimulated L02 hepatocyte. Acta Biochim Biophys Sin Shanghai 46:31–39
Liu R, Li J, Song J et al (2014b) Pinocembrin protects human brain microvascular endothelial cells against fibrillar amyloid-ß1 40 injury by suppressing the MAPK/NF-κB inflammatory pathways. Biomed Res Int 2014:1–14
Liu B, Xu C, Wu X et al (2015) Icariin exerts an antidepressant effect in an unpredictable chronic mild stress model of depression in rats and is associated with the regulation of hippocampal neuroinflammation. Neuroscience 294:193–205
Liu D, Perkins JT, Hennig B (2016) EGCG prevents PCB 126-induced endothelial cell inflammation via epigenetic modifications of NF-κB target genes in human endothelial cells. J Nutr Biochem 28:164–170
Lu YC, Yeh WC, Ohashi PS (2008) LPS/TLR4 signal transduction pathway. Cytokine 42(2):145–151
Ma MM, Li Y, Liu XY et al (2015) Cyanidin-3-O-Glucoside ameliorates lipopolysaccharide-induced injury both *in vivo* and *in vitro* suppression of NF-κB and MAPK pathways. Inflammation 38:1669–1682
Mackert JD, McIntosh MK (2016) Combination of the anthocyanidins malvidin and peonidin attenuates lipopolysaccharide-mediated inflammatory gene expression in primary human adipocytes. Nutr Res 36(12):1353–1360.
Maher P (2015) How fisetin reduces the impact of age and disease on CNS function. Front Biosci (Schol Ed) 7:58–82
Malik S, Suchal K, Khan S et al (2017) Apigenin ameliorates streptozotocin-induced diabetic nephropathy in rats via MAPK-NF-κB-TNF-α and TGF-β1-MAPK-fibronectin pathways. Am J Physiol Renal Physiol 313(2):F414–F422
Manigandan K, Manimaran D, Jayaraj RL et al (2015) Taxifolin curbs NF-κB-mediated Wnt/β-catenin signaling via up-regulating Nrf2 pathway in experimental colon carcinogenesis. Biochimie 119:103–112
Manna K, Das U, Das D et al (2015) Naringin inhibits gamma radiation-induced oxidative DNA damage and inflammation, by modulating p53 and NF-κB signaling pathways in murine splenocytes. Free Radic Res 49:422–439
Marín L, Miguélez EM, Villar CJ et al (2015). Bioavailability of dietary polyphenols and gut microbiota metabolism: antimicrobial properties. Biomed Res Int 2015:905215
Middleton E, Kandaswami CH (1994) The impact of plant flavonoids on mammalian biology: implications for immunity, inflammation and cancer. In: Harbone JB (ed) The flavonoids. Advances in research since 1986. Chapman and Hall, London, p 619
Min G, Ku SK, Park MS et al (2016) Anti-septic effects of pelargonidin in HMGB1-induced inflammatory responses *in vitro* and *in vivo*. Arch Pharm Res 39:1726–1738
Nathan C (1992) Nitric oxide as a secretory product of mammalian cells. FASEB J 6(12):3051–3064
Owuor ED, Kong AN (2002) Antioxidants and oxidants regulated signal transduction pathways. Biochem Pharmacol 64(5–6):765–770
Paixão J, Dinis TC, Almeida LM (2012) Malvidin-3-glucoside protects endothelial cells up-regulating endothelial NO synthase and inhibiting peroxynitrite-induced NF-κB activation. Chem Biol Interact 199(3):192–200
Pal HC, Athar M, Elmets CA et al (2015) Fisetin inhibits UVB-induced cutaneous inflammation and activation of PI3K/Akt/NF-κB signaling pathways in SKH-1 hairless mice. Photochem Photobiol 91:225–234

Palmieri D, Perego P, Palombo D (2012) Apigenin inhibits the TNF-α-induced expression of eNOS and MMP-9 via modulating Akt signalling through oestrogen receptor engagement. Mol Cell Biochem 371(1–2):129–136

Park SE, Sapkota K, Kim S et al (2011) Kaempferol acts through mitogen-activated protein kinases and protein kinase B/Akt to elicit protection in a model of neuroinflammation in BV2 microglial cells. Br J Pharmacol 164:1008–1025

Pietta PG (1998) Natural-antioxidants in nutrition, health and disease. Paper presented at the 2nd international conference on natural antioxidants and anticarcinogens, Helsinki, 24–27 June 1998

Pietta PG (2000) Flavonoids as antioxidants. J Nat Prod 63(7):1035–1042

Pratheeshkumar P, Son YO, Divya SP et al (2014) Luteolin inhibits Cr(VI)-induced malignant cell transformation of human lung epithelial cells by targeting ROS mediated multiple cell signaling pathways. Toxicol Appl Pharmacol 281:230–241

Pubchem (2017) Naringin – compound summary. https://pubchem.ncbi.nlm.nih.gov/compound/naringin#section=Top. Accessed 23 Nov 2017

Qi S, Xin Y, Guo Y et al (2012) Ampelopsin reduces endotoxic inflammation via repressing ROS-mediated activation of PI3K/Akt/NF-κB signaling pathways. Int Immunopharmacol 12(1):278–287

Qi Z, Xu Y, Liang Z et al (2015) Naringin ameliorates cognitive deficits via oxidative stress, proinflammatory factors and the PPAR-γ signaling pathway in a type 2 diabetic rat model. Mol Med Rep 12:7093–7101

Rabinovich GA, Zwirner NW, Toscano MA (2011) Regulación de la expresión génica en el sistema inmunitario. In: Fainboim L, Geffner J (eds) Introducción a la inmunología humana, 6th edn. Editorial Médica Panamericana, Buenos Aires, pp 219–239

Rangel-Huerta OD, Pastor-Villaescusa B, Aguilera CM et al (2015) A systematic review of the efficacy of bioactive compounds in cardiovascular disease: phenolic compounds. Nutrients 7(7):5177–5216

Rani N, Bharti S, Bhatia J et al (2016) Chrysin, a PPAR-γ agonist improves myocardial injury in diabetic rats through inhibiting AGE-RAGE mediated oxidative stress and inflammation. Chem Biol Interact 250:59–67

Ren X, Shi Y, Zhao D et al (2016) Naringin protects ultraviolet B-induced skin damage by regulating p38 MAPK signal pathway. J Dermatol Sci 82:106–114

Ribeiro D, Freitas M, Lima JL et al (2015) Proinflammatory pathways: the modulation by flavonoids. Med Res Rev 35(5):877–936

Rücker H, Al-Rifai N, Rascle A et al (2015) Enhancing the anti-inflammatory activity of chalcones by tuning the Michael acceptor site. Org Biomol Chem 13:3040–3047

Sakamoto Y, Kanatsu J, Toh M et al (2016) The dietary isoflavone daidzein reduces expression of pro-inflammatory genes through PPARα/γ and JNK pathways in adipocyte and macrophage co-cultures. PLoS One 11(2):e0149676

Sakthivel KM, Guruvayoorappan C (2013) Amentoflavone inhibits iNOS, COX-2 expression and modulates cytokine profile, NF-κB signal transduction pathways in rats with ulcerative colitis. Int Immunopharmacol 17(3):907–916

Santangelo C, Varì R, Scazzocchio B et al (2007) Polyphenols, intracellular signalling and inflammation. Ann Inst Super Sanita 43:394–405

Serafini M, Peluso I, Raguzzini A (2010) Flavonoids as anti-inflammatory agents. Proc Nutr Soc 69(3):273–278

Shalini V, Bhaskar S, Kumar KS et al (2012) Molecular mechanisms of anti-inflammatory action of the flavonoid, tricin from Njavara rice (*Oryza sativa* L.) in human peripheral blood mononuclear cells: possible role in the inflammatory signaling. Int Immunopharmacol 14(1):32–38

Shaulian E, Karin M (2001) AP-1 in cell proliferation and survival. Oncogene 20(19):2390–2400

Shin HJ, Shon DH, Youn HS (2013) Isobavachalcone suppresses expression of inducible nitric oxide synthase induced by Toll-like receptor agonists. Int Immunopharmacol 15(1):38–41

Smith WL, Langenbach R (2001) Why there are two cyclooxygenase isozymes. J Clin Invest 107:1491–1495
Song X, Chen Y, Sun Y et al (2012) Oroxylin A, a classical natural product, shows a novel inhibitory effect on angiogenesis induced by lipopolysaccharide. Pharmacol Rep 64(5):1189–1199
Spencer JP (2010) The impact of fruit flavonoids on memory and cognition. Br J Nutr 104(3):S40–S47
Spencer JPE, Schroeter H, Rechner AR et al (2001) Bioavailability of flavan-3-ols and procyanidins: gastrointestinal flavonoids tract influences and their relevance to bioactive forms in vivo. Antioxid Redox Signal 3:1023–1039
Spencer JPE, Srai SK, Rice-Evans C (2003) Metabolism in the small intestine and gastrointestinal tract. In: Rice-Evans C, Packer L (eds) Flavonoids in health and disease. Marcel Dekker, New York, pp 363–390
Spencer JP, Abd-el-Mohsen MM, Rice-Evans C (2004) Cellular uptake and metabolism of flavonoids and their metabolites: implications for their bioactivity. Arch Biochem Biophys 423:148–161
Tang NP, Zhou B, Wang B et al (2009) Flavonoids intake and risk of lung Cancer: a meta-analysis. Jpn J Clin Oncol 39(6):352–359
Tuñón MJ, García-Mediavilla MV, Sánchez-Campos S et al (2009) Potential of flavonoids as anti-inflammatory agents: modulation of pro-inflammatory gene expression and signal transduction pathways. Curr Drug Metab 10(3):256–271
Vauzour D, Rodriguez-Mateos A, Corona G et al (2010) Polyphenols and human health: prevention of disease and mechanisms of action. Nutrients 2(11):1106–1131
Wagner EF (2001) AP-1 – introductory remarks. Oncogene 20(19):2334–2335
Wang L, Lee IM, Zhang SM et al (2009) Dietary intake of selected flavonols, flavones, and flavonoid-rich foods and risk of cancer in middle-aged and older women. Am J Clin Nutr 89(3):905–912
Wang J, Zhang T, Ma C et al (2015a) Puerarin attenuates airway inflammation by regulation of eotaxin-3. Immunol Lett 163:173–178
Wang Y, Wang B, Du F et al (2015b) Epigallocatechin-3-gallate attenuates oxidative stress and inflammation in obstructive nephropathy via NF-κB and Nrf2/HO-1 signalling pathway regulation. Basic Clin Pharmacol Toxicol 117:164–172
Wang Y, Lu P, Zhang W et al (2016) GEN-27, a newly synthetic isoflavonoid, inhibits the proliferation of colon cancer cells in inflammation microenvironment by suppressing NF-κB pathway. Mediat Inflamm 2016:1–17
Wang F, Yin J, Ma Y, Jiang H, Li Y (2017) Vitexin alleviates lipopolysaccharide-induced islet cell injury by inhibiting HMGB1 release. Mol Med Rep 15(3):1079–1086.
Wells TN, Power CA, Shaw JP et al (2006) Chemokine blockers-therapeutics in the making? Trends Pharmacol Sci 27:41–47
Williams RJ, Spencer JP, Rice-Evans C (2004) Flavonoids: antioxidants or signalling molecules? Free Radic Biol Med 36(7):838–849
Williamson G (2017) The role of polyphenols in modern nutrition. Nutr Bull 42(3):226–235
Wu J, Zhou J, Chen X et al (2012) Attenuation of LPS-induced inflammation by ICT, a derivate of icariin, via inhibition of the CD14/TLR4 signaling pathway in human monocytes. Int Immunopharmacol 12(1):74–79
Wu LH, Lin C, Lin HY et al (2016) Naringenin suppresses Neuroinflammatory responses through inducing suppressor of cytokine signaling 3 expression. Mol Neurobiol 53:1080–1091
Wu KC, Huang SS, Kuo YH et al (2017) Ugonin M, a *Helminthostachys zeylanica* constituent, prevents LPS-induced acute lung injury through TLR4-mediated MAPK and NF-κB signaling pathways. Molecules22(4):pii:E573
Wun ZY, Lin CF, Huang WC et al (2013) Anti-inflammatory effect of sophoraflavanone G isolated from *Sophora flavescens* in lipopolysaccharide-stimulated mouse macrophages. Food Chem Toxico 62:255–261

Xiao JB (2017) Dietary flavonoid aglycones and their glycosides: what show better biological benefits? Crit Rev Food Sci Nutr 57(9):1874–1905

Xiao J, Ho CT, Liong EC et al (2014) Epigallocatechin gallate attenuates fibrosis, oxidative stress, and inflammation in non-alcoholic fatty liver disease rat model through TGF/SMAD, PI3 K/Akt/FoxO1, and NF-κB pathways. Eur J Nutr 53:187–199

Xie C, Kang J, Li Z et al (2012) The açaí flavonoid velutin is a potent anti-inflammatory agent: blockade of LPS-mediated TNF-α and IL-6 production through inhibiting NF-κB activation and MAPK pathway. J Nutr Biochem 23(9):1184–1191

Xie H, Sun J, Chen Y et al (2015) Epigallocatechin-3-gallate attenuates uric acid-induced inflammatory responses and oxidative stress by modulating notch pathway. Oxidative Med Cell Longev 2015:1–10

Xu CQ, Liu BJ, Wu JF et al (2010) Icariin attenuates LPS-induced acute inflammatory responses: involvement of PI3 K/Akt and NF-κB signaling pathway. Eur J Pharmacol 642(1–3):146–153

Yang G, Ham I, Choi HY (2013) Anti-inflammatory effect of prunetin via the suppression of NF-κB pathway. Food Chem Toxicol 58:124–132

Yao Y, Chen L, Xiao J et al (2014) Chrysin protects against focal cerebral ischemia/reperfusion injury in mice through attenuation of oxidative stress and inflammation. Int J Mol Sci 15(11):20913–20926

Yao J, Jiang M, Zhang Y et al (2016) Chrysin alleviates allergic inflammation and airway remodeling in a murine model of chronic asthma. Int Immunopharmacol 32:24–31

Ye T, Zhen J, Du Y et al (2015) Green tea polyphenol (-)-epigallocatechin-3-gallate restores Nrf2 activity and ameliorates crescentic glomerulonephritis. PLoS One 10(2):e0119543

Yoo H, Ku SK, Baek YD (2013) Anti-inflammatory effects of rutin on HMGB1-induced inflammatory responses *in vitro* and *in vivo*. Inflamm Res 63(3):197–206

Yoo H, Ku SK, Han MS et al (2014) Anti-septic effects of fisetin *in vitro* and *in vivo*. Inflammation 37(5):1560–1574

You OH, Shin EA, Lee H et al (2017) Apoptotic effect of astragalin in melanoma skin cancers via activation of caspases and inhibition of sry-related HMg-Box Gene 10. Phyther Res 31(10):1614–1620

Yu DH, Ma CH, Yue ZQ et al (2014) Protective effect of naringenin against lipopolysaccharide-induced injury in normal human bronchial epithelium via suppression of MAPK signaling. Inflammation 38(1):195–204

Zhang X, Liu T, Huang Y et al (2014) Icariin: does it have an osteoinductive potential for bone tissue engineering? Phyther Res 28(4):498–509

Zhang JX, Xing JG, Wang LL et al (2017) Luteolin inhibits fibrillary β-amyloid1-40-induced inflammation in a human blood-brain barrier model by suppressing the p38 MAPK-mediated NF-κB signaling pathways. Molecules 22(3):pii:E334

Zhou X, Yuan L, Zhao X et al (2014) Genistein antagonizes inflammatory damage induced by β-amyloid peptide in microglia through TLR4 and NF-κB. Nutrition 30(1):90–95

Zhou CH, Wang CX, Xie G et al (2015a) Fisetin alleviates early brain injury following experimental subarachnoid hemorrhage in rats possibly by suppressing TLR 4/NF-κB signaling pathway. Brain Res 1629:250–259

Zhou LT, Wang KJ, Li L et al (2015b) Pinocembrin inhibits lipopolysaccharide-induced inflammatory mediators production in BV2 microglial cells through suppression of PI3K/Akt/NF-κB pathway. Eur J Pharmacol 761:211–216

Zurier RB (2013) Prostaglandins, leukotrienes, and related compounds. In: Firestein GS, Budd RC, Gabriel SE, McInnes IB, O'Dell JR (eds) Kelley's textbook of rheumatology, 9th edn. Elsevier Saunders, Philadelphia, pp 340–357

Current Approaches to the Isolation and Structural Elucidation of Active Compounds from Natural Products

Alice L. Perez

1 State of the Art

In 1999 the Nobel Prize winner Elias J. Corey, in his editorial review for the publication of the Comprehensive Natural Products Chemistry, wrote: "the wonders of biological evolution are manifest not only in the countless living species found in nature but also in the vast numbers of organic natural products generated by them. Natural products have been of central importance to the field of organic chemistry as engines of its development and as links to the domain of biology" (Corey 1999). Time has passed since then, but his words are still valid almost 20 years later.

According to Newman and Cragg (2016), natural products continue to play a fundamental role in drug discovery and are "still alive and well" in the pharmaceutical industry. Depending on which analysis one uses (Newman and Cragg 2016), at the present time more than two thirds of prescription drugs on the market are natural products or natural product-based derivatives.

Molinski (2014), from the perspective of the impact and advance of organic chemistry as a field, considered that the study of natural products was the most important. This impact can be seen in its contributions in the development of synthetic methodology, in various theoretical calculations, in the biochemistry of primary and secondary metabolism, and in drug development, among other fields. Molinski (2014) also highlights the consequences that these studies have had on the development of different analytical methods. He writes that "without natural products, the study of conformational analysis, developments in circular dichroism (CD), nuclear magnetic resonance (NMR), mass spectrometry (MS), "the art of organic synthesis" and, of course, logical frameworks for total synthesis of natural products of high molecular complexity may have followed very different paths."

A. L. Perez (✉)
School of Chemistry and Natural Products Research Center, University of Costa Rica, San Jose, Costa Rica
e-mail: alice.perez@ucr.ac.cr

V. Cechinel Filho (ed.), *Natural Products as Source of Molecules with Therapeutic Potential*, https://doi.org/10.1007/978-3-030-00545-0_6

Nicolaou (2014), in his assay on drug discovery and development, summarized what he considered the elements of the actual process of achieving new drugs (Fig. 1). Drug discovery, as indicated in the figure, now has become a more complex venture than before. Even though modern technology (e.g., high-throughput screening (HTS), combinatorial chemistry) has added more expedience to this work (Koehn 2008; Singh and Pelaez 2008), going back to basics is in mode once again, as it is becoming "basic" in the study of natural products. "It might be true (Nicolaou 2014) that natural products chemistry requires longer-term plans and higher initial investments, but, in the long run, the endeavor pays off as demonstrated by its rich and glorious history." Previous to the clinical stages, the conjunction of several disciplines is becoming critical to better understand and guide the discovery and synthesis of new and better medicines as is showed in Fig. 1 under the name of "auxiliary arms."

Therefore, high-tech, multi- and transdisciplinary studies, re-exploring and investigating the vast biodiversity of our planet and the wide range of biological targets and screening tests of new or revisited natural products, promise a challenging but bright future in the field.

The combination of areas of study, theory, and techniques (in addition to a classical methodology to identify metabolites derived from different organisms or symbionts) is today a more reliable approach. In 2003, Bleicher et al. (Fig. 2) suggested a scheme that summarizes the increasing trends in drug discovery including the natural product approach, a full cycle of interactions that does not focus necessarily in one single entry.

Moreover, different reports in the literature add new applications and advancements in hyphenated spectroscopic techniques to aid the discovery of natural product entities. The use of metabolomic profiling and dereplication approaches to

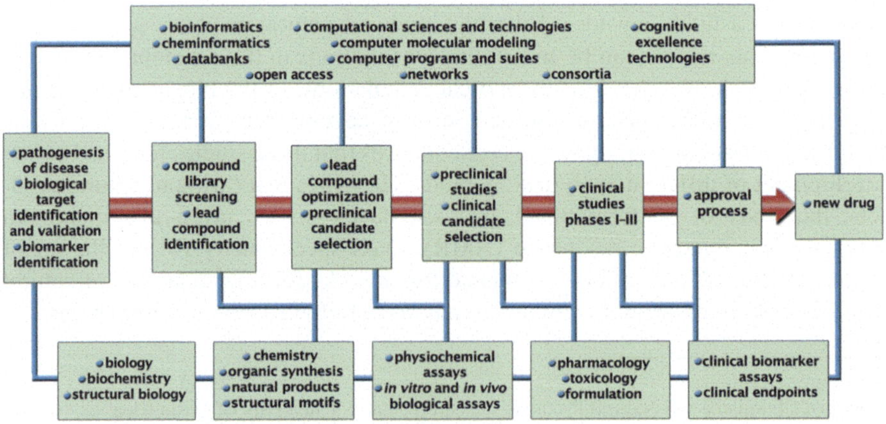

Fig. 1 Nicolaou's description of the drug discovery and development process and its auxiliary arms (top and bottom). (Taken from Nicolaou 2014)

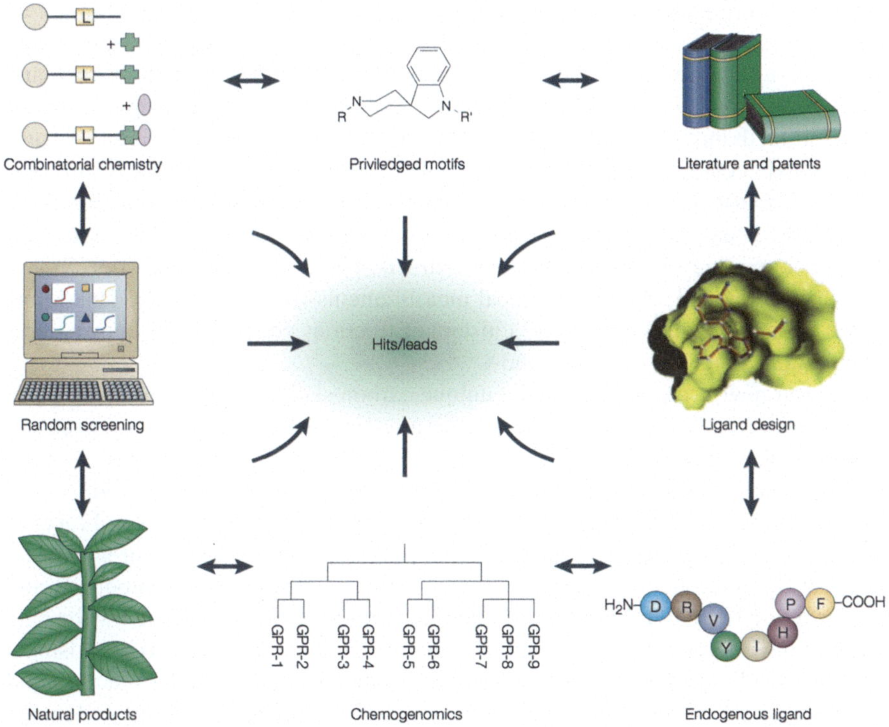

Fig. 2 Hit-identification strategy proposed by Bleicher et al. (2003). (Taken from Bleicher et al. 2003)

natural product extracts open new windows of opportunity for more thorough research (Dias et al. 2012).

A recent quantitative dataset analysis of all published microbial and marine-derived natural products, performed by Pye et al. (2017), led these researchers to conclude that even though most of the compounds analyzed presented structural similarities to previously published ones and although it seems that there are a finite number of scaffolds available in nature, there are still a good number of new structural features to be discovered. This fact continues to encourage the pursuit of finding new compounds and their examination for different biological activities. New approaches to the screening of extracts of natural product libraries may help in the prediction of the identity and mode of action of bioactive molecules. Hovarth et al. (2016) discussed the development of new, cell-based assay technologies in the field and suggested that these kinds of assays may enhance the conventional process of drug discovery.

Two examples of these new approaches have been reported recently. The first example is by Kurita et al. (2015), who integrates image-based phenotypic screening in HeLa cells with high-resolution untargeted metabolomics (database built with liquid chromatography-mass spectrometry (LC-MS) spectra of libraries of extracts) to create "compound activity mapping" (Fig. 3). The extract library was composed of fractionated extracts (extractions using 1:1 methanol/dichloromethane and later fractionated on a reverse-phase C_{18} column with an eluotropic series of water and methanol [20%, 40%, 60%, 80%, and 100% (vol/vol) methanol in water and an ethyl acetate wash], Kurita et al. 2015) obtained from bacterial strains isolated from marine sediments collected from coastal areas of Panama that were grown under standard fermentation conditions. This sequence allowed them to build a library containing 234 natural products extracts. They later combined 10,977 mass spectral features and 58,032 biological measurements from the library. This information allowed them to identify 13 clusters of fractions containing 11 known compound families and 4 new compounds.

The four new compounds posed a unique carbon skeleton that were named quinocinnolinomycins. These compounds were associated with endoplasmic reticulum stress (Fig. 4, Kurita et al. 2015).

The second approach by Kremb and Voolstra (2017) uses broad-spectrum phenotypic profiling. In this case, they used the Sigma Life Science's Library of Pharmacologically Active Compounds (LOPAC) (see sigmaaldrich.com web page, details at the reference), selecting 720 single molecule reference compounds and another set of 124 natural products purchased from Specs (see Specs.net web page, details at the reference), that the authors classified as poorly characterized. Using a panel of 14 cellular markers and cytological results, they aimed to have access and to achieve a better understanding on physiology, mechanisms of actions, and multilevel toxicity of diverse group of molecules, highlighting the easy setup of this technology for laboratories in charge of natural products research.

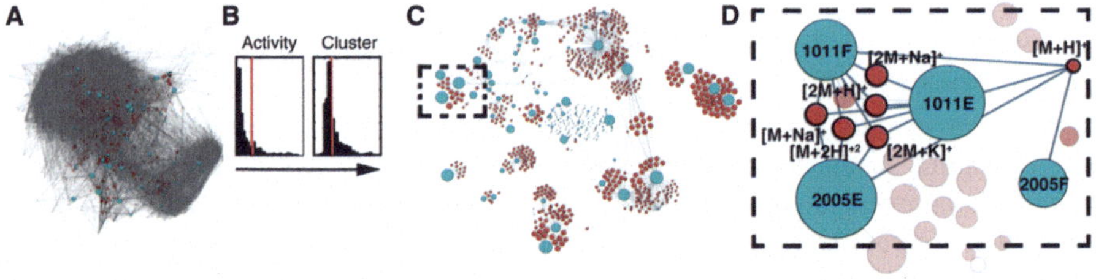

Fig. 3 Compound activity map constructed by data gather from bacterial extracts. (Taken from Kurita et al. 2015)

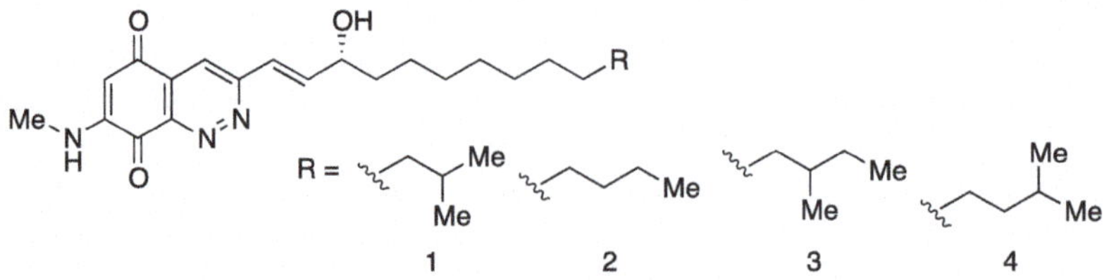

Fig. 4 Quinocinnolinomycins, a new natural product isolated and identified by Kurita et al. (2015)

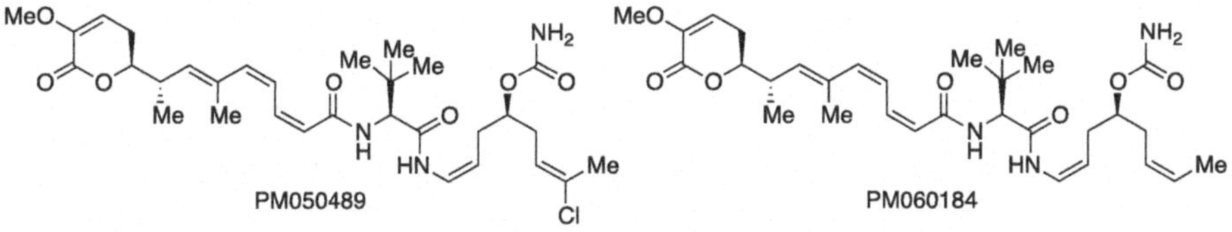

Fig. 5 New marine metabolites discovered by Cuevas' group from PharmaMar

At the same time that these new approaches were being developed, an example of the importance of natural products research in drug discovery was exemplified by the identification of molecules like PM0500489 and PM060184 (Fig. 5) by the group of Cuevas from PharmaMar in 2013 (Martin et al. 2013; Pera et al. 2013; Martínez-Diez et al. 2014).

PM0500489 and PM06018 inhibit the growth of different cancer cell lines at the sub-nanomolar level, and these molecules are considered as highly potent microtubule inhibitors that disrupt mitosis with a distinct molecular mechanism. Molinski (2014) considered this "once in a decade" discovery a landmark, putting the spotlight on natural products in drug discovery one more time.

2 Dark Chemical Matter as a Source of New Active Compounds

The total mass energy of the universe is about 4.9% and is called ordinary matter (which comprises everything that can be observed with instruments). The other ~95%, which cannot be seen, is the sum of dark energy (68%) and dark matter (27%). In astrophysics and cosmology, dark matter and dark energy are important issues, even though they cannot be defined completely. More is unknown than is known in this subject. Even experts consider it as a complete mystery (National Aeronautics and Space Administration, 2018).

By analogy, Wassermann et al. (2015) introduced the equivalent term "dark chemical matter" (DCM) to group and categorized molecules that belong to a particular library and have shown no biological activity. The criteria established by the authors in order to build the DCM was that the compounds presented no activity by a minimum of 100 assays. They used, for the purpose of their work, the compounds contained in the Novartis and NIH Molecular Libraries Program screening collections and 234 (Novartis) and 429 (PubChem) assays. They analyzed 803,990 compounds from the Novartis database and found 112,872 (14%) molecules that satisfied the chosen criterion, whereas from the PubChem-NIH database, they ran through 363,598 molecules and found 131,726 (36.2%) DCM compounds. Also, they analyzed molecules from the Novartis natural product library and found 294 compounds that can be classified as DCM; some examples are shown in Fig. 6.

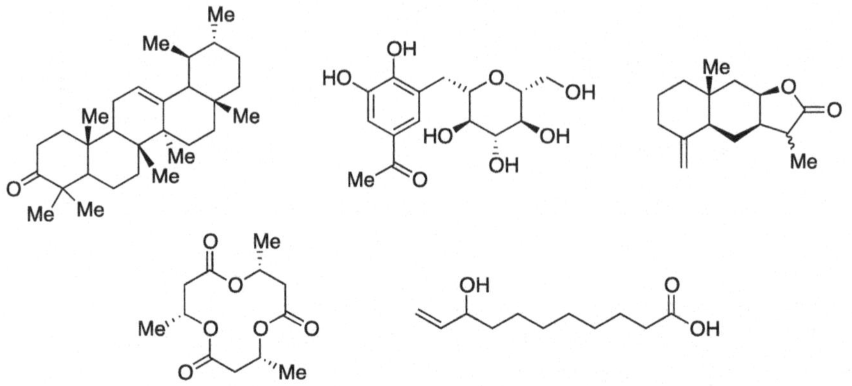

Fig. 6 Examples of natural products found in the Novartis database as DCM. (Taken from Wassermann et al. 2015)

They analyzed this information using different methods and algorithms that allowed them to assign the "dark matter in chemical space," easing the selection of the chemical motif. They were able to identify dark matter hotspots (using multidimensional scaling) to cluster the different compounds. One of the most important conclusions from this work is that "DCM compounds have active neighbors and are structurally not very distinct from biologically active chemical matter," and therefore DCM should not be considered biologically inert and should be considered a new and potentially valuable resource for hit finding and lead optimization. (Wassermann et al. 2015). This important paper has attracted the attention of the pharmaceutical world as it was commented on by Derek Lowe in his blog published in *Science Translational Medicine* in 2015. Lowe addresses an issue long discussed in natural product research and the results coming from HTS, as he put it: "this paper made me think about screening collections from an angle that I really never had before, and I really appreciate the substantial time and effort that went into it."

Since then, new papers have been published evaluating and outlining the proposal of Wassermann et al. (2015). Muegge and Mukherjee (2016) presented their results analyzing the total data of 203 high-throughput screening campaigns done at Boehringer Ingelheim (BI), (a datasets comprising between 406,000 and 985,000 compounds) that were conducted between the years 1999 and 2012. They claimed to use a rigorous statistical analysis in order to improve selection and eliminate bias. They concluded, similar to Wassermann, that "screening DCM may generate hits with higher selectivity and perhaps more attractive properties such as higher polarity, lower molecular weight, and higher 3D character compared to previously active compounds and with lower tendency to generating false positives in HTS assays" (Muegge and Mukherjee 2016). Jasial and Bajorath (2017) made their contribution on DCM using the PubChem BioAssay database. They identified 367,557 screening compounds that were tested in at least 100 primary assays. For these compounds, 81,597 met the DCM requirement of being inactive in all primary and confirmatory assays in which they were tested. The researchers then did a systematic search for bioactive analogs to pair DCM using the ChEMBL database (Gaulton et al., 2012). This pairing allowed them to propose an analogous series containing DCM and known bioactive compounds and to derive target hypotheses for more than 8000 extensively assayed DCM molecules. In addition to this contribution, Wassermann, Tudo, and Glick (2017) ran the DCM criteria with the Merck & Co. database. The major conclusion from this analysis is that "the definition of DCM is highly dependent on the assay panel and also the number of times that a compound has to be screened in order to be considered dark." According to the authors, this result suggested that there are targets and phenotypes that are uniquely investigated by each group and therefore some overlap or none can be found, making the definition of DMC a little bit more complex than previously thought. However, all the authors cited above agree on the usefulness of DCM as a source of new leads in drug discovery, not only for synthetic compounds but also for natural product-based entities.

This sporadic indiscriminate/promiscuous behavior found in these analyses had been previously explained in a different way from the results suggested by DCM researchers. Schulze et al. (2013) proposed, after analyzing a training set of commercial compounds of known mechanism and comparing these profiles to those obtained from natural product library components, the creation of "function-first" approaches to natural products discovery, annotating natural products extracts based on the mode of action (MOA).

More recently, the issue of addressing the existence of compounds with reported bioactivity for almost any biological/pharmacological test [invalid metabolic panaceas (IMPs, Fig. 7), pan-assay interference compounds (PAINS, Fig. 8)], has been studied. These studies use different approaches to the same problem: how to identify and select new molecules with distinct biological activity without false positives or false negatives. Caution is recommended when blind filters are used during the

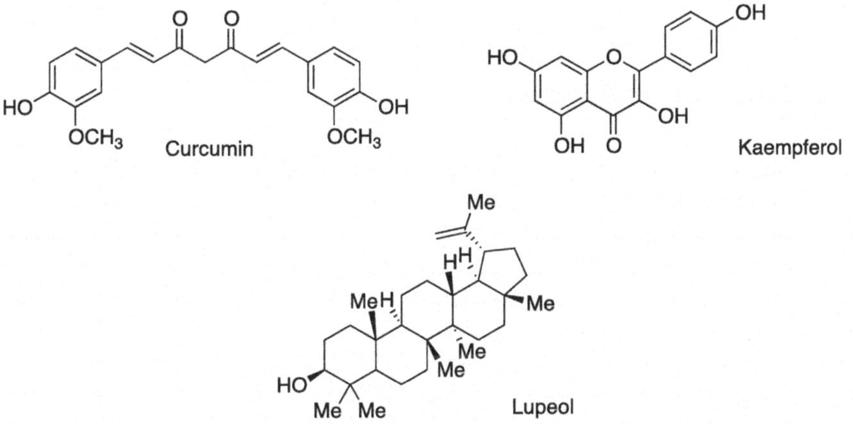

Fig. 7 Examples of the top reported compounds using the NAPRALERT database that represented the natural products most reported by occurrence, activity, and distinct activity (Bisson et al. 2016)

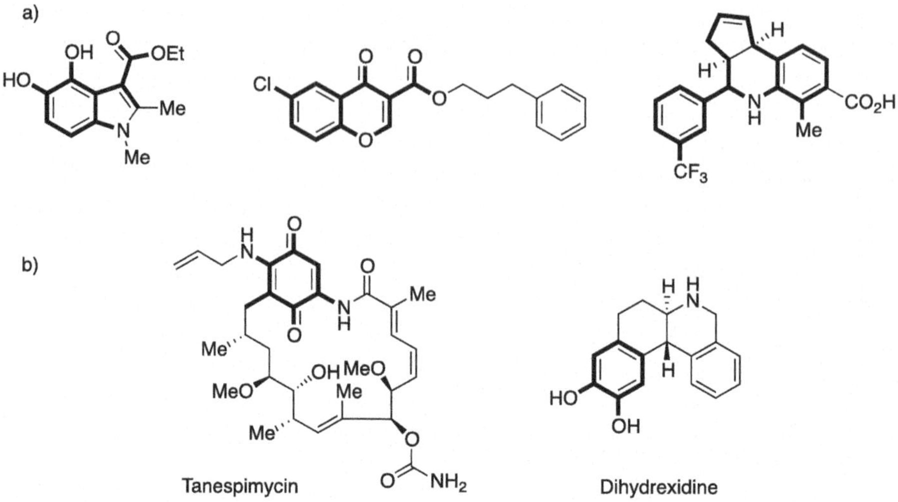

Fig. 8 (**a**) Highlight substructures of compounds considered as PAINS according to Baell and Holloway (2010). (**b**) Two active compounds having PAINS motifs (Capuzzi et al. 2017)

triage of compounds. Misleading information could produce false negatives and, therefore, potentially an important molecular hit may be overlooked [Baell and Holloway (2010), Bisson et al. (2016), Capuzzi et al. (2017)].

The typical concentrations used in several HTS runs could be one of the key factors for missing certain biological active hits. Siramshetty and Preissner (2018) argue that from the original DCM criteria, a more reasonable number should be used, one that takes into account the inactivity in multiple confirmatory screening assays (not only 100 as an arbitrary number) at multiple concentrations (preferably higher than the typical screening concentrations).

3 Labile Natural Products

The occurrence of artifacts during the process of isolation of natural products is a risk that any researcher may encounter on a daily basis. Some molecules are highly sensitive to the extraction conditions (e.g., solvents, temperature, chromatographic procedures) and may decompose, rearrange, or even degrade during some of these steps, making the process of metabolite identification more difficult or misleading. Moreover, this new metabolite or artifact (artificially formed during processing) may be deprived of its original biological activity. Thus, even though the word "artifact" is a term well-known and widely used by the practitioners in the field, there is no consistent definition reported. Recently, DeHaven and collaborators (2012) define an artifact as "any chemical whose presence can be attributed to sample handling and processing and not originating from the biological sample. Artifacts can include releasing agents and softeners present in plastic sample vials and tubing, solvent contaminants, etc."

Maltese et al. (2009), highlight six problems that are complicated by artifact formation:

1. Formation of new compounds, which cannot be identified as natural products.
2. Loss of activity of active components.
3. Formation of active compounds from inactive natural products (false positives).
4. Loss in total yield of important chemicals during isolation.
5. Generation of toxic components from inactive substances.
6. Difficulty in the reproducibility of the method.

Considering the multiple steps and risk factors leading to metabolite isolation, rigorous scrutiny should always be a major concern; this also includes sample preparation during all metabolomics studies (Kim and Verpoorte 2009). What happens when there is a labile metabolite present in a sample? How to proceed? Ignoring labile molecules may eliminate deeper understanding of a mode of action or even may overlook a new bioactive compound. Wakimoto and Abe (2012) review five of these successful encounters of labile products: (a) the red pigments from hippopotamus (hipposudoric acid), (b) labile polyketides from filamentous fungus (lambertellols), (c) cyclopropene carboxylic acid from a mushroom, (d) an unusual aziridine amino acid from a mushroom (pleurocybellaziridine), and (e) a furan fatty acid from a mussel (Fig. 9).

In all the original cases presented by the authors, the compounds were elusive during "normal" isolation procedures. It took careful technique and a better understanding of the chemical nature of these particular metabolites in order to proceed with a proper extraction protocol and later structural elucidation.

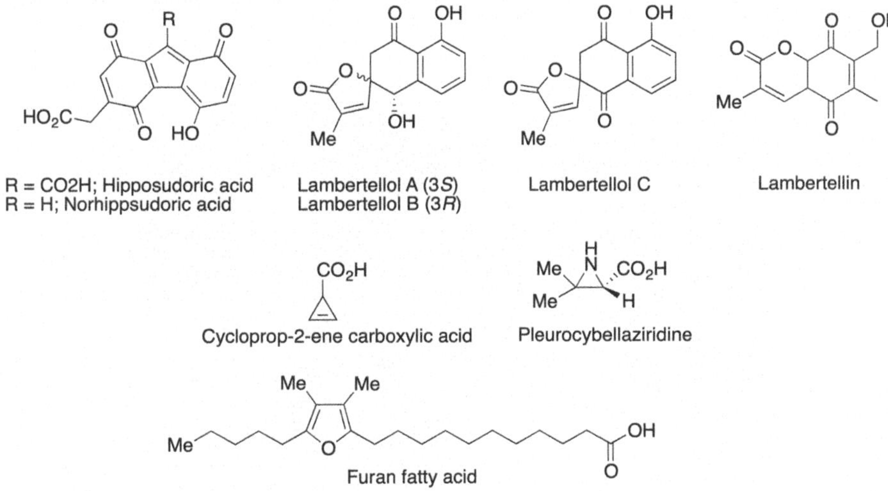

Fig. 9 Some examples of labile natural products (Wakimoto and Abe 2012)

4 Metabolomics and Dereplication

4.1 Metabolomics and Natural Products

A rapid search on Google Scholar (http://scholar.google.com) of the word "metabolomics" through the years 2005 and 2018 (excluding citations) leads to 50,800 entries. Narrowing the search to include "natural products," the number of hits went down to 26,200 entries (about 2000 entries per year). This exercise indicates the rapid growth of this field associated not only to profiling complex mixtures but also analysis leading to metabolite identification. This tendency is also recognized by Haug et al. (2017) in their analysis of the years 1995–2015. Even though this article deals with the need for openness of the databases generated through years of research on this field and the central point that the authors make for data sharing, the statistics presented in the article disclose the increasing amount of work developed in this area. The authors foresee that if more data is deposited in public repositories, more re-analyses and information about metabolites present in different matrixes would be possible, therefore extending the scope and contributions developed from this research. All the data appear to indicate that along with the more "traditional" methods for structural elucidation, metabolomics will be a standard protocol in natural products studies in the near future.

Regarding the actual applications of metabolomics analyses, Liu and Locasale (2017) reported a wide range of areas where the metabolomics analysis had found a niche, in which drug discovery comprises 13.2% of the total analyzed published papers (Fig. 10).

Concerning natural products and metabolite profiling, Wolfender et al. (2015) compiled a review paper considering the main analytical techniques used for generic and comprehensive profiling of secondary metabolites in natural extracts. Three levels of metabolite analysis are reviewed: (a) metabolite fingerprinting, (b)

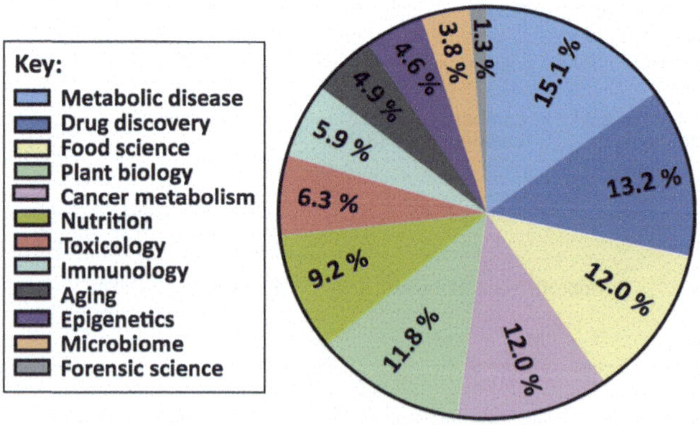

Fig. 10 Distribution of recent publications on applications of metabolomics by area. (Taken from Liu and Locasale 2017)

metabolite profiling, and (c) metabolite target analysis. In each level, a different degree of accuracy on the determination of the chemical composition of a given biological matrix or extract can be achieved, and, therefore, it depends on the researcher's criteria and expertise to decide which level and when to use it. It is definitively a growing field contributing to the chemical structural elucidation of natural products.

NMR and MS (alone or as hyphenated systems) are the usual techniques used for metabolomics studies. The criteria to choose one over the other or its complimentary information depends of the type of study to be pursued. General aspects like sensitivity, high throughput, the number of molecules to be measured, the complexity of the biological sample, and the purpose of the analysis (among other facts) determine which technique should be used (Wolfender et al. 2015; Robinette et al. 2012; Margueritte et al. 2018; Ganzera and Sturm 2018; Gomes et al. 2018; Dias et al. 2016; Kruk et al. 2017).

Some criticism has arisen recently over the increased use of untargeted metabolite profiling and the poor understanding of the data obtained, as pointed out in a paper by Mahieu and Patti (2017). They refer to the uncertainty in the number of unique metabolites being profiled in general and how to differentiate among all types of possible metabolites (known, unknown and new metabolites, impurities, artifacts, and degeneracies) when the untargeted method is used due to the lack of a comprehensive evaluation. The authors present an example of the analysis of LC-MS data on *Escherichia coli* samples analyzed with one untargeted metabolomics method. The researchers initially found 25,230 targets (using LC-MS) that were then narrowed to no more than 892 (a 90% reduction in data). This differentiation required utilizing annotation data, the use of better databases including accurate mass, retention time, and MS/MS fragmentation data and annotations of all credentialed features plus ^{13}C-NMR. This research is clarion call for metabolomics practitioners to choose suitable methodology that also includes statistical methods and a reliable database analysis.

Dührkop and co-workers (2015) previously attempted to address this same issue of how to analyze the profiling results of untargeted metabolites. They approached the identification problem through the creation of a searching method using tandem MS data of the metabolite in a database of molecular structures, originating what is known today as Compound Structure Identification: FingerID (CSI:FingerID). It is based on fragmentation trees that are predicted from MS/MS data by an automated computational method. The peaks in the MS/MS spectrum are annotated with molecular formulas of the corresponding fragments, and fragments are connected via assumed losses leading eventually to a single molecule. Da Silva et al. (2015) commented on this approach: CSI:FingerID as a tool to "begin to illuminate more of the chemical dark matter." This profiling is an ongoing process; more information about it can be found in https://bio.informatik.uni-jena.de/software/sirius/.

Literature reviews, for example, those of Wolfender et al. (2015) and Liu and Locasale (2017), are a good start for someone that wants to understand the different aspects that a metabolomic profile implies and, more importantly, what kind of information and pitfalls can be found if the focus is the structural elucidation of natural products (Fig. 11). Throughout these reviews guidelines are drawn, review-

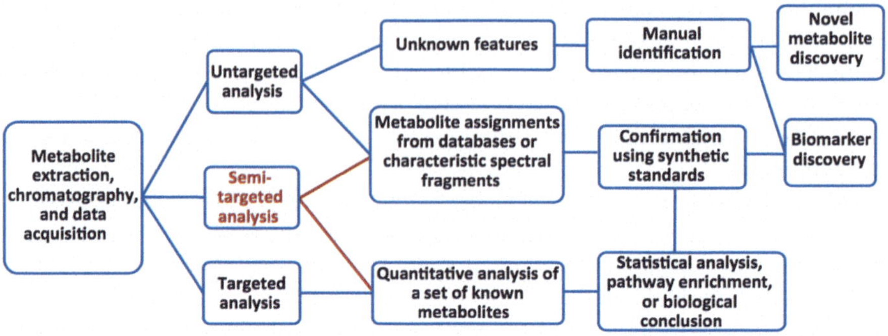

Fig. 11 Targeted, semi-targeted, and untargeted analysis used in metabolomics studies. (Taken from Liu and Locasale 2017)

ing also the state of the art regarding extraction protocols and analytical techniques used frequently in metabolomics studies.

4.2 Dereplication

Based on data from 2016, it is estimated that more than 250,000 natural products have been structurally characterized from plants, animals, and microorganisms. This number continues to increase each year (Mohamed et al. 2016). This fact indicates an opportunity of more discoveries but a challenge of originality: the time-consuming process of the re-isolation and reidentification of known compounds when the search of novel or new ones is intended. Dereplication has been fostered as a promise to help in this endeavor. Classically, any dereplication study begins in the same way as any other standard natural product research. Dereplication is based on the comparison of data originating from chromatographic and spectroscopic techniques. The process diverges at the spectral measurements. Techniques such as NMR, LC-MS, LC-MS/MS, and GC-MS, and structure fragments present in standard compounds libraries are used for comparison and as searching tools (Gaudencio and Pereira 2015; Mohamed et al. 2016, Fig. 12). Therefore, the need to go through all the structural elucidation process is not required in order to identify (or eliminate from different samples) known compounds, leaving the effort to concentrate on the search of novel chemical structures (Carnevale Neto et al. 2016).

The concept of dereplication was introduced in 1978 by Hanka and collaborators (1978). These researchers proposed this methodology as a way to avoid the "rediscovery" of known anticancer antibiotics from bioactive extracts derived from microorganisms during triage tests. The simplest definition is presented by Carter (2017), as the "rapid identification of a compound (or class of compounds)"; the identification in this case (different from the usual process of natural products structural elucidation) refers as an expected or unwanted (a nuisance) compound. In this line of thought, dereplication would be the process of using the information already reported of a compound that has been previously characterized (chemical structure known) to identify it in any new sample being analyzed.

Since 1978, the definition and focus has changed and diversified; the number of reported studies using this method has grown rapidly. Dereplication has, as well as metabolomics, increased its presence in the world of natural products. Recent review articles (Gaudencio and Pereira 2015; Mohamed et al. 2016; Hubert et al. 2017) accounts for this fact. An excellent statistical analysis on this topic was done by Gaudencio and Pereira (2015) based on a literature search indexed in databases (mainstream and nonmainstream). As can be seen (Fig. 13), the increase on the number of publications is evident and represents an 89% growth since 1993 according to the authors.

As a consequence, the original (or simplest) definition of dereplication seemed to experience changes through time. After a literature analysis, Hubert et al. (2017) summarized at least five definitions that the authors identified according to the nature of the sample under examination and the goal of the study (Table 1).

From all these reports arising from the literature, it is clear when performing any dereplication study to have access to advanced spectroscopic equipment (MS and NMR). Also necessary are reliable databases and software tools to compare and process all the spectra and to carry out correlations to any other available data (e.g.,

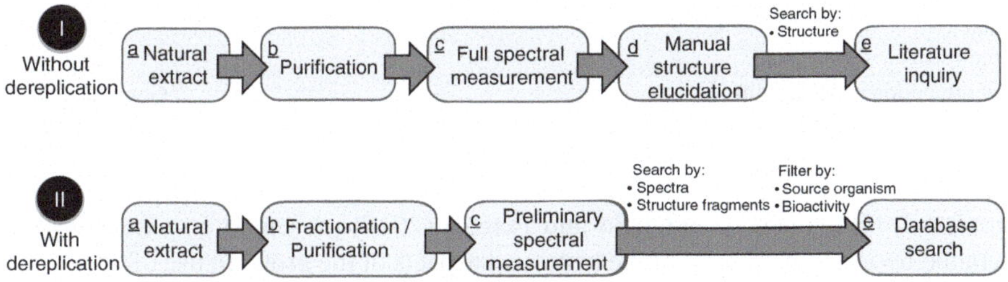

Fig. 12 Compound identification in natural products without and with dereplication. (Taken from Mohamed et al. 2016)

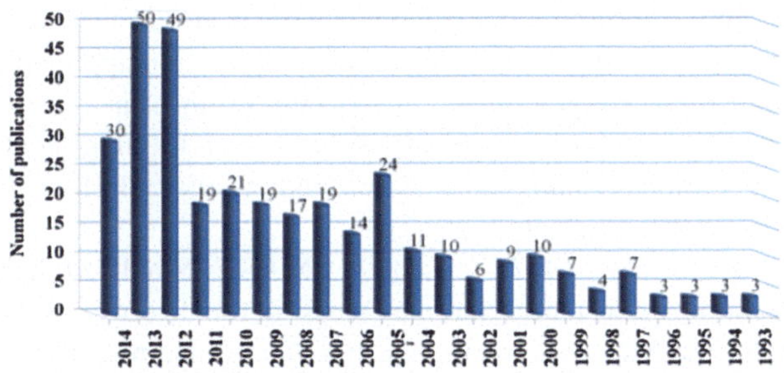

Fig. 13 Number of publications per year on dereplication. (Taken from Gaudencio and Pereira 2015)

Table 1 Natural products dereplication categories according to Hubert and collaborators. (Taken from Hubert et al. 2017)

Goal	DEREP 1 Identification of the major compounds in a single extract	DEREP 2 Acceleration of activity-guided fractionation	DEREP 3 Chemical profiling of crude extract collections	DEREP 4 Chemical profiling of target compounds	DEREP 5 Taxonomoic identification of microbial strains
Targeted chemical class	No			Yes	No
Biological assays	Independent	Yes, systematically	Independent		
Samples	Single extract		Extract collection	Single extract or collection	Extract collection
Fractionation step	In most cases yes	Yes	No (direct sample analysis)	In most cases no	No (direct sample analysis)
Analytical tools	LC/MS, GC/MS and/or NMR			Mostly LC/MS	Gene-sequencing
Computer tool and statistics for data treatments	May include		Yes. Systematically	May include	Yes. Systematically
Identification	Metabolite database				Gene database

structural, biological). Gaudencio and Pereira (2015) presented an extensive and detailed review, which hints at the historical aspects of the growing use of dereplication, parallel to the development of the analytical capabilities of the associated spectroscopic techniques required for any meaningful study. This correlation is very important in natural products research. The dereplication analyses expand their presence as the development of better detectors and spectroscopy equipment has been made. Computer programs in statistical analysis are also a must-have. In this sense, the need of technology still limits many dereplication applications.

In order to aid these studies, different working groups have put together their expertise in creating databases, new algorithms, and training neural networks, among other applications (Zhang et al. 2016, Nothias et al. 2018; Zani and Carroll 2017; Allard et al. 2017; Chen et al. 2017).

Zhang and co-workers (2017) presented SMART, Small Molecule Accurate Recognition Technology, as a tool that may assist in the discovery efforts in natural products research. SMART uses a combination of a trained neural network (convolutional neural networks CNN, Nothias et al. 2018), databases, and NMR data (non-uniform sampling 2D NMR) for this purpose. The authors point out that even though various algorithms have been explored for comparing NMR data applied to dereplication studies, the presence of spectroscopic artifacts, solvent effects, and functional group diversity, among other facts, make it difficult to compare chemical shifts. To test their algorithm, a database containing 2054 HSQC spectra as a training set was put together and used to "automatically locate" a previously isolated new depsipeptide, Viequeamide A and related compounds (Boudreau et al. 2012). SMART was able to pick out the structure and indicated the occurrence of closely related compounds. The working flowchart for SMART is shown in Fig. 14, and, according to the authors, this protocol may not only help in dereplication studies but also in the structural determination of new compounds. The potential for this or any other tool as a method to ease the structural elucidation problem is something we have to look forward to in the near future.

5 Natural Products Isolation and Chemical Structure Elucidation: A More Traditional View

A useful review, covering up to the year 2012, regarding different approaches for the isolation of secondary metabolites can be found in the third edition of the book *Natural Products Isolation* (Sarker and Nahar 2012).

A year later, Bucar and collaborators (2013) published a review on the same subject comprising the years 2008–2012. This review covers general aspects of the authentication and preparation of plant material and marine organisms, employing morphological and anatomical analysis through analytical techniques such as TLC/HPTLC, HPLC and GC analyses, spectroscopic methods (NMR, MS, NIR, FT-IR), and molecular biological methods and postharvest changes in plant material. A various array of extraction protocols (general extraction methods, uses of classical solvent procedures, microwave and ultrasound-assisted extraction, use of ionic liquids, supercritical fluid extraction, distillation, and solid-phase methods) are also reviewed. The paper ends up with a section on isolation procedures (liquid-solid chromatography, preparative planar chromatography, column chromatographic methods, vacuum liquid chromatography, flash chromatography, low-pressure liquid chromatography, medium-pressure liquid chromatography, high-performance liquid chromatography, chiral chromatographic methods, and preparative gas chromatography).

Ciesla and Moaddel (2016) review this topic with different criteria. In their case, they focus on pairing hyphenated analytical techniques with identification of bioactive compounds. This approach combines the metabolomics and dereplication studies previously discussed. They highlight the use of an integrated UHPLC-UV-MS-SPE-NMR system for phytochemical analysis (Fig. 15). Several examples spotlighting the success of this system is presented by the authors. However, as they mention, this procedure has at this stage some limitations: the required access to the equipment, the protocol is labor-intensive and time-consuming, and, more importantly, for this methodology to work efficiently, the dependence on the concentration of the secondary metabolites. Without a reasonable quantity of sample, it is possible that the identified metabolites would only correspond to those in higher concentration but not necessarily the most active.

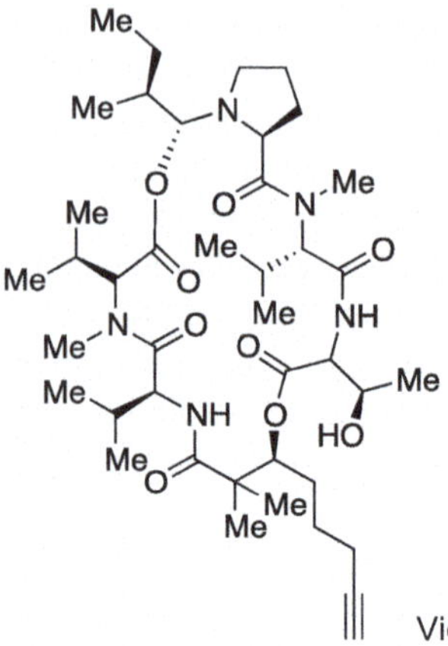

Viequeamide A

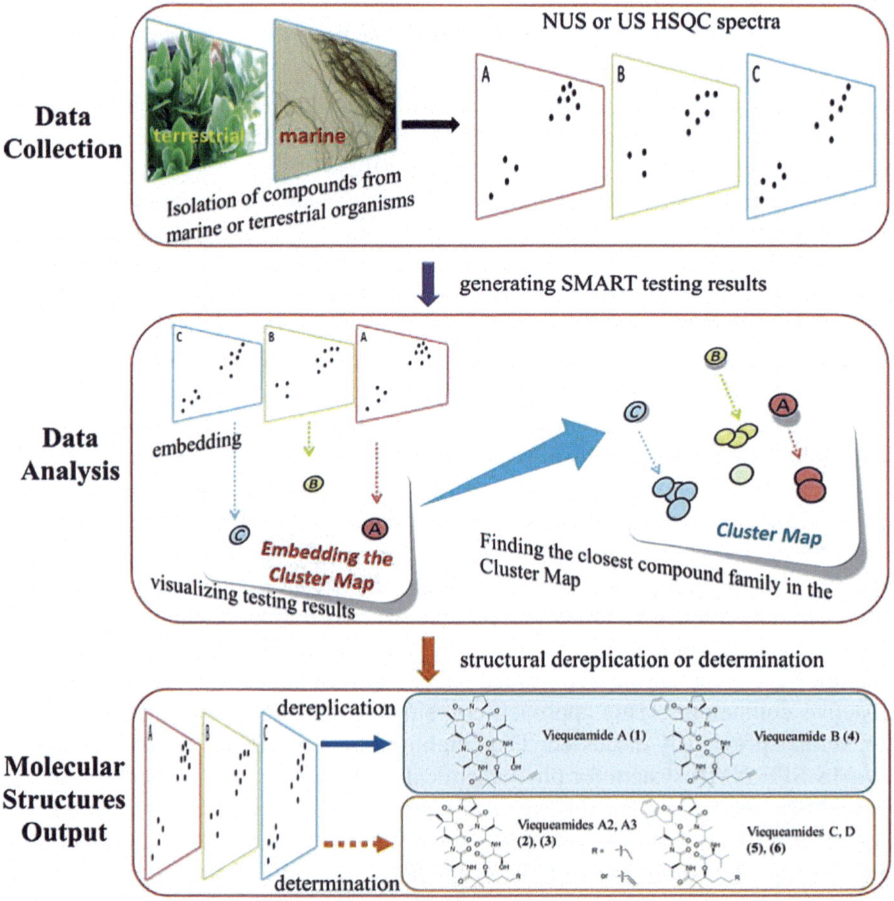

Fig. 14 Workflow for SMART. (Taken form Zhang et al. 2017)

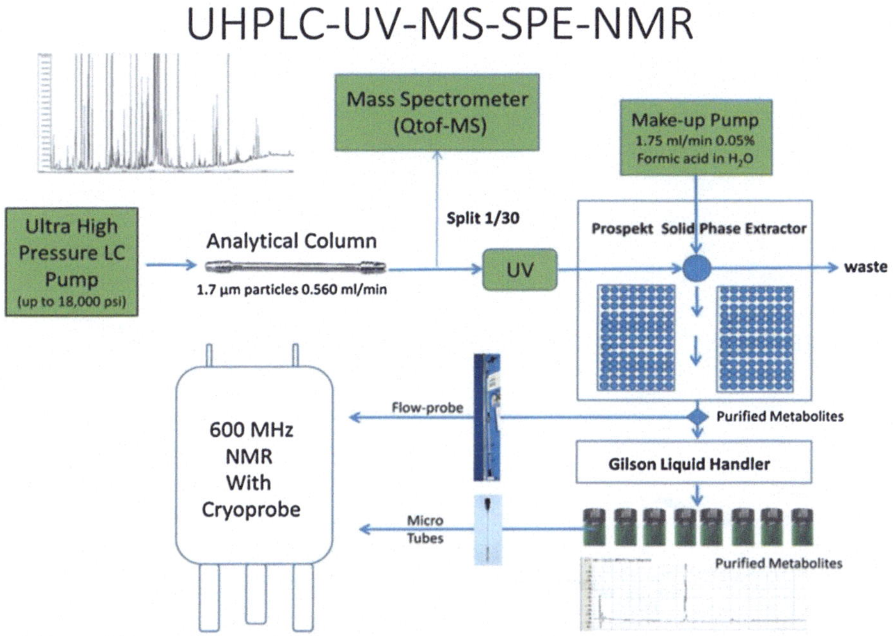

Fig. 15 Integrated UHPLC-UV-MS-SPE-NMR system for phytochemical analysis. (Taken from Ciesla and Moaddel 2016)

These authors also review recent developments as alternative methods (ultrafiltration, bioaffinity chromatography, cellular membrane affinity chromatography, and ligand fishing) to the traditional-guided fractionation. The more widespread use of these new techniques is something to look forward to in the near future. This popularity would imply not only access to the proper instrumentation but also to intervention from the different research groups.

Along this line, Wubshet and collaborators (2015), using the magnetic ligand fishing technique coupled with HPLC-HRMS-SPE-NMR analysis, enable the identification of four α-glucosidase inhibitory ligands, six new alkyl resorcinol glycosides, and five known flavonoids from *Eugenia catharinae* (Fig. 16), showing a very promising application of this technique.

Brusotti et al. (2014) review the chemical structure elucidation process from an even more traditional way using ethnopharmacological prospecting. Considering this approach, the authors analyzed, step-by-step, what to expect based on the classic phytochemical flowchart showed in Fig. 17. having as a guideline the information of the uses of herbal medicines or information coming from traditional medicine.

The authors also underlined the importance of “coming back to the future”: the traditional solid-liquid extraction method (as seen in Fig. 18). This protocol includes water maceration and/or decoction as representative of what is the customary procedure performed by native healers. Clearly Brusotti et al. (2014) call attention to

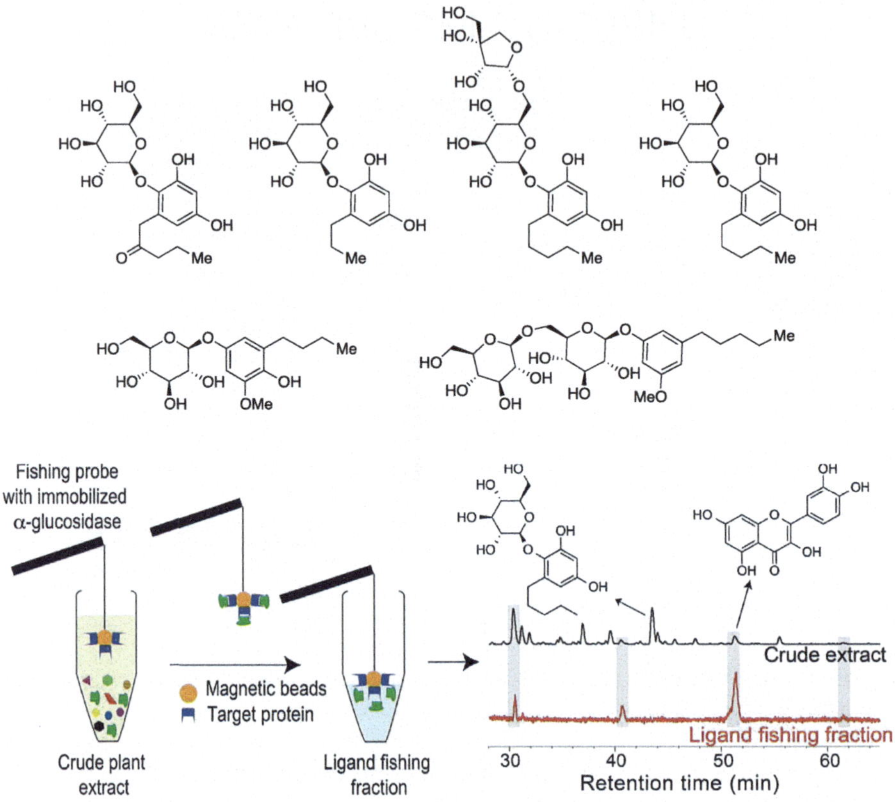

Fig. 16 Ligand fishing strategy and the structure of six new alky resorcinol glycosides found in *Eugenia catharinae*. (Taken from Wubshet and collaborators 2015)

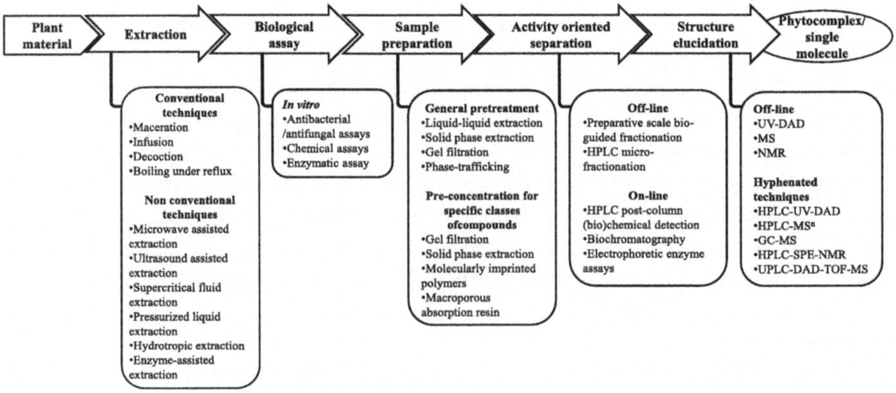

Fig. 17 Phytochemical approach using ethnopharmacological information. (Taken from Brusotti et al. 2014)

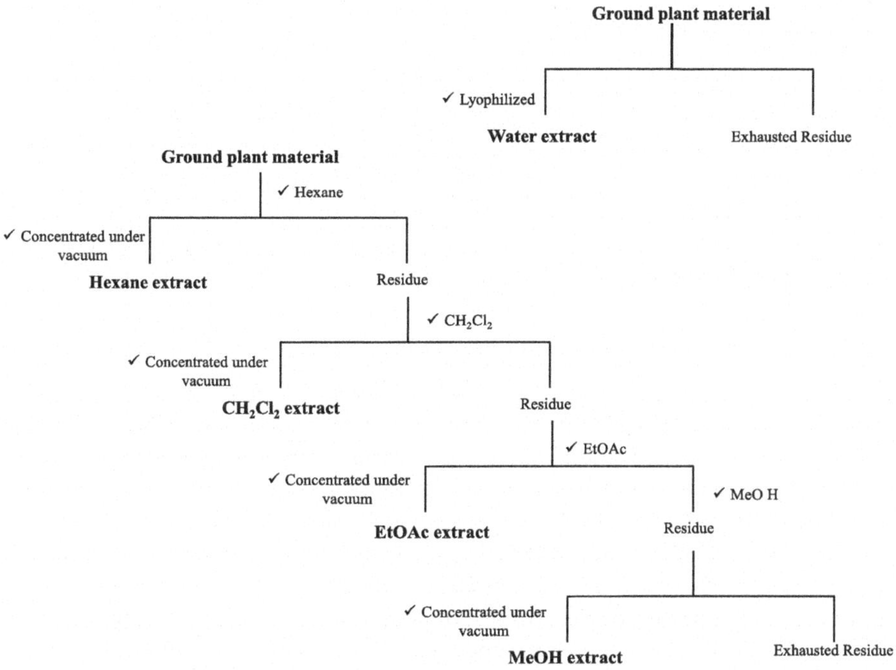

Fig. 18 Conventional solid-liquid extraction protocol. (Taken from Brusotti et al. 2014)

the fact that even though we have experienced a fast development of analytical tools in the past years, good guidelines and laboratory practices have to be kept in mind when dealing with any natural products research if we want the results to be meaningful and successful.

More recently, Buenz and co-workers (2018) have argued to rethink the ethnopharmacological consideration in drug discovery. Traditional knowledge, say the authors, "is a resource that has been underappreciated in the past and the contemporary contribution of ethnopharmacology to drug discovery has generally been intermittent. The goal moving forward, particularly in the context of the resources available today, is to institutionalize the use of ethnopharmacology to augment the drug discovery process and accelerate the identification of new therapeutics."

6 Concluding Remarks

What is the future of research in natural products? As mentioned at the beginning of this chapter, despite the ups and downs reported throughout the history of this field, the future is promising. This area faces the great challenge of the discovery process: new organisms from very diverse sources to investigate, knowledge and traditional uses to decipher, both from the molecular and biological point of

Table 1 | Selection of natural product databases.

Database	Natural product entries	Content	Website
Dictionary of Natural Products	210,213	Natural products described in the literature	http://dnp.chemnetbase.com
Traditional Chinese Medicine	32,364	Natural products from herbs and animals, as listed in Chinese medical texts and dictionaries	http://tcm.cmu.edu.tw
SuperNatural	325,508	Natural products with toxicity and target prediction	http://bioinformatics.charite.de/supernatural/
ChEMBL	>75,000	Chemical structures from more than 48,000 publications	http://www.ebi.ac.uk/chembl/
MarinLit	26,490	Marine natural products, compiled from journal articles	http://pubs.rsc.org/marinlit/

Fig. 19 Total number of register natural products in different available databases. (Taken from Rodrigues and collaborators 2016)

view. The arrival of new technologies, the development, and refinement of techniques of structural analysis, increasingly open doors and new possibilities in the process of structural elucidation and understanding of the mechanisms of action of many of these substances. This allows you to lighten the investigation process like never before. As presented by Rodrigues and collaborators, "natural products and their intricate molecular frameworks offer medicinal chemists a range of uncharted chemo types for the discovery of chemical probes and drugs." It is well-known, in the area of drug design, the use of molecular scaffolds and pharmacophores from isolated and identified compounds from diverse natural sources, which are considered privileged fragments (Rodrigues et al. 2016). Together with the well-known assembly of natural products collected in different databases (Fig. 19), it now joins the development of computational tools that allow, with the appropriate algorithms, the calculation, prediction and design of structures, and better understanding of biological properties.

Polypharmacoloy is another approach to pursue the study of natural products (Anighoro et al., 2014, Moya-García et al., 2017, Ho et al., 2018), deviating from the paradigm of "magic bullet". This view change could allow a re-analysis of many structures and sub-structures of natural products in relation to their biological activity, further enhancing the structural richness and their contribution to the design of new and better drugs, this is undoubtedly a new stage with a promising future and that sustains and motivates the studies in this area.

References

http://www.specs.net/snpage.php?snpageid=home

https://www.sigmaaldrich.com/life-science/cell-biology/bioactive-small-molecules/lopac1280-navigator.html

Allard P-M, Genta-Jouve G, Wolfender JL (2017) Deep metabolome annotation in natural products research: towards a virtuous cycle in metabolite identification. Curr Opin Chem Biol 36:40–49

Anighoro A, Bajorath J, Rastelli G (2014) Polypharmacology: challenges and opportunities in drug discovery. J Med Chem 57:7874–7887

Baell JB, Holloway GA (2010) New substructure filters for removal of pan assay interference compounds (PAINS) from screening libraries and for their exclusion in bioassays. J Med Chem 53:2719–2740

Bisson J, McAlpine JB, Friesen JB, Chen S.-N, Graham J, Pauli GF (2016) Can Invalid Bioactives Undermine Natural Product-Based Drug Discovery? J Med Chem 59:1671–1690

Bleicher KH, Böhm H-J, Müller K, Alanine AI (2003) Hit and lead generation: beyond high-throughput screening. Nat Rev Drug Discov 2:369–378

Boudreau PD, Byrum T, Liu WT, Dorrestein PC, Gerwick WH (2012) Viequeamide A, a cytotoxic member of the kulolide superfamily of cyclic depsipeptides from a marine button cyanobacterium. J Nat Prod 75:1560–1570

Brusotti G, Cesari I, Dentamaro A, Caccialanza G, Massolini G (2014) Isolation and characterization of bioactive compounds from plant resources: the role of analysis in the ethnopharmacological approach. J Pharm Biomed Anal 87:218–228

Bucar F, Wube A, Schmid M (2013) Natural product isolation-how to get from biological material to pure compounds. Nat Prod Rep 30:525–545

Buenz EJ, Verpoorte R, Bauer BA (2018) The ethnopharmacologic contribution to bioprospecting natural products. Annu Rev Pharmacol Toxicol 58:509–530

Capuzzi SJ et al (2017) Phantom PAINS: problems with the utility of alerts for Pan-Assay INterference CompoundS. J Chem Inf Model 57:417–427

Carnevale Neto F, Pilon AC, Selegato DM, Freire RT, Gu H, Raftery D, Lopes NP, Castro-Gamboa I (2016) Dereplication of natural products using GC-TOF mass spectrometry: improved metabolite identification by spectral deconvolution ratio analysis. Front Mol Biosci 3:59. https://doi.org/10.3389/fmolb.2016.00059

Carter GT (2017) Natural products in drug discovery. In: Strømgaard K, Krogsaard-Larson P, Madsen U (eds) Textbook of drug design and discovery, 5th edn. CRC Press, chapter 7, Boca Raton

Chen Y, de Bruyn Kops C, Kirchmair J (2017) Data resources for the computer-guided discovery of bioactive natural products. J Chem Inf Model 57:2099–2111

Ciesla L, Moaddel R (2016) Comparison of analytical techniques for the identification of bioactive compounds from natural products. Nat Prod Rep 33:1131–1145

Corey EJ (1999) Editorial review for comprehensive natural products chemistry. In: Barton DHR, Nakanishi K, Meth-Cohn O (eds) Comprehensive natural products chemistry. Elsevier, Amsterdam/New York

da Silva RR, Dorrestein PC, Quinn RA (2015) Illuminating the dark matter in metabolomics. Proc Natl Sci Acad 112(41):12549–12550

DeHaven CD, Evans AM, Dai H, Lawton KA (2012) Software techniques for enabling high-throughput analysis of metabolomic datasets. In: Roshner U (ed) Metabolomics. Intech. isbn:978-953-51-0046-1. Available from http://www.intechopen.com/books/metabolomics/software-techniques-for-enabling-high-throughput-analysis-on-metabolomic-datasets. Chapter 7, 167–92

Dias DA, Urban S, Roessner U (2012) A historical overview of natural products in drug discovery. Meta 2:303–336

Dias DA, Jones OAH, Beale DJ, Boughton BA, Benheim D, Kouremenos KA, Wolfender J-L, Wishart DS (2016) Current and future perspectives on the structural identification of small molecules in biological systems. Meta 6(4):46. https://doi.org/10.3390/metabo6040046

Dührkop K, Shen H, Meusel M, Rousu J, Böcker S (2015) Searching molecular structure databases with tandem mass spectra using CSI: FingerID. Proc Natl Sci Acad 112(41):12580–12585

Ganzera M, Sturm S (2018) Recent advances on HPLC/MS in medicinal plant analysis – an update covering 2011–2016. J Pharm Biomed Anal 147:211–233

Gaudencio SP, Pereira F (2015) Dereplication: racing to speed up the natural products discovery process. Nat Prod Rep 32:779–810

Gaulton A, Bellis LJ, Bento AP, Chambers J, Davies M, Hersey A, Light Y, McGlinchey S, Michalovich D, Al-Lazikani B, Overington JP (2012) ChEMBL: a large-scale bioactivity database for drug discovery. Nucleic Acids Res 40:D1100–D1107

Gomes NGM, Pereira DM, Valentão P, Andrade PB (2018) Hybrid MS/NMR methods on the prioritization of natural products: applications in drug discovery. J Pharm Biomed Anal 147:234–249

Hanka LJ, Kuentzel SL, Martin DG, Wiley PF, Neil GL (1978) Detection and assay of antitumor antibiotics. In: Carter SK, Umezawa H, Douros J, Sakurai Y (eds) Antitumor antibiotics. Recent results in cancer research/Fortschritte der Krebsforschung/Progrès dans les recherches sur le cancer, vol 63. Springer, Berlin/Heidelberg

Haug K, Salek RM, Steinbeck C (2017) Global open data management in metabolomics. Curr Opin Chem Biol 36:58–63

Ho TT, Tran QTN, Chai CLL (2018) The polypharmacology of natural products. Future Med Chem 10:1361–1368

Hovarth P et al (2016) Screening out irrelevant cell-based models of disease. Nat Rev Drug Discov 15:751–769

Hubert J, Nuzillard J-M, Renault J-H (2017) Dereplication strategies in natural product research: how many tools and methodologies behind the same concept? Phytochem Rev 16:55–95

Jasial S, Bajorath J (2017) Dark chemical matter in public screening assays and derivation of target hypotheses. Med Chem Commun 8:2100–2104

Kim HK, Verpoorte R (2009) Sample preparation for plant metabolomics. Phytochem Anal 21:4–13

Koehn F (2008) High impact technologies for natural products screening. In: Petersen F, Amstutz R (eds) Progress in drug research, vol 65; Natural products as drugs, vol I. Birkhäuser Verlag AG, Basel, pp 177–210

Kremb S, Voolstra CR (2017) High-resolution phenotypic profiling of natural products- induced effects on the single-cell level. Sci Rep 7:44472

Kruk J, Doskocz M, Jodłowska E, Zacharzewska A, Łakomiec J, Czaja K, Kujawski J (2017) NMR techniques in metabolomic studies: a quick overview on examples of utilization. Appl Magn Reson 48:1–21

Kurita KL, Glassey E, Linington RG (2015) Integration of high-content screening and untargeted metabolomics for comprehensive functional annotation of natural product libraries. Proc Natl Sci Acad 112(39):11999–12004

Liu X, Locasale JW (2017) Metabolomics: a primer. Trends Biochem Sci 42:274–284

Lowe, D (2015) The “Dark Matter” of a compound collection. From http://blogs.sciencemag.org/pipeline/archives/2015/10/20/the-dark-matter-of-a-compound-collection. Accessed 26 Feb 2018

Mahieu N, Patti GJ (2017) Systems-level annotation of a metabolomics data set reduces 25 000 features to fewer than 1000 unique metabolites. Anal Chem 89:10397–10406

Maltese F, van der Kooy F, Verpoorte R (2009) Solvent derived artifacts in natural products chemistry. Nat Prod Commun 4(3):447–454

Margueritte L, Markov P, Chiron L, Starck J-P, Vonthron-Sénécheau C, Bourjot M, Delsuc M-A (2018) Automatic differential analysis of NMR experiments in complex samples. Magn Reson Chem 56:1–11

Martin MJ et al (2013) Isolation and first total synthesis of PM050489 and PM060184, two new marine anticancer compounds. J Am Chem Soc 135(27):10164–10171

Martínez-Diez et al (2014) PM060184, a new tubulin binding agent with potent antitumor activity including P-glycoprotein over-expressing tumors. Biochem Pharmacol 88(3):291–302

Mohamed A, Nguyen CH, Mamitsuka H (2016) Current status and prospects of computational resources for natural product dereplication: a review. Brief Bioinform 17(2):309–321

Molinski TF (2014) All natural: the renaissance of natural products chemistry. Org Lett 16(15):3849–3855

Moya-García A, Adeyelu T, Kruger FA, Dawson NL, Lees JG, Overington JP, Orengo C, Ranea JAG (2017) Structural and functional view of polypharmacology. Sci Rep 7:10102. https://doi.org/10.1038/s41598-017-10012-x

Muegge I, Mukherjee P (2016) Performance of dark chemical matter in high throughput screening. J Med Chem 59:9806–9813

National Aeronautics and Space Administration (2018) Dark energy, dark matter. Available from https://science.nasa.gov/astrophysics/focus-areas/what-is-dark-energy. Accessed 6 Feb 2018

Newman DJ, Cragg GM (2016) Natural products as sources of new drugs from 1981 to 2014. J Nat Prod 79:629–661

Nicolaou KC (2014) Advancing the drug discovery and development process. Angew Chem Int Ed 53:9128–9140. https://doi.org/10.1002/anie.201404761

Nothias L-F, Nothias-Esposito M, da Silva R, Wang M, Protsyuk I, Zhang Z, Sarvepalli A, Leyssen P, Touboul D, Costa J, Paolini J, Alexandrov T, Litaudon M, Dorrestein PC (2018) Bioactivity-based molecular networking for the discovery of drug leads in natural product bioassay-guided fractionation. J Nat Prod 81:758–767

Pera B et al (2013) New interfacial microtubule inhibitors of marine origin, PM050489/PM060184, with potent antitumor activity and a distinct mechanism. ACS Chem Biol 8(9):2084–2094

Pye CR, Bertin MJ, Lokey RS, Gerwick WH, Linington RG (2017) Retrospective analysis of natural products provides insights for future discovery trends. Proc Natl Sci Acad 114(22):5601–5606

Robinette SL, Schweiler RB, Schroeder FC, Edison AS (2012) NMR in metabolomics and natural products research: two sides of the same coin. Acc Chem Res 45(2):288–297

Rodrigues T, Reker D, Schneider P, Schneider G (2016) Counting on natural products for drug design. Nat Chem 8:531–541

Sarker SD, Nahar F (eds) (2012) Natural products in drug discovery, 3rd edn. Humana Press/Springer, New York

Schulze CJ et al (2013) "Function-first" lead discovery: mode of action profiling of natural product libraries using image-based screening. Chem Biol 20:285–295

Singh SB, Pelaez F (2008) Biodiversity, chemical diversity and drug discovery. In: Petersen F, Amstutz R (eds) Progress in drug research, vol 65; Natural products as drugs, vol I. Birkhäuser Verlag AG, Basel, pp 142–174

Siramshetty V, Preissner R (2018) Drugs as habitable planets in the space of dark chemical matter. Drug Discov Today https://doi.org/10.1016/j.drudis.2017.07.003 23:481

Wakimoto T, Abe I (2012) Labile natural products. Med Chem Commun 3:866–870

Wassermann AM et al (2015) Dark chemical matter as a promising starting point for drug lead discovery. Nat Chem Biol 11:958–966

Wassermann AM, Tudor M, Glick M (2017) Deorphanization strategies for dark chemical matter. Drug Discov Today Technol 23:69–74

Wolfender J-L, Marti G, Thomas A, Bertrand S (2015) Current approaches and challenges for the metabolite profiling of complex natural extracts. J Chromatogr A 1382:136–164

Wubshet SG, Brighente IMC, Moaddel R, Staerk D (2015) Magnetic ligand fishing as a targeting tool for HPLC-HRMS-SPE- NMR: α-glucosidase inhibitory ligands and alkylresorcinol glycosides from *Eugenia catharinae*. J Nat Prod 78:2657–2665

Zani CL, Carroll AR (2017) Database for rapid Dereplication of known natural products using Data from MS and fast NMR experiments. J Nat Prod 80:1758–1766

Zhang O, Fu Y, Huang C, Zheng C, Wu Z, Zhang W, Yang X, Gong F, Li Y, Chen X, Gao S, Chen X, Li Y, Lu A, Wang Y (2016) New strategy for drug discovery by large-scale association analysis of molecular networks of different species. Sci Rep 6:21872. https://doi.org/10.1038/srep21872

Zhang C, Idelbayev Y, Roberts N, Tao Y, Nannapaneni Y, Duggan BM, Min J, Lin EC, Gerwick EC, Cottrell GW, Gerwick WH (2017) Small molecule accurate recognition technology (SMART) to enhance natural products research. Sci Rep 7:14243. https://doi.org/10.1038/s41598-017-13923-x

follow their intrinsic biological activity (Křen 2011). On the other hand, C-glycosidic compounds feature metabolic stability (desirable for new drug candidates and biochemical probes, among others), which makes them similar to natural pyran-containing polyethers and macrocycles, which are the main focus of the review. Finally, praising synthetic approaches to material needs encountered in natural product inspired medicinal chemistry and drug discovery, we use the term "total synthesis" in a rather general sense – when presenting syntheses of such complex target molecules as marine toxins, chiral pool is amply represented, with many examples of derivatives obtained from natural carbohydrates. The same concerns new generation of therapeutics, like SGLT2 inhibitor antidiabetics discussed in this chapter, which are developed based on natural product glycosides.

2 Synthesis of Six-Membered Oxygen Heterocycles Involving Cyclization

Numerous classes of natural compounds isolated from plant material, marine organisms, and animals contain oxa-heterocyclic structures. Many of the compounds (Fig. 2) bearing five-membered cyclic units (furans) and six-membered oxygen heterocycles (pyrans) exhibit biological activity. Based on these properties, a large number of pharmaceutical agents containing oxacyclic compounds have been developed (Wilson and Danishewsky 2006). Intensive investigations of their biological activity are hampered by the limited availability of the natural products. The significant biological and pharmaceutical activity of oxygen heterocyclic compounds has

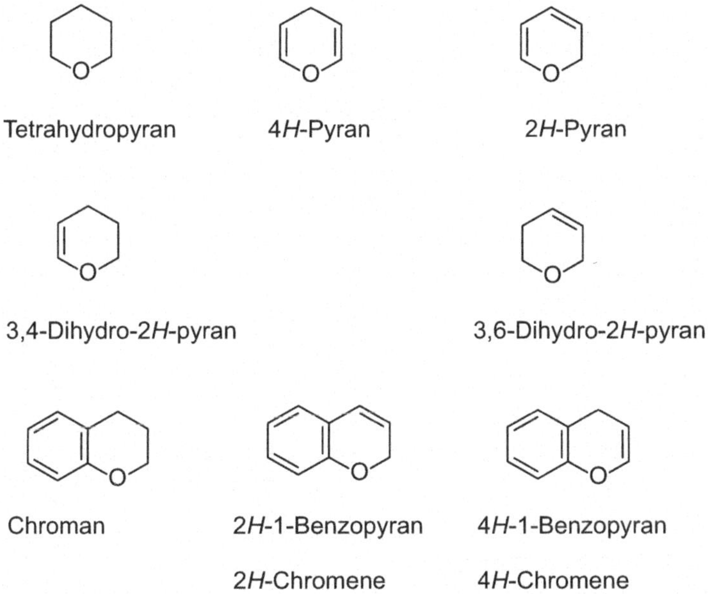

Fig. 2 The structure of six-membered oxygen heterocycles widespread in natural products

stimulated study on the synthesis of cyclic compounds, especially pyrans. The literature covering oxygen heterocycles' synthesis is extensive, as can be seen from the compilation referenced here (Nishiwaki 2014; Caberele and Reiser 2016; Perry et al. 2014; Ylijoki and Stryker 2012). The hetero-Diels-Alder reaction (HDA reaction) is frequently used at an early stage of a synthesis of a complex natural product to establish a structural heterocyclic core, which can be elaborated to the more complex target structure. The design of the synthesis includes its analysis in terms of potentially available dienophile or diene components, variants on the structure of the intermediate for Diels-Alder disconnection, tactics for ensuring stereocontrol and/or position control in the Diels-Alder addition, and possible chiral control elements for an enantioselective Diels-Alder reaction. Several syntheses of 2*H*-pyrans are based on hetero-Diels-Alder reactions (HDA reactions) of dienophiles with dienes, or α,β-unsaturated aldehydes, ketones, acids, esters, and acetals. As the chirality of a natural product is crucial to its bioactivity (Mori 2011), the development of a synthetic strategy that can provide both enantiomers in adequate amounts is clearly important for the further investigation of its biological activity. In some cases, α,β-unsaturated carbonyl compounds react with electron-rich 1,2-dialkoxy-substituted alkenes to generate dihydropyrans, and multiple stereogenic centers are formed in a single step. In the strategies outlined for the synthesis of complex biologically active compounds, the dihydropyran products could be converted to the desired compounds by C-C bond formation, reduction, oxidation, and other functional group transformations. Selected effective methods for the preparation of natural biologically active compounds using HDA reactions are discussed.

Several synthesis of dihydropyrans are based on the hetero-Diels-Alder reaction (HDA reaction) of electron-rich dienophiles with α,β-unsaturated carbonyl compounds, mainly aldehydes, ketones, acids, esters, and acetals. In the total synthesis of natural compounds, the electron-rich 1,3-dialkoxy substituted dienes designed by Danishewsky (Danishewsky et al. 1982) are conventionally used as substrates. The hetero-Diels-Alder (HDA) reaction between 1-methoxy-3-trimethylsilyloxy-1,3-butadiene (Danishewsky's diene) and aldehydes is a very important method for synthesizing dihydropyrones (Danishewsky 1986). The first HDA reaction of Danishewsky's diene and aldehydes was achieved with Lewis acid catalysts, leading to a mixture of stereoisomers (Danishewsky et al. 1982) (Fig. 3).

Early efforts to control the stereoselectivity in HDA reactions focused on the use of chiral aldehydes (Danishewsky et al. 1985) in substrate-controlled diastereoselective reactions (Fig. 4).

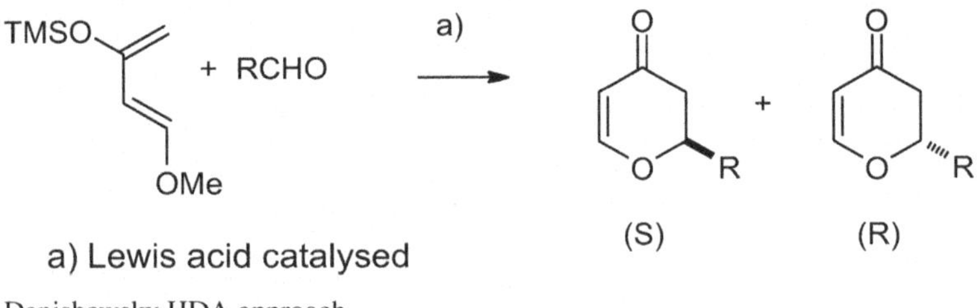

Fig. 3 Danishewsky HDA approach

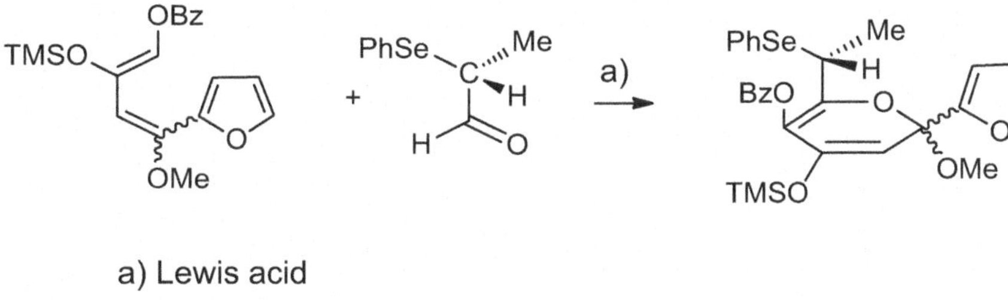

Fig. 4 Substrate-controlled diastereoselective HDA reaction

In this application, the hetero-Diels-Alder reaction occurred via a BF_3-catalyzed cycloaddition between the diene and aldehyde (as a chiral acrolein equivalent). Subsequent post-Diels-Alder transformation was used to install the carbohydrate functionality into the initial Diels-Alder adduct, leading to the synthesis of rare sugars. The development of a method of chiral substrate control of the HDA reaction and asymmetric approaches to various hexoses synthesis has been published (Danishewsky and De Ninno 1987). Asymmetric catalysis is an efficient methodology for the construction of a wide variety of enantiomerically enriched carbon skeletons as well as for the enantioselective installation of the desired functional groups (List 2010; Büschleb et al. 2016). Intensive efforts in this field have led to the emergence of numerous asymmetric catalysts, and now a number of reactions can be performed with high stereoselectivity (Kumagai and Shibasaki 2011; Fan et al. 2002).

Attention has been given to the discovery of chiral catalysts to effect enantioselective HDA reactions of Danishewsky's diene **1** and related electron-rich dienes (Jiang and Wang 2013; Cherney et al. 2015). Especially noteworthy are the results of a condensation reaction carried out with the participation of organometallic reagents. The Jacobsen research group has developed two related catalyst systems for effecting enantioselective HDA reactions. The (salen)Cr(III)-BF_4 complex was found to be a reactive and effective catalyst for the HDA reaction between **1** and achiral aldehydes (Guy et al. 2002). To establish the viability of chiral catalyst-controlled doubly diastereoselective HDA reactions, the reaction between diene and optically active chiral lactic aldehyde was investigated as a model reaction (Fig. 5). Jacobsen established that chiral (salen)-Cr(III) complexes could effectively catalyze Danishewsky's diene hetero-Diels-Alder/elimination reaction in excellent yields and enantiomeric excesses.

This methodology provides selective access to dihydropyranone stereoisomers by the controlled use of aldehyde and catalyst enantiomers (Guy et al. 2002). Suitable dihydropyranone products are not readily accessible using chiral substrate-controlled diastereoselective reactions. To improve the asymmetric HDA reaction of Danishewsky's diene with aldehydes, various chiral Lewis acids, chiral complexes of 1,1'-bi-2-naphthol (BINOL), such as its Ti-BINOL, Zr-BINOL, and Zn-BINOL complexes, have been employed for this type of reaction, in some cases with medium stereoselectivity (Cherney et al. 2015). It was found that the chiral zinc complex obtained from diethyl zinc and 3,3'-dibromo-1,1'-bi-2-naphthol [3,3'-Br_2-BINOL-Zn] was an efficient catalyst in various HDA reactions (Du et al. 2002) (Fig. 6).

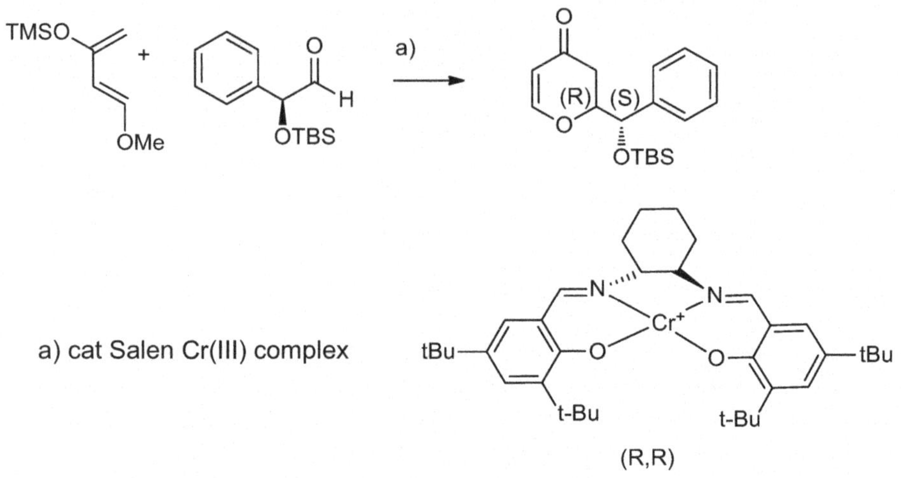

Fig. 5 The catalyst-controlled doubly diastereoselective HDA reaction

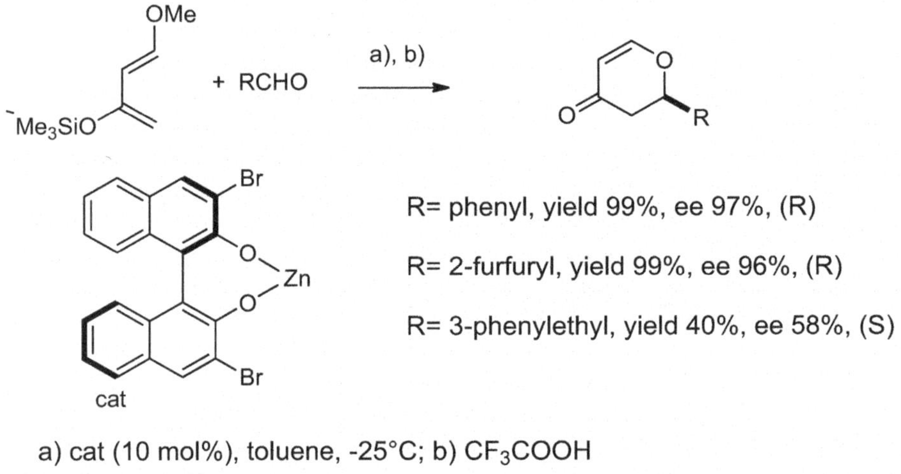

Fig. 6 The enantioselective HDA reaction of Danishewsky's diene and aldehydes

The Jørgensen group reported the first highly enantioselective HDA reaction of Danishewsky's diene with ketones catalyzed by a chiral copper(II) complex (Yao et al. 1998). Recently, an efficient catalytic enantioselective hetero-Diels-Alder reaction of Danishewsky's dienes with α-carbonyl esters using a chiral In(III)-pybox complex has been presented (Zhao and Loh 2013) (Fig. 7).

This protocol offers several advantages, including mild reaction conditions, relatively low catalyst loading, and good to excellent enantioselectivities. The highly enantioselective hetero-Diels-Alder reaction of Danishewsky's diene with glyoxals was carried out with the participation of a readily accessible chiral copper catalyst

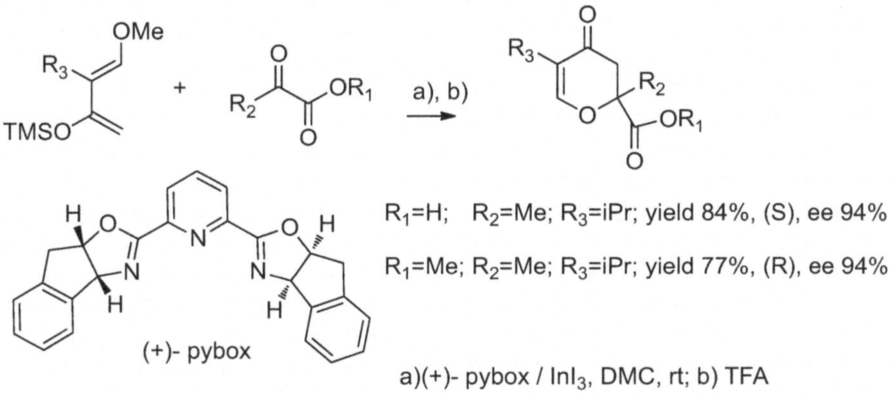

Fig. 7 Asymmetric HDA reactions of Danishewsky diene with α-ketoesters

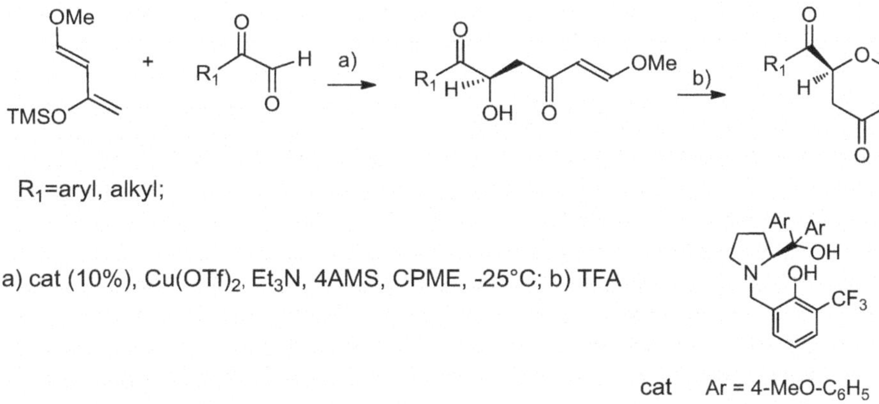

Fig. 8 Asymmetric HDA reactions of Danishewsky diene with α-ketoaldehyde

(Li et al. 2016). This efficient transformation provided a facile and scalable access to a wide range of biologically active dihydropyrones with a high level of enantioselectivity (Fig. 8). Moreover, the substrate scope of this reaction could be extended to isatins with this catalytic system.

Chiral 5,6-dihydropyran-2-one or α,β-unsaturated δ-lactone derivatives are key structural subunits of natural products with a wide range of biological activity (Bindseil and Zeeck 1993), such as antifungal and antitumor properties. Thus, the synthetic methodology of δ-lactones has been an area of intense research efforts. From a synthetic viewpoint, one of the most convenient ways to δ-lactones is based on an HDA reaction of Brassard's diene (Savard and Brassard 1979) with suitable aldehydes or ketones (Fig. 9).

Ding's group (Du et al. 2004) has used TADDOL to catalyze the reaction of aldehydes and Brassard's diene **1** (route A), thus preparing lactones (45–85% yield,

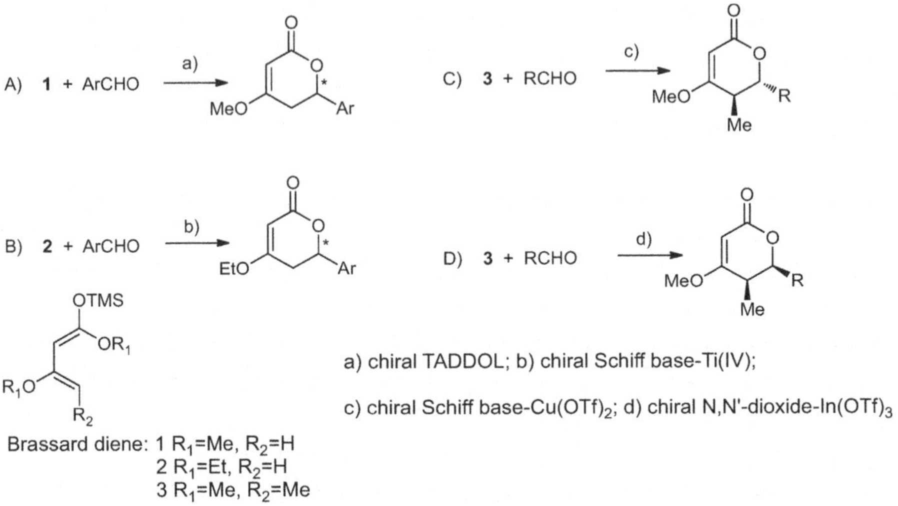

Fig. 9 Synthesis of δ- lactones by reaction of Brassard dienes with aldehydes

68–91% ee). However, when the system was employed for the reaction of diene **3** with benzaldehyde, disappointing results were obtained (yield 50%, ee 2%). Feng and co-workers described (Fan et al. 2004) a highly enantioselective synthesis of δ-lactones by an HDA reaction of Brassard's diene **2** with aromatic aldehydes catalyzed by titanium (IV) tridentate chiral Schiff base complexes (route B). Only $Ti(OiPr)_4$ gave a high ee, up to 93%. At the same time, the HDA reaction of the Brassard-type diene **2** with aromatic aldehydes was also achieved by Cu(II)/Schiff base complexes (route B) (Lin et al. 2006). Efficient chiral Schiff base copper(II) complexes were developed for the highly enantio- and diastereoselective HDA reaction of the Brassard-type diene **3** with aldehydes (route C) to afford the corresponding α,β-unsaturated δ-lactone derivatives in moderate yields, high enantioselectivities (up to 99% ee), and excellent diastereoselectivities (up to 99:1 anti/syn) (Lin et al. 2006). A highly diastereo- and enantioselective hetero-Diels-Alder reaction of the Brassard-type diene **3** with aliphatic aldehydes catalyzed by the chiral N,N'-dioxide/$In(OTf)_3$ complex has been developed (route D). The corresponding β-methoxy-γ-methyl-α,β-unsaturated-δ-lactones were obtained in good yields (83%) as well as high dr and ee values (up to 97:3 syn/anti and 94% ee) (Lin et al. 2011). Such adducts could be easily transformed into the building blocks existing in many natural products by hydrogenation.

An excellent example of the synthetic application of Brassard-type dienes is the preparation of kavain and analogous natural products, which are interesting for drug discovery against a variety of cellular targets, including P-glycoprotein (Pgp), cytochrome P450, and cyclooxygenase (COX) enzymes, among others (Rowe et al. 2011). An efficient catalytic asymmetric HDA reaction of Brassard's diene with aliphatic aldehydes led to kavain (Fig. 10). The catalyst, which was generated from (R)-BINOL, $Ti(i\text{-}PrO)_4$, and 4-picolyl chloride hydrochloride, promoted the reaction smoothly to afford the corresponding α,β-unsaturated δ-lactone derivatives in

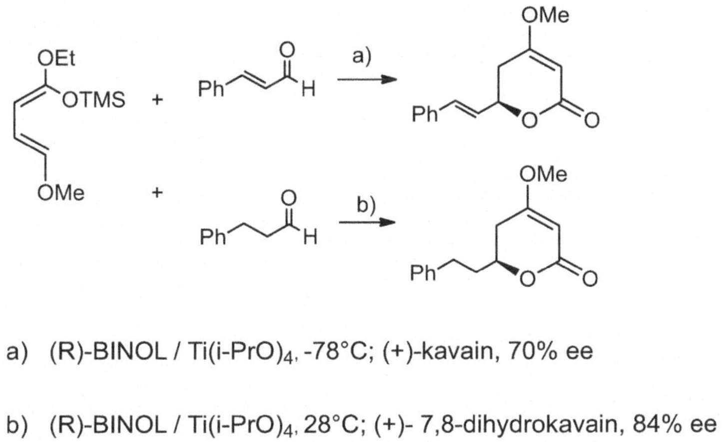

Fig. 10 One-pot synthesis of (+)-kavain and (+)-7,8-dihydrokavain

moderate to good yields (46–79%) with high enantioselectivities (up to 88% ee). By using this methodology, the natural products (R)-(+)-kavain (70% ee, >99% ee after a single recrystallization) and (S)-(+)-dihydrokavain (84% ee) were prepared in one step starting from cinnamaldehyde and 3-phenylpropionaldehyde (Lin et al. 2008).

In various studies on the preparation of biologically active natural products, attention was focused on chiral acids, which are nonmetal catalysts. Stereoselective variants have been achieved by means of different types of chiral catalysts, including phosphoric, carboxylic, and sulfonic acids (El-Sepelgy et al. 2014; Momiyama et al. 2009). Examples of their catalytic activity are found in various papers on the preparation of natural products of plant origin. The synthesis of paeoveitol is a good example. The root of *Paeonia veitchii*, known as Chuan Chi Shao in China, is an important crude drug in traditional Chinese medicine and is used as a sedative, analgesic, and cardiovascular agent (Ruan et al. 2017). Biologically active compounds, paeoveitols, have been isolated from the root. However, investigations into the biological activity of paeoveitol are hampered by the limited availability of this natural product, and the development of a synthetic strategy that can provide both enantiomers in adequate amounts is clearly important for further investigation of its biological activity. Brønsted acid catalysts have been exploited to obtain a diversity of chiral dihydropyrones, and the first catalytic asymmetric total synthesis of (+)- and (−)-paeoveitol has been accomplished in 42% overall yield via a biomimetic HDA reaction (Li et al. 2017). The chiral phosphoric acid-catalyzed HDA reaction showed excellent diastereo- and enantioselectivity (>99:1 dr and 90% ee); two rings and three stereocenters were constructed in a single step to produce (−)-paeoveitol on a scale of 452 mg. The highly stereoselective synthesis of both enantiomers has been accomplished by a catalytic transformation involving chiral acids as catalysts (Fig. 11).

The current status of organic synthesis is hampered by costly protecting group strategies and lengthy purification procedures after each synthetic step. To circumvent these problems, the synthetic potential of multicomponent domino reactions has

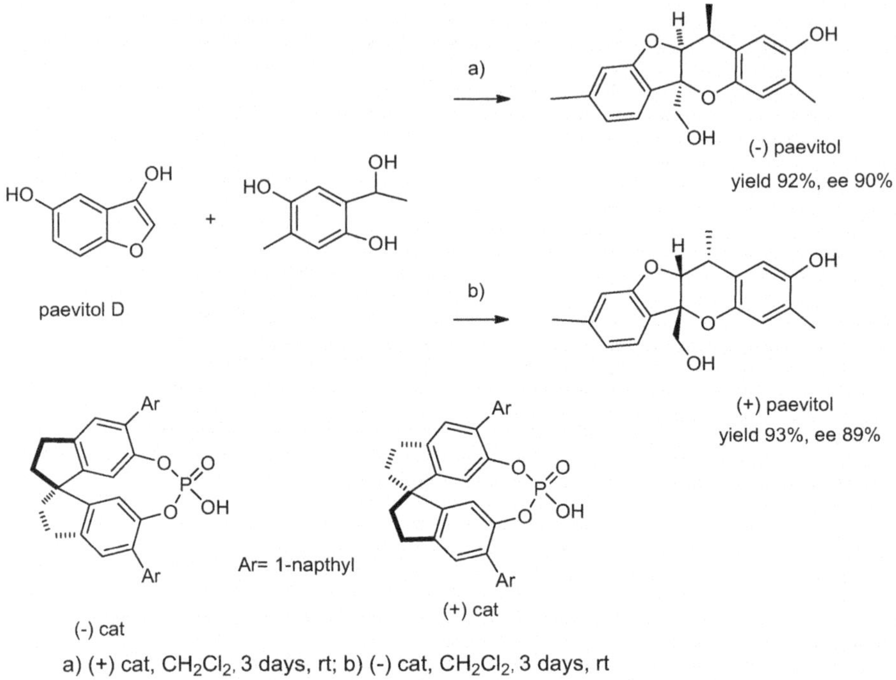

Fig. 11 Stereoselective synthesis of (+)-paevitol

been utilized for the efficient and stereoselective construction of complex molecules from simple precursors in a single process (Witczak and Bielski (eds) 2016). A domino reaction has been defined by Tietze as a "reaction involving two or more bond-forming transformations that take place under the same reaction conditions, without adding additional reagents and catalysts, and in which the subsequent reactions result as a consequence of the functionality formed by bond formation or fragmentation in the previous step" (Tietze 2014; Tietze and Düfert 2012). Domino reactions avoid time-consuming and costly protection/deprotection processes, as well as the purification of intermediates. They often proceed with excellent stereoselectivities and are environmentally friendly. The efficiency of asymmetric domino reactions can be judged by the number of bonds formed, the number of newly created stereocenters, and the increase in molecular complexity. In particular, asymmetric domino sequences involving cycloaddition reactions are highly effective processes for the rapid elaboration of complex polycyclic systems, since each cycloaddition event generates a new ring and two new covalent bonds (Tietze et al. 2006). In particular, domino reactions mediated by organocatalysts are in a way biomimetic, as this principle is used very efficiently in the biosynthesis of complex natural products starting from the natural products scaffolds (List 2010). As a natural product scaffold, the benzoquinone core is ubiquitous in many bioactive natural products and pharmaceuticals (Enders et al. 2007). An example is embelin (1), a naturally occurring alkyl substituted hydroxybenzoquinone and a major constituent of the Andean medicinal

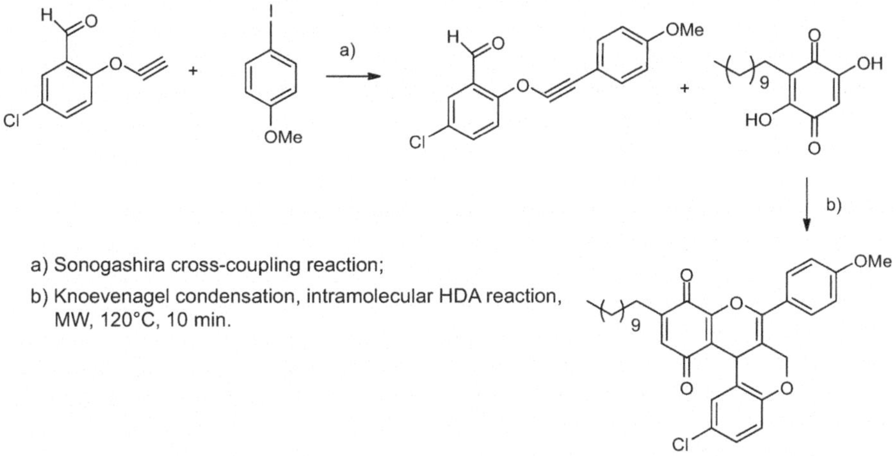

Fig. 12 Synthesis of embelin analogs

plant *Oxalis erythrorhiza*, belonging to the *Oxalidaceae* family (Naik et al. 2013). It has been reported that embelin possesses a number of therapeutic properties, including antidepressant, antitumor, anti-inflammatory, and antibacterial properties (Deshmukh and Gupta 2013; Harvey 2008). All these activities make embelin an interesting scaffold for synthesizing new and more selective therapeutic agents. A highly efficient and regioselective approach to new polycyclic embelin derivatives through a domino Knoevenagel condensation/intramolecular hetero-Diels-Alder reaction using O-(arylpropynyloxy)-salicylaldehydes in the presence of ethylenediamine diacetate (EDDA) has been reported (Martín-Acosta et al. 2016). This organocatalyzed protocol is compatible toward a wide range of aryl-substituted alkynyl ethers with electron-donating and electron-withdrawing groups. A microwave-assisted intramolecular approach for the synthesis of a set of new angular tetracyclic embelin derivatives was proposed (Fig. 12). The presence of the phenyl group is essential in order to favor the intramolecular cycloaddition reaction. Diversely substituted embelin adducts could be prepared in good yields from a large variety of aryl-substituted alkynyl ethers with electron-donating and electron-withdrawing groups.

This efficient organocatalyzed protocol was successfully applied to a variety of active methylene compounds and for de novo synthesis of rare sugars. An example is the preparation of forosamine, an integral part of the *Saccharopolyspora spinosa* metabolite spinosyn (Kühne and Benson 1965), which shows high antibiotic and insecticidal activity.

The sugar scaffold was constructed (Tietze et al. 2009) by a new domino-Knoevenagel-HDA reaction of nitroaldehyde, formaldehyde, and ethyl vinyl ether (Fig. 13).

Selective reduction of the double bond in pyran led to the nitrosugars which was followed by isomerization, reduction of the nitro group, and successive Cbz-protection. Chromatographic resolution, deprotection of the amino group with

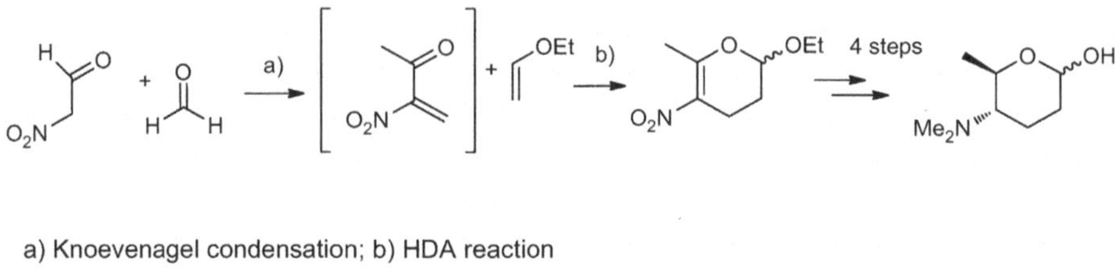

Fig. 13 Synthesis of (+)-D-forosoamine

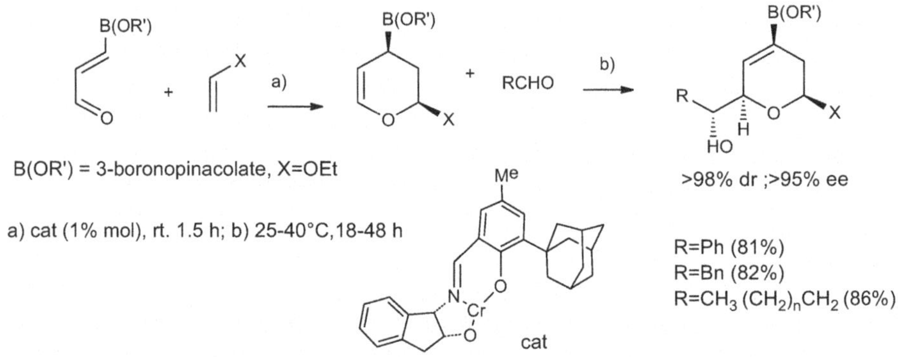

Fig. 14 Tandem hetero-[4 + 2] cycloaddition/allylboration reaction

simultaneous demethylation, and removal of the anomeric protecting group led to (+)-D-forosamine in an enantiomeric excess of up to 95%.

Multicomponent reactions (MCR), transformations employing three or more simple substrates in a single and highly atom-economical operation, are very attractive in both natural product synthesis and the diversity-oriented synthesis of drug-like molecules (Zhu et al. 2015; Touré and Hall 2009; Hall et al. 2016). Several popular multicomponent reactions have been designed by combining two well-established individual reactions that utilize mutually compatible substrates. In this regard, the Diels-Alder cycloaddition and carbonyl allylboration can produce a powerful and highly versatile tandem MCR process. This methodology was applied to the synthesis of α-hydroxyalkyl pyrans, subunits of a number of natural products (Cioc et al. 2014). These substances display a broad range of biological properties, including antibiotic and anticancer activity. They are attractive tools in the diversity-oriented synthesis of drug-like molecules (Hulme and Gore 2003). An effective method of pyran synthesis (Hall et al. 2012) is oxa [4 + 2] cycloaddition/allylboration (Fig. 14). In these HDA [4 + 2] cycloadditions, a boronodiene was reacted with an alkene bearing an EDG group, followed by the addition of aldehydes to the cyclic allylboronates in a separate step, leading to α-hydroxyalkylated carbocycles 4 via a highly diastereoselective nucleophilic allylation.

Multicomponent reactions (MCR), transformations employing three or more simple substrates in a single and highly atom-economical operation, are very attractive in both natural product synthesis and diversity oriented synthesis of drug like molecules (Zhu et al. 2015; Touré and Hall 2009; Hall et al. 2016). Several popular multicomponent reactions were designed by combining two well-established individual reactions that utilize mutually compatible substrates. In this regard, the Diels-Alder cycloaddition and carbonyl allylboration would produce a powerful and highly versatile tandem MCR process. This methodology was applied in the synthesis of α-hydroxyalkyl pyrans, subunits of a number of natural products (Cioc et al. 2014). These substances display a broad range of biological properties, including antibiotic and anticancer activity. They are attractive tools in diversity-oriented synthesis of drug-like molecules (Hulme and Gore 2003). The effective method of synthesis pyrans is oxa [4 + 2] cycloaddition/allylboration. The Diels-Alder cycloaddition and carbonyl allylboration (Hall et al. 2012) would produce a powerful and versatile tandem MCR process (Fig. 14). In HDA [4 + 2] cycloadditions, boronodiene was reacted with alkene-bearing EDG group, followed by addition of aldehydes to the cyclic allylboronates, in a separate step, leading to α-hydroxyalkylated carbocycles via a highly diastereoselective nucleophilic allylation.

Consequently, this oxa [4 + 2] cycloaddition/allylboration sequence can be executed in a one-pot three-component procedure from a boronodiene by simple cycloaddition with a suitable alkene, catalyzed by a chiral Jacobsen chromium catalyst (Gademann et al. 2002), and the addition of another aldehyde after the completion of the cycloaddition stage. Simple aldehydes react at a relatively low temperature (40°C) in neat vinyl ethyl ether to afford α-hydroxy dihydropyrans as a single diastereomer. Suitable aldehydes include both aromatic ones, with various electronic characteristics, and aliphatic ones. The resulting products were obtained in good to excellent yields and >95% ee.

Application of the oxa [4 + 2] cycloaddition/allylboration to the use of chiral α-substituted aldehydes gave a convincing demonstration of utility featuring the synthesis of the cytotoxic antitumor natural product, goniodiol, and its natural analog 8-methoxygoniodiol (Carreaux et al. 2006; Favre et al. 2008) (Fig. 15). To this end, use of (2*R*)-methoxy(phenyl)-acetaldehyde in the double-diastereoselective MCR led to product as a single stereoisomer. Four more steps were needed to complete the

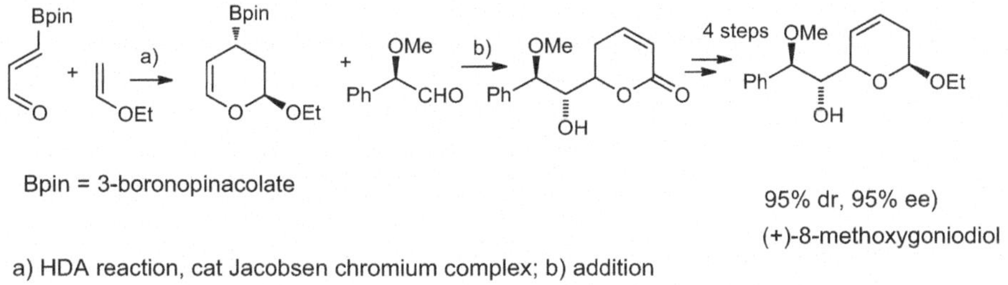

Fig. 15 Synthesis of (+)-8-methoxygoniodiol

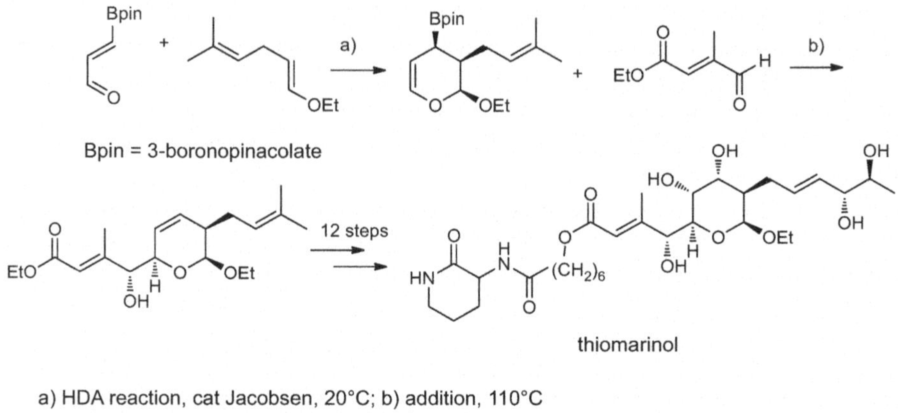

Fig. 16 Synthesis of thiomarinol

synthesis of desired compound: alcohol protection, followed by oxidation of the acetal to the corresponding lactone, then alkene migration, and fluorodesilylation. A similar sequence was exploited in the syntheses of (+)-goniotriol, (−)-goniofupyrone, (+)-altholactone, and (+)-iso-exo-brevicomin (Favre et al. 2008).

Hall and co-workers used an example of the oxa [4 + 2] cycloaddition/allylboration MCR in their concise total synthesis of the thiomarinol family of antibiotics (Marion et al. 2009; Rybak and Hall 2015) related to the commercial antibacterial agent, mupirocin. The key MCR step of this synthesis aimed at preparing the dihydropyran (Fig. 16). A boronodiene and a 3:1 Z/E mixture of enol ether had to be employed in the [4 + 2] cycloaddition. Fortunately, the Jacobsen chromium catalyst promoted the cycloaddition of the requisite Z-enol ether faster than that of the corresponding E isomer, thus affording the cyclic allylboronate as the sole stereoisomer. The latter added stereoselectively to the unsaturated aldehyde in a "one-pot" sequential process that delivered the targeted product in good yield. The remarkable selectivity of this MCR enabled an expedient enantio-controlled synthesis of thiomarinol, which was obtained in 22% overall yield.

This method allowed the design of a series of thiomarinol analogs that were evaluated for their antibacterial activity (Marion et al. 2009). As recently demonstrated with a short enantioselective synthesis of diospongin B (Rybak and Hall 2015), merging hetero [4 + 2] cycloadditions with transition metal-catalyzed cross-coupling reactions may open further opportunities for designing new MCRs.

In recent years, transition metal-catalyzed metathesis has become one of the most efficient methods for the formation of carbon-carbon double bonds (Grubbs et al. 2014; Grela 2014). This reaction has become the method of choice in the formation of medium to large rings and the functionalization of intermediate compounds in the synthesis of complex natural products and pharmaceuticals (Higman et al. 2016). In this methodology, Ru-dichloro compounds fulfil a special role as catalysts, due to their ability to promote the reactions of hindered alkenes bearing a neighboring hydroxyl, carbonyl, or carboxylic group. The most significant input into laboratory practice was the discovery by Grubbs and co-workers (Grubbs et al.

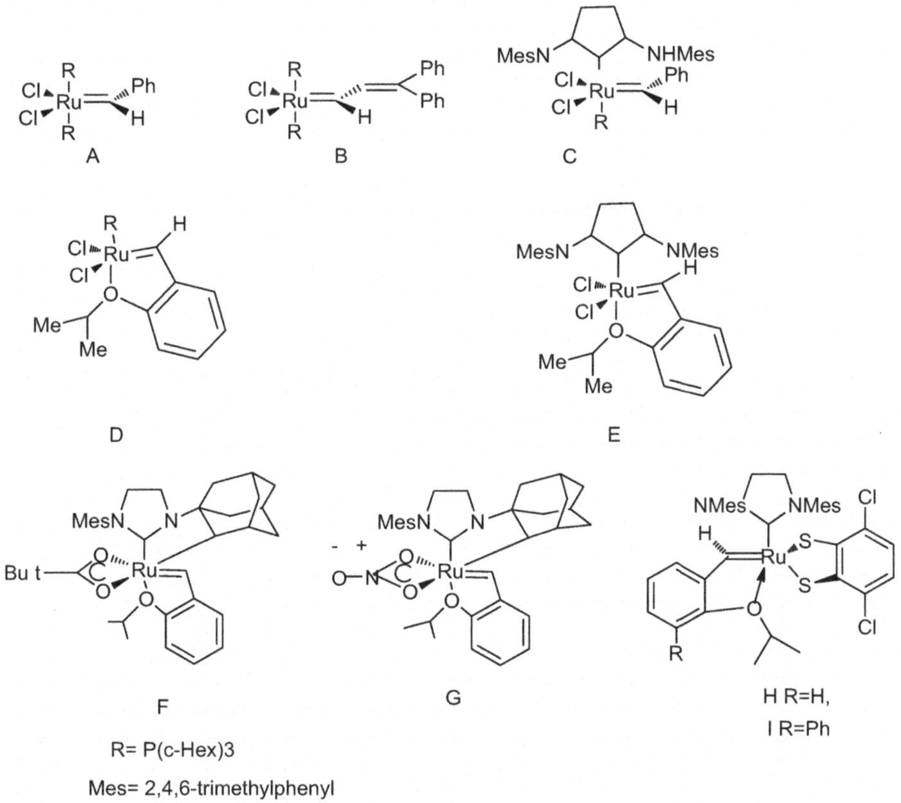

Fig. 17 The selected ruthenium-based metathesis catalysts

2014) of the stable benzylidene complexes, which are known as first-generation Grubbs catalysts (Fig. 17 cat A, cat B). As the methods for performing metathesis transformations have evolved and increased, a great deal of effort has been applied to the design of catalysts with increased lifetime and functionality. Hoveyda and co-workers introduced the O-isopropoxybenzylidene group (Hoveyda 2014), improving the stability of the catalysts (Fig. 17 cat E, cat F). An important improvement in catalyst effectiveness was a phosphine ligand's replacement by N-heterocyclic carbenes (Fig. 17 cat C), providing increased stability, activity, and stereoselectivity (Grela 2014). These Ru-based complexes exhibit high reactivity, show little sensitivity to air and moisture, and can be stored in the air without decomposition; they can be recycled by chromatography. Achieving high selectivity for 1,2-disubstituted olefinic products with a particular stereochemistry (E or Z) was a long-standing challenge in olefin metathesis. Grubbs and co-workers reported that kinetic control of Z-selectivity (Keitz et al. 2012) can be achieved by substituting anionic chlorides with an alkyl and an oxo ligand (Fig. 17 G, H, I).

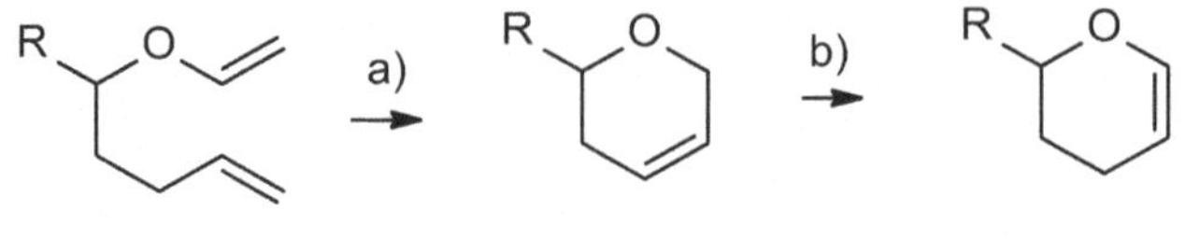

R= Ph ,$PhCH_2CH_2$, PhCH=CH

A a) Grubbs' cat A (10 mol%), benzene, D; b) cat rutenium hydride,H_2

B a) Grubbs' cat A (5 mol%), toluene, rt.; b) NaH or $NaBH_4$, 110°C

Fig. 18 RCM synthesis of dihydropyrans

Hoveyda and co-workers disclosed the design of a sterically and electronically distinct bidentate disulfide ligand (cat H), which may be used for efficient catalytic Z-selective olefin cross-metathesis and macrocyclic ring-closing metathesis (Khan et al. 2014). Recently, Hoveyda and co-workers reported an excellent methodology for the highly stereoselective formation of E isomers (Xu et al. 2017) by incorporation of Z-butene together with readily accessible Ru-based dithiolate catalysts. The first demonstration of highly kinetic trans-selective transition metal-catalyzed olefin cross metathesis was reported last year, with the unexpected discovery that ruthenium-based catalysts bearing chelated dithiolate ligands are able to perform cross-metathesis between two trans olefins or between a trans olefin and a terminal olefin to generate products with high trans selectivity (Ahmed and Grubbs 2017). His result significantly expands the applicability of the CM reaction at the functionalization stage of complex natural compounds.

The possible applications of metathesis in the synthesis of simple and complex combinations comprising the pyran structure are illustrated by the following examples. Sturino was able to use Grubbs' catalyst **A** to catalyze the ring-closing metathesis (RCM) reactions of a variety of vinyl ethers (Sturino and Wong 1998) (Fig. 18, route A). Schmidt reported a similar approach for preparing 3,4-dihydropyrans and other five- and seven-membered cyclic enol ethers from allyl homoallylic ethers (Schmidt 2003) (Fig. 18, route B). In this case, the catalyst was activated in a second step using NaH or $NaBH_4$, thereby making the method a two-step, one-pot procedure. The dienes first underwent RCM reaction with **A** for 20–60 min at room temperature, NaH or $NaBH_4$ was then added, and the reaction mixture was heated to 110°C to form 3,4-dihydro-2*H*-pyrans. This procedure appears to give products in high yields, although the harsh conditions for the second step might be a disadvantage for sensitive substrates.

The stereoselectivity of the RCM reactions of chiral olefins was reported by Schmidt and Wildemann (Schmidt and Wildemann 2000). Depending on the steric demand of the oxo substituent of the divinyl carbinol moiety (either unprotected OH, TBDMS, or benzyl ether), different diastereomers are preferably formed upon

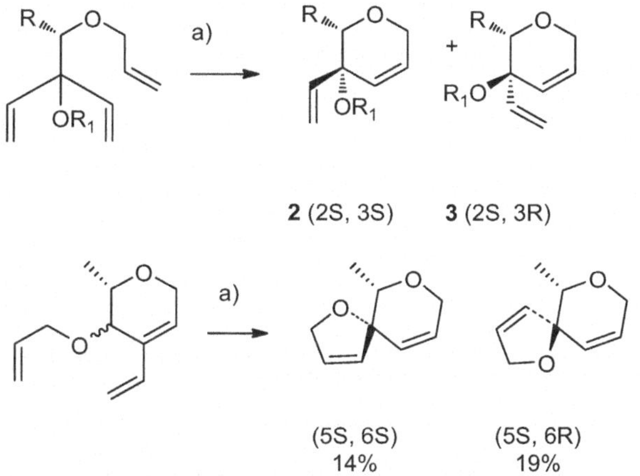

R	R1	2/3	yield(%)
Me	H	3/1	66
Me	Bn	1/3	45
Me	TBDMS	1/2	91
Ph	Bn	1/10	62
Ph	TBDMS	1/6	87

a) Grubbs' catalyst A, $Cl_2(Cy_3P)_2Ru{=}CHPh$ (3 mol %)

Fig. 19 Asymmetric synthesis of 2*H* Pyrans by RCM reaction

ring-closing metathesis (Fig. 19). Stereoinduction is more efficient for the phenyl derivatives due to the higher steric demand of the phenyl substituent compared to the methyl group. Finally, a two-step procedure to synthesize the spirocyclic system was proposed. Allylation of the alcohol mixture (2*S*,3*S*):(2*S*,3*R*) = 3:1 led to dihydropyran. Ring-closing metathesis of this intermediate proceeded smoothly in the presence of Grubbs' catalyst to produce a 3:1 mixture of spirocycles.

Six-membered lactone rings constitute a structural feature common to numerous biologically active natural products (Hoffmann and Rabe 1985), many of which exhibit antitumor properties. A number of lactones have been synthesized using the RCM reaction, and an example is found in Honda's synthesis of malyngolide (Mizutani et al. 2002). Strategies that were developed for preparing valerolactone-derived natural products applied RCM reactions wherein esters of acrylic acid were treated with Grubbs' catalyst (sometimes in the presence of a Lewis acid such as titanium tetraisopropoxide) to yield 2,3-dihydro-4*H*-pyran-4-one derivatives (Fig. 20). When the RCM substrate was treated with Grubbs' catalyst **A**, the lactone was obtained in low yield. However, use of 1 mol% of the more reactive and recyclable Hoveyda catalyst **E** afforded the desired product in very good yield. The synthesis of malyngolide was then completed in two steps, deprotection followed by hydrogenation.

It has been claimed that multi-RCM reactions constitute a powerful tool for the rapid assembly of complex polycyclic ethers. Such strategies have been used to develop approaches to natural products. An interesting application is illustrated by the two-directional approach toward the precursor of the C(20)–C(36) subunit of halichondrin B. This fragment has been prepared in a one-pot operation (Burke et al. 1998) involving a ring opening and a double ring closing metathesis reaction of ethers employing Schrock's catalyst (Fig. 21).

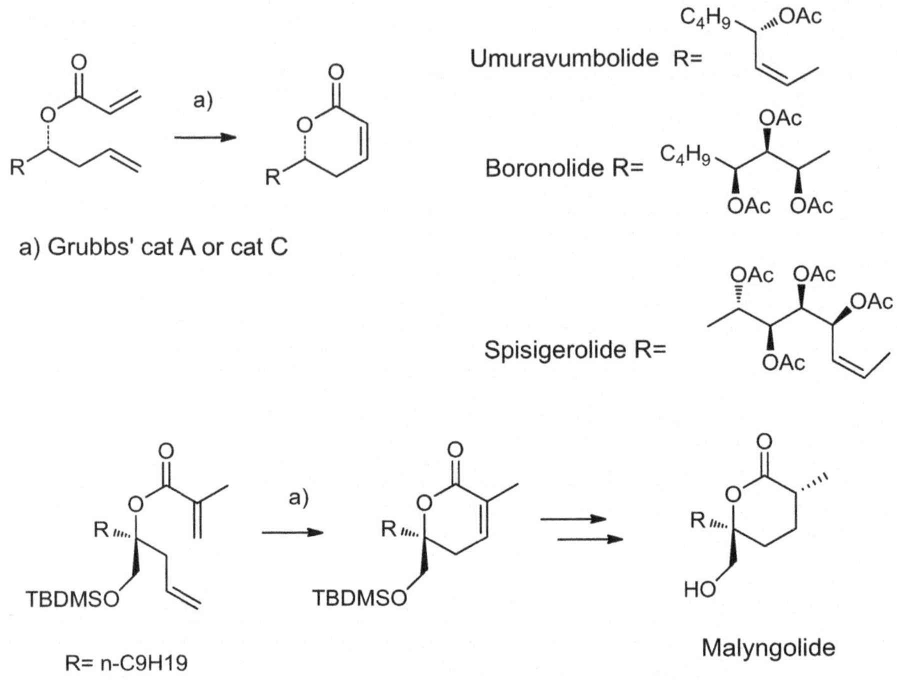

a) Grubbs' cat E (5 mol%), benzene ,70°C

Fig. 20 RCM synthesis of δ-lactones

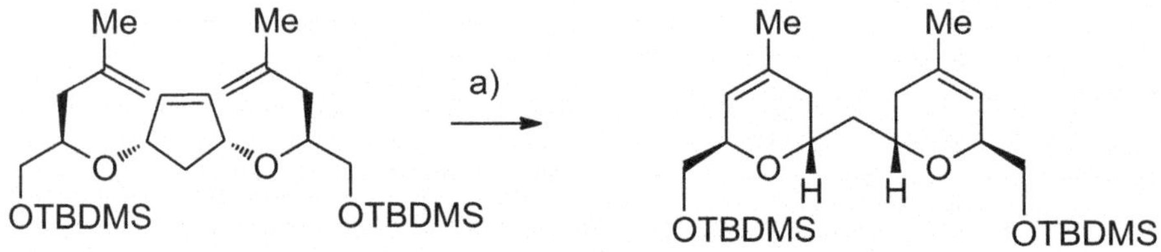

a) Grubbs' cat F (25 mol%), benzene, 60°C

Fig. 21 Synthesis of the halichondrin B subunit

Benzo-fused oxabicyclic compounds are found in a variety of important natural products and hence are interesting scaffolds for drug design. Of these, one prominent class of oxygen heterocycles are the chromenes (Schweizer and Meeder-Nycz 1977). Grubbs reported a facile route to 2*H*-chromenes that features a RCM reaction as the key transformation (Chang and Grubbs 1998). The necessary dienes were

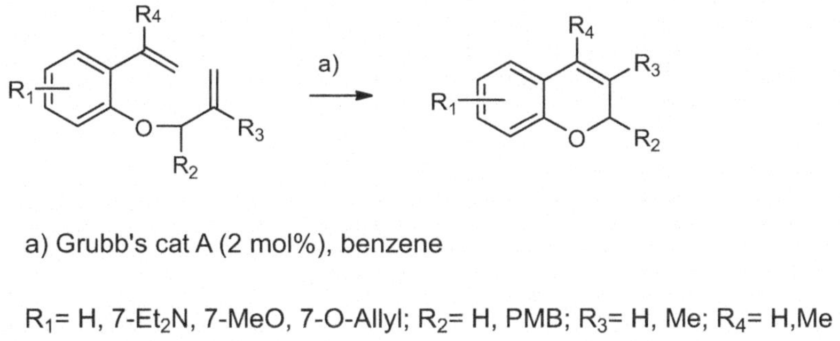

Fig. 22 RCM synthesis of chromenes

prepared in either one step by a Mitsunobu coupling of an *o*-hydroxystyrene with a secondary alcohol or in two steps by an allylation of an *o*-hydroxybenzaldehyde followed by a Wittig olefination (Fig. 22). The RCM reactions of the dienes generally proceeded smoothly with catalyst **A** to furnish the 2*H*-chromenes in high yields. The electronic properties of the ring substituents R_1 had little influence on the cyclization, and the catalyst **A** was tolerant of a variety of functional groups. However, when R_3 was a methyl group, the reaction required elevated temperatures, and when R_2 was either aryl or methyl group, the reaction times were longer, and 5 mol% of **A** was required.

Many polycyclic xanthone natural products exhibit potent antibacterial and anticancer activity. A general strategy for the synthesis of polycyclic xanthone natural products has been proposed by Martin and co-workers (Nichols et al. 2012). The utility of this method is illustrated by its application to a synthesis of a pentacyclic precursor of IB-00208, a glycoside which displays strong antibiotic activity against Gram-positive bacteria and potent anticancer activity against several cancer cell lines (Rodriguez et al. 2003). The key steps of this methodology are preparing angularly fused allyl benzocyclobutenones (a), which underwent ring-closing metathesis in the presence of Grubbs II catalyst (b) to provide angularly fused benzocyclobutenones (Knueppel et al. 2015) (Fig. 23). The next stage was piece together the benzocyclobutenone with the trisubstituted aldehyde (c), and Moore rearrangement (Foland et al. 1989) upon heating (d) to give the tetracyclic intermediate. In the next steps, the aglycone of IB-00208 and its tautomer were obtained (Knueppel et al. 2015). This is a general strategy for the synthesis of polycyclic xanthone natural products.

Significant efforts have been devoted to the development of efficient methodologies for the synthesis of tetrahydropyran derivatives (Núñez et al. 2010), useful building intermediates for the synthesis of biologically active compounds. Among a number of proposed methodologies, the addition of carbon nucleophiles to conjugate acceptor systems is one of the most effective methods, and intramolecular oxa-Michael conjugate cyclizations are the key transformations in a number of total syntheses of complex natural products. Very often it is the method of choice for the asymmetric synthesis of heterocycles (Nising and Bräse 2012). An advantage of this reaction is the formation of a carbon-heteroatom bond in a single step with high

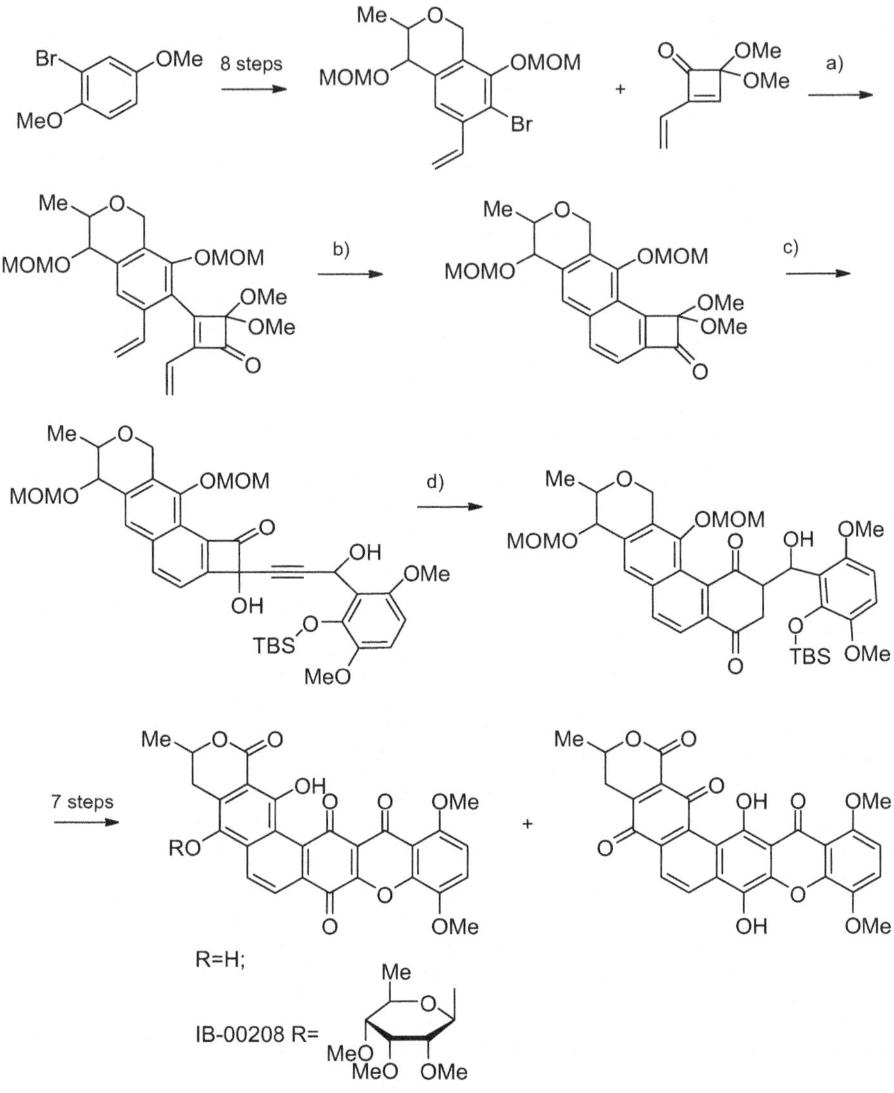

a) t-BuLi; $(CF_3CO)_2O$; b) cros-methathesis, Grubb's cat B;
c) Moore rearrangement; d) thermal cyclization

Fig. 23 Synthesis of polycyclic xanthone, aglycone of IB-00208

atom economy. The most convenient methodology is based on the organocatalytic activation of the carbonyl group toward an electrophilic or nucleophilic cycloaddition reaction (Jensen et al. 2012). Activation of a carbonyl-conjugated C-C double bond through an iminium intermediate has proven to be a particularly powerful strategy, with frequent applications in asymmetric organocatalysis. Among the

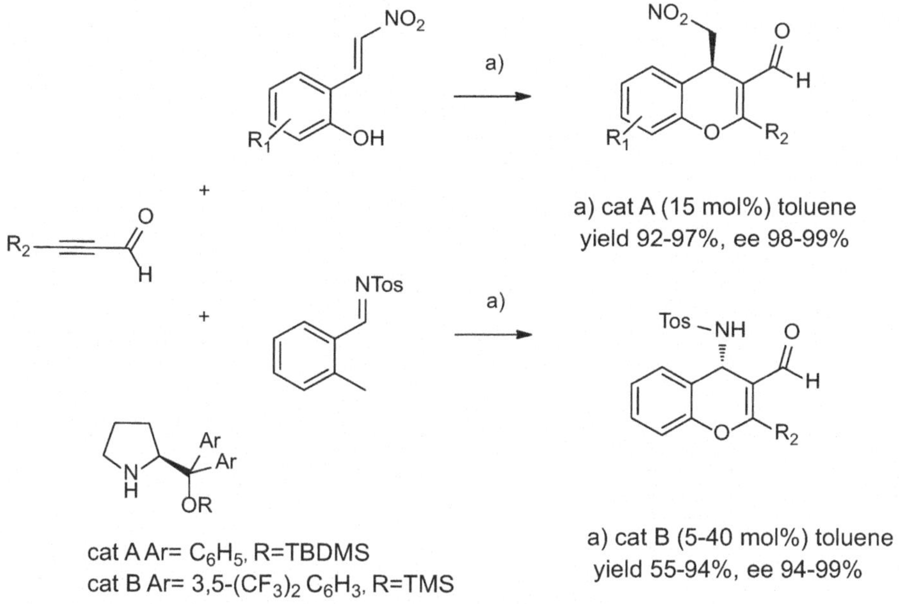

Fig. 24 Synthesis of 4*H*-chromenes

organocatalysts used for covalent activation, chiral secondary amines play a central role, and the diarylprolinol silyl ethers have proved to be the most effective catalysts. The organocatalytic oxa-Michael reaction has been used for the asymmetric synthesis of 4*H*-chromenes and analogous natural products (Alemán et al. 2010; Xu et al. 2008). With 2-(E)-(2-nitrovinyl)phenols as substrate, the reaction leads to 4*H*-chromenes featuring one newly generated stereogenic center, as well as synthetically versatile aldehyde and nitro functionalities (Fig. 24). After optimization of the reaction conditions and catalyst screening, diarylprolinol silyl ethers turned out to be the most effective catalysts, with toluene as the optimal solvent. It was found that alkynals bearing aromatic or aliphatic substituents could be transformed with high yields and enantiomeric excesses. In these reactions a wide range of substitution in the aromatic ring were accepted.

Hong and co-workers (2010) applied an organocatalytic oxa-Michael/Michael/Michael/aldol-reaction to the direct total synthesis of the marine metabolite (+)-conicol (Fig. 25).

In a key step, an organocatalyzed oxa-Michael/Michael reaction between a suitably substituted 2-(E)-(2-nitrovinyl)-phenol and 3-methylbut-2-enal led to the chromene in excellent yield and enantiomeric excess. This compound could then be submitted to a second organocatalyzed domino reaction using the same catalyst as before. In this step, a domino Michael/aldol reaction between the chromene and 4,4-dimethoxy-but-2-enal led to the complete carbon skeleton of conicol. Importantly, the abovementioned reaction steps could be combined in one pot, with a similar yield and enantiomeric excess. The ability of the diarylprolinol silyl ether catalysts to activate aldehydes and α,β-unsaturated aldehydes via enamine and

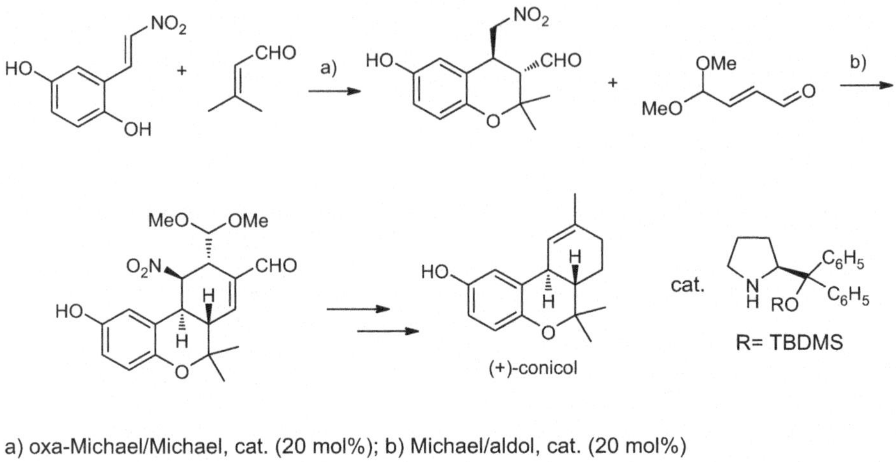

Fig. 25 Synthesis of (+)-conicol

iminium-ion formation, respectively, makes these catalysts ideal for employment in cascade reactions with α,β-disubstituted aldehydes having at least two stereocenters (Jensen et al. 2012). The Michael adducts obtained in reaction of aromatic enals and dialkyl malonate using the diarylprolinol silyl ether catalyst can serve as versatile chiral building blocks for the synthesis of lactams and lactones (Brandau et al. 2006) (Fig. 26), as well as for various pharmacologically active compounds, such as (-)-paroxetine and (+)-femoxetine. Following this initial report, the catalyst was shown to promote various chemo- and stereoselective additions of different 1,3-dicarbonyl compounds to α,β-unsaturated aldehydes. For instance, when 1,3-diketones were employed as Michael donors, the originally formed Michael adducts cyclized to give stable hemiacetal products (Franke et al. 2008; Rueping et al. 2008).

The use of allenes in pyran synthesis has been presented (Zhou et al. 2017). The formal [3 + 3] annulations of δ-acetoxy allenoates have been reported, using 6'-deoxy-6'-perfluorobenzamido-quinine as a catalyst, which provides rapid access to 4*H*-pyrans with excellent enantioselectivity (Fig. 27).

A variety of allenates with different types of substituents at the δ-position, including an aromatic ring, alkene, and alkyl groups, were applicable to the reaction. The products were obtained in good yields and high enantioselectivity. This process was extended to 1,3-dicarbonyl compounds. Cyclohexane-1,3-dione and substituted derivatives were found to be capable of engaging in [3 + 3] annulations with various allenoates (Fig. 27), and products were prepared with excellent ee.

Additionally, acyclic 1,3-dicarbonyl compounds were found to be suitable substrates, and similar levels of reaction performance were observed. Notably, β-carbonyl esters also reacted well with allenoate. The compatibility of a 1,3-dicarbonyl substrate was further illustrated by the reactions of lactones with allenoates. To demonstrate the synthetic utility of this [3 + 3] annulation, an analog of the antiproliferative natural product calyxin I (Gewali et al. 1999) was obtained.

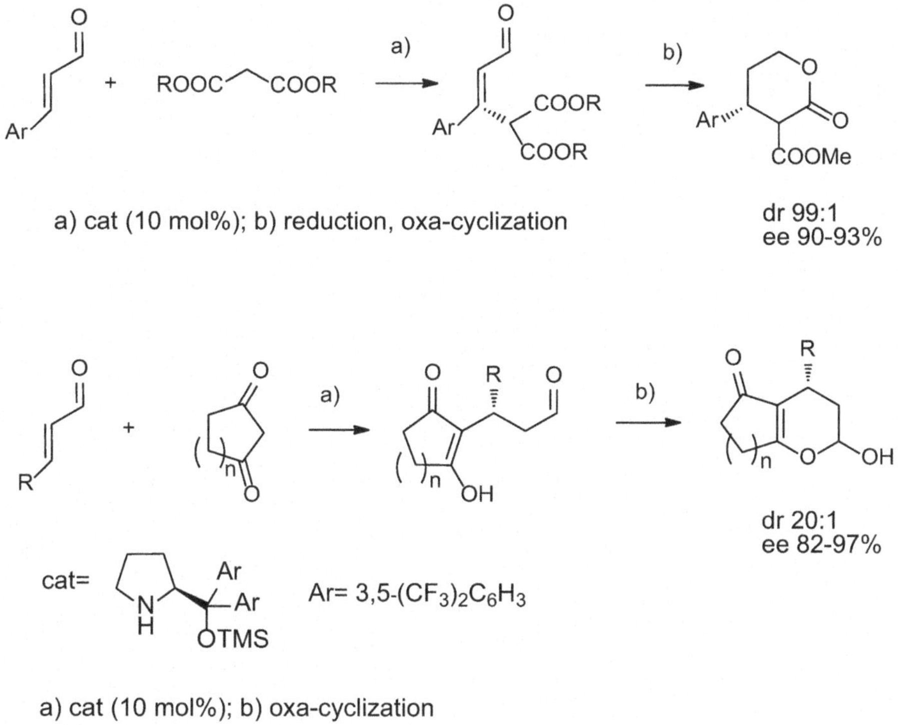

Fig. 26 Domino process for the synthesis of 4*H*-chromenes

Treatment of allene with keto ester led to pyran (a) as a single isomer in 80% yield. After hydrogenation and deprotection of the MOMO group, calyxin I analog was obtained via lactonization with the help of TFA. While the configuration of C3 is opposite to that of calyxin I, the epimerization was achieved via the isomerization/lactone opening and relactonization processes, giving compound desired compound (90% ee) (Zhou et al. 2017) (Fig. 28).

A bifunctional aminoboronic acid has been used to facilitate for the first time the intramolecular aza- and oxa-Michael reactions of α,β-unsaturated carboxylic acids (Azuma et al. 2014). The combination of an arylboronic acid with a chiral aminothiourea allowed for these reactions to proceed successfully in an enantioselective manner to afford the desired heterocycles in high yields and enantioselectivity.

The overall utility of this dual catalytic system was demonstrated by a one-pot enantioselective synthesis of (+)-erythrococcamide B, which proceeded via sequential Michael and amidation reactions (Fig. 29). The asymmetric oxa- Michael reaction of the α,β-unsaturated carboxylic acid proceeded to completion within 24 h under the optimized conditions to give the corresponding carboxylic acid in 94% yield and 94% ee. A one-pot amidation with amine proceeded smoothly to give highly enantio-enriched (+)-erythrococcamide B. It is noteworthy that no racemization was observed during the one-pot reaction process. Another example demon-

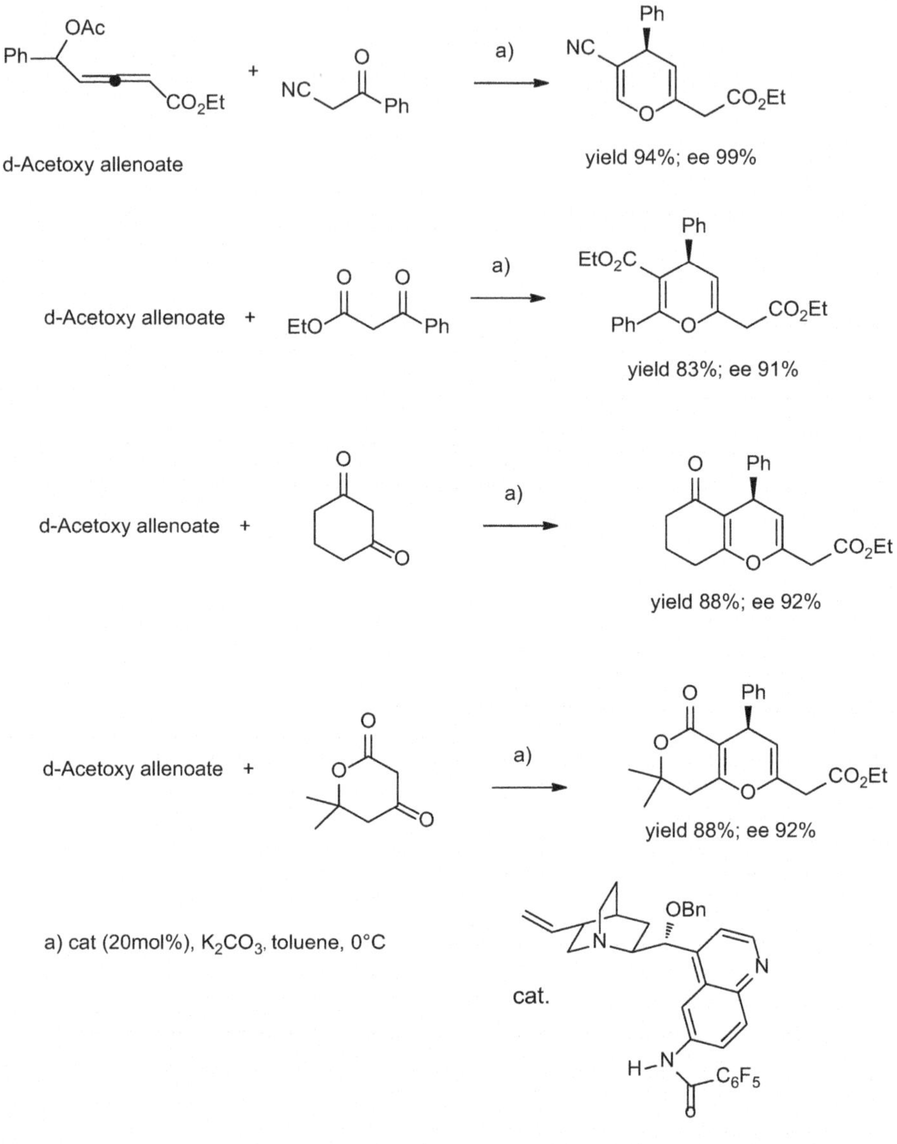

Fig. 27 Enantioselective synthesis of 4*H*-pyran via amine-catalyzed formal [3 + 3] annulation

strates the usefulness of the oxa-cyclization method in the synthesis of complex macrolides containing a tetrahydropyran block. The marine macrolide Leucascandrolide A was isolated from the calcareous sponge *Leucascandra caveolata* by Pietra and co-workers (D'Ambrosio et al. 1996). Leucascandrolide A is an extremely potent inhibitor of tumor cell proliferation. A facile synthetic route for the biologically active natural product Leucascandrolide A was developed (Lee et al. 2011). In this stereoselective synthesis of the macrolide lactone, the key stages

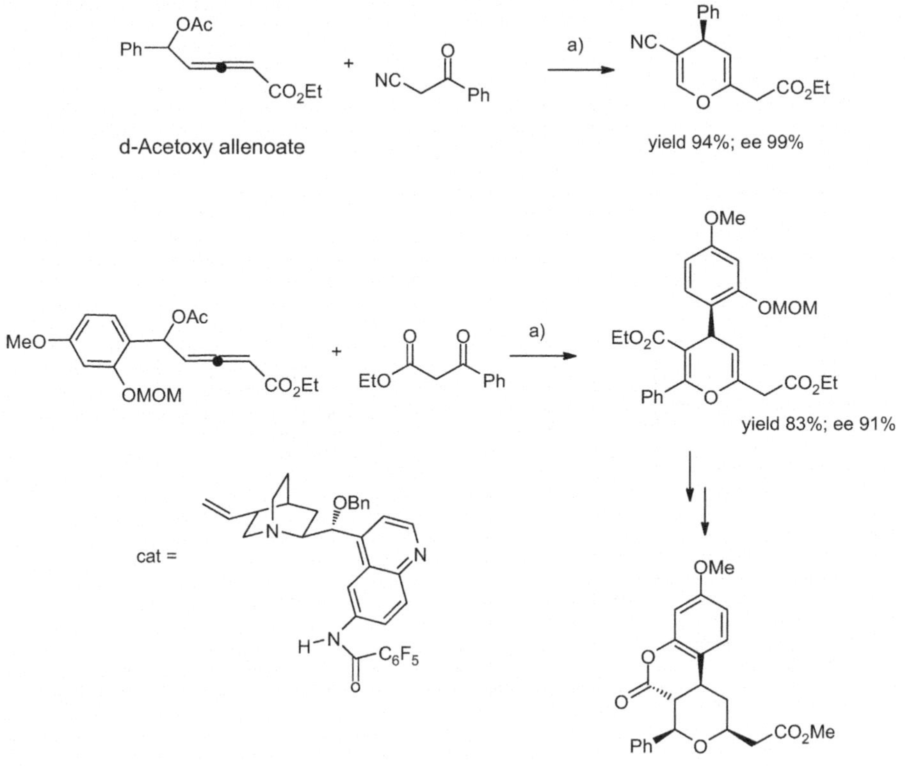

Fig. 28 Synthesis of calyxin I analog

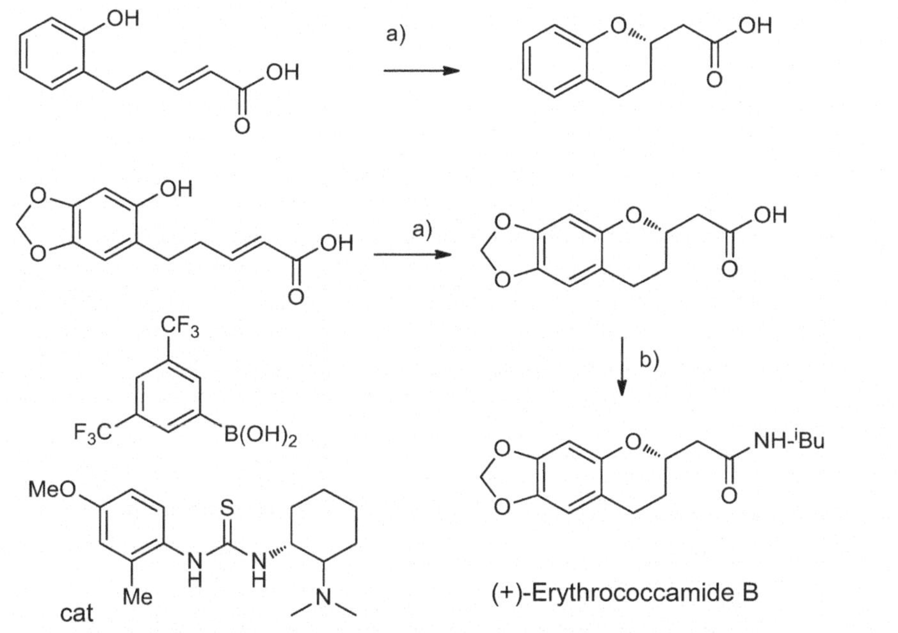

a) oxa-Michael reaction, boronic acid (20 mol%), cat (20 mol%); b) ⁱbutylamine

Fig. 29 Total synthesis of (+)-erythrococcamide B

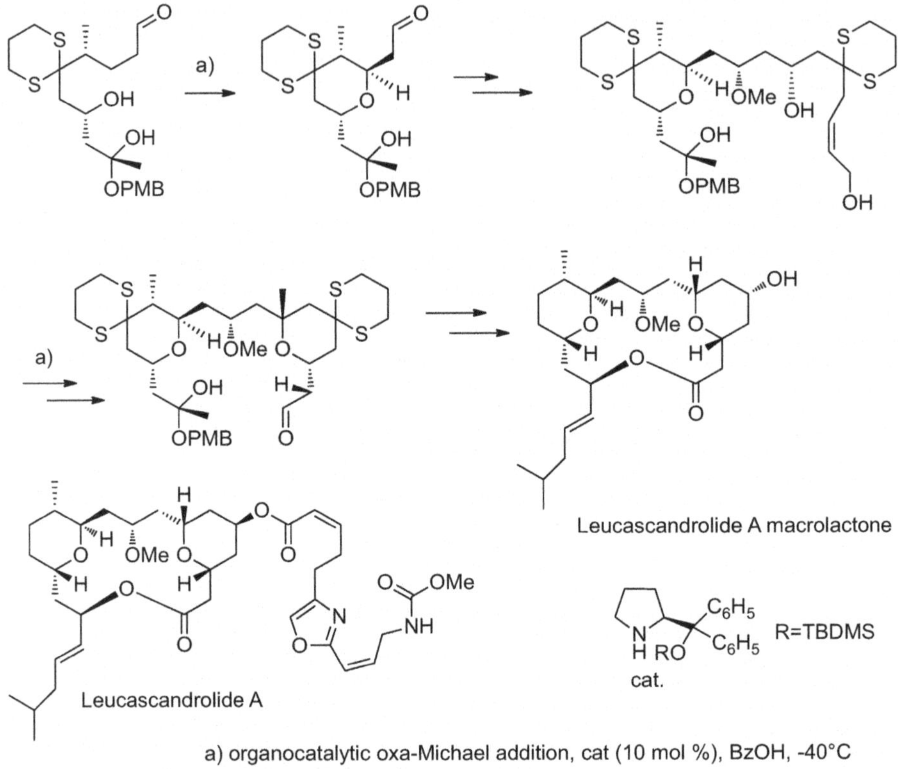

Fig. 30 Synthesis of macrolide lactone incorporating tetrahydropyran rings

are the tandem and organocatalytic oxa-Michael reactions in conjunction with the dithiane coupling reaction (Fig. 30). When the aldehyde was treated with (S)-diphenyl-(trimethylsilyl)prolinate, the organocatalytic oxa-Michael reaction proceeded smoothly to provide tetrahydropyran derivatives with excellent stereoselectivity and yield (dr > 20:1, 98%). The next step was the functionalization of the 2,3-*trans*-2,6-*trans*-tetrahydropyran obtained and the installation of the dithiane moiety. The final stage in the synthesis of Leucascandrolide A lactone was oxidation of the hydroxyl group followed by oxa-Michael cyclization. The presented examples demonstrate that the oxa-Michael reaction is an extremely powerful and versatile tool for the rapid construction of cyclic oxygen-containing building blocks.

3 Approaches to C-Glycosyl Compounds and Synthesis of Gliflozins

It follows, from general knowledge of glycosylation reaction, for which vast theoretical knowledge and practical experience is being accumulated since Michael discovery of chemical conjugation of acetobromoglucose with phenol in 1879 (Michael

1879), that various glycosylating agents, initially designed for functionalization of simple and complex alcohols, can also react with other nucleophiles: thiols (as well as Se and Te analogs), nitrogen (and phosphorus) compounds, and also electron-rich carbon centers of various organic substrates, leading to glycoside analogs in which anomeric oxygen is replaced by a heteroatom (or carbon atom) (Zhu and Schmidt 2009; Yu et al. 2012; Li and Zhu 2016; Levy and Tang 1995). Carbon analogs of O-glycosides are of particular interest as metabolically stable mimics of natural compounds with recognized biological activity (Křen 2008). Moreover, many C-glycosidic compounds were found in nature, and there is a close structural connection between these unmistakably sugar derivatives and many other naturally occurring pyrans of noncarbohydrate origin, which nevertheless may formally be considered 1,5-C-substituted deoxypyranoses. Since the synthetic methods by which C-glycosidic compounds can be obtained are extensively studied and well covered in literature, only a short methodical survey is presented in this review, with focus on aryl and heteroaryl aglycons, which are of particular interest in medicinal chemistry, pharmacology, and pharmaceutical industry. Classical studies on the subject were summarized in principal references (Demchenko 2008; Mydock and Demchenko 2010; Das and Mukhopadhyay 2016) and are not going to be discussed here in detail. Approximately a couple of decades later, entirely new state of art in creation of C–C bond emerged, as a consequence of application transition metal-catalyzed reactions, such as Suzuki-Miyaura, Heck, Stille, Negishi, and Sonogashira protocols (Johansson Seechurn 2012). Accordingly, new summaries on C-glycosylation appeared, which contain a selection of novel solutions to the problem of stereoselective mounting of carbon-carbon bonds at the anomeric position (Yang and Yu 2017; Bokor et al. 2017; Martin 2017; Zhu et al. 2017). In general, glycosyl donors used for O-glycosylation are practically not effective for C-C bond formation with aromatic substrates, unless the latter are metalated. However, unsaturated pyranoses feature somewhat separate reactivity which stems from more profound stabilization of intermediate electrophilic oxocarbonium ions. This can be illustrated by excerpts from glycal chemistry, with either Lewis acid promoted (Ferrier rearrangement) or palladium catalyzed (Heck-type or Tsuji-Trost-type reactions, including O'Doherty glycosylations), in which various C-nucleophilic reagents can be used. As illustrated on the Fig. 31, each of these transformations results in placement of a nucleophilic C-substituent at the anomeric position (Ansari et al. 2013; Gomez et al. 2015).

As can be expected, geometry of corresponding transition states which govern a stereochemical outcome of the reaction is strongly dependent on spatial and electronic factors of participating substituents. A nice example of the Heck-type reaction (Bai et al. 2013), in which very high α-stereoselectivity was observed features acetylated glycols and phenylhydrazines reacted in acetic acid in the presence of oxygen and palladium diacetate (Fig. 32).

More typically arylboronic acids or aryl halides are used in such reactions. The catalytic reaction cycle involves syn-addition of catalyst-ligand fragments to the double bond, which is followed by either β-hydride or β-heteroatom containing group elimination (Fig. 33). Consequently, three types of products could be expected, depending on experimental conditions (Ansari et al. 2013).

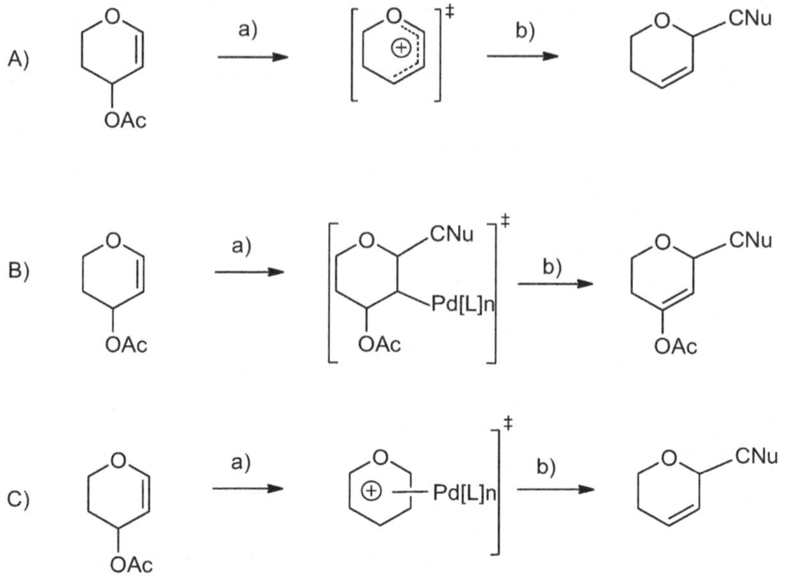

Route A a) Lewis acid, b) carbon nucleophile, CNu or $^{-}$CNu
Route B Heck type glycosulation a) syn-addition, CNu, cat PdLn b) 2,3-elimination
Route C Tsui-Trost glycosylation a) cat PdLn b) $^{-}$CNu

Fig. 31 Syntheses of C-glycosyl 2,3-unsaturated pyranoses from glycals

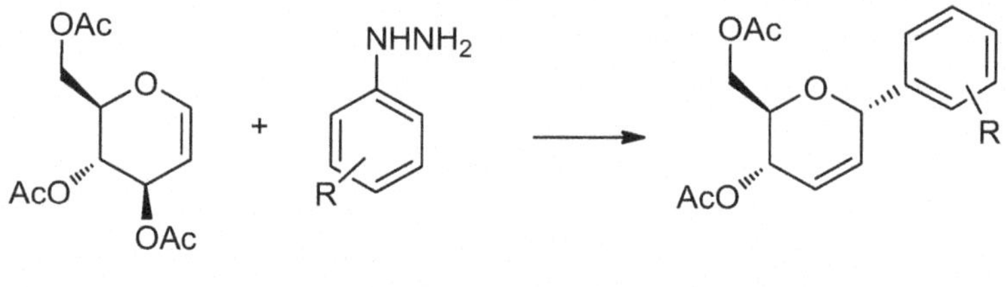

a) cat $Pd(OAc)_2$ / Ligand, O_2, AcOH

Fig. 32 Stereoselective Heck-type C-glycosylation of arylhydrazines with D-glucal acetate

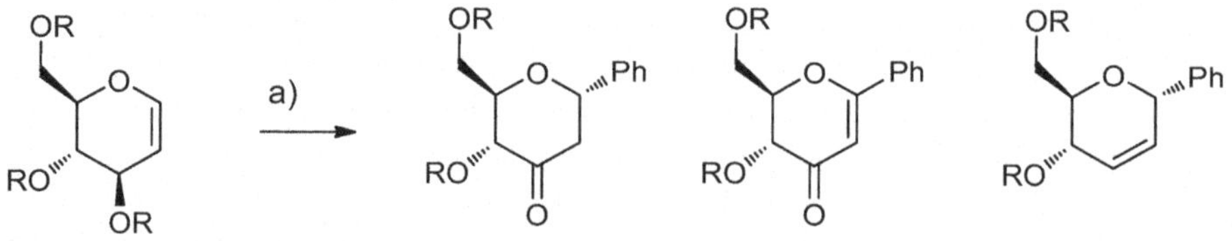

a) $Pd(OAc)_2$, $PhB(OH)_2$

Fig. 33 Three type of C-glycosyl products obtainable from Heck reaction with protected D-glucal

An ability of unsaturated pyranosides to alkylate electron-rich aromatic substrates under Friedel-Crafts reaction conditions was recorded long before an avalanche of reports on application of transition metal-catalyzed coupling reactions commenced (Grynkiewicz and Zamojski 1980; Grynkiewicz and BeMiller 1982). Ferrier rearrangement, which started as an observation of a glycal reactivity toward hydroxylic substrates in 1964, has by now accumulated a vast amount of literature, among which C-glycosylation reactions are well represented (Ferrier and Hoberg 2003; Fraser-Reid and Lopez 2009; Gomez et al. 2013; Grynkiewicz et al. 2014; Gomez et al. 2015). Initial C-aryl glycoside syntheses carried out by reacting electrophilic glycosylation agents with methylated aryl compounds are well covered in basic monographs on sugar reactivity (Levy and Fügedi 2006; Nishikawa et al. 2008). Examples quoted include also wider engagement of an anomeric reactivity, including heteroaromatic substituted sugars, stabilized anomeric carbanions, and anomeric radicals. In recent years, new conditions were also found for efficient coupling of metalated aryl reagents with halogenoses (Lemaire et al. 2012; Adak et al. 2017). Conversely (Zhu et al. 2016), benzylated 1-tributyltin substituted pyranoses smoothly react with aromatic bromides and iodides, with replacement of stannate stereoselectively and in good yield (Fig. 34).

Gluconic acid lactone addition, followed by the reduction of resulting ketol with a silane in the presence of Lewis acid (Kishi reduction), is a fairly popular approach to aromatic C-glycosides (Fig. 35). Such sequence of reaction is fairly selective, affording mainly, if not exclusively the beta anomer, similar to direct C-glycosylations (Czernecki and Perlat 1991; Bokor et al. 2017).

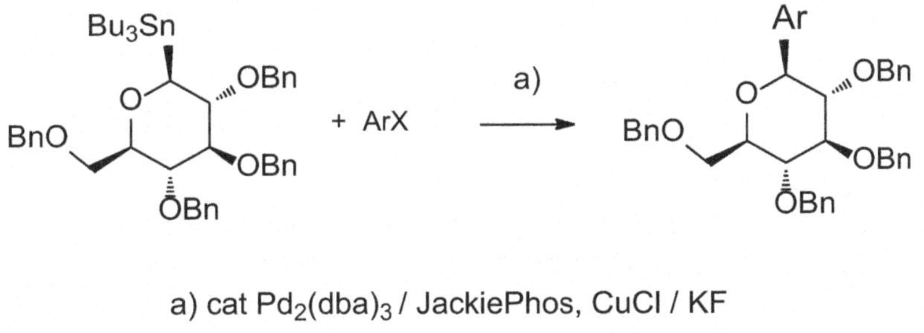

Fig. 34 Stereospecific benzylated pyranosyl stannate coupling with an aryl halide

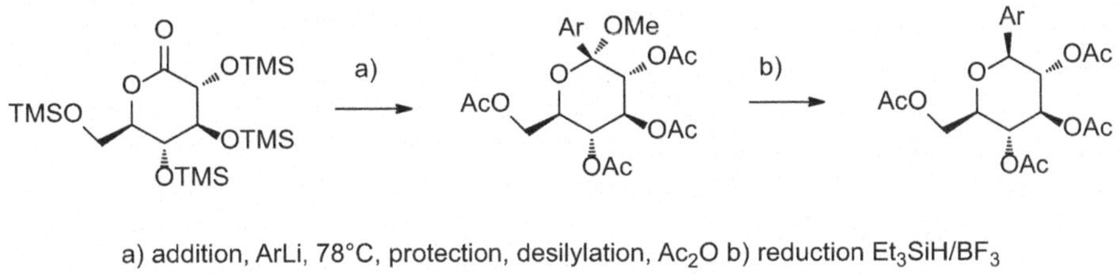

Fig. 35 Synthesis of β-C-glycosides

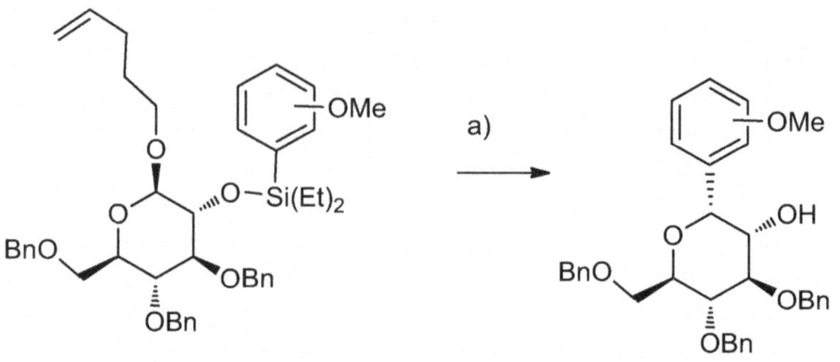

a) IDCP, TBAF

Fig. 36 Stereoselective synthesis of aryl α-C-glycosides by intramolecular delivery of an aromatic group

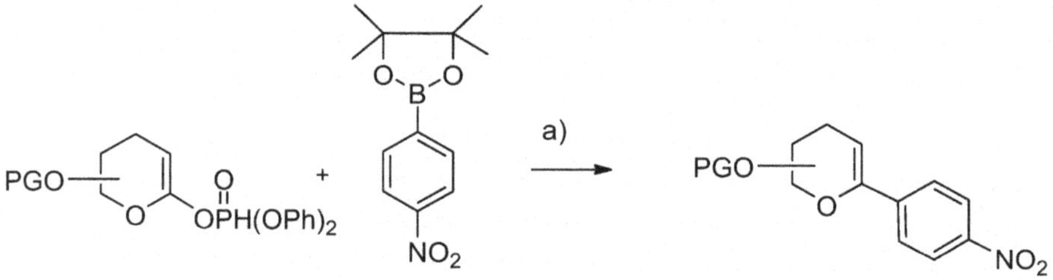

PG protecting group

a) catalyst Pd(II)

Fig. 37 Synthesis of C-aryl glycals

The opposite stereoselectivity can be achieved via intramolecular C-arylation as depicted on the Fig. 36 (Rousseau and Martin 2003).

Ketene acetal 1-O-phosphates, easily obtainable from properly protected 2-deoxy aldonolactones, undergo Suzuki-Miyaura cross-coupling with arylboronic acids, (Leidy et al. 2013), affording in the presence of Pd(II) catalysts C-aryl glycals (Fig. 37).

Interestingly, different kinds of C-glycosyl products can be obtained from benzylated glycals when they are reacted with aryl bromides, in the presence of a base and a phase transfer catalyst, under Pd(II) catalysis, when heated in dry DMF during microwave irradiation (Lei et al. 2009). Yet another type of coupling reactivity under similar conditions was observed during decarboxylative arylation of glycals with substituted benzoic acids (Xiang et al. 2011). Vinylic C–H borylation in the presence of Iridium (I) complexes has also been demonstrated (Kikuchi et al. 2008). 1,2-anhydro- and 1,6-anhydropyranoses are also known to be susceptible to a nucleophilic attack with formation of a new bond attaching an anomeric substituent, and C-glycosyl bonds can also be easily installed that way, in case of glycal epoxides with either anti- or syn- mode, depending on the reaction conditions (Rainier and Cox 2000) (Fig. 38).

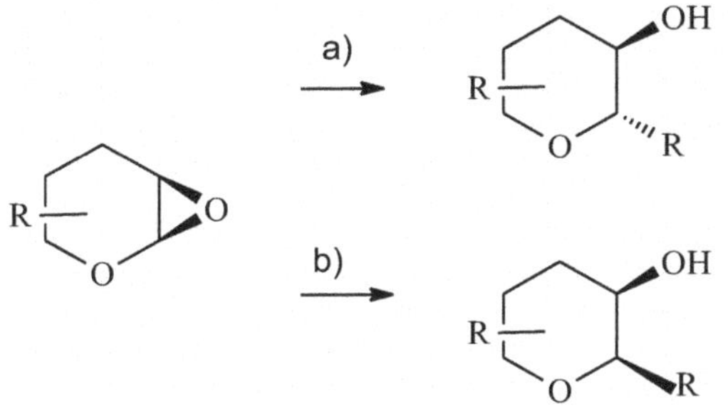

a) RMe: Me=Mg, Cu, Li, Sn; b) R_3Me: Me=Al, B

Fig. 38 Glycosylation of 1,2-anhydropyranoses

The cause for a considerable revival of an interest in practical procedures for highly stereoselective methods of C-glycosylation recently can be explained by pressing needs for syntheses in two areas of natural products of prospective (and largely proven) application in pharmacy and medicine. The first area is that of fused polycyclic ethers, known as antibiotics of microbial origin and marine toxins. Syntheses connected to this subject have been presented in earlier parts of this chapter. The second concerns a new group of antidiabetic medicines, which evolved as natural plant glycoside analogs recently and already became an indispensable part of the metabolic syndrome treatment, globally.

Gliflozins

Phlorizin, β-D-glucopyranoside of dihydrochalcone-phloretin, is among the first neutral compounds of plant origin obtained in pure chemical state at the beginning of the nineteenth century, after organic acids and alkaloids, whose isolation could be efficiently assisted by salt formation (Ehrenkranz et al. 2005). Plant phenolics and their glycosides were of interest for practical reasons, since they were recognized as useful dyes (carminic acid, alizarins, flavonoids) and medicines (cardiac glycosides, salicin). First isolated from apple tree root bark in 1835, phlorizin was assumed to have antipyretic properties, by analogy to salicin, but that activity was not confirmed experimentally. Instead, it was demonstrated that the glycoside applied per os to mice, produced glycosuria, which turned attention to possible application of phlorizin in diabetes mellitus (Ehrenkranz et al. 2005). Contemporary pharmacology includes phlorizin into a group of natural nonselective inhibitors of sodium-dependent glucose-transporting proteins. Its phytochemical function is that of a growth regulator and phytoalexin, hence its presence in all parts of apple tree including fruits and consequently in human diet. Apple pulp contains 4–20 mg of the glycoside per kilogram, which also occurs in cider (3–16 mg/100 ml), while apple peel content can reach more than 400 mg/kg. Finding that phlorizin blocks renal glucose resorption and improves insulin sensitivity made it a drug candidate, and Japanese company Tanabe Seiyaku developed its semisynthetic derivative (T-1095), which performed well as a blood glucose level-lowering agent in preclinical tests (Ehrenkranz et al. 2005; Neumiller et al. 2010; Pathania et al. 2016). Similar effort in GSK resulted in development of sergliflozin (Fig. 39). Eventually, serious drawbacks of phlorizin,

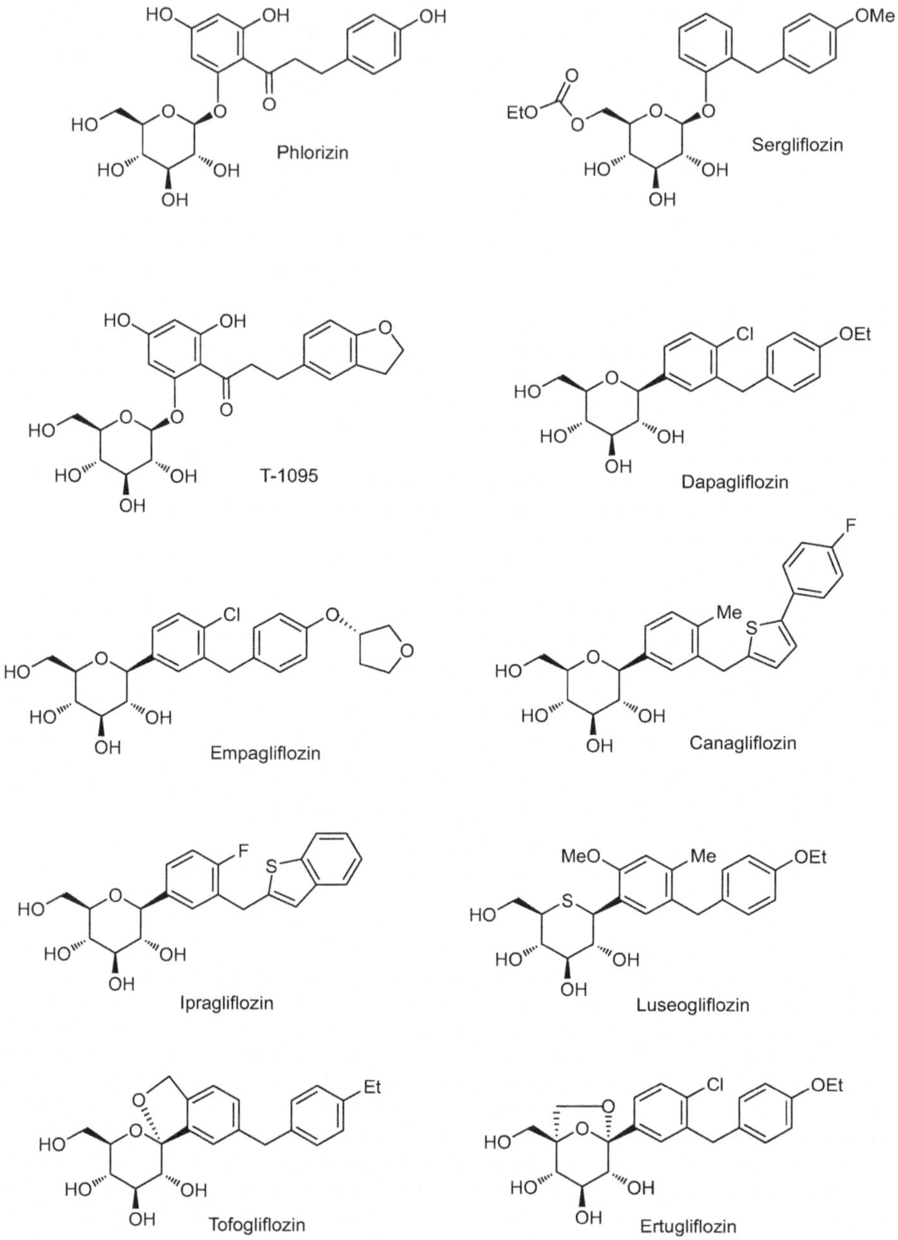

Fig. 39 Synthetic gliflozins, used as antidiabetic therapeutics, and their O-glycosyl precursors

and their first-generation analogs, such as low bioavailability due to metabolic instability and lack of distinction between SGLT1 and SGLT2 cotransporters, have caused termination of the projects, in favor of development of the second-generation mimics featuring C-glycosyl linkage between aglycon and sugar moiety. Dapagliflozin

(Farxiga) was the first synthetic C-aryl glycosyl analog of phlorizin, approved as a therapeutic SGLT2 inhibitor in 2012 (EU) and (USA) upon application of BMS/Astra Zeneca; structures of natural phlorizin and its O-glycosyl analogs, sergliflozin and T-1095, along with the next five gliflozins approved to date (canagliflozin (Invokana; Janssen Pharmaceuticals), empagliflozin (Jardince; Beringher Ingelheim/Eli Lilly), ipragliflozin (Suglat; Astellas/Kotobuki Jp), tofogliflozin (Deberza; Chugai Jp), luseogliflozin (Lusefi; Taisho Jp), and advanced drug candidate ertugliflozin) are presented on Fig. 39. Many other analogs of registered drugs form SGLT2 inhibitor group are in pipelines of pharmaceutical companies. Detailed accounts on this massive drug design and synthetic effort are given in recent book chapters (Lemaire and Schils 2015; Mishra et al. 2016; Hoang et al. 2017).

As can be noticed from Fig. 39, gliflozins share some characteristic structural features, such as β-C-glycosyl linkage between D-glucopyranosyl moiety and the aromatic aglycone, diarylmethane scaffold with possible substitutions at the distal ring, and metatype substitution of the sugar and linker at the proximal aromatic ring. As a consequence, synthetic methods initially applied for preparation of active pharmaceutical ingredients (API) in this group of therapeutics are common and as a rule share a protocol for C-glycosylation, in which suitably metalated aglycone is added to protected pyranosyl form of D-gluco-aldonolactone. Such approach is illustrated in Fig. 40, which depicts synthesis of dapagliflozin (Deshpande et al. 2007; Washburn 2012a; Braem et al. 2014).

Addition of the lithioaromatic reagent to silylated D-gluconolactone at low temperature is followed by MsOH-catalyzed reaction with methanol and then acetylation with Ac_2O. Intermediate-acetylated glycoside is purified by crystallization and then subjected to Kishi reduction, which affords pure C-glycosyl product in acety-

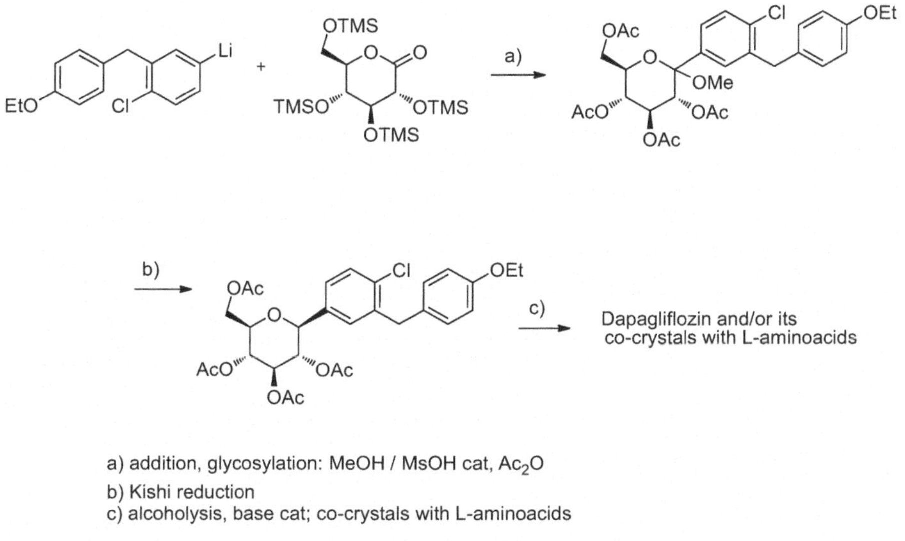

Fig. 40 The key chemical steps in technical process of dapagliflozin API manufacturing

lated form. Base-catalyzed deacetylation gives dapagliflozin, further co-crystallized with L-proline, for final formulation. Along the same line, synthesis of empagliflozin was completed, by application of a bi-arylmethane aglycone, which was synthesized from commercially available halogenated benzoic acid (four steps; 64% overall yield). In a process elaborated by Boehringer Ingelheim, iodoaromatic reagent was applied in addition to silylated aldonolactone, in the presence of isopropylmagnesiumchloride and lithium chloride, which allowed to run the reaction in a comfortable range of temperature (−20°C to +10°C). The lactol intermediate thus obtained had tendency to form furanosyl-mixed methyl ketal under HCl/MeOH treatment but could be equilibrated to expected pyranoside, which secured formation of desired C-glycosyl compound under modified Kishi reduction reaction, which proceeded smoothly without use of protecting groups, when run in the presence of aluminum trichloride (Henschke et al. 2015b) (Fig. 41).

More recently, an alternative to aryllithium-gluconolactone addition, with better promise of stereoselective placement of aryl aglycone, based on 1,6-anhydropyranose ring opening appeared. In this approach the β-face selectivity of the aryl residue is secured by intramolecular addition (Henschke et al. 2015a), as depicted in Fig. 42.

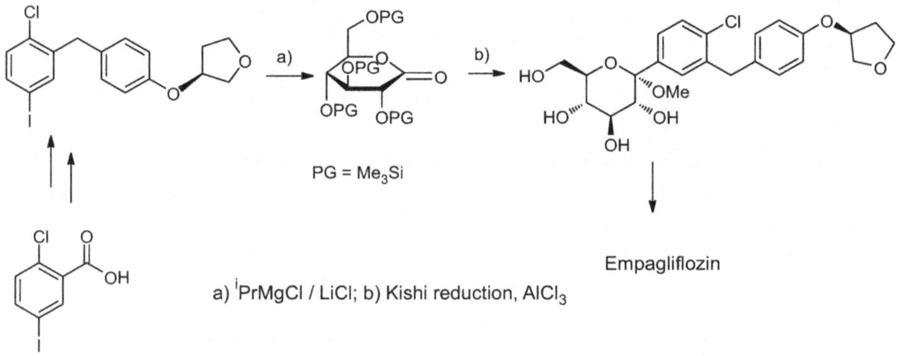

Fig. 41 Key steps in synthesis of empagliflozin

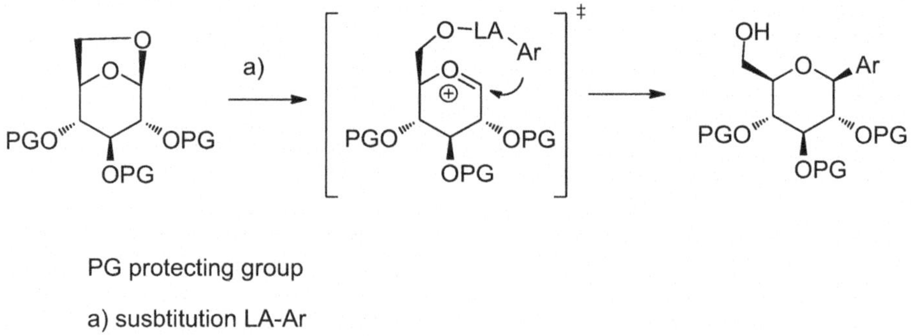

Fig. 42 C-glycosylation by intramolecular delivery of the Ar group

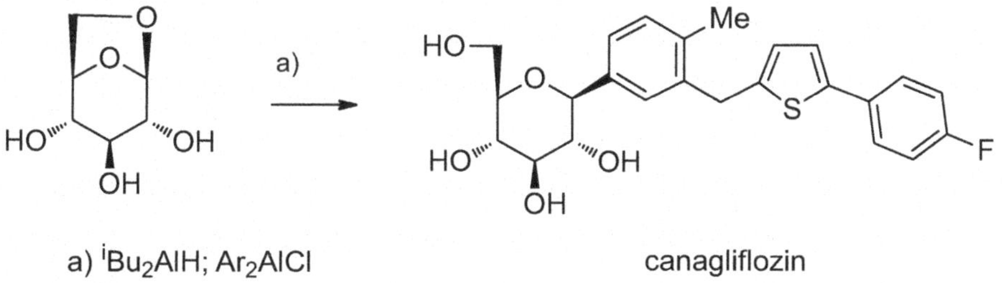

Fig. 43 C-glycosylation via arylaluminum reagents

In a follow-up study, Henschke has used the intramolecular aryl delivery principle for elaboration of a simplified synthesis of canagliflozin which allows for application of 1,6-anhydroglucose as the sugar moiety component, without hydroxyl groups protection (Fig. 43). Stepwise application of diisobutylaluminum hydride and diarylated aluminum chloride resulted in efficient and stereoselective C-glycosylation (Henschke et al. 2015b; Washburn 2014).

Since several SGLT2 inhibitors achieved status of registered drugs, their API are manufactured in validated processes on a technical scale. It is fortunate that some of them were described in scientific literature, apart from earlier patent disclosures. The technical details, which are important from a process engineering point of view, concern choice of protecting groups for the sugar synthons, conditions for C-glycosylation and the following lactol reduction, and efficiency of the aglycon assembly. Nevertheless, the main synthetic drawbacks, like necessity for cryogenic conditions in carbon-carbon bond formation steps, or use of expensive or toxic reagents as an alternative, still need to be addressed. An example of the two scalable syntheses of tofogliflozin hydrate is shown below to illustrate some notoriously difficult steps (Ohtake et al. 2016; Yang et al. 2016) (Fig. 44).

Supplementing antidiabetic drugs armamentarium with SGLT2 inhibitors has undoubtedly met with great success, in the opinion of medical experts (Neumiller et al. 2010; Washburn 2012b; Bokor et al. 2017). Therefore, after some period of strict adherence to basic pharmacophore hypothesis, many researchers started testing gliflozin analogs, which are modified not only in aglycon part but also in sugar moiety. The 6-deoxy analogs, which seem to be natural choice, were already reported, and ertugliflozin – the promising drug candidate from Pfizer – seems to start a new trend in that direction. Ertugliflozin represents the first example from dioxabicyclo [3.2.1] octane class, which is derived from unnatural 5,5-dihydroxymethyl pyranoses (Mascitti et al. 2013). Resulting C-aryl glycosids have stiff bicyclic framework in place of relatively flexible pyranosyl moiety of natural hexoses. Thus it appears that gliflozin family is going to continue to expand in many directions, in search for promising new chemical entities which mimic SGLT2 drugs known to date.

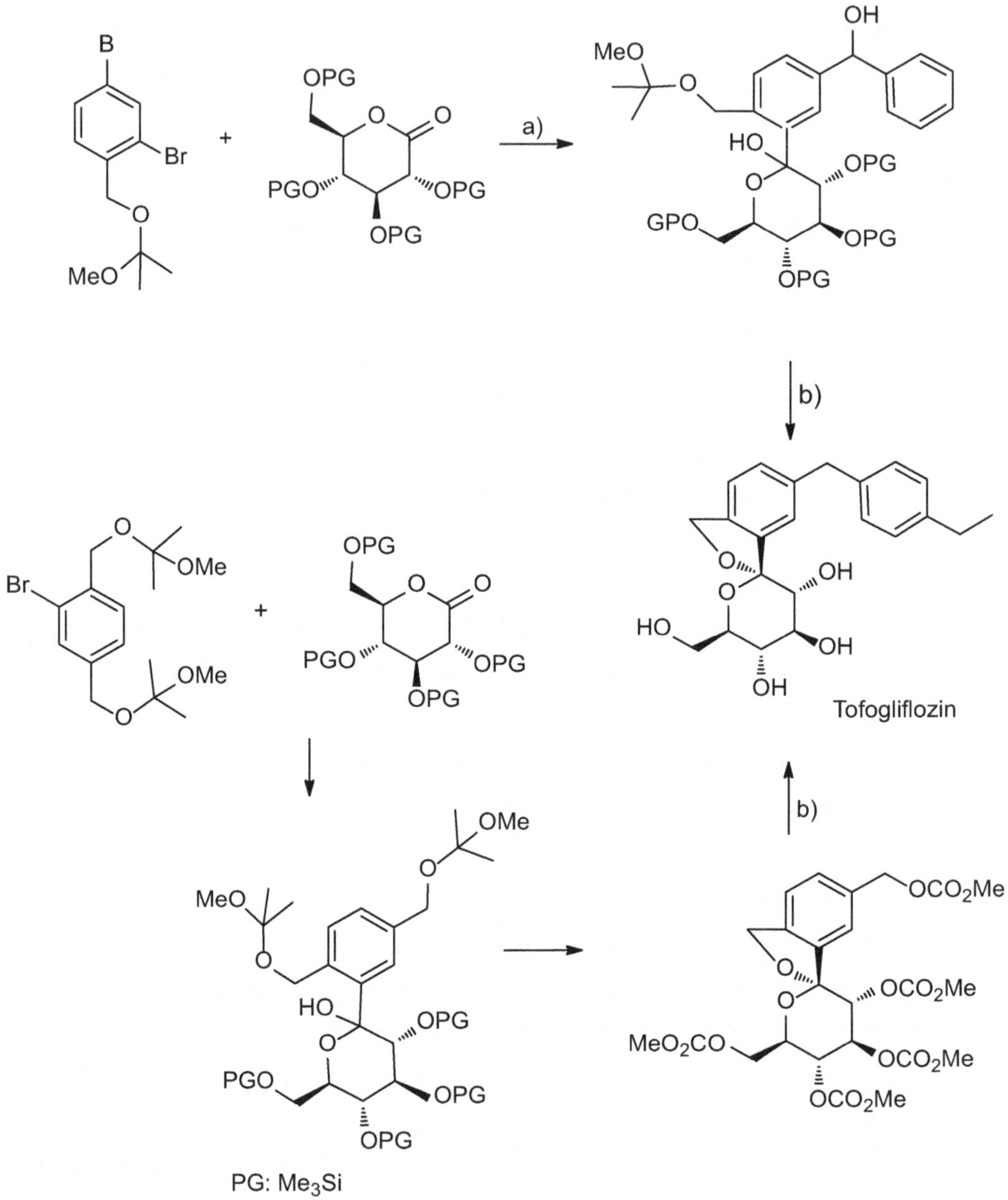

a) 4-ethylbenzaldehyde b) Pd(II) / 4-Et-Ph-B(OH)$_2$

Fig. 44 Comparing sequence of C–C bond formation in scalable syntheses of tofogliflozin

4 Preparation of Pyran Derivatives from Furylcarbinols: Applications of Achmatowicz Rearrangement (AR)

Biomass is gaining increased appreciation as a renewable resource of energy and materials, in consideration of inevitable depletion of geological deposits of carbon. Being mainly of plant origin, it is known to be a complex polymeric and polycondensate matter composed chiefly of oxygen heterocyclic rings (cellulose, hemicelluloses, lignins, etc.), from which individual monomers, like glucose, are not easily recoverable. Consequently the first-generation biomass utilization programs focused primarily on transformation into various forms of a biofuel like solid fuels, biogas, ethanol, or biodiesel (Van Putten et al. 2013; Dusselier et al. 2014; Chatterjee et al. 2015). Dehydration reactions of saccharide materials, in which pyranoses are converted into furan derivatives, are known since the nineteenth century and for a long time were neglected by practitioners of academic synthetic chemistry on obvious grounds of structural complexity and chirality loss along the transformation. Modern heterocyclic chemistry, supported by novel means in control of stereoselectivity of basic C–O, C–C, and C–heteroatom bond formation reactions, changed this view radically. In particular, furan derivative oxidation reactions became an inspiration for new syntheses of pyran compounds, including natural and modified carbohydrates (Zamojski et al. 1982; Vogel and Robina 2008; Gomez et al. 2013). Currently, availability of furan starting materials increases greatly, as scalable chemical processes are being developed, aimed at platform chemicals, which can replace petrochemicals as raw materials for preparation of chemical intermediates and substrates for blended polymers, used in such manufacturing sectors as civil engineering, automobile industry, agrochemicals, pharmaceuticals, etc. In particular, cellulosic biomass is of interest as a starting point for group of furan derivatives, represented by 5-hydroxymethylfurfural (HMF), which can be considered as straightforward dehydration products of D-fructofuranose (Van Putten et al. 2013) (Fig. 45).

A more substantial part of five or six carbon-containing furan derivatives, obtainable from biomass and already functioning as platform chemicals, is presented in Fig. 46, which summarizes manufacturing processes starting with dehydration of polysaccharide monomers with furfual F and hydroxymethylfurfural HMF as the key intermediates (Chheda et al. 2007; Binder and Raines 2009; Serrano-Ruiz et al. 2011).

Naturally, structural variety of available furans and their versatile reactivity suggested their use as achiral substrates for total synthesis of monosaccharides (Achmatowicz et al. 1971; Aljahdali et al. 2013). Chemical means for carrying out

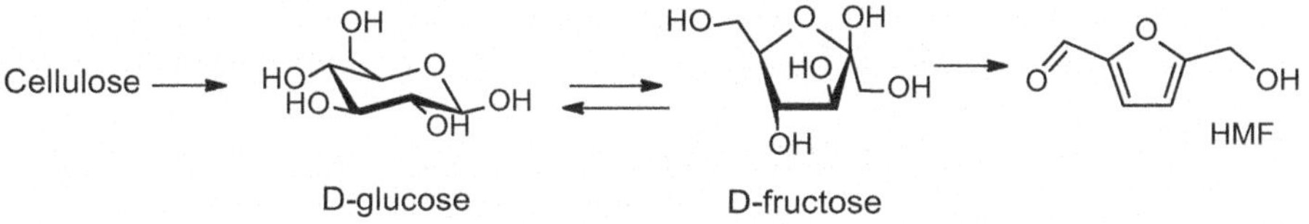

Fig. 45 Summary of hydrolytic and dehydrative processes in which plant cellulose is converted into 5-hydroxymethyl furfural

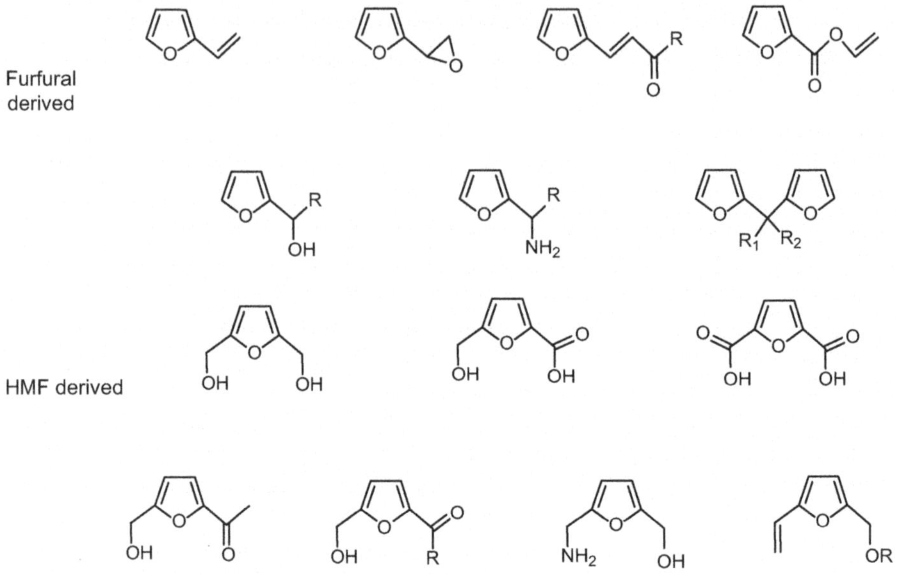

Fig. 46 Furan derivative products of biomass transformation

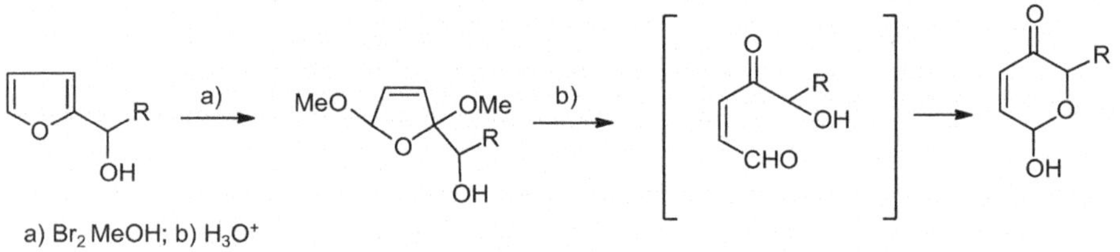

Fig. 47 Achmatowicz rearrangement

a stepwise reversal transformation to the dehydrative process pictured on Fig. 1 started to emerge around the 1950s. The Clauson-Kaas discovery that furans can be easily transformed into corresponding 2,5-dialcoxy compounds provided entry to masked 1,4-dicarbonyl synthons, which soon found use in preparation of variety of heterocyclic compounds (Elming 1960). It was reasonable to assume that such synthons, when generated from furylcarbinols, would have strong tendency for intramolecular acetalization, particularly under acidic conditions (Fig. 47), and this assumption served as the cornerstone of extensive program aimed at total synthesis of monosaccharides (Achmatowicz et al. 1971; Achmatowicz 1981; Zamojski et al. 1982) (Fig. 43).

For the initial experiments, 2-furylcarbinol and 2-furyletanol were selected, and their stereoisomeric 2,5-methoxylated derivatives were successfully transformed into desired pyranos-4-uloses, which already resembled some simple deoxy monosaccharides of antibiotic origin (Fig. 48). 2-Furylethandiol has followed, after adoption of proper measures for the primary group contemporary protection (Merino et al. 2007).

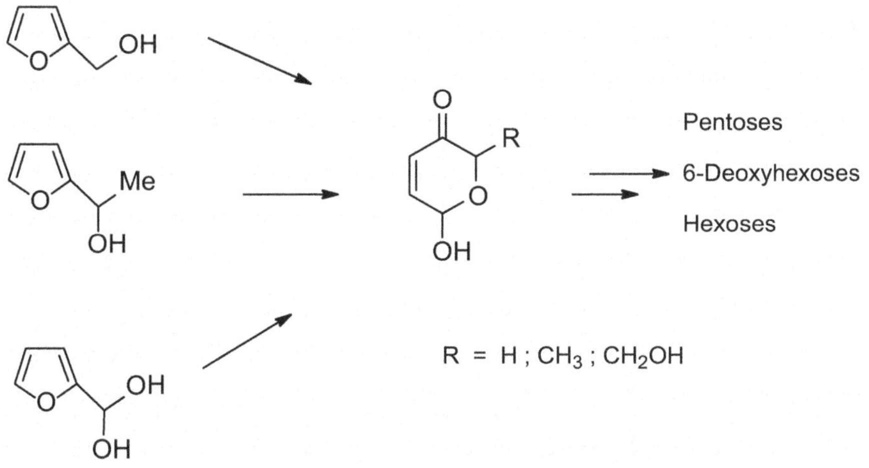

Fig. 48 Conversion of furylcarbinols via AR to enuloses on the way to monosaccharide syntheses

The proof of concept for the furylcarbinols rearrangement (Achmatowicz rearrangement, AR) as key step toward versatile pyranosulose synthons from which great variety of pyranosides including natural pentoses and hexoses can be obtained, was provided on racemic substrates, but correlation of C-5 configuration followed shortly on two independent pathways: by use of homochiral furan starting materials and by conversion of D-glucose, via corresponding glucal, into 2-furylethandiol (Hauser and Ellenberger 1986). Nevertheless, it was the scarce availability of furylcarbinol pure enantiomers, which hampered advancement of AR application as a useful transformation in total synthesis of heterocyclic natural products. A few decades have passed before satisfactory efficient methods of stereoselective catalysis, such as enantioselective dihydroxylation, epoxidation, and carbonyl group reduction, became routine tools of organic synthesis (Noyori 2002; Koskinen 2012; Babu et al. 2012; Yu and O'Doherty 2008). While the first total synthesis of D-glucoside achieved by application of AR involved a racemic furylcarbinol resolution via fractional crystallization of diastereoisomeric esters, contemporary syntheses are carried out practically without loss of a preselected prochiral starting material, as demonstrated on the Fig. 49, which depicts total synthesis of papulacandin D (Balachari and O'Doherty 2000).

Thus, it is obvious that the stereogenic center of the starting furylcarbinol is preserved throughout further steps of preparation. Consequently, for the purpose of natural products of total synthesis, homochirality of the furan starting materials is presently a crucial requirement. The methods for their preparation mentioned before should be supplemented by such possibilities as efficient enantioselective catalytic Friedel-Crafts alkylation of furans with activated carbonyl compounds (Majer et al. 2008), as well as stereoselective allylation of furfural or furylalcohol, in the presence of dedicated transition metal catalysts complexed with chiral ligands (Fig. 50).

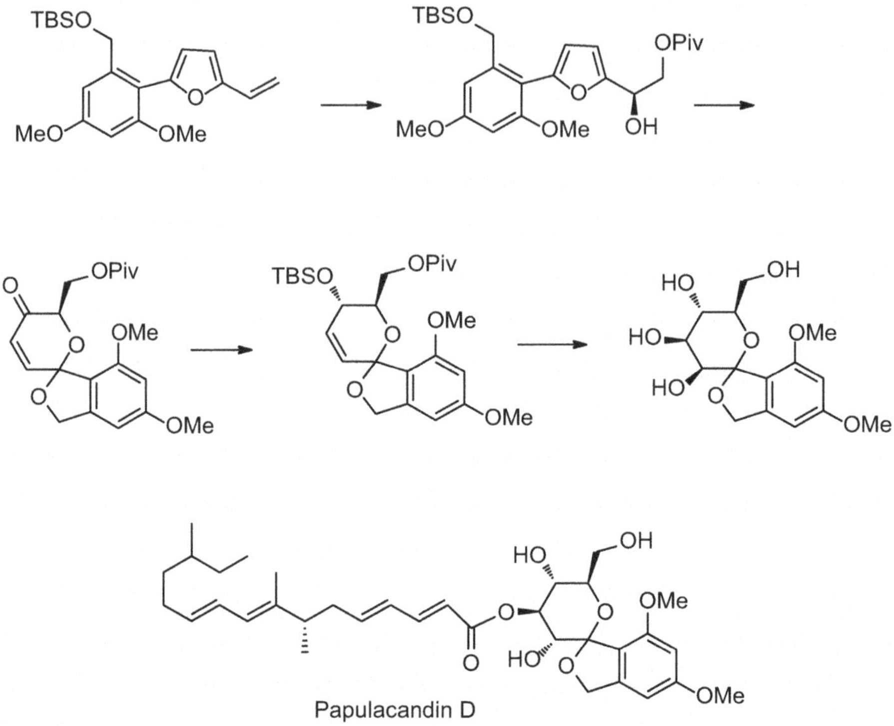

Fig. 49 Papulacandin D: synthesis of its tricyclic spiroketal nucleus

Several enantioselective methods of furylcarbinols preparation were scaled up and became validated for purpose of a large-scale processing (Merino 2015). Biocatalysts are useful for racemate resolution (e.g., lipase for esterification or ester hydrolysis), and similar efficiency was observed during exposition of racemic furylcarbinols to chiral epoxidation under Sharpless conditions (Smith et al. 1990). Simple furan derivatives are thus well suited for generating small molecule libraries helpful in diversification of selected natural lead structures (Burke et al. 2004; Yu et al. 2013; Asta et al. 2013).

Many other examples of dynamic asymmetric transformations are known, which can be applied to furylcarbinols (Bandini et al. 2004; Bhat et al. 2017). These involve crystallization-induced separations, ester hydrolysis, and acylations (Jakubec et al. 2006; Mahajan et al. 2017; Wang et al. 2017). The Sharpless epoxidation reagent catalyzed racemate resolution depicted below remains among the most popular methods of the AR substrates to enantioselective total syntheses (Fig. 51).

Conditions initially applied for AR assumed a two-step procedure, starting with stoichiometric bromine promoted alkoxylation of the starting material and isolation of the isomeric 2,5-dimethoxy intermediates, followed by a protic acid-catalyzed aqueous hydrolysis. That procedure was gradually replaced by various modifications leading to more convenient one-step preparations carried out in

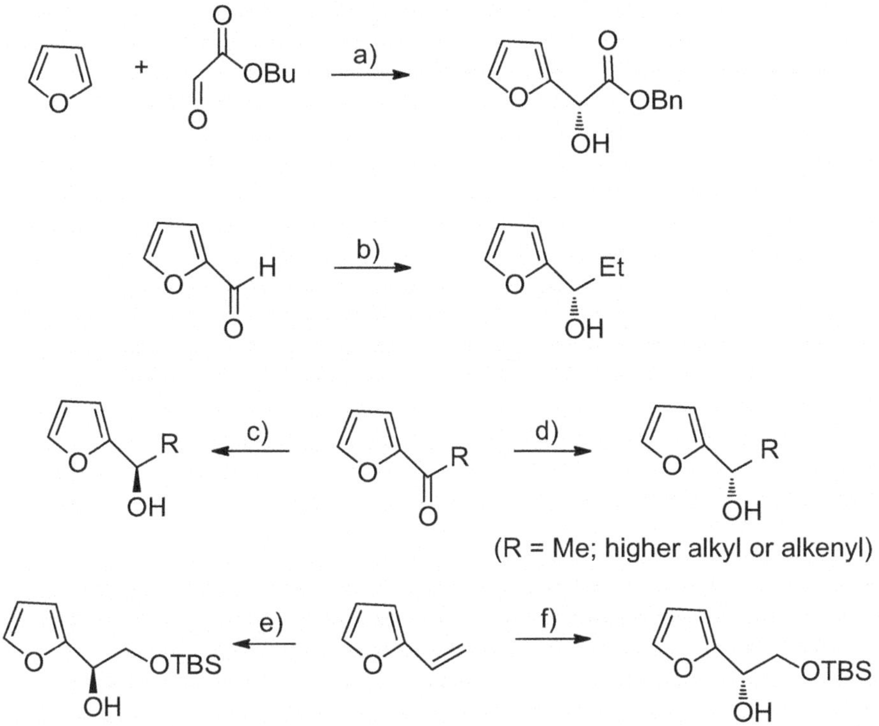

a) (R)-6,6'-Br_2-BINOL (2-5 mol%); b) $ZnEt_2$ / MIB (4 mol%); c) Noyori (R,R); d) Noyori (S, S); e) AD mix alpha; f) AD mix beta

Fig. 50 Catalytic asymmetric syntheses of chiral 2-furylalcohols

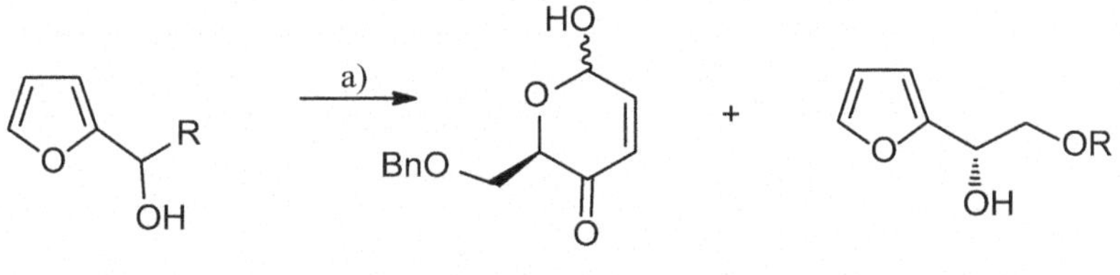

a) L-(+)-DIPT (15 mol %) / Ti $(O^iPr)_4$ (10 mol %), TBHP (0.7 eqv.)

Fig. 51 Kinetic resolution of racemic furylcarbinols upon action of Sharpless epoxidation reagent

mixed solvents, with less aggressive oxidants, with preference for applications of catalytic cycles. A list of AR-promoting reagents includes PCBA, other peroxyacids and their salts, NBS, oxone, PCC, CAN, and $VO(acac)_2$ – $t$$BuO_2H$ (Georgiadis et al. 1992; Merino 2015; Li and Zhu 2016). Oxidative enzymes (glucose peroxi-

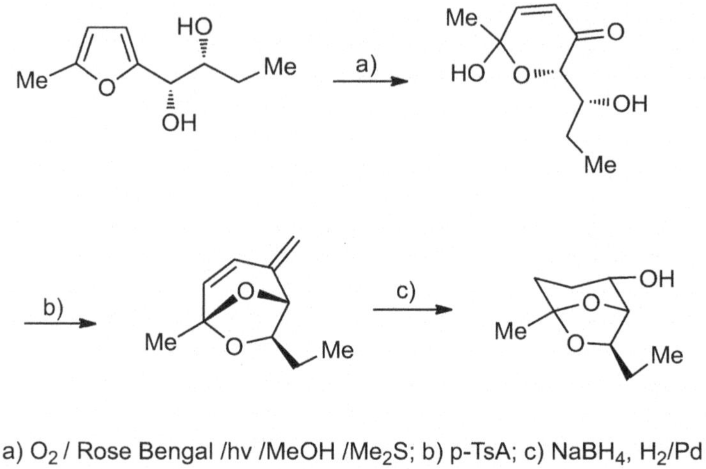

Fig. 52 Singlet oxygen promoted synthesis of 2-hydroxy-exo-brevicomin via AR

dase; laccase) under proper aeration conditions were found to carry out AR of various furylcarbinols in very good yields (Thiel et al. 2014; Deska et al. 2015; Harris et al. 2004). Special attention was given to application of singlet oxygen for AR, which starts with formation of the endoperoxide and can lead to a cascade of a spiroketal-directed cyclizations when applied to multifunctional substrates, as exemplified by preparation of 2-hydroxy-*exo*-brevicomin (Fig. 52) (Montagnon et al. 2008; Ghogare and Greer 2016; Noutsias et al. 2011).

Following studies on pyranosulose chemistry, stemmed from observation of versatile reactivity of the highly polarized enulose system, which is adjacent to an activated anomeric center. This arrangement can promote both: anomeric substituent exchange as well as its elimination. Examples of such reactivity are presented on the Fig. 53. Both types of bicyclic compounds, formed by intermolecular dipolar addition or intramolecular ketalization, found application as intermediates in a variety of natural products, including pyran secondary metabolites (Pellissieri 2011).

Similarly to furylalcohols, enuloses obtainable by AR are also amenable to racemate resolution via catalytic, kinetically controlled processes (Jones and Krische 2009). In an example presented above, (*S*)-pentenulose pivalate was obtained in 92% ee and 70% isolated yield in the reaction of racemic anomeric ester with 55 molar % amount of nitrobenzyl alcohol, catalyzed by bidentate chiral phosphine and palladium salt. Chiral pivalate thus obtained was subjected to phosphine-catalyzed cyclization with allenic carboxylate ester, to afford a bicyclic intermediate, from which iridoid genipine and its D-glucoside – geniposide – was obtained in few simple steps (Fig. 54).

It is obvious that reactivity of enuloses produced in AR, and their congeners obtained by follow-up transformations, strongly resembles chemistry of unsaturated compounds obtained from natural pyranoses in Fraser-Reid, Ferrier, and other studies (Fraser-Reid and Lopez 2009; Ferrier and Hoberg 2003; Ferrier and Zubkov 2003; Goel and Ram 2009). Recent accounts on unsaturated sugar chemistry clearly

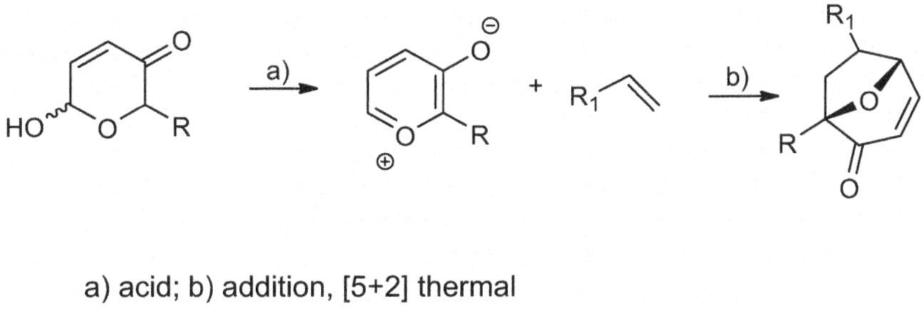

Fig. 53 General figure of addition of AR-derived oxonium derivative to an olefinic dipolarophile

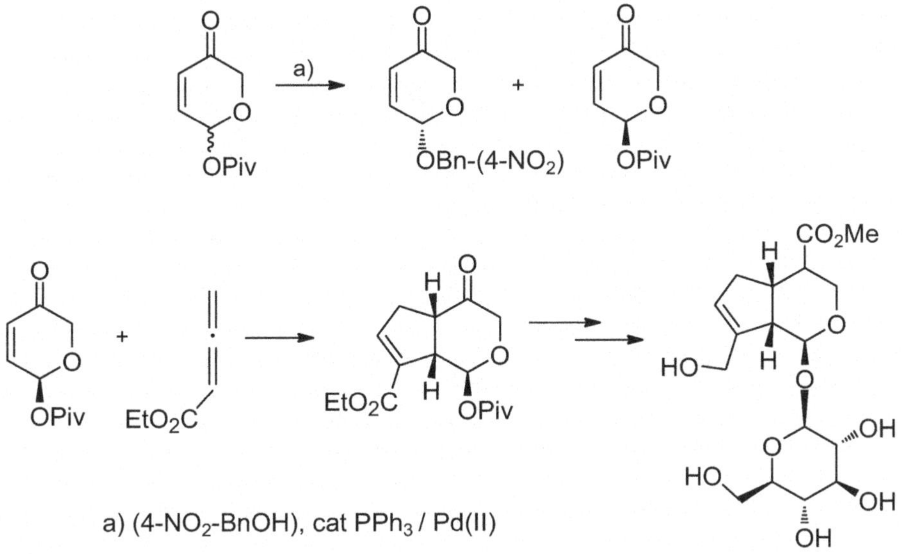

Fig. 54 Synthesis of geniposide

indicate that combination of hexenose stereoelectronic versatility with the ligand bonding and exchange capability of transition metal complexes offers remarkable synthetic possibilities in pyranose ring functionalization, in which well-established rules of stereoselection can be applied for chirality transfer (Li and Tong 2016; Mahajan et al. 2017). As an example, preparation of a stale, stoichiometric chiral enulose complexes, by stereoselection shown by a transition metal complex, can be quoted (Coombs et al. 2008; Li and Zhu 2016) (Fig. 55).

Although the ample amount of new chemistry, which followed AR, was delivered along during first two decades after its discovery (Georgiadis et al. 1992), it undoubtedly owes recent renaissance of interest to a wide accessibility of homochiral furylcarbinols, resulting from selection of new enantioselective methods of aromatic alkylation, catalytic carbonyl reduction (Noyori 2002), olefin dihydroxylation,

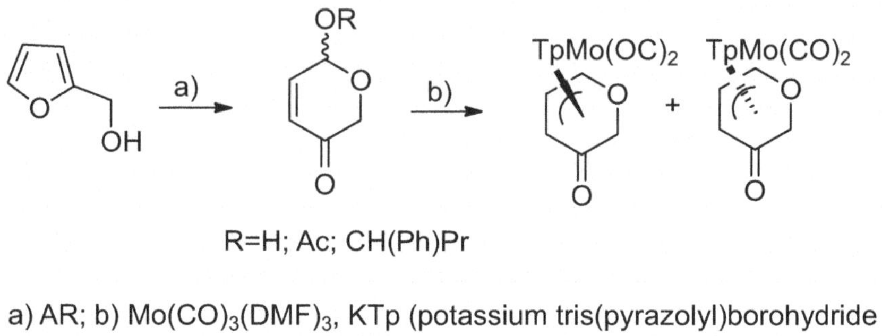

Fig. 55 Preparation of chiral oxopyranyl scaffolds

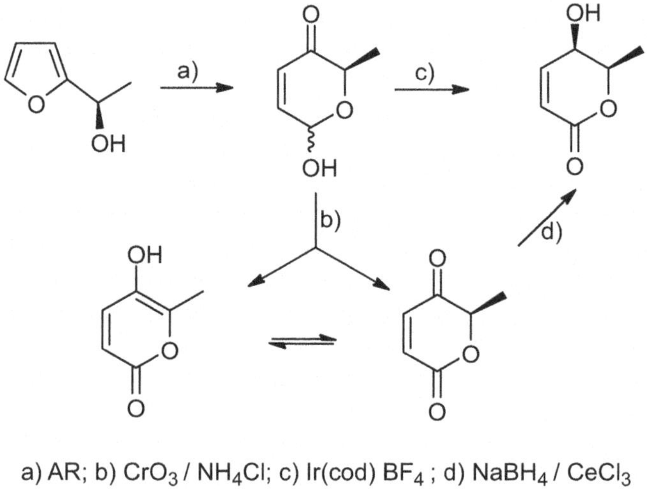

Fig. 56 Isomerization of pyranosulose lactols and lactones

and epoxidation (Koskinen 2012; Ghosh and Brindisi 2016), which makes them preferred synthons for asymmetrically substituted pyrans. Presently, 2-acyl furans (e.g., 2-acetylfuran) and 2-alkenyl furans (e.g., 2-vinylfuran) can be easily converted in a catalytic process into furylcarbinols of preferred absolute configuration (Merino 2015; Li and Tong 2016). Similarly, selected homochiral furylcarbinols can be obtained in catalytic Friedel-Crafts-type alkylation of furans with appropriate carbonyl electrophiles (Wang et al. 2015a, b; Bandini et al. 2004) (Fig. 56).

Englerins were discovered in 2009, as sesquiterpenoid constituents of extracts obtained from East African shrub of Euphorbiaceae family – *Phyllanthus engleri* – well recognized as a traditional medicinal plant of the region but poorly explored from the point of view of modern pharmacognosy. Initial screen against NCI-60 panel of human cancer lines of *P. engleri* extracts indicated remarkable activity against five out of eight renal cancer lines present in the panel, which prompted

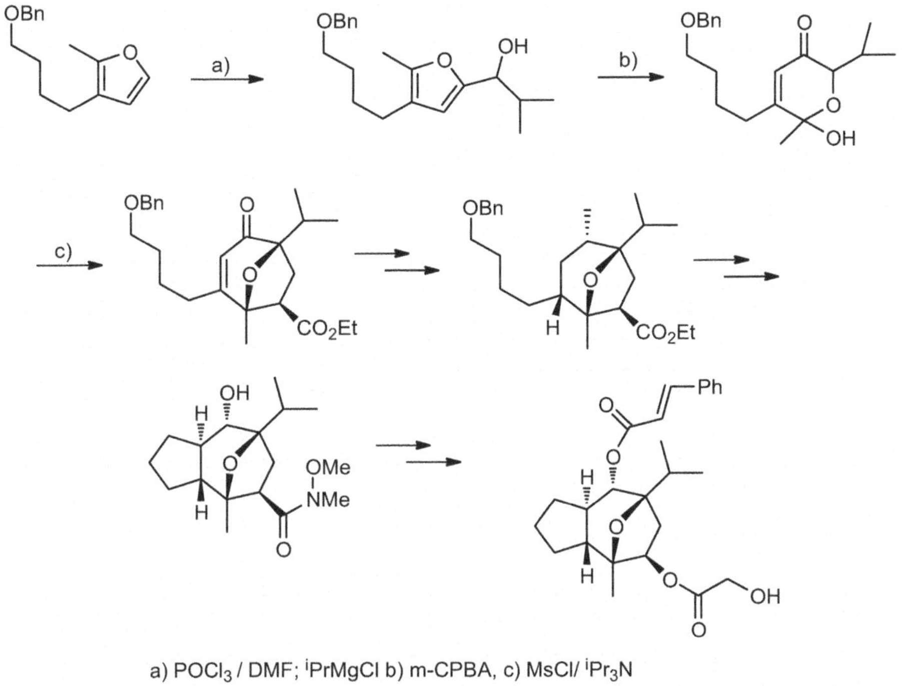

Fig. 57 Synthesis of englerin A

identification of two guaiane analogs named englerin A and B, responsible for the activity (Pouwer et al. 2012; Wu et al. 2017). Englerins instantly became a target of numerous synthetic efforts; current review lists fifteen successful total syntheses (two formal and next two ending with racemic mixtures) majority of them aimed at (−)-englerin A, while its enantiomer is also available synthetically. While these efforts differ greatly in their design, number of steps (8–25), and achieved product yield (2.9–20%), it should be noted that the key advanced bicyclic intermediate in englerin preparation can be obtained in a good yield by application of AR. The synthesis depicted below (Fig. 57) starts with assembling appropriate en-yn-ol substrate from which suitably substituted furan intermediate is obtained by base-catalyzed cyclization. Required furylcarbinol is obtained by Vilsmeier-Haack formylation, followed by addition of isopropylmagnesium chloride. AR is carried out by action of slight excess of m-CPBA; thus obtained pyranosulose is subjected to thermal [5 + 2] cycloaddition with ethyl acrylate used in excess. Resulting racemic oxabicycloenone is obtained as a mixture of C-9 epimers (β:α = 8:1) in which the needed diastereosomer prevails. From this point, the next 13 steps are needed to complete synthesis of racemic englerin A, with terminal olefin Wacker oxidation followed by carbocyclizarion, transformation of carboethoxy substituent into hydroxyl group and couple of delicate esterification (Wu et al. 2017; Nicolaou et al. 2012).

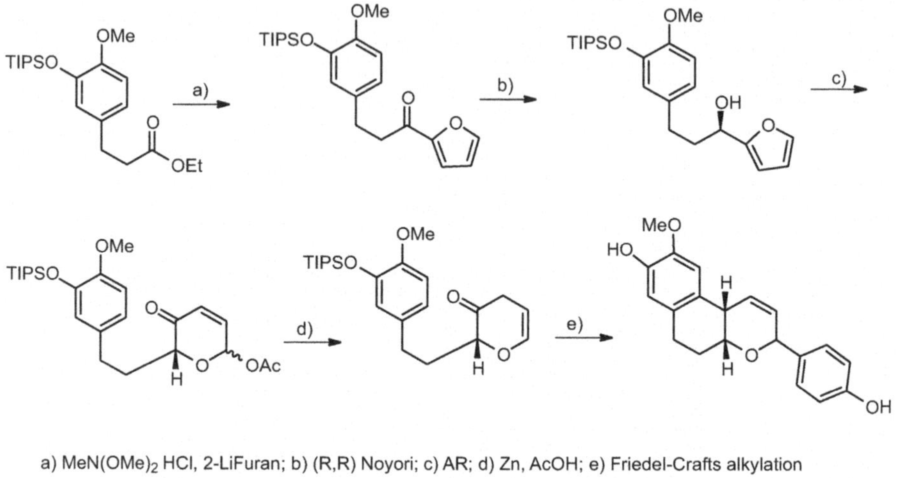

Fig. 58 Total synthesis of musellarin A comprising AR step

Musellarins are a small group of pyran-containing secondary plant metabolites, often classified as part of family of diarylheptanoid natural products, isolated from various sources and characterized by a potent biological activity, such as antioxidant, anticancer, antifungal, and antibacterial. After the first report of occurrence in Peruvian plant (2002), musellarin A was isolated in China (2011), from *Musella lasiocarpa*, in only 0.0006% yield, which spurred synthetic activity toward the compound. Linking the aromatic synthon derived from isovanillin with furfural gave after AR step pyranoid (enulose) intermediate, in which anomeric ester group was reductively eliminated to form glycal-type compound, suitable for arylation with a diazonium salt via Heck-type coupling. The final carbocyclic ring closure was carried out as intramolecular Friedel-Crafts alkylation with secondary cyclic triflate ester (Li et al. 2014, 2015) (Fig. 58).

This particular method of arylation proved useful also for asymmetric total syntheses of another diarylheptanoids from Asian herbs, named hedycoropyrans, differing from musellarins in a degree of condensation. The synthetic plan of their assembly is depicted on the figure showing p-substituted arylethanol, furfural, and aryldiazonium salt, which contain all carbon atoms present in the target molecule. Formation of substituted furylcarbinol leads to enulose synthon from which the cyclic vinyl ether needed for arylation is obtained by zinc-acetic acid reduction. The entire synthesis takes close to 20 steps including protection/deprotection, and the overall yield of the products, which is a mixture of anomeric C-glycosides – hedycoropyrans A and B – is little over 5% (Li and Tong 2017) (Fig. 59).

Perhaps the best illustration of the general AR utility for assembling pyran rings, including their spiroketal and condensed ring arrangements, comes from its multiple application in syntheses of very complex marine toxins, such as halichondrin and maitotoxin (Nicolaou et al. 2012). Some examples of a stereoselective assembly of condensed pyran ring fragments involving AR are presented below (Fig. 60).

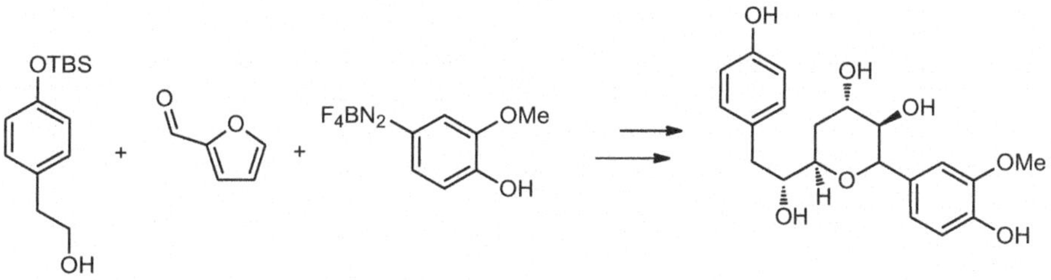

Fig. 59 General plan of synthesis for hedycoropyrans

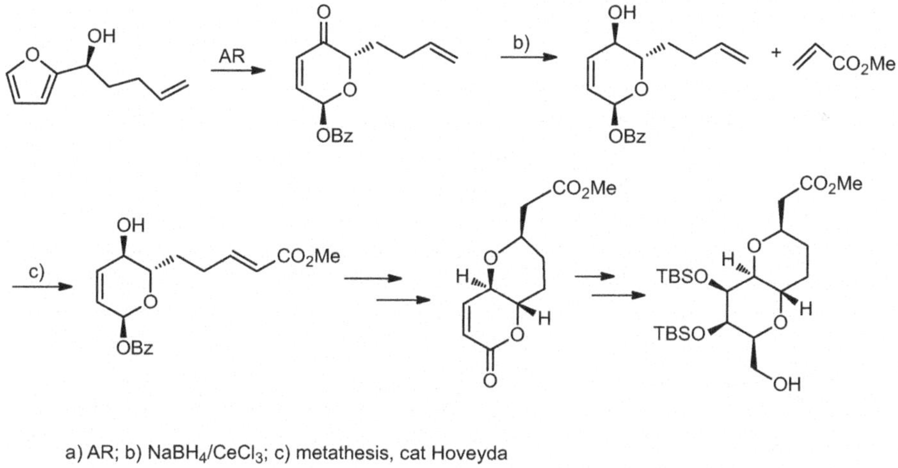

Fig. 60 Synthesis of bicyclic dipyranyl synthon for further functionalization into polycyclic system

The highly substituted bicyclic product at the end of the sequence can be easily converted into the pentacyclic structure in which five-membered ketals form the arrangement characteristic for marine toxin – halichondrin (Nicolaou et al. 2012; Mahajan et al. 2017). AR is currently applied for various types of heterocyclic intermediates in both: classical pyranosulose forms particularly useful for total syntheses of sugars and other pyrans of natural origin, as well as in aza-version, which became indispensible tool in preparation of some simple N-heterocycles, also useful for complex alkaloids preparation (Guo and O'Doherty 2008; Mahajan et al. 2017). Naturally, its wide application in an over four-decade span resulted in multitude of preparative procedures, some even validated for large scale, which claim optimization toward green chemistry, process safety, materials sustainability, and energy transmission efficiency. AR which has started as a stoichiometric two-step redox reaction evolved into an array of catalytic transformations, chemically, electrochemically, photochemically, or enzymatically driven, offering technical means for preparation of versatile intermediates, which are energetically friendly and environmentally benign. At the enulose product step, remarkable side enantiodifferentiation is possible with aid of transition metal carbonyl complexes containing chiral nucleophilic ligands (Wang et al. 2015a, b, 2017). Racemic furylcarbinols, as well as the primary AR products, can be subjected to kinetic resolution by action of Sharpless epoxidation reagents or other chiral selectors, which greatly facilitate application of the furan-pyran transformations in the total syntheses of heterocyclic natural products (Merino 2015; Li and Zhu 2016; Mahajan et al. 2017).

5 Synthesis of Selected Pyran Derivatives

5.1 Synthesis of KDO

3-Deoxy-D-manno-2-octulosonic acid (KDO) is expressed in Gram-negative bacteria, in higher plants, and in green algae. It is a vital component of the outer lipopolysaccharide membrane and is an essential substance for bacterial replication (Unger 1981; Angata and Varki 2002). The main step in the biosynthesis of KDO is the condensation of D-arabinose-5-phosphate and phosphoenolpyruvate to yield KDO-8-phosphate, providing the donor for KDO incorporation into oligosaccharides (Ghalambor et al. 1966; Ghalambor and Heath 1966). In lipopolysaccharide biosynthesis, the incorporation of KDO consists of the formation of CMP-KDO (Ghalambor and Heath 1966) by CMP-KDO synthetase (EC 2.7.7.38) followed by coupling with lipid-A precursor (Munson et al. 1978). The incorporation of KDO is an important step in LPS biosynthesis and in the growth of Gram-negative bacteria. Since the rate-limiting step is the activation of the KDO moiety, inhibitors of CMP-KDO synthetase are potentially useful as antibacterial agents. The most logical and simple synthetic approaches to 3-deoxy-2-ulosonic acids in general are those based on biomimetic pathways, coupling the aldehyde sugar with an equivalent of pyruvate unit (Unger 1981). The carbon-chain elongation strategy of pentose or hexose into octulose derivatives has been the subject of many papers, in which both conventional chemical and biochemical methods have been presented (Branchaud and Meier 1989). A chemoenzymatic methodology, involving the functionalization of aldoses by aldol condensation reactions catalyzed by enzymes, has been described by Wong and co-workers (Gijsen et al. 1996). The stereoselectivity of the aldol reaction is in general controlled by the enzyme and does not depend on the structure or stereochemistry of the substrate, making the stereochemistry of the products highly predictable (Gijsen et al. 1996; Brovetto et al. 2011). It is considered that the enzymatic aldol reactions of pyruvate with D-arabinose and its analogs catalyzed by KDO aldolase under nearly neutral and mild conditions may be useful. KDO aldolase (EC 4.1.2.23) from *Aureobacterium barkerei*, strain KDO-37-2, accepted an even wider variety of aldose substrates, including hexoses, pentoses, tetroses, and trioses; among these, the pentoses and tetroses were the best substrates. The isolated enzyme catalyzes the reversible condensation of pyruvate with D-arabinose to form KDO (Fig. 61). The enzyme was found to be specific for substrates having a (R)-configuration at C-3. In all the cases observed, the pyruvate attacked the aldose

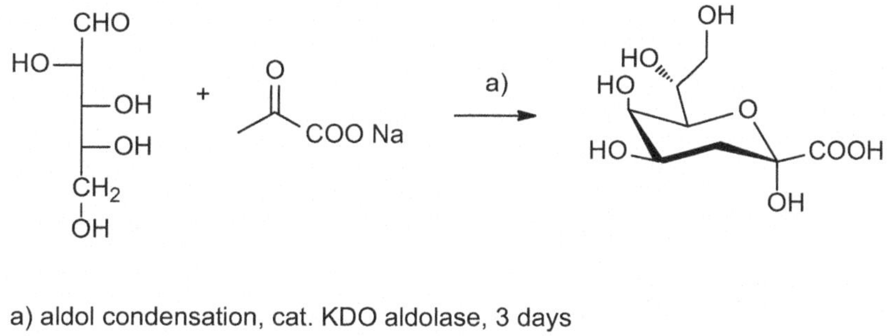

Fig. 61 Chemo enzymatic synthesis of KDO

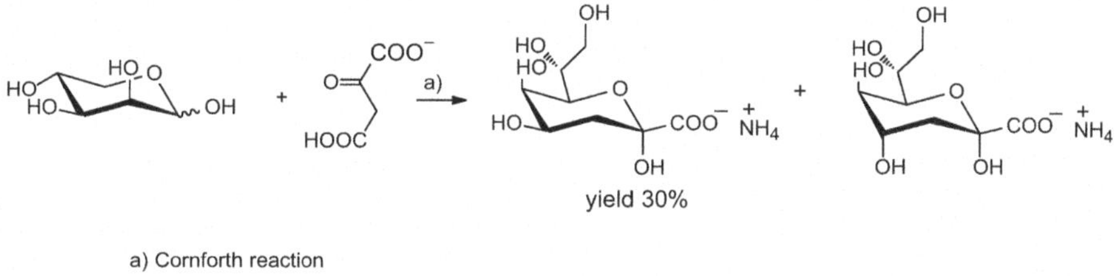

Fig. 62 Chemical synthesis of KDO

substrate stereoselectively on the re face of the carbonyl group. The enzymatic synthesis of KDO on multi-mmol scales using 10 molar excess of pyruvate worked well, and KDO was obtained in 67% yield (Sugai et al. 1993).

Of many chemical methods presented (Feng et al. 2015; Hekking et al. 2006; Martin and Zinke 1991), Cornforth's reaction of an aldehyde sugar and oxalacetic acid has been considered the most simple (Cornforth et al. 1958). Early chemical syntheses of KDO of and its derivatives have been developed, mainly using D-arabinose, D-mannose as substrates, and monosaccharides with C-2, C-3, and C-4 carbon configurations identical to C-4, C-5, and C-6 configuration of KDO. The condensation of oxalacetic acid with D-arabinose (Unger 1981; McNicholas et al. 1986; Shirai and Ogura 1989) has been efficiently applied to the preparation of KDO ammonium salt in crystalline form. As the carbonyl group of the sugar is diastereotopic, the *manno* and the *gluco* diastereoisomers are formed (Fig. 62). However, KDO (the *manno* configuration) is a main product. This method provides a short route to KDO, but due to the formation of a mixture of epimers, and difficulties with their separation, yields are always low.

A similar methodology of KDO synthesis by condensation of di-tert-butyl oxaloacetate with D-arabinose was proposed (Hershberger et al. 1968; Hershberger and Binkley 1968). A diastereomeric mixture of lactones was formed, and the epimers were separated by crystallization. The lactone with the D-*manno* configuration was converted into ammonium KDO by treatment with aqueous ammonia (Fig. 63).

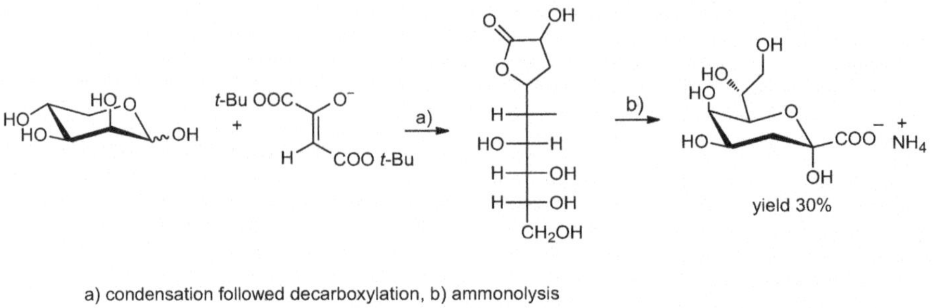

Fig. 63 Synthesis of ammonium KDO

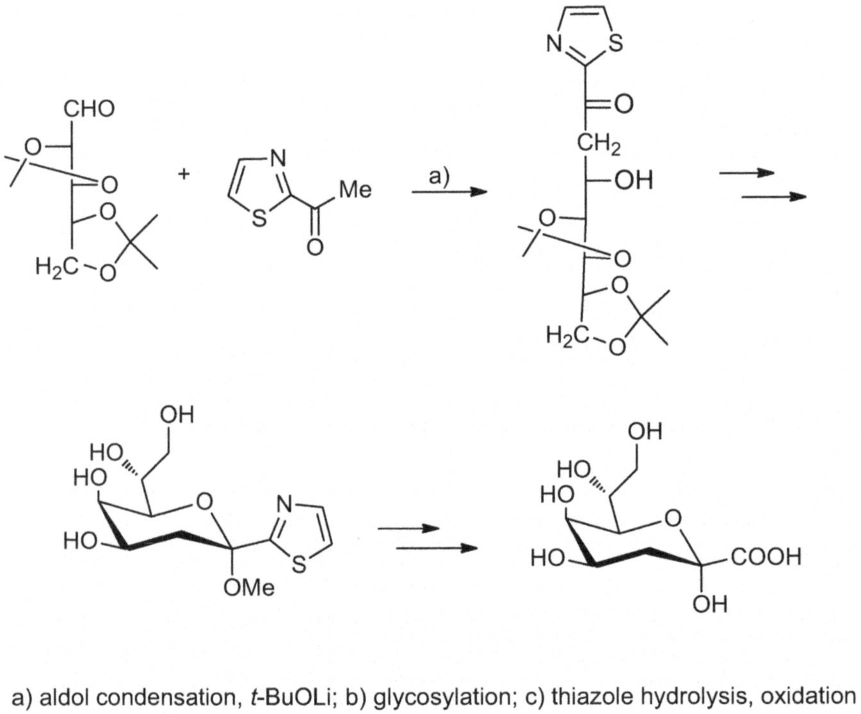

Fig. 64 Synthesis of KDO by D-arabinose elongation

The Dondoni group developed synthetic routes leading to higher sugars based on the thiazole-aldehyde synthesis. Crucial to the viability of these routes was ready access to *R*- or *S*-configured β-hydroxy or β-alkoxy ketones (Dondoni and Marra 2004). The synthesis of KDO was carried out starting from the 2,3:4,5-di-O-isopropylidene-D-arabinose (Dondoni and Merino 1991) (Fig. 64). These routes employed the thiazolyl ketone as a synthetic equivalent of pyruvaldehyde. The acetyl group of 2-acetylthiazole was selectively metallated by treatment with lithium tert-butoxide. The lithium enolate, formed under conditions of kinetic control,

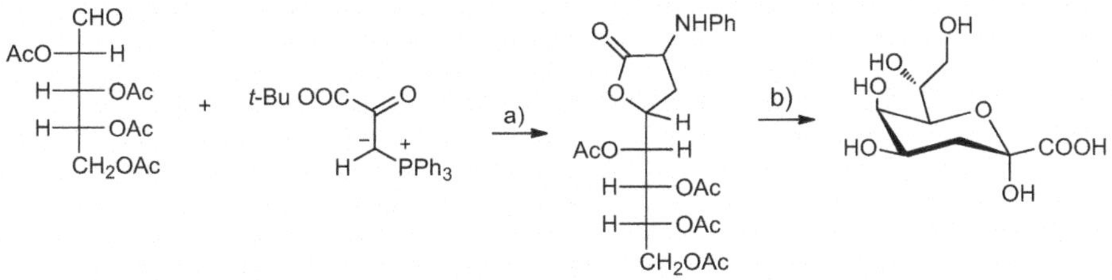

a) condensation, ester hydrolysis followed aminolysis; b) methanolysis,acid catalysed hydrolysis

Fig. 65 KDO synthesis via Wittig reaction

reacted with an aldehyde sugar to give the β-hydroxyketone. The reaction of the lithium enolate with the aldehyde arabinose occurred with a high degree of anti-diastereoselectivity (90%). The synthesis was continued: key steps in it were the glycosylation by treatment with a methanol solution of HCl, protection of the methyl glycoside, conversion to aldosulose by hydrolytic cleavage of the thiazole ring, oxidation, and deprotection, followed by oxidation leading to the desired product. The application of this three-carbon-chain elongation to protected D-arabinose afforded the octulosonic acid KDO in 6.8% overall yield (Dondoni and Merino 1991).

Kochetkov and co-workers (Kochetkov et al. 1967) used the Wittig reaction to build the skeleton of KDO. The reaction of 2,3,4,5-tetra-O-acetyl-*aldehydo*-D-arabinose with a triphenylphosphorane gave the Wittig adduct (Fig. 65). KDO was prepared by a sequence of reactions: hydrolysis of the ester, followed by treatment of the keto sugar with aniline, gave a mixture of D-*manno* and D-*gluco* diastereomeric enaminolactones. After separation of the diastereoisomers, methanolysis of the D *manno*-lactone and hydrolysis, the desired KDO was obtained.

A total synthesis of (+)-KDO was reported by Hu and co-workers (Hu et al. 1998); their method relied on the highly stereoselective HDA addition between an oxy-diene and ethyl glyoxalate (Fig. 66). The key step was the Wittig reaction of 2,3-O-isopropylidene-R-glyceraldehyde with triphenylphosphorane (route a). The KDO precursor was prepared by a highly stereoselective two-carbon-chain elongation via HDA addition of the diene to ethyl glyoxalate catalyzed by the (salen)Co^{II} complex (route c). The HDA addition provided a mixture of substituted pyrans, and the major product was separated. Hydroboration led to the tetrahydropyran derivative prepared in an enantiomerically pure form (route d). The next step required a change of the C4 carbon's configuration by oxidation of the hydroxyl group with subsequent reduction. Finally, the desired product was obtained by a series of conventional functionalizations.

In the total synthesis of natural compounds, the dienes designed by Danishewsky are conventionally used as substrates (Danishewsky et al. 1988). Early efforts to control stereoselectivity in these reactions focused on the use of chiral aldehydes in substrate-controlled diastereoselective reactions. An example of this approach can be seen in Danishewsky's synthesis of KDO (Fig. 67). Danishewsky's approach to KDO began with the synthesis of the diene (as a mixture of *E,Z*-isomers). In this

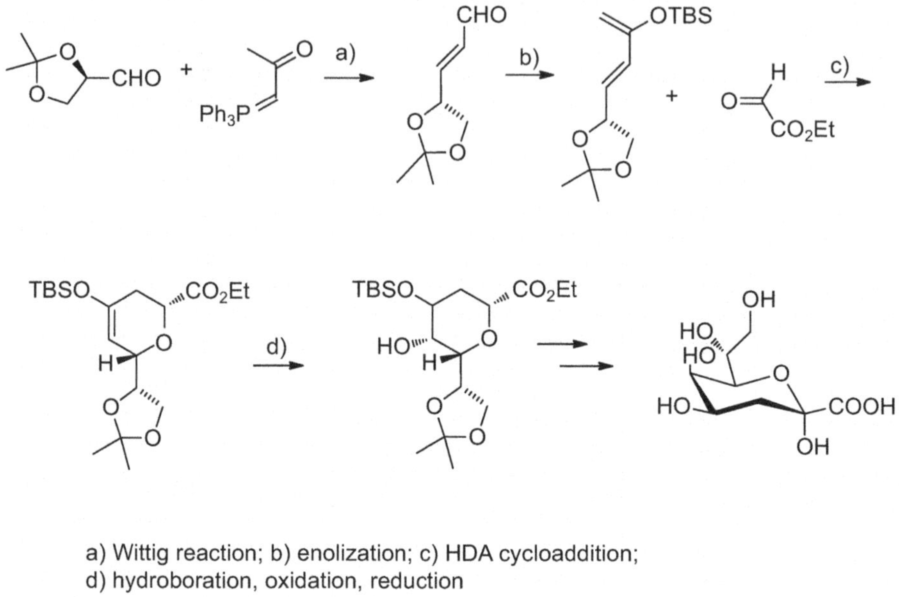

Fig. 66 Total synthesis of KDO

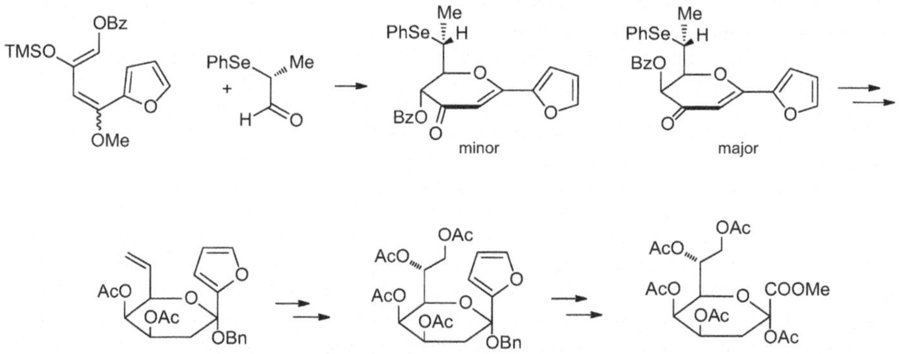

Fig. 67 Danishewsky's HDA synthesis of KDO

application, the hetero-Diels-Alder reaction occurred via a BF_3-catalyzed cycloaddition between the diene and aldehyde. The adduct mixture, on treatment with trifluoroacetic acid, delivered a mixture of dihydropyrones. Subsequent transformation of the major product by stereoselective Luche reduction of the carbonyl group and protection of hydroxyl group led to the glycal. Glycosylation followed oxidative elimination of the phenylseleno group gave the alkene; subsequent diastereoselective dihydroxylation and acylation were used to install the C-7/8 acetoxy groups. An oxidative cleavage of the electron-rich furan ring was used to form a carboxylic acid, which was then converted to a methyl ester with diazomethane. Finally, hydrogenolysis and acylation reactions were used to give the peracylated KDO sugar.

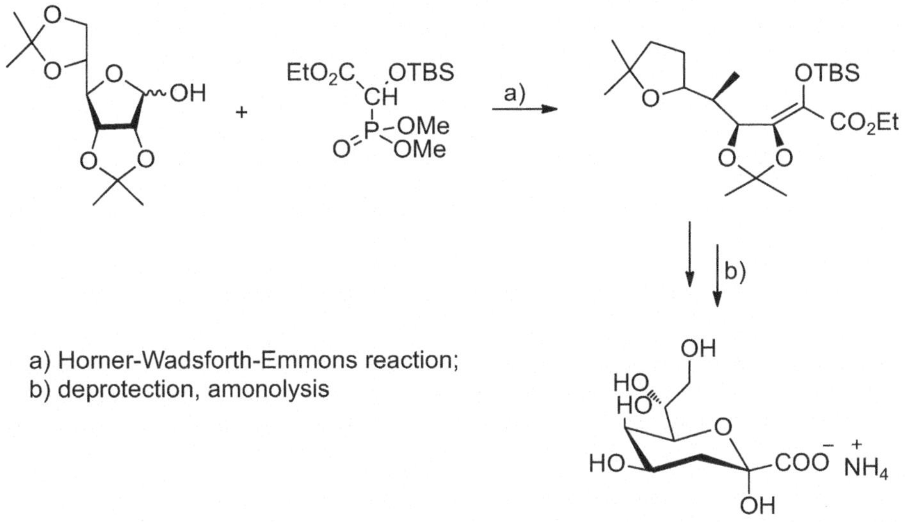

Fig. 68 Large-scale synthesis of KDO

An efficient method to synthesize KDO on a large scale was developed by Chai and co-workers (Feng et al. 2015) (Fig. 68). The key step was the Horner-Wadsworth-Emmons reaction of the commercially available phosphate ester with readily prepared 1,2:5,6-di-O-isopropylidene-D-mannose. The intermediate ester was obtained on a gram scale in 87% yield when t-BuOLi was used as the base and THF as the solvent. Deprotection and ammonolysis of KDO ester provided the ammonium salt in high yield.

5.2 *An Antitumor Drug Derived from Marine Toxins: Eribulin Story*

In 1985 Uemura et al. described isolation of a new polyether macrolides, which they named halichondrins, isolated from the marine sponge *Halichondria okadai* Kadota, the source known from earlier elaboration of okadaic acid structure (Uemura et al. 1985; Hirata and Uemura 1986). Their rather unusual structure, elucidated by combined molecular spectroscopy, became an instant target of retro-analytical approaches to various designs of the total synthetic design effort. Discovery of its potent anticancer activity has spurred a large synthetic effort initiatives in several academic centers, which eventually led to development of a new experimental drug (NSC 707389; E7389) and eventually to commercial therapeutic (Eribulin), launched under registered name of Halaven (Jackson et al. 2009). The prime lead for the drug development was the structure of halichondrin B, shown below (Fig. 69), which was confirmed by total synthesis completed by Y. Kishi, as early as 1992 (Aicher et al. 1992). By this time, halichondrins were detected in some unrelated species of

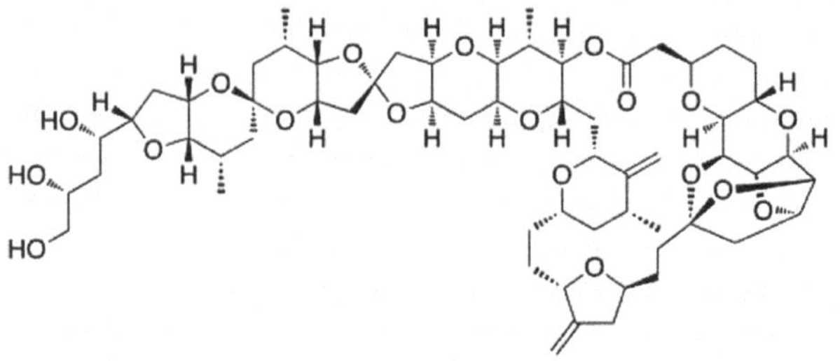

Fig. 69 Structural formula of halichondrin B, a marine toxin lead compound

sponge (*Raspalia agminata, Axinella*, and *Lissodendoryx* sp.) in total amounts varying from 0.1 to approx. 1 mg/kg of the sponge mass. As a result of surprisingly positive results in the halichondrins antitumor activity screening, the National Cancer Institute supported a deepwater sponge harvesting, which afforded less than 100 mg of halichondrins mixture (eight constituents identified) from over 600 kg of the biomass. Norhalichondrin A (35 mg) was the most abundant ($5.8 \times 10^{-8}\%$), while halichondrin B (HB), which is 50 times more active in standard antitumor tests, was isolated in $2.1 \times 10^{-8}\%$ yield (12.5 mg) (Allred et al. 2017).

The minute amount of the natural metabolite proved sufficient for an initial preclinical pharmacology studies, which revealed two important findings: (a) HB is an inhibitor of tubulin dynamics and organization, which results in halting cells in G2 phase and directing them toward apoptotic pathway, and (b) general toxicity of HB against normal cell is low enough to foster regular drug development program. (Jackson et al. 2009). Additionally, it has been established that tubulin-binding domain for HD is similarly located to the dimeric indole alkaloid (vinblastine, vinorelbine) domain, but its mechanism of action is distinct enough to recognize HB as a first-in-class non-taxane tubulin inhibitor of good promise for therapeutic significance. Under the circumstance of dramatic shortage of the substance for continuation of pharmacological studies, even the total synthesis of HB, then comprising 90 chemical synthetic steps, was welcomed as a temporary support for the initial drug development effort. While synthetic studies in the Y. Kishi group at Harvard University was initially focused on demonstrating the utility of a new, stereoselective method of C-C bond formation (presently known as Nozaki-Hiyama-Kishi reaction) (Hargaden and Guiry 2007), it soon produced substantial pool of HB synthetic analogs, together with an important discovery that promising antitumor activity resides exclusively in the right part of the macrocyclic molecule. Then, Kishi methods for HB intermediates synthesis and the final synthetic target assembly were licensed out to Japanese Eisai Pharmaceuticals and its US East Coast subsidiary – Eisai Research Institute (ERI) – where the new drug candidate Eribulin (Fig. 70) was designed and scalable methods for its synthetic process were developed and implemented (Towle et al. 2001; Seletsky et al. 2004; Zheng et al. 2004).

As can be seen by comparison of the above figures, the three dramatic structural changes undertaken consisted of exchange of macrolide ester function for a more stable ketone, truncation of the left-hand side cyclic polyether fragment, and

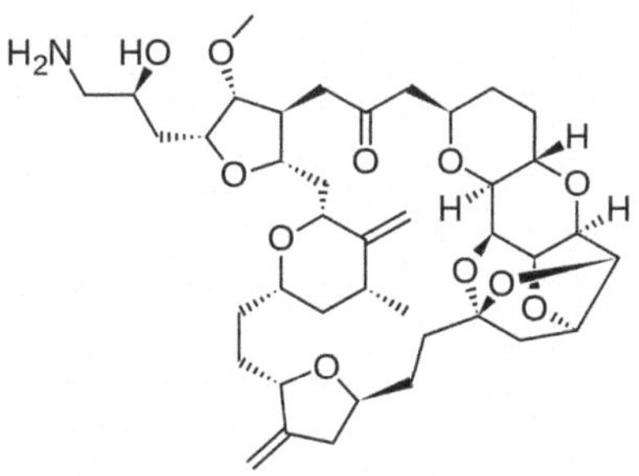

Fig. 70 Structure of eribulin (used as anticancer drug in the form of methanesulfonate salt)

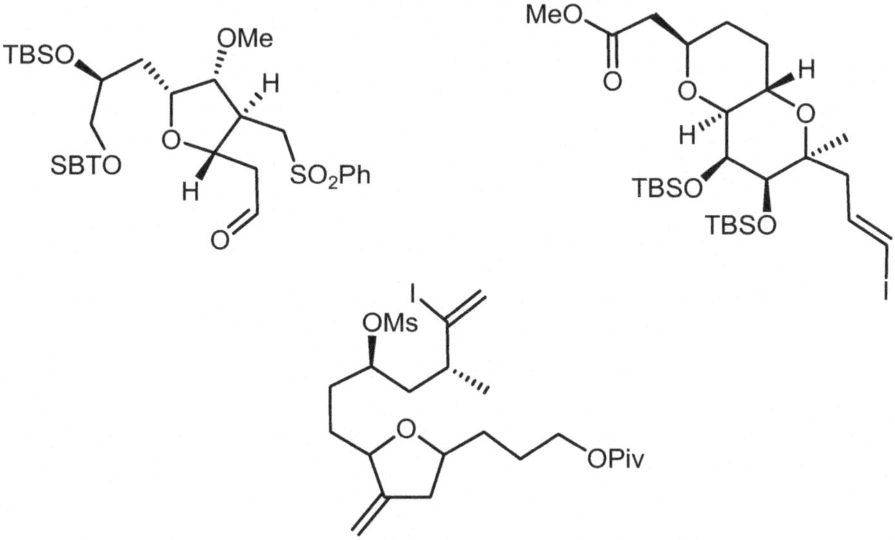

Fig. 71 Principal synthons for eribulin preparation deduced from retrosynthetic analysis

substitution of the four carbon triols containing side chain, with terminal primary amine as a part of aminopropandiol residue. Notably, these changes in molecular composition from $C_{54}H\ O_{21}$ to $C_{40}H_{59}NO_{11}$ (MW 729.9) have made the task of total synthesis more realistic, although 19 out of the original 32 stereogenic centers were still present in the target molecule. The general retrosynthetic approach to eribulin in ERI was based on earlier studies of HB synthesis by Kishi, making use of three advanced intermediates, shown below (Fig. 71), obtained by application of the chiral pool strategy.

In the technical process aimed studies at ERI, special care was taken to secure high stereoselectivity of the transformations used and to choose protecting and functional groups which give a fair chance of intermediate isolation and purification by crystallization. Optimization, side products elimination, and process reproducibility were attempted at every step on the way to multikilogram scale of synthesis. In the real world process, used for eribulin API manufacturing in multikilogram scale, the three crucial synthons have the structures shown below (Fig. 72)

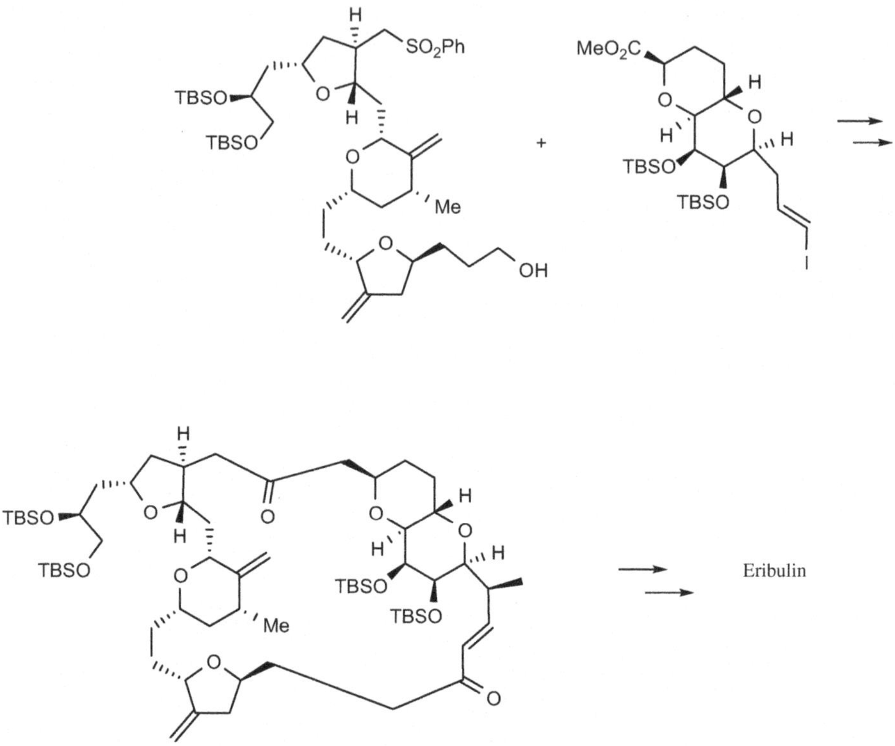

Fig. 72 Final steps of industrial total synthesis of eribulin from chiral pool-derived building blocks

(Fukuyama et al. 2016; Hasebe et al. 2013). The two tetrahydrofurans are linked together first, to prepare the intermediate which is suitable for making crucial ketone linkage with bicyclic tetrahydropyran synthons obtained from D-gulono-1,4-lactone, assembling all subunits of final eribulin macrocycle via sulfone-aldehyde coupling. Then, SmI_2 promoted sulfone group removal takes place, and the final C-C bond can be introduced, by intramolecular coupling of vinyl iodide with aldehyde group, catalyzed by $CrCl_2/NiCl_2$ according to Nozaki-Hiyama-Kishi procedure. The macrocyclic intermediate then undergoes hydroxyl group deprotection, with concomitant intramolecular ketalization, to afford the final skeletal arrangement. Last steps convert in the standard way terminal primary hydroxyl group into amine and its sulfonate salts, affording eribulin monomethylsulfonate, the active ingredient of Halaven drug.

It is evident from the presented short quotients of extensive synthetic figures describing several approaches to the total synthesis of eribulin in literature that the main challenge of the polycyclic molecule assembly consists of carbon-carbon bond formation between carefully composed chiral synthons. From the three tetrahydropyran (THP) rings among eight oxygen heterocycles present in the molecule, condensed assembly is derived from a natural sugar synthons, while isolated

THP ring is formed during base promoted intramolecular cyclization, which follows NHK coupling of tetrahydrofuran synthons shown in Fig. 7. Thus, pyran ring construction can hardly be considered a highlight of this over 60-step synthesis. Nevertheless eribulin synthesis deserves mentioning as an unprecedented effort, translated directly from academic research and undertaken during apparent pharmaceutical industry crisis, against all rules of the profit-driven enterprises (Austad et al. 2013). After its microtubule inhibiting activity was firmly established in some preclinical models (Wang et al. 2007; Smith et al. 2010), patient-related study started, with investigation of dose-limiting toxicity and pharmacokinetics, following eribulin mesylate solution infusion. Maximum-tolerated dose was found to be 1 mg per square meter of body area, on which further clinical trial schedules were based. In 2010, EM was successfully launched under propriety name Halaven® with indication for breast cancer patients, which were previously pretreated without positive result, and in the following year, the same therapy was approved by EMA in Europe (Twelves et al. 2010; Shablak 2013). Presently Halaven is registered and used in oncological clinical practice in over 40 countries.

5.3 Total Synthesis of Bryostatin

The bryostatins are a class of highly oxygenated macrolides originally isolated by the Pettit group (Pettit et al. 1982) under the NCI program designed to discover novel antitumor agents from natural sources. A marine bryozoan is the source of the bryostatins, a family of macrocyclic lactones. Bryostatins are the bacterial products found in *Bugula neritina*, which harbors the uncultivated gamma proteobacterial symbiont "*Candidatus Endobugula sertula*." Bryostatins are complex polyketides similar to bacterial secondary metabolites synthesized by modular type I polyketide synthases (PKS-I). Bryostatins show excellent potential as therapeutic agents that act through protein kinase C (PKC) signal transduction to alter cellular activity. A number of bryostatins have been isolated and characterized. The biological properties of the bryostatins stimulated syntheses of these compounds (Fig. 73) (Wender et al. 2015).

Enantioselective total synthesis of bryostatins have been reported: bryostatin 7 was achieved by Masamune's group (Kageyama et al. 1990), Lu and co-workers (2011), bryostatin 2 by Evans and co-workers (1999), bryostatin 3 by Yamamura and co-workers (Ohmori et al. 2000), and bryostatin 16 by Trost and Dong (2010), and the synthesis of bryostatin 9 was achieved by Wender and Schrier (2011). None of these methods, however, are viable for the synthesis of any of the bryostatins for medical applications. Particular attention has been paid to the synthesis of bryostatin 1, a lead compound which shows useful pharmacological activity (Kazuhiro et al. 2014) and can be the starting natural product for drug design. It is recognized that at least one mechanism for the function of this agent involves activity on protein kinase C (PKC) isozymes (Schwartz and Shah 2005; Etcheberrigaray et al. 2004). Bryostatin 1 binds to the cysteine rich domains of PKC, resulting in its activation and translocation to the cell membrane. These signalling proteins are known to reg-

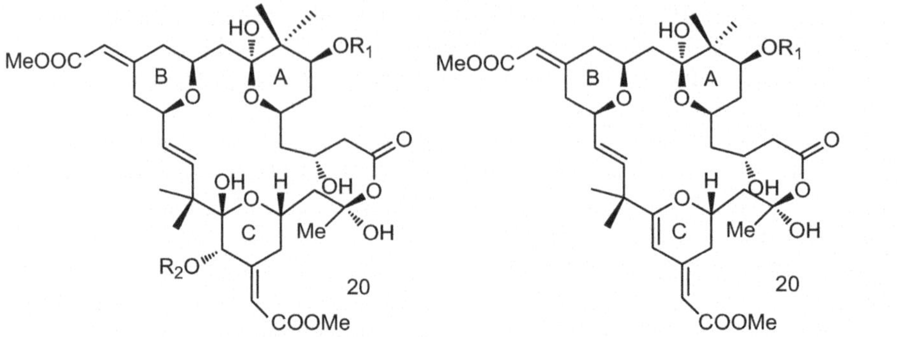

Bryostatins:1,2,7,9

Bryostatin 16: R1 = Me_3CCO

1: $R_1 = CH_3CO$, $R_2 = CH_3CH_2CH_2(CH = CH)_2CO$
2: $R_1 = H$, $R_2 = CH_3CH_2CH_2(CH{=}CH)_2CO$
7: $R_1 = R_2 = CH_3CO$
9: $R_1 = CH_3CO$, $R_2 = CH_3CH_2CH_2CO$

Fig. 73 Structure of bryostatins

ulate some of the most critical cellular processes and properties (Griner and Kazanietz 2007; Ali et al. 2016). Bryostatin 1 has shown activity against a range of cancers (Maier 2015; Gomes et al. 2016; Kazuhiro et al. 2014) and has also shown synergism with established oncolytic agents such as Taxol (Schwartz and Shah 2005). Bryostatin 1 is in clinical development directed at HIV/AIDS eradication, cancer therapy, and also neurological and cardiovascular diseases (Wender et al. 2017). The extension of clinical trials is limited due to the availability of the compound. The concentration of bryostatins in the marine organism is low, and the extracts contain the active principle in small quantities. Chemical synthesis is an alternative to extraction for marine-derived natural products. The work has been carried out in many laboratories and is the subject of many review publications, some of the recent ones we report (Maier 2015; Kazuhiro et al. 2014; Wender et al. 2007; Wender et al. 2011; Hale and Manaviazan 2010; Manaviazar and Hale 2011). The elegant first total synthesis of this high complexity molecule was presented by Keck and co-workers (2011). In the illustrated method, involving complex derivatives of pyran, the building blocks ring A and ring C were joined by "pyran annulation" with concomitant formation of ring B. The desired macrolide was synthesized from the intermediate by macrolactonization and functionalization. The synthetic route presented was a highly convergent one, in which further elaborations of the resulting very highly functionalized intermediate led to a family of the biologically active analogs. The synthesis of the ring A intermediate is presented below (Fig. 74).

The allylstannate, one of the coupling partners, was synthesized in eight overall steps (Keck et al. 2006) from commercially available α,β-unsaturated aldehyde. The key step of the synthesis was the catalytic asymmetric allylation to afford the protected homoallylic alcohol in high enantioselectivity (route a). The total synthesis of the intermediate compound was accomplished in the following way: reduction of

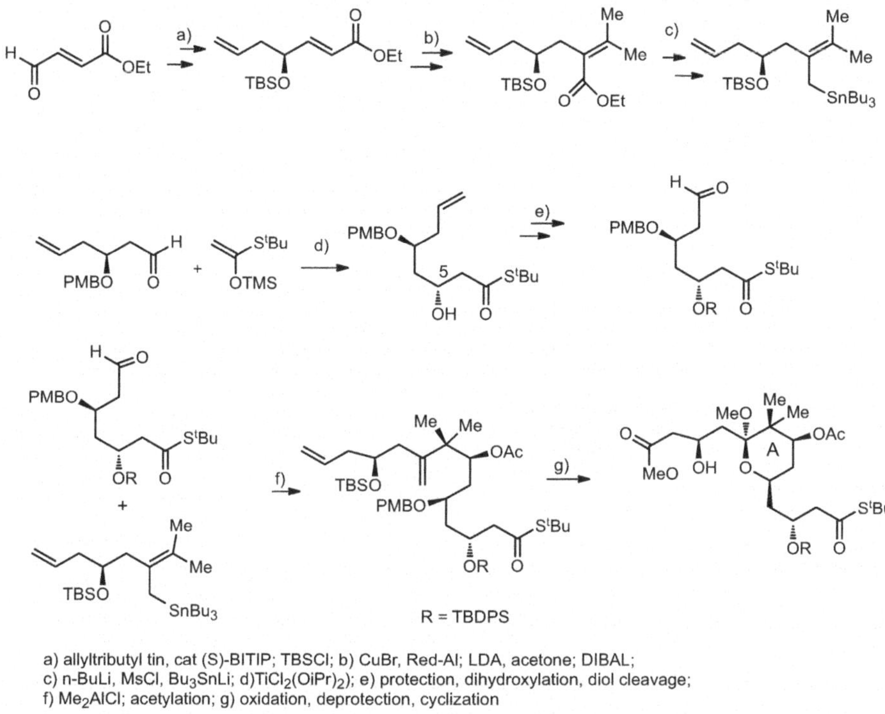

Fig. 74 Synthesis of functionalized ring A subunit

the double bond, condensation of the ester with acetone to provide a tertiary alcohol and introduction of the dimethyl moiety, elimination of the hydroxyl group, reduction of the α,β-unsaturated ester, mesylation of the alcohol, and Bu_3SnLi displacement providing the intermediate stannane in 37% overall yield. An important step in the synthesis of the next intermediate was the Mukaiyama aldol reaction. Optimization of the reaction conditions afforded the aldol adduct as a 41:1 mixture of diastereomers, and the obtained alcohol precursor had the desired configuration at the C-5 stereocenter (route d). Coupling of the aldehyde and stannate, followed by functionalization of the intermediate, gave the C_1-C_{13} A ring subunit.

The next step was the synthesis of the fully functionalized C-ring fragment (Fig. 75). The homoallylic alcohol was prepared from (R)-isobutyl lactate in six steps (Keck and Truong 2005). Esterification of the alcohol with carboxylic acid, followed by oxidation of the double bond and Wittig chain extension, afforded the olefin (route a) (Keck et al. 2011). Transformation of the carbonyl group to a methylene group and a Rainier metathesis reaction of the olefin led to the glycal derivative (route b) (Iyer and Rainier 2007). Oxidative functionalization of the glycal afforded the methoxyketone; this was followed by aldol condensation with methyl glyoxylate leading to the enoate (Keck et al. 2008) (route c). Luche reduction of carbonyl group gave an intermediate alcohol that was acetylated immediately. The desired aldehyde was obtained by desilylation and Ley oxidation using TPAP and NMO (Keck et al. 2011) (route d).

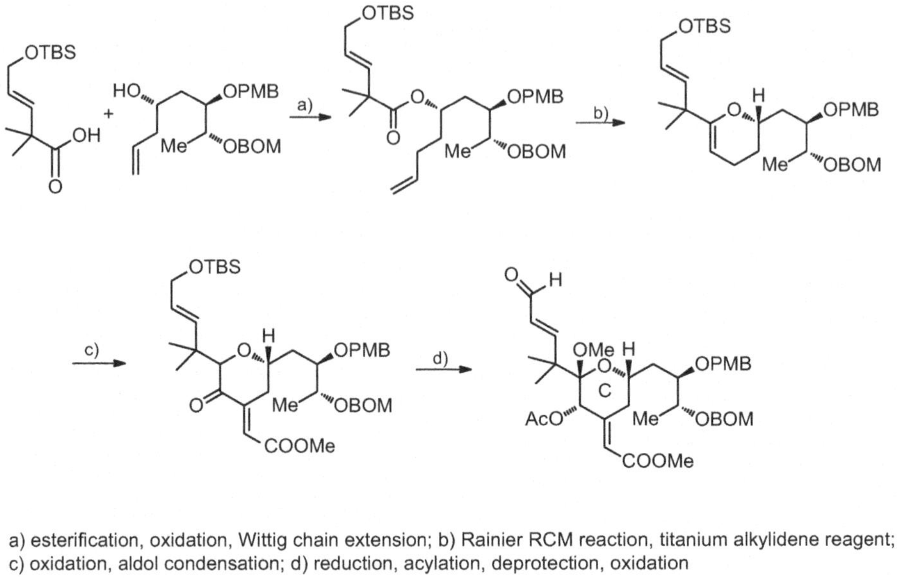

Fig. 75 Synthesis of functionalized C-ring fragment

Attempts were made to obtain bryostatin 1 in enantiomeric form by cycloaddition of the ring A and ring C portions (Fig. 76). The crucial step was the pyran annulation reaction between the A-ring hydroxyallylsilane and the C-ring aldehyde, which provided the tricyclic intermediate (route a). Hydrolysis of the thiolester to reveal the carboxylic acid and removal of the PMB protecting group, followed by Yamaguchi macrolactonization (Inanaga et al. 1979), afforded the macrolactone (route b). The regioselective oxidative cleavage of the olefin by Sharpless asymmetric dihydroxylation (Sharpless et al. 1992) followed by periodate oxidation provided the ketone in good yield (route c). An asymmetric Horner-Wadsworth-Emmons reaction on the ketone using Fuji's chiral BINOL phosphonate furnished a 4:1 Z/E mixture of α,β-unsaturated methyl esters (Tanaka et al. 1993). The main isomer was separated, regioselective methanolysis of the acetate providing the alcohol, which was immediately esterified with anhydride; desilylation then provided bryostatin 1. This first total synthesis of bryostatin 1 was thus accomplished in 30 steps for the longest linear sequence from commercially available (R)-isobutyl lactate; however, the long linear sequence and number of total steps have so far restrained this synthetic route from serving as a practical supply source for this natural product. Recently, Wender and his co-workers presented a scalable synthesis of bryostatin 1 and its analogs (Wender et al. 2017). This convenient synthesis proceeded in 29 steps with 4.8% overall yield. The synthetic strategy was based on proven methodologies for the synthesis of bryostatin 9 (Wender and Schrier 2011). The key steps of the synthesis of ring A (Fig. 77) and ring B (Fig. 78) are presented.

An efficient route to the ring A subunit consisted in Claisen condensation of t-butyl acetate and diethoxy propionate, followed by the highly stereoselective reduction of the keto group in the presence of Noyori catalyst to the alcohol with the correct C-3

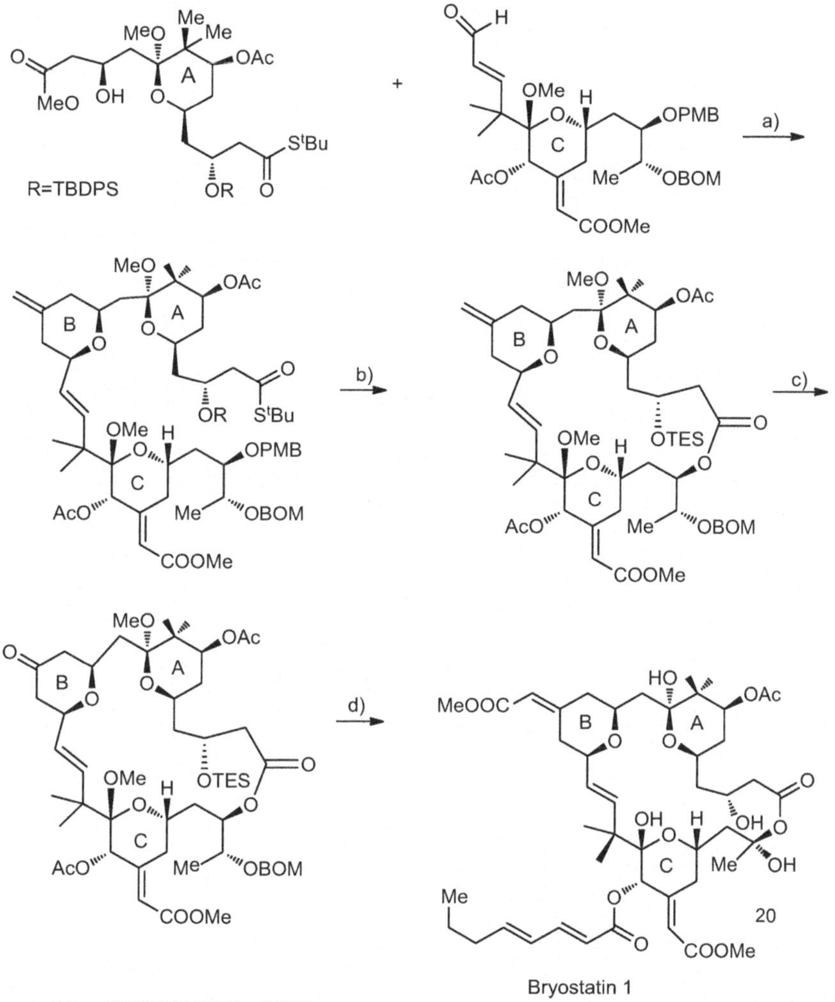

a) Pyranannulation (TMSOTf, Et2O, -780C);

b) deprotection thioester, cleavage PMB ether, macro lactonization

c) dihydroxylation, periodate oxidation,

d) Horner-Wadsworth-Emmons reaction ketone with BINOL phosphate, regioselective methanolysis, esterification, and final deprotection

Fig. 76 Total synthesis of bryostatin 1

configuration (route a). The boron-aldol condensation of the ester and diketone led to the product with the correct configuration at the stereogenic center (route b). In this way, the synthesis of a substrate having the desired number of carbon atoms as well as the necessary functional groups was achieved. The next steps consisted of the chemo- and diastereoselective reduction of the C7 carbonyl group and ketalization to afford the desired pyran derivative (route c). The A ring subunit having a defined configuration at the stereogenic centers and the correct functional groups was readily available in a series of transformations in 13% overall yield.

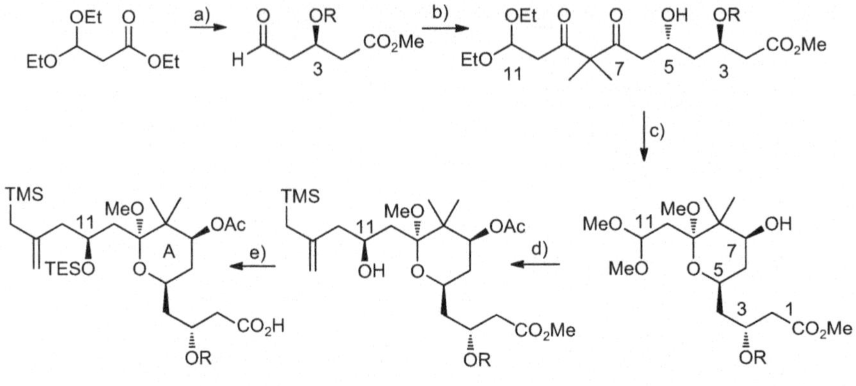

R = TBDPS

a) condensation,$MeCO_2{}^tBu$; (R)-BINAP-$RuCl_2$, H_2; b) boron-aldol reaction, 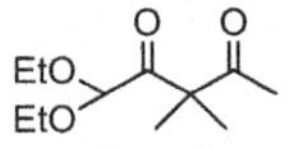 o

c) ketalization (trimethylortoformate) d) acylation, deprotection,Leighton allylation(Cl_3Si $CH_2C(=)CH_2TMS$);
e) protection,hydrolysis

Fig. 77 Synthesis of ring A subunit

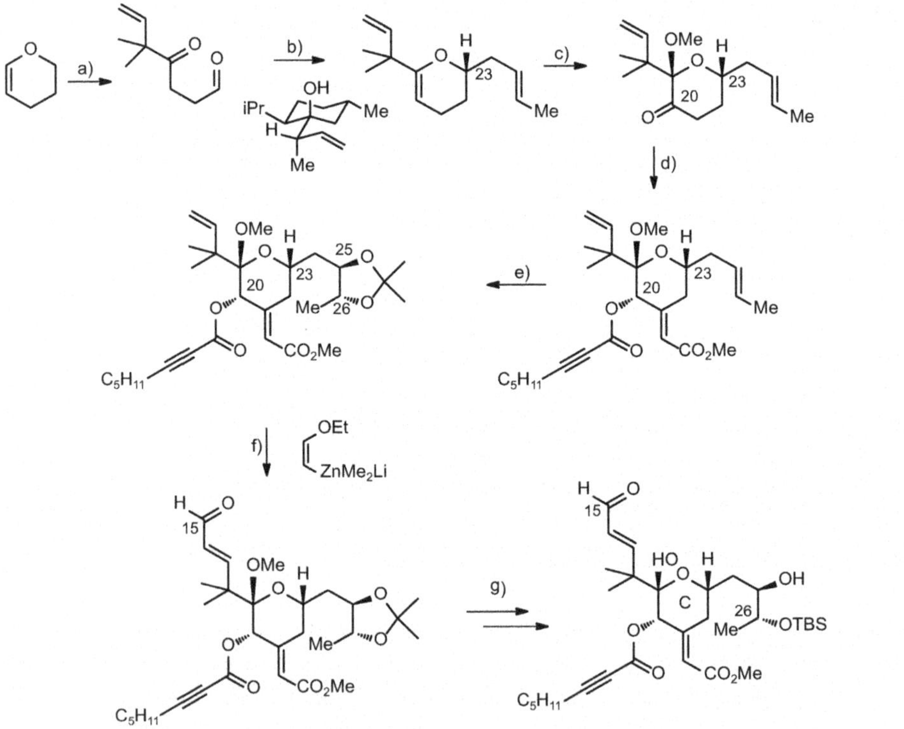

a) hydrolysis, prenylation, Swern oxidation; b) Nokami crotylation, cyclodehydratation
c) epoxidation, Dess-Martin oxidation; d) aldol condensation-methyl glyoxalate, reduction, esterification;
e) Sharpless oxidation, ketalization; f) ozonolysis, homologation of aldehyde; g) functionalization

Fig. 78 Synthesis of C ring subunit

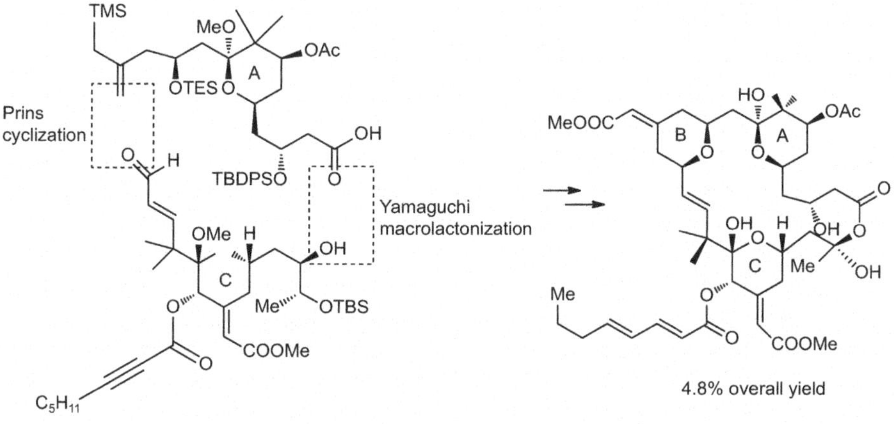

Fig. 79 Scalable total synthesis of bryostatin 1

Dihydropyran was used for a highly satisfactory synthesis of the ring C subunit (Fig. 78). This substrate led to the aldehyde via a sequence of reactions (route a). Nokami crotylation of the aldehyde with crotyl transfer reagent and in situ cyclodehydration proceeded with high chemo- and enantioselectivity to set up the C-23 stereochemistry (route b). In order to explain the correct configuration at C-20, the formation of the intermediate ketone was assumed to occur by Dess-Martin oxidation of the glycoside and by enantioselective reduction (route d). Highly stereoselective introduction of the hydroxyl groups at C-25 and C-26 was possible by Sharpless hydroxylation of the double bond (route e). The introduction of the enal at C-15 was performed by ozonolysis of the terminal double bond and homologation of the intermediate aldehyde with vinyl zinc reagent (route f). Functionalization of the intermediate compound afforded the bryostatin C-ring subunit (13 steps, 16% overall yield). Yamaguchi macrolactonization (Inanaga et al. 1979) then joined the A- and C-ring fragments in high yields. The next key step was a Prins cyclization catalyzed by pyridinium p-toluenesulfonate (PPTS) in methanol; this process formed bryostatin's B ring while closing the macrocycle (Fig. 79). Functionalization of the macrocycle led to the desired bryostatin. The key functionalization step was the diastereoselective transformation of the triple bond to conjugate double bonds using Fuji's phosphonate. The B-ring enolate was installed by a Horner-Wadsworth-Emmons reaction using Fuji's phosphonate.

This synthetic product was >99.5% pure after high-performance liquid chromatography purification and crystallization. This study opens practical, gram-scale access to bryostatin 1 and a wide range of a simpler analogs synthesis from late-stage intermediates. Research on design and synthesis of analogs carried out in the Wender team led to a series of biologically active compounds (Wender et al. 2011; Wender et al. 2017).

6 Conclusions

Natural products (secondary metabolites) continue to remain the central inspiration for biological activity oriented research, as exemplified by current trends in medicinal chemistry and in drug discovery and development efforts. Pyran ring containing natural products are fairly common in many categories of secondary metabolites, which include acetogenins, products of shikimic acid biogenetic pathway, chromones, flavonoids, coumarins, iridoids, terpenoids, etc. Greatly, many of these compounds proved useful as molecular probes in all sorts of research related to mammal biology, and a fair proportion advanced to a stage of drug leads and candidates for pharmaceutical development, with some significant examples of registered therapeutics. A newly composed list of pyran-containing privileged structures, with focus on condensed aromatic systems, was added quite recently (Kumar et al. 2017). Currently, the needs for majority of complex natural products, which feature pyran scaffold, are satisfied by chemical synthesis. The present state of total synthetic approaches to scarcely available natural products relies on methods described as enantioselective catalysis, which use sophisticated chiral catalysts for efficient chirality transfer and amplification. Principal methods of the pyran ring assembly were surveyed, with intention to indicate the approaches which combine chemical efficiency with effective control of stereoselection, without discriminating a priori any source of chirality. In the real world processes of target-oriented chemical synthesis, great significant achievements of stereoselective total synthesis well harmonize with the chiral pool application, as indicated in examples devoted to preparation of cyclic polyether marine toxins, which require assembly of several units often differing in substitution patterns and stereochemistry (Marmsäter and West 2002; Donner et al. 2004; Guidotti and Coelho 2015).

References

Achmatowicz O (1981) An approach to the synthesis of higher-carbon sugars. In: Trost BM, Hutchinson CR (eds) Organic chemistry today and tomorrow. Pergamon, Oxford/New York, pp 307–318

Achmatowicz O, Szechner B, Zwierzchowska Z et al (1971) Synthesis of methyl 2,3-dideoxy-DL-alk-2-enopyranosides from furan compounds: a general approach to the total synthesis of monosaccharides. Tetrahedron 27:1973–1976

Adak L, Kawamura S, Toma G et al (2017) Synthesis of aryl C-glycosides via iron-catalyzed cross coupling of halosugars: stereoselective anomeric arylation of glycosyl radicals. J Am Chem Soc 139:10693–10701

Ahmed TS, Grubbs RH (2017) Fast-initiating, ruthenium-based catalysts for improved activity in highly E-selective cross metathesis. J Am Chem Soc 139:1532–1537

Aicher TD, Buszek KR, Fang FG et al (1992) Total synthesis of halichondrin B and norhalichondrin B. J Am Chem Soc 114:3162–3164

Alemán J, Núñez A, Marzo L et al (2010) Asymmetric synthesis of 4-Amino-4*H*-Chromenes by organocatalytic oxa-Michael/Aza-Baylis–Hillman tandem reactions. Chem Eur J 16:9453–9456

Ali ES, Hua J, Wilson CH et al (2016) The glucagon-like peptide-1 analogue exendin-4 reverses impaired intracellular Ca2+ signalling in steatotic hepatocytes. BBA-Mol Cell Res 1863:2135–2146

Aljahdali AZ, Shi P, Zhong Y, et al (2013) De novo asymmetric synthesis of the pyranoses: from monosaccharides to oligosaccharides. Adv Carbohydr Chem Biochem 69:55–123

Allred TK, Manoni F, Harran PG (2017) Exploring the boundaries of "practical": de novo syntheses of complex natural product-based drug candidates. Chem Rev 117:11994–12051

Angata T, Varki A (2002) Chemical diversity in the sialic acids and related α-keto acids: an evolutionary perspective. Chem Rev 102:439–469

Ansari AA, Lahiri R, Vankar YD (2013) The carbon-Ferrier rearrangement: an approach towards the synthesis of C-glycosides. ARKIVOC 11:316–362

Asta C, Schmidt D, Conrad J et al (2013) The first enzymatic Achmatowicz reaction: selective laccase-catalyzed synthesis of 6-hydroxy-(2H)-pyran-3(6H)-ones and (2H).-pyran-2,5(6H)-diones. RSC Adv 3:19259–19263

Austad BC, Calkins TL, Chase CE et al (2013) Commercial manufacture of Halaven®: chemoselective transformations en route to structurally complex macrocyclic ketones. Synlett 24:333–337

Azuma T, Murata A, Kobayashi Y et al (2014) Arylboronic acid–aminothiourea catalytic system for the asymmetric intramolecular hetero-Michael reaction of α,β-unsaturated carboxylic acids. Org Lett 16:4256–4259

Babu RS, Chen Q, Kang S-W et al (2012) GA. *De novo* asymmetric synthesis of all D- all L- and D-/L-oligosaccharides using atom-less protecting groups. J Am Chem Soc 134:11952–11955

Bai Y, Kim LMH, Liao H et al (2013) Oxidative Heck reaction of glycals and aryl hydrazines: a palladium-catalyzed C-glycosylation. J Org Chem 78:8821–8825

Balachari D, O'Doherty GA (2000) Enantioselective synthesis of the papulacandin ring system: conversion of the mannose diastereoisomer into a glucose stereoisomer. Org Lett 2:4033–4036

Bandini M, Melloni A, Umani-Ronchi A (2004) New catalytic approaches in the stereoselective Friedel-Crafts alkylation reaction. Angew Chem Int Ed 43:550–556

Bennett CS (2017) Selective glycosylations: synthetic methods and catalysts. VCH, Weinheim

Bhat V, Welin ER, Guo X et al (2017) Advances in stereoconvergent catalysis from 2005 to 2015: transition-metal-mediated stereo ablative reactions, dynamic kinetic resolutions, and dynamic kinetic asymmetric transformations. Chem Rev 117:4528–4561

Binder JB, Raines RT (2009) Simple chemical transformation of lignocellulosic biomass into furans for fuel and chemicals. J Am Chem Soc 131:1979–1985

Bindseil KU, Zeeck A (1993) Metabolic products of microorganisms. Part 265 Prelactones C and B, oligoketides from *streptomyces* producing concanamycins and bafilomycins. Helv Chim Acta 76:150–157

Bokor E, Kun S, Goyard D et al (2017) C-glycosylarenes and hetarenes: synthetic methods and bioactivity focused on antidiabetic potential. Chem Rev 117:1687–1764

Braem A, Deshpande PP, Ellsworth BA et al (2014) Discovery and development of selective renal sodium-dependent glucose cotransporter 2 (SGLT2) dapagliflozin for the treatment of type 2 diabetes. Top Med Chem 12:73–94

Branchaud BP, Meier MS (1989) A novel strategy for the synthesis of ammonium 3-deoxy-D-manno-2-octulosonate (ammonium KDO). from lower monosaccharides. C–C bond construction of C6 of D-mannose via cobaloxime-mediated radical alkyl-alkenyl cross coupling. J Org Chem 54:1320–1326

Brandau S, Landa A, Franzen J et al (2006) Organocatalytic conjugate addition of malonates to α,β-unsaturated aldehydes: asymmetric formal synthesis of (-).-Paroxetine, chiral lactams, and lactones. Angew Chem Int Ed 45:4305–4309

Brovetto M, Gamenara D, Saenz Méndez P et al (2011) C–C bond-forming lyases in organic synthesis. Chem Rev 111:4346–4403

Burke SD, Quinn KJ, Chen VJ (1998) Synthesis of a C(22).-C(34). Halichondrin B precursor via ring opening-double ring closing metathesis. J Org Chem 63:8626–8627

Burke MD, Berger EM, Schreiber SL (2004) A synthesis strategy yielding skeletally diverse small molecules combinatorially. J Am Chem Soc 126:14095–14104

Büschleb M, Dorich S, Hanessian S et al (2016) Synthetic strategies toward natural products containing contiguous stereogenic quaternary carbon atoms. Angew Chem Int Ed 55:4156–4186

Caberele C, Reiser O (2016) The modern face of synthetic heterocyclic chemistry. J Org Chem 81:10109–10125

Carreaux F, Favre A, Carboni B et al (2006) First synthesis of (+).-8-Methoxygoniodiol and its analogue, 8-Deoxygoniodiol, using a three component strategy. Tetrahedron Lett 47:4545–4548
Chang S, Grubbs RH (1998) A highly efficient and practical synthesis of chromene derivatives using ring-closing olefin metathesis. J Org Chem 63:864–866
Chatterjee C, Pong F, Sen A (2015) Chemical conversion pathways for carbohydrates. Green Chem 7:40–71
Cherney AH, Kadunce NT, Reisman SE (2015) Enantioselective and enantiospecific transition-metal-catalyzed cross-coupling reactions of organometallic reagents to construct C–C bonds. Chem Rev 115:9587–9652
Chheda JN, Huber GW, Dumesic JA (2007) Liquid-phase catalytic processing of biomass-derived oxygenated hydrocarbons to fuels and chemicals. Angew Chem Int Ed 46:7164–7183
Cioc RC, Ruijter E, Orru RVA (2014) Multicomponent reactions: advanced tools for sustainable organic synthesis. Green Chem 16:2958–2975
Coombs TC, Lee MD, Wong H et al (2008) Practical, scalable, high-throughput approaches to η^3-Pyranyl and η^3-Pyridinyl organometallic enantiomeric scaffolds using the Achmatowicz reaction. J Org Chem 73:822–888
Corey EJ, Chelg X-M (1995) The logic of chemical synthesis. Wiley, New York
Cornforth JW, Firth ME, Gottschalk A (1958) The synthesis of N-acetylneuraminic acid. Biochem J 68:57–61
Czernecki S, Perlat MC (1991) C-glycosides 9. Stereospeific synhesis of C-glycosidic spiroketal of the papulacandins. J Org Chem 56:6289–6292
D'Ambrosio M, Guerriero A, Debitus C et al (1996) Leucascandrolide A, a new type of macrolide: the first powerfully bioactive metabolite of calcareous sponges *Leucascandra caveolata*, a new genus from the coral sea. Helv Chim Acta 79:51–60
Danishewsky SJ (1986) Reflection on organic synthesis: the evolution of a general strategy for stereoselective construction of polyoxygenated natural compounds. Aldrichimica Acta 19:59–65
Danishewsky SJ, De Ninno MP (1987) Totally synthetic routes to the higher monosaccharides. Angew Chem Int Ed Engl 26:15–23
Danishewsky SJ, Kerwin JF, Kobayashi S (1982) Lewis acid catalyzed cyclocondensations of functionalized dienes with aldehydes. J Am Chem Soc 104:358–360
Danishewsky SJ, Pearson WH, Segmuller BE (1985) Total synthesis of (+ -).-3-Deoxy-D-manno-2-octulopyranosate (KDO). J Am Chem Soc 107:1280–1285
Danishewsky SJ, DeNinno M, Chen S (1988) Stereoselective total syntheses of the naturally occurring enantiomers of N-acetylneuraminic acid and 3-deoxy-D-manno-2-octulosonic acid. A new and stereospecific approach to sialo and 3-deoxy-D-manno-2-octulosonic acid conjugates. J Am Chem Soc 110:3929–3940
Das R, Mukhopadhyay B (2016) Chemical O-glycosylations: an overview. ChemistryOpen 5:401–433
Demchenko AV (2008) Handbook of chemical glycosylation: advances in stereoselectivity and therapeutic relevance. Wiley VCH, Weinheim
Deshmukh PT, Gupta VB (2013) Embelin accelerates cutaneous wound healing in diabetic rats. J Asian Nat Prod Res 15:158–165
Deshpande PP, Ellsworth BA, Buono FG et al (2007) Remarkable β-selectivity in the synthesis of β-1-C- arylglucosides: stereoselective reduction of acetyl-protected methyl 1C-arylglucosides without acetoxy-group participation. J Org Chem 72:9746–9749
Deska J, Thiel D, Gianolio E (2015) The Achmatowicz rearrangement – oxidative ring expansion of furfuryl alcohols. Synthesis 47:3435–3450
Dondoni A, Marra A (2004) Thiazole-mediated synthetic methodology. Chem Rev 104:2557–2599
Dondoni A, Merino P (1991) Chemistry of the enolates of 2-acetylthiazole: aldol reactions with chiral aldehydes to give 3-deoxy aldos-2-uloses and 3-deoxy 2-ulosonic acids a short total synthesis of 3-deoxy-d-manno-2-octulosonic acid (KDO). J Org Chem 56:5294–5301
Donner CD, Gill M, Tewierik LM (2004) Synthesis of pyran and pyranone natural products. Molecules 9:498–512
Du H, Long J, Hu J et al (2002) 3,3'-Br_2-BINOL-Zn complex: a highly efficient catalyst for the enantioselective hetero-Diels-Alder reaction. Org Lett 4:4349–4352

Du H-F, Zhao D-B, Ding K-L (2004) Enantioselective catalysis of the hetero-Diels–Alder reaction between Brassard's diene and aldehydes by hydrogen-bonding activation: a one-step synthesis of (S)-(+)-dihydrokawain. Chem Eur J 10:5964–5970

Dusselier M, Mascal M, Sels BF (2014) Top chemical opportunities from carbohydrate biomass: a chemist's view of the biorefinery. Top Curr Chem 353:1–40

Ehrenkranz JRL, Lewis NG, Kahn CR et al (2005) Phlorizin: a review. Diabetes Metab Res Rev 21:31–38

El-Sepelgy O, Haseloff S, Alamsetti SK, Schneider C (2014) Brønsted acid catalyzed, conjugate addition of β-dicarbonyls to in situ generated ortho-quinone methides--enantioselective synthesis of 4-aryl-4H-chromenes. Angew Chem Int Ed 53:7923–7927

Elming N (1960) Dialkoxydihydrofurans and diacyloxydihydrofurans as synthetic intermediates. Adv Org Chem 2:67–115

Enders D, Grondal C, Hüttl MRM (2007) Asymmetric organocatalytic domino reactions. Angew Chem Int Ed 46:1570–1581

Etcheberrigaray R, Tan M, Dewachter I et al (2004) Therapeutic effects of PKC activators in Alzheimer's disease transgenic mice. Proc Natl Acad Sci U S A 101:11141–11146

Evans DA, Carter PH, Carreira EM et al (1999) Total synthesis of bryostatin 2. J Am Chem Soc 121:7540–7552

Fan Q-H, Li Y-M, Chan ASC (2002) Recoverable catalysts for asymmetric organic synthesis. Chem Rev 102:3385–3466

Fan Q, Lin L, Liu J et al (2004) Highly enantioselective hetero-Diels-Alder reaction of Brassard diene with aromatic aldehydes. Org Lett 6:2185–2188

Favre A, Carreaux F, Deligny M et al (2008) Stereoselective synthesis of (+).-Goniodiol, (+).-Goniotriol, (−).-Goniofupyrone, and (+).-Altholactone using a catalytic asymmetric hetero Diels−Alder/Allylboration approach. Eur J Org Chem 29:4900–4907

Feng Y, Dong J, Xu F (2015) Efficient large scale syntheses of 3-deoxy-D-manno-2-octulosonic acid (Kdo) and its derivatives. Org Lett 17:2388–2391

Ferrier RJ, Hoberg JO (2003) Synthesis and reactions of unsaturated sugars. Adv Carbohydr Chem Biochem 58:55–111

Ferrier RJ, Zubkov OA (2003) Transformation of glycals into 2,3-unsaturated glycosyl derivatives. In: Overman LE (ed) Organic reactions, vol 62. Wiley, New York, pp 569–736

Foland LD, Karlsson JO, Perri ST et al (1989) Rearrangement of 4-alkynylcyclobutenones. A new synthesis of 1,4-benzoquinones. J Am Chem Soc 111:975–989

Franke PT, Richter B, Jørgensen KA (2008) Organocatalytic asymmetric synthesis of functionalized 3,4-dihydropyran derivatives. Chem Eur J 14:6317–6321

Fraser-Reid B, Lopez JC (2009) Unsaturated sugars: a rich platform for methodological and synthetic studies. Curr Org Chem 13:532–553

Fukuyama T, Chiba H, Kuroda H et al (2016) Application of continuous flow for DIBAL-H reduction and N-BuLi mediated coupling reaction in the synthesis of eribulin mesylate. Org Process Res Dev 20:503–509

Gademann K, Chavez DE, Jacobsen EN (2002) Highly enantioselective inverse-electron-demand hetero-Diels−Alder reactions of α,β-unsaturated aldehydes. Angew Chem Int Ed 41:3059–3061

Georgiadis MP, Albizati KF, Georgiadis TM (1992) Oxidative rearrangement of furylcarbinols to 6-hydroxy-2H-pyran-3(6H).-ones, a useful synthon for the preparation of a variety of heterocyclic compounds. A review. Org Prep Proced Int 24:95–118

Gerard B, Marie J-C, Pandya BA et al (2011) Large-scale synthesis of all stereoisomers of a 2,3-unsaturated C-glycoside scaffold. J Org Chem 76:1898–1901

Gewali MB, Tezuka Y, Banskota AH et al (1999) Epicalyxin F and calyxin I: two novel antiproliferative diarylheptanoids from the seeds of *Alpinia blepharocalyx*. Org Lett 1:1733–1736

Ghalambor MA, Heath EC (1966) The biosynthesis of cell wall lipopolysaccharide in *Escherichia coli*. IV. Purification and properties of cytidine monophosphate 3-deoxy-d-*manno*-octulosonate synthetase. J Biol Chem 241:3216–3221

Ghalambor MA, Levine EM, Heath EC (1966) The biosynthesis of cell wall lipopolysaccharide in *Escherichia coli*. III. The isolation and characterization of 3-deoxyoctulosonic acid. J Biol Chem 241:3207–3215

Ghogare GA, Greer A (2016) Using singlet oxygen to synthesize natural products and drugs. Chem Rev 116:9994–10034
Ghosh AK, Brindisi M (2016) Achmatowicz reaction and its application in the syntheses of bioactive molecules. RSC Adv 6:111564–111598
Gijsen HJM, Qiao L, Fitz W et al (1996) Recent advances in the chemoenzymatic synthesis of carbohydrates and carbohydrate mimetics. Chem Rev 96:443–473
Goel A, Ram VJ (2009) Natural and synthetic 2H-pyran-2-ones and their versatility in organic synthesis. Tetrahedron 65:7865–7913
Gomes NGM, Dasari R, Chandra S et al (2016) Marine invertebrate metabolites with anticancer activities: solutions to the "supply problem". Mar Drugs 14:98
Gomez AM, Lobo F, Uriel C et al (2013) Recent developments in the Ferrier rearrangement. Eur J Org Chem 2013:7221–7262
Gomez AM, Lobo F, Miranda S et al (2015) A survey of recent synthetic applications of 2,3-dideo xy-hex-2-enopyranosides. Molecules 20:8357–8394
Grela K (2014) Olefin metathesis: theory and practice. Wiley, Hoboken
Griner EM, Kazanietz MG (2007) Protein kinase C and other diacylglycerol effectors in cancer. Nat Rev Cancer 7:281–294
Grubbs RH, Wenzel AG, O'Leary DJ et al (2014) Handbook of metathesis. Wiley-VCH, Weinheim
Grynkiewicz G, BeMiller JN (1982) Use of unsaturated sugars as alkylating agents. Addition of enol esters and ethers to glycals. J Carbohydr Chem 1:121–127
Grynkiewicz G, Zamojski A (1980) Electrophilic substitution of aromatic compounds by unsaturated sugar derivatives. Z Naturforsch 35b:1024–1027
Grynkiewicz G, Szeja W, Krzeczyński P et al (2014) Hexenoses in design of glycoconjugates – from chemistry to function. Chem Biol Interface 4:301–320
Guidotti BB, Coelho F (2015) Sequential Morita-Baylis-Hillman/Achmatowich reactions: an expeditious access to pyran 3(6H).-ones with a unique substitution pattern. Tetrahedron Lett 56:6356–6359
Guo H, O'Doherty GA (2008) De novo asymmetric syntheses of D-, L- and 8-epi-D-swainsonine. Tetrahedron 64:304–313
Guy D, Joly GD, Jacobsen EN (2002) Catalyst-controlled diastereoselective hetero-Diels-Alder reactions. Org Lett 4:1795–1798
Hale KJ, Manaviazan S (2010) New approaches to the total synthesis of the bryostatin antitumor macrolides. Chem Asian J 5:704–754
Hall DG, Lachance H, Denmark SE (2012) Allylboration of carbonyl compounds. Org React 73:1–596
Hall DG, Rybak T, Verdelet T (2016) Multicomponent hetero-[4 + 2] cycloaddition/allylboration reaction: from natural product synthesis to drug discovery. Acc Chem Res 49:2489–2500
Hanessian S (1993) Reflections on the total synthesis of natural products: art, craft, logic, and the Chiron approach. Pure Appl Chem 65:1189–1204
Hargaden GC, Guiry PJ (2007) The development of the asymmetric Nozaki-Hiyama-Kishi reaction. Adv Synth Catal 349:2407–2424
Harris JM, Li M, Scott JG et al (2004) Achmatowicz approach to 5,6-dihydro-2H-pyran-2-one containing natural products. In: Harmata M (ed) Strategies and tactics in organic synthesis, vol 5. Elsevier, Amsterdam, pp 221–253
Harvey AL (2008) Natural products in drug discovery. Drug Discov Today 13:894–901
Hasebe T, Hoshino Y, Ishizuka H et al (2013) Commercial manufacture of Halaven®: chemoselective transformations en route to structurally complex macrocyclic ketones. Synlett 24:333–337
Hauser FM, Ellenberger SR (1986) Syntheses of 2,3,6,-trideoxy-3-amino and 2,3,6,-trideoxy-3-nitrohexoses. Chem Rev 86:35–67
Hekking KFW, Moelands MAH, Van Delft FL et al (2006) An in-depth study on ring-closing metathesis of carbohydrate-derived α-alkoxyacrylates: efficient syntheses of DAH, KDO, and 2-Deoxy-β-KDO. J Org Chem 71:6444–6450
Henschke JP, Lin C-W, Wu P-Y et al (2015a) β-Selective C-arylation of diisobutylaluminum hydride modified 1,6-anhydroglucose: synthesis of canagliflozin without recourse to conventional protecting groups. J Org Chem 80:5189–5195

Henschke JP, Wu PY, Lin CW et al (2015b) β-Selective C-arylation of silyl protected 1,6-anhydroglucose with arylalanes: the synthesis of SGLT2 inhibitors. J Org Chem 80:2295–2309

Hershberger C, Binkley SS (1968) Chemistry and metabolism of 3-deoxy-d-mannooctulosonic acid: I. Stereochemical determination. J Biol Chem 243:1578–1584

Hershberger C, Davis M, Binkley SB (1968) Chemistry and metabolism of 3-deoxy-D-*manno*octulosonic acid. II. Practical synthesis and stability. J Biol Chem 243:1585–1608

Higman CS, Lummiss JAM, Fogg DE (2016) Olefin metathesis at the dawn of implementation in pharmaceutical and specialty-chemicals manufacturing. Angew Chem Int Ed 55:3552–3565

Hirata Y, Uemura D (1986) Halichondrins – antitumor polyether macrolides from a marine sponge. Pure Appl Chem 58:701–710

Hoang KLM, Leng W-L, Tan Y-J et al (2017) Stereoselective C-glycosylation from glycal scaffolds. In: Bennett CS (ed) Selective glycosylations: synthetic methods and catalysts. Wiley-VCH, Weinheim, pp 137–153

Hoffmann HMR, Rabe J (1985) Synthesis and biological activity of α-methylene gamma-butyrolactones. Angew Chem Int Ed Engl 24:94–104

Honda T (2012) Investigation of innovative synthesis of biologically active compounds on the basis of newly developed reactions. Chem Pharm Bull 60:687–705

Hong BC, Kotame P, Tsai CW et al (2010) Enantioselective total synthesis of (+).-Conicol via cascade three-component organocatalysis. Org Lett 12:776–779

Hoveyda AH (2014) Evolution of catalytic stereoselective olefin metathesis: from ancillary transformation to purveyor of stereochemical identity. J Org Chem 79:4763–4792

Hu YJ, Huang XD, Yao ZI (1998) Formal synthesis of 3-deoxy-d-manno-2-octulosonic acid (KDO). via a highly double-stereoselective hetero Diels–Alder reaction directed by a (salen) Co^{II} catalyst and chiral diene. J Org Chem 63:2456–2461

Hulme C, Gore V (2003) Multi-component reactions: emerging chemistry in drug discovery, from Xylocain to Crixivan. Curr Med Chem 10:51–80

Inanaga J, Hirata K, Saeki H et al (1979) A rapid esterification by means of mixed anhydride and its application to large-ring lactonization. Bull Chem Soc Jpn 52:1989–1993

Iyer K, Rainier JD (2007) Olefinic ester and diene ring-closing metathesis using a reduced titanium alkylidene. J Am Chem Soc 129:12604–12605

Jackson KL, Henderson JA, Phillips AJ (2009) The halichondrins and E7389. Chem Rev 109:3044–3079

Jakubec P, Berkes D, Siska R et al (2006) Crystallization induced asymmetric transformation (CIAT) in the synthesis of furyloalanines and furylcarbinols. Tetrahedron Asymmetry 17:1629–1637

Jensen KI, Dickmeiss G, Jiang H, Albrecht L, Jørgensen KA (2012) The diarylprolinol silyl ether system: a general organocatalyst. Acc Chem Res 45:248–264

Jiang X, Wang R (2013) Recent developments in catalytic asymmetric inverse-electron-demand Diels–Alder reaction. Chem Rev 113:5515–5546

Johansson Seechurn CCC, Litching MO et al (2012) Palladium-catalyzed cross-coupling: a historical contextual perspective to the 2010 Nobel Prize. Angew Chem Int Ed 51:5062–5085

Jones RA, Krische MJ (2009) Asymmetric total synthesis of the iridoid β-glucoside (+).-geniposide via phosphine organocatalysis. Org Lett 11:1849–1851

Kageyama M, Tamura T, Nantz MH, Roberts JC, Somfai P, Whritenour DC, Masamune S (1990) Synthesis of bryostatin 7. J Am Chem Soc 112:7407–7408

Kazuhiro IK, Ryo C, Yanagita RC (2014) Synthesis and biological activities of the simplified analogs of natural PKC ligands, bryostatin-1 and aplysiatoxin. Chem Rec 14:251–267

Keck GE, Truong AP (2005) Synthetic studies on the bryostatins: preparation of a truncated BC-ring intermediate by pyran annulation. Org Lett 7:2149–2152

Keck GE, Welch DS, Vivian PK (2006) Synthetic studies toward the bryostatins: a substrate-controlled approach to the A-ring. Org Lett 8:3667–3670

Keck GE, Kraft MB, Truong AP et al (2008) Convergent assembly of highly potent analogues of bryostatin 1 via pyran annulation: bryostatin look-alikes that mimic phorbol ester function. J Am Chem Soc 130:6660–6661

Keck GE, Poudel YB, Cummins TJ et al (2011) Total synthesis of bryostatin 1. J Am Chem Soc 133:744–747
Keitz BK, Endo K, Patel PR et al (2012) Improved ruthenium catalysts for Z-selective olefin metathesis. J Am Chem Soc 134:693–699
Khan SH, O'Neill RA (1996) Modern methods in carbohydrate synthesis. Harwood Academic, Amsterdam
Khan RKM, Torker S, Hoveyda AH (2014) Reactivity and selectivity differences between catecholate and catechothiolate Ru complexes. Implications regarding design of stereoselective olefin metathesis catalysts. J Am Chem Soc 136:14337–14340
Kikuchi T, Takagi J, Isou H et al (2008) Vinylic C–H borylation of cyclic vinyl ethers with bis(pinacolato) diboron catalyzed by an Iridium(I)-dtbpy complex. Chem Asian J 3:2082–2090
Knueppel D, Yang J, Cheng B et al (2015) Total synthesis of the aglycone of IB-00208. Tetrahedron 71:5741–5757
Kochetkov NK, Dmitriev BA, Backinowsky LV (1967) A new approach to the synthesis of higher 3-deoxyglyculosonic acids. Carbohydr Res 5:399–405
Koskinen AMP (2012) Asymmetric synthesis of natural products, 2nd edn. Wiley, Chichester
Křen V (2008) Glycoside vs aglycon: the role of glycosidic residue in biological activity. In: Fraser-Reid B, Tatsuta K, Thiem J (eds) Glycoscience: chemistry and chemical biology III. Springer, Berlin, pp 2589–2644
Křen V (2011) Chemical biology and biomedicine of glycosylated natural compounds. In: Fraser-Reid BO, Tatsuta K, Thiem J (eds) Glycoscience: chemistry and chemical biology I–III. Springer, Berlin, pp 2471–2529
Kühne ME, Benson BW (1965) The structures of the spiramycins and magnamycin. J Am Chem Soc 87:4660–4662
Kumagai N, Shibasaki M (2011) Recent advances in direct catalytic asymmetric transformations under proton-transfer conditions. Angew Chem Int Ed 50:4760–4772
Kumar D, Sharma P, Singh H et al (2017) The value of pyrans as anticancer scaffolds in medicinal chemistry. RSC Adv 7:36977–36999
Kunz H (2002) Emil Fischer – unequalled classicist, master of organic chemistry research, and inspired trailblazer of biological chemistry. Angew Chem Int Ed 41:4439–4451
Lee K, Kim H, Hong JA (2011) Stereoselective formal synthesis of leucascandrolide A. Org Lett 13:2722–2725
Lei M, Gao L, Yang J-S (2009) Microwave-assisted palladium catalyzed cross-coupling reactions between pyranoid glycals and aryl bromides. Synthesis of 2'-deoxy C-aryl-β-D-glycopyranosides. Tetrahedron Lett 50:5135–5138
Leidy MR, Hoffman JM, Pongdee R (2013) Preparation of C-arylglycals via Suzuki-Miyaura cross coupling of dihydropyranylphosphates. Tetrahedron Lett 54:6889–6891
Lemaire S, Schils D (2015) Development of efficient routes to access C-glycosides as SGLT-2 inhibitors for the treatment of type 2 diabetes. In: Časar Z, (ed), Synthesis of heterocycles in contemporary medicinal chemistry. Springer, Top Heterocycl Chem 44:29–50
Lemaire S, Houpis IN, Xiao T et al (2012) Stereoselective C-glycosylation reactions with arylzinc reagents. Org Lett 14:1480–1483
Levy DE (2006) Strategies towards C-glycosides. In: Levy DE, Fügedi P (eds) The organic chemistry of sugars. CRC Taylor & Francis, Boca Raton, pp 269–348
Levy DE, Fügedi P (2006) The organic chemistry of sugars. CRC Taylor & Francis, Boca Raton
Levy DE, Tang C (1995) The chemistry of C-glycosides. Pergamon, Oxford
Li Z, Tong R (2016) A catalytic environment-friendly protocol for Achmatowicz rearrangement. J Org Chem 81:4847–4855
Li Z, Tong R (2017) Asymmetric total syntheses of (-)-hedycoropyrans A and B. J Org Chem 82:1127–1135
Li X, Zhu J (2016) Glycosylation via transition-metal catalysis: challenges and opportunities. Eur J Org Chem 2016:4724–4767
Li Z, Leung TF, Tong R (2014) Total syntheses of (±)-musellarins A-C. Chem Commun 50:10990–10993

Li Z, Ip FCP, Ip NY, Tong R (2015) Highly trans-selective arylation of Achmatowicz rearrangement products by reductive γ-deoxygenation and Heck-Matsuda reaction: asymmetric total synthesis of (-)-musellarins A-C and their analogs. Chem Eur J 21:11152–11157
Li Y, Hu Y, Zhang S, Sun J et al (2016) Copper-catalyzed enantioselective hetero-Diels−Alder reaction of Danishewsky's diene with glyoxals. J Org Chem 81:2993–2999
Li TZ, Geng C-A, Yin XJ et al (2017) Catalytic asymmetric total synthesis of (+)- and (−)-paeoveitol via a hetero-Diels−Alder reaction. Org Lett 19:429–431
Lin L, Fan Q, Qin B et al (2006) Highly enantio- and diastereoselective Brassard type hetero-Diels-Alder approach to 5-methyl-containing α,β-unsaturated δ-lactones. J Org Chem 71:4141–4146
Lin L, Chen Z, Yang X et al (2008) Efficient enantioselective hetero-Diels-Alder reaction of Brassard's diene with aliphatic aldehydes: a one-step synthesis of (R)-(+)-kavain and (S)-(+)-dihydrokavain. Org Lett 10:1311–1314
Lin L, Kuang Y, Liu X (2011) Indium(III)-catalyzed asymmetric hetero-Diels Alder reaction of Brassard-type diene with aliphatic aldehydes. Org Lett 13:3868–3871
List B (2010) Asymmetric organocatalysis, Topics in current chemistry. Springer, Heidelberg
Lu Y, Woo SK, Krische MJ (2011) Total synthesis of bryostatin 7 via C-C bond-forming hydrogenation. J Am Chem Soc 133:13876–13879
Mahajan PS, Humne VT, Mhaske SB (2017) Achmatowicz reaction: a versatile tool in bioactive natural products synthesis. Curr Org Chem 21:503–545
Maier ME (2015) Design and synthesis of analogues of natural products. Org Biomol Chem 13:5302–5343
Majer J, Kwiatkowski P, Jurczak J (2008) Highly enantioselective synthesis of 2-furanylhydroxyacetates from furans via the Friedel-Crafts reaction. Org Lett 10:2955–2958
Manaviazar S, Hale KJ (2011) Total synthesis of bryostatin 1: a short route. Angew Chem Int Ed 50:8786–8789
Marion O, Gao X, Marcus S, Hall DG (2009) Synthesis and preliminary antibacterial evaluation of simplified thiomarinol analogs. Bioorg Med Chem 17:1006–1017
Marmsäter FP, West FG (2002) New efficient iterative approaches to polycyclic ethers. Chem Eur J 8:4346–4353
Martin SF (2017) Natural products and their mimics as targets of opportunity for discovery. J Org Chem 82:10757–10794
Martin SF, Zinke PW (1991) The furan approach to oxygenated natural products. Total synthesis of (+).-KDO. J Org Chem 56:6600–6601
Martín-Acosta P, Feresin G, Tapia A et al (2016) Microwave-assisted organocatalytic intramolecular Knoevenagel/hetero Diels−Alder reaction with O-(arylpropynyloxy).-salicylaldehydes: synthesis of polycyclic embelin derivatives. J Org Chem 81:9738–9756
Mascitti V, Thuma BA, Smith AC et al (2013) On the importance of synthetic organic chemistry in drug discovery: reflections on the discovery of antidiabetic agent ertugliflozin. Med Chem Comm 4:101–111
McNicholas PA, Batley M, Redmond JW (1986) Synthesis of methyl pyranosides and furanosides of 3-deoxy-d-*manno*-oct-2-ulosonic acid (KDO). by acid-catalysed solvolysis of the acetylated derivatives. Carbohydr Res 146:219–231
Merino P (2015) Oxidative cleavage of furans. Org React 87:1–256
Merino P, Merino P, Tejero T et al (2007) Furan oxidations in organic synthesis: recent advances and applications. Curr Org Chem 11:1076–1091
Michael A (1879) On the synthesis of helicon and phenolglucoside. Am Chem J 1:305–312
Mishra S, Singh AS, Mishra N et al (2016) Carbohydrate-based antidiabetic agents from nature. In: Brachmachari G (ed) Discovery and development of antidiabetic agents from natural products: natural products drug discovery. Elsevier, Amsterdam, pp 147–184
Mizutani H, Watanabe M, Honda T (2002) Enantioselective total synthesis of δ -lactonic marine natural products, (1)-tanikolide and (2)-malyngolide, via RCM reaction. Tetrahedron 58:8929–8936
Momiyama N, Tabuse H, Terada M (2009) Chiral phosphoric acid-governed anti-diastereoselective and enantioselective hetero-Diels-Alder reaction of Glyoxylate. J Am Chem Soc 131:12882–12883

Montagnon T, Tofi M, Vassilikogiannakis G (2008) Using singlet oxygen to synthesize polyoxygenated natural products from furans. Acc Chem Res 41:1001–1011
Mori K (2011) Bioactive natural products and chirality. Chirality 23:449–462
Munson RS, Rasmussen NS, Osborn MJ (1978) Biosynthesis of lipid A. Enzymatic incorporation of 3-deoxy-d-*manno*octulosonate into a precursor of lipid A in Salmonella typhimurium. J Biol Chem 252:1503–1511
Mydock LK, Demchenko AV (2010) Mechanism of chemical glycosylation: from early studies to recent discoveries. Org Biomol Chem 8:497–510
Naik SR, Niture NT, Ansari AA et al (2013) Anti-diabetic activity of embelin: involvement of cellular inflammatory mediators, oxidative stress and other biomarkers. Phytomedicine 20:797–804
Neumiller JJ, White JR Jr, Campbell RK (2010) Sodium-glucose co-transport inhibitors progress and therapeutic potential in type 2 diabetes mellitus. Drugs 70:377–385
Nichols AL, Zhang P, Martin SF (2012) Concise approach to 1,4-dioxygenated xanthones via novel application of the Moore rearrangement. Tetrahedron 68:7591–7597
Nicolaou KC, Mitchell HJ (2001) Adventures in carbohydrates chemistry: new synthetic technologies, chemical synthesis, molecular design, and chemical biology. Angew Chem Int Ed 40:1576–1624
Nicolaou KC, Hale CRH, Nilewski C et al (2012) Constructing molecular complexity and diversity: total synthesis of natural products of biological and medicinal importance. Chem Soc Rev 41:5185–5238
Nishikawa T, Adachi M, Isobe M (2008) C-Glycosylation. In: Fraser-Reid BO, Tatsuta K, Thiem J (eds) Glycoscience – chemistry and chemical biology. Springer, New York, pp 755–811
Nishiwaki N (2014) Methods and applications of cycloaddition reactions in organic syntheses. Wiley, Hoboken
Nising CF, Bräse S (2012) Recent developments in the field of oxa-Michael reactions. Chem Soc Rev 41:988–999
Noutsias D, Kouridaki A, Vassilikogiannakis G (2011) Scope and limitations of the photooxidations of 2-(α-Hydroxyalkyl). furans: synthesis of 2-hydroxy-exo-brevicomin. Org Lett 13:1166–1169
Noyori R (2002) Asymmetric catalysis: science and opportunities. Angew Chem Int Ed 41:2008–2022
Núñez MG, García P, Moro RF et al (2010) Asymmetric organocatalytic synthesis of six-membered oxygenated heterocycles. Tetrahedron 66:2089–2109
Ohmori K, Ogawa Y, Obitsu T et al (2000) Total synthesis of bryostatin 3. Angew Chem Int Ed 39:2290–2294
Ohtake Y, Sato T, Kobayashi T et al (2012) Discovery of tofogliflozin, a novel C-arylglucoside with an O- spiroketal ring system, as a highly selective sodium glucose cotransporter 2 (SGLT2) inhibitor for the treatment of type 2 diabetes. J Med Chem 55:7828–7840
Ohtake Y, Emura T, Nishimoto M et al (2016) Development of a scalable synthesis of tofogliflozin. J Org Chem 81:2148–2153
Pathania S, Kashyap N, Singh V et al (2016) Development and recent advancement of SGLT2 inhibitors for the treatment regime of T2DM. J Biomed 1(4):e10052
Pellissieri H (2011) Recent developments in the [5+ 2] cycloaddition. Adv Synth Catal 353:189–218
Perry MA, Rychnovsky SD, Sizemore N (2014) Synthesis of saturated tetrahydropyrans. Springer, Berlin
Pettit GR, Herald CL, Doubek DL et al (1982) Isolation and structure of bryostatin 1. J Am Chem Soc 104:6846–6848
Pouwer RH, Richard JA, Tseng CC et al (2012) Chemical synthesis of the englerins. Chem Asian J 7:22–35
Rainier JD, Cox JM (2000) Aluminum- and boron-mediated C-glycoside synthesis from 1,2-anhydroglycosides. Org Lett 2:2707–2709
Rodriguez JC, Fernandez-Puentes JL, Baz JP et al (2003) IB-00208, a new cytotoxic polycyclic xanthone produdced by a marine-derived *actinomadura*. J Antibiot 56:318–321
Rousseau C, Martin OR (2003) Stereodirected synthesis of aryl α-C-glycosides from 2-O-arylsilylglucopyranosides. Org Lett 5:3763–3766

Rowe A, Narlawar R, Groundwater P et al (2011) Kavalactone pharmacophores for major cellular drug targets. Mini Rev Med Chem 11:79–83

Ruan J, Zheng C, Liu Y et al (2017) Chemical and biological research on herbal medicines rich in xanthones. Molecules 22:1698

Rueping M, Sugiono E, Merino E (2008) Asymmetric organocatalysis: an efficient enantioselective access to benzopyranes and chromenes. Chem Eur J 14:6329–6332

Rybak T, Hall DG (2015) Stereoselective and regiodivergent allylic Suzuki-Miyaura cross-coupling of 2-Ethoxydihydropyranyl boronates: synthesis and confirmation of absolute stereochemistry of diospongin B. Org Lett 17:4156–4159

Savard J, Brassard P (1979) Regiospecific syntheses of quinones using vinyl ketene acetals derived from unsaturated esters. Tetrahedron Lett 20:4911–4914

Schmidt B (2003) An olefin metathesis/double bond isomerization sequence catalyzed by an in situ generated ruthenium hydride species. Eur J Org Chem 2003:816–819

Schmidt B, Wildemann H (2000) Diastereoselective ring-closing metathesis in the synthesis of dihydropyrans. J Org Chem 65:5817–5822

Schwartz GK, Shah MA (2005) Targeting the cell cycle: a new approach to cancer therapy. J Clin Oncol 23:9408–9421

Schweizer EE, Meeder-Nycz D (1977) 2*H*- and 4*H*-1-benzopyrans. In: Ellis GP (ed) The chemistry of heterocyclic compounds; 31; chromenes, chromanones, and chromones. Wiley, New York, pp 11–139

Seletsky BM, Wang Y, Hawkins LD et al (2004) Structurally simplified macrolactone analogues of halichondrin B. Bioorg Med Chem Lett 14:5547–5550

Serrano-Ruiz JC, Luque R, Sepulveda-Escribano A (2011) Transformations of biomass-derived platform molecules: from high added-value chemicals to fuels via aqueous-phase processing. Chem Soc Rev 40:5266–5281

Shablak A (2013) Eribulin for advanced breast cancer: a drug evaluation. J Breast Cancer 16:12–15

Sharpless KB, Amberg W, Bennani YL et al (1992) The osmium-catalyzed asymmetric dihydroxylation: a new ligand class and a process improvement. J Org Chem 57:2768–2771

Shirai R, Ogura H (1989) Improved syntheses of two 3-deoxyald-2-ulosonic acids (KDN, KDO). by condensation of oxalacetic acid with aldoses followed by Ni^{2+} catalyzed decarboxylation. Tetrahedron Lett 30:2263–2264

Smith DB, Wang Z, Schreiber SL (1990) The asymmetric epoxidation of divinyl carbinols: theory and applications. Tetrahedron 46:4793–4808

Smith JA, Wilson L, Azarenko O et al (2010) Eribulin binds at microtubule ends to a single site on tubulin to suppress dynamic instability. Biochemistry 49:1331–1337

Sturino CF, Wong JCK (1998) The ring-closing metathesis of vinyl ethers with Grubbs' catalyst for the synthesis of dihydropyrans. Tetrahedron Lett 39:9623–9626

Sugai T, Shen G-J, Ichikawa Y, Wong C-H (1993) Synthesis of 3-deoxy-D-manno-2-octulosonic acid (KDO) and its analogs based on KDO aldolase-catalyzed reactions. J Am Chem Soc 115:413–421

Tanaka K, Ohta Y, Fuji K (1993) Differentiation of enantiotopic carbonyl groups by the Horner-Wadsworth-Emmons reaction. Tetrahedron Lett 34:4071–4074

Thiel D, Doknic D, Deska J (2014) Enzymatic aerobic ring rearrangement of optically active furylcarbinols. Nat Commun 5:5278

Tietze LF (2014) Domino reactions: concepts for efficient organic synthesis. Wiley-VCH, Weinheim

Tietze LF, Düfert A (2012) Modern tools for the synthesis of complex bioactive molecules. Wiley, Hoboken

Tietze LF, Brasche G, Gericke KM (2006) Domino reactions in organic synthesis. Wiley-VCH, Weinheim

Tietze LF, Böhnke N, Dietz S (2009) Synthesis of the deoxyaminosugar (+).-D-forosamine via a novel domino-Knoevenagel-hetero-Diels-Alder reaction. Org Lett 11:2948–2950

Touré B, Hall DG (2009) Natural product synthesis using multicomponent reaction strategies. Chem Rev 109:4439–4486

Towle MJ, Salvato KA, Budrow J et al (2001) In vitro and in vivo anticancer activities of synthetic macrocyclic ketone analogues of halichondrin B. Cancer Res 61:1013–1021

Trost BM, Dong G (2010) Total synthesis of bryostatin 16 using a Pd-catalyzed diyne coupling as macrocyclization method and synthesis of C20-*epi*-bryostatin 7 as a potent anticancer agent. J Am Chem Soc 132:16403–16416

Twelves C, Cortes J, Vahdat LT et al (2010) Phase III trials of eribulin mesylate (E7389) in extensively pretreated patients with locally recurrent or metastatic breast cancer. Clin Breast Cancer 10:160–163

Uemura D, Takahashi K, Yamamoto T et al (1985) Norhalichondrin A: an antitumor polyether macrolide from a marine sponge. J Am Chem Soc 107:4796–4798

Unger FM (1981) The chemistry and biological significance of 3-deoxy-D-*manno*-2-octulosonic acid (KDO). Adv Carbohydr Chem Biochem 38:323–388

Van Putten RJ, Van der Waal JC, de Jong E et al (2013) Hydroxymethylfurfural, a versatile platform chemical made from renewable resources. Chem Rev 113:1499–1597

Vogel P, Robina I (2008) De novo synthesis of monosaccharides. In: Fraser-Reid B, Tatsuta K, Thiem J (eds) Glycoscience. Springer, Berlin, pp 861–956

Wang Y, Serradell N, Bolos J et al (2007) Eribulin mesilate. Drugs Future 32:681–698

Wang H-Y, Yang K, Yin D et al (2015a) Chiral catalyst-directed dynamic kinetic diastereoselective acylation of lactols for de novo synthesis of carbohydrate. Org Lett 17:5272–5275

Wang HY, Yang K, Bennett SR et al (2015b) Iridium-catalyzed dynamic kinetic isomerization: expedient synthesis of carbohydrates from Achmatowicz rearrangement products. Angew Chem 127:8880–8883

Wang H-Y, Simmons CJ, Zhang Y et al (2017) Chiral catalyst-directed dynamic kinetic diastereoselective acylation of anomeric hydroxyl groups and a controlled reduction of the glycosyl ester products. Org Lett 19:508–511

Washburn WN (2012a) SGLT2 Inhibitors in development. In: Jones RM (ed) New therapeutic strategies for type 2 diabetes: small molecule approaches. Royal Society of Chemistry, UK, pp 29–87

Washburn WN (2012b) Sodium glucose Co-transporter 2 (Sglt2). inhibitors: novel antidiabetic agents. Expert Opin Ther Pat 22:483–494

Washburn WN (2014) Case history: Forxiga (dapagliflozin) a potent selective SGLT2 inhibitor for treatment of diabetes. Annu Rep Med Chem 49:363–382

Wender PA, Schrier AJ (2011) Total synthesis of bryostatin 9. J Am Chem Soc 133:9228–9231

Wender PA, Baryza JL, Hilinski MK et al (2007) Beyond natural products: synthetic analogues of bryostatin 1. In: Huang Z (ed) Drug discovery research: new frontiers in the post-genomic era. Wiley, Hoboken, pp 127–162

Wender PA, Loy BA, Schrier AJ (2011) Translating nature's library: the bryostatins and function-oriented synthesis. Isr J Chem 51:453–472

Wender PA, Ryan V, Quiroz RV et al (2015) Function through synthesis-informed design. Acc Chem Res 48:752–760

Wender PA, Hardman CT, Ho S et al (2017) Scalable synthesis of bryostatin 1 and analogs, adjuvant leads against latent HIV. Science 358:218–223

Wilson RM, Danishewsky SJ (2006) Small molecule natural products in the discovery of therapeutic agents: the synthesis connection. J Org Chem 71:8329–8351

Witczak ZJ, Bielski R (2016) Domino and intramolecular rearrangement reactions as advanced synthetic methods in glycoscience. Wiley, Hoboken

Witczak ZJ, Tatsuta K (2003) Carbohydrate synthons in natural products chemistry; synthesis, functionalization, and applications. ACS, Washington, DC

Wu Z, Zhao S, Fash DM et al (2017) Englerins: a comprehensive review. J Nat Prod 80:771–781

Xiang S, Cai S, Zeng J et al (2011) Regio- and stereoselective synthesis of 2-deoxy-C-aryl glycosides via palladium catalyzed decarboxylative reactions. Org Lett 13:4608–4611

Xu DQ, Wang YF, Luo SP et al (2008) A novel enantioselective catalytic tandem oxa-Michael-Henry reaction: One-pot organocatalytic asymmetric synthesis of 3-Nitro-2H-chromenes. Adv Synth Catal 350:2610–2616

Xu C, Shen X, Hoveyda AH (2017) In situ methylene capping: a general strategy for efficient stereoretentive catalytic olefin metathesis. The concept, methodological implications, and applications to synthesis of biologically active compounds. J Am Chem Soc 139:10919–10928

Yang Y, Yu B (2017) Recent advances in the chemical synthesis of C-glycosides. Chem Rev 117:12281–12356

Yang XD, Pan ZX, Li DJ et al (2016) A scalable synthesis of tofogliflozin hydrate. Org Process Res Dev 20:1821–1827

Yao S, Johannsen M, Audrain H et al (1998) Catalytic asymmetric hetero-Diels-Alder reactions of ketones: chemzymatic reactions. J Am Chem Soc 120:8599–8605

Ylijoki KEO, Stryker JM (2012) [5+ 2] Cycloaddition reactions in organic and natural product synthesis. Chem Rev 113:2244–2266

Yu X, O'Doherty GA (2008) De novo synthesis in carbohydrate chemistry: from furans to monosaccharides and oligosaccharides. Chapter 1. In: Chen X (ed) Chemical glycobiology. ACS, Washington DC, pp 3–28

Yu B, Sun J, Yang X (2012) Assembly of naturally occurring glycosides, evolved tactics and glycosylation methods. Acc Chem Res 45:1227–1236

Yu MJ, Zheng W, Seletsky BM (2013) From micrograms to grams: scale-up synthesis of eribulin mesylate. Nat Prod Rep 30:1158–1164

Zamojski A, Banaszek A, Grynkiewicz G (1982) The synthesis of sugars from non-carbohydrate substrates. Adv Carbohydr Chem Biochem 40:1–129

Zhao B, Loh TP (2013) Asymmetric hetero-Diels_Alder reaction of Danishewsky's dienes with α-carbonyl esters catalyzed by an Indium(III).-PyBox complex. Org Lett 15:2914–2917

Zheng W, Seletsky BM, Palme MH et al (2004) Macrocyclic ketone analogues of halichondrin B. Bioorg Med Chem Lett 14:5551–5554

Zhou W, Ni C, Chen J et al (2017) Enantioselective synthesis of 4H-Pyran via amine-catalyzed formal (3 + 3) annulation of δ-acetoxy allenoate. Org Lett 19:1890–1893

Zhu R, Schmidt RR (2009) New principles for glycoside-bond formation. Angew Chem Int Ed 48:1900–1934

Zhu J, Wang Q, Wang M (2015) Multicomponent reactions in organic synthesis. Wiley-VCH & Co. KGaA, Weinheim

Zhu F, Rourke MJ, Yang T et al (2016) Highly stereospecific cross-coupling reactions of anomeric stannates for the synthesis of C-aryl glycosides. J Am Chem Soc 138:12049–12052

Zhu F, Yang T, Walczak MA (2017) Glycosyl Stille cross-coupling with anomeric nucleophiles – a general solution to a long standing problem of stereocontrolled synthesis of C-glycosides. Synlett 28:1510–1516

Natural Products as Sources of Anticancer Agents: Current Approaches and Perspectives

Gordon M. Cragg and David J. Newman

1 Introduction

More than 60% of current anticancer drugs have their origin in one way or another from natural sources (Newman and Cragg 2016b), though the absolute figure varies by a percentage or two depending upon the years considered, as shown in earlier reviews by the same authors (Newman and Cragg 2012; Newman 2008; Newman et al. 2000). Nature continues to be the most prolific source of biologically active and diverse chemotypes. Though relatively few of the compounds initially isolated from natural sources advance to become clinically effective drugs in their own right, these structurally unique and complex molecules frequently serve as model scaffolds for the preparation of more efficacious analogues and prodrugs using chemical methodologies for semi- or total syntheses of closely related structures or, with the advent of rapid genomic analyses, the manipulation of biosynthetic pathways. In addition to these, the use of formulations based on nanoparticle or liposomal techniques has led to more efficient drug delivery.

The essential role played by natural products in the discovery and development of effective anticancer agents and, in particular, the importance of multidisciplinary collaboration in the generation and optimization of novel molecular leads from natural product sources has been the subject of several recent reviews (Giddings and Newman 2015, 2017; Bebbington 2017; Newman and Cragg 2014, 2015, 2016a, b, 2017; Newman 2016, 2017; Cragg and Pezzuto 2016; Newman et al. 2015; Basmadjian et al. 2014).

Within the last few years, old compounds from a variety of sources, which are too toxic to use as drug candidates in their own right (usually found from failed

G. M. Cragg
Gaithersburg, MD, USA

D. J. Newman (✉)
Wayne, PA, USA

V. Cechinel Filho (ed.), *Natural Products as Source of Molecules with Therapeutic Potential*, https://doi.org/10.1007/978-3-030-00545-0_8

clinical trials), are now gaining a "new lease on life" as warheads linked to monoclonal antibodies and, in some recent cases, to polymeric carriers. These antibodies and/or polymeric carriers are specifically targeted to epitopes expressed by tumors of interest and have led to the development of approved targeted therapies, with significant numbers in earlier phases of clinical trials. These will be discussed in a separate section as the methodologies are similar, but "warheads" are sourced from all natural product sources.

In addition to updating the data on compounds that are natural products or derived therefrom, the following sections will also discuss the development of new anticancer agents through the synthesis of analogues and prodrugs of established drugs or recently discovered active lead compounds and the development of improved formulations and methods of delivery of established drugs and their active analogues.

Discussions will be limited to new agents currently in some stage of clinical development, including some which have been approved for commercial use in the last few years. Links to completed and ongoing clinical trials have been accessed via www.clinicaltrials.gov or the European Clinical Trials Register via www.clinicaltrialsregister.eu/ctr-search/search?, and the information is accurate as of November 2017.

2 Plant-Sourced Antitumor Agents

2.1 Introduction

In an earlier chapter in this series (Cragg and Newman 2014), we extensively discussed many plant-sourced products (e.g., taxanes, vinca alkaloids, camptothecins, and ingenol derivatives); however, it currently appears that new agents from plant sources are not being reported, but there are still significant advances in modification of the underlying structures, different formulations to overcome the problems associated with toxicity in the earlier formulations, and recently, modification of the base structure for use as "warheads" in monoclonal and/or other methods of targeted delivery. Their use as warheads will be discussed in a separate section so that the methodologies used can be discussed in fair detail.

2.2 Vinca Alkaloid Derivatives

Vintafolide (EC-145 or MK-8109; Fig. 1) was placed into a number of Phase I through Phase III trials, but the Phase III trial under Merck in Europe (the PROCEED trial; NCT01170650), where patients with platinum-resistant ovarian cancer were treated in conjunction with pegylated liposomal doxorubicin, was terminated due to

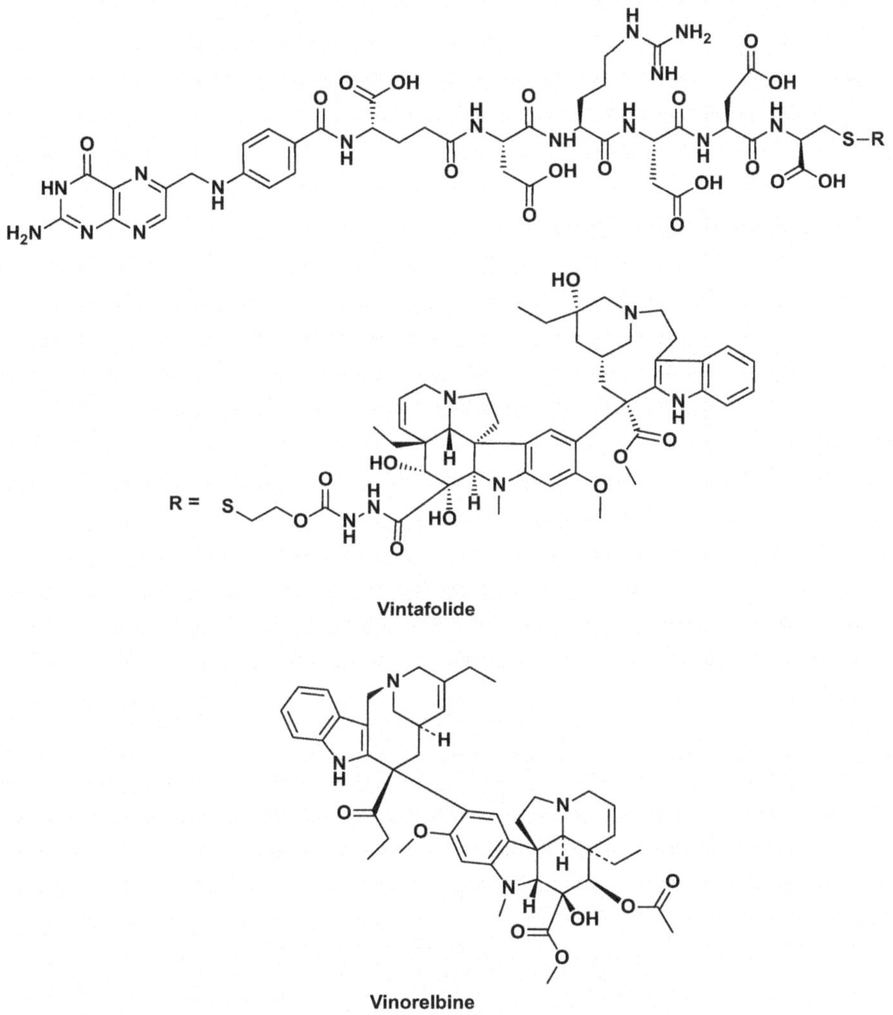

Fig. 1 Vinca alkaloid derivatives

a lack of efficacy when compared to the doxorubicin arm. Merck then returned the agent to Endocyte. However, Endocyte have continued with Phase II trials, with an extension study for patients who showed efficacy against solid tumors (mainly lung) under the trial number NCT01002924. No details have yet been published on this trial, but data from the Phase III trial referred to above indicated that data from a Phase II trial (PRECEDENT) that led to the Phase III trial demonstrated that the folate receptor status of a patient did not influence any adverse effects seen in treated patients (Herzog et al. 2016). The future of this and similar folate-linked drug entities was recently discussed by Graybill and Coleman, and this paper should be consulted for further details (Graybill and Coleman 2016).

Liposomal vinorelbine (TLC178; Fig. 1) is currently in Phase I/Phase IIa trials under the aegis of Taiwan Liposome Company (NCT02925000), and the compounded mixture was given an orphan drug classification in October 2016 by the US FDA for the treatment of cutaneous T-cell lymphoma.

2.3 Camptothecin Derivatives

Irinotecan-loaded beads (Fig. 2). What might almost be considered a "stealth" camptothecin derivative is a composition known as DEBIRI, DC-loaded beads or CM-BC2, where irinotecan hydrochloride is delivered by "drug-eluting beads." This was missed by Newman and Cragg in their reviews, primarily as it was never approved in the USA or Canada, and the early EMEA listings are rather difficult to search, as they go by year, though it was approved in November 2007. However, the construct has been in, or is currently in, a number of clinical trials listed in the NIH trials listing (Phase II, NCT01839877; Phase I, NCT02350400; Phase I/II, NCT02481960; Phase II/III, NCT03175016 {in the PRC}), plus one early trial result (Phase I/II, NCT00844233, PARAGON-II) has recently been published by investigators from Biocompatibles in the UK where the beads were used for pre-treatment prior to resection of colorectal cancers, demonstrating efficacy (Jones et al. 2016).

Karenitecin (Fig. 2) is a lipophilic silicon-substituted camptothecin derivative that reached Phase III studies against platinum-resistant ovarian cancer patients in a trial in 2007 (NCT0047782). Although no data has been reported on this trial, even after 10 years from its commencement, a paper in 2014 expressed significant concerns over exactly what is a "platinum-resistant ovarian cancer" (Davis et al. 2014), and the questions raised in that paper might well be the reason(s) why nothing further has been reported on this compound.

Namitecan (ST1968; Fig. 2) is a hydrophilic camptothecin derivative that has had one Phase I trial listing (NCT01748019) with a report in 2015 giving PK-PD data that suggested that the optimal dosing strategy was flat dosing every 3 weeks (Joerger et al. 2015). Although this paper refers to data from two Phase I trials, we can only find the one listed in the NCT database. The EU "EudraCT" database does not show any other trial.

Etirinotecan pegol (NKTR-102; Fig. 2) is a pegylated version of irinotecan with four molecules of irinotecan linked via a pegylated framework. An MAA was submitted in 2016 to the EU from data from a Phase III clinical trial (NCT01492101), but in July 2017, a negative opinion was given by the EMEA. There have been more than 13 trials reported in the EudraCT and NCT databases, though some overlap has occurred in these reports. The current approval status is not known.

Irinotecan/HM30181A (Oratecan; Fig. 2) is a very interesting combination of this well-known camptothecin derivative and the P-glycoprotein inhibitor that prevents a cell from eliminating the drug. Currently this combination is in Phase I

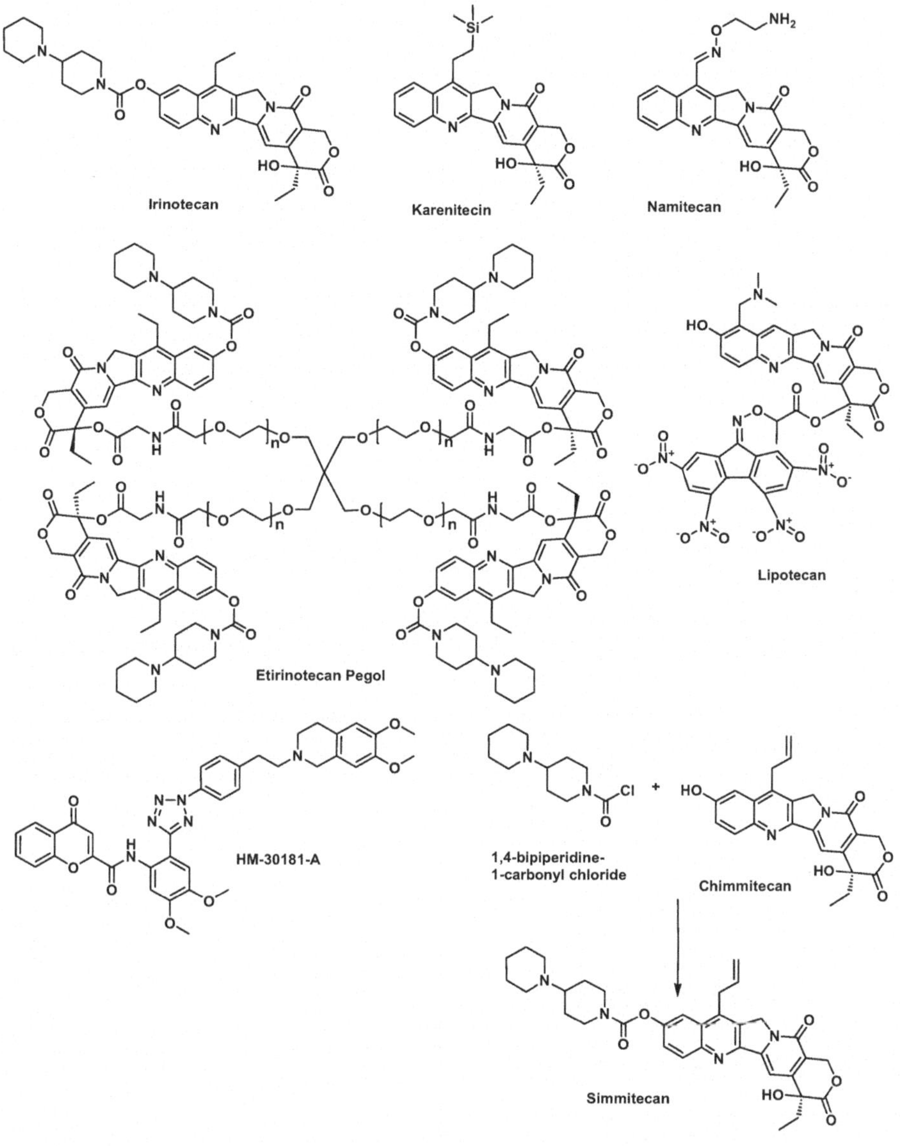

Fig. 2 Camptothecin derivatives

clinical trials, with two completed (NCT00986843; NCT01463982) and one recruiting (NCT02250157).

Lipotecan (Fig. 2) is a chemically modified camptothecin that is designed to also have radiosensitizing capability (Huang et al. 2010). It is currently in a Phase I/II trial in Taiwan under NCT03035006.

Simmitecan (Fig. 2) is a modified 9-allylcamptothecin synthesized from chimmitecan (Huang et al. 2007) in the one-step reaction as shown in Fig. 2 and is in Phase I clinical trials in China with one completed (NCT01832298) and the other recruiting (NCT02870036).

Liposomal irinotecan (Onivyde®), a nanoparticle formulation that consists of irinotecan sucrosofate encapsulated into anion-stabilized liposomes, was approved by the US FDA in October 2015 and then by the EMEA a year later.

Currently there are a number of camptothecin-derived agents using various liposomal, peptide-linked, or nanoparticle systems for delivery, but none have yet gone beyond Phase I. In addition to the materials listed above, there are also some antibody-drug conjugates (ADCs) using camptothecin derivatives as warheads. These will be discussed later in the chapter.

2.4 *Taxol Derivatives*

Currently there are a number of variations on Taxol that are in Phase III trials or have completed them.

Probably the oldest is CT-2103 (Fig. 3) which is a polyglutamic acid where some of the gamma carboxylates are linked via esterification to a hydroxyl group on the Taxol side chain and has been in at least 24 trials as shown in the NIH clinical trials in database, with a number of Phase III trials that have not had any updates for a significant time. One Phase III trial, NCT00108745, which dates from April 2005, is still ongoing but not recruiting, and of the remaining five, four have not had any

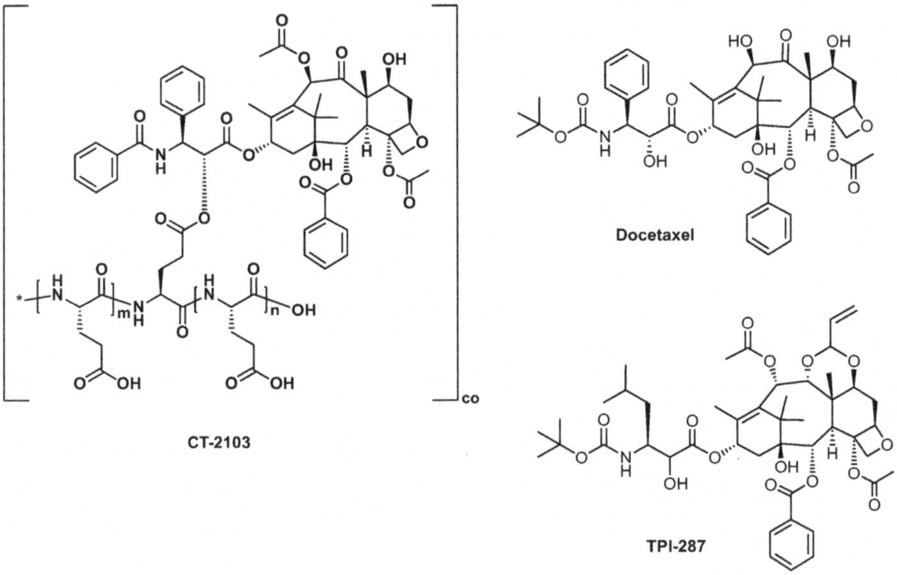

Fig. 3 Taxol derivatives

data for at least the last 2 years, and one was terminated. Thus their current status is unknown at this time.

Following along with a similar effort that was used in Oratecan, Athenex Pharma and Hanmi have combined the P-glycoprotein inhibitor HM30181A (see structure in Fig. 2) with Taxol (under the name Oraxol) and have the combination in a Phase III clinical trial (NCT02594371) that is actively recruiting. Recently, in 2015 a report from one of the Phase I/II studies was published demonstrating some efficacy as a chemotherapeutic against gastric cancer (Lee et al. 2015).

Similar to Oratecan and Oraxol, Athenex has commenced studying the combination of docetaxel (Fig. 3) and HM30181A (see structure in Fig. 2) under the name Oradoxel, with a Phase I trial under NCT02963168 that was first listed in November 2016.

EndoTAG-1 is a formulation of Taxol in cationic liposomes that can target endothelial cells and has been in six reported clinical trials with two Phase III trials now underway; one (NCT03002103) is actively recruiting and the other (NCT03126435) is approved but not yet recruiting. Recently the results from a Phase II trial (EudraCT 2006-002221-23 and NCT00448305) were reported demonstrating antitumor efficacy (Awada et al. 2014).

NK-105 is a nanoparticle formulation of Taxol, and there is one record of a completed Phase III trial (NCT01644890), but no formal results appear to have been reported other than a press release from the president of Nippon Kayaku in July of 2016 reporting that the desired endpoints were not achieved in that trial. The current status is not known.

TPI-287 (Fig. 3) an analogue of Taxol has been in a variety of clinical trials, but currently only one, a Phase I/Phase II (NCT01933815), is currently recruiting, with one (NCT01966666) looking at the effects in Alzheimer's disease which is active but not recruiting. The compound does cross the blood-brain barrier, so the results of the Alzheimer's trial may well be of interest.

In earlier versions of this chapter (Cragg and Newman 2014), a number of Taxol analogues were shown as being in clinical trials in 2014, such as taxopressin (7-DHA-paclitaxel), larotaxel, tesetaxel, and ortataxel. None of these appear to have progressed any further toward approval as new drug entities in the last 3 years, so we are not considering them any further.

Finally, with respect to this class of molecules, there are other nanoparticle/liposome variations using taxanes based on the Taxol structure, but all are at the Phase I level at this moment. These will not be listed in this chapter since from experience, very few will actually become drugs in their own right.

2.5 Homoharringtonine (Fig. 4)

Although this compound was discussed at length following its approval by the FDA in 2012 in the 2014 version of this chapter (Cragg and Newman 2014), as is customary with approved drug, it is still in clinical trials as a method of extending its pharmacological activities and potential usage. Currently there are five trials at the

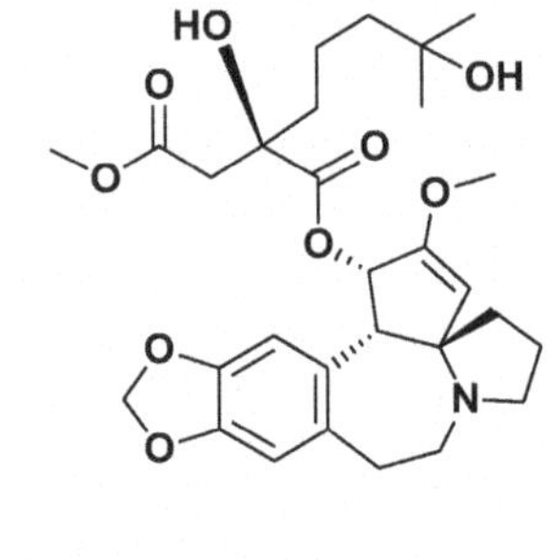

Fig. 4 Homoharringtonine

Phase II level shown in the NCT database that are recruiting or approved and not yet recruiting (NCT03170895, NCT03135054 {in Hong Kong}, NCT02078960, NCT02159872, NCT01873495, NCT02440588 {all in the USA}) and six Phase II trials listed in the EudraCT database that date from the early 2000s that still have at least one country where the trials are not shown as completed.

3 Terrestrial Microbial-Derived Agents

3.1 Staurosporine Derivatives

Although there has been a dearth of recent agents isolated from terrestrial microbes (including fungi) approved in the last few years, the first staurosporine derivative, midostaurin (the very simple synthesis from staurosporine is shown in Fig. 5), was approved by the US FDA in April of 2017 and was rapidly followed by the EMEA with approval in September 2017. Both approvals were for systemic mastocytosis, mast cell leukemia, and acute myeloid leukemia. Currently there are 20 trials listed in the NCT database that are active and/or recruiting at the Phase I to III levels, and there is little doubt that there will be many more as these progress, expanding the "reach" of this molecule and related compounds. The first full paper on this particular derivative under the code number CGP 41251 (a Ciba-Geigy number in the pre-Novartis days) was published by Andrejauskas-Buchdunger and Regenass in 1992 (Andrejauskas-Buchdunger and Reganass 1992). Thus it took close to 25 years for this initial staurosporine-derived compound to be approved as a drug.

There are two other modified staurosporine derivatives in advanced clinical trials related to cancer. In one case enzastaurin (Fig. 5), the sugar moiety has been completely removed, and one indole nitrogen has a methyl substituent with the other having a piperidine-pyridine substituent. Currently the compound is in a Phase III trial utilizing a biomarker (DGM1) in diffuse large B-cell lymphoma and compared with the standard "R-CHOP" protocol under NCT03263026. The second molecule, sotrastaurin (Fig. 5), is altered quite differently with one indole and the sugar being

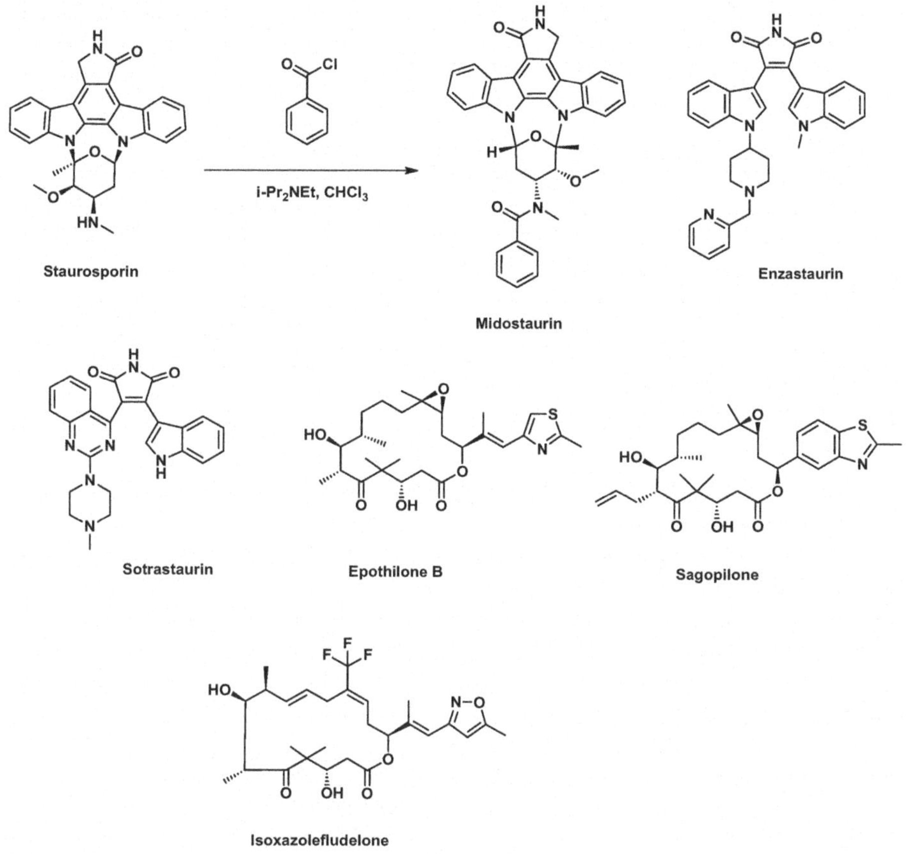

Fig. 5 Staurosporine and epothilone derivatives

replaced with quinazoline and a piperazine. This molecule has been in 20 clinical trials at the Phase I/Phase II levels and is currently only active in 1 Phase I trial (NCT02273219) though a Phase Ib/Phase II study with patients with metastatic uveal melanoma (NCT01801358) was terminated in early 2015 by the sponsor for "scientific reasons."

3.2 *Epothilones*

Epothilone B (Fig. 5) was the first epothilone to enter clinical trials, and it is a natural product. Although it has 68 listings in the NCT database up through Phase III, only 7 trials are still active but not recruiting, with three (NCT00159484; 00877500 and 1168232) at the Phase II level, but in 2010 a Phase III trial did not meet endpoints in ovarian cancer. Thus the drop back to Phase II.

Sagopilone (Fig. 5) is a synthetic derivative of epothilone that reached multiple Phase II trials against ovarian, small-cell lung, prostate, and melanoma carcinomas, but no recent studies are listed in the Bayer AG listing under {pharma.bayer.com/en/innovation-partnering/development-pipeline/} or when searched under Bayer AG and sagopilone, but a recent interesting chemistry paper shows how biocatalysis was used in the large-scale synthesis of this compound (Gottfried et al. 2015).

Isoxazolefludelone (Fig. 5) is a derivative of the basic epothilone structure that has been in a Phase I clinical trial (NCT01379287) since 2011 at Memorial Sloan Kettering in New York. Currently the trial is ongoing but no longer recruiting patients.

3.3 Anthracyclines

Doxorubicin is the classical model for this class of antitumor drugs, but it has significant problems with cardiac toxicity, in spite of which, it is still one of the first-line treatments for breast cancer, and a liposomal formulation was approved in 2001, which helped with the toxicity problem but was not a panacea. Variations on this theme including a liposomal variant with a pegylated doxorubicin were subsequently approved, and currently there are a number of different variations, including nanoparticle formulations in clinical trials with this agent. A recent review article covering cancer nanomedicine should be read in conjunction with these examples (Shi et al. 2017).

Currently, ThermoDox® from Celsion, a heat-activated doxorubicin liposomal preparation, has one Phase III trial completed (NCT00617981), and one more at the same level (the OPTIMA trial) is recruiting (NCT02112656), with the recommendation in August of 2017 to continue following an initial assessment of safety after the first 50% of patients had enrolled. Other trials at Phases I and II are also underway.

A nanoparticle formulation of doxorubicin that is stated to be a lyophilized doxorubicin preparation on polyisohexylcyanoacrylate nanoparticles (the Transdrug™ nanotechnology) is currently in a Phase III trial that is not currently recruiting, and there are press reports from Onxeo SA, the company with the drug candidate, that the current Phase III trial (NCT01655693) has problems due to the other arm having a much greater than expected survival rate.

A preparation of doxorubicin using liposomal coated with glutathione-conjugated polyethylene glycol has completed a Phase I/Phase II study under the code name 2B3-101 (NCT01286580), and a second Phase II trial (NCT01818713) was shown to have been prematurely ended under the EudraCT number, 2011-001119-30, but the NCT trial mentioned above simply states "no current information."

An interesting variation on liposomal preparations is to coat the liposome with immunoglobulin fragments. A group at the Universitatsspital Basel (USB) made such a preparation by covalently linking Fab fragments from cetuximab to pegylated liposomes containing doxorubicin. The methodology involved in producing this material under cGMP conditions was described in 2015 (Wicki et al. 2015).

Currently there is a Phase II clinical trial recruiting (NCT02833766) that is specifically directed toward patients with triple-negative breast cancer.

Aldoxorubicin (Fig. 6) which is a prodrug of doxorubicin completed a Phase III trial (NCT02029905) and was set for an NDA when the agent was put on hold in late 2014 due to a death under a compassionate use protocol. The partial on hold was lifted 2 months later. Currently, it appears from Cytex press releases that the FDA has agreed to an NDA under what is known as a 505(b)(2) submission protocol with a potential approval in 2018. The compound also has orphan drug status in both the USA and the EU.

Liposomal annamycin (Fig. 6) is a preparation with a checkered history with earlier Phase I and Phase II trials under different companies. The base molecule was designed to overcome the toxicity of doxorubicin, and as mentioned, it was developed by different companies as interest waxed and waned. However, it now has a new IND submitted by Moleculin and currently has orphan drug status from the FDA.

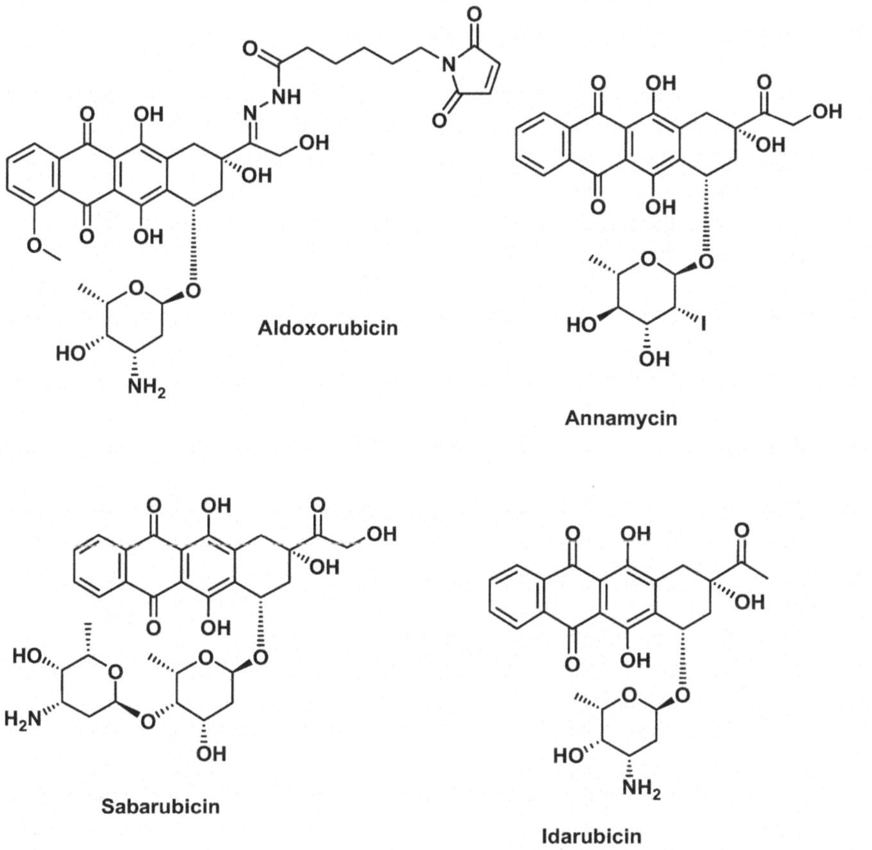

Fig. 6 Anthracyclines

Sabarubicin (Fig. 6) is a disaccharide derivative of doxorubicin that is reported to be in Phase II/Phase III studies against small-cell lung cancer. This is from the Menarini press release in the middle of 2016, but to date, neither of the main clinical trial databases show any new trials. If the company is correct, then the trial of this agent will be interesting.

Idarubicin-eluting beads (Fig. 6). This technology was mentioned earlier under both camptothecin and Taxol, and currently a Phase II trial (NCT02185768) of "IDASPHERE" is recruiting in France.

4 Nucleoside-Derived Agents

As covered extensively in the 2014 review (Cragg and Newman 2014), the influence of the discoveries by Bergmann of bioactive arabinose derivatives isolated from sponges in the Caribbean effectively pointed the way for subsequent work on nucleoside derivatives such as cytosine arabinoside (Ara-C) as antitumor agents, and at the same time, the group also isolated the ribose nucleoside spongiosine (Fig. 7). We therefore are considering nucleoside-derived antitumor agents to be "descended" from these initial discoveries. What is also of significant interest is that there is now definitive evidence that a substantial number of marine-derived metabolites isolated from marine invertebrates are in fact produced by microbes. Some of these "producers" are cultivatable, but the majority cannot yet be cultured but can be "interrogated" by genetic techniques as shown by the seminal work of the Piel group in Switzerland (Wakimoto et al. 2014; Wilson et al. 2014).

Further evidence of microbe(s) being involved in the original reports by Bergmann came from a report by the Gerwick laboratory in 2015 where they demonstrated the production of spongiosine by a *Vibrio harveyi* isolate from the sponge (Bertin et al. 2015). The data as yet does not show production of the arabinose derivatives by any isolated microbe; therefore one cannot rule out the possibility that the sponge plays a role in the production of these arabinose derivatives, but work is ongoing.

Uridine triacetate (Fig. 7) is an orally active prodrug of uridine and was approved by the FDA in 2015. The approval was not for antitumor activity but for emergency treatment of chemotherapeutic overdoses. However, it was in a Phase III trial in conjunction with 5-fluorouracil for treatment of pancreatic cancer under NCT00024427, but no details have been published as yet.

Sapacitabine (Fig. 7) is an orally bioavailable nucleoside analogue prodrug that shows a unique mechanism of action, causing single-strand breaks (SSBs) after incorporation into DNA, which are converted into double-strand breaks (DSBs) when cells enter a second S-phase. The active metabolite of sapacitabine is CNDAC (2′-C-cyano-2′-deoxy-1-β-D-arabino-pentofuranosyl-cytosine), and DSBs caused by CNDAC are largely repaired through homologous recombination (HR). Thus, cells deficient in HR components should be greatly sensitized to CNDAC, implying that sapacitabine could specifically target malignancies defective in HR (Liu et al.

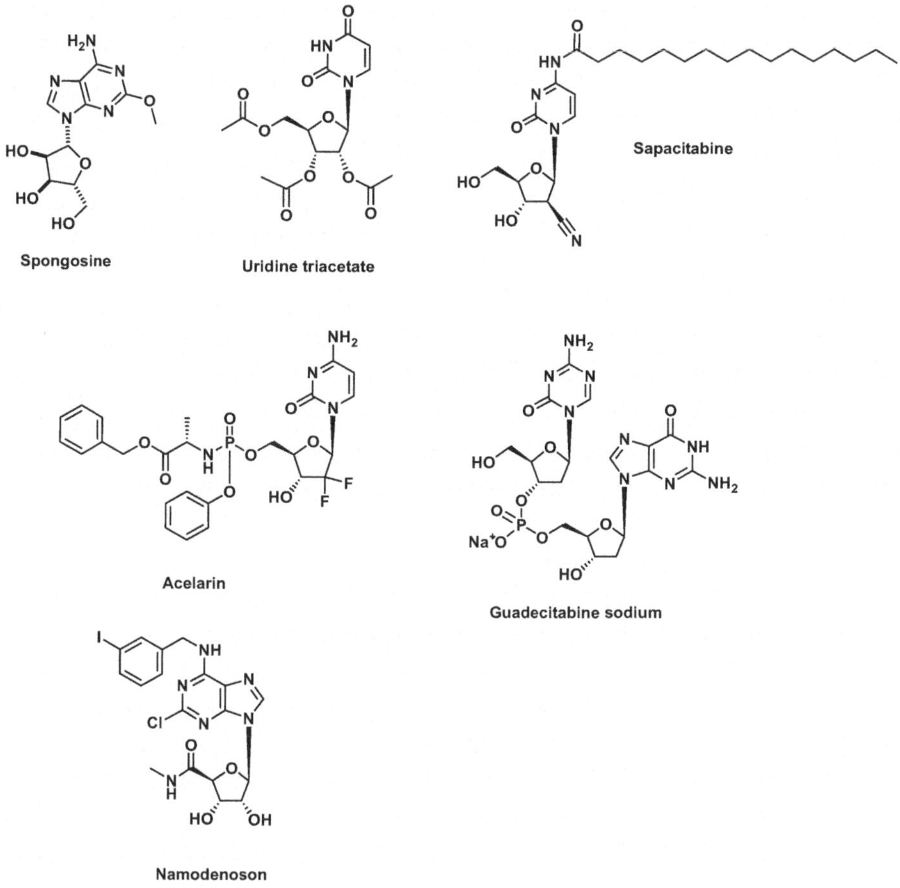

Fig. 7 Substituted nucleosides

2012). The results from a pivotal Phase III study (NCT01303796) that involved 110 sites and close to 500 patients did not meet the primary endpoint of a statistically significant increase in overall survival, and is now being analyzed on the basis of subgroup responses. The company, Cyclacel, is now concentrating on DNA damage responses and regulation of transcription by this compound.

Acelarin (Fig. 7) is a substituted difluorocytosine derivative that has been in Phase I clinical trials, and the Phase II trial (NCT03146663) shows "recruiting status" for patients with platinum-resistant ovarian cancer. However, the company developing the compound is based in the UK, and a check of the EudraCT database shows that a Phase III trial commenced in April of 2015 under the trial number 2014-004653-14 known as the ACELARATE trial, and compares acelarin with gemcitabine in patients with metastatic pancreatic carcinoma. To date, no results have been shown.

Guadecitabine (Fig. 7) is a specific targeted inhibitor against DNA methyl transferase that is in three Phase III clinical trials at the moment (NCT02920008; 02907369; 02348489), and three Phase I trials involving immune-oncologic treatments where a chemotherapeutic agent is administered with a specific antibody (NCT03206047; 02892318; 03085849). The results from all of these trials will definitely be of interest since they involve the use of an epigenetic modifier under different conditions, and cover carcinomas from small-cell lung to ovarian to acute myeloid leukemia. A discussion of these types of treatment was published in the journal *Leukemia* and should be consulted by interested readers (Montalban-Bravo and Garcia-Manero 2015).

Namodenoson (Fig. 7) is a highly substituted adenine derivative and is a high-affinity and selective adenosine A3 receptor agonist that is in Phase II clinical trials against hepatocellular carcinoma. One (NCT00790218) has been completed and one (NCT02128958) that is actively recruiting, together with one listed in the EU database (EudraCT 2014-000489-23) that is also ongoing at the Phase II level in Bulgaria. In addition to this activity, it also protects against cardiotoxicity of doxorubicin (Galal et al. 2016).

5 Marine Derived Non-nucleoside Agents

Although many compounds from marine sources (from invertebrates through to free-living microbes) have been reported to have "anticancer" activities in the scientific literature, it should be pointed out that in most cases, these are simply of the form "compound X is cytotoxic against 'cell-line Y' at 'Z micromolar levels'." However, there are agents direct from nature that are in clinical trials at advanced levels, as shown below.

Aplidine (Fig. 8) can be considered an oxidized version of didemnin B (Fig. 8) which was the first "direct from the sea" compound to go into humans as a prospective antitumor compound, in that the proline on the side chain has an N-pyruvyl substituent in place of the lactyl group in didemnin B. The compound is now before the EMEA for registration in the EU as a treatment for multiple myeloma in conjunction with dexamethasone, and as is customary, it is also in a variety of earlier trials in conjunction with different approved agents against other hematological cancers. (Aplidine was rejected by the EMA in March 2018 and is now back at the Phase II level).

Lurbinectedin (Fig. 8) is a derivative of PharmaMar's first approved agent, ET743, which was described in our earlier contribution to this series, and it is currently in Phase III trials against small-cell lung cancer and ovarian cancer, plus other cancers at Phases I and II. It is an inhibitor of the transcription process in three ways. (1) It binds to CG-rich sequences which are mainly located around promotors of protein-coding genes. (2) It irreversibly stalls elongating RNA polymerase II (Pol II) on the DNA template, which triggers degradation of the phosphorylated Pol II. (3) It generates DNA breaks and subsequent apoptosis (Santamaría et al. 2016).

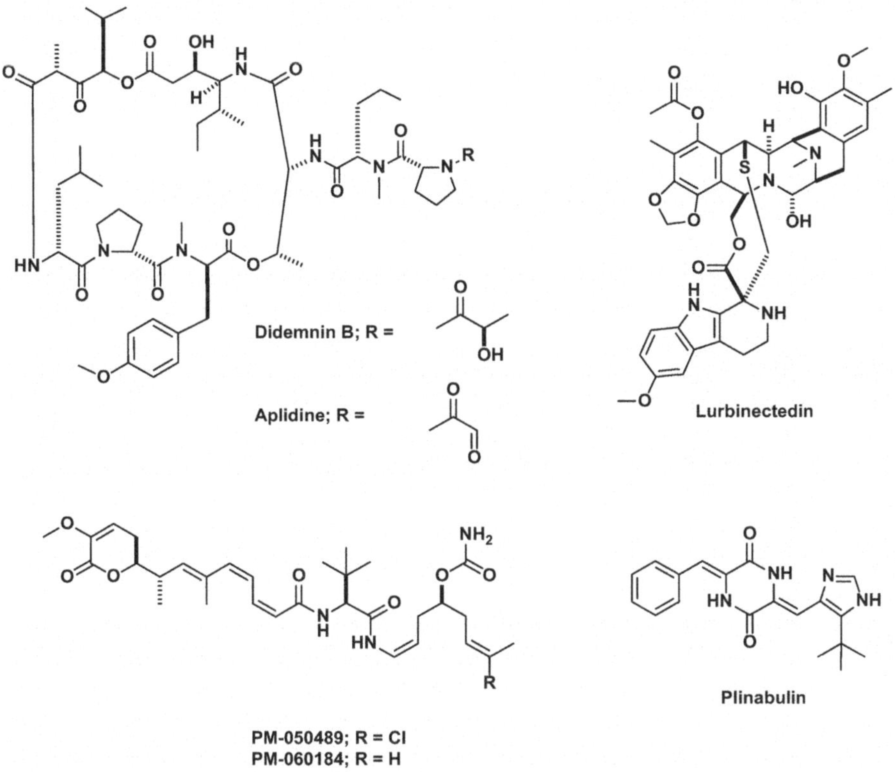

Fig. 8 Marine-sourced antitumor agents

Plocabulin (PM184; Fig. 8) and its chloro-derivative (Fig. 8; PM-050489) were isolated by PharmaMar scientists from the sponge *Lithoplocamia lithistoides*, but due to the low levels in the extract, both compounds were synthesized, and their mechanisms of action were shown to be tubulin inhibition but at a novel site as shown by resistance studies in *Aspergillus nidulans* (Martín et al. 2013; Pera et al. 2013). Currently this agent is in Phase II trials according to the PharmaMar web site, but the only trial shown in the NCT database is NCT02533674 (which refers to a Phase I trial), and there are no records shown in the EU database (http://www.eudrapharm.eu/eudrapharm/clinicaltrials.do) accessed on November 29, 2017. The use of the chloro-derivative as a warhead will be covered later in this chapter.

Plinabulin (Fig. 8) is a compound originally from Nereus that was a variation on the fungal product halimide, and though originally considered as an anticancer agent (Phase II against lung cancer under Nereus), it is now in Phase II/Phase III clinical trials in the USA (NCT03294577) comparing it against pegfilgrastim as a treatment for the neutropenia seen in breast cancer patients using the "TAC" therapy regimen (taxane, Adriamycin, cyclophosphamide). This series of trials is in addition to trials against lung cancer at the Phase I level in conjunction with nivolumab under NCT02812667 and a Phase II trial under NCT02846792 with another series of lung cancers. There is also a press release from BeyondSpring Pharmaceuticals that the first patients in a Phase III trial in the People's Republic of China against lung cancer have been enrolled.

6 Natural Product-Based Compounds as "Warheads"

As alluded to earlier in this chapter, significant numbers of compounds have been isolated from marine sources that have biological activity in a variety of pharmacological screens, but very few have become drug candidates as single agents. However, the advent of monoclonal antibodies directed toward a particular epitope expressed by a cancerous cell has led to a "rethinking" of what these highly toxic agents can be used for. The success of Mylotarg® where the highly toxic terrestrial microbial product, calicheamicin, was linked to an anti-CD33 humanized antibody, and launched in 2005, has led to a significant number of constructs entering preclinical and clinical trials as antitumor agents, using as warhead molecules originally sourced from nature, plant, and microbial and "nominally" marine sources.

The following tables show the current status of agents (as early November 2017) that are in or have been in clinical trials. Structures of the warhead are shown in figures related to each table, with the linkers if necessary. No preclinical agents are listed, though there are some marine-sourced materials in this stage that are potential candidates for Phase I trials, including derivatives of hemiasterlin and the chloro-derivative of PM184 shown in Fig. 8, but due to the difficulty of establishing where agents such as these are in the development process, we have not given any further information until they are in a listed clinical trial.

The tables are ordered by the now known source, rather than the one(s) from which the prototype was isolated. Relevant references will be given under each heading.

7 Microbial (Marine and Terrestrial)

Dolastatin-combinations (Newman and Cragg 2017) (Fig. 9).

Warhead	Approved	Phase III	Phase II	Phase I	Dropped
MMAE	1		9	10	4
MMAF			3	2	3
Amberstatin 269				2	
Auristatin W and analogue				2	
Auristatin 0101				2	
BMS-986148			1		

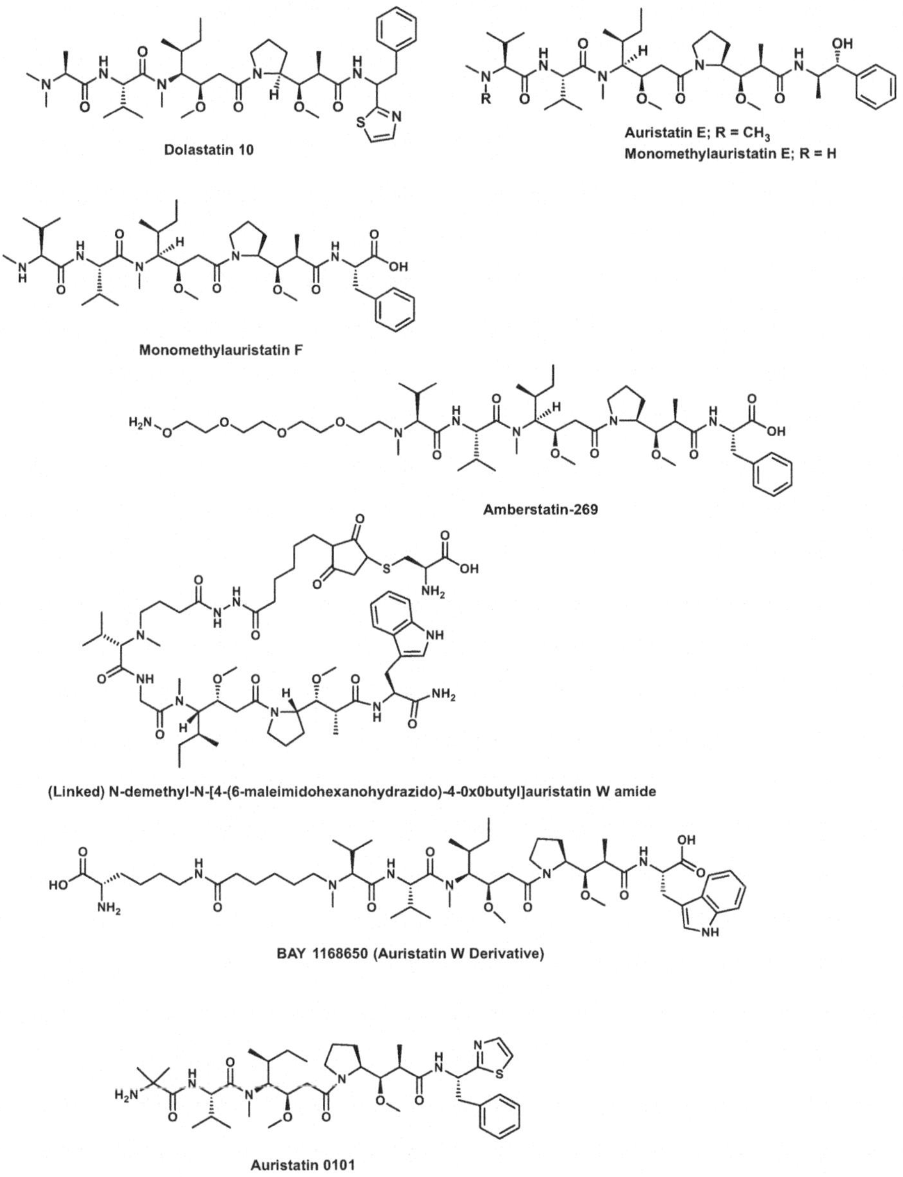

Fig. 9 Dolastatin-derived warheads

Maytansine combinations (Lambert 2012; Kusari et al 2014, 2016).

Warhead	Approved	Phase III	Phase II	Phase I	Dropped
DM-1	1		2	2	7
DM-4		1	5	3	4

Calicheamicin combinations (Chan et al. 2003; Damelin et al. 2015).

Warhead	Approved	Phase III	Phase II	Phase I	Dropped
Ozogamicin[a]	2			1	2

[a]All warheads are the same, a linked iodinated calicheamicin, only the mAb alters

Duocarmycin- combinations (Black et al. 2016; Elgersma et al. 2015) (Fig. 10).

Warhead	Approved	Phase III	Phase II	Phase I	Dropped
vc-seco-DUBA		1			
mb-vc-MGBA					1

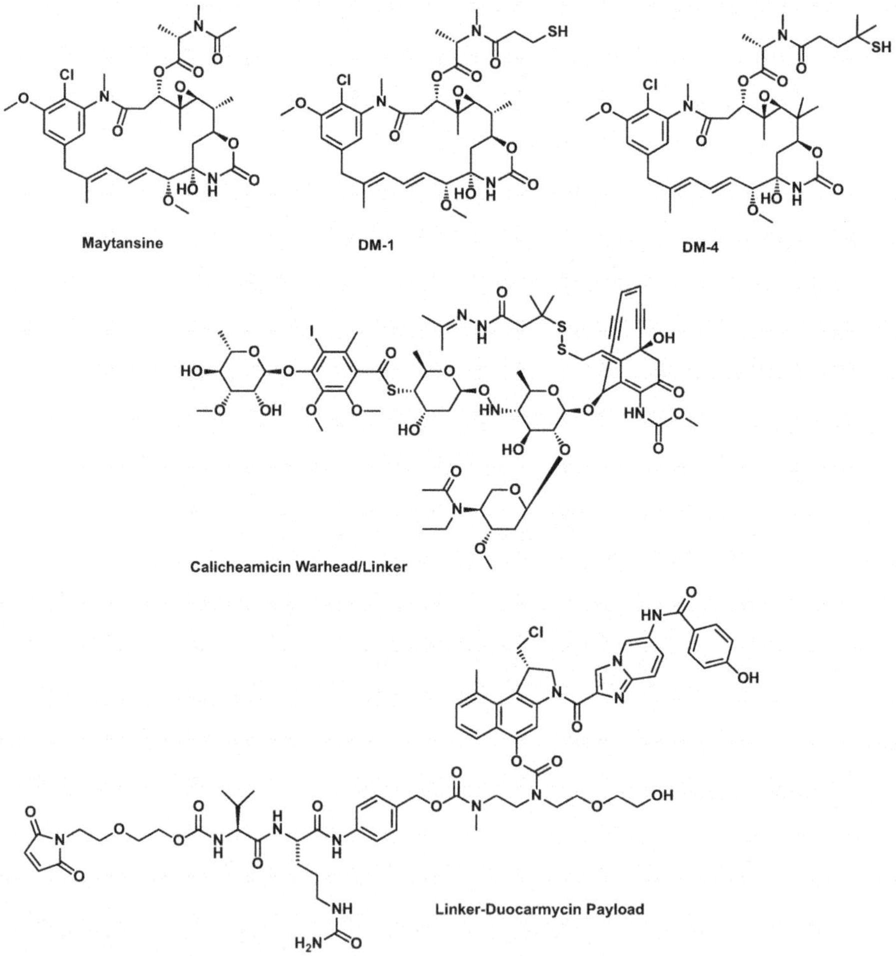

Fig. 10 Maytansine-, calicheamicin-, and duocarmycin warheads

PBD combinations (Mantaj et al. 2017).

Warhead	Approved	Phase III	Phase II	Phase I	Dropped
va-SGD1882				3	2
PEG8-va-SG3199		1		3	
Undisclosed				2	

Tubulysin-combination (Li et al. 2016) (Fig. 11).

Warhead	Approved	Phase III	Phase II	Phase I	Dropped
AZ13599185			1		

8 Plant-Sourced

Camptothecin combinations (Agatsuma 2017; Nakada et al. 2016) (Fig. 12).

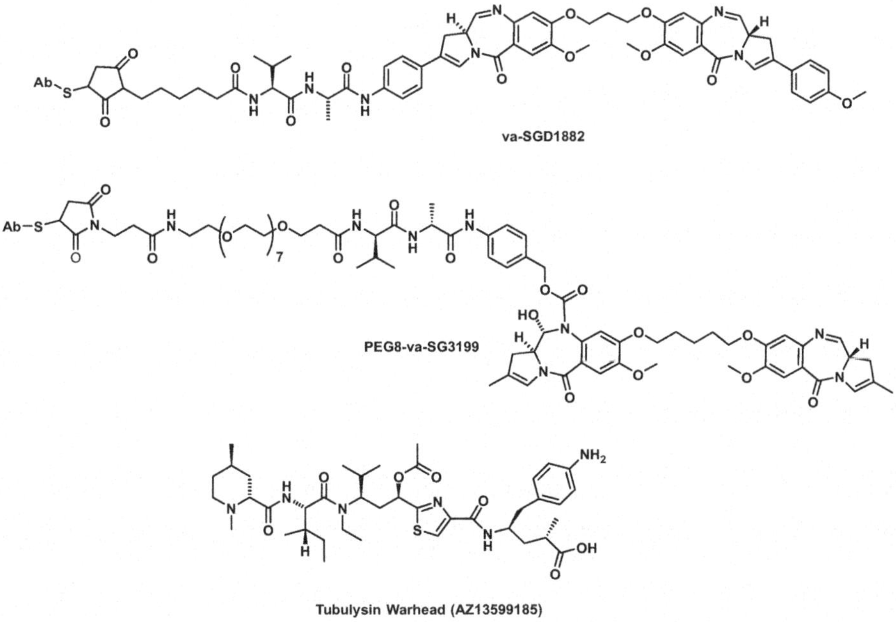

Fig. 11 Pyrrolobenzodiazepine- and tubulysin warheads

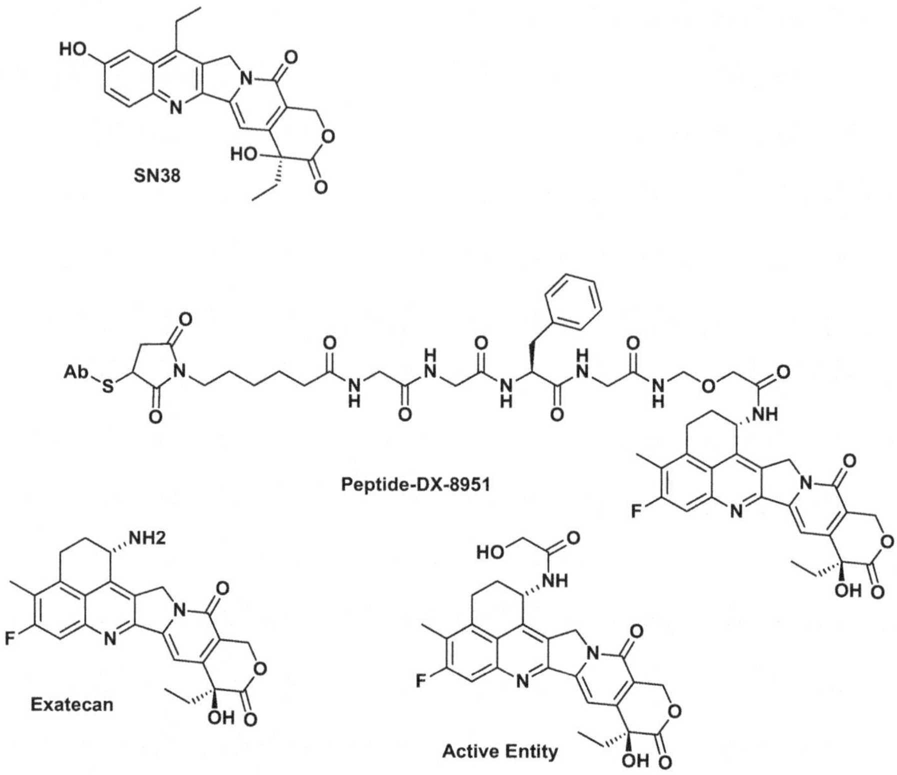

Fig. 12 Camptothecin-derived warheads

Warhead	Approved	Phase III	Phase II	Phase I	Dropped
CL2A-SN38[a]		1	1		
Peptide-DX-8951[b]			2		

[a]SN38 is the active principle of irinotecan
[b]Modified exatecan

9 Conclusions

Although the direct line from nature to an anticancer drug is probably not functional any longer in the case of true plant-sourced molecules, their scaffolds are still useable as leads to new active materials. In contrast, in the case of marine- and microbial-sourced compounds, there are still some compounds that are potential drug candidates in their own right.

All of these sources, however, have now entered what might be considered the equivalent of the early days of antibiotic discovery, in that subtle (and in some cases, not so subtle) modifications to the base structures have led to their use as warheads on mAbs, as can be seen in the tables above. By using the exquisite targeting abilities of humanized mAbs and careful control of linkage chemistries, extremely toxic agents (e.g., the tubulysins and calicheamicins) can be "tamed" for use as antitumor agents.

Does this mean that we should switch over to only looking for microbial-sourced agents? There is a case to be made for this, but even with today's advances in "omics," finding an active agent irrespective of its "true source" is just the start of any development, so novel agents from any "nominal source" are necessary to start the development process. If the source is then shown to be microbial (single-celled organism) later on, then the advances in "omics" can be brought into the development program, but first one has to find that novel structure.

References

Agatsuma T (2017) Development of new ADC technology with topoisomerase I inhibitor. Yakugaku Zasshi 137:545–550

Andrejauskas-Buchdunger E, Reganass U (1992) Differential inhibition of the epidermal growth factor-, platelet derived growth factor-, and protein kinase C-mediated signal transduction pathways by the staurosporine derivative CGP 41251. Cancer Res 52:5353–5358

Awada A, Bondarenko IN, Bonneterre J et al (2014) A randomized controlled phase II trial of a novel composition of paclitaxel embedded into neutral and cationic lipids targeting tumor endothelial cells in advanced triple-negative breast cancer (TNBC). Ann Oncol 25:824–831

Basmadjian C, Zhao Q, Djehal A et al (2014) Cancer wars: natural products strike back. Front Chem 2:20

Bebbington MWP (2017) Natural product analogues: towards a blueprint for analogue-focused synthesis. Chem Soc Rev 46:5059–5109

Bertin MJ, Schwartz SL, Lee J et al (2015) Spongosine production by a *Vibrio harveyi* strain associated with the sponge *Tectitethya crypta*. J Nat Prod 78:493–499

Black J, Menderes G, Bellone S et al (2016) SYD985, a novel duocarmycin-based HER2-targeting antibody-drug conjugate, shows antitumor activity in uterine serous carcinoma with HER2/Neu expression. Mol Cancer Ther 15:1900–1909

Chan SY, Gordon AN, Coleman RE et al (2003) A phase 2 study of the cytotoxic immunoconjugate CMB-401 (hCTM01-calicheamicin) in patients with platinum-sensitive recurrent epithelial ovarian carcinoma. Cancer Immunol Immunother 52:243–248

Cragg GM, Newman DJ (2014) Natural products as sources of new anticancer agents. In: Feliciano AS, Filho VC (eds) Descoberta, Desenho e Desenvolvivmento de Novos Agentes Anticancer no Ambito do Programa Iberoamericano CYTED. Editoria Univali, Itajai, pp 67–118

Cragg GM, Pezzuto JM (2016) Natural products as a vital source for the discovery of cancer chemotherapeutic and chemopreventive agents. Med Princ Pract 25:41–59

Damelin M, Bankovich A, Park A et al (2015) Anti-EFNA4 calicheamicin conjugates effectively target triple-negative breast and ovarian tumor-initiating cells to result in sustained tumor regressions. Clin Cancer Res 21:4165–4173

Davis AM, Tinker AV, Friedlander M (2014) "Platinum resistant" ovarian cancer: what is it, who to treat and how to measure benefit? Gynecol Oncol 133:624–631

Elgersma RC, Coumans RG, Huijbregts T et al (2015) Design, synthesis, and evaluation of linker-duocarmycin payloads: toward selection of HER2-targeting antibody-drug conjugate SYD985. Mol Pharm 12:1813–1835

Galal A, El-Bakly WM, El-Demedash E (2016) Selective A3 adenosine receptor agonist protects against doxorubicin-induced cardiotoxicity. Cancer Chemother Pharmacol 77:309–322

Giddings LA, Newman DJ (2015) Bioactive compounds from extremophiles, genomic studies, biosynthetic gene clusters, and new dereplication methods. In: Tiquia-Arashiro SM, Mormile M (eds) Extremophilic bacteria, Springer briefs in microbiology. Springer, Heidelberg, pp 1–58

Giddings LA, Newman DJ (2017) Microbial involvement in the production of natural products by plants, marine invertebrates and other organisms. In: Atta-ur-Rahman (ed) Frontiers in natural product chemistry, vol 3. Bentham, Karachi, pp 1–64

Gottfried K, Klar U, Platzek J et al (2015) Biocatalysis at work: applications in the development of Sagopilone. ChemMedChem 10:1240–1248

Graybill WS, Coleman RL (2016) Folate receptor-targeted therapeutics for ovarian cancer. Drugs Future 41:137–143

Herzog TJ, Kutarska E, Bidzińsk M et al (2016) Adverse event profile by folate receptor status for vintafolide and pegylated liposomal doxorubicin in combination, versus pegylated liposomal doxorubicin alone, in platinum-resistant ovarian cancer: Exploratory analysis of the Phase II PRECEDENT trial. Int J Gynecol Cancer 26:1580–1585

Huang M, Gao H, Chen Y et al (2007) Chimmitecan, a novel 9-substituted camptothecin, with improved anticancer pharmacologic profiles *in vitro* and *in vivo*. Clin Cancer Res 13:1298–1307

Huang G, Wang H, Yang LX (2010) Enhancement of radiation-induced DNA damage and inhibition of its repair by a novel camptothecin analog. Anticancer Res 30:937–944

Joerger M, Hess D, Delmonte A et al (2015) Integrative population pharmacokinetic and pharmacodynamic dose finding approach of the new camptothecin compound namitecan (ST1968). Br J Clin Pharmacol 80:128–138

Jones RP, Malik HZ, Fenwick SW et al (2016) PARAGON II – a single arm multicentre phase II study of neoadjuvant therapy using irinotecan bead in patients with resectable liver metastases from colorectal cancer. Eur J Surg Oncol 42:1866–1872

Kusari S, Lamsho M, Kusari P et al (2014) Endophytes are hidden producers of maytansine in *Putterlickia* roots. J Nat Prod 77:2577–2584

Kusari P, Kusari S, Eckelmann D et al (2016) Cross-species biosynthesis of maytansine in *Maytenus serrata*. RSC Adv 6:10011–10016

Lambert JM (2012) Drug-conjugated antibodies for the treatment of cancer. Br J Clin Pharmacol 76:248–262

Lee KW, Lee KH, Zang DY et al (2015) Phase I/II study of weekly Oraxol for the second-line treatment of patients with metastatic or recurrent gastric cancer. Oncologist 20(8):896–897

Li JY, Perry SR, Muniz-Medina V et al (2016) A biparatopic HER2-targeting antibody-drug conjugate induces tumor regression in primary models refractory to or ineligible for HER2-targeted therapy. Cancer Cell 29(1):117–129

Liu X, Kantarjian H, Plunkett W (2012) Sapacitabine for cancer. Expert Opin Investig Drugs 21(4):541–555

Mantaj J, Jackson PJM, Rahman KM et al (2017) From anthramycin to pyrrolobenzodiazepine (PBD) containing antibody-drug conjugates (ADCs). Angew Chem Int Ed Engl 56(2):462–488

Martín MJ, Coello L, Fernández R et al (2013) Isolation and first total synthesis of PM050489 and PM060184, two new marine anticancer compounds. J Am Chem Soc 135(27):10164–10171

Montalban-Bravo G, Garcia-Manero G (2015) Novel drugs for older patients with acute myeloid leukemia. Leukemia 29(4):760–769

Nakada T, Masuda T, Naito H et al (2016) Novel antibody drug conjugates containing exatecan derivative-based cytotoxic payloads. Bioorg Med Chem Lett 26:1542–1545

Newman DJ (2008) Natural products as leads to potential drugs: An old process or the new hope for drug discovery? J Med Chem 51:2589–2599

Newman DJ (2016) Predominately uncultured microbes as sources of bioactive agents. Front Microbiol 7:1832

Newman DJ (2017) Recent advances in screening and identification of novel biologically active natural compounds. F1000 Fac Rev 6:783

Newman DJ, Cragg GM (2012) Natural products as sources of new drugs over the 30 years from 1981 to 2010. J Nat Prod 75(3):311–335

Newman DJ, Cragg GM (2014) Making sense of structures by utilizing Mother Nature's chemical libraries as leads to potential drugs. In: Osbourn A, Goss RJ (eds) Natural products: discourse, diversity and design. Wiley, New York, pp 397–411

Newman DJ, Cragg GM (2015) Endophytic and epiphytic microbes as "sources" of bioactive agents. Front Chem 3:34

Newman DJ, Cragg GM (2016a) Natural product scaffolds of value in medicinal chemistry. In: Brase S (ed) Privileged scaffolds in medicinal chemistry. Royal Society of Chemistry, London, pp 348–378

Newman DJ, Cragg GM (2016b) Natural products as sources of new drugs from 1981 to 2014. J Nat Prod 79(3):629–661

Newman DJ, Cragg GM (2017) Current status of marine-derived compounds as warheads in anti-tumor drug candidates. Mar Drugs 15(4) 99.

Newman DJ, Cragg GM, Snader KM (2000) The influence of natural products upon drug discovery. Nat Prod Rep 17(3):215–234

Newman DJ, Cragg GM, Kingston DGI (2015) Natural products as pharmaceuticals and sources for lead structures. In: Aldous D, Rognan D, Raboisson P et al (eds) The practice of medicinal chemistry, 4th edn. Elsevier, Amsterdam, pp 102–138

Pera B, Barasoain I, Pantazopoulou A et al (2013) New interfacial microtubule inhibitors of marine origin, PM050489/PM060184, with potent antitumor activity and a distinct mechanism. ACS Chem Biol 8(9):2084–2094

Santamaría NG, Robles CM, Giraudon C et al (2016) Lurbinectedin specifically triggers the degradation of phosphorylated RNA polymerase II and the formation of DNA breaks in cancer cells. Mol Cancer Ther 15(10):2399–2412

Shi J, Kantoff PW, Wooster R et al (2017) Cancer nanomedicine: Progress, challenges and opportunities. Nat Rev Cancer 17(1):20–37

Wakimoto T, Egami Y, Nakashima Y et al (2014) Calyculin biogenesis from a pyrophosphate protoxin produced by a sponge symbiont. Nat Chem Biol 10(8):648–655

Wicki A, Ritschard R, Loesch U et al (2015) Large-scale manufacturing of GMP-compliant anti-EGFR targeted nanocarriers: Production of doxorubicin-loaded ant-EGFR-immunoliposomes for a first-in-man clinical trial. Int J Pharm 484(1–2):8–15

Wilson MC, Mori T, Ruckert C et al (2014) An environmental bacterial taxon with a large and distinct metabolic repertoire. Nature 506(7486):58–62

Virtual Screening for the Discovery of Active Principles from Natural Products

Benjamin Kirchweger and Judith M. Rollinger

1 Computational Tools in Drug Discovery

Computer-aided drug design is of emerging importance and usefulness for the development of therapeutically relevant small molecules. With recent advances in structural determination, e.g., cryo-electron microscopy (Fernandez-Leiro and Scheres 2016), we face a growing number of possible drug targets and binding positions. Likewise, combinatorial chemistry and high-throughput screening performed excessively in the last 20 years have led to an explosion in the number of small molecules and in vitro data (Gaulton et al. 2012; Yan et al. 2006; Lo et al. 2018). Although not as dramatic as in synthetic chemistry, the number of small molecules from nature and their occupied chemical space is constantly increasing by the discovery of new sources (e.g., marine and microbial organisms) and the diligent exploitation of already known material (Pye et al. 2017; Harvey et al. 2015). From this already existing large quantity of data, learning and making predictions offer the chance of rationalizing research. With the help of computational tools, extrapolation from this huge volume of data enables the prediction of new events, such as a molecule's putative ligand-target interaction, biological activity, or its properties including metabolism, toxicology, and pharmacokinetics. The implementation of computational methods aims to concentrate the capacities for experimental testing on less but more encouraging subjects. It can thus help to focus time and money by streamlining experimental efforts (Sliwoski et al. 2014). In iterative research processes, e.g. lead optimization and bioassay-guided fractionation, it can guide research projects. Recent advances in machine learning algorithms, molecular dynamics, more accurate ADMET predictions, and the fading of the earlier computational power bottleneck brought computer-aided drug

B. Kirchweger · J. M. Rollinger (✉)
Department of Pharmacognosy, University of Vienna, Vienna, Austria
e-mail: benjamin.kirchweger@univie.ac.at; judith.rollinger@univie.ac.at

V. Cechinel Filho (ed.), *Natural Products as Source of Molecules with Therapeutic Potential*, https://doi.org/10.1007/978-3-030-00545-0_9

discovery (CADD) into the spotlight of research interest. Especially in the field of NP, which is generally regarded as an expensive endeavor (Strohl 2000), in silico tools can help to overcome key difficulties and fast-forward investigations as recently reviewed (Rollinger et al. 2006a, b; Rollinger 2009, Rollinger et al. 2009; Rollinger and Wolber 2011, Kaserer et al. 2018). Although delayed, their application in NP research has led to outstanding findings. However, the interface of computer-aided drug design and NP research (two multidisciplinary disciplines on their own) is highly interdisciplinary as it embraces organic chemistry and phytochemistry, informatics, structural biology, genomics, biochemistry, pharmacology, mathematics, biophysics, and medicinal chemistry besides other fields. Furthermore, there is an abundance of CADD tools suited for different applications and problems (Schneider 2010; Gasteiger 2016).

This survey shall give a general overview to students and researchers, who will step in this emerging and exciting field. It gives a brief introduction into the field of cheminformatics and further presents and explains different virtual screening (VS) approaches with a focus on opportunities and obstacles in general and in combination with NP. Finally, the chapter presents various studies embracing different in silico strategies to exemplarily show the great diversity of VS in NP drug discovery.

2 A Brief Introduction to Cheminformatics

2.1 Definition of Cheminformatics

In 1998 F. K. Brown (1998) made the striking definition: "Cheminformatics is the mixing of those information resources to transform data into information and information into knowledge for the intended purpose of making better decisions faster in the area of drug lead identification and optimization." Since then the cheminformatics field has extended and is today better and more general defined as broad field of solving chemical problems with computational methods (Gasteiger and Engel 2003). They can be of great assistance, but it is fundamental that one understands the basic concepts and fundaments of cheminformatics: the representation of chemical structures, the calculation of molecular descriptors and chemical fingerprints, and the analysis of chemical space.

2.2 Chemical Structure Formats

All cheminformatics tools are based on the representation and storage of chemical structures in a format accessible to software tools. As computers can only manage bits of 0 and 1, languages have been sought which are readable for computers and operators. They should be fast processible and have a small memory footprint but a maximum of information content. Similar to human chemical description, there are also several complexity levels for computer chemical structure exchange formats depicted in Fig. 1. For each hierarchy level, there have been plenty of language formats developed, most specialized to particular software tools and incorporating specific information. The interconversion from one to another is often difficult, sometimes impossible to perform without loss of information (Kirchmair et al. 2008). However, there are some standard structure exchange formats implemented by nearly all software applications:

1. SMILES codes, developed by Daylight (O'Boyle 2012; Weininger 1988), are simple line notations of molecules, enabling memory saving storage. The atoms of a molecule are simply notated by their connections. The SMILES smiles code for cyclohexane, C1CCCCC1, is simple; for larger molecules like piperine (C1CCN(CC1)C(=O)C=CC=CC2=CC3=C(C=C2)OCO3) or strychnine (C1CN2CC3=CCO[C@@H]4CC(=O)N5[C@@H]6[C@@H]4[C@H]3C[C@@H]2C61C7=CC=CC=), it gets more complex, and some additional rules are necessary, but they still need few memory. Because of its linguistic construction, operators can easily learn this language. Moreover there are rules for canonical notations, making the canonical SMILES synonymous with a molecule. Isomeric SMILES can incorporate information about double bond geometry and chirality and should therefore be used for three-dimensional (3D) uses.
2. The connection table formats (Dalby et al. 1992) mol and sdf are more memory intensive and although still text based less comprehensible for researchers (Table 1). They can incorporate additional information on the compound and support 3D information. Sdf files can store series of molfiles joined together and

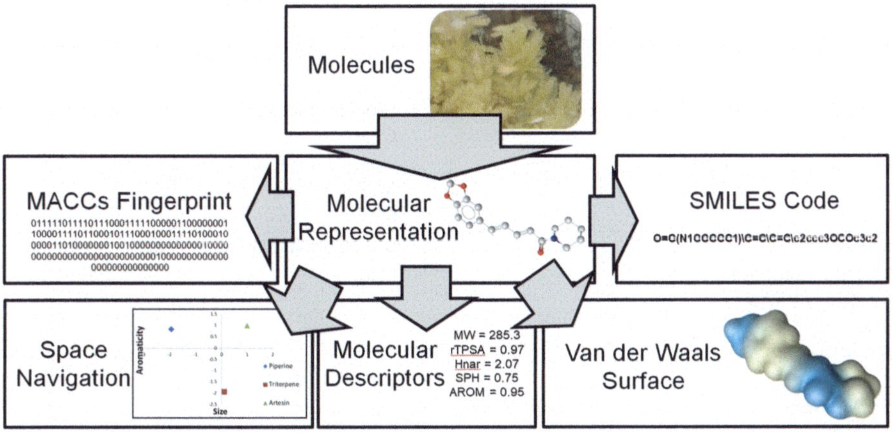

Fig. 1 Data processing in cheminformatics on the example piperine: Besides different molecular structure formats, like 3D sdf and SMILES line notation formats, it is possible to calculate molecular fingerprints, different descriptors like the 3D Van der Waals surface, and principal components for space navigation

Table 1 Structure of the exemplary molfile of anthranilic acid

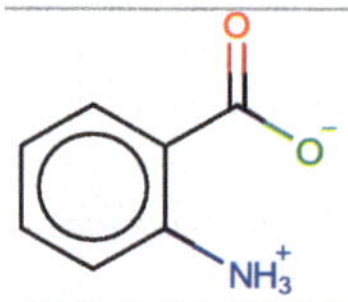

Anthranilic acid Chemdraw02161816332D			Name Used software	Header block (3 lines)
Structure 1			Comment line	
10 10 0 0 0 0 0 0 0 0 0999 V2000			Counts line (10 heavy atoms and 10 bonds)	Connection table
−1.4289 −1.4289 −0.7145 −0.0000 −0.0000 −0.7145 0.7145 0.7145 1.4289 0.7145	0.0000 −0.8250 −1.2375 −0.8250 0.0000 0.4125 0.4125 1.2375 0.0000 −1.2375	0.0000 C 0 0 0 0 0 0 0 0 0 0 0 0 0.0000 C 0 0 0 0 0 0 0 0 0 0 0 0 0.0000 C 0 0 0 0 0 0 0 0 0 0 0 0 0.0000 C 0 0 0 0 0 0 0 0 0 0 0 0 0.0000 C 0 0 0 0 0 0 0 0 0 0 0 0 0.0000 C 0 0 0 0 0 0 0 0 0 0 0 0 0.0000 C 0 0 0 0 0 0 0 0 0 0 0 0 0.0000 O 0 0 0 0 0 0 0 0 0 0 0 0 0.0000 O 0 5 0 0 0 0 0 0 0 0 0 0 0.0000 N 0 3 0 0 0 0 0 0 0 0 0 0	Atom block: A line for each heavy atom with position in a x, y, and z coordination system is given (in Angstrom). In this example, the third column of the atom table is always 0, because the given file only describes a 2D representation. Right to the coordinate specification, there is information stored on elements (C, O, N, etc.), isotopes, or charges.	
1 2 4 0 2 3 4 0 3 4 4 0 4 5 4 0 5 6 4 0 1 6 4 0 5 7 1 0 7 8 2 0 7 9 1 0 4 10 1 0			Bond block: Specifies the connection between the atoms; atom 1 is connected with atom 2 by an aromatic bond (2 for double bond, 3 for triple, 4 for aromatic) stereospecificity annotation 0 (1, up; 0, no stereo; 6, down). Atom 2 is connected to atom 3 by an aromatic bond without stereospecification, etc.	
M CHG 2 9 −1 10 1 M END			Property block: (CHG = Charge) 2 charges, atom 9 charge −1, atom 10 charge +1 M END: Ends connection table	
> <Origin> Mentha x piperita > < Rel. TPSA> 3.99 > < # Rings> 1 > <MolWt> 137.14			> < Associated data subject> Data	Associated data
$$$$			Structure data files (sdf) can store more than one conformation or multiple compounds. They are delimited by lines consisting of four dollar signs	Delimiting line

therefore be used for exchanging libraries of structure data, together with annotated metadata.

3. The mol2 file format is also a text-based connection table format. It can store 3D small molecule representations but also large proteins and nucleic acids.
4. The Protein Data Bank (PDB) format was developed for the 3D storage of biological macromolecular structures (proteins and nucleic acids). It is a text-based format with rather complicated ordering and formatting of data sections and records (record = line of information), atomic coordinates, assignment of secondary structures, and side-chain rotamers. Bonds are not specified, which makes the correct reconstruction of structures error-prone. For the atoms of amino acid standard residues, only 3D coordinates are stored, because the connectivity of the atoms is heuristic and can be looked up in implemented databases. The pdb format is the standard exchange format for proteins and other macromolecules (Berman et al. 2003; Henrick et al. 2008).

The broader basis of chemical structure formats and their processing would by far exceed this chapter, and interested ones are referred to literature (Gasteiger and Engel 2003).

2.3 *Molecular Descriptors*

The chemical information incorporated in a molecular structure format can be transformed into quantitative molecular descriptors, which play a fundamental role in cheminformatics (Fig. 1) (Corwin et al. 1995; Danishuddin and Khan 2016). Numerical molecular descriptors can be scalar (one-dimensional), e.g., heavy atom count and molecular weight. Two-dimensional (2D) chemical descriptors include topological indices or molecular profiles, and 3D descriptors extract their content from 3D coordinate representations, e.g., surface/volume descriptors and pharmacophore descriptors. Four-dimensional descriptors are 3D descriptors considering multiple conformations (Bajorath 2001; Karelson et al. 1996; Sliwoski et al. 2016) (Todeschini and Consonni 2008; Lo et al. 2018). Some descriptors can be obtained by standardized experimental measurements (physicochemical properties) but most of them by a mathematical calculation, which transforms chemical information present in the molecular structure into a useful, sometimes purely abstract number. The number of numerical descriptors is growing, e.g., the commercial DRAGON system can generate up to 5000 different descriptors (Sawada et al. 2014; Chavan et al. 2014). They can be computed by several software and online tools or easily defined by simple scripts. Tetko reviews some resources for molecular descriptors and tools to calculate them (Tetko 2003). The most important application for molecular descriptors is quantitative structure activity relationship (QSAR) and quantitative structure property relationship (QSPR) but also chemical similarity analysis.

2.4 Molecular Fingerprints

Molecular fingerprints can be an abstract but useful way to encode structural features of molecules. Basically they are bit strings or high-dimensional vectors, which are generated by hashing functions. The most widespread molecular fingerprints are the binary Molecular ACCes System (MACCS) keys. Binary means they only use two digits (0 and 1). For the presence or absence of one of the 166 substructures in a molecule, zeros and ones are appointed (Fig. 1). This bit set on the one hand quite well defines a molecule and on the other hand enables easier computation than a "bulky" molecular representation. The easiness of bit set fingerprints compared to complex molecular representations is the most important factor to use them for similarity searches. PubChem, e.g., uses their own PubChem substructure fingerprint with 881 fragments for similarity searching. Next to such 2D substructure fingerprints, there exists a broad palette of other fingerprints including topological (connectivity or spatial distribution of fragments), pharmacophore, text-based, protein-ligand interaction, and hybrid fingerprints. Next to similarity searching, they are useful tools for VS.

2.5 Chemical Space Analysis

Characterization of molecules with quantitative descriptors enables their arrangement in a multidimensional space. Placement of small molecules in this chemical space enables their classification and comparison. A general approach would be to calculate numerical descriptors (e.g., physicochemical or structural descriptors) and perform a principal component analysis to reduce the descriptor vector to a 2D or 3D space which can be plotted (Singh et al. 2009; Wetzel et al. 2007). Chemical space examination has grown to major relevance, since it was shown that biologically relevant small molecules occupy only small regions of the possible chemical space. While in theory 10^{60} small organic molecules are thinkable (Reymond et al. 2010), most of the chemical space they occupy can be seen as useless for drug discovery (Payne et al. 2006; Macarron 2006). Good news for NP researchers is the fact that NP occupy exactly this biologically important space, which is in accordance with the understanding of NP as privileged structures (Gu et al. 2013; Harvey et al. 2015). Since there is great interest in compound libraries which are focused on this drug-relevant space (Akella and DeCaprio 2010), it is aimed that screening libraries with compounds similar to NP are created (Cordier et al. 2008) and their quality is assessed with a NP likeness score (Jayaseelan et al. 2012). Chemical space analysis has high applicability, e.g., for the quality control of VS libraries, for the comparison of query molecules, or for the selection of virtual hits. A useful and comprehensible resource is the free chemGPS-NP online tool (Larsson et al. 2007). For uploaded molecules, 35 descriptors converted to 8 principal components are calculated and returned. The principal components are accountable as they apply properties like a molecule's size, shape, or flexibility. An iterative process for the definition of PCA space avoids outliers. The chemical space analysis of chemGPS compares all molecules from a reference set to a test set (each dot stands for a molecule) (Larsson et al. 2007). If it is more appropriate to compare each molecule of a test set against the mean of a reference set, multi-fusion similarity mapping based on molecular fingerprints is a viable alternative (Medina-Franco et al. 2007; Singh et al. 2009).

2.6 *Chemical Similarity Analysis*

The basic principle of chemical similarity analysis is the assumption that compounds with similar structures in the narrower or the wider sense have similar biological properties, which is often but not always the case (Hu et al. 2013; Stumpfe et al. 2014; Bajorath 2017). Chemical similarity is quantified by distance/similarity metrics between fingerprints derived from molecular features (e.g., topological, physicochemical, or pharmacophore features). The most important similarity metric is probably the Tanimoto coefficient (TC). The higher the TC score, which ranges from 0 to 1, the higher the similarity between two molecules (Bajusz et al. 2015). Chemical similarity search can be used for VS, e.g., 3D shape and pharmacophore matching. Moreover, molecules can be clustered into groups based on their similarity in chemical reference space. Molecules in the same cluster are similar to each other; molecules in different clusters are thought to be different from each other. Several cluster analysis methods are available, e.g., Jarvis-Patrick (nonhierarchical) or Ward's (hierarchical) methods; there is however no universal solution for all problems, as there is no single measure of similarity. With similarity clustering, target fishing can be sufficiently performed, e.g., features of query compound fragments are compared to pre-calculated drug compound clusters (Reker et al. 2014; Rodrigues et al. 2016). Besides this astonishing works, similarity measurements and clustering techniques are of utmost importance in cheminformatics. For additional reading, one is referred to Nikolova and Jaworska (2003).

3 Virtual Screening for Hit Generation from Natural Products

3.1 *Definition*

Similar to physical high-throughput screening, where a large number of compounds is tested in wet lab in any assay to identify those compounds which exert biological activities, VS is an in silico technique to virtually probe molecular libraries for those structures, which are most promising to putatively exert an activity on a focused target.

It is more or less the mining of hypothetical molecule piles to identify the most promising candidates to possess a desired property (i.e., activity on a target) (Rester 2008). For evaluation of the predictive power of the used filtering tool (model), and primarily for the identification of hit compounds, the experimental testing of predicted hits is an indispensable component in the drug discovery process. VS is always a heavily knowledge-driven process and depends on the information already available for the system under investigation. The quality and amount of information and its preceding selection and preparation is imperative for successful experiments (Sichao et al. 2013). Moreover, the subjective expertise of an operator or working group should not be underestimated (Ban et al. 2017). Generally, one can classify VS into structure-based approaches with information from experimental protein structures, either from X-ray crystallography, NMR, or computational homology models, and ligand-based approaches with information on known ligands. Although usually less reliable than the structure-based techniques, the latter approach is still the method of choice for membrane-bound G-protein-coupled receptors and ion channels (Evers et al. 2005; Seidel et al. 2010). A comprehensive list of resources for VS is provided at http://www.click2drug.org.

3.2 Molecular Docking

Molecular docking is regarded by many as the central technology for VS; therefore, the programs are under intense development and evaluation (Kitchen et al. 2004; Hauser and Windshügel 2016; Cavasotto and Orry 2007; Grosdidier et al. 2011). They can filter out promising hits in a virtual database but can also give answers to related problems like prediction of binding pose and affinity (Anderson 2003; Jain and Nicholls 2008). Molecular docking is a computational approach, which aims to first virtually predict the binding of a ligand (in a specific conformation) to a target. Secondly, binding affinities of the predictions are approximated. The binding modes of the virtually screened molecules are usually ranked from estimated most active to inactive ones. Prediction of (a) the correct binding pose with search algorithms and (b) the correct estimations of binding affinities, termed as scoring, is a nontrivial task. This is why more than 60 different docking tools and programs with different search algorithms and scoring functions have been developed (Pagadala et al. 2017). As they may incorporate their own benefits and shortcomings, their performance has been evaluated several times, but the claims of superiority differ largely. Commercial software not necessarily has superior performance over open-source tools (Wang et al. 2016; Huang et al. 2010; Warren et al. 2006). According to Chen (2015), the three most frequently used programs are the freeware AutoDock (Osterberg et al. 2002) and the commercial programs GOLD (Jones et al. 1997) and Glide (Friesner et al. 2006). Whereas AutoDock uses stochastic search algorithms, GOLD is based on genetic and Glide on systematic ones (Taylor et al. 2002). Other commonly used programs are FlexX, Surflex, LigandFit, Dock, and AutoDock Vina, besides different web services like SwissDock (Grosdidier et al. 2011). It has to be emphasized that the choice of software and algorithm(s) strongly depends on the focus of the VS project. Chen has recently given a comprehensive overview on different applications and their benefits (Chen 2015). Although molecular docking is illustrative and has led to outstanding findings in drug design (Shoichet et al. 2002; Claude Cohen 2007), shortcomings independent of the software are pervasive. While docking algorithms perform quite well in sampling correct binding poses, it is still impossible to calculate the solvation effects and entropic parts of ligand binding energy, causing inaccurate scoring. Another shortcoming is the time-consuming screening as high calculation demands are necessary causing programs to balance accuracy and speed. Issues like side-chain flexibility, explicit water, as well as solvent effects and backbone movements are currently under research. Nevertheless the long list of unclear issues can impede success by causing high numbers of false-positive hits (Pagadala et al. 2017; Warren et al. 2006; El-Houri et al. 2015). Moreover, some tools are empirically trained to molecules diverging from NP, which are more flexible and have a higher molecular weight (Wetzel et al. 2007), which questions their suitability to such exercises (Rollinger and Wolber 2011). However, molecular docking bears some distinct advantages over other VS techniques, e.g., incorporation of structural and mechanistic information, which gives the possibility to not only identify novel binders and scaffolds but also novel modes of binding (Ma et al. 2011).

With the exponential increase in computer processing power, advances in elucidation of macromolecule structures, consensus and machine learning-based scoring methods (Oda et al. 2006; Wójcikowski et al. 2017), and growing understanding of intermolecular interactions that take part in the protein-ligand molecular interaction (Ren et al. 2014; Yang et al. 2015), docking will continue to play its important role in drug discovery and optimization (Wang and Zhu 2016).

3.3 Target-Based Pharmacophores

The "pharmacophore" concept is approximately 50 years old and is today defined by the International Union of Pure and Applied Chemistry as "the ensemble of steric and electronic features that is necessary to ensure the optimal supramolecular interactions with a specific biological target structure and to trigger (or to block) its biological response" (Wermuth et al. 1998). The concept states that the complexity of ligand-target interactions can be reduced to a distinct abstract blueprint, a 3D arrangement of the most important interaction types and their proximity and angle between each other. Ligands that comprise similar combinations of pharmacophore features in similar spatial orientation are likely to have similar activity towards a biological macromolecule. That is why query pharmacophore models can act as proficient VS filters (Van Drie 2010; Langer and Wolber 2004). Figure 2 shows an illustrating example of such pharmacophore models. The interaction types are commonly named pharmacophore features and comprise the chemical characteristics of functional groups participating in the ligand-target interactions. Different types of

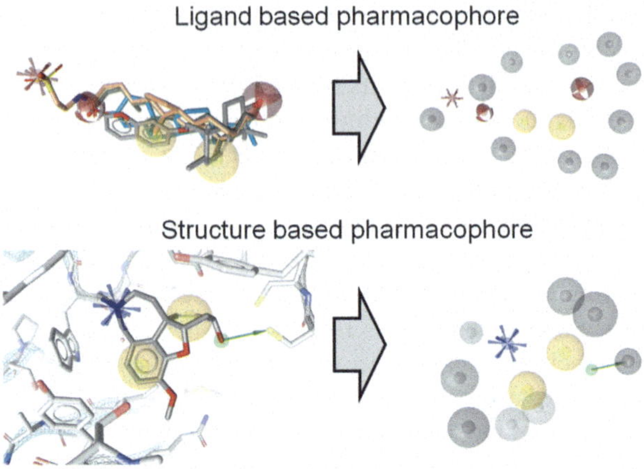

Fig. 2 Creation of a ligand-based pharmacophore for the GPBAR1 receptor (top) and a target-based pharmacophore for the ACh-binding protein (bottom). While a ligand-based pharmacophore is generated based on common patterns observed in aligned ligand sets, a structure-based pharmacophore uses the information of co-crystallized structures. The features are visualized as colored spheres (yellow, lipophilic feature; stars, ionic feature; green arrow, hydrogen bond donoring feature; red sphere, hydrogen bond accepting feature). Gray spheres model steric exclusion volumes

occurring interactions are classified into a set of a few pharmacophore features like charged group features for ionic interactions, hydrogen bond donor and acceptor features for hydrogen bonds, aromatic features for π-π and π-cation interactions, and lipophilic features for Van der Waal's interactions between receptor and ligand. Steric constraints can be simulated by negative features, so-called exclusion volumes. Other pharmacophore features including metal complexing, ring, covalent, and halogen bonding features may be implemented additionally in some tools. There are several software packages for pharmacophore modeling and VS, most importantly LigandScout, Phase, Catalyst, and MOE (Seidel et al. 2010). As there is no golden algorithm for alignment, a previously performed comparative VS experiment recognized a high variance between the hit lists obtained from Phase, Catalyst, and MOE (Spitzer et al. 2010). Query pharmacophore models can be created by two approaches. With a target-based approach, the pharmacophore is extrapolated from experimental structures of protein targets or homology models. If there is no target structure available, the ligand-based approach can still represent a reasonable method by extracting information on known ligands. Because the target-based approach is built on observed complementarities between ligand and target, they better incorporate directionality of binding-site interactions like hydrogen bonds. Furthermore, a quite rationale modeling of steric hindrances in the binding pocket can be simulated by placing exclusion volume spheres. As soon as a pharmacophore model is generated, it is of utmost importance that it is theoretically validated by prospective VS of a set of known ligands and inactive molecules. In the optimum case, the model should discard inactive molecules and find all of the active molecules of the training set. Model refinement can be achieved by adding or deleting features and changing their tolerance or weight. Nevertheless, when generation of a single restrictive model is not possible, it is preferable to perform the VS experiment with a set of highly specific local models rather than with one global model. This approach usually results in lower false-positive hit rates. A single model would only have the ability to describe the whole ligand diversity, if it is clear that all ligands bind at the same binding site with the same molecular interactions, whereby crystal water and dynamics are negligible (Schuster et al. 2010). The pharmacophore model may always suffer from a bias toward its input information or its validation set. The only way to assess its predictive power is to validate it in vitro, preferably by target or binding assays. Cellular assays can also be suitable (sometimes they are the only possible way), but possible off-target effects may impede the significance of the experimental results and in turn their validity for model validation.

3.4 Ligand-Based Pharmacophores

As mentioned before, proficient pharmacophore models can be created without information on the target structure. More than half of all small molecule drugs act on G-protein-coupled receptors or ion channels (Santos et al. 2016), where protein structures are rare (Hauser et al. 2017), although the number is growing (Pándy-Szekeres et al. 2018). Only around one third of all projects in pharmaceutical research can rely on X-ray structural target information and another third on homology structural target information (Scior et al. 2012). It is therefore an approach with great potential and a wide range of success stories (Vuorinen et al. 2014; Acharya et al. 2011; Ha et al. 2015; Evers et al. 2005; Kratz et al. 2014; Kirchweger et al. 2018). The principle of this attempt is the alignment of different conformations of active molecules in order to visualize common electrochemical features, which might be necessary for target inhibition/activation assuming that suggested groups within the molecules trigger their biological action. The most crucial step of the method is therefore the selection of the training set. These molecules should be potent, small, rigid and most importantly should all have the same binding position. Without an experimental target structure, such information is normally absent. Therefore, it is a good way to cluster molecule sets based on their superficial pharmacophores. LigandScout's (http://www.inteligand.com) implemented pharmacophore radial distribution function (RDF) code similarity clustering tool is an unsupervised, fast, and efficient tool to do so (Goldmann et al. 2015). Still, also an operator can quickly get an idea of the pharmacophore feature patterns in a dataset and investigate it by conformational sampling and alignment. From different alignments of the molecule set (molecular superpositioning), geometrically overlapping features can then be extrapolated as a pharmacophore model. This is achieved by an operator or automatically by different software solutions. The conformer generation (the calculation of possible binding conformations from 2D represented molecules) of the training set is a further crucible step. Just thinking of a supposed molecule with three rotatable bonds: If every rotatable bond is sampled in 10° intervals, it would give rise to 46.656 conformations. Some software tools are able to reliably suggest conformations, e.g., as experimentally observed in co-crystals, and are therefore recommended for this purpose (Ebejer et al. 2012; Friedrich et al. 2017). When the pharmacophore model (set) performs well in the theoretical validation, it is used as a query for VS. A recent study by Karaboga and coworkers highlighted the predictive power of pharmacophore-based VS by comparing different VS approaches (Karaboga et al. 2013). The biggest advantages can be summarized as:

1. Their capacity of reducing the immense complexity of ligand-target interactions to an uncomplicated but illustrating model.
2. By just aligning pre-generated conformational libraries to the models, fast VS is achieved.
3. The simplicity of the models allows the identification of ligands with deviating structures from the training molecules. So-called scaffold hopping can be achieved.

Next to the establishment of the models as VS queries, pharmacophores can be used as descriptors for QSAR studies, as similarity metrics for machine learning approaches (Schneider and Schneider 2017) or to pre-filter large compound libraries for molecular docking.

3.5 Molecular Dynamic Simulations in Structure-Based Virtual Screening

The main problem of the structure-based VS approach is the assumption of a rigid lock-and-key binding theory mainly derived from X-ray co-crystal data, e.g., from the Protein Data Bank (PDB) (Berman et al. 2000, 2003). Although reflecting an experimental finding, it completely ignores flexibility and dynamics of proteins and protein-ligand complexes. A co-crystallized X-ray structure is not more than a snapshot of a ligand-protein system, taken at unphysiological conditions, and it therefore does not depict reality. To some extent, this may be one reason why PDB structure-based pharmacophore approaches are likely to fail in four of ten cases (Wieder et al. 2017). The new paradigm in our understanding of ligand-target interaction is the dynamic induced fit of a ligand to a target. It is aimed to implement this into structure-based VS tools and to introduce such dynamics in structure-based pharmacophore models (Sperandio et al. 2010; Bock et al. 2016; Sohn et al. 2013; Spyrakis et al. 2015) and molecular docking (Makeneni et al. 2018; Campbell et al. 2014; Sabbadin et al. 2014; Liu and Kokubo 2017). Molecular dynamic (MD) simulations are hereby a valuable method to link protein structure and dynamics and help to analyze conformational changes and allosteric modulations. Basically MD is a physics-based method for studying the interaction and motion of atoms according to Newton's law of motion. This allows the atoms and molecules to interact for a fixed period of time, giving a view on the dynamic evolution of the system. The simulation of this system of interacting particles is defined by numerically computed Newton's equations of motions. Involved forces and potential energies are calculated by force fields. In theory one could simply simulate the association of a ligand with a protein (Adcock and McCammon 2006). Indeed there are attempts to do so, but usually computational constraints only allow simulations on time scales in the low nanosecond scale, with specialized hardware even milliseconds (Feig and Sugita 2013). However, making a molecular dynamics simulation with a co-crystallized complex can show side-chain rotations and frequently occurring hydrogen bonds or visualize the stability of the complex. How the information obtained from those trajectories can be implemented into models and what else can be learned are a topic of current interest well outlined in a recent review by De Vivo and coworkers (De Vivo et al. 2016). MD can be used to score free energies of protein-ligand complexes (Gumbart et al. 2013; Rastelli et al. 2009), study the role of bridging water molecules in ligand binding (Sabbadin et al. 2014), visualize ligand unbinding to estimate mean residence times (Mollica et al. 2015), or discover new hidden binding pockets for allosteric activation (Bowman et al. 2015). Further improvements in VS experiments can be achieved, e.g., by ensemble docking. Thereby ligands are docked into several simulated but rigid protein conformations, and the results from single screens are then merged. This approach is especially helpful for the identification of ligands of proteins, which function is strongly connected to structural flexibility (Pang and Kozikowski 1994; Tarcsay et al. 2013; Tian et al. 2014). Pharmacophores can be created using snapshots of the molecular dynamic trajectory or by extracting feature densities (dynophores) (Mortier et al. 2017; Bock et al. 2016). Another unbiased method is the common hits approach, where the whole dataset is screened against a large number of single molecular dynamic derived pharmacophores. The virtual hits are then ranked based on their hit count without user intersection (Wieder et al. 2017).

3.6 *Shape-Based Screening*

A VS technique based purely on ligand information is shape-based screening. The 3D Gaussian shape of a set of molecules is calculated and compared to the shape of a known active query molecule and ranked according to the 3D shape similarity (Fig. 3). This can be achieved by finding and quantifying the maximal

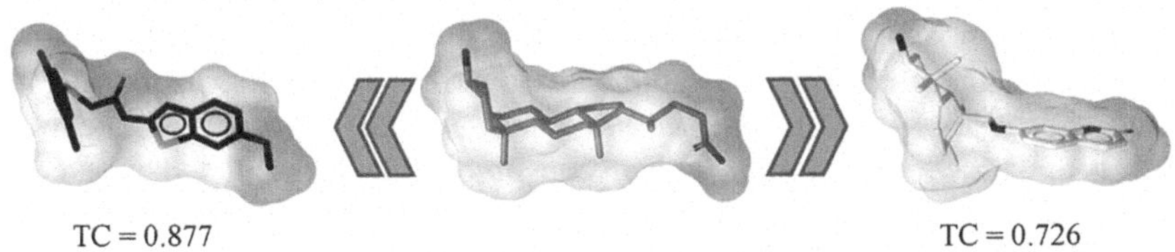

TC = 0.877 TC = 0.726

Fig. 3 Shape focused comparison of a bile acid (middle) with a synthetic ligand (left) and the polycyclic sesquiterpene coumarin microlobiden (right). As shown by relatively high TC scores, they share similar shape and electrostatic features overlap of the query molecule's volume with that of the screened molecules (Rush et al. 2005; Hawkins et al. 2007). The similarity of two shapes is usually quantified with Tanimoto or Tversky scores. ROCS, one of the most powerful shape-matching tools (Shin et al. 2015), offers combinations of shape and pharmacophore features for similarity assessment. Molecular shape is a fundamentally important feature of small molecules, which can not only predict activities on certain targets but also absorption, distribution, metabolism, and excretion (Kortagere et al. 2009). Shape comparison solely has proven to be a useful method for certain projects (Grienke et al. 2014). The advantages are the ease of use, the traceability of results, the likelihood of scaffold hopping (although pharmacophores are hereby the gold standard (Hessler and Baringhaus 2010)), the independence of available target structures, and the rapid screening speed. These advantages reveal it as a well-established method for virtual high-throughput screening. However, shape-based screening has several pitfalls, which should not be neglected: Typically, the bioactive conformation is not known without a co-crystallized target complex. We also know from X-ray structures of different molecules in the active site of the same target that the shape overlap between them is not always as high as assumed by shape-based VS tools. Similar to ligand-based pharmacophores, different ligands may bind to different regions in the same protein or in the same binding site. This causes quite high uncertainty, and VS entirely based on shape comparison does not perform well, since physicochemical and pharmacophore properties are neglected. However the method can be combined with other screening methods, e.g., using the shape-matching score for rescoring hit lists obtained from molecular docking or pharmacophore-based VS. Moreover 3D shape descriptors as similarity metrics are powerful descriptors for QSAR, QSPR, and machine learning models not only for predicting biological activity but also metabolism, e.g., drug metabolism (Kirchmair et al. 2015).

3.7 Virtual Parallel Screening

Due to the fast alignment-based procedure of pharmacophore screening, individual (low-energetic) conformers of a molecule but also libraries of molecule conformers can be matched with a set of pharmacophores, which represent different targets or protein isoforms (Steindl et al. 2006a, b) Thus, this approach can be used for:

1. **Virtual target fishing**: Conformers of isolated NP are screened against a set of validated pharmacophore models representing drug targets in order to determine those promising for testing. It is applied to identify bioactivities for recently isolated and novel NP (with resolved structure). Targets for observed phenotypic effects of extracts with multiple known constituents can be predicted by screening against a set of target models potentially influential to the assay outcomes. This may guide the targeted isolation and testing of specific compounds on predicted drug targets (Rollinger 2009; Rollinger et al. 2009).
2. **Bringing traditionally used herbal remedies on a molecular basis**: Herbal remedies usually contain hundreds of constituents. The challenge consists in

identifying those compounds mainly contributing to a beneficial effect of the used remedy. On the other hand, the aim is to identify the involved targets and underlying mechanisms of action. Virtual parallel screening may be tremendously helpful to explore yet undiscovered biological actions in this framework (Grienke et al. 2015).

3. **Prediction of side effects:** Potential side effects like cardiotoxicity, unfavorable cytochrome metabolism, nausea, and psychotic symptoms or anticholinergic effects of herbal remedies, dietary supplements, or single molecules can be predicted and assessed in a rationale manner (Klabunde and Evers 2005; Adhami et al. 2012; Kratz et al. 2014; Kratz et al. 2017; Hochleitner et al. 2017) .
4. **Polypharmacological profiling**: A specific modulation profile is crucial for the safety and therapeutic usefulness of drug groups like kinase inhibitors or dopamine receptor modulators. Therefore a broad set of selective inhibition/activation models for the related targets is applied for virtual parallel screening. In this way the obtained predictions contribute to a fast and rationale selection of molecules for testing (Malo et al. 2010).

Independent of the approach, a few prerequisites should however be considered:

1. Usage of published, experimentally validated models is recommended since most pharmacophore models are only validated theoretically on molecule sets lacking diversity. Their performance on novel molecules is therefore doubtful.
2. Depending on the intentions of virtual parallel screening, the specificity of the models should be appropriate. For an off-target, like the hERG potassium channel, models with high selectivity are favorable as a higher number of false positives are acceptable in order to not miss real actives. Vice versa for target fishing, models with high specificity are preferable.
3. It is important to have some information on the models. What was their enrichment factor in the theoretical and in the experimental validation? Are the applied models local or global? Which software was used to generate the models? Which conformer generator was used for the query molecules; is it the same as library generation? What was the query crystal structure or was it ligand based? All this information may be important for the interpretation of the results.

3D pharmacophore modeling is one of the most important virtual parallel screening approaches as it combines comprehensive and transparent results with fast screening. Virtual parallel screening can also be performed with molecular docking (Chen et al. 2003; Wang et al. 2012). This reverse docking approach is illustrated in Fig. 4. However for large-scale screening including side-chain flexibility, a high-performance computing cluster may be necessary. The fact that target identification for novel NP is a common obstacle, difficult to overcome with other methods, highlights the importance of this approach. Moreover we currently experience a paradigm shift from a one-drug-one-target to a polypharmacologic one-drug-many-target model for drug discovery. This is in particular the case for chronic diseases, such as chronic inflammation (Koeberle and Werz 2014). That NP bear great potential for the development of curing agents in these disease areas can be presumed.

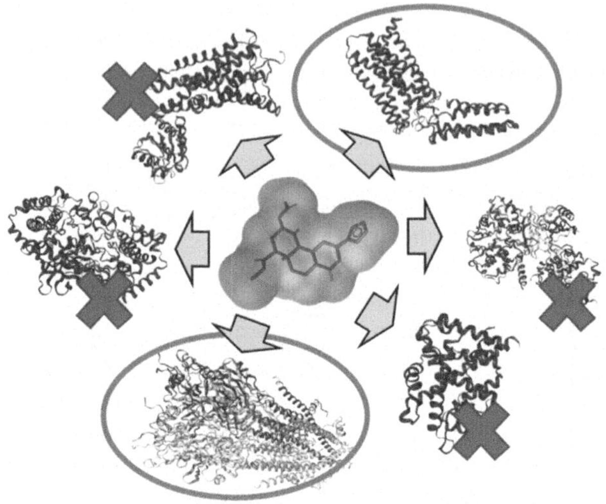

Fig. 4 Principle of virtual target fishing with a reverse docking approach. New molecular entities (middle) are docked into binding sites of different high-quality crystal structures. Those achieving high scores (highlighted in circles) give clues about potentially addressed targets to be prioritized for experimental testing

3.8 *Machine Learning*

Machine learning is a broad term for computer programs, which mine useful knowledge for drug discovery from molecular structures. It is one of the most dynamic topics in computer-aided drug discovery (Varnek and Baskin 2012). One cause is the onset of big data in cheminformatics. Different to previous methods, machine learning can be easily scaled up to big datasets. They aim to identify patterns in empirical datasets to generate mathematical relationships which can be extrapolated to predict properties of novel compounds. One important application is QSAR and QSPR. Artificial intelligence is hereby used to predict how chemical modifications might influence activity or biological properties like toxicity, carcinogenesis, or metabolism. QSAR is based on the pioneering works of Hansch and Free-Wilson, who used multivariate regression models to create mathematical formulas which correlate activity to molecular properties like lipophilicity (Hansch et al. 1962; Free and Wilson 1964). There exists a broad range of methods: multiple regression analysis, support vector machines, principal component analysis, hierarchical cluster analysis, decision trees, random forest, and k-nearest neighbor besides artificial neural networks just to name few of them. They can be classified into supervised and unsupervised learning. The workflow can be broken down to the extraction of molecular features (e.g., topological descriptors, physicochemical properties, pharmacophores) from a training set, creation of molecular fingerprints from those features, similarity comparison for shared/different features (supervised or unsupervised), generation of models based on observations in the training set, and finally validation of the obtained models with test sets (Lavecchia 2015). Resulting models can be powerful tools: Reker and coworkers (Reker et al. 2014) as well as Schneider and Schneider (2017) used the neural network technique of self-organizing maps to make target predictions for new molecular entities with good results. With the advent of big data and deep learning, there is also reemerging interest in neural network algorithms for solving chemoinformatic tasks. Although shallow neural network approaches are used in drug research for a long time, newer advances with stunning performance have prompted many to consider it as a game-changing technology for VS (Gawehn et al. 2016; Pereira et al. 2016; Schneider 2017; Balaban 1997).

4 Limitations and Caveats in VS

Although there are abundant reports on successful VS application examples, also for the discovery of novel ligands from nature, awareness should be prevalent that virtual hits are just predictions and no method regardless of complexity and rationality is bullet proof. Moreover, there are certain caveats and limitations pervasive. Mistakes can happen to researchers, originally not coming from a computational chemistry-related field, which are sometimes easily circumvented as far as awareness is given. A review of Scior and coworkers addresses these "pitfalls" and is highly recommended for further reading on this topic (Scior et al. 2012). This chapter will give a brief overview on the ten most important caveats:

1. An in silico experiment strongly depends on its input information. The quality should therefore be controlled and prepared in detail. Macromolecular and ligand structures from the PDB entries (Berman et al. 2000, 2003) are not error-free. They can be partially incomplete with missing atoms and residues (Brandt et al. 2008), binding pockets and ligands may not fit to their experimentally determined electron density, asparagine and glutamine rotamers can be incorrectly assigned, and irregular protonation states of ionizable residues and ligands can occur. Open-source databases like PubChem (Kim et al. 2016) and ChEMBL (Gaulton et al. 2012; Bento et al. 2014) offer a large quantity of information, but it should be corrected with the original literature for correct conformations, annotated bioactivities, and other errors.
2. Generally, there is only limited information available on inactive molecules. Although lots of inhibitors/activators have been reported, only few inactives are published for certain targets. In such cases, it may be possible to generate decoys for model validation. In contrast to inactives, which are experimentally confirmed to be inactive, decoys are drug-like molecules sharing similarity with the active molecules but have never been tested. The DUDe web service (Mysinger et al. 2012) appoints decoys with similar physicochemical but deviant topological properties for active molecules. The topological differences

should minimize the possibility of retrieving real active molecules as decoys. Similar physicochemical properties guarantee challenging decoys for the validation of molecular docking settings and pharmacophore models.

3. As described in Chap. 5, there are several molecules, which give frequently false-positive results in bioactivity measurements. If such a molecule is employed as training compound or part of the validation set, it may destroy the predictive power of the query. Applying substructure filters to the training and validation sets as well as checking original publications for inaccuracies can help prevent this pitfall.
4. The data obtained from literature can be difficult to compare, as it is generated with different assays, cell lines, and working groups or deviates in protocol settings (incubation time, readout, etc.). Highly potent activators for model validation or model generation are not always easy to identify in such noise. In such cases, it is the best approach to simply discard all molecules endowed with unreliability and only focus on compounds reported frequently as they are acknowledged positive controls, clinical candidates, or drugs. In some cases, subjective picking can be the only way to classify the molecules.
5. Assessing the performances and comparison of several screening methods or models is crucial to identify the most suitable. Comparison of VS protocols and models is difficult. There are general quality metrics used for VS like specificity, sensitivity, accuracy, enrichment of actives, ROCS curves, area under the curve, and BEDROC curves. Most of these metrics are dependent on the molecule sets for validation, especially on the ratio of active to inactive molecules (Kirchmair et al. 2009; Sheridan 2008). Truchon and Bayly gave practical recommendations on their use (Truchon and Bayly 2007).
6. Most 3D VS methods are dependent on conformer generators with the ability to identify bioactive molecule conformations. The performance of several conformer generators was recently assessed, and the conformer generators Omega, ConfGenX, and ICon were recommended. The best performing open-source tool according to the study of Friedrich et al. (2017) is RDKit.
7. There is an abundance of different structure formats, and it is necessary to read in, read out, or interconvert one to another. But because of format incompatibilities, information can get distorted. The correctness of atomic coordinates, handling of aromatic moieties, chirality, hybridization and protonation can cause problems and make manual correction necessary (Kirchmair et al. 2008).
8. Ligand-based but also structure-based modeling is always partly based on assumptions. There are multiple or allosteric binding pockets. Selected hits may be active in the experimental confirmation, but they may also bind to alternative binding pockets. It is generally rare to address the question, whether the hits identified by VS really bind in the pose predicted by docking or in the ligand binding interaction mimicked by a pharmacophore model; or it was simply a discovery by serendipity, stating that the method is not as proficient as supposed.
9. Ligand-based VS, whether shape-based or shape-focused screening and to less extent pharmacophores, assume a large overlap of ligands in the binding pocket.

However the observed overlap in X-ray crystal structures is not always as large as one would assume. Proper theoretical model validation and a set of local models instead of a global one can avoid this pitfall (Schuster et al. 2010; Scior et al. 2012).

10. New highly potent lead-like compounds are strived for. Hits identified by VS are generally less active than the query molecules. Accordingly operators (and reviewers) should therefore not expect activities of the virtual hits, which are higher than those of the query structures. The main goal should rather be to generate several novel and structural diverse hits as starting points for further research. Zhu and coworkers critically reviewed the literature and gave recommendations among others on hit identification criteria (Zhu et al. 2013).

However if the obstacles and recommendations are considered, VS can be a powerful tool. As an example, Doman and coworkers reported a random screening hit rate of 0.02% for protein tyrosine phosphatase inhibitors, while a screening of virtually predicted hits yielded a hit rate of 34.8% (Doman et al. 2002).

5 Special Considerations and Application Examples of VS of Natural Products

The straightforward applicability of many computational tools prompted NP scientists to implement them into their research. Vice versa the privileged structures and their metabolite likeness caused computational chemists to screen and work with NP libraries. Obstacles but also opportunities arise on the interface of these two scientific disciplines and should be considered.

5.1 Quality and Availability of Resources

The availability of virtual databases comprising structurally and stereochemically well-defined compounds is necessary. Next to proper commercial NP databases like the dictionary of natural products (http://dnp.chemnetbase.com), several partially smaller open-source libraries exist. However, most virtual hits from VS studies are not physically available or affordable. Only approximately 10% of all molecules in NP libraries can be obtained commercially. The resources for the computer-guided discovery of bioactive NP have been reviewed recently (Chen et al. 2017). Virtual databases of in-house compound libraries can moreover be of great value. Physically available compounds for testing can be easily drawn and compiled to small databases in usable formats.

The quality of some virtual libraries may be uncertain. Wrong structure determination or erroneous appointment of conformations can be apparent and should be considered. Proper work-up of the virtual hit list and comparison of hits with the literature can clear up reservations. When creating an own virtual library, special attention should be given to the stereochemistry. Grienke et al. created their own virtual library of 279 constituents of *Ganoderma lucidum* to predict the molecular mechanism of the antiviral and metabolic activity of this TCM drug (Grienke et al. 2015).

Computational methodologies strongly rely on the experience from studies on synthetic molecules, which were used to train algorithms. NP generally have a deviating architecture, e.g., more unsaturated bonds, higher flexibility, and oxygen content but less halogens and more fused rings. Proper theoretical validation of workflows and models prior to their application should be seen as mandatory (Rollinger and Wolber 2011). However, as discussed before, it is exactly their fascinating molecular architecture, which determines them as bioactive agents and researchers can be comforted to work with them.

5.2 Postprocessing of a NP Virtual Screening Hit List

As a result of a prospective VS, a hit list comprised of dozens to thousands of small molecules (which are according to the VS filter likely to exert a biological response) is retrieved. The prioritization and selection of hits for experimental evaluation is a vital part of the VS process, especially for hit generation from NP, where pure compounds are precious, since they are rarely available (or affordable) from commercial suppliers (Fig. 5). The pure substances may have to be isolated from natural starting material with enormous efforts, and the obtainment issue is not the only challenge faced with a natural product hit list. However, classical pharmacognostic know-how (i.e., on the traditional use of herbal preparations from which the compounds were originated) can assist the selection process and may increase the success rate of finding bioactive molecules. Several considerations and tools exist, which can assist this process. Screening hit selection of NP has to be performed with utmost care and can be a complex process. On the one hand, medicinal chemistry (scaffold diversity, prediction of assay interference) as well as pharmacognostic considerations should influence the decision. For achieving a chemical diversity, hit lists can be clustered into groups (e.g., hierarchical cluster analysis). Another approach is to prioritize the virtual hits based on QSAR models or similarity (shape comparison) to query molecules. Substructure, physicochemical, and machine learning filters can be applied to scale down the hit list and clear out drug unlike molecules and pan-assay interfering substances (PAINS). The best example for a physicochemical property filter is the Lipinski's rule of five (Lipinski et al. 2001). Next to Lipinski's rule of five, the Veber rules (Veber et al. 2002), Ghose filter (Ghose et al. 1999), as well as more specific rules like the blood-brain barrier rule (Pardridge 2005) should be taken into account depending on the target to be reached. Especially for testing in-cell targets, cell permeability should be guaranteed. Although the downsizing of a hit list in such a way is a very rational approach, it should be considered that many agent classes such as antibiotics or molecules targeting protein-protein interactions routinely fall

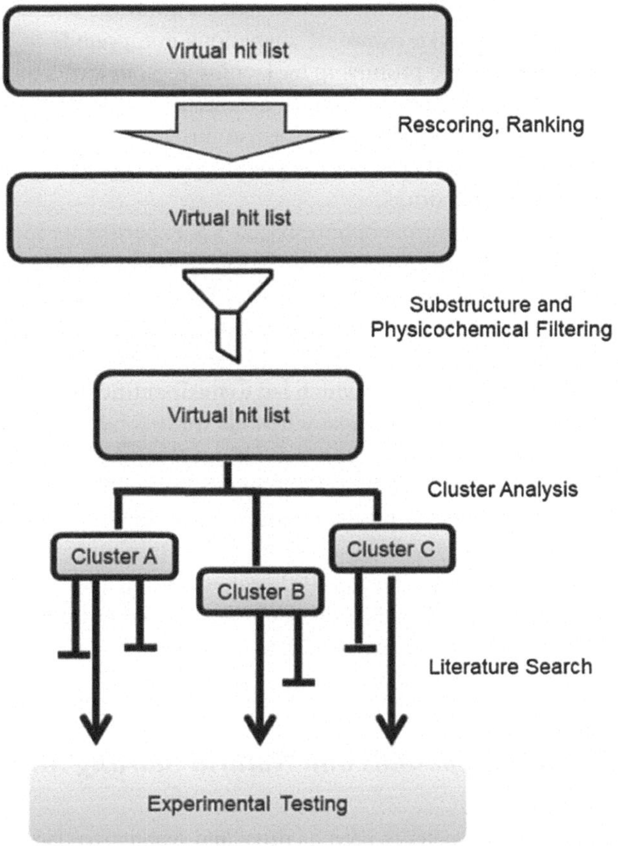

Fig. 5 Schematic overview of an exemplary postprocessing of a VS hit list. Rescoring and ranking should guarantee that virtual hits, which are more likely to exert a bioactive effect, are ranked higher. Filtering can clean the hit list from PAINS and compounds without drug-likeness. The cluster analysis then guarantees a high molecular diversity in the final hit selection. Literature search can give valuable hints for the final selection

outside the constraint of these rules. Moreover NP-derived small molecules are elaborated in aqueous solution and are perceived as drug-like all along. Therefore, it can be worth looking outside the scope of such general drug-likeness. Next to drug-likeness, NP often contain reactive groups and have high lipophilicity, autofluorescence, or other undesirable functionalities making them prone to in vitro assay interference and, in turn, false-positive results. Broadly distributed PAINS motifs in NP are catechol, hydroquinones, epoxide and peroxide bridges, phenolic Mannich bases, and others (Baell 2016). Although several filters are widespread for sorting out PAINS and are able to process thousands of compounds within seconds, such black box treatment is under criticism. Control experiments to check for, e.g., aggregation or fluorescence interferences, are then mandatory for proof of concept.

Substructure filters as, for example, contained in the Faf4drugs online tool (Lagorce et al. 2017) are useful to make aware of suspicious compounds in your hit lists. Nevertheless, if virtual hits are promising molecules, e.g., in terms of ranking, scaffold diversity, ethnobotanical use, or simply availability, a PAINS alert should not be a no-go as long as one is prepared to perform control experiments. Mitoxantrone, rifampicin, cephalosporin, and artemisinin would have never passed a PAINS filter but turned out to be valuable drugs.

Kirchweger et al. used pharmacophore-based VS experiments to identify novel ligands for the GPBAR1 (Kirchweger et al. 2018). The obtained hit list with more than 1000 compounds was re-ranked with the TC score obtained from shape-focused screening. Diversity was obtained by physicochemical clustering using principal components obtained by chemGPS. In the final selection, substructure and PAINS filters were included flexible, which led to the identification of sesquiterpene coumarins and triterpenes as high-efficacy ligands for this bile acid receptor.

Su and coworker performed a VS approach for novel inhibitors of the Rho kinase. Prior to the VS, they cleaned and focused their library with fingerprint clustering and drug-likeness filters. The VS hit list obtained by molecular docking was scaled down by re-docking and a QSAR-based scoring function, specially developed to kinase inhibitors. Only 6 from the top 100 ranked virtual hits were subjected to experimental validation, which led to the identification of phloretin and baicalein showing IC_{50}s in the nanomolar range (Su et al. 2015).

5.3 *Selection of Compounds and Natural Starting Material*

The screening of herbal extracts is seen as dirty and expensive. Isolation, fractionation, dereplication, and characterization are labor- and time-intensive. Complexity of multicomponent mixtures, aggregation, assay interference, and instability of possible constituents make it questionable if the active principles can be identified. Computational methods can bypass many of these steps in testing specific compounds in silico and guide the selection of compounds or proper natural starting material for experimental investigation. After computational prediction and post-processing of the hit list, several questions have to be answered. Reports from medicinal usage, phenotypic or in vivo experiments without molecular modes of action, are apparent for some NP and herbal preparations. Several successful projects have previously shown that extracting that knowledge and using it for decision-making enrich the outcomes (Rollinger et al. 2004, 2006b, 2008; Kratz et al. 2016; Waltenberger et al. 2016). The virtual hit itself may be obtained in sufficient purity from commercial suppliers, or the secondary metabolite has to be re-isolated from a reported natural source. In this case literature on its isolation should be available. The natural starting material should be accessible and legally available for collection/acquisition considering issues on bioprospecting, intellectual property rights, and transfer of natural material from outside (Nagoya protocol, (Matthias and Clare 2011)).

Considerations on detection, dereplication, and targeted isolation methods of the virtual hit and analogs from an extract can influence decision-making. Finally, virtual hits and congeners must be able to get isolated in an adequate time, amount, and purity.

In their attempt to assess the cardiotoxic risk by hERG channel blockage of commonly consumed NP, Kratz et al. screened a 3D multiconformational NP database comprising 130,000 molecules against a validated pharmacophore model set (Kratz et al. 2014). The majority of virtual hits have identified as constituents of 12 often used medicinal plant genera. Small-scale lead-like enhanced extracts from these plant materials were prepared and tested in a patch clamp assay. Thereof, four plant extracts showed potent hERG inhibition, among them Ipecacuanhae radix. Preparations of this antiemetic traditional medicine are easily available OTC products and underline the necessity for systematic risk assessment by antitarget identification (Kratz et al. 2016).

For comparative analysis of the performance of different bioactivity detection tools, Rollinger et al. (2005) used both, pharmacophore-based VS and a classical bioassay-guided strategy for the identification of new cyclooxygenase inhibitors. Since different Diels-Alder adducts from Sang Bai Pi, the root bark of *Morus alba*, were predicted by pharmacophore-based VS, the methanolic extract of this traditionally used herbal drug was examined. Bioassay-guided isolation then led to the isolation and identification of nine COX-inhibitors, the majority of them have been predicted in the VS approach (Rollinger et al. 2004).

For the discovery of natural FXR activators, a set of validated pharmacophore models (Schuster et al. 2011) were used as queries for the VS of a Chinese Herbal Medicine (CHM) database comprising 10,000 compounds from traditional Chinese medicine. Because several triterpenes known from the fruit body of *Ganoderma lucidum* were found in the virtual hit list, extracts of this medicinal mushroom were prepared and proved to be active in the experimental setup. Phytochemical work-up then led to the identification of five lanostane triterpenes which induced FXR activation in the low micromolar range (Grienke et al. 2011).

5.4 Find Molecular Targets for Novel Compounds

The approach-inherent incapability to find novel compounds constitutes a clear limitation of VS studies. This however can be circumvented, when prior to VS a phytochemical investigation is performed to eventually isolate probably new compounds from given natural starting material. VS is then applied to all (also new) isolated and structurally identified molecular entities for the elucidation of potentially hit targets (Rollinger 2009). NP are often multi-target compounds, and the assessment of the whole target ensemble is an almost impossible task. Although only hypothetical and limited to known and structurally defined target proteins, VS may give a clue for finding the involved molecular mechanism by in silico target prediction.

This strategy was previously exemplified for 16 constituents isolated from the aerial parts of the medicinal plant *Ruta graveolens* (Rollinger et al. 2009). The small compound collection was virtually screened against a set of 2208 pharmacophore models, which helped to identify novel inhibitors of acetylcholinesterase, the human rhinovirus coat protein, and the cannabinoid receptor type 2.

Schneider and Schneider (2017), for example, used their Target Interference Generator (TIGER) software to find protein targets for the new molecular entity marinopyrrole A. TIGER uses topological pharmacophore similarities between the subject and a set of reference compounds with 331 known targets. Four of six predicted and tested targets were experimentally validated, among them the glucocorticoid receptor with a KB of 0.7 μM.

Gong et al. (2014) used a reverse docking approach against 211 cancer-related targets to unravel the phenotypic cytotoxic effect of two novel sponge isolates. Only the ten best ranked targets were tested and led to the discovery of novel h(p300) inhibitors.

5.5 Elucidation of Natural Product Molecular Binding Mechanism

Docking and MD simulations besides other methods can accurately predict the binding mode of NP to their respective targets offering valuable support for the understanding of bioactivities on a molecular level.

For example, Fu and coworkers identified the binding mode of xanthohumol to the anti-inflammatory target myeloid differentiation protein 2 with docking experiments. They proved their in silico prediction by MD simulation of the proposed complex. The authors verified their hypothesis in vitro by surface plasmon resonance experiments and an enzyme-linked immunosorbent assay with MD-2 mutants, whose predicted key residues for hydrogen bond formation were transformed into alanine (Fu et al. 2016).

Atanasov et al. (2013) found polyacetylenes from *Notopterygium incisum* as PPARγ ligands using an ethnobotanical screening. To elucidate their binding mode, docking experiments against a magnolol-bound PPARγ structure was performed. Although not experimentally validated, the results suggested a similar binding mode and highlighted key residues in the ligand-target recognition.

DNA intercalation is a common mode of action for NP, and Mulholland and Wu (2016) performed an extensive research on the dynamics of this process. They investigated in silico the binding mechanism of telomestatin, a *Streptomyces* isolate which induces apoptosis in cancer cells, to telomeric G-quadruplex DNA. Although previous docking studies revealed a plausible binding mode, they gave no detailed information on the process of binding. One millisecond molecular binding simulations showed the formation of three stable binding poses. Next to these binding poses, the authors also showed the dynamics of DNA intercalation and observed interconversion of one to another pose.

6 Conclusions

The here presented studies can only give a limited insight into research applications, which have been published in the last years. Although NP scientists all over the world strive through thousands and thousands of extracts and their isolates, many NP remain to be discovered. In particular, the search for new sources, e.g., in the field of microbes and marine organisms, is a renewed area of interest in NP research and may provide new chemical scaffolds. These explorations are mandatory in the light of enriching our pool of NP diversity. Additionally, historical information from traditional medicine and findings from observational studies on the one hand, and the increasing knowledge we observe in structural and biological data from new chemical entities, macromolecular targets and their physiological role in humans on the other hand provide an infinite source of data. Combining information derived from all these heterogeneous sources, structuring big data, and not getting lost within it will be a future challenge in our society. VS experiments can derive a maximum benefit from this increase of life science data and thereby strengthen its way in NP drug discovery. However, awareness concerning data reliability and a critical view on and an unbiased attitude toward predicted results are indispensable prerequisites for successful projects. Considering its limits and pitfalls and exploiting its potential, VS will successfully guide future studies and thereby augment our knowledge on bioactive natural lead structures.

References

Acharya C, Coop A, Polli JE et al (2011) Recent advances in ligand-based drug design: relevance and utility of the conformationally sampled pharmacophore approach. Curr Comput Aided Drug Des 7:10–22

Adcock SA, Mccammon JA (2006) Molecular dynamics: survey of methods for simulating the activity of proteins. Chem Rev 106:1589–1615

Adhami HR, Linder T, Kaehlig H et al (2012) Catechol alkenyls from Semecarpus anacardium: acetylcholinesterase inhibition and binding mode predictions. J Ethnopharmacol 139(1):142–8

Akella LB, Decaprio D (2010) Cheminformatics approaches to analyze diversity in compound screening libraries. Curr Opin Chem Biol 14:325–330

Anderson AC (2003) The process of structure-based drug design. Chem Biol 10:787–797

Atanasov AG, Blunder M, Fakhrudin N et al (2013) Polyacetylenes from Notopterygium incisum – new selective partial agonists of peroxisome proliferator-activated receptor-gamma. PLoS One 8:e61755

Baell JB (2016) Feeling nature's PAINS: natural products, natural product drugs, and pan assay interference compounds (PAINS). J Nat Prod 79:616–628

Bajorath J (2001) Selected concepts and investigations in compound classification, molecular descriptor analysis, and virtual screening. J Chem Inf Comput Sci 41:233–245

Bajorath J (2017) Molecular similarity concepts for informatics applications. Methods Mol Biol 1526:231–245

Bajusz D, Rácz A, Héberger K (2015) Why is Tanimoto index an appropriate choice for fingerprint-based similarity calculations? J Cheminform 7:20

Balaban AT (1997) Neural networks in QSAR and drug design. In: J Devillers (ed) vol. 2 in the series: principles of QSAR and drug design. J Chem Inf Comput Sci 37:628–629

Ban F, Dalal K, Li H et al (2017) Best practices of computer-aided drug discovery: lessons learned from the development of a preclinical candidate for prostate cancer with a new mechanism of action. J Chem Inf Model 57:1018–1028

Bento AP, Gaulton A, Hersey A et al (2014) The ChEMBL bioactivity database: an update. Nucleic Acids Res 42:D1083–D1090

Berman HM, Westbrook J, Feng Z et al (2000) The protein data bank. Nucleic Acids Res 28:235–242

Berman H, Henrick K, Nakamura H (2003) Announcing the worldwide protein data bank. Nat Struct Mol Biol 10:980

Bock A, Bermudez M, Krebs F et al (2016) Ligand binding ensembles determine graded agonist efficacies at a G protein-coupled receptor. J Biol Chem 291:16375–16389

Bowman GR, Bolin ER, Hart KM et al (2015) Discovery of multiple hidden allosteric sites by combining Markov state models and experiments. Proc Natl Acad Sci U S A 112:2734–2739

Brandt BW, Heringa J, Leunissen JA (2008) SEQATOMS: a web tool for identifying missing regions in PDB in sequence context. Nucleic Acids Res 36:W255–W259

Brown FK (1998) Chapter 35 – Chemoinformatics: what is it and how does it impact drug discovery. Annu Rep Med Chem 33:375–384

Campbell AJ, Lamb ML, Joseph-Mccarthy D (2014) Ensemble-based docking using biased molecular dynamics. J Chem Inf Model 54:2127–2138

Cavasotto CN, Orry AJ (2007) Ligand docking and structure-based virtual screening in drug discovery. Curr Top Med Chem 7:1006–1014

Chavan S, Nicholls IA, Karlsson BC et al (2014) Towards global QSAR model building for acute toxicity: Munro database case study. Int J Mol Sci 15:18162–18174

Chen YC (2015) Beware of docking! Trends Pharmacol Sci 36:78–95

Chen X, Ung CY, Chen Y (2003) Can an *in silico* drug-target search method be used to probe potential mechanisms of medicinal plant ingredients? Nat Prod Rep 20:432–444

Chen Y, De Bruyn Kops C, Kirchmair J (2017) Data resources for the computer-guided discovery of bioactive natural products. J Chem Inf Model 57:2099–2111

Claude Cohen N (2007) Medicine pipeline: structure-based drug design and the discovery of aliskiren (Tekturna®): perseverance and creativity to overcome a R&D pipeline challenge. Chem Biol Drug Des 70:557–565

Cordier C, Morton D, Murrison S et al (2008) Natural products as an inspiration in the diversity-oriented synthesis of bioactive compound libraries. Nat Prod Rep 25:719–737

Corwin HA, Leo DH, Hoekman D (1995) Exploring QSAR: fundamentals and applications in chemistry and biology. Hydrophobic, electronic, and steric constants. American Chemical Society, Washington, DC

Dalby A, Nourse JG, Hounshell WD et al (1992) Description of several chemical structure file formats used by computer programs developed at molecular design limited. J Chem Inf Comput Sci 32:244–255

Danishuddin KAU (2016) Descriptors and their selection methods in QSAR analysis: paradigm for drug design. Drug Discov Today 21:1291–1302

De vivo M, Masetti M, Bottegoni G et al (2016) Role of molecular dynamics and related methods in drug discovery. J Med Chem 59:4035–4061

Doman TN, Mcgovern SL, Witherbee BJ et al (2002) Molecular docking and high-throughput screening for novel inhibitors of protein tyrosine phosphatase-1B. J Med Chem 45:2213–2221

Ebejer JP, Morris GM, Deane CM (2012) Freely available conformer generation methods: how good are they? J Chem Inf Model 52:1146–1158

El-Houri RB, Mortier J, Murgueitio MS et al (2015) Identification of PPARγ agonists from natural sources using different *in silico* approaches. Planta Med 81:488–494

Evers A, Hessler G, Matter H et al (2005) Virtual screening of biogenic amine-binding G-protein coupled receptors: comparative evaluation of protein- and ligand-based virtual screening protocols. J Med Chem 48:5448–5465

Feig M, Sugita Y (2013) Reaching new levels of realism in modeling biological macromolecules in cellular environments. J Mol Graph Model 45:144–156
Fernandez-Leiro R, Scheres SHW (2016) Unravelling the structures of biological macromolecules by cryo-EM. Nature 537:339–346
Free SM, Wilson JW (1964) A mathematical contribution to structure-activity studies. J Med Chem 7:395–399
Friedrich NO, De Bruyn Kops C, Flachsenberg F et al (2017) Benchmarking commercial conformer ensemble generators. J Chem Inf Model 57:2719–2728
Friesner RA, Murphy RB, Repasky MP et al (2006) Extra precision glide: docking and scoring incorporating a model of hydrophobic enclosure for protein-ligand complexes. J Med Chem 49:6177–6196
Fu W, Chen L, Wang Z et al (2016) Determination of the binding mode for anti-inflammatory natural product xanthohumol with myeloid differentiation protein 2. Drug Des Devel Ther 10:455–463
Gasteiger J, Engel T (2003) Chemoinformatics: a textbook. Wiley-VCH, Weinheim
Gasteiger J (2016) Chemoinformatics: achievements and challenges, a personal view. Molecules 21:151
Gaulton A, Bellis LJ, Bento AP et al (2012) ChEMBL: a large-scale bioactivity database for drug discovery. Nucleic Acids Res 40:D1100–D1107
Gawehn E, Hiss JA, Schneider G (2016) Deep learning in drug discovery. Mol Inform 35:3–14
Ghose AK, Viswanadhan VN, Wendoloski JJ (1999) A knowledge-based approach in designing combinatorial or medicinal chemistry libraries for drug discovery. 1. A qualitative and quantitative characterization of known drug databases. J Comb Chem 1:55–68
Goldmann D, Pakfeifer P, Hering S et al (2015) Novel scaffolds for modulation of TRPV1 identified with pharmacophore modeling and virtual screening. Future Med Chem 7:243–256
Gong J, Sun P, Jiang N et al (2014) New steroids with a rearranged skeleton as (h)P300 inhibitors from the sponge theonella swinhoei. Org Lett 16:2224–2227
Grienke U, Mihaly-Bison J, Schuster D et al (2011) Pharmacophore-based discovery of FXR-agonists. Part II: identification of bioactive triterpenes from *Ganoderma lucidum*. Bioorg Med Chem 19(22):6779–6791
Grienke U, Braun H, Seidel N et al (2014) Computer-guided approach to access the anti-influenza activity of licorice constituents. J Nat Prod 77:563–570
Grienke U, Kaserer T, Pfluger F et al (2015) Accessing biological actions of Ganoderma secondary metabolites by in silico profiling. Phytochemistry 114:114–124
Grosdidier A, Zoete V, Michielin O (2011) SwissDock, a protein-small molecule docking web service based on EADock DSS. Nucleic Acids Res 39:W270–W277
Gu J, Gui Y, Chen L et al (2013) Use of natural products as chemical library for drug discovery and network pharmacology. PLoS One 8:e62839
Gumbart JC, Roux B, Chipot C (2013) Standard binding free energies from computer simulations: what is the best strategy? J Chem Theory Comput 9:794–802
Ha H, Debnath B, Odde S (2015) Discovery of novel CXCR2 inhibitors using ligand-based pharmacophore models. J Chem Inf Model 55:1720–1738
Hansch C, Maloney PP, Fujita T et al (1962) Correlation of biological activity of phenoxyacetic acids with hammett substituent constants and partition coefficients. Nature 194:178–180
Harvey AL, Edrada-Ebel R, Quinn RJ (2015) The re-emergence of natural products for drug discovery in the genomics era. Nat Rev Drug Discov 14:111–129
Hauser AS, Windshügel B (2016) LEADS-PEP: a benchmark data set for assessment of peptide docking performance. J Chem Inf Model 56:188–200
Hauser AS, Attwood MM, Rask-Andersen M et al (2017) Trends in GPCR drug discovery: new agents, targets and indications. Nat Rev Drug Discov 16:829
Hawkins PC, Skillman AG, Nicholls A (2007) Comparison of shape-matching and docking as virtual screening tools. J Med Chem 50:74–82
Henrick K, Feng Z, Bluhm WF et al (2008) Remediation of the protein data bank archive. Nucleic Acids Res 36:D426–D433

Hessler G, Baringhaus KH (2010) The scaffold hopping potential of pharmacophores. Drug Discov Today Technol 7:e263–e269
Hochleitner J, Akram M, Ueberall M et al (2017) A combinatorial approach for the discovery of cytochrome P450 2D6 inhibitors from nature. Sci Rep 7:8071
Hu Y, Stumpfe D, Bajorath J (2013) Advancing the activity cliff concept. F1000Research 2:199
Huang SY, Grinter SZ, Zou X (2010) Scoring functions and their evaluation methods for protein-ligand docking: recent advances and future directions. Phys Chem Chem Phys 12:12899–12908
Jain AN, Nicholls A (2008) Recommendations for evaluation of computational methods. J Comput Aided Mol Des 22:133–139
Jayaseelan KV, Moreno P, Truszkowski A et al (2012) Natural product-likeness score revisited: an open-source, open-data implementation. BMC Bioinformatics 13:106
Jones G, Willett P, Glen RC et al (1997) Development and validation of a genetic algorithm for flexible docking. J Mol Biol 267:727–748
Karaboga AS, Planesas JM, Petronin F et al (2013) Highly specific and sensitive pharmacophore model for identifying CXCR4 antagonists. Comparison with docking and shape-matching virtual screening performance. J Chem Inf Model 53:1043–1056
Karelson M, Lobanov VS, Katritzky AR (1996) Quantum-chemical descriptors in QSAR/QSPR studies. Chem Rev 96:1027–1044
Kaserer T, Schuster D, Rollinger JM (2018) Chapter 6.3. Chemoinformatics in natural product research. In: Applied chemoinformatics: achievements and future opportunities. Wiley-VCH, Weinheim
Kim S, Thiessen PA, Bolton EE et al (2016) PubChem substance and compound databases. Nucleic Acids Res 44:D1202–D1213
Kirchmair J, Markt P, Distinto S et al (2008) Evaluation of the performance of 3D virtual screening protocols: RMSD comparisons, enrichment assessments, and decoy selection – what can we learn from earlier mistakes? J Comput Aided Mol Des 22:213–228
Kirchmair J, Distinto S, Markt P et al (2009) How to optimize shape-based virtual screening: choosing the right query and including chemical information. J Chem Inf Model 49:678–692
Kirchmair J, Goller AH, Lang D et al (2015) Predicting drug metabolism: experiment and/or computation? Nat Rev Drug Discov 14:387–404
Kirchweger B, Kratz JM, Ladurner A et al (2018) *In silico* workflow for the identification of natural products targeting GPBAR1. Front Chem 6:242
Kitchen DB, Decornez H, Furr JR (2004) Docking and scoring in virtual screening for drug discovery: methods and applications. Nat Rev Drug Discov 3:935–949
Klabunde T, Evers A (2005) GPCR antitarget modeling: pharmacophore models for biogenic amine binding GPCRs to avoid GPCR-mediated side effects. Chembiochem 6:876–889
Koeberle A, Werz O (2014) Multi-target approach for natural products in inflammation. Drug Discov Today 19:1871–1882
Kortagere S, Krasowski MD, Ekins S (2009) The importance of discerning shape in molecular pharmacology. Trends Pharmacol Sci 30:138–147
Kratz JM, Schuster D, Edtbauer M et al (2014) Experimentally validated HERG pharmacophore models as cardiotoxicity prediction tools. J Chem Inf Model 54:2887–2901
Kratz JM, Mair CE, Oettl SK et al (2016) hERG channel blocking ipecac alkaloids identified by combined *in silico – in vitro* screening. Planta Med 82(11–12):1009–1015
Kratz JM, Grienke U, Scheel O et al (2017) Natural products modulating the hERG channel: heartaches and hope. Nat Prod Rep 34:957–980
Lagorce D, Bouslama L, Becot J et al (2017) FAF-Drugs4: free ADME-tox filtering computations for chemical biology and early stages drug discovery. Bioinformatics 33:3658–3660
Langer T, Wolber G (2004) Pharmacophore definition and 3D searches. Drug Discov Today Technol 1:203–207
Larsson J, Gottfries J, Muresan S et al (2007) ChemGPS-NP: tuned for navigation in biologically relevant chemical space. J Nat Prod 70:789–794
Lavecchia A (2015) Machine-learning approaches in drug discovery: methods and applications. Drug Discov Today 20:318–331

Lipinski CA, Lombardo F, Dominy BW et al (2001) Experimental and computational approaches to estimate solubility and permeability in drug discovery and development settings. Adv Drug Deliv Rev 46:3–26
Liu K, Kokubo H (2017) Exploring the stability of ligand binding modes to proteins by molecular dynamics simulations: a cross-docking study. J Chem Inf Model 57:2514–2522
Lo YC, Rensi SE, Torng W et al (2018) Machine learning in chemoinformatics and drug discovery. Drug Discov Today S1359-6446(17):30469–30465
Ma DL, Chan DSH, Leung CH (2011) Molecular docking for virtual screening of natural product databases. Chem Sci 2:1656–1665
Macarron R (2006) Critical review of the role of HTS in drug discovery. Drug Discov Today 11:277–279
Makeneni S, Thieker DF, Woods RJ (2018) Applying pose clustering and md simulations to eliminate false positives in molecular docking. J Chem Inf Model 58:605–614
Malo M, Brive L, Luthman K et al (2010) Selective pharmacophore models of dopamine D(1) and D(2) full agonists based on extended pharmacophore features. ChemMedChem 5:232–246
Matthias B, Clare H (2011) The Nagoya protocol on access to genetic resources and the fair and equitable sharing of benefits arising from their utilization to the convention on biological diversity. R.E.C.I.E.L. 20:47–61
Medina-Franco JL, Maggiora GM, Giulianotti MA et al (2007) A similarity-based data-fusion approach to the visual characterization and comparison of compound databases. Chem Biol Drug Des 70:393–412
Mollica L, Decherchi S, Zia SR et al (2015) Kinetics of protein-ligand unbinding via smoothed potential molecular dynamics simulations. Sci Rep 5:11539
Mortier J, Prévost JRC, Sydow D et al (2017) Arginase structure and inhibition: catalytic site plasticity reveals new modulation possibilities. Sci Rep 7:13616
Mulholland K, Wu C (2016) Binding of telomestatin to a telomeric g-quadruplex dna probed by all-atom molecular dynamics simulations with explicit solvent. J Chem Inf Model 56:2093–2102
Mysinger MM, Carchia M, Irwin JJ et al (2012) Directory of useful decoys, enhanced (dud-e): better ligands and decoys for better benchmarking. J Med Chem 55:6582–6594
Nikolova N, Jaworska J (2003) Approaches to measure chemical similarity – a review. QSAR Comb Sci 22:1006–1026
O'Boyle NM (2012) Towards a Universal SMILES representation – a standard method to generate canonical SMILES based on the InChI. J Cheminform 4:22–22
Oda A, Tsuchida K, Takakura T et al (2006) Comparison of consensus scoring strategies for evaluating computational models of protein-ligand complexes. J Chem Inf Model 46:380–391
Osterberg F, Morris GM, Sanner MF et al (2002) Automated docking to multiple target structures: incorporation of protein mobility and structural water heterogeneity in AutoDock. Proteins 46:34–40
Pagadala NS, Syed K, Tuszynski J (2017) Software for molecular docking: a review. Biophys Rev 9:91–102
Pándy-Szekeres G, Munk C, Tsonkov TM et al (2018) GPCRdb in 2018: adding GPCR structure models and ligands. Nucleic Acids Res 46:D440–D446
Pang YP, Kozikowski AP (1994) Prediction of the binding sites of huperzine A in acetylcholinesterase by docking studies. J Comput Aided Mol Des 8:669–681
Pardridge WM (2005) The blood-brain barrier: bottleneck in brain drug development. NeuroRx 2:3–14
Payne DJ, Gwynn MN, Holmes DJ et al (2006) Drugs for bad bugs: confronting the challenges of antibacterial discovery. Nat Rev Drug Discov 6:29
Pereira JC, Caffarena ER, Dos Santos CN (2016) boosting docking-based virtual screening with deep learning. J Chem Inf Model 56:2495–2506
Pye CR, Bertin MJ, Lokey RS et al (2017) Retrospective analysis of natural products provides insights for future discovery trends. Proc Natl Acad Sci U S A 114:5601–5606
Rastelli G, Degliesposti G, Del Rio A et al (2009) Binding estimation after refinement, a new automated procedure for the refinement and rescoring of docked ligands in virtual screening. Chem Biol Drug Des 73:283–286

Reker D, Perna AM, Rodrigues T et al (2014) Revealing the macromolecular targets of complex natural products. Nat Chem 6:1072–1078

Ren J, He Y, Chen W et al (2014) Thermodynamic and structural characterization of halogen bonding in protein-ligand interactions: a case study of PDE5 and its inhibitors. J Med Chem 57:3588–3593

Rester U (2008) From virtuality to reality – virtual screening in lead discovery and lead optimization: a medicinal chemistry perspective. Curr Opin Drug Discov Devel 11:559–568

Reymond JL, Van Deursen R, Blum LC et al (2010) Chemical space as a source for new drugs. Med Chem Comm 1:30–38

Rodrigues T, Sieglitz F, Somovilla VJ et al (2016) Unveiling (−)-Englerin A as a modulator of l-type calcium channels. Angewandte Chemie (International Ed in English) 55:11077–11081

Rollinger JM (2009) Accessing target information by virtual parallel screening – the impact on natural product research. Phytochem Lett 2:53–58

Rollinger JM, Wolber G (2011) Computational approaches for the discovery of natural lead structures. In: Bioactive compounds from natural sources, Natural products as lead compounds in drug discovery, 2nd edn. CRC Press, Boca Raton, pp 167–186

Rollinger JM, Haupt S, Stuppner H et al (2004) Combining ethnopharmacology and virtual screening for lead structure discovery: COX-inhibitors as application example. J Chem Inf Comput Sci 44:480–488

Rollinger JM, Bodensieck A, Seger A et al (2005) Discovering COX-inhibiting constituents of Morus root bark: activity-guided versus computer-aided methods. Planta Med 71:399–405

Rollinger JM, Langer T, Stuppner H (2006a) Integrated *in silico* tools for exploiting the natural products' bioactivity. Planta Med 72:671–678

Rollinger JM, Langer T, Stuppner H (2006b) Strategies for efficient lead structure discovery from natural products. Curr Med Chem 13:1491–1507

Rollinger JM, Steindl TM, Schuster D et al (2008) Structure-based virtual screening for the discovery of natural inhibitors for human rhinovirus coat protein. J Med Chem 51:842–851

Rollinger JM, Schuster D, Danzl B et al (2009) In *silico* target fishing for rationalized ligand discovery exemplified on constituents of *Ruta graveolens*. Planta Med 75:195–204

Rush TS, Grant JA, Mosyak L et al (2005) A shape-based 3-D scaffold hopping method and its application to a bacterial protein–protein interaction. J Med Chem 48:1489–1495

Sabbadin D, Ciancetta A, Moro S (2014) Bridging molecular docking to membrane molecular dynamics to investigate GPCR-Ligand recognition: the human A2A adenosine receptor as a key study. J Chem Inf Model 54:169–183

Santos R, Ursu O, Gaulton A (2016) A comprehensive map of molecular drug targets. Nat Rev Drug Discov 16:19

Sawada R, Kotera M, Yamanishi Y (2014) Benchmarking a wide range of chemical descriptors for drug-target interaction prediction using a chemogenomic approach. Mol Inform 33:719–731

Schneider G (2010) Virtual screening: an endless staircase? Nat Rev Drug Discov 9:273–276

Schneider G (2017) Automating drug discovery. Nat Rev Drug Discov 17:97

Schneider P, Schneider G (2017) A computational method for unveiling the target promiscuity of pharmacologically active compounds. Angew Chem Int Ed Engl 56:11520–11524

Schuster D, Waltenberger B, Kirchmair J et al (2010) Predicting cyclooxygenase inhibition by three-dimensional pharmacophoric profiling. Part I: model generation, validation and applicability in ethnopharmacology. Mol Inform 29:75–86

Schuster D, Markt P, Grienke U et al (2011) Pharmacophore-based discovery of FXR agonists. Part I: model development and experimental validation. Bioorganic Med Chem 19:7168–7180

Scior T, Bender A, Tresadern G et al (2012) Recognizing pitfalls in virtual screening: a critical review. J Chem Inf Model 52:867–881

Seidel T, Ibis G, Bendix F et al (2010) Strategies for 3D pharmacophore-based virtual screening. Drug Discov Today Technol 7:e221–e228

Sheridan RP (2008) Alternative global goodness metrics and sensitivity analysis: heuristics to check the robustness of conclusions from studies comparing virtual screening methods. J Chem Inf Model 48:426–433

Shin WH, Zhu X, Bures MG et al (2015) Three-dimensional compound comparison methods and their application in drug discovery. Molecules 20:12841–12862
Shoichet BK, Mcgovern SL, Wei B et al (2002) Lead discovery using molecular docking. Curr Opin Chem Biol 6:439–446
Sichao W, Youyon GL, Lei X et al (2013) Recent developments in computational prediction of hERG blockage. Curr Top Med Chem 13:1317–1326
Singh N, Guha R, Giulianotti MA et al (2009) Chemoinformatic analysis of combinatorial libraries, drugs, natural products, and molecular libraries small molecule repository. J Chem Inf Model 49:1010–1024
Sliwoski G, Kothiwale S, Meiler J et al (2014) Computational methods in drug discovery. Pharmacol Rev 66:334–395
Sliwoski G, Mendenhall J, Meiler J (2016) Autocorrelation descriptor improvements for QSAR: 2DA_Sign and 3DA_Sign. JCAMD 30:209–217
Sohn YS, Park C, Lee Y et al (2013) Multi-conformation dynamic pharmacophore modeling of the peroxisome proliferator-activated receptor γ for the discovery of novel agonists. J Mol Graph Model 46:1–9
Sperandio O, Mouawad L, Pinto E et al (2010) How to choose relevant multiple receptor conformations for virtual screening: a test case of Cdk2 and normal mode analysis. Eur Biophys J 39:1365–1372
Spitzer GM, Heiss M, Mangold M et al (2010) One Concept, three implementations of 3D pharmacophore-based virtual screening: distinct coverage of chemical search space. J Chem Inf Model 50:1241–1247
Spyrakis F, Benedetti P, Decherchi S et al (2015) A Pipeline to enhance ligand virtual screening: integrating molecular dynamics and fingerprints for ligand and proteins. J Chem Inf Model 55:2256–2274
Steindl TM, Schuster D, Laggner C et al (2006a) Parallel screening: a novel concept in pharmacophore modeling and virtual screening. J Chem Inf Model 46:2146–2157
Steindl TM, Schuster D, Wolber G et al (2006b) High-throughput structure-based pharmacophore modelling as a basis for successful parallel virtual screening. JCAMD 20:703–715
Strohl WR (2000) The role of natural products in a modern drug discovery program. Drug Discov Today 5:39–41
Stumpfe D, De La Vega De Leon A, Dimova D et al (2014) Advancing the activity cliff concept, part II. F1000Res 3:75
Su H, Yan J, Xu J et al (2015) Stepwise high-throughput virtual screening of Rho kinase inhibitors from natural product library and potential therapeutics for pulmonary hypertension. Pharm Biol 53:1201–1206
Tarcsay A, Paragi G, Vass M et al (2013) The impact of molecular dynamics sampling on the performance of virtual screening against GPCRs. J Chem Inf Model 53:2990–2999
Taylor RD, Jewsbury PJ, Essex JW (2002) A review of protein-small molecule docking methods. JCAMD 16:151–166
Tetko IV (2003) The WWW as a tool to obtain molecular parameters. Mini Rev Med Chem 3:809–820
Tian S, Sun H, Pan P et al (2014) Assessing an ensemble docking-based virtual screening strategy for kinase targets by considering protein flexibility. J Chem Inf Model 54:2664–2679
Todeschini R, Consonni V (2008) Handbook of molecular descriptors. Wiley-VCH, Weinheim
Truchon JF, Bayly CI (2007) Evaluating virtual screening methods: good and bad metrics for the "early recognition" problem. J Chem Inf Model 47:488–508
Van Drie JH (2010) History of 3D pharmacophore searching: commercial, academic and open-source tools. Drug Discov Today Technol 7:e255–e262
Varnek A, Baskin I (2012) Machine learning methods for property prediction in chemoinformatics: quo vadis? J Chem Inf Model 52:1413–1437
Veber DF, Johnson SR, Cheng HY et al (2002) Molecular properties that influence the oral bioavailability of drug candidates. J Med Chem 45:2615–2623

Vuorinen A, Engeli R, Meyer A et al (2014) Ligand-based pharmacophore modeling and virtual screening for the discovery of novel 17β-Hydroxysteroid dehydrogenase 2 inhibitors. J Med Chem 57:5995–6007

Waltenberger B, Atanasov AG, Heiss EH et al (2016) Drugs from nature targeting inflammation (DNTI): a successful Austrian interdisciplinary network project. Monatsh Chem 47:479–491

Wang G, Zhu W (2016) Molecular docking for drug discovery and development: a widely used approach but far from perfect. Future Med Chem 8:1707–1710

Wang JC, Chu PY, Chen CM et al (2012) idTarget: a web server for identifying protein targets of small chemical molecules with robust scoring functions and a divide-and-conquer docking approach. Nucleic Acids Res 40:W393–W399

Wang Z, Sun H, Yao X et al (2016) Comprehensive evaluation of ten docking programs on a diverse set of protein-ligand complexes: the prediction accuracy of sampling power and scoring power. Phys Chem Chem Phys 18:12964–12975

Warren GL, Andrews CW, Capelli AM et al (2006) A critical assessment of docking programs and scoring functions. J Med Chem 49:5912–5931

Weininger D (1988) SMILES, a chemical language and information system. 1. Introduction to methodology and encoding rules. J Chem Inf Model 28:31–36

Wermuth CG, Ganellin CR, Lindberg P et al (1998) Glossary of terms used in medicinal chemistry (IUPAC recommendations 1998). Pure Appl Chem 70:1129

Wetzel S, Schuffenhauer A, Roggo S et al (2007) Cheminformatic analysis of natural products and their chemical space. Chimia 61:355–360

Wieder M, Garon A, Perricone U et al (2017) Common Hits approach: combining pharmacophore modeling and molecular dynamics simulations. J Chem Inf Model 57:365–385

Wójcikowski M, Ballester PJ, Siedlecki P (2017) Performance of machine-learning scoring functions in structure-based virtual screening. Sci Rep 7:46710

Yan SF, King FJ, He Y et al (2006) Learning from the data: mining of large high-throughput screening databases. J Chem Inf Model 46:2381–2395

Yang Y, Xu Z, Zhang Z et al (2015) Like-charge guanidinium pairing between ligand and receptor: an unusual interaction for drug discovery and design? J Phys Chem B 119:11988–11997

Zhu T, Cao S, Su PC et al (2013) Hit identification and optimization in virtual screening: practical recommendations based on a critical literature analysis. J Med Chem 56:6560–6572

Current Regulatory Environment of Herbal Medicinal Products in the European Union

Werner Knöss

Abbreviations

CTD	Common Technical Document
DCP	Decentralized procedure
EC	European Community
EDQM	European Directorate for Quality of Medicines & Health Care
EMA	European Medicines Agency
GCP	Good Clinical Practice
HMPC	Committee on Herbal Medicinal Products
MLWP	Working Party on Community Monographs and List Entries
MRP	Mutual Recognition Procedure
ORGAM DG	Organizational Matters Drafting Group
Q DG	Quality Drafting Group
TCM	Traditional Chinese medicine

1 Introduction

Medicinal plants have been used all over the world since ancient times. The earliest documents about medicinal plants and their usage in Europe were written in Greece more than 2000 years ago. The systematic knowledge was extended in the medieval age and famous textbooks were created also in central Europe. During the last centuries of the second millennium, the natural constituents of medicinal plants became more and more subject of research and education, especially in pharmacy and medicine.

Legal regulation of herbal medicinal products in Europe has been developed since the second half of the twentieth century. Legal documents have been established

W. Knöss (✉)
Federal Institute for Drugs and Medical Devices, Bonn, Germany
e-mail: werner.knoess@bfarm.de

V. Cechinel Filho (ed.), *Natural Products as Source of Molecules with Therapeutic Potential*, https://doi.org/10.1007/978-3-030-00545-0_10

to ensure quality, efficacy and safety of herbal medicinal products. In the Member States of the European Union, there were cultural differences with respect to usage of medicinal plants. Accordingly, different approaches were followed at the national level. A harmonized legal framework was enforced in the European Union to establish a harmonized assessment of herbal medicinal products and to facilitate access to the market in different Member States of the European Union. The legal framework provided in a series of European Union directives consists of basic definitions, different options to grant access to the market and common standards and requirements for herbal medicinal products. A specific scientific Committee on Herbal Medicinal Products (HMPC) was established at the European Medicines Agency (EMA) in 2004 (Regulation (EC) 726/2004). This expert committee is following defined legal tasks. The most important task is the development of so-called European Union monographs. These monographs provide harmonized standards for safety and efficacy of herbal substances and preparations thereof, thus representing a basic recommendation for decisions by the national competent authorities of the Member States of the European Union. The monographs are based on assessment of public data and non-confidential knowledge of the authorities. Following the legal requirements, the existing level of evidence could result in a well-established use monograph, which is the precondition for a marketing authorization as herbal medicinal product. If the scientific evidence is not sufficient, the European legal framework offers the option for a monograph addressing the traditional use. This may be the base for a registration as a traditional herbal medicinal product. The HMPC is an excellent model for scientific evaluation of herbal medicines at a multi-national level. The result is a common harmonized and science-based standard to ensure public health. At the same time, resources are shared and access to the market in different Member States is facilitated.

2 Globalization of Herbal Medicines

The ongoing globalization of traditional medicines has led to a broad diversity of regulatory systems in different countries and regions. Until now, there is a lack of internationally accepted definitions and standard requirements for quality, safety and efficacy. Different concepts have been established to consider the particular characteristics of traditional medicines. As a consequence, companies face great challenges when trying to gain access to different markets for herbal medicines. An international dialogue about scientific and regulatory issues is necessary to develop reasonable and adequate requirements. Such a conversation should also address topics such as translating indications into another cultural context or therapeutic environment (e.g. an additional diet or a parallel physical treatment), using material of non-herbal origin and classifying herbal products.

The European legislation was primarily designed to deal with traditional herbal medicinal products with a well-known origin in Europe. However, the existence of therapeutic systems and products from Traditional Chinese Medicine (TCM) or Ayurvedic medicine within Europe has prompted the HMPC to address issues related to non-European traditional medicines. A question and answer document was released in the spring of 2014 that explained the European regulatory framework and the options and limitations for traditional products originating from non-European regions. In addition, the HMPC has started a pilot project to create monographs for the herbal substances used in Asian traditional medicines, such as TCM and Ayurvedic medicine. With respect to quality standards for herbal substances, the European Pharmacopoeia (European Pharmacopeia 2016) has established a particular expert group to elaborate Pharmacopoeia monographs for herbal substances originating in TCM. Until spring 2018 this expert group has prepared 73 monographs for TCM herbal substances for adoption by the European Pharmacopoeia Commission.

3 Basic Legal Definitions and Access to the Market

Worldwide there is no unique definition for "herbal medicines". Regulation in Canada is using the term "natural health products"; in the USA the term "botanicals" is applied. In Asia TCM, Ayurvedic medicine or Kampo medicine provides particular classifications and definitions. In the European Union, the basic definitions are laid down in Directive 2001/83/EC (Consolidated Directive 2001). European regulatory framework provides in Article 1 definitions for herbal medicinal products, traditional herbal medicinal products, herbal substances and herbal preparations.

3.1 Herbal Medicinal Products

Any medicinal product exclusively containing as active ingredients one or more herbal substances or one or more herbal preparations or one or more such herbal substances in combination with one or more such herbal preparations.

3.2 Traditional Herbal Medicinal Products

A herbal medicinal product that fulfils the conditions laid down in Article 16a(1) of Directive 2001/83/EC (Consolidated Directive 2001). *Vitamins and minerals may be added if their action is ancillary to the herbal constituent(s).* As this is the original basic definition, no further explanation is given here, but the approach and criteria are described in more detail in part N.4.

Herbal substances (synonym "herbal drug" according to the European Pharmacopoeia (European Pharmacopeia 2016))

All mainly whole, fragmented or cut plants, plant parts, algae, fungi, lichen in an unprocessed, usually dried, form, but sometimes fresh. Certain exudates that have not been subjected to a specific treatment are also considered to be herbal substances. Herbal substances are precisely defined by the plant part used and the botanical name according to the binomial system (genus, species, variety and author).

Herbal preparations (synonym "herbal drug preparation" according to the European Pharmacopoeia (European Pharmacopeia 2016)).

Preparations obtained by subjecting herbal substances to treatments such as extraction, distillation, expression, fractionation, purification, concentration or fermentation. These include comminuted or powdered herbal substances, tinctures, extracts, essential oils, expressed juices and processed exudates.

The basic approach in the European Union is to assess quality, efficacy and safety of herbal medicinal products before they have access to the market. An application for marketing authorization or registration has to be submitted via a distinct procedure. The following procedures are legally established (Consolidated Directive 2001; 'Marketing Authorisation' n.d.):

- *Centralized procedure*: This procedure for marketing authorization is directed to the EMA and is linked to an assessment coordinated by EMA. If the marketing authorization is granted, a medicinal product can be marketed in all Member States of the European Union. This procedure is foreseen for a defined set of indications (e.g. oncological or neurological indications) or medicinal products of special importance for public health.

  Until now, this procedure has been applied only twice for herbal medicinal products.
- *Decentralized procedure* (DCP): This procedure for marketing authorization or registration is directed to a subset of Member States. A Reference Member State is taking the lead for the assessment, and the other Member States involved (Concerned Member States) are mainly checking the assessment of the Reference Member State. At the end of a successful procedure, a marketing authorization or registration is granted in the Member States participating.

  During the last years, there has been growing experience with DCP. The application of this procedure is a driving force of harmonization at product level.
- *Mutual recognition procedure* (MRP): If a medicinal product is already authorized or registrated in one Member States, a procedure may be started which is built up on the existing assessment. At the end of a successful procedure, a marketing authorization or registration is granted in the Member States participating.
- *National procedure*: An application can be directed to a single national competent authority and finally only a marketing authorization or registration for one Member State is granted.

The national procedure is still substantially used by applicants.

The application has to be made in a distinct format, the so-called Common Technical Document (CTD). The content is depending on the type of procedure as described above and on the type of application, marketing authorization or registration (Fig. 1):

Marketing authorization

- Full application – for new herbal medicinal products
- Bibliographic application – for known herbal medicinal products with well-established use
- Hybrid forms may be used

Registration

- Bibliographic application with additional data on safety if necessary – for traditional herbal medicinal products.

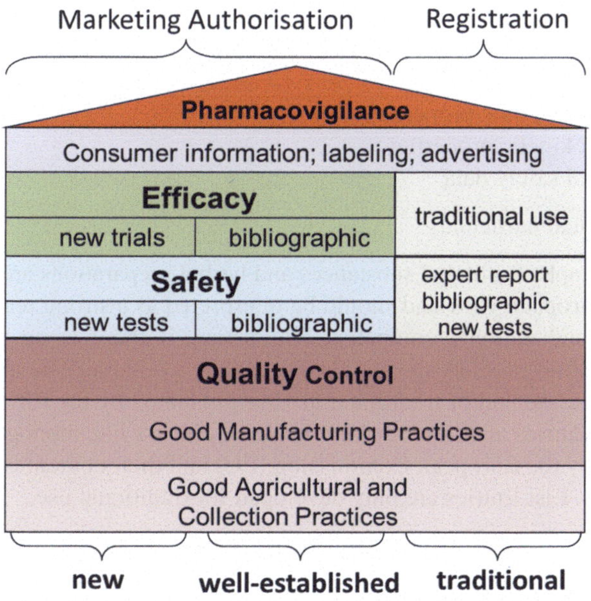

Fig. 1 Overview on requirements for marketing authorization and registration of herbal medicinal products and traditional herbal medicinal products in the European Union

4 Monographs and List Entries: Well-Established Use and Traditional Use

The European Union monographs of the HMPC are intended to facilitate marketing authorization of herbal medicinal products and registration of traditional herbal medicinal products in the Member States of the European Union. They basically follow the structure of a summary of product characteristics in order to give the national competent authorities the backbone for a product-specific assessment.

Sections of Community Monographs and List Entries

- Qualitative and quantitative composition
- Pharmaceutical form
- Clinical particulars
  - Therapeutic indications
  - Posology, method of administration
  - Contraindications
  - Special warnings and precautions for use
  - Interactions
  - Pregnancy and lactation, fertility
  - Effects on the ability to drive and use machines
  - Undesirable effects
  - Overdose
- Pharmacological properties
  - Pharmacodynamic properties
  - Pharmacokinetic properties
  - Preclinical safety data
- Pharmaceutical particulars

The monographs for herbal substances and herbal preparations are reflecting the harmonized European view and should be interpreted as a strong recommendation to applicants and national competent authorities, if there is no new scientific knowledge or if no product-specific data are made available. The monographs on well-established use and/or traditional use are published by the EMA. In contrast, so-called List Entries are developed by the same process like monographs but they are published by the European Commission (EC) and their content is binding to all Member States. List Entries are only developed for traditional use.

4.1 Well-Established Use

The concept of well-established use was implemented in the European legislation not only for herbal medicinal products but also for other medicinal products which have been already on the market when the legislation was developed. The basic idea of the concept was to facilitate applications. The applicant shall *not* be required to provide the results of toxicological and pharmacological tests or the results of clinical trials if it can be demonstrated that the constituent or constituents of the medicinal product have a well-established medicinal use with recognized efficacy and an acceptable level of safety, by means of a detailed scientific bibliography. This option should help to avoid unnecessary tests and trials. Non-clinical and clinical characteristics shall be addressed in a detailed scientific bibliography of published scientific literature which is discussed in the dossier for application by an expert.

The time of accepted medicinal use in the European Union must be at least 10 years. The HMPC agreed that it is important to consider as well quantitative aspects of the use of the active substance, the degree of scientific interest in the use of the substance and the coherence of scientific assessments and published scientific literature. Basic reflections were laid down in the guideline on the assessment of clinical safety and efficacy (EMEA/HMPC/104613/05-rev 1). In the development of monographs for well-established use, a systematic review of all clinical data is performed, taking into account the quality of the clinical trials (e.g. sufficient number of patients, GCP, etc.). According to the guideline, at least one controlled clinical study (clinical trial, post-marketing study, epidemiological study) of good quality is required to substantiate efficacy for a well-established use monograph.

4.2 *Traditional Use*

When experiences with evaluation of well-established use of herbal medicinal products were reflected in Europe, it became obvious that the requirements could probably not be fulfilled for many herbal medicines with a tradition in the European market. Accordingly, a new legislation was established – Directive 2004/24/EC (Directive 2004/24/EC of the European Parliament 2004) which was amending the overall Directive 2001/83/EC (Consolidated Directive 2001) – in order to create an option for a simplified registration for those products which had a long tradition of usage which could be accepted as a substitute for the data especially on safety and efficacy. The time to establish a tradition was set to 30 years, at least 15 of which a product should have been in medicinal use in the European Union. The second part of 15 years of medicinal use could be as well in the European Union but could also be demonstrated for any other part of the world. In Article 16 (1) of Directive 2001/83/EC (Consolidated Directive 2001) as amended, the following inclusion criteria were defined as a precondition for traditional use:

- Indication(s) appropriate to traditional herbal medicinal products
- Use without the supervision of a medical practitioner for diagnosis, prescription or monitoring of treatment
- Specified strength/posology
- Only oral use, external use and inhalation
- Sufficient data on traditional use of the product (to demonstrate safety)
- Pharmacological effects/efficacy plausible on the basis of long-standing use and experience

The objective of these inclusion criteria was to assure that only safe traditional herbal medicinal products are subject to a registration. If necessary, a national competent authority could ask for additional data on safety. The quality of traditional herbal medicinal products must meet the same criteria as any other herbal medicinal products. The legislation demands a specific labelling for traditional herbal medicinal products. The package leaflet must include a statement that the product is a traditional herbal medicinal product for use in specified indications exclusively based upon long-standing use, and the user should consult a doctor or qualified health-care practitioner if the symptoms persist during the use of the product or if adverse effects not mentioned in the package leaflet occur.

5 HMPC: Establishment and Working Structure

The HMPC was established as one of the seven scientific expert committees of the European Medicines Agency in 2004:

Table:	Scientific committees at the EMA
CHMP	Committee on Medicinal Products for Human Use
COMP	Committee on Orphan Medicinal Products
PDCO	Paediatric Committee
HMPC	**Committee on Herbal Medicinal Products**
CAT	Committee on Advanced Therapies
CVMP	Committee on Medicinal Products for Veterinary Use
PRAC	Pharmacovigilance Risk Assessment Committee

Further details were laid down in directives and regulations of the European Union (Regulation (EC) 726/2004; Consolidated Directive 2001; Directive 2004/24/EC of the European Parliament 2004). The European Medicines Agency was built to coordinate the network of the national competent authorities for medicinal products of the Member States of the European Union.

Ten years before the HMPC started, its work there had been a working party on herbal medicinal products which was elaborating first steps towards harmonized evaluation of herbal medicinal products. However, although the working party played a substantial role in initiating exchange between the Member States in the field of herbal medicines, the impact of the HMPC was much stronger because the tasks were legally defined. Moreover, the decisions and scientific opinions of the HMPC were intended to create a common standard, which due to the specific procedure could be legally binding or should be considered as a strong recommendation.

The legal tasks of the HMPC are defined as follows. The HMPC shall:

- Prepare European Union herbal monographs on herbal substances or herbal preparations that may be used for full marketing authorizations of well-established herbal medicinal products or simplified registrations
- Establish a list of traditional herbal substances/preparations/combinations

- Draw up an opinion on the adequacy of the evidence of the long-standing use at the request of a Member State
- After referral of a Member State, draw up a European Union herbal monograph on traditional herbal products used less than 15 years within the community
- Be responsible for arbitration/referral procedures originating from different views among Member States on registered traditional herbal medicinal products
- Give an opinion on other medicinal products containing herbal substances for human use referred to the EMA

Each Member State of the European Union is nominating one delegate and one alternate member of the HMPC. The competence is complemented by up to five so-called coopted members who are elected by the HMPC and who are representing specific fields of expertise. Currently, the following subjects are covered by the coopted members: paediatrics, toxicology, pharmacology, clinical pharmacology and general medicine. Decisions of the HMPC should strive for a consensus but can be also based on a majority. Delegates and coopted members are allowed to vote (i.e. currently 28 delegates and 5 coopted members, April 2018). The regular plenary meetings of the HMPC are also including delegates from Norway and Iceland, which are members of the European Economic Association and observers, e. g. from Switzerland. A secretariat is established at EMA, which is supporting the work administratively as well as scientifically. In the same way, EMA is providing legal or regulatory advice to the HMPC, if necessary, and ensures an adequate coordination with other committees established at EMA. Plenary meetings of HMPC are usually held six times a year every other month. The meetings have been scheduled so far for 1 or 1.5 day.

There are three subgroups which have been established to support the work of the HMPC:

- Organizational Matters Drafting Group (ORGAM DG)
- Quality Drafting Group (Q DG)
- Working Party on Monographs and List Entries (MLWP)

The ORGAM DG is composed of few experts who are elected by the HMPC. The task of this drafting group is to develop suitable procedures to organize the work, to provide appropriate templates to facilitate the work and to give advice to the HMPC and its subgroups whenever questions address procedural or organizational topics. Meetings are usually held four times per year. Whereas initially all meetings were face-to-face meetings, the work is meanwhile performed during virtual meetings. The Q DG is bringing together the knowledge of ten experts in the field of quality of herbal medicines. For all issues identified by the HMPC related to the quality of herbal medicines, the Q DG is preparing draft decisions or comments for further consideration and discussion at the HMPC. Meetings are as well organized routinely four times per year; the Q DG meetings are performed either as face-to-face meetings or as virtual meetings. The third subgroup, MLWP, has been established as a permanent subgroup. The contribution is of high importance for the core task of the HMPC because the MLWP is drafting monographs and related documents for final discussion and decision by the HMPC. Without any doubt a major part of scientific discussion and evaluation of herbal substances and preparations derived thereof is taking place at the MLWP.

6 Procedure to Establish Monographs

HMPC established a priority list of herbal substances, for which a monograph and/or a list entry should be established. Important parameters for prioritization were interests from the Member States of the European Union, suggestions from interested parties and also inclusion in other sets of monographs. The process of development of a monograph is started by approval of a rapporteur which is suggested by MLWP to HMPC. At the same time, a peer-reviewer is nominated who is responsible to cross-check the documents at specific steps of the process in order to achieve an appropriate quality and consistency. EMA is publishing a general call for scientific data, which are, e. g. provided by interested parties, and the rapporteur is requesting data on existing medicinal products from all Member States. Moreover, the rapporteur is carefully collecting all scientific data which are available in the public domain. Subsequently, the rapporteur elaborates a draft assessment report and a draft reference list, from which a monograph is derived. The draft documents are discussed at the MLWP; the documents are improved and rediscussed until the MLWP decides to forward the draft monograph and the accompanying documents to the HMPC. Before discussion at the HMPC, the peer-reviewer is controlling quality of the documents and consistency with decisions made so far. If the HMPC agrees with the draft suggested by MLWP, it is adopted for public consultation. Monograph, assessment report and reference list are published at the website of EMA for a period of 3 months to enable interested parties and the public in general to submit comments. At the end of the consultation period, the rapporteur compiles an overview of comments, and based on this document, MLWP discusses whether a modification of the monograph is justified. When the whole package of monograph, assessment report, list of references and overview of comments is finalized by MLWP, the documents are peer-reviewed again and then forwarded to HMPC for final adoption. After final adoption of a monograph, all the documents for a herbal substance are published at the website of EMA (www.ema.europa.eu) (European Medicines Agency n.d.) and are available to the public by just three mouse clicks.

When assessing the data for development of a monograph, it is checked in parallel whether a list entry could be drafted. However, in praxis there are always some data missing (e.g. data on genotoxicity). Therefore, only a limited number of list entries have been released so far. In case that the HMPC is adopting a list entry, the final decision and publication are up to the European Commission.

The HMPC also experienced that it may not be possible to develop a monograph. A simple reason may be overall lack of sufficient data, and routinely the projects are put on hold in order to avoid investment of additional resources. In some cases there may be legal reasons, which do not allow establishment of a monograph, or there may be some concerns on the safety of a herbal substance. In this situation the process is finalized by releasing a public statement which is explaining the reasons that hindered establishment of a monograph. Until April 2018 the HMPC released 154 monographs, 12 list entries and 21 public statements on herbal substances. The majority of monographs resulted in an assignment of traditional use, 13 monographs concluded well-established and traditional use for different herbal preparations of a herbal substance and 13 monographs concluded for well-established use only. An overview about the results of the projects finished is given in Tables 1, 2 and 3.

After the HMPC initiated the process in the first years after its establishment, the committee could meanwhile finish about 20 monographs per year. About 90% of herbal substances of current economic value in the European Union have been assessed. An important step to guarantee sustainability of the system was the start of a revision process. Each monograph will be regularly updated and modified according to the needs of current scientific knowledge every 5 years. Following this approach the set of European Union monographs will be a valuable and official standard in the Member States of the European Union for the next decades.

In comparison with other sets of monographs, e. g. by ESCOP (ESCOP Monographs 2009) and WHO (WHO Monographs 2009), the process is more advanced with respect to transparency and public availability. At the end of the twentieth century in Germany and other countries, the monographs of the Commission E (Blumenthal et al. 1998) were regarded as a gold standard, but this historic set is outdated because the monographs had not been updated to current knowledge.

7 Guidance on Quality, Efficacy and Safety: Coordination

Traditional and herbal medicinal products are defined by the manufacturing procedure and a set of specifications. The reproducible quality is a precondition to assure safe and effective therapeutic use of these products. The quality must be demonstrated at all steps of the manufacturing process:

- Harvest or collection of the plant material
- Herbal substance
- Herbal preparation
- Finished herbal medicinal product or finished traditional herbal medicinal product

Whereas the European Pharmacopoeia provides standards on methodology in general monographs and basic quality requirements for herbal substances and selected herbal preparations in specific monographs, the quality guidance of the HMPC is addressing quality issues which have to be regarded when providing a dossier for application. The HMPC installed a Quality Drafting Group which is supporting the HMPC in establishing harmonized positions and guidance with respect to requirements and assessments linked to applications for marketing authorizations or registrations. For example, guidelines are addressing quality requirements, specifications,

Table 1 Herbal substances evaluated by the HMPC resulting in a monograph or a list entry

	Herbal substance (Latin name)	Botanical name of plant	Common name
T	Millefolii herba	*Achillea millefolium L.*	Yarrow
T	Millefolii flos	*Achillea millefolium L.*	Yarrow flower
T	Hippocastani cortex	*Aesculus hippocastanum L.*	Horse chestnut bark
T, W	Hippocastani semen	*Aesculus hippocastanum L.*	Horse chestnut seed
T	Agrimoniae herba	*Agrimonia eupatoria L.*	Agrimony
T	Agropyri repentis rhizoma	*Agropyron repens (L.) P. Beauv.*	Couch grass rhizome
T	Allii sativi bulbus	*Allium sativum L.*	Garlic
W	Aloes folii succus siccatus	*Aloe barbadensis Mill. and Aloe (various species, mainly Aloe ferox Mill. and its hybrids)*	Aloes
T	Althaeae radix	*Althaea officinalis L.*	Marshmallow root
T	Arctii radix	*Arctium lappa L.*	Burdock root
T	Uvae ursi folium	*Arctostaphylos uva-ursi (L.) Spreng.*	Bearberry leaf
T	Arnicae flos	*Arnica montana L.*	Arnica flower
T	Absinthii herba	*Artemisia absinthium L.*	Wormwood
T	Avenae fructus	*Avena sativa L.*	Oat fruit
T	Avenae herba	*Avena sativa L.*	Oat herb
T	Betulae folium	*Betula pendula Roth / Betula pubescens Ehrh.*	Birch leaf
T, LE	Calendulae flos	*Calendula officinalis L.*	Calendula flower
T	Camelliae sinensis non fermentatum folium	*Camellia sinensis (L.) Kuntze*	Green tea leaf
T	Bursae pastoris herba	*Capsella bursa-pastoris (L.) Medikus*	Shepherd's purse
W	Capsici fructus	*Capsicum annuum L. var. minimum (Miller) Heiser*	Capsicum
T	Carvi aetheroleum	*Carum carvi L.*	Caraway oil
T	Carvi fructus	*Carum carvi L.*	Caraway fruit
W	Sennae fructus	*Senna alexandrina Mill.*	Senna pods
W	Sennae folium	*Senna alexandrina Mill.*	Senna leaf
T	Centaurii herba	*Centaurium erythraea Rafn. s.l.*	Centaury
T	Lichen islandicus	*Cetraria islandica (L.) Acharius s.l.*	Iceland moss
T	Chamomillae romanae flos	*Chamaemelum nobile (L.) All.*	Roman chamomile flower
T	Cichorii intybi radix	*Cichorium intybus L.*	Chicory root
W	Cimicifugae rhizoma	*Cimicifuga racemosa (L.) Nutt.*	Black cohosh
T	Cinnamomi cortex	*Cinnamomum verum J. S. Presl, (Cinnamomum zeylanicum Nees)*	Cinnamon
T	Cinnamomi corticis aetheroleum	*Cinnamomum verum J. S. Presl (Cinnamomum zeylanicum Nees)*	Cinnamon bark oil

(continued)

Table 1 (continued)

	Herbal substance (Latin name)	Botanical name of plant	Common name
T	Colae semen	*Cola nitida (Vent.) Schott et Endl. and its varieties and Cola acuminata (P. Beauv.) Schott et Endl.*	Cola
T,W	Combination: Valerianae radix and Lupuli flos	*Combination: Valeriana officinalis L. and Humulus lupulus L.*	Valerian root and hop strobile
T	Myrrha, gummi-resina	*Commiphora molmol Engler*	Myrrh
T	Crataegi folium cum flore	*Crataegus spp.*	Hawthorn leaf and flower
T	Cucurbitae semen	*Cucurbita pepo L.*	Pumpkin seed
T	Curcumae xanthorrhizae rhizoma	*Curcuma xanthorrhiza Roxb. (C. xanthorrhiza D. Dietrich).*	Javanese turmeric
T	Curcumae longae rhizoma	*Curcuma longa L.*	Turmeric
T	Cynarae folium	*Cynara scolymus L.*	Artichoke leaf
T, W, LE	Echinaceae purpureae herba	*Echinacea purpurea (L.) Moench*	Purple coneflower herb
T	Echinaceae angustifoliae radix	*Echinacea angustifolia DC.*	Narrow-leaved coneflower root
T	Echinaceae purpureae radix	*Echinacea purpurea (L.) Moench.*	Purple coneflower root
T	Echinaceae pallidae radix	*Echinacea pallida (Nutt.) Nutt.*	Pale coneflower root
T, LE	Eleutherococci radix	*Eleutherococcus senticosus (Rupr. et Maxim.) Maxim.*	Eleutherococcus root
T	Epilobii herba	*Epilobium angustifolium L. and/or Epilobium parviflorum Schreb.*	Willow herb
T	Equiseti herba	*Equisetum arvense L.*	Horsetail herb
T	Eschscholziae herba	*Eschscholzia californica Cham.*	California poppy
T	Eucalypti folium	*Eucalyptus globulus Labill.*	Eucalyptus leaf
T	Eucalypti aetheroleum	*Eucalyptus globulus Labill.; Eucalyptus polybractea R.T. Baker; Eucalyptus smithii R.T. Baker.*	Eucalyptus oil
T	Filipendulae ulmariae flos	*Filipendula ulmaria (L.) Maxim. (= Spiraea ulmaria L.).*	Meadowsweet flower
T	Filipendulae ulmariae herba	*Filipendula ulmaria (L.) Maxim. (= Spiraea ulmaria L.).*	Meadowsweet
T, LE	Foeniculi amari fructus	*Foeniculum vulgare Miller subsp. vulgare var. vulgare*	Bitter fennel
T, LE	Foeniculi dulcis fructus	*Foeniculum vulgare Miller subsp. vulgare var. dulce (Miller) Thellung.*	Sweet fennel
T	Foeniculi amari fructus aetheroleum	*Foeniculum vulgare Miller subsp. vulgare var. vulgare*	Bitter fennel fruit oil

(continued)

Table 1 (continued)

	Herbal substance (Latin name)	Botanical name of plant	Common name
T	Fraxini folium	*Fraxinus excelsior L. or Fraxinus angustifolia Vahl*	Ash leaf
T	Fucus vesiculosus, thallus	*Fucus vesiculosus L.*	Bladderwrack
T	Fumariae herba	*Fumaria officinalis L.*	Fumitory
T	Gentianae radix	*Gentiana lutea L.*	Gentian root
T, W	Ginkgo folium	*Ginkgo biloba L.*	Ginkgo leaf
T	Soiae oleum raffinatum	*Glycine max (L.) Merr.*	Soya bean oil, refined
T	Lecithinum ex soya	*Glycine max (L.) Merr.*	Soya bean lecithin
T	Liquiritiae radix	*Glycyrrhiza glabra L. and/or Glycyrrhiza inflata Bat. and/or Glycyrrhiza uralensis Fisch.*	Liquorice root
T	Grindeliae herba	*Grindelia robusta Nutt., Grindelia squarrosa (Pursh) Dunal, Grindelia humilis Hook. et Arn., Grindelia camporum Greene*	Gumweed herb
T	Hamamelidis cortex	*Hamamelis virginiana L.*	Hamamelis bark
T	Hamamelidis folium	*Hamamelis virginiana L.*	Hamamelis leaf
T, LE	Hamamelidis folium et cortex aut ramunculus destillatum	*Hamamelis virginiana L.*	Hamamelis leaf and bark distillate
T	Harpagophyti radix	*Harpagophytum procumbens DC.; Harpagophytum zeyheri Decne*	Devil's claw root
W	Hederae helicis folium	*Hedera helix L.*	Ivy leaf
T	Pilosellae herba cum radice	*Hieracium pilosella L.*	Mouse-ear-hawkweed
T	Lupuli flos	*Humulus lupulus L.*	Hop strobile
T, W	Hyperici herba	*Hypericum perforatum L.*	St. John's wort
T	Mate folium	*Ilex paraguariensis St. Hilaire*	Maté leaf
T	Juglandis folium	*Juglans regia L.*	Walnut leaf
T	Juniperi pseudo-fructus	*Juniperus communis L.*	Juniper berry
T	Juniperi aetheroleum	*Juniperus communis L.*	Juniper oil
T	Lavandulae aetheroleum	*Lavandula angustifolia Mill.*	Lavender oil
T	Lavandulae flos	*Lavandula angustifolia Mill.*	Lavender flower
T	Leonuri cardiacae herba	*Leonurus cardiaca L.*	Motherwort
T	Levistici radix	*Levisticum officinale Koch.*	Lovage root
T, W	Lini semen	*Linum usitatissimum L.*	Linseed
T	Marrubii herba	*Marrubium vulgare L.*	White horehound
T	Matricariae aetheroleum	*Matricaria recutita L.*	Matricaria oil
T	Matricariae flos	*Matricaria recutita L.*	Matricaria flower

(continued)

Table 1 (continued)

	Herbal substance (Latin name)	Botanical name of plant	Common name
T, LE	Melaleucae aetheroleum	*Melaleuca alternifolia (Maiden and Betch) Cheel, M. linariifolia Smith, M. dissitiflora F. Mueller and/or other species of Melaleuca*	Tea tree oil
T	Meliloti herba	*Melilotus officinalis (L.) Lam.*	Melilot
T	Melissae folium	*Melissa officinalis L.*	Melissa leaf
T,	Menthae piperitae folium	*Mentha x piperita L.*	Peppermint leaf
T, W, LE	Menthae piperitae aetheroleum	*Mentha x piperita L.*	Peppermint oil
T	Oenotherae biennis oleum	*Oenothera biennis L.; Oenothera lamarckiana L.*	Evening primrose oil
T	Oleae folium	*Olea europaea L.*	Olive leaf
T	Ononidis radix	*Ononis spinosa L.*	Restharrow root
T	Origani dictamni herba	*Origanum dictamnus L.*	Dittany of Crete herb
T	Origani majoranae herba	*Origanum majorana L.*	Majoram
T	Orthosiphonis folium	*Orthosiphon stamineus Benth.*	Java tea
T	Ginseng radix	*Panax ginseng C. A. Meyer.*	Ginseng root
T	Passiflorae herba	*Passiflora incarnata L.*	Passion flower
T	Paulliniae semen	*Paullinia cupana Kunth ex H.B.K. var sorblis (Mart.) Ducke*	Guarana seed
T	Pelargonii radix	*Pelargonium sidoides DC, Pelargonium reniforme Curt.*	Pelargonium root
T	Boldi folium	*Peumus boldus Molina*	Boldo leaf
T	Phaseoli fructus (sine semine)	*Phaseolus vulgaris L.*	Green bean pod
T	Anisi aetheroleum	*Pimpinella anisum L.*	Anise oil
T, LE	Anisi fructus	*Pimpinella anisum L.*	Aniseed
T	Pistacia lentiscus, resinum (mastix)	*Pistacia lentiscus L.*	Mastic tree resin
W	Plantaginis ovatae semen	*Plantago ovata Forssk.*	Ispaghula seed
W	Psyllii semen	*Plantago afra L.; Plantago indica L.*	Psyllium seed
W	Plantaginis ovatae seminis tegumentum	*Plantago ovata Forssk.*	Ispaghula husk
T	Plantaginis lanceolatae folium	*Plantago lanceolata L.*	Ribwort plantain
T	Polygoni avicularis herba	*Polygonum aviculare L.*	Knotgrass herb
T	Polypodii rhizoma	*Polypodium vulgare L.*	Polypody rhizome
T	Tormentillae rhizoma	*Potentilla erecta (L.) Raeusch.*	Tormentil

(continued)

Table 1 (continued)

	Herbal substance (Latin name)	Botanical name of plant	Common name
T	Primulae radix	*Primula veris L.; Primula elatior (L.) Hill*	Primula root
T	Primulae flos	*Primula veris L.; Primula elatior (L.) Hill*	Primula flower
T	Pruni africanae cortex	*Prunus africana (Hook f.) Kalkm.*	Pygeum africanum bark
T	Quercus cortex	*Quercus robur L.; Quercus petraea (Matt.) Liebl.; Quercus pubescens Willd.*	Oak bark
W	Rhamni purshianae cortex	*Rhamnus purshianus D.C.*	Cascara
W	Frangulae cortex	*Rhamnus frangula L.*	Frangula bark
W	Rhei radix	*Rheum palmatum L.; Rheum officinale Baillon*	Rhubarb
T	Rhodiolae roseae rhizoma et radix	*Rhodiola rosea L.*	Arctic root
T	Ribis nigri folium	*Ribes nigrum L.*	Blackcurrant leaf
W	Ricini oleum	*Ricinus communis L.*	Castor oil
T	Rosae flos	*Rosa centifolia L.; Rosa gallica L.; Rosa damascena Mill.*	Rose flower
T	Rosmarini folium	*Rosmarinus officinalis L.*	Rosemary leaf
T	Rosmarini aetheroleum	*Rosmarinus officinalis L.*	Rosemary oil
T	Rubi idaei folium	*Rubus idaeus L.*	Raspberry leaf
T	Rusci aculeati rhizoma	*Ruscus aculeatus L.*	Butcher's broom
T, W	Salicis cortex	*Salix [various species including S. purpurea L.; S. daphnoides Vill.; S. fragilis L.]*	Willow bark
T	Salviae officinalis folium	*Salvia officinalis L.*	Sage leaf
T	Sambuci flos	*Sambucus nigra L.*	Elder flower
T, W	Sabalis serrulatae fructus	*Serenoa repens (W. Bartram) Small*	Saw palmetto fruit
T, LE	Sideritis herba	*Sideritis scardica Griseb.; Sideritis clandestina (Bory & Chaub.) Hayek; Sideritis raeseri Boiss. & Heldr.; Sideritis syriaca L.*	Ironwort
T	Sisymbrii officinalis herba	*Sisymbrium officinale (L.) Scop.*	Hedge mustard
T	Solani dulcamarae stipites	*Solanum dulcamara L.*	Woody nightshade stem
T	Solidaginis virgaureae herba	*Solidago virgaurea L.*	European goldenrod
T	Symphyti radix	*Symphytum officinale L.*	Comfrey root
T	Caryophylii floris aetheroleum	*Syzygium aromaticum (L.) Merill et L. M. Perry*	Clove oil

(continued)

Table 1 (continued)

	Herbal substance (Latin name)	Botanical name of plant	Common name
T	Tanaceti parthenii herba	*Tanacetum parthenium (L.) Schultz Bip.*	Feverfew
T	Taraxaci radix cum herba	*Taraxacum officinale Weber ex Wigg.*	Dandelion root with herb
T	Taraxaci folium	*Taraxacum officinale Weber ex Wigg.*	Dandelion leaf
T, LE	Thymi aetheroleum	*Thymus vulgaris L.; Thymus zygis Loefl. ex L.*	Thyme oil
T	Thymi herba	*Thymus vulgaris L.; Thymus zygis L.*	Thyme
T	Tiliae flos	*Tilia cordata Miller, Tilia platyphyllos Scop., Tilia x vulgaris Heyne or their mixtures*	Lime flower
T	Trigonellae foenugraeci semen	*Trigonella foenum-graecum L.*	Fenugreek
T	Urticae radix	*Urtica dioica L.; Urtica urens L.*	Nettle root
T	Urticae folium	*Urtica dioica L.; Urtica urens L.*	Nettle leaf
T	Urticae herba	*Urtica dioica L.; Urtica urens L.*	Nettle herb
T	Myrtilli fructus siccus	*Vaccinium myrtillus L.*	Dried bilberry fruit
T	Myrtilli fructus recens	*Vaccinium myrtillus L.*	Fresh bilberry fruit
T, W, LE	Valerianae radix	*Valeriana officinalis L.*	Valerian root
T	Valerianae aetheroleum	*Valeriana officinalis L.*	Valerian essential oil
T	Verbasci flos	*Verbascum thapsus L.; V. densiflorum Bertol. (V. thapsiforme Schrad); V. phlomoides L.*	Mullein flower
T	Violae tricoloris herba cum flore	*Viola tricolor L. and/or subspecies Viola arvensis Murray (Gaud); Viola vulgaris Koch (Oborny)*	Wild pansy, heartsease
T, W	Agni casti fructus	*Vitex agnus-castus L.*	Agnus castus fruit
T, W, LE	Vitis viniferae folium	*Vitis vinifera L.*	Grapevine leaf
T, W	Zingiberis rhizoma	*Zingiber officinale Roscoe*	Ginger
	Combination: Thymi herba and Primulae radix	*Combination: Thymus vulgaris L., Thymus zygis L. and Primula veris L., Primula elatior (L.) Hill*	Thyme and primula root

T traditional use monograph, *W* well-established use monograph, *LE* list entry

stability testing and labelling. Meanwhile, the harmonization is also targeting very specific and tricky issues, for instance, by publishing guidance about the level of purification of herbal preparations or the application of marker concepts. In the future the HMPC will include a statement in the assessment report of a monograph to clarify the classification of herbal preparations (standardized, quantified or other extract). HMPC and EDQM have established a very valuable coordination of their activities in order to complement the requirements and to offer a complete set of guidance to applicants and national competent authorities.

Table 2 Herbal substances evaluated by the HMPC resulting in a public statement (PS)

	Latin name of herbal substance	Botanical name of plant	English common name of herbal substance
PS	Adhatodae vasicae folium	*Adhatoda vasica Nees*	Malabar nut leaf
PS	Allii cepae bulbus	*Allium cepa L.*	Onion
PS	Andrographidis paniculatae folium	*Andrographis paniculata Nees*	Kalmegh
PS	Angelicae sinensis radix	*Angelica sinensis (Oliv.) Diels*	Angelica sinensis root
PS	Centellae asiaticae herba	*Centella asiatica L. Urban*	Centella
PS	Chelidonii herba	*Chelidonium majus L.*	Greater celandine
PS	Citri bergamia aetheroleum	*Citrus bergamia Risso & Poiteau*	Bergamot oil
PS	Euphrasiae herba	*Euphrasia officinalis L. and Euphrasia rostkoviana Hayne*	Eyebright
PS	Balsamum peruvianum	*Myroxylon balsamum (L.) Harms var. perierae (Royle) Harms*	Peru balsam
PS	Paeoniae radix rubra	*Paeonia lactiflora Pall. or Paeonia veitchii Lynch*	Red Peony root
PS	Paeoniae radix alba	*Paeonia lactiflora Pallas*	White Peony root
PS	Picrorhizae kurroae rhizoma et radix	*Picrorhiza kurroa Royle ex. Benth.*	Katula
PS	Piperis methystici rhizoma	*Piper methysticum G. Forst.*	Kava Kava
PS	Salviae fruticosae folium	*Salvia fruticosa, Mill.*	Three-lobed sage leaf
PS	Salviae officinalis aetheroleum	*Salvia officinalis L.*	Sage oil
PS	Sambuci fructus	*Sambucus nigra L.*	Elderberry
PS	Caryophylli flos	*Syzygium aromaticum (L.) Merill et L. M. Perry*	Clove
PS	Tiliae tomentosae flos	*Tilia tomentosa Moench*	Silver lime flower
PS	Uncariae tomentosae cortex	*Uncaria tomentosa (Willd. ex Schult.) DC.*	Cat's claw
PS	Visci albi herba	*Viscum album L.*	Mistletoe
PS	Withaniae somniferae radix	*Withania somnifera (L.) Dunal*	Winter cherry root

The guidance on non-clinical efficacy and safety is addressing on one hand the evaluation criteria for establishing monographs. On the other hand, topics of specific and multidisciplinary interest are addressed. These may refer to safety concerns (e. g. recommendations for thresholds for the levels of thujone) or cover a special field like genotoxicity. Because traditional knowledge is not suitable to substitute data on cancerogenicity and genotoxicity, there was an approach to have a basic investigation of genotoxicity for traditional herbal medicinal products by means of an Ames test. If there is no concern from literature and the Ames test is

Table 3 Selected examples of documents on quality of traditional and herbal medicinal products established by the HMPC

Guidelines
Declaration of herbal substances and herbal preparations in herbal medicinal products/traditional herbal medicinal products, EMA/HMPC/CHMP/CVMP/287539/2005 rev.1
Good agricultural and collection practice for starting materials of herbal origin, EMEA/HMPC/246816/2005
Quality of combination herbal medicinal products/traditional herbal medicinal products, EMEA/HMPC/CHMP/CVMP/214869/2006
Quality of herbal medicinal products/traditional herbal medicinal products, EMA/HMPC/201116/2005 rev. 2
Specifications: test procedures and acceptance criteria for herbal substances, herbal preparations and herbal medicinal products/traditional herbal medicinal products, EMA/HMPC/162241/2005 rev. 2
Questions and answers
Questions and answers on quality of herbal medicinal products/traditional herbal medicinal products, EMA/HMPC/41500/2010 rev.5
Reflection papers
Level of purification of extracts to be considered as herbal preparations, EMA/HMPC/186645/2008
Markers used for quantitative and qualitative analysis of herbal medicinal products and traditional herbal medicinal products, EMEA/HMPC/253629/2007
Microbiological aspects of herbal medicinal products, medicinal products and traditional herbal medicinal products, EMA/HMPC/95714/2013
Quality of essential oils as active substances in herbal medicinal products/traditional herbal medicinal products, EMA/HMPC/84789/2013
Stability testing of herbal medicinal products and traditional herbal medicinal products, EMA/HMPC/3626/2009
Use of fumigants, EMEA/HMPC/125562/2006
Use of recovered/recycled solvents in the manufacture of herbal preparations for use in herbal medicinal products/traditional herbal medicinal products, EMA/HMPC/453258/2013
Concept papers
Development of a reflection paper on new analytical methods/technologies in the quality control of herbal medicinal products, EMA/HMPC/541422/2017

negative, then no further data are required. In case of concerns or a positive Ames test, the investigations have to follow a decision tree (Tables 4 and 5).

An intrinsic part of the European regulatory framework is to include all medicinal products in a pharmacovigilance system which is surveying the market after marketing authorization or registration in order to detect signals from recording adverse events. Herbal and traditional herbal medicinal products are embedded in this system, but the legal provisions are following an approach to take into account the particular characteristics of these products. For example, periodic safety update reports are only requested if there is a distinct safety concern.

Manufacturing site inspections are principally possible. Responsibilities are attributed following different concepts (centralized or federal), but basic requirements are also valid for traditional and herbal medicinal products.

Table 4 Selected examples of guidance documents on non-clinical and clinical topics related to traditional and herbal medicinal products established by HMPC

Guidelines
Assessment of genotoxicity of herbal substances/preparations, EMEA/HMPC/107079/2007
Non-clinical documentation for herbal medicinal products in applications for marketing authorization, EMEA/HMPC/32116/2005
Selection of test materials for genotoxicity testing for traditional herbal medicinal products/herbal medicinal products, EMEA/HMPC/67644/2009
Assessment of clinical safety and efficacy in the preparation of EU herbal monographs for well-established and traditional herbal medicinal products, EMA/HMPC/104613/2005–rev.1
Clinical assessment of fixed combinations of herbal substances/herbal preparations, EMEA/HMPC/166326/2005
Reflection papers
Adaptogenic concept, EMEA/HMPC/102655/2007
Necessity of initiatives to stimulate the conduct of clinical studies with herbal medicinal products in the paediatric population, EMA/HMPC/833398/2009

Table 5 Selected examples of public statements related to safety of traditional and herbal medicinal products released by the HMPC

Allergenic potency of herbal medicinal products containing soya or peanut protein, EMEA/HMPC/138139/2005
Capsicum/capsaicin containing herbal medicinal products, EMEA/HMPC/138379/2005
Chamomilla containing herbal medicinal products, EMEA/HMPC/138309/2005
Contamination of herbal medicinal products/traditional herbal medicinal products with pyrrolizidine alkaloids, EMA/HMPC/328782/2016
CPMP list of herbal drugs with serious risks, dated 1992, EMEA/HMPC/246736/2005
Environmental risk assessment of herbal medicinal products, EMA/HMPC/121934/2010
Herbal medicinal products containing Cimicifugae racemosae rhizoma – serious hepatic reactions, EMEA/269259/2006
Risks associated with the use of herbal products containing *Aristolochia* species, EMEA/HMPC/138381/2005
Use of herbal medicinal products containing asarone, EMEA/HMPC/139215/2005
Use of herbal medicinal products containing estragole, EMEA/HMPC/137212/2005
Use of herbal medicinal products containing methyleugenol, EMEA/HMPC/138363/2005
Use of herbal medicinal products containing pulegone and menthofuran, EMA/HMPC/138386/2005 *Rev. 1*
Use of herbal medicinal products containing thujone, EMA/HMPC/732886/2010 rev.1
Use of herbal medicinal products containing toxic, unsaturated pyrrolizidine alkaloids (PAs), EMA/HMPC/893108/2011

8 Usage and Acceptance of Monographs

Until April 2018 more than 1900 registrations for traditional herbal medicinal products have been granted by the Member States of the European Union since 1004. About 700 of these registrations were combination products. About 800 marketing authorizations for herbal medicinal products based on well-established use have been granted, about 200 being combination products (European Medicines Agency n.d.). Many of these applications are based on European Union monographs. In most of the Member States of the European Union, the standards are accepted and applied when decisions are made at the national level. The pharmaceutical companies are increasingly exploring the benefits of European procedures addressing a selection of Member States, DCP and MRP. If a harmonized decision is predictable, a company needs only to prepare one application file in the format of the Common Technical Document (CTD), and the national competent authorities share their resources used for assessment. More and more examples are demonstrating that the system is working appropriately.

The indications which are granted for registered traditional herbal medicinal products are rather representative for therapeutic use of medicinal plants in European phytotherapy (see listing below). Nevertheless, some indications are still under debate, e. g. the majority of Member States does not regard cardiovascular indication acceptable for traditional herbal medicinal products because of safety considerations linked to an obligate diagnosis. In Austria and Germany, this type of indications is still granted for traditional herbal medicinal products due to a long-standing tradition.

Therapeutic areas of major importance for traditional herbal medicinal products and herbal medicinal products approved (European Medicines Agency n.d.):

- Cough and cold
- Mental stress and mood disorders
- Gastrointestinal disorders
- Urinary tract and gynaecology disorders
- Sleep disorders and temporary insomnia
- Pain and inflammation
- Skin disorders and minor wounds
- Fatigue and weakness
- Mouth and throat disorders
- Venous circulatory disorders (Germany and Austria)
- Loss of appetite

The following herbal substances were most frequently approved as single active substance traditional herbal medicinal products (European Medicines Agency n.d.):

- Harpagophyti radix
- Pelargonii radix
- Valerianae radix
- Hyperici herba
- Thymi herba
- Passiflorae herba
- Ginseng radix
- Salviae officinalis folium
- Echinaceae purpureae radix
- Silybi Mariani fructus

The following herbal substances were most frequently approved as single active substance herbal medicinal products based on well-established use (European Medicines Agency n.d.):

- Hederae helicis folium
- Ginkgo folium
- Valerianae radix
- Hyperici herba
- Silybi Mariani fructus
- Cimicifugae rhizoma
- Echinaceae purpureae herba
- Vitis viniferae folium
- Glycine semen
- Pelargonii radix

9 Future Visions

The legal framework for herbal medicinal products in the European Union has been established to set up a harmonized approach to offer European citizens herbal medicinal products with appropriate quality, safety and efficacy. Pharmaceutical companies can follow defined requirements to develop successful applications. Despite the long tradition of medicinal use, data on many herbal medicinal products are still limited. The current framework in the European Union was set up to avoid disappearance of all these products due to regulation. Such products, provided they are very safe, can be marketed as traditional herbal medicinal products. However, their use is restricted, and mostly the usage is not recommended for special groups of patients like pregnant or lactating women or younger children. These patients are under special protection of European legislation. There were even incentives in legislation to improve the availability of medicinal products to the paediatric population (Regulation (EC) 1901/2006). However, in most European Union monographs, these special patient groups cannot be particularly addressed due to lack of data.

The set of European Union monographs on herbal substances and preparations derived thereof is defining a unique standard. It is an excellent model of multinational harmonization of scientific assessment among a large set of countries with different traditional backgrounds in application of herbal medicines. As the development of monographs is taking into account comments from the scientific community and interested parties from pharmaceutical industry the standards provided are quite robust. There are still some issues for which Member States of the European Union have a divergent opinion, but these are made public together with the Monographs and accompanying documents. Together with the basic quality requirements defined by the European Pharmacopeia the European Union Monographs of the HMPC provide a complete system for regulation of herbal medicinal products in the market.

As plants are not only used as herbal medicines but also for other purposes, there is still a problem of classification of products. There is an existing borderline area where products derived from plants may be marketed also as food (especially food supplements), medical devices or cosmetics. These different categories of products are covered by different legal frameworks, and until now, the final classification is not harmonized but is still left to the responsibility of the individual Member States. This situation is not satisfying, and there is a need for better definitions to minimize problems and confusion of citizens with borderline products.

Besides the approach of a harmonized European legislation, the market of herbal products has also been challenged by globalization during the past decades. European herbal medicinal products have been exported worldwide, and vice versa herbal and traditional medicines from other parts of the world were brought to the European Union. HMPC initiated a project within its work program which is exploring options and limitations of the European regulatory framework to handle herbal traditional medicines of non-European origin. Reflections about this issue and an overview of relevant parts of legislation were published in a question and answer document. Globally, there is no unique definition of herbal medicines or related products. Without any doubt there is a similar approach by regulators to assess quality, efficacy and safety, but criteria for evaluation and requirements are divergent. Obviously there is a need to discuss regulation of herbal and traditional medicines at a global level to improve exchange and availability of safe products of reasonable quality all over the world. Moreover, the scientific community should increase current knowledge with research initiatives including new methodology and technology. Better knowledge about multi-target mode of actions, synergies and interactions and availability in the human body will be a precondition for future use of traditional and herbal medicinal products.

Whereas people have used medicinal plants and traditional medicines worldwide for thousands of years, regulation and evaluation have evolved regionally. Thus, today there are different legal environments for similar herbal products. Some legislations focus on medicinal products, others refer to products classified as food or address specific traditional therapies. A unique terminology is missing – "herbal medicinal products", "natural health products" and "botanical medicines" are only a few examples. Responsible authorities have developed different perspectives towards assessment, risk-benefit-evaluation, risk management, information of patients and communication. In parallel, based on cultural habits, there are differences between countries in classification and acceptance of herbal and traditional medicinal products. In times of globalization, it should be realized that a global market for herbal and traditional medicinal products or similar products with alternative classification already exists.

Despite all differences there is a common sense approach on some key elements of regulatory systems for medicinal products. Basically, quality, efficacy and safety are to be addressed. The depth of an evaluation is defined by risk management strategies which should ensure public health. Assessment of products must be based on scientific principles. There is also a consensus that appropriate information about herbal and traditional medicinal products should be communicated to the patients.

Regulation should be made as efficient as possible considering the different roles of patients, authorities and industry.

Without any doubt regulation of herbal and traditional medicinal products has the overall objective of safeguarding public health. The assessment of these products is science-driven. Consequently and taken into account objectives and methodology, there is an option to strive for a common practice. Of course it is not realistic to start thinking about a worldwide identical legislation. Nevertheless, a stepwise improvement is realistic and may offer mid- and long-term perspectives to identify adequate practices and to save resources. The first challenge is communication and knowledge about the different regulatory systems. This will lead to a mutual understanding and might help to select topics which especially deserve further discussion. Major obstacles and weaknesses should be identified as well as already existing overlaps. In a second stage, common standards – at least partially – could be established. Accordingly, it could be checked which existing data are complying with such standards and might support mutual acceptance. Overall, the process will have better chances when adequate data for herbal and traditional medicinal products are provided.

Further progress depends on support at the political level in order to adjust existing regulatory systems. Adequate definitions of the diverse product categories containing material of botanical origin could facilitate regulation. A best practice in the field of herbal and traditional medicinal products cannot be reduced only to regulation of the medicinal products as such but must also address the therapeutic approach including the health-care professionals involved. Further elements which must be taken into account are e.g. marketing status, advertising and public information.

On a global scale, a harmonized regulatory approach applying a unique best practice and mutual acceptance is an extraordinary challenge. However, there are already regional regulatory networks between selected authorities, which have been established at different continents, e.g. Europe, Asia and South America. The regulatory network within the European Union based on a common legal framework may be the most advanced system with respect to the multilateral acceptance. A global platform for communication among regulatory agencies is provided by WHO by the International Regulatory Conference on Herbals (IRCH). Thus, the process as such has already been initiated. Science and continuous exchange will be the key elements which will determine the degree of future development. Overall, a convergence of the diverse regulatory systems might save resources and lead to an adequate availability of herbal and traditional medicinal products for patients without neglecting public health.

10 Conclusions

A common legal framework for herbal and traditional herbal medicinal products has been established in the European Union. This harmonized platform provides agreed standards and guidance, contributes to work sharing and facilitates access to the market for herbal medicinal products or traditional herbal medicinal products.

Patients have a choice for treatment with products, which are classified as medicines and which have been evaluated before access to the market. Monographs on quality (EDQM) and monographs on safety and efficacy (HMPC, EMA) have been developed for nearly all herbal substances of major importance for the European market. Projects to include herbal substances with a tradition outside of the European Union are ongoing. Future developments should take into account globalization, new methodology and communication with all interested stakeholders including the scientific community.

Conflict of Interest The views expressed in this article are the views of the authors and may not be understood or quoted as being made on behalf of or reflecting the position of the European Medicines Agency or one of its committees or working parties. There is no conflict of interest.

The data and figures provided are based on data available in April 2018.

References

Blumenthal M, Busse WR, Goldberg A, Gruenwald J et al (eds) (1998) The complete German Commission E monographs. American Botanical Council, Austin

Consolidated Directive 2001/83/EC of the European Parliament and of the Council of 6 November 2001 on the Community code relating to medicinal products for human use as amended by Directive 2002/98/EC, Directive 2003/63/EC, Directive 2004/24/EC, Directive 2004/27/EC and Directive 2008/29/EC

Directive 2004/24/EC of the European Parliament and of the Council of 31 March 2004 amending, as regards traditional herbal medicinal products, Directive 2001/83/EC on the Community code relating to medicinal products for human use

ESCOP Monographs.1st and 2nd edition and supplements (2009). European Scientific Cooperative on Phytotherapy, editor. Georg Thieme, Stuttgart

European Medicines Agency: http://www.ema.europa.eu – Regulatory – Human medicines

European Pharmacopeia 9th edition (2016). EDQM, Strasbourg

'Marketing Authorisation', the Rules governing medicinal products in the European Community, notice to applicants, Volume 2A, Chapter 1; EU Pharmaceutical legislation – EudraLex Volume 1

Regulation (EC) no 726/2004 of the European Parliament and of the council of 31 March 2004 laying down community procedures for the authorisation and supervision of medicinal products for human and veterinary use and establishing a European medicines agency

Regulation (EC) Nr. 1901/2006 of the European Parliament and of the council of 12 December 2006 on medicinal products for paediatric use

WHO monographs on selected medicinal plants. Volumes 1, 2, 3 and 4 (2009). World Health Organisation, Geneva

Development and Use of Polymeric Nanoparticles for the Encapsulation and Administration of Plant Extracts

Ashik Mosaddik, Vijaya Ravinayagam, Silviya Elaanthikkal, Hatem Fessi, Waisudin Badri, and Abdelhamid Elaissari

1 Introduction

Plants as a natural source of different molecules possess various biological activities, which are known since long times. The treatment of diverse diseases was taken place by these molecules (WHO 2015). Nowadays, plant extracts are employing for the therapy of different health problems within approximately three quarter of people in the world (Fabricant and Farnsworth 2001). Natural products cover essential oils, plant extracts, tea, salves, and so on. Principally, natural extracts, which are the mixture of chemicals having biological activities, are obtained from the medicinal plants' leaves, stems, fruits, or roots. In fact, antifungal, antioxidants, antibiotic, antiparasitic, anticancer, hypoglycemic, and antihypertensive are from the most noticeable biological activities that are presented by the plant extracts (Clark 1996; Butler and Buss 2006; Surya et al. 2014; Memvanga et al. 2015; Chakraborty et al. 2014; Njimoh et al. 2015; Patten et al. 2016). Encapsulated plant extracts are also applied in the fields of cosmetics, food technology, and phytotherapy (Armendáriz-Barragán et al. 2016). Currently, the innovation of pharmaceutical products is being evolved thanks to the advances of scientific technologies. These advanced technologies caused conventional therapies to be gradually supplemented by further flexible and well-developed dosage forms. The undertaking of traditional drug delivery-related limitations attracted

A. Mosaddik
Clinical Pharmacy Department, Institute for Research & Medical Consultations, Imam Abdulrahman Bin Faisal University, Dammam, Saudi Arabia

V. Ravinayagam · S. Elaanthikkal
Deanship of Scientific Research, Department of Nanomedicine, Institute for Research & Medical Consultations, Imam Abdulrahman Bin Faisal University, Dammam, Saudi Arabia

H. Fessi · W. Badri · A. Elaissari (✉)
University of Lyon, Lyon, France
e-mail: abdelhamid.elaissari@univ-lyon1.fr

V. Cechinel Filho (ed.), *Natural Products as Source of Molecules with Therapeutic Potential*, https://doi.org/10.1007/978-3-030-00545-0_11

a special attention. Low bioavailability, bitter taste, poor stability, and disagreeable odor of several active ingredients are from the most commonly faced challenges. However, these obstacles toward dosage form design can be dealt with through the drug encapsulation strategy. The encapsulation approach could avoid the early degradation of active molecules such as proteins and peptides.

Moreover, controlled and targeted drug delivery systems obtaining could be supported by the encapsulation technologies. Colloidal carriers get the wide applications in the field of biomedicine and biotechnology. Dendrimers, block ionomer complexes, polymer-based biodegradable nanoparticles (NPs), polymer-based micelles, liposomes (Naseer et al. 2014; Laouini et al. 2012), nanotubes, nanorods (Wang et al. 2012), and quantum rods are the different types of colloids used in medicine (Iqbal et al. 2015). The utilization of particulate carriers creates the opportunities for more progress in the fields of biotechnology and biomedicine. These colloidal carriers were employed in both in vivo and in vitro investigations. In comparison with the simple solution of active molecules, particulate carriers have their own advantages including active ingredient protection from degradation or inactivation (by enzyme or light) and reduction of toxicity. The unpleasant taste and odor accompanying some active molecules can be masked via encapsulation of drugs. Colloidal carriers could be absorbed on the membrane and target the tissues for pharmacotherapeutic action better than drug solution.

Therefore, active ingredient reproducible and prolonged release is obtained (Cintra e Silva et al. 2012; Levchenko et al. 2012; Poletto et al. 2012; Wang et al. 2012; Cenni et al. 2008; Sahoo et al. 2007; Miladi et al. 2013). In addition, as upon encapsulation of drugs, their biodistribution no longer relates to the drugs' physicochemical characteristics but to carrier's ones; therefore, the therapeutic efficacy of active ingredients is improved (Gagliardi et al. 2012; Heneweer et al. 2012; Herrero et al. 2012; Mora-Huertas et al. 2010). The usage of carriers in biomedical application is progressively being increased (Ahmad 2013; Soares 2013; Miladi et al. 2014).

However, complex composition and toxicity to the organism are the two factors that limited therapeutic usage of plant extracts. In addition, organic solvents such as methanol, ethanol, hexane, dichloromethane, ethyl acetate, etc. are commonly used for obtaining plant extracts. Therefore, these vehicle presences by which plant extracts are obtained avoid plant extracts' direct application on the organisms. Furthermore, plant extract conservation, targeted delivery to the tissues, and protection are another obstacle that should be solved for their usage in diseases treatment (Rubió et al. 2013). Polymer-based nanoparticles are one of the most recent and modern approaches of plant extract application that decreases the already mentioned restrictions. Recently, researches are progressively concentrated on the formulation design that includes polymer-based nanoparticles and plant extracts in order to associate plant extract biological activities and polymeric nanoparticle advantages. Cosmetics, medicine, and food technology are the fields in which these formulations would be potentially applied.

Nowadays, plant extract field researchers are more focused on the issues such as encapsulation method standardization to obtain nanoparticles, drug molecule encapsulation (encapsulation efficiency), formulation stability investigation, drug release kinetic of the carriers embedding plant extracts, free and encapsulated plant extract biological evaluations throughout in vitro and in vivo models, and nanoparticle physicochemical characterization (Armendáriz-Barragán et al. 2016).

2 History and Development of Herbal Medicine from Plant Extracts

Natural products, such as plants, have been the fundamental of human disease treatment. The main concept of the development of modern medicine can be traced to traditional medicine and therapies (Sharma et al. 2011; Patwardhan et al. 2004, 47). In several parts of the world like Africa, America, China, Egypt, and India, plants had been employed for medicinal use long before recorded history. Chemical analysis first became accessible in the early nineteenth century, which begins the extraction and modification of herbal extract (Patwardhan et al. 2004; Zuckerman and Bielory 2002). For a long duration of time, herbal medicines were not recognized for development as a novel formulation due to the lack of scientific proof, characterization, and processing, such as extraction, standardization, and identification of individual drug constituents in complex polyherbal systems.

However, modern phytopharmaceutical research dealt with the scientific requirements for herbal medicines as in modern medicine, which proffer a way for fabricating novel nanocarrier such as nanoparticles, nanoliposomes, matrix systems, microemulsions, solid dispersions, and SLNs. Nanomicellar system (Bisht et al. 2007) colloidal nanogels and nanotubes (Zheng and Song 2009) have been developed for curcumin or in combination with several other chemotherapeutic agents such as paclitaxel (Yadav et al. 2011).

3 Plant as a Source of Bioactive Compounds

Typically, bioactive compounds of plants are produced as secondary metabolites (Bernhoft 2010). The production of secondary metabolites in different species is mainly selected through the course of evaluation and the need of that species. Among secondary metabolites, some of these substances have an effect on biological systems which are considered as bioactive. Thus, a simple definition of bioactive compounds in plants is secondary plant metabolites eliciting pharmacological or toxicological effects in human and animals (Bernhoft 2010).

According to Croteau et al. (2000), bioactive compounds of plants are divided into three main categories: (a) terpenes and terpenoids (approximately 25,000 types), (b) alkaloids (approximately 12,000 types), and (c) phenolic compounds (approximately 8000 types). The plant extract's bioactivities are also related with compounds like fiber, vitamins, phytosterols, sulfur-containing compounds, carotenoids, and organic acid anions together with polyphenolics (Manach et al. 2005). Figure 1 presents some of the common bioactive compounds present in plant extracts.

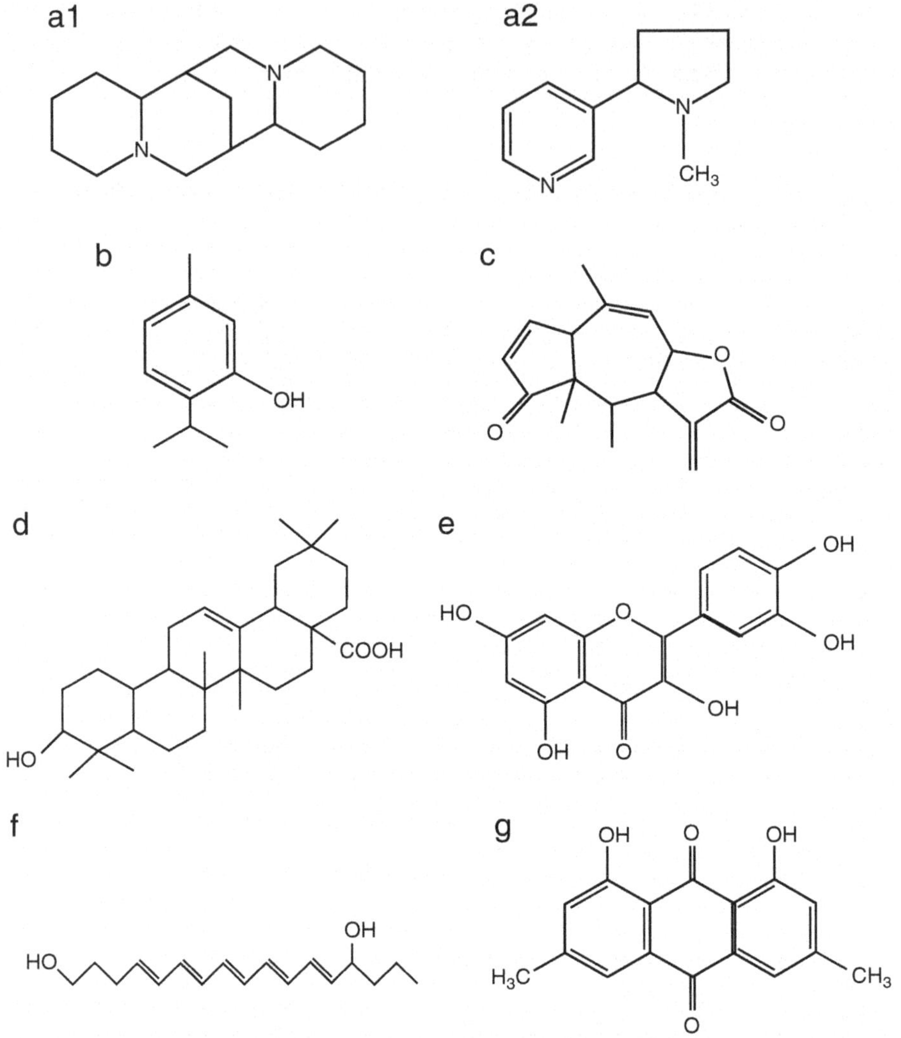

Fig. 1 General structures of different catagories of plant bioactive compounds, alkaloids (**a1, a2**), monoterpenes (**b**), sesqueterpenes (**c**), triterpenes, saponins, steroids (**d**), flavonoids (**e**), polyacetylenes (**f**), and polyketides (**g**). (Adopted from Wink 2003)

4 Different Extraction Process Employed for Nanoparticle Plant Extracts

Different techniques, many of them remaining almost the same through hundreds of years, can also be used to extract bioactive compounds. All these techniques have some common objectives:

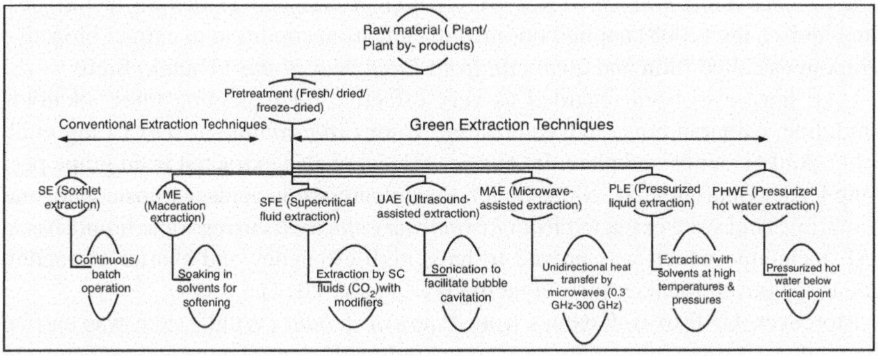

Fig. 2 Conventional and modern extraction methods for plant bioactives. (Adopted from Ameer et al. 2017)

- To extract targeted bioactive compounds from complex plant sample
- To increase selectivity of analytical methods
- To increase sensitivity of bioassay by increasing the concentration of targeted compounds
- To convert the bioactive compounds into a more suitable form for detection and separation
- To provide a strong and reproducible method that is independent of variations in the sample matrix (Smith 2003)

Some of the most promising techniques are ultrasound-assisted extraction, enzyme-assisted extraction, microwave-assisted extraction, pulsed electric field-assisted extraction, supercritical fluid extraction, and pressurized liquid extraction. These techniques are also considered as "green techniques" (Fig. 2) as they comply with standards set by the Environmental Protection Agency, USA. (http://www.epa.gov/greenchemistry/pubs/about_gc.html).

4.1 Ultrasound-Assisted Extraction (UAE)

The main benefit of UAE can be observed in solid plant sample because ultrasound energy facilitates organic and inorganic compounds leaching from plant matrix (Herrera and Luque de Castro 2005). UAE has emerged as a promising technique that fulfills the required criteria as an inexpensive green extraction technique. Rostagno et al. (2003) showed extraction efficiency of four isoflavone derivatives, namely, daidzin, genistin, glycitin, and malonyl genistin, from soybean. Herrera and Luque de Castro (2004) extracted phenolic compounds such as rutin, naringin, naringenin, quercetin, ellagic acid, and kaempferol from strawberries. Li et al. (2005) found better recovery of chlorogenic acid from fresh leaves, fresh bark, and dried bark of *Eucommia ulmoides* Oliv. by UAE than classical extraction techniques. Yang and Zhang (2008) applied optimized sonication condition to extract bioactive compounds called rutin and quercetin from *Euonymus alatus* (Thund.) Sieb.

UAE have also been regarded as very effective for extracting three alkaloids (vindoline, catharanthine, and vinblastine) from *Catharanthus roseus* (Yang et al. 2011). Anthocyanins and phenolic compounds were also extracted from grape peel using UAE (Ghafoor et al. 2009, 2011). Phenolcarboxylic acids, carnosic acid, and rosmarinic acid were extracted from *Rosmarinus officinalis* using Ionic liquid-based UAE technique which was proved to have high efficiency and shorter extraction time than conventional extraction methods (Zu et al. 2012).

Moreover, UAE of isoflavones from *Pueraria lobata* (Willd.) stem was carried out, and extraction efficiency was compared with that of conventional solvent extraction (CSE) (Huaneng et al. 2007). Table 1 presents a comparative overview of the UAE of polyphenol from various plant matrices with benefit and use.

4.2 Microwave-Assisted Extraction (MAE)

The microwave-assisted extraction is also considered as a novel method for extracting soluble products into a fluid from a wide range of materials using microwave energy (Paré et al. 1994). MAE can extract bioactive compounds more rapidly, and a better recovery is possible than conventional extraction processes. It is a selective technique to extract organic and organometallic compounds that are more intact. MAE is also recognized as a green technology because it reduces the use of organic solvent (Alupului et al. 2012).

For polyphenols and caffeine extraction from green tea leaves, MAE achieved higher extraction yield at 4 min than any extraction methods at room temperature for 20 h (Pan et al. 2003). Ginsenosides extraction yield from ginseng root obtained by 15 min using focused MAE technique was better than conventional solvent extraction for 10 h (Shu et al. 2003). Dhobi et al. (2009) showed increased extraction efficiency of MAE by extracting a flavolignin and silybinin from *Silybum marianum* compared with the conventional extraction techniques like Soxhlet and maceration. Asghari et al. (2011) extracted some bioactive compounds (E- and Z-guggulsterone, cinnamaldehyde, and tannin) from various plants under optimum conditions. MAE was applied to release bound phenolic acids from bran and flour fractions of sorghum and maize of different hardness by Chiremba et al. (2012).

Other biomolecules such as terpenoids, alkaloids, and saponins have also been recovered utilizing MAE (Zhang et al. 2011b). Higher yields and higher antioxidant activity were obtained in peel extracts of citrus mandarin (Hayat et al. 2009), tomatoes (Li et al. 2011a), and onions (Zill-e-Huma et al. 2011) as compared to rotary extraction. MAE has been exploited for the extraction of health-promoting flavonoids from artichoke herb (*Cynara scolymus* L.) leaves (Alupului et al. 2012).

Table 1 Reported application of ultrasonic-assisted extractions (UAE) for extracting bioactive molecules (polyphenols) from plant extract and their possible therapeutic uses

Plant source with used parts	Bioactive compounds	Benefit in application	Possible therapeutic use	References
Spirulina platensis alga	Beta-carotene	Application for pharmaceuticals	Protect against cancer, diabetes, and other chronic diseases	Dey and Rathod (2013)
Forsythia suspensa plant	Phillyrin		Anti-inflammatory, antioxidant, antiviral, and vasorelaxant	Xia et al. (2011)
Penggan peel	Hesperidin	Applicable in food and pharmaceutical industries	Antioxidant, anti-inflammatory, and antiallergic	Ma et al. (2008)
Prunella vulgaris L. plant	Flavonoids		Against sore throat, reducing fever, and accelerating wound healing	Zhang et al. (2011a)
Nannochloropsis oculata alga	Lipids		Feedstock for biodiesel production	Adam et al. (2012)
Hawthorn seeds	Flavonoids		Used in coronary heart diseases	Pan et al. (2011)
Litchi seeds	Polysaccharides	Applicable in food andpharmaceutical industries	Antitumoral and antioxidant and hypoglycemic properties	Chen et al. (2011)
Chilean papaya seeds	Isothiocyanates, phenolic acids, and flavanols	Rapid and enhanced extraction process	Antioxidant and antimicrobial	Briones-Labarca et al. (2015)
Citrus peel	Flavonoids	Better yield than conventional extraction	Food supplements	Londóno-Londóno et al. (2010)
Red raspberry fruit	Anthocyanins	Better yield than conventional extraction	Antioxidant	Chen et al. (2007)
Grapes fruit	Flavonoids	Better yield than conventional extraction	Cancer, diabetes, food and cosmetic industries	Carrera et al. (2012)
Orange peel	Flavonoids	Rapid extraction and better recovery of compounds	Cancer, diabetes, food and cosmetic industries	Khan et al. (2010)
Jabuticaba skin	Anthocyanins	Rapid extraction and better recovery of compounds	Food and cosmetic industries	Santos et al. (2012)

(continued)

Table 1 (continued)

Plant source with used parts	Bioactive compounds	Benefit in application	Possible therapeutic use	References
Olive leaves	Flavonoids	Lower extraction time	Health-promoting effects	Sáhin and Sámli (2013)
Wheat bran	Flavonoids	Rapid extraction and better recovery of compounds	Health-promoting effects	Wang et al. (2008c)

Similarly, Routray and Orsat (2014) identified highbush blueberry (*Vaccinium corymbosum*) as a potent source of flavonoids, particularly chlorogenic acids and anthocyanins with the help of MAE (Table 2).

4.3 Enzyme-Assisted Extraction (EAE)

Enzymatic pre-treatment has been considered as a novel and an effective way to release bounded compounds and increase overall yield (Rosenthal et al. 1996). Some enzymes such as cellulase, β-glucosidase, xylanase, β-gluconase, and pectinase help to degrade cell wall structure and depolymerize plant cell wall polysaccharides, facilitating the release of linked compounds (Moore et al. 2006). There are two approaches for enzyme-assisted extraction: (1) enzyme-assisted aqueous extraction (EAAE) and (2) enzyme-assisted cold pressing (EACP) (Latif and Anwar 2009).

Bhattacharjee et al. (2006) described EACP as an ideal alternate for extracting bioactive components from oilseed, because of its nontoxic and noninflammable properties. The oil extracted by enzyme-assisted methods was found to contain higher amount of free fatty acids and phosphorus contents than traditional hexane extracted oil (Dominguez et al. 1995). EAE of phenolic antioxidants from grape pomace during wine production was tested by Meyer et al. (1998). Chandini et al. (2011) employed the enzymes tannase and pectinase independently to improve the quality of black tea extracts, and the maximum level of polyphenol extraction was observed when tannase was used alone.

Enzyme-assisted extraction was also used to improve the antioxidant composition of black carrot juice and, more recently, to obtain vegetable oils (Khandare et al. 2010; Szydlowska-Czerniak et al. 2010). Landbo and Meyer (2001) showed improved release of phenolic compounds from *Ribes nigrum* pomace using various enzymes. Maier et al. (2008) used mixture of pectinolytic and cellulolytic enzyme in the ratio of 2:1 to extract bioactive compounds (phenolic acids, non-anthocyanin flavonoids, and anthocyanins) from grape pomace. Extraction of phenolic antioxidant from raspberry solid wastes was increased by application of enzyme in hydroalcoholic extraction compared with nonenzymatic control (Laroze et al. 2010). Gómez-García et al. (2012) extracted phenolic compounds from grape waste using different types of enzymes. Other examples of applying EAE for extracting bioactive compounds are present in Table 3 with the therapeutic uses.

Table 2 List of some other applications of microwave-assisted extractions (MAE) in extracting bioactive molecules from plant extract and their possible therapeutic uses

Plant source with used parts	Bioactive compounds	Benefit in application	Possible therapeutic use	References
Pigeonpea leaves	Cajaninstilbene acid and pinostrobin	Rapid and enhanced extraction process	Postmenopausal osteoporosis, hypocholesterolemic and hypoglycemic effects	Kong et al. (2010)
Fucus vesiculosus alga	Sulfated polysaccharides	Rapid and enhanced extraction process	Anticoagulant, antithrombotic, antitumor, antiviral, contraceptive	Rodriguez-Jasso et al. (2011)
Peanut skins	Phenolic compounds	Applicable in food and pharmaceutical industries	Health-promoting compounds including cancer prevention	Ballard et al. (2010)
Green coffee beans	Chlorogenic acid, caffeine, and polyphenols		Used as functional foods	Upadhyay et al. (2012)
Dunaliella tertiolecta and Cylindrotheca closterium microalga	Chlorophyll a and b and β-carotene and fucoxanthin	Better yield than conventional extraction	Food, health, and biotechnological applications	Pasquet et al. (2011)
Uncaria sinensis herb	Catechin, caffeic acid, epicatechin, and rhynchophylline	Rapid and enhanced extraction process	Fears and nervous disorders	Tan et al. (2011)
Rosemary leaves	Phenolic compounds, rosmarinic and carnosic acids	Applicable in food and pharmaceutical industries	Antioxidants for the food industry	Rodríguez-Rojo et al. (2012)
Grape seeds	Polyphenols	Rapid and enhanced extraction process	Pharmaceutical, cosmetic, and food industry	Li et al. (2011b)
Huang qi (*Radix astragali*) root	Flavonoids	High yield and recovery rate		Xiao et al. (2008b)
Sea buckthorn fruit, leaves, and seeds	Flavonoids	Better yield than conventional method		Périno-Issartier et al. (2011)
Soybean beans	Isoflavone	Better yield than conventional method		Rostagno et al. (2007)
Cortex fraxini Bark	Coumarins, Flavones	96% recovery rate with high yield		Zhou et al. (2011)

(continued)

Table 2 (continued)

Plant source with used parts	Bioactive compounds	Benefit in application	Possible therapeutic use	References
Hu Zhang (*Rhizoma polygoni Cuspidati*) leaves	Resveratrol	Rapid extraction and better recovery of compounds		Wang et al. (2008b)
Red bayberry Leaves	Myricetin	Lower extraction time		Wang and Weller (2006)
Apple	Flavonoids	Rapid extraction and better recovery		Bai et al. (2010)
Rosemary leaves	Phenolic acids and flavonoids	Lower extraction time	Antioxidants for the food industry	Svarc-Gajić et al. (2013)
Dittany of crete stem	Phenolic compounds	Rapid extraction and better recovery of compounds	Pharmaceutical, cosmetic, and food industry	Proestos and Komaitis (2008)
Guava (*Psidium guajava*) leaves	Flavonoids	Lower extraction time		Du et al. (2009)

4.4 Pressurized Liquid Extraction (PLE)

This method is now known by several names, pressurized fluid extraction (PFE), accelerated fluid extraction (ASE), enhanced solvent extraction (ESE), and high-pressure solvent extraction (HSPE) (Nieto et al. 2010). Nowadays, for extraction of polar compounds, PLE is considered as a potential alternative technique to supercritical fluid extraction (Kaufmann and Christen 2002). PLE has also been used for the extraction of bioactive compounds from marine sp. (Oszmianski et al. 2011). Flavonoids extracted from spinach by PLE using a mixture of ethanol and water (70:30) solvent at 50–150°C were more effective (Howard and Pandjaitan 2008). Phenolic compounds such as gallocatechin (GCT), catechin, epicatechin gallate, caffeic acid, chlorogenic acid, and myricetin and total phenolic contents were also recovered from various parts of *Anatolia propolis* using PLE at optimum condition (Erdogan et al. 2011).

PLE has also been successfully employed for extraction of anthocyanins from freeze-dried red grape skin (Ju and Howard 2003). Other examples of applying EAE for extracting bioactive compounds are present in Table 4. Despite the advantages over conventional methods, this method is not found to be suitable for thermolabile compounds as high temperature can have deleterious effects on their structure and functional activity (Ajila et al. 2011).

Table 3 Reported some other applications of enzyme-assisted extractions (EAE) in extracting bioactive molecules from plant extract and their possible therapeutic uses

Plant source	Bioactive compounds	Enzyme used in extraction	Possible therapeutic use	References
Pigeonpea leaves	Flavones: luteolin and apigenin	Pectinase, cellulose, and beta-glucosidase	Anti-inflammatory, antiallergic, antiproliferative	Chen et al. (2010a)
Turmeric (*Curcuma longa L.*) spice	Oleoresin	Alpha-amylase and glucoamylase	Food formulations for the prevention of cancer	Fu et al. (2008)
Citrus peels: Yen Ben and Meyer lemon, grapefruit, mandarin, and orange	Total phenolics	Cellulase® MX, Cellulase® CL, and Kleerase ® AFP	Antioxidant and free radical scavenging activities. Implications in human health	Kurmudle et al. (2010)
Rapeseed	Phenolics, tocopherols, and phospholipids	ROHALASE® OS and ROHAPECT® PTE (cellulase, glucanase, and xylanase activity)	Prevention and treatment of chronic diseases: heart and neurodegenerative diseases, aging, cancer, and rheumatoid arthritis	Li et al. (2006a)
Pomace	Polyphenols	Pectinex XXL and Pectinex Ultra SPL (pectolytic enzymes)	Effectiveness against colon cancer	Munoz et al. (2004)
Grape skin from three varieties	Anthocyanins	Pectinex B3-L, Vinozym EC, and Vinozym G	Food additives providing health benefits	Oszmianski et al. (2011)
Ginkgo biloba leaves	Flavonoids: quercetin, kaempferol, and isorhamnetin	Cellulase from *Trichoderma reesei*, pectinase from *A. niger* and *P. decumbens* cellulose.	Physiological activities in therapies for inflammations, heart diseases, and cancer	Parrado et al. (2006)
Rice bran	Enzymatic extract	Endoprotease mixture	Prevention of diseases including cancer, fatty liver, hypercalciuria, kidney stones, and so on	Wang et al. (2010)

4.5 Supercritical Fluid Extraction (SFE)

Supercritical fluid technique has attracted wide scientific interest, and it was successfully used in environmental, pharmaceutical and polymer applications, and food analysis (Zougagh et al. 2004). Supercritical fluid possesses gas-like properties of diffusion, viscosity, and surface tension and liquid-like density and solvation power. These properties make it suitable for extracting compounds in a short time with higher yields (Sihvonen et al. 1999).

Table 4 Reported some other applications of pressure liquid extractions (PLE) and pressure hot water extraction (PHWE) in extracting bioactive molecules from plant extract

Plant source with parts	Bioactive compounds	Condition used	Benefit in extraction	References
Apple pulp and peel	Catechins, flavonols (quercetin), and anthocyanins	99% methanol (solvent), 313.15 K temperature, 7 MPa pressure	Comparable recovery with reduced solvent amount, handling, and time required than CSE	Alonso-Salces et al. (2001)
Jabuticaba (*Plinia cauliflora*) skin	Anthocyanins	Extractor conditions: pressure (5 MPa) and 553 K temperature	Higher recovery: 2.15-fold anthocyanins (13%) and 1.66-fold (8%) total phenolic compounds than CSE at lower temperature	Santos et al. (2012)
Parsley (*Petroselinum crispum*)	Flakes (glycone of apiin and melonyl-apiin	Temperature (313.15 K), pressure (7 MPa), particle size (<850 μm), S/F ratio (250), and 75% flush volume	Improved recovery of six phenolic compounds with wider solvent choice without any thermal degradation	Luthria (2008)
Cabbage (red) leaves (sample)	Anthocyanins 2.5 g	372.15 K at 5 MPa solvent ratios: water/ethanol/formic acid (94:5:1, v/v/v)	Fast recovery and identification of polyphenols coupling with HPLC-DAD	Arapitsas and Turner (2008)
Bitter melon (*Momordica charantia*)	Fruit chlorogenic acid, genistic acidand catechin	Methanol as solvent at 5 MPa and 423.15–473.15 K with 2 mL/min of flow rate 120 min	Faster and high yield in short time (2 h)	Budrat and Shotipruk (2009)
Citrus (*Citrus unshiu*) peel	Flavanones (hesperidin and narirutin)	One cycle of PHWE at 433.15 K temperature and 10.13 MPa pressure	High-yield hesperidin (3.2-fold) and narirutin (3.7-fold) than CSE	Cheigh et al. (2012)

Saldaña et al. (1999) extracted purine alkaloids (caffeine, theobromine, and theophylline) from *Ilex paraguariensis* (herbal mate tea) using SFE. Supercritical CO_2 modified with ethanol (15 wt.%) gave higher extraction yields of naringin (flavonoid) from *Citrus paradise* (Giannuzzo et al. 2003). Polyphenols and procyanidins were extracted from grape seeds using SFE, where methanol was used as modifier (Khorassani and Taylor 2004). Verma et al. (2008) used optimized condition of SFE to extract indole alkaloids from *Catharanthus roseus* leaves. Zuo et al. (2008) extracted the soybean isoflavones (predominantly daidzein, genistein, and daidzin) from soybean meal. Similarly, Kong et al. (2009) reported extraction of cajaninstilbene acid (CSA) and pinostrobin (PI) are, respectively, a stilbene and flavonone from pigeonpea leaves. Hassas-Roudsari et al. (2009) studied the extraction of antioxidant compounds from canola seed meal using subcritical water and ethanolic and hot water extraction. Other vegetable matrices that have been used to extract bioactive compounds by SFE from *Citrus pomaces* (Kim et al. 2009a) and oregano (Rodríguez-Meizoso et al. 2006), as well as some microalgae (Herrero et al. 2006). Plaza et al. (2010b, c) studied neoformation of antioxidants during SFE extraction of different natural compounds, including microalgae (*Chlorella vulgaris*), algae (*Sargassum vulgare, Sargassum muticum, Porphyra* spp., *Cystoseira abies-marina, Undaria pinnatifida*, and *Halopitys incurvus*), and plants (rosemary, *Rosmarinus officinalis* L.; thyme, *Thymus vulgaris*; and verbena, *Verbena officinalis*). Other examples of applying EAE for extracting bioactive compounds are present in Table 5 with the therapeutic uses.

5 Need for Novel Drug Delivery System (DDS) "Nanoparticles"

Before getting to the bloodstream, many phytochemicals of the plant extracts or herbal remedies will be decomposed plying through the highly acidic pH of the stomach and liver enzymatic action. Thus, the optimum amount of the plant extracts may not get to the blood. If an optimum quantity of the active molecules does not reach the infected region at "minimum effective level," then it will not be potent enough to elucidate a therapeutic effect. Nanodrug delivery for herbal remedies or plant extract can carry an optimum amount of the drug to their site of action circumventing all the barriers such as the stomach acidic pH, metabolism of the liver, and an increase in the systemic drug circulation due to their small size (Yadav et al. 2011; Bairwa et al. 2010).

Nanodrug delivery system is a novel technique to reduce the shortcomings of the traditional DDS. Nano-sized delivery system was chosen because they can deliver high concentrations of drugs to disease sites due to their unique size and high loading efficiency, they deliver the drug in an enclosed small particle size that improves the entire surface area of the drugs allowing fast distribution when they get to the bloodstream, and their concentration persists at the sites for a longer period (controlled delivery). In addition to the above-listed advantages, they show EPR (enhanced permeation and retention) effect, i.e., enhanced permeation via barriers, because of their small size and retention due to weak lymphatic drainage such in cancer. They also exhibit passive disease targeting without the attachment of any other specific ligand moiety, decrease in the side effects, and decrease in the drug formulation dose (Namdaria et al. 2017).

6 Benefits of Nanoformulation (Encapsulation)

Encapsulation of the plant extracts and/or herbal formulation into nanocarrier systems have certain added merit, such as their bulk dosing and smaller absorption which can be overcome, which poses a major problem, attracting the attention of

Table 5 Reported some other applications of supercritical fluid extraction (SFE) for extracting bioactive compounds from plant extracts and their possible therapeutic use

Plant source with used parts	Bioactive compounds	Benefit in application	Possible therapeutic use	References
Basil and oregano herbs	Terpenes: alpha-pinene, limonene, camphor, citronellol, and carvacrol	Stability of molecules increased	Anti-inflammatory and antioxidant	Yang et al. (2007)
Centella asatica herb	Asiatic acid and asiaticoside	Better yield than conventional extraction	Antibacterial and fungicidal, colon and breast cancer	Kim et al. (2009b)
Mahkota dewa fruit	Mangiferin	Better yield than conventional extraction	Antidiabetic, anti-HIV, anticancer, immunomodulatory, and antioxidant	Kim et al. (2010)
Morinda citrifolia fruit	Damnacanthal	Better yield than conventional extraction	Anticancer	Anekpankul et al. (2007)
Terminalia chebula Retz. fruit	Gallic acid, ellagic acid, and corilagin		Anticancer, antimicrobial, and anti-inflammatory	Rangsriwong et al. (2009)
Apple and peach pomaces fruit	Polyphenols		Health-promoting effects	Adil et al. (2007)
Ground red paprika fruits	Carotenoids	Better yield than conventional extraction	Protective against cancer, heart and eye diseases	Rutkowska and Stolyhwo (2009)
Olive leaves	Mannitol		Used in pharmaceutical and diabetic food products	Ghoreishi and Shahrestani (2009)
Rosemary, thyme, and verbena leaves	Phenols, protein, amino acids, and sugars		Functionals food, nutraceuticals, and antioxidant	Plaza et al. (2010a)
Quisqualis indica L. flower	Essential oil		Germicide against skin	Rout et al. (2008)
Haematococcus pluvialis	Vit. E, phenolic compounds		Antioxidant and antimicrobial	Rodríguez-Meizoso et al. (2010)
Rice bran biomass	Phenolic compounds		Cancer, diabetes, food, and cosmetic	Pourali et al. 2010
Winery grape seed	Catechins and proanthocyanidins	Lower extraction time	Pharmaceutical and food industries	García-Marino et al. (2006)

(continued)

Table 5 (continued)

Plant source with used parts	Bioactive compounds	Benefit in application	Possible therapeutic use	References
Bovine bones	Hydroxyapatite		Use as osteoconductive, osteoinductive, and bioceramics	Barakat et al. (2009)
Onion skin	Quercetin	Lower extraction time	Anticancer, antivirus, and anti-inflammatory	Ko et al. (2011)
Oregano leaves	Flavone, flavonone, and flavonols		Food ingredient	Cavero et al. (2006)
Rosemary leaves	Volatiles oil	Better yield than conventional extraction	Functional foods, antioxidant, and anticancer	Carvalho (2005)
Pfaffia paniculata and *Pfaffia glomerata* plant	Ginseng	Better yield than conventional extraction	Dermatological and cosmetic compositions	Leal et al. (2010)
Pitanga leaves	Polyphenolic compounds and flavonoids	Rapid extraction and better recovery of compounds	Antibacterial, anticarcinogenic, antiviral, and anti-inflammatory	Martinez-Correa et al. (2010)
Patrinia villosa Juss herb	Volatiles		Antiviral and antibacterial	Xie et al. (2008)
Cherry fruit	Phenols perillyl alcohol		Antioxidants, anticancer agent for the colon, skin, and lung cancer	Serra et al. (2010)
Tomato juice	Lycopene	Better yield than conventional extraction	Coloring agent	Egydio et al. (2010)
Sargassum muticum, *S. vulgare*, *Hypnea spinella*, *Porphyra* spp., *Chondrus crispus*, *Halopytis incurvus*, *Spongiochloris spongiosa*, *Scenedesmus*, and Nostoc 7 (*Cyanobacteria*)	Isoflavones	Lower extraction time	Functional foods and pharmaceuticals industries.	Klejdus et al. (2010)

(continued)

Table 5 (continued)

Plant source with used parts	Bioactive compounds	Benefit in application	Possible therapeutic use	References
Synechococcus spp. (microalga)	Carotenoids and chlorophyll	Rapid extraction and better recovery of compounds	Functional foods and in drinks and ice creams	Maćıas-Śanchez et al. (2007)
Nannochloropsis gaditana, *Dunaliella salina*, and *Synechococcus* spp.	Carotenoids	Rapid extraction and better recovery of compounds	Food additives	Maćıas-Śanchez et al. (2009)
Scenedesmus almeriensis (microalga)	Beta-carotene and lutein		Antioxidants and food coloring agent, preventer of cataracts, atherosclerosis and some type of cancer	Maćıas-Śanchez et al. (2010)
Nannochloropsis oculata (microalga)	Carotenoids and lipids	Better yield than conventional extraction	Food supplements and functional foods	Liau et al. (2010)
Chorella pyrenoidosa (alga)	Antioxidants		Dietary supplements.	Hu et al. (2007)
Grape seed oil	Triacylglycerides		Antioxidants	Passos et al. (2010)
Tomato skins	Lycopene	Better yield than conventional extraction	Protective against cardiovascular, coronary heart diseases and cancer	Yi et al. (2009)
Roasted wheat germ	Phenolic compounds and tocopherols		Pharmaceutical, foods, cosmetic formulation, and bioinsecticide	Gelmez et al. (2009)
Coriander seeds	Antioxidant fractions	Better yield than conventional extraction	Dietary supplements	Yepez et al. (2002)
Mangosteen pericarp	Xanthones		Inhibition of lipid peroxidation, antioxidant, neuroprotective, and inhibitor of HIV-1 protease	Zarena and Udaya Sankar (2009)
Grape seed	Proanthocyanidins		Anticarcinogenic, antiviral, and anticancer	Yilmaz et al. (2010)

(continued)

Table 5 (continued)

Plant source with used parts	Bioactive compounds	Benefit in application	Possible therapeutic use	References
Hibiscus cannabinus L. seed	Edible oil	Better yield than conventional extraction	Functional foods	Chan and Ismail (2009)
Strawberry fruits	Phenolic compounds		Cancer, diabetes, food and cosmetic industries	Akay et al. (2011)
Canola	Hydroxycinnamic acid		Antibacterial, anticarcinogenic, antiviral, and anti-inflammatory	Li et al. (2010)
Wine grapes seeds	Flavonoids and phenolic acids		Antioxidant and antimicrobial	Prado et al. (2012)
Maritime pine bark	Flavonoids (catechin and epicatechin)		Health-promoting effects	Braga et al. (2008)
Grape bagasse stems, skin, and seed	Anthocyanins, catechins, and glycosides of flavonols	Rapid extraction and better recovery of compounds	Cancer, diabetes, food and cosmetic industries	Farías-Campomanes et al. (2013)
Coffee grounds and husk	Phenolic compounds	Better yield than conventional extraction	Antioxidant and antimicrobial	Andrade et al. (2012)

big pharmaceutical corporations (Chaturvedi et al. 2011). Herbal medicine activity relies on overall function of a several active components, as all the phytochemicals they presented provide synergistic action thereby enhancing the therapeutic value. Each active phytochemical plays a crucial role, and they are all connected to each other.

However, majority of the herbal origin drugs are insoluble leading to a lower bioavailability and elevated systemic clearance, which requires repeated administration or a higher dose, making the drug to be less potent for therapeutic use. In phytochemical formulation studies, developing nano-dosage forms (polymeric nanoparticles [nanospheres and nanocapsules], proliposomes, liposomes, Nanoemulsion, etc.) has a great number of merit for plant extracts that are illustrated in Fig. 3.Thus, the nanoformulation of herbal drugs or plant extracts has a probable future for improving the activity and defeating problems related to plant extracts (Table 6).

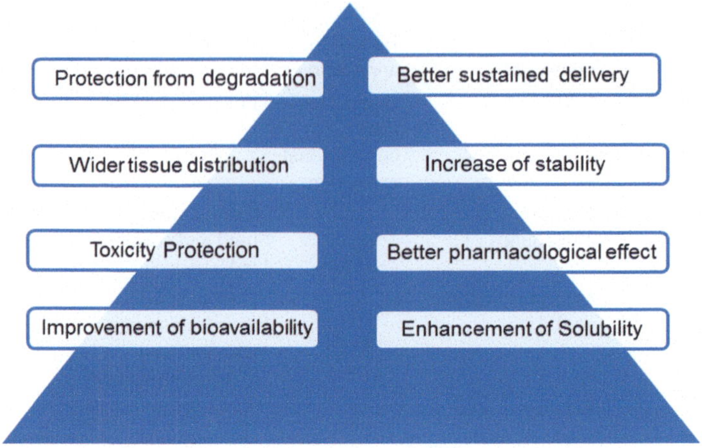

Fig. 3 Different advantages in nanoformulation with plant extract for better therapeutic use

7 General Encapsulation Processes

The process of entrapment of one substance (active agent) within another substance (wall materials) is called encapsulation, which can prepare nanospheres or nanocapsules as shown in Fig. 4 (Nedovic et al. 2011; Mora-Huertas et al. 2010). In addition, the active substance can be adsorbed on the surface of the nanoparticle (Miladi et al. 2016). Newly, the biodegradable particles are progressively prepared from the polymers that might be obtained from natural source, as gelatin, chitosan, albumin, etc. or polymers can be synthetic, like methacrylates and so on. In addition to the local therapeutic effects, drug molecule, genes, and vaccines can be delivered to the target organs by such particles (Jahanshahi 2008; Zafar et al. 2014). Polymer selection criteria are the toxicity and final application of the polymer. Indeed, polymers are used in regenerative medicines and tissue engineering as well. The used biodegradable polymers for drug encapsulation indicate excellent characteristics such as nontoxicity and stability in the blood circulation.

The physicochemical characteristics (e.g., zeta potential, hydrophobicity), drug release profile (e.g., triggered, delayed, prolonged), and biological behavior (e.g., enhanced cellular uptake, bioadhesion) modification of nanoparticles are obtained by the polymeric materials (Galindo-Rodriguez et al. 2005; Kumari et al. 2010; Rieux des et al. 2006). Poly(lactic acid), poly(glycolic acid), and poly(lactic-co-glycolic acid) (PLGA) are the common employed biodegradable polymers for the encapsulation of active ingredients. In fact, release profile of encapsulated drug within the nanoparticles is governed principally by the used polymer degradation kinetics, which consider as an advantage of biodegradable polymers. Drug molecules comprising paclitaxel, 9-nitrocamptothecin, cisplatin, insulin, dexamethasone, estradiol, progesterone, tamoxifen, tyrphostins, and haloperidol

Table 6 Examples of some very recent (2015–2017) encapsulation studies involving various plant extracts for better biological activity

Type of study	Support	Bioactive	% efficiency/loading dose	Advantages/benefit	References
Encapsulation of lycopene from plant extract	Alginate gelatin	Lycopene	79 ± 3% (in olive oil extract)	The stability against isomerization and release effect (diffusion coefficient) of lycopene was enhanced	Calvo et al. (2017)
Encapsulation of pantothenic acid	Liposome, alginate, or alginate-pectin mixture	Pantothenic acid	~80% loading in liposomes	The efficiency and stability at acidic pH (4.0) improved	Ota et al. (2018)
Synthesis and characterization of polyphenols extracted from fresh strawberry fruits	Polymer chitosan	Polyphenols	36% (loading)	Functional amino group is reported to enhance the loading of negatively charged polyphenols and improve bioavailability and sustained release	Pulicharla et al. (2016)
Microencapsulation of lutein, an extract from marigold flowers	Maltodextrin (polysaccharide base) and copovidone (polyvinyl pyrrolidone)	Lutein	95%	Improve the bioavailability, antioxidant ability, and stability	Nalawade and Gajjar (2016)
Tragacanth gum for peppermint encapsulation	Natural polysaccharide tragacanth gum	Peppermint	16.7%	The peppermint encapsulation showed better antimicrobial action over *C. albicans* than *S. aureus* and *E. coli*	Ghayempour et al. (2015)
Aloe vera extract encapsulation into natural polysaccharide tragacanth gum	Tragacanth gum	Aloe Vera	16.1%	*Aloe vera* extract reported as effective wound healer due to controlled release	Ghayempour et al. (2016)
Encapsulation of grape seed	Polylactide	Grape seed	38 wt% of proanthocyanidins extract	The proanthocyanidins are reported to enhance sustained release and dental matrix stability	Yourdkhani et al. (2017)
The encapsulation of bitter melon using spray-drying technology for antioxidant activity	Mixture of maltodextrin and arabic gum	Bitter melon	71.4 ± 1.4% (yield after spray drying powder)	The infusion of bitter melon has been reported to improve antioxidant performance of about ≥ 87.9 ± 2.6%	Tan et al. (2015)

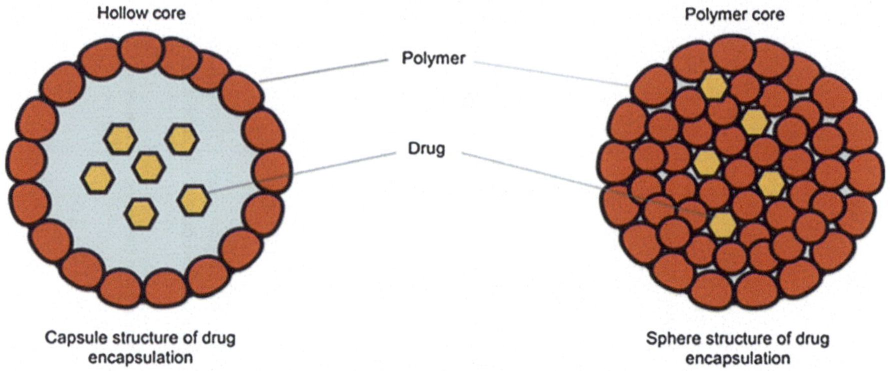

Fig. 4 Drug encapsulation structures in a depictive schema (Rivas et al. 2017)

have successfully encapsulated within different natural and synthetic polymers (Kumari et al. 2010).

Active molecule encapsulation techniques are classified into two main categories: (i) chemistry-related processes (polymerization of monomers) and (ii) physicochemical characteristic-based process (dispersion of preformed polymers). Preformed polymer methods comprise solvent evaporation, nanoprecipitation, solvent diffusion, and dialysis, while polymerization of monomers consists of processes like radical polymerization, miniemulsion, interfacial polymerization, and microemulsion (Armendáriz-Barragán et al. 2016). Such methods are used for the preparation of the various carriers such as microparticles, nanoparticles, and liposomes (Miladi et al. 2013).

The principles of encapsulation techniques or the active molecule nature that is to be encapsulated is the criterion, which differentiates these methods from each other. To design a formulation with proper characteristics for in vitro and in vivo uses, it is very crucial to select correctly the encapsulation method (Armendáriz-Barragán et al. 2016).

To obtain controlled release formulations in pharmaceutical industries, microencapsulation via solvent evaporation is frequently employed for which different methods are available to use. The choice of suitable encapsulation technique is based on the hydrophilicity and hydrophobicity characteristics of drug molecules (Mora-Huertas et al. 2010).To prepare particulate carriers, various polymers possessing different physicochemical properties are used. Generally, these polymers are biodegradable and biocompatible. Polymeric encapsulation is an approach to make drugs nontoxic, noninflammatory, and stable in blood.

Moreover, polymeric nanoparticles as a drug formulation have certain advantages as follows (Armendáriz-Barragán et al. 2016):

- Encapsulation of different chemicals having various properties within a formulation
- Drug molecule protection from environment, enzyme, etc. via encapsulation

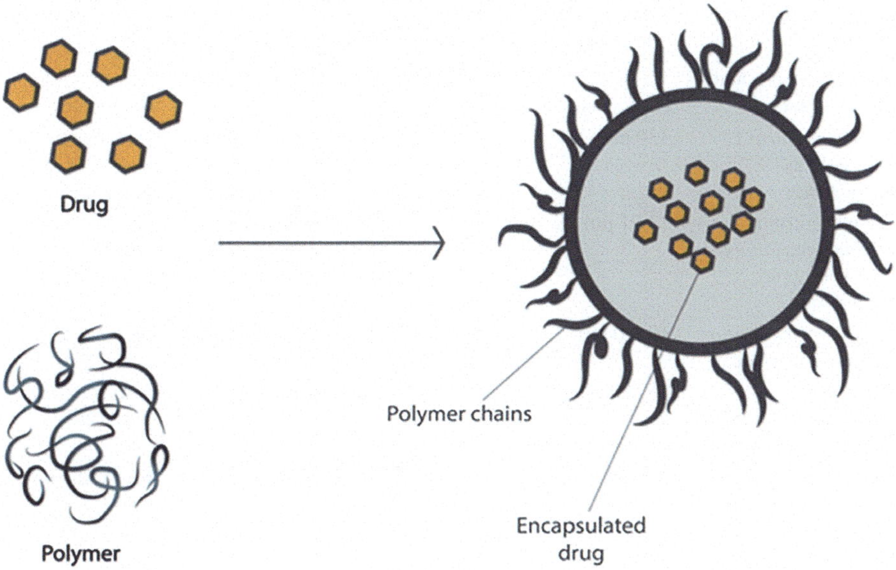

Fig. 5 Drug polymeric encapsulation schematic representation. (Adopted from Rivas et al. 2017)

- Organic solvent's easy elimination throughout the nanoparticles elaboration
- Encapsulated drug release profile control
- Particular tissue or organ targeting

As depicted in the Fig. 5, the coating of particles by substances, such as polyethylene glycol (PEG) that modulates their surface, can avoid particles uptake through macrophages (Zafar et al. 2014) (Table 7).

7.1 Nanoprecipitation

Nanoprecipitation, that is, likewise named solvent displacement or interfacial deposition, was firstly developed by Fessi et al. Nanoprecipitation is taken into account as an early developed active ingredient in encapsulation method, which was firstly employed (Fessi et al. 1989). Basically, the solvent phase can be provided by the dissolving of a film-forming substance such as a polymer, active ingredient, oil, lipophilic surfactant, and in case of need active ingredient solvent or oil solvent in a solvent or in a solvents mixture (Fig. 6). Furthermore, the non-solvent phase would be obtained from the film-forming substance non-solvent or a mixture of non-solvents, complemented with one or more naturally occurring or synthetic tensioactive (Mora-Huertas et al. 2010). In fact, nanoprecipitation method needs two miscible phases (Miladi et al. 2016). According to the study performed by Lince et al.

Table 7 List of reported polymeric nanoparticles/nanostructured (complex and precipitation) formulations using plant extract/bioactive molecules and their possible therapeutic use

Bioactive compounds/ plant extract	Particle size	Encapsulation efficiency	Possible therapeutic use	References
Euphorbia hirta L. with gold as nanocarrier	530 nm		Antibacterial and antifungal	Annamalai et al. (2013)
Silymarin from *Silybum marianum* with solid lipid formulation	22.91 μm		Antioxidant and hepatoprotectant	Pitsiree et al. (2013)
Argemone mexicana L. with iron oxide	10–30 nm		Diuretic and purgative	Arokiyaraj et al. (2013)
Indigofera aspalathoides Vahl with silver as carrier	45 and 69 nm		Hepatoprotective	Arunachalam et al. (2013)
Vincarosea from *Catharanthus roseus* Linn. G. Donn with silver as carrier	28 nm		Antimicrobial activity	Kotakadi et al. (2013)
Aloe from *Aloe* leaf extract with silver as carrier	15.2 ± 4.2 nm		Carcinogenic activity	Zhang et al. (2013a)
β-sitosterol from green tea extract with solid lipid formulation			Antioxidant activity	Lacatusu et al. (2012)
Safflower from *Carthamus tinctorius* with gold as carrier	40–200 nm			Nagaraj et al. (2012)
Ajwain and opium poppy seed from *Trachyspermum ammi and Papaver somniferum* with silver	3.2 and 7.6 μm		Antispasmodic	Vijayaraghavan et al. (2012)
English ivy from Hedera helix with nanoparticles formulation	60–85 nm		Antiaging	Burris et al. (2012)
Triptolide from *Tripterygium wilfordii* Hook F with nanoparticle formulation			Anti-inflammatory	Xue et al. (2012)
Ginger from *Zingiber officinale* rhizome with nanoparticle formulation	453.1–551.7 nm		Anti-inflammatory	Ratcharin and Indranupakorn (2012)

(continued)

Table 7 (continued)

Bioactive compounds/ plant extract	Particle size	Encapsulation efficiency	Possible therapeutic use	References
Tripterygium with nanoparticle formulation			Male reproductivetoxicity in rats	Xue et al. (2011)
Triptolide obtained from *Tripterygium wilfordii* Hook with poly(DL-lactic acid) as nanocarrier	149.7 nm	85.7%	Autoimmune diseases, especially rheumatoid arthritis, psoriasis, leukemia, and antineoplastic	Liu et al. (2005)
Curcumin from the root of *Curcuma longa* with Methoxy poly(ethylene glycol)-palmitate as carrier	41.43 nm	100%	Antitumor, antioxidant, antiamylodin, antiplatelet aggregation, and anti-inflammatory	Sahu et al. (2008)
Camptothecin from bark and stem of the oriental tree *Camptotheca acuminate* with glycol chitosan	280–330 nm	80%	Gastric, rectum, bladder, colon, lung, breast, and ovarian cancer	Min et al. (2008)
Hypericin from *Hypericum perforatum* with polylactic acid/ polylactic-co-glycolic acid as nanoparticles	200–300 nm	70%	Photosensitizer used in photochemotherapy	Labouebe et al. (2006)
Cuscuta chinensis (active constituents – flavonoids and lignans such as quercetin, kaempferol) obtained from *Cuscuta chinensis* Lam.	267 nm	90%	Used as tonic and to improve sexual function, prevent senescence, and regulate immune system. Also for anticancer, antiaging, and immunostimulatory effects	Yen et al. (2008)
Catechins (active constituents – (+)-catechin, (−)-epicatechin, (−)-epigallocatechin-3-gallate) from the *Camellia sinensis* with chitosan as nanoparticles	1.97–6.83 μm	27.9–40.12%	Chemopreventive, anticarcinogenic, antiviral, anti-oxidative, anti-obesity, anti-inflammatory, antidiabetic, antimutagenic, antiangiogenic, antibacterial, and antiaging activities	Zhang and Kosaraju (2007)

(continued)

Table 7 (continued)

Bioactive compounds/ plant extract	Particle size	Encapsulation efficiency	Possible therapeutic use	References
Plant extract of *Ziziphus mauritiana* with chitosan			Immunomodulatory activity	Bhatia et al. (2011)
Flavonoids and lignans from *Cuscuta chinensis*		90%	Hepatoprotective and antioxidant effect	Feng-Lin et al. (2008)
Triptolide			Anti-inflammatory	Zhinan et al. (2005)
Artemisinin		90–93%	Anticancer	Youfang et al. (2009)
R. salvia miltiorrhiza		96.68%	Coronary heart diseases, anginapectoris, and myocardial infarction	Su et al. (2008)
Taxel		99.44%	Anticancer	Fu et al. (2006)
Berberine		65.40 ± 0.70%	Anticancer	Lin et al. (2007)
Silibinin		95.64%	Hepatoprotective	Li et al. (2007b)
Tetrandrine		84%	Lung	Xiaoyan et al. (2008)
Glycyrrhizic acid		91.76%	Anti-inflammatory and antihypertensive	Hou and Zhou (2008)
Quercetin		over 99%	Antioxidant	Tzu-Hui et al. (2008)
Breviscapine		93.1%	Cardiovascular and cerebrovascular	Liu et al. (2008)
Zedoary turmeric oil		1.62 ± 0.15%	Hepatoprotection Anticancer and antibacterial	Lertsutthiwong et al. (2008)
Naringenin			Hepatoprotective	Feng-Lin et al. (2009)
Curcuminoids		70%	Anticancer and antioxidant	Mukerjee and Vishwanatha (2009)
Camptothecin		80%	Anticancer	Min et al. (2008)
Ginkgo biloba extract			Brain function activation	Shimada (2008)
Root extract of *Phytolacca decandra* with PLGA	25 gm/kg		Lung cancer	Das et al. (2012)
Leave extract of *Ocimum sanctum* with alginate chitosan			Antimicrobial	Rajendran et al. (2013)

(continued)

Table 7 (continued)

Bioactive compounds/ plant extract	Particle size	Encapsulation efficiency	Possible therapeutic use	References
Curcumin from *Curcuma longa*	50 nm		Pancreatic tumor cell	Bisht et al. (2007)
Curcumin-loaded polymeric nanoparticles	2–40 nm		Antibacterial activity	Basniwal et al. (2011)
Honokial from *Magnolia officinalis*			Anti-inflammatory, antithrombotic, antirheumatic, antioxidant, anxiolytic, CNS depressant, muscle relaxant, and antitumor activity	Zheng et al. (2010)
Coumarin from *Gelsemium sempervirens*			Antitumoral activity	Khuda-Bukhsh et al. (2010)
Ethanolic extract (flavonoids) of *Harungana madagascariensis* with PLG	1000 mg/ml		Antibacterial, antifungal, and antiviral	Moulari et al. (2005)
Ethyl acetate extract of *Harungana madagascariensis* with PLG	500 mg/ml		Dental caries and gingivitis due to bacteria	Moulari et al. (2006)
Quercetin with PVA		99%	Anti-inflammatory, antioxidant, and hepatoprotective	Wu et al. (2008)
Naringenin with PVA as carriers	100 mg/kg		Hepatoprotective in vivo	Yen et al. (2009)
Ethanolic extract of *Polygala senega* with PLGA			Anticancer	Paul et al. (2011)
Aq. extracts of *Plectranthus ecklonii*	1 mg/ml	About 100%	Antioxidant activity	Rijo et al. (2014)

particles within nanoprecipitation method formed throughout three steps of nucleation, growth, and aggregation (Lince et al. 2008).

Nanoprecipitation is almost restricted to the encapsulation of hydrophobic actives. However, a modified nanoprecipitation technique has been designed by Bilati et al. (2005), in order to encapsulate the hydrophilic drug molecules (Bilati et al. 2005). Biodegradable polyesters, principally poly-e-caprolactone (PCL), poly(lactide) (PLA), and poly(lactide-co-glicolide) (PLGA), are usually used as a polymer in nanoprecipitation method. In fact, greater purity and improved reproducibility can be provided by synthetic polymers in comparison with natural polymers

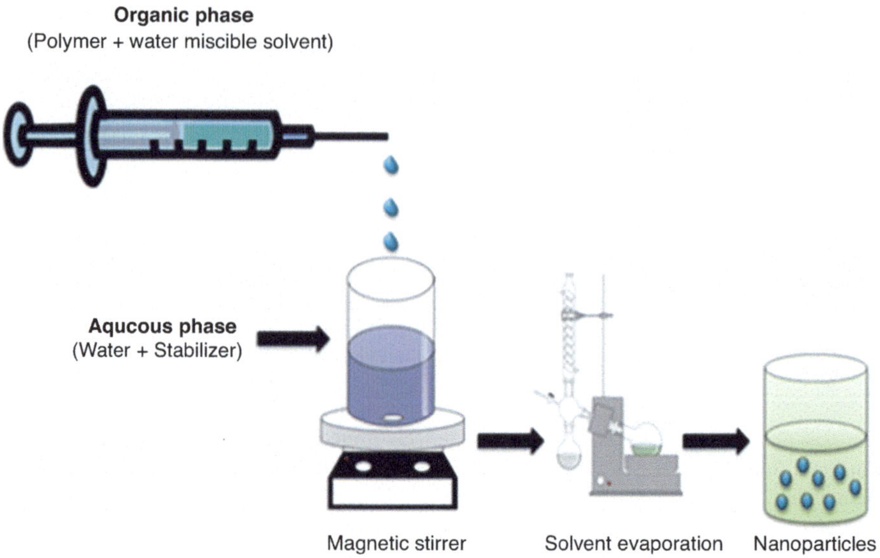

Fig. 6 Schematic representation of nanoprecipitation method. (Adopted from Badri et al. 2017)

(Khoee and Yaghoobian 2009; Mora-Huertas et al. 2010). Mostly, in this technique acetone is used as a polymer solvent. For the dissolution of active substance and oil, another solvent like ethanol is also employed. As non-solvent and stabilizer, water or buffer solutions and poloxamer 188 or polysorbate 80 are, respectively, used. Polymer's interfacial deposition after displacement of semipolar solvent miscible with water is the mechanism in which the nanoprecipitation method is based (Fessi et al. 1989). Nanoprecipitation due to its advantages that are cited in the Table 10 is a widely used method for the nanoparticle preparation (Miladi et al. 2016). The most commonly used administration route and targeted organs for obtained nanoparticles by nanoprecipitation method are shown in Fig. 7 (Table 8).

7.2 *Emulsification Process*

Generally, an emulsion is made up of at least two immiscible liquids where one acts as a dispersed phase and the other acts as a continuous phase, water in oil emulsion or oil in water emulsion. Emulsion can also be a multiphase water/oil/water or oil/water/oil. An emulsifier can also be used to stabilise the emulsion. The particle size of the emulsion depends on the type and amount of emulsifier used and the emulsification technique. Encapsulation efficiency depends on the particle size. During emulsification, the bioactive compound will be embedded in the continuous phase, and thus the active compound is protected from degradation and thermo-oxidation. The morphology of the encapsulated material depends on the dispersed phase, its distribution in the continuous phase, suspension viscosity, size of the droplets,

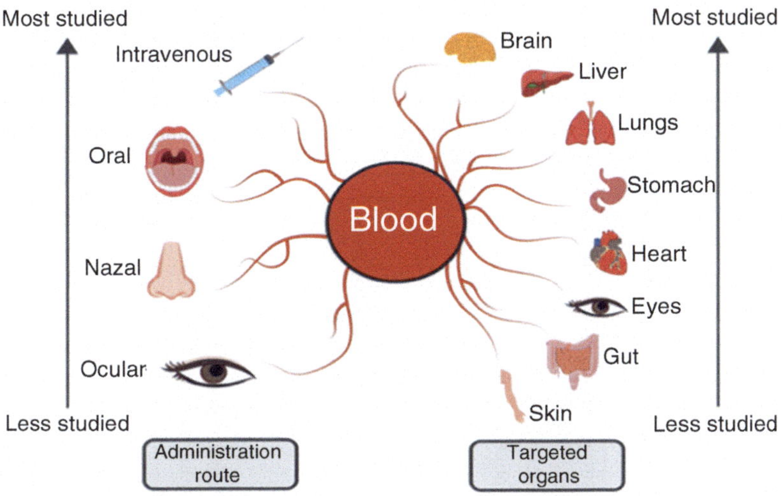

Fig. 7 The most common targeted organs and administration routes for nanoparticles designed by nanoprecipitation technique. (Adopted from Rivas et al. 2017)

processing conditions, etc. Colloidal dispersions with particle size less than 100 nm are called nanoemulsion.

For preparing nanoemulsion, small instrument like ultrasonicator, homogenizer usually used. Mohammadi et al. (2016) used nanoemulsion method to encapsulate olive leaves extract in soybean oil, and this nanoencapsulated olive oil extract exhibited better antioxidant activity. Ghayempour et al. (2016) used microemulsion technique to encapsulate *Aloe vera* extract in tragacanth gum and produced a natural wound healing product.

7.2.1 Emulsification: Solvent Evaporation

For the preparation of nanocapsules, an emulsion is first prepared by mixing the organic polymer solution with aqueous phase. An organic solution of the polymer is mixed with bioactive oil in a non-solvent to form a suspension. Then the solvent is evaporated, and nanocapsules are formed by entrapping the active component inside the polymer matrix. Solvent evaporation is an easy and commonly used method.

Yourdkhani et al. (2017) used a combination of double emulsion and solvent evaporation technique to encapsulate grape seed extract in polylactide and produced polynuclear microcapsules with an average diameter of 1.38 μm and loading efficiency of 38% weight. Microencapsulation resulted in the preservation of the bioactivity of the extract. Pink pepper is an important medicinal plant which possesses antitumor activity, anti-inflammatory property, and antioxidant property. Andrade et al. (2017) encapsulated pink pepper extract in polylactic acid by emulsification and subsequent solvent extraction (Table 9).

Table 8 List of reported microsphere-encapsulated (nanoprecipitation) formulations using plant extract/bioactive molecules and their possible therapeutic use

Bioactive compounds/ plant extract	Benefit in application	Route of admiration	Size	Possible therapeutic use	References
Jaboticaba	Applied in food	In vitro	Uniform with few wrinkles and smooth surfaces	Antioxidant activity	Silva et al. (2013)
Locust bean gum	Applied as thickening and gelling agent in food technology	Oral	734.18–293.17 μm	Hypocholesterolemic activity	Kaity et al. (2013)
Soy protein isolate	Applied as a good gelling, emulsifying, fat-absorbing, and water-binding agent	In vitro	5.5–9.3 μm	Estrogen beta-agonist (decreases serum testosterone levels inhealthy men)	Nesterenko et al. (2012)
Rosemary extract	Value addition in noninvasive drug delivery systems	In vitro	254.5 nm	Antiproliferative	Yesil-Celiktas and Cetin-Uyanikgil (2012)
Rutin	Applied for standardization of Chinese traditional medicines	In vitro	Small and uniform	Antioxidant activities	Zeng et al. (2012)
Chelerythrine	Enhance delivery of drug to a tumor organ	In vitro	12.18 μm	Antimicrobial, anti-inflammatory, antitumor, and antiplaque effect	Li et al. (2011d)
Thymol, clove, origanum, and camphor white oil	Applied for use in pest control	In vitro	5 μm to over 300 μm	Larvicidal activity	Glenn et al. (2010)
Rutin	Cerebrovascular region and cardiocascular targeting	In vitro	165–195 μm	Cerebrovascular and cardiovascular	Xiao et al. (2008a)
Zedoary oil	Sustained release and higher bioavailability	Oral	100–600 μm	Hepatoprotective	You et al. (2006)
Camptothecin	Prolonged release of camptothecin	Intraperitoneally andintravenously	10 μm	Anticancer	Machida et al. (2000)
Quercetin	Significantly decreases the dose size	In vitro	6 μm	Anticancer	Chao et al. (2010)
Cynara scolymus extract	Controlled release of neutraceuticals	Oral	6–7 μm	Nutritional supplement	Gavini et al. (2005)

Table 9 List of reported emulsion (solvent evaporation and solvent extraction method) encapsulated formulations using plant extract/bioactive molecules and their possible therapeutic use

Bioactive compounds	Benefit in application	Route of administration	Size	Possible therapeutic use	References
Triptolide	Enhance the penetration of drugs through the stratum corneum by increased hydration	Topical	Less than 100 nm	Anti-inflammatory	Zhinan et al. (2003)
Zedoary turmeric oil	Improved aqueous dispersibility, stability, and oral bioavailability	Oral	68.3 ±1.6 nm	Hepatoprotection, anticancer, and antibacterial	Zhao et al. (2010)
Docetaxel	Improve residence time	Intravenous	166 nm	Anticancer high pressure	Li et al. (2007a)
Berberine	Improve residence time and absorption	Oral	56.80 nm	Anticancer	Sun and Ouyang (2007)
Silybin	Sustained release formulation	Intramuscular	21.20 nm	Hepatoprotective	Song et al. (2005)
Quercetin	Enhance penetration into the *stratum corneum* and epidermis	Topical	10–100 nm	Antioxidant	Fabiana et al. (2008)
Triptolide from *Tripterygium wilfordii* Hook F	Reduction in toxicity of triptolide following transdermal delivery		18–20 nm	Used in treatment of autoimmune diseases, especially rheumatoid arthritis, psoriasis, leukemia, and antineoplastic activity	Chen et al. (2004)
Furocoumarin (psoralen) from seed of *Psoralea corylifolia*	Enhanced anti-inflammatory effects			Treatment of skin diseases characterized by hyperproliferation such as psoriasis	Ali et al. (2008)
Curcumin isolated from root of *Curcuma longa*	Enhanced anti-inflammatory effects		61.8–79.5 nm	Antitumor, antioxidant, anti amyloid, antiplatelet aggregation, and anti-inflammatory	Wang et al. (2008a)

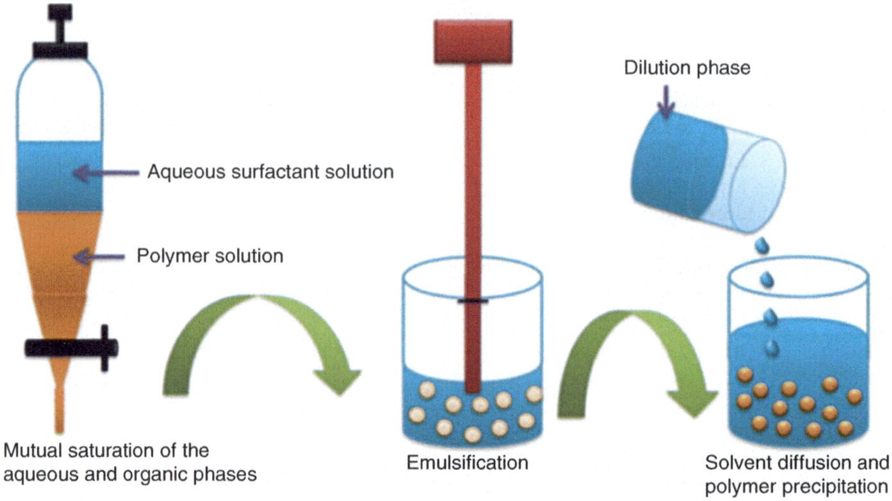

Fig. 8 Illustration of emulsion solvent diffusion method setup. (Adopted from Armendáriz-Barragán et al. 2016)

7.2.2 Emulsion-Diffusion Method

Quintanar-Guerrero and Fessi (Quintanar-Guerrero et al. 1996) for the first time developed the emulsion-diffusion method in order to prepare polylactide (PLA) nanoparticles. Both lipophilic and hydrophilic active ingredients can be nanoencapsulated by the emulsion-diffusion technique. However, emulsion-diffusion method is mainly used for the encapsulation of hydrophobic drug molecules. In emulsion-diffusion method three liquid phases, namely, organic phase, aqueous phase, and dilution phase, are required (Miladi et al. 2014). The organic, aqueous, and dilution phases might be achieved via preforming the experimental procedure. The organic phase including polymer, active ingredient, oil, and an organic solvent (partially miscible with water) has to be water-saturated, while a lipophilic active molecule is intended to be nanoencapsulated (Mora-Huertas et al. 2010). The size of the prepared particles in emulsion-diffusion method can be affected by the operating conditions such as rate of emulsification stirring, diluting water temperature and volume, concentration of polymer, and stabilizer amount and ratio of phase (Quintanar-Guerrero et al. 1996; Mora-Huertas et al. 2010). Based on the study taken place on the homogenization and sonication effect on the size of the particles, sonication is more crucial than homogenization to the particles size (Miladi et al. 2014) (Fig. 8).

For various components in this method, organic phase plays the role of solvent. In case of need, it is also possible that active ingredient solvent or oil solvent be included in the organic phase. Usually, the dilution phase is a large volume of water, while the aqueous phase forms from the stabilizing agent aqueous dispersion, which is obtained employing solvent-saturated water. Eudragit®, PCL, and PLA are the polymers, which are usually used in this technique (Mora-Huertas et al. 2010).

7.2.3 Emulsification-Ionic Gelation

Emulsification-ionic gelation can be applied in charged polymers like chitosan and alginate. In this method, charged polymer chain interacts with oppositely charged medium to form particles. The charged medium acts as a cross-linking agent. Lertsutthiwong et al. (2008) encapsulated turmeric oil by this method. Turmeric oil is emulsified in sodium alginate aqueous solution and then subjected to gelification with chitosan and calcium chloride and subsequent solvent evaporation

7.2.4 Double Emulsion Method

Double emulsion (DE) complex systems are named emulsion of emulsion as well (Garti and Bisperink 1998). Commonly double emulsions are classified into two groups of water-oil-water (w/o/w) and oil-water-oil (o/w/o). Usually double emulsions are prepared in two-step process; the size of droplets is mostly polydispersed in double emulsion. Double emulsion technique includes aqueous phase dispersion into a nonmiscible organic solvent (in the presence of short-time low-power sonication or high-shear homogenization) to obtain the first emulsion (W1/O). Once prepared emulsion is dispersed in a second aqueous phase that includes hydrophilic emulsifier, then homogenization or sonication step is conducted (Fig. 9). This step of homogenization can be repeated under the same condition. To avoid the first emulsion breaking in case if sonication is used, it should be carried out in short time and at low power. The volatile organic solvent evaporation at ambient temperature or under low pressure (via rotary evaporator) after the multiple emulsion formation permits the reparation colloidal dispersion or particulate carrier (Giri et al. 2013). In fact, small droplets of double emulsion are containing one or few droplets, while sometimes droplets are too large that each drop encompasses certain small

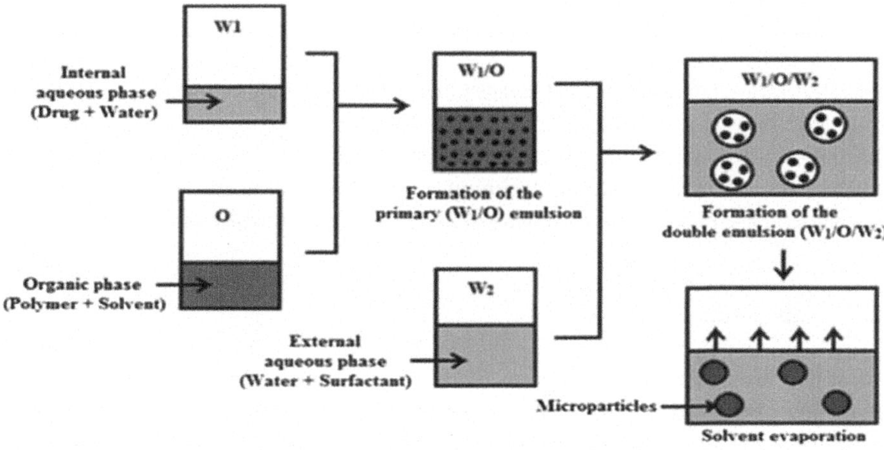

Fig. 9 Double emulsion solvent evaporation method schemes. (Adopted from Giri et al. 2013)

compartments with 50–100 droplets (Garti and Bisperink 1998; Schuch et al. 2013). Double emulsion is an appropriate method to encapsulate the hydrophilic drug molecules (Miladi et al. 2014). Prepared carrier stability and release profile are possible to be significantly enhanced via alteration of used stabilizer type and amount within the system. Cancer can be efficiently treated thanks to the targeted drug delivery and prolonged drug release, which are associated with the usage of double emulsion systems. Indeed, double emulsions possess numerous advantages such as biocompatibility, biodegradability, and versatility.

In addition, the encapsulation and protection of both hydrophobic and hydrophilic types of actives molecules can be take place by double emulsion method. Nevertheless, multiple emulsions are encountered several challenges such as vulnerability to physical and chemical degradation, formulation trouble, and bulky. To tackle the stability problems of multiple emulsions, various efforts (e.g., surfactant amount variation, polymerization gelling, steric stabilization, pro-emulsion, addition of excipients, and interfacial complexation) were made (Garti 1997; Garti and Aserin 1996; Hino et al. 2000). In order to improve the efficacy of cosmetics, through double emulsion, incompatible substances may combine within the same formulation. The disadvantages of double emulsion method are, namely, process complexity, thermodynamic instability, and production of comparatively heterogeneous and size-sensitive (sensitive to different double emulsion method-related parameter) nanoparticles. In comparison with other encapsulation methods, commonly double emulsion provides polydisperse particles. Usually, main advantages and disadvantages of drug encapsulation methods are shown (Table 10). Encapsulation of both hydrophilic and lipophilic active ingredients is the unique advantage of double emulsion method. The characteristics of nanoparticles provided by double emulsion technique can be influenced by parameters as speed of evaporation (khoee et al. 2012), external phase composition (Péan et al. 1998; Tse et al. 2009), relative ratio of phases (Khoee et al. 2012), polymer concentration, nature and molecular weight (Zambaux et al. 1998; Péan et al. 1998; Van de Ven et al. 2011), surfactants nature and amount (Zhao et al. 2007; Khoee and Yaghoobian 2009; Dhanaraju et al. 2004), and speed of homogenization (Eley and Mathew 2007; Basarkar et al. 2007).

Furthermore, encapsulation efficiency could be considerably influenced by operating condition as well (Billon et al. 2005). Nanoprecipitation, emulsion-diffusion technique, microemulsion, phase inversion temperature technique, and high-pressure homogenization are the techniques, which do not need the use of organic solvents or toxic solvents for the encapsulation of drug molecules (Iqbal et al. 2015).

7.2.5 Liposomes

Liposomes are defined as the systems made by one or several phospholipid bilayers describing one or several aqueous compartments (core) (Gulati et al. 1998; Walde and Ichikawa 2001). Cholesterol and phospholipids are the principal constituents of liposomes. Liposomes that are spherical-shaped vesicles could be categorized in

Table 10 Advantages and drawbacks of different encapsulation methods

Methods	Advantages	Disadvantages	References
Nanoprecipitation	No need for high-shear stress Monodispersed particle preparation Fast and simple Does not require highly toxic solvent High reproducibility Easy to scale-up	Prepared nanoparticle size is mainly related to the polymer concentration Commonly, restricted to the encapsulation of hydrophobicdrug molecules	Katara and Majumdar (2013), Siqueira-Moura et al. (2013), Han et al. (2013), Seremeta et al. (2013), Miladi et al. (2016)
Double emulsion method	Encapsulation of both hydrophobic and hydrophilic drug molecules	Long process (two steps) Hard to scale-up Preparation of polydispersed particles Hydrophilic drug molecule leakage into external aqueous phase	Ibraheem et al. (2013), Bitar et al. (2015), Zakeri-Milani et al. (2013)
Emulsion-diffusion method	Easy to scale-up Good reproducibility Nontoxic solvents usage Narrow particles size distribution and mean particle size reduction Thermosensitive drug incorporation Lipophilic drugs high entrapment	To prepare nanoparticles, larger volume of water is required Hydrophilic drugs poor encapsulation Final formulation ingredients concentration is needed Final formulation may contain organic solvent residues Need long emulsion agitation	Campos et al. (2013), Hao et al. (2013), Souguir et al. (2013)

multivesicular, multilamellar, oligolamellar, and unilamellar classes (Fig. 10) (Zhai and Zhai 2014).

Liposomes are broadly employed such as carrier for hydrophobic and hydrophilic drug molecules (Yoshida et al. 2010; Detoni et al. 2012). Liposomes as a nonimmunogenic, nontoxic, biocompatible, and biodegradable drug carriers have certain advantages including drug biodistribution and pharmacokinetic improvement, toxicity reduction, and specific target drug delivery design (Drulis-Kawa and Dorotkiewicz-Jach 2010; Voinea and Simionescu 2002). Thanks to the liposomes structure, they can encapsulate hydrophobic, hydrophilic, and amphiphilic active ingredients (Fig. 11) (Yoshida et al. 2010).

Usually plant extracts are susceptible to oxygen, heat, and light degradation, which have restricted their usage in medicine. Liposomes as an attractive encapsulation approach can overcome the challenges that encountered plant extracts like low water solubility-associated decreased bioavailability, problems of stability (volatility and oxygen, light, temperature sensibility), and toxicity (Detoni et al. 2012). In addition, liposomes can improve tissue targeting and enhance the biological activity

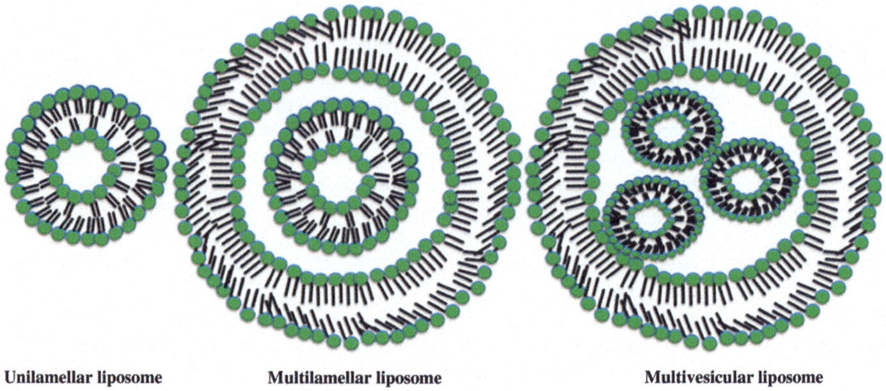

Fig. 10 Types of liposomes. (Adopted from Badri et al. 2016)

Fig. 11 The structure of liposome. (Adopted from Badri et al. 2016)

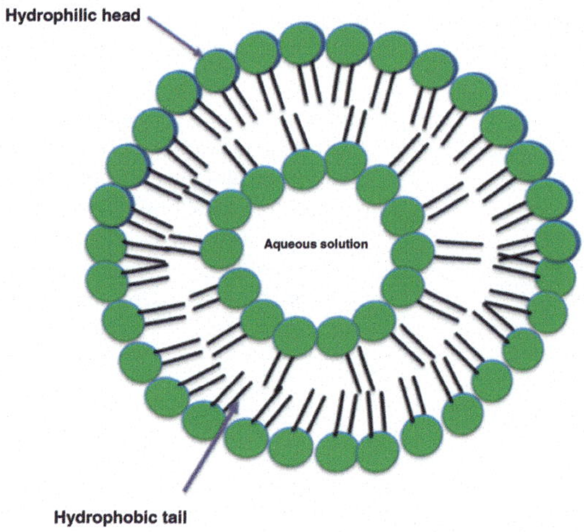

of plant extracts due to the modification of plant extract physicochemical property modification. To encapsulate plant extracts, various techniques have been employed (Detoni et al. 2012; Asbahani et al. 2015) and presented in Table 11.

7.2.6 Niosomes

Niosomes are microscopic vesicles that are composed from admixture of cholesterol and alkyl or dialkyl polyglycerol ether class of nonionic surfactants, which are successively hydrated in the aqueous media. The nonionic surfactant as Span 60 is used as vesicles forming amphiphile in noisome. For the aim of stabilization, cholesterol and small quantity of anionic surfactant like dicetyl phosphate are commonly added (Makeshwar and Wasankar 2013; Buckton 2000). Niosome's main

Table 11 List of reported liposomal formulations using plant extract/bioactive molecules with their possible therapeutic use

Bioactive compounds/plant extract	Benefit in formulations	Route of admiration	Encapsulation efficiency	Possible therapeutic use	References
Silymarin	Hepatic targeting capability for delivering silymarin to the liver	Parenteral administration	60%	Antioxidant activity	El-Mowafy et al. (2013)
Quercetin	Used to treat UVB-irradiation	Topically	80.41 ± 4.22%	Antioxidant activity	Liu et al. (2013)
Actein	To prevent and treat breast cancer	in vitro		Anticancer activity	Einbond et al. (2013)
Artemisia princeps Pampanini	Enhanced transdermal delivery	in vitro	51.96 ± 0.01	Anti-infective anti-inflammatory	Yang et al. (2013)
Persicae Semen and Carthami Flos	The bioavailability of iron, manganese, and zinc was significantly improved	in vitro	Well encapsulated	Immunosuppressive and chemotherapeutic activity	Zheng et al. (2013)
Propolis	Immunological activity of propolis flavonoids enhanced with liposome encapsulation	in vitro	Well encapsulated	Anti-inflammatory, anti-oxidative, hepatoprotective	Yuan et al. (2012)
TOH, GTE, epicatechin (EC), and catechin (C)	Protect oxidation through enhancement of the activity for endogenous antioxidants	in vitro	Well encapsulated	Antioxidant	Yin et al. (2012)
Fisetin	Suitable for in vivo administration	in vitro	73%	Antioxidant, anticarcinogenic, antiangiogenesis	Mignet et al. (2012)
Berberine and palmatine	Proton delivery to model lipid membranes as well as in isolated mitochondria	in vitro	Well encapsulated	Antioxidants rotary evaporation	Pustovidko et al. (2012)
Ginseng	Effectively suppress the depolarization of mitochondrial membrane	in vitro	234.1 ± 13.9%	Antioxidant activity	Tsai et al. (2012)
Ammonium glycyrrhizinate	Improve the drug anti-inflammatory activity in mice	Subcutaneously	28.8%	Anti-inflammatory	Marianecci et al. (2012)
Radix Salviae Miltiorrh.	Standardization of traditional Chinese medicines	Orally		Immunoreactivity	Chen et al. (2012)

(continued)

Table 11 (continued)

Bioactive compounds/plant extract	Benefit in formulations	Route of admiration	Encapsulation efficiency	Possible therapeutic use	References
Resveratrol	Protect the dopaminergic neurons in Parkinson's disease rats	Intragastrically	73.54%	Antioxidant	Wang et al. (2011)
α-Tocopherol and ascorbic acid	Orange juice-mixed liposomal formulation exhibited stable microbiological results	in vitro	99%	Antioxidant	Marsanasco et al. (2011)
Carotenoid	Successful delivery of highly lipophilic enzymatic substrate in aqueous media	in vitro	Well entrapped	Antioxidant	Nacke and Schrader (2011)
Phenolic compounds	Liposome formulation fails topical delivery of antioxidant phenolic compounds	Topical	33.03 ± 3.84%	Antioxidant	González-Paredes et al. (2011)
Quercetin	Reduced dose, enhance penetration in blood-brain barrier	Intranasal	60%	Antioxidant, anticancer	Aroonsri et al. (2008)
Silymarin	Improve bioavailability	Buccal	69.22 ± 0.6%	Hepatoprotective	El-Samaligy et al. (2006a)
Artemisia arborescens essential oil	Targeting of essential oils to cells, enhance penetration into cytoplasmatic barrier	in vitro	60–74%	Antiviral	Chiara et al. (2005)
Ampelopsin	Increase efficiency	in vitro	62.30%	Anticancer	He et al. (2008)
Paclitaxel	High entrapment efficiency and PH sensitive	in vitro	94%	Anticancer	Rane and Prabhakar (2009)
Curcumin	Long circulating with high entrapment efficiency	in vitro	88.27 ± 2.16%	Anticancer	Hong et al. (2008)
Garlicin	Increase efficiency		90.77%	Lungs	Sun et al. (2009)
Quercetin and rutin	Binding of flavonoids with Hb is enhanced	in vitro		Hemoglobin	Juqun and Rong (2007)

Usnea	Increase solubility and localization with prolonged releaseprofile	in vitro	99.5%	Antimycobacterial	Lira et al. (2009)
Wogonin	Sustained release effect	in vivo	81.20 ± 4.20%	Anticancer	Ke et al. (2007)
Colchicine	Enhance skin accumulation, prolong drug release andimprove site specificity	Topical	66.3 ± 2.2%	Antigout	Godin and Touitou (2004)
Catechins	Increased permeation through the skin	Transdermal	93.0 ± 0.1%	Antioxidant and chemopreventive	Fang et al. (2006)
Breviscapine	Sustained delivery of breviscapine	Intramuscular	87.9 ± 3.1%	Cardiovascular	Zhong et al. (2005)
Curcumin	Photoaging attenuation (demonstration in mice)	Oral		Antioxidant, Anti-inflammatory,	Agrawal and Kaur (2010)
Resveratrol	Improvement of the cellular oxidative stress via rapid and potent cellular internalization	in vitro	>70%	Antioxidant Photo-protector	Kristl et al. (2009)
Resveratrol	Nano-sized vesicles, inclusion of resveratrol retarded drug release in vitro	In vitro and in vivo Intraperitoneal injection	≈70%	Cardiovascular protector	Hung et al. (2006)
Quercetin	Reduced anxiety and cognitive functions, dose administered decrease, increase in circulation time, vectorization, increase in brain penetration efficiency	Nasal	60%	Antioxidant, anticancer	Tong-Un et al. (2010)
Quercetin	Biodisponibility increased, vectorization, hepatic membrane penetration efficiency greatly improved	Transdermic		Hepatoprotector	Mandal and Das (2005)
Myrtle (*Myrtuscommunis*) extract	Antioxidant and antimicrobial activities superior to free forms	in vitro		Antioxidant, antimicrobial	Gortzi et al. (2008)

(continued)

Table 11 (continued)

Bioactive compounds/plant extract	Benefit in formulations	Route of admiration	Encapsulation efficiency	Possible therapeutic use	References
Catechin, (-)-epicatechin and EGCG	Liposomes may influence drug deposition in tumor tissues	Topical application and intratumoral		Antioxidant, anticancer	Fang et al. (2005)
Curcumin isolated from the root of *Curcuma longa*	Improved intravenous delivery of curcumin to tissue macrophages			Antitumor, antioxidant, antiplatelet, and aggregation	Sou et al. (2008)
Silymarin (silybin, taxifolin, isosilybin, silydianin, silychristin) obtained from fruits of *Silybum marianum*	Improved permeation and stability of silymarin		70%	Hepatoprotective agent	El-Samaligy et al. (2006b)
Lectin from seeds of *Cratylia Mollis*	Production of immunoglobulins in human and causes antibacterial			Immunostimulatory action	Andrade et al. (2004)
Camptothecin isolated from *Camptotheca acuminate* Decne	Overcome insolubility and stability	IV		Antitumor	Watanabe et al. (2008)
Essential oil from *Artemisia arborescens*	Effective to treat against HSV-1 and HSV-2	tropically		Antiviral	Sinico et al. (2005)
Triptolide from *Tripterygium wilfordii* Hook	Improve permeation	in vitro		Anti-inflammatory, immunosuppressive, and antifertility	Mei et al. (2003)
Extracts from fruit of *Syagrus romanzoffiana*	Improve permeation	in vitro		Antioxidant	Mezadri (2010)

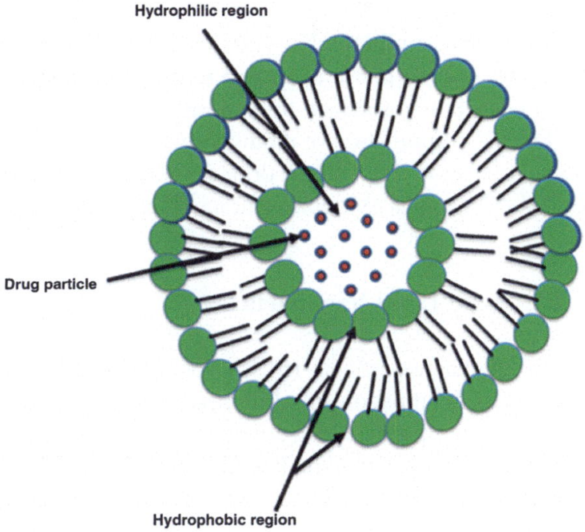

Fig. 12 Niosome structure

components are non-ionic surfactant, cholesterol, and charged molecules. Niosomes are classified based on their size into three groups of (a) small unilamellar vesicles (SUV) with a size range of 0.025–0.05 μm, (b) multilamellar vesicles (MLV) with a size of larger than 0.05 μm, and (c) large unilamellar vesicles (LUV) with a size of larger than 0.10 μm (Makeshwar and Wasankar 2013; Moghassemi and Hadjizadeh 2014).

Comparison of Liposomes with Niosomes

1. Liposome's ingredients such as phospholipids are not stable due to degradation associated with oxidative predisposition; natural phospholipids are different in terms of purity grade. Liposome's storage and handling need special care and conditions; meanwhile liposomes are expensive.
2. The properties of liposomes and niosomes are different, since double-chain phospholipids (charged or neutral) are used for the preparation of liposomes, while cholesterol and uncharged single-chain surfactants are employed for the preparation of niosomes (Moghassemi and Hadjizadeh 2014) (Fig. 12).

Niosomes Preparation

According to the wanted double-layer number, sizes, and size distribution of niosomes, vesicle membrane permeability, and aqueous phase entrapment efficiency, niosomes can be obtained through different techniques. Thus, sonication and microfluidization methods are used for the preparation of small unilamellar vesicles, while handshaking technique (thin-film hydration method) and transmembrane pH gradient (inside acidic) drug uptake process (remote loading) are employed for the preparation of multilamellar vesicles. In addition, large unilamellar vesicles are prepared by reverse phase evaporation method (REV) and ether injection technique. Factors such as resistance to osmotic stress, structure, nature, type, and amount of surfactant, composition of membrane, hydration temperature, niosomes preparation technique, and encapsulated drug molecule nature affect the physicochemical properties of niosomes (Moghassemi and Hadjizadeh 2014).

7.2.7 Solid Lipid Nanoparticles

Solid lipid nanoparticles (SLNs) are developed in 1990, which were taken into account as the most outstanding lipid-based carriers. SLNs have certain advantages such as scale-up and sterilization feasibility, low toxicity, and good biocompatibility (Ram et al. 2012; Dolatabadi et al. 2014). Indeed, to prepare SLNs, solid lipid components (physiologically tolerated) are used. In addition to the protection of drugs within SLNs, nanoparticle's solid matrix may adjust drug release behavior as well (Pallerla and Prabhakar 2013). SLNs were the promising transdermal drug delivery system in the promptly progressive era of nanotechnology due to their structural similarity with lipids of skin epidermis layer. SLNs since several previous years are used in cosmetics (Müller et al. 2002; Wissing et al. 2004). Based on the report, SLNs as the occlusive can prevent skin water loss that consequently boosts skin moisturization. Moreover, according to the previously performed claim, UV absorbance into the skin would be enhanced through the SLNs, which is too important from the cosmetics industry point of view.

However, SLNs are not utilized within the commercialized sunscreen products until now that is most probably attributed to the SLNs pretty complex manufacturing processes, like high-pressure and high-temperature homogenization (Fang et al. 2008). The prospect is limited because of the UV sunscreen low loading in SLN final colloidal systems (Zhai and Zhai 2014). SLNs thanks to their advantages including drug release rate well control and sterilization feasibility have a constant role in the local drug delivery. SLN properties such as small size and fine size distribution make easy the penetration of drug into deeper regions of the skin (Uner and Yener 2007). Table 12 represents the different examples of different plant extracts used as solid lipid nanoparticles (Fig. 13).

7.2.8 Coacervation

There are two types of coacervation methods, simple coacervation in which one type of polymer only used and complex coacervations in which two types of polymers are used. In simple coacervation, a poor solvent is added to a colloidal solution, and two phases are formed, one rich in colloidal particles and the other colloid-free solution phase. For encapsulating the plant extract, the active

Table 12 List of reported phytosomal (solid lipid nanoparticles) formulations using plant extract/bioactive molecules and their possible therapeutic use

Bioactive compounds/plant extract	Benefit in application	Route of administration	Dose	Possible therapeutic use	References
Bacopa	Enhance the antiamnesic activity	Oral	40 mg/kg	Antiamnesic	Habbu et al. (2013)
Curcumin	Used as a sustained delivery system	Oral	100 mg/kg	Antioxidant, anti-inflammatory, antimicrobial, anti-amyloid, and antitumor activities	Zhang et al. (2013b)
Rutin	Solubility enhancement of poorly soluble rutin	in vitro		Antioxidant	Singh et al. (2012)
Trichosanthes cucumerina and *Abrus precatorius* aqueous extract	Hair growth promoters	Topical	2%	Hair growth promoter	Sandhya et al. (2012)
Catechin	Systemic absorption	in vitro		Anti-inflammatory, antioxidant, antitumour, and hepatoprotective	Semalty et al. (2012)
Gallic acid	Effective against carbon tetrachloride induced liver and kidney damage	Oral	45 mg/kg	Antibacterial, antiviral, analgesic, and anti-apoptotic activities	Shyam et al. (2012)
Quercetin, kaempferol, and isorhamnetin	Bioavailability enhancement	Oral	20.3 mg/kg	Antioxidant phospholipid complexation	Chen et al. (2010b)
Flavonoids	Flavonoids of GBP stabilize the ROS	Subcutaneous	100/200 mg/kg	Cardioprotective, antioxidant activity	Panda and Naik (2008)
Flavonoids	Inhibits lipid peroxidation (LPO), stabilize the ROS	Oral	25/50 mg/kg	Hepatoprotective, antioxidant	Naik and Panda (2008)
Flavonoids	Absorption of silybin phytosome® from silybin is approximately seven times greater	Oral	120 mg	Hepatoprotective, antioxidant	Yanyu et al. (2006)
Ginsenosides	Increase absorption nutraceutical	Oral	150 mg	Immunomodulator	Bhattacharya (2009)

(continued)

Table 12 (continued)

Bioactive compounds/plant extract	Benefit in application	Route of admiration	Dose	Possible therapeutic use	References
Epigallocatechin	Increase absorption systemic	Oral	50–100 mg	Antioxidant, anticancer	Bhattacharya (2009)
Procyanidins	A controlled elevation in the blood total radical trapping antioxidant parameter were observed	Oral	50–100 mg	Systemic antioxidant, cardioprotective	Bhattacharya (2009)
Flavonoids	To enhance the absorption rate	Oral	100 mg	Cardioprotective and antihypertensive	Bhattacharya (2009)
Quercetin	Used to get better therapeutic effect	Oral		Antioxidant, anticancer	Maiti et al. (2005)
Curcumin	Increases the bioavailability by showing antioxidant activity	Oral	360 mg/kg	Antioxidant, anticancer	Maiti et al. (2007)
Naringenin	Prolonged duration of action	Oral	100 mg/kg	Antioxidant	Maiti et al. (2006)
Flavonoids	Stabilize the ROS cardioprotective	Subcutaneous	100/200 mg/kg	Antioxidant activity	Vandana and Suresh (2008)
Flavonoids	Inhibits lipid peroxidation (LPO), stabilize the ROS	Oral	25/50 mg/kg	Hepatoprotective, antioxidant	Suresh and Vandana (2008)
Flavonoids	Absorption of silybin phytosome from silybin is approximately seven times greater	Oral	120 mg	Hepatoprotective, antioxidant for the liver and skin	Yanyu et al. (2006)
Curcuminoids	Enhanced stability of curcuminoids			Antitumor, antioxidant, antiamylodin, antiplatelet aggregation, and anti-inflammatory	Tiyaboonchai et al. (2007)
Tetrandrine	Enhanced solubility and encapsulation of tetrandrine			Anti-inflammatory, antiplatelet aggregation, and free radical scavenging activity	Li et al. (2006b)

Triptolide	Enhanced anti-inflammatory and transdermal delivery of triptolide			Used in autoimmune diseases, rheumatoid arthritis, psoriasis, leukemia and antineoplastic	Mei et al. (2003)
Podophyllotoxin (active constituent podophyllin	Reduction of adverse effects of podophyllotoxin			Antivirus in the treatment of warts through topical application and anticancer activity	Chen et al. (2006)
Cryptotanshinone	Enhancement of bioavailability of cryptotanshinone			Anti-inflammatory, cytotoxic, antibacterial, antiparasitic, anti-angiogenic, and anti-oxidative	Hu et al. (2010)
Quercetin	Enhancement of bioavailability more than five times greater	in vitro		Antioxidant, anticancer	Li et al. (2009)
Quercetin	Promote permeation in the epidermis and dermis	in vitro		Antioxidant and anti-inflammatory	Guo et al. (2012)
Curcumin	Improve bioavailability and prolonged drug release	in vitro	1, 12.5, 25, and 50 mg/kg	Antioxidant, anticancer	Kakkar et al. (2011)

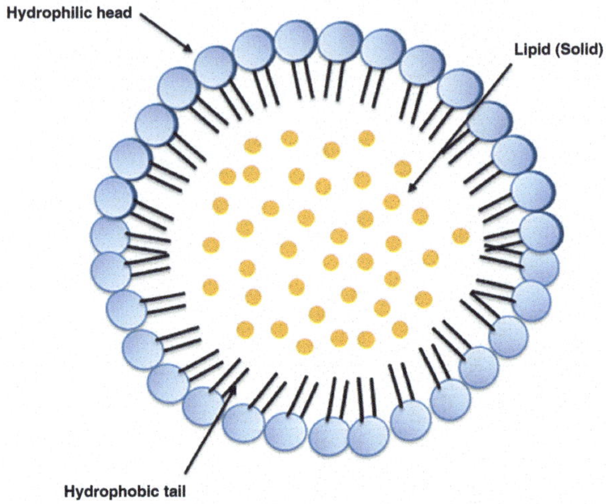

Fig. 13 Solid lipid nanoparticle schematic representation. (Adopted from Badri et al. 2016)

component is dispersed in a polymer solution, and a desolvation agent is added for phase separation. Polymer coating is deposited on the active fluidized-bed drying or component and is stabilized and hardened. The obtained microcapsules can be dried by spray drying. These microcapsules possess more than 50% encapsulation efficiency. The drawback of this technique is that they are stable only in a narrow range of pH and temperature.

In complex coacervation, complexation occurs between two oppositely charged polymers. At first, the active material (oil) is dispersed in to a polymer solution (cationic, e.g., gelatin); a second anionic polymer solution (Arabic gum) is then added to form dispersion. The two oppositely charged polymers undergo complexation and deposition of shell on the active material occurs. Prepared microcapsules can be stabilized by thermal treatment, desolvation, or chemical cross-linking. Researchers have used this method to encapsulate different oils. Patchouli oil is highly volatile, unstable oil with strong smell, which possess medicinal properties. In order to reduce the volatility and strong smell and to prevent the oxidation, Han et al. (2013) encapsulated this oil using complex coacervation technique.

7.2.9 Sol-Gel Encapsulation

Sol-gel encapsulation can be used to trap hydrophobic agents inside a spherical shell of amorphous silica. The hydrophobic material is solubilized in the silicon phases such as tetramethoxy silane or tetraethoxy silane, and oil in water emulsion is formed. Silica droplets are hydrolyzed and condensed at the oil water interface to form hard silica shell in which the hydrophobic agents are entrapped.

7.2.10 Spray Drying

Spray drying is a widely used method for encapsulation because of its low cost and simplicity. In this process, the suspension or dispersion of the active material (core) and shell material is sprayed into a hot drying chamber. During spraying, the solvent will be evaporated, and shell material gets deposited on the active component. Microparticles obtained by this method are small and spherical with homogeneous distribution. Advantages of this process are the following: it can be operated continuously, and it is a quick process. However, the use of air at high temperatures may affect the biological activity of the component. These materials are susceptible to easy oxidation; therefore, the shelf life of spray-dried products is less. Carvalho et al. (2014) used spray-drying technology to encapsulate green coffee oil. Green coffee oil has cosmetic properties like emollient, antioxidant, and UV absorption capacity. The shell material used in this method includes arabic gum, maltodextrin, and modified starch. In spray congealing method, the protective coating material is applied as melt. The core material is dispersed in the melt of the coating polymer. Solidification of the melt is achieved by passing through hot air to cold air stream.

7.2.11 Freeze-Drying

In freeze-drying, the suspension or dispersion of the active component and shell is frozen, and then solvent is evaporated via sublimation under high vacuum. Compared to spray-dried products, freeze-dried products are superior in quality and more stable to oxidation. For heat-sensitive products, freeze-drying is the best method. The drawbacks of freeze-drying process are high energy consumption and long processing time. Since the freeze-dried particles are not homogenous and possess irregular shapes, their encapsulation efficiency is less compared to spray-dried particles.

Tao et al. (2017) used freeze-drying method for encapsulating blueberry anthocyanin extract. A mixture of whey protein isolates, β-cycodextrin, maltodextrin, and gum arabic was used as the matrix material. Homogeneous mixture was prepared by magnetic stirring followed by ultrasonication. The obtained dispersion was dried using freeze dryer. In another study, tomato oleoresins, pumpkin oleoresins, and wheat bran oleoresins were encapsulated in α-cyclodextrin by freeze-drying method (Durante et al. 2016). Supercritical carbon dioxide technology was adopted to extract oleoresins.

7.2.12 Fluidized Bed Coating or Air Suspension

In this method, active agents (core material) to be coated are suspended in a stream of air, and then the coating material (polymer solution) is sprayed onto the moving particles. The solvent gets evaporated, and an outer layer is formed over the particles. Desired thickness and weight of the product can be achieved by repeating the process. Fluid bed coater can be top spray type, bottom spray type, or tangential spray type.

7.2.13 Polymer Encapsulation by Rapid Expansion of Supercritical Fluids

Super critical fluid – highly compressed gases (CO_2, N_2O) – containing the shell material and core material maintained at high pressure is released at atmospheric pressure through a small nozzle. The sudden pressure drops cause desolvation, and the shell material gets coated over the active component (core). Coating material used in this method includes polyethylene glycol and paraffin wax. The main condition in this technique is that both the active component and the coating material should be soluble in supercritical fluid.

8 Drug Release Profiles vs Administration Route of Encapsulated Plant Extracts

The presence of different types of bioactive components (polyphenols and flavonoids) is traditionally used to treat various diseases. However, the direct oral administrations of such natural bioactive molecules degrade during the course of the administration and absorption period leading to the significant loss of bioactivity and therapeutic efficiency. The nanotherapeutics are expected to subvert the limitation of current drug therapy (conventional), which includes less target-oriented drug release, less bioavailability, and therapeutic index. In the case of drug delivery system, the biocompatible polymeric template is expected to take advantage of engineering capability to reduce burst release and improve target efficiency at the action site. Compared to conventional therapeutic approaches, the well-designed nano-based drug delivery system is expected to stabilize the phytocomponents and improve the nano delivery as shown in Fig. 14.

9 Drug Release Pattern

9.1 Effect of pH on Drug Release

The polyphenol (epigallocatechin-3-gallate) present in green tea has been encapsulated into the biocompatible polymer and used as anticancer drug. However, the bioavailability decreases, and degradation increases due to structural complication of these types of polyphenols. For instance, the presence of a number of hydroxyl and gallolyl groups has the difficulty to diffuse through the intestinal epithelium (Li et al. 2011c). The presence of acidic pH environment in human gastrointestinal

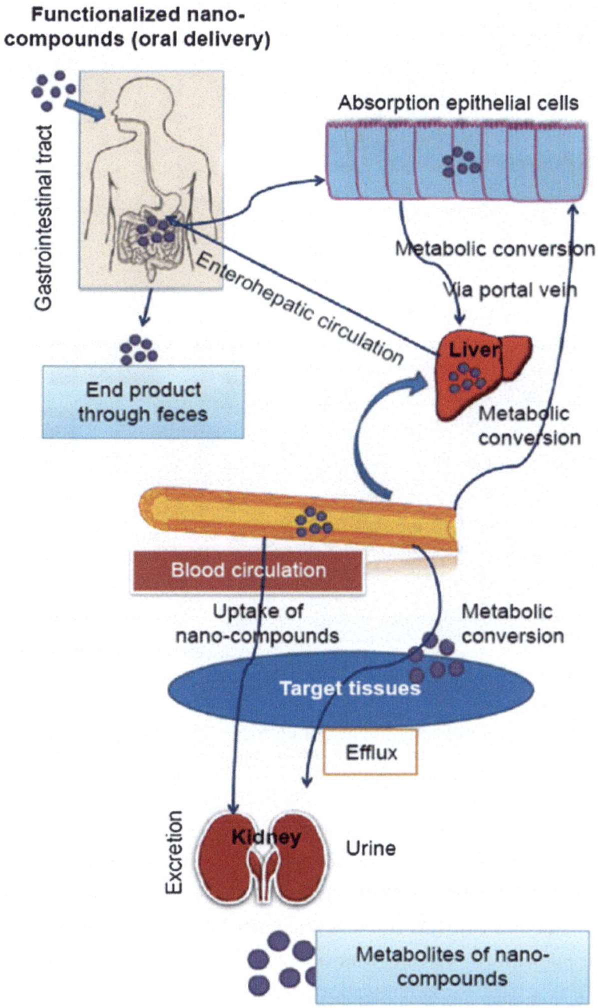

Fig. 14 Nano-based phytobioactive molecule bioavailability route through oral delivery in humans. (Adopted from Ganesan et al. 2017)

tract also reported to degrade the content through formation of dimerization process (Lun Su et al. 2003). Therefore, encapsulation into nanomaterials are gaining importance, which is reported to favor the detainment of the plant extract property through stabilization and therefore enhancing their medicinal property availability (biocompatibility) for a longer time.

The effect of pH was observed when the synthesis and characterization of polyphenols extracted from fresh strawberry fruits were assessed through chitosan encapsulation. The presence of positively charged functional amino group is reported to enhance the loading of negatively charged polyphenols and improve bioavailability and sustained release. The trapping of polyphenols was reported to

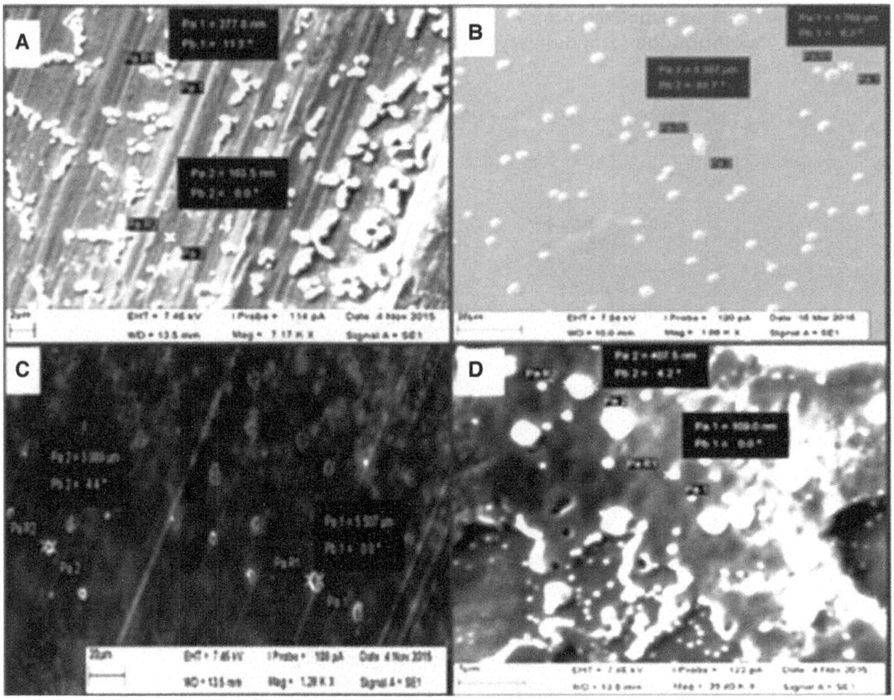

Fig. 15 Scanning electron microscopy images of chitosan-TPP NPs before and after release, (**a**) before release, (**b**) after release at pH 1.4, (**c**) after release at pH 7.4, and (**d**) after release at pH 10.4. (Adopted from Pulicharla et al. 2016)

be 58%, while optimum-loading capability was found to be 36%. The release profile was found to be pH dependent by studying the drug release at pH 1.4, pH 7.4, and 10.4, respectively. An initial burst release profile for polyphenols was observed at pH 7.4, while sustained release was observed at pH 1.4 (Pulicharla et al. 2016) (Fig. 15).

The release profiles of polyphenols are corroborated with particle size distributions using SEM images (Fig. 14). The polyphenol's sustained release at pH 1.4 shows the presence of small-sized particle in sustained manner, while increased release of polyphenols with large particle sizes are observed at high pH condition (Pulicharla et al. 2016).

9.2 Food Intake and Body Weight Influence the Drug Release

Microencapsulation of shrub-based *Catha edulis* termed as Khat using gelation was reported to be effective against obesity. The slow release rate of Khat through subcutaneous injection route in controlled fashion is studied on food intake (FI),

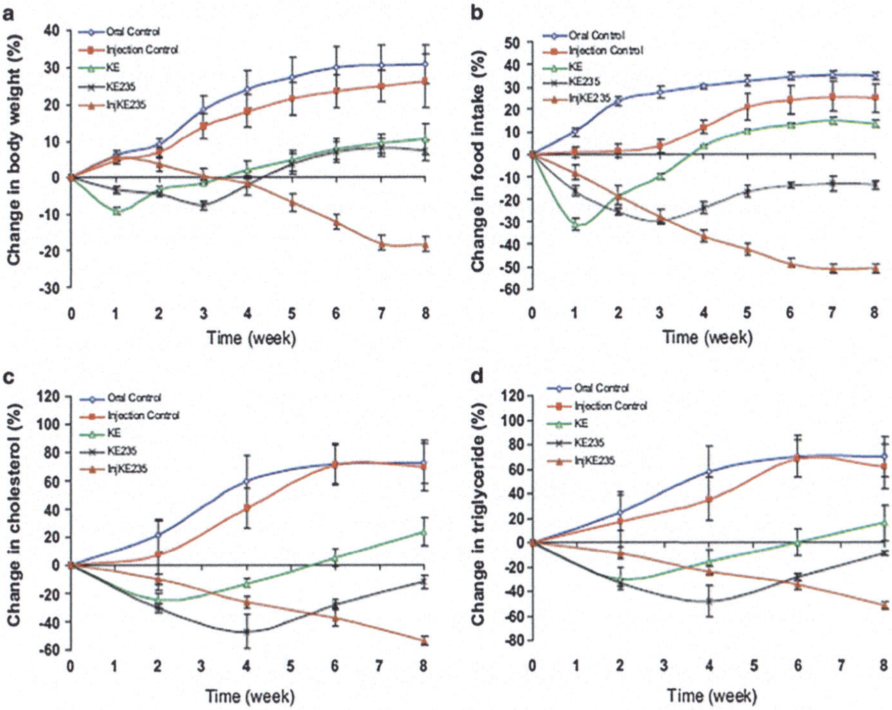

Fig. 16 The change in (**a**) body weight, (**b**) food intake, (**c**) cholesterol, and (**d**) triglyceride levels (%). Mean ± SEM, N = 12. (Adopted from Aziz et al. 2011)

body weight (BW), cholesterol (CS), and triglyceride (TG) levels. The study showed correlation between $T_{50\%}$ and reduction of BW, CS, and TG (Aziz et al. 2011) (Fig. 16).

9.3 Route of Administration Influence the Drug Release

The solubility of plant extract camptothecin derived from *Camptotheca acuminate* Decne was reported to improve and shown to exert anticancer effect through intravenous injection route. The hydrophobically designed glycol-based chitosan with about 80% camptothecin loading through dialysis technique has shown to target intracellular topoisomerase. The injection amount of 10 mg per kg and 30 mg per kg is reported to be an effective dose compared to free camptothecin with 30 mg per kg. Primarily the activity was attributed due to enhanced solubility and longer blood circulation at the targeted tumor region (Min et al. 2008).

Moreover, intravenous injection route of natural polyphenol resveratrol-loaded lipid was reported to be effective for Alzheimer disease. The low solubility of flavonoid subdues the biological advantage of resveratrol that further tends to isomerize and degrade during environmental exposures (pH, temperature, and light). However, the encapsulation of resveratrol into lipid core hydrophobic structure is reported to bypass the filtration by the liver and spleen and also helps to cross endothelial cells of the blood-brain barrier. The lipid functionalization with antibody, in particular anti-transferrin receptor monoclonal antibody, was reported to help the transfer of natural component to the targeted brain (Loureiro et al. 2017). In addition, solid-based lipids like glyceryl dilaurate, stearic acid, hydrine, cetyl alcohol, and glyceryl monostearate and liquid-based lipids (glyceryl monodicaprylate, oleic acid, and capric acid) are found to be effective for dermal administration route. The lipids tend to enhance the encapsulation efficiency of drugs by about 70%, and the nanoform increases the dermal penetration by improving contact with stratum corneum (Santos et al. 2013). Quercetin-lipid nanoformulations with particle size in the range of 215.2 nm and entrapment ability of 89.95% are reported to be effective for dermal delivery route. The flavonoid-rich quercetin was reported to improve the traverse through stratum corneum and exert anti-inflammatory action (Guo et al. 2012). Some other recent developments of route of administration in case of encapsulation drug efficiency can be summarized in Table 13.

10 Commercially Available Plant Extract/Herbal Formulations

Nano-phytomedicines are prepared from plant extracts or with their therapeutically active constituents. Nano-drug delivery systems help in better bioavailability that decreases side effects and toxicity. Nowadays, there are many companies involved to market nano-herbal formulation. Among them, two companies dominate the market, viz., Cosmetochem and Indena. For herbal drug delivery, Cosmetochem launches Herbasec® technology in the market which is actually a liposomal preparation of various herbal constituents like extracts of white tea, green tea, white hibiscus, guarana, and aloe vera. These extracts are used in cosmetics because of their antioxidant effects for prevention of aging. Indena patented the technology of Phytosomes® and launched many products in the market under this having diverse therapeutic benefits. Indena commercializes the plant constituents/extracts of liquorice (18ß-glycyrrhetinic acid), *Ammi visnaga* (visnadin), *Centella asiatica* (triterpenes), *Ginkgo biloba* (ginkgo flavone glucosides, ginkgolides, bilobalide), hawthorn flower (vitexin-2″-O-rhamnoside), milk thistle (silymarin and silybin), horse chestnut (escin ß-sitosterol), *Terminalia sericea* (sericoside), *Panax* ginseng (ginsenosides), grape seed (polyphenols), green tea (polyphenols), etc. (Devi et al. 2010; Pinto 2010). Table 14 presents some of the marketed nano-plant extract or herbal medicines.

Table 13 Summary of recent encapsulation studies involving various plant extracts and administration routes

Study type	Nanocarrier	Bioactives	% efficiency	Finding/benefits	References
Anthocyanin encapsulation using liposomes in supercritical CO_2	Liposomes	Anthocyanin	50.6% (EE)	The flavonoid release from liposome carrier was found to be a sustained release (≤3509%) in simulated intestinal fluid condition	Zhao et al. (2017)
In vitro and in vivo study of curcumin-mixed micelles (mPEG-PLA/TPGS) for oral route	Mixed micelles Poly(ethylene glycol)-poly (lactide)	Curcumin	16.1% (curcumin loading)	The in vitro study in simulated gastrointestinal solution showed the controlled curcumin release and improved bioavailability	Duan et al. (2016)
Oral administration of Soluplus-*Angelica gigas* Nakai nanocombination extract fabricated through electrohydro dynamic technique	Soluplus (polyethylene glycol 6000)	*Angelica gigas* Nakai	Mean entrapment efficiency (EE) of decursin and decursinol angelate to be 100% and 85.4%	The study showed that herb loaded over nanocomposite (AGN/SP2 NC) can be efficiently used for oral delivery that showed high exposure effect and high concentration in plasma	Lee et al. (2016)
Antioxidant derived from red grapes in combination with micelle chitosan for oral route study	Quaternary ammonium (thiolated or non-thiolated) and chitosan	Grape seed extract (antioxidant)	15–20% loading (80% EE)	The technique improved the loading of antioxidant (15–20%) and internalization by endothelial progenitor cells	Fabiano et al. (2016)
Curcumin/microparticles for ulcerative colitis through oral route administration	Eudragit, polymer poly(lactide-co-glycolide) (PLGA)	Curcumin	80% (curcumin loading)	Curcumin in the micron-sized particles ranging 1.52–1.91 μm was shown to be effective for ulcerative colitis	Xiao et al. (2015)
Oral administration of quercetin on a mouse model of Alzheimer's disease	Zein	Quercetin	70 ± 1.3 μg/mg payload (EE of 81.2 ± 1.3%)	Quercetin flavonoid has been loaded over zein (a natural type of polymer) and treated for Alzheimer's disease through oral absorption	Moreno et al. (2017)

(continued)

Table 13 (continued)

Study type	Nanocarrier	Bioactives	% efficiency	Finding/benefits	References
Enhancing the bioavailability of *Silybum marianum* dry extract	Coground (Gelucire)	*Silybum marianum*	Production yield of 94–95 by wt and EE of 92–98%	Mechanochemical and spray drying of dry extract of *Silybum marianum* along with activated coground (Gelucire) improve bioavailability	Passerini et al. (2012)
Self-assembly of green tea catechin derivatives in nanoparticles for oral lycopene delivery	Green tea catechin derivatives	Lycopene	9% (lycopene loading), EE (89%)	The in vivo study in mice showed an improved pharmacokinetics and can be potential oral drug delivery system	Li et al. (2017)
Self-assembly of green tea catechin derivatives in nanoparticles for oral lycopene delivery	Green tea-derived oligomerized (-)-epigallocatechin-3-O-gallate	Lycopene	9% (lycopene loading), 89% (EE)	Green tea-based nanoparticle-chitosan polymer was used to load lycopene (natural antioxidant). The composite showed an improved pharmacokinetics and can be potential oral drug delivery system	Li et al. (2017)
Oridonin-mixed micelles (Soluplus-Pluronic P105) for oral administration study	Soluplus and Pluronic P105	Oridonin, a diterpenoid compound ($C_{20}H_{28}O_6$)	15.08 ± 0.38% (loading), 90.48 ± 1.85% (EE)	The optimized drug formulation showed improved bioavailability (210.55%) and can be a potential drug therapy for cancer treatment	Ke et al. (2017)
Proliposomes containing a bile salt for oral delivery of *Ginkgo biloba* extract: formulation optimization, characterization, oral bioavailability, and tissue distribution in rats	Proliposomes	*Ginkgo biloba*	Up to 88% (encapsulation efficiency)	The nanoformulation enhanced the absorption in gastrointestinal tract and subsequently reduced elimination	Zheng et al. (2015)

Table 14 Commercially available herbal formulations with plant extract in pharmaceutical and cosmetic industry

Product name with photo	Plant extract with active ingredients	Formulation	Route of administration	Therapeutic use	References
	*Cuscuta chinensis* A/I: flavonoids and lignans	Nano-suspension method	Oral	Hepatoprotective and antioxidants effect	Yen et al. (2008)
	*Artemisinin* A/I: artemisinin	Self-assembly procedure	IV	Anticancer	Youfang et al. (2009)
	*Radix salvia miltiorrhiza A/C: R. salvia miltiorrhiza extracts*	Spray-drying technique	IV	Coronary heart diseases, angina pectoris, and myocardial infarction	Su et al. (2008)
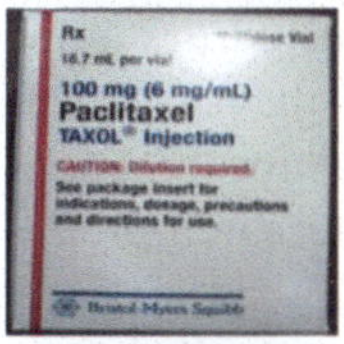	Taxel-loaded nanoparticles A/I: taxel	Emulsion solvent evaporation method	IV	Anticancer	Fu et al. (2006)
	Berberine-loaded nanoparticles A/I: berberine	Ionic gelation method	IV	Anticancer	Lin et al. (2007)
	Sunscreens A/I: ultraviolet filters	Nano-form	Topical	UV protection	Online data

Table 14 (continued)

Product name with photo	Plant extract with active ingredients	Formulation	Route of administration	Therapeutic use	References
	Breast cream *Pueraria mirifica* A/I: St. herb	Niosomes	Topical	Increased size	Online data
	Hair care Nettle leaf extract, black elderberry extract, hamomile combined with citrus and mint oils	Nanoceutical-shampoo	Topical	Diminish dandruff and increase hair volume and shine	Online data

11 Future Prospective

Throughout the whole world, research is ongoing on plant extract remedies and natural products. Herbal formulation development is being carried out in a number of institutes at the basic and clinical trial levels (Namdaria et al. 2017). The only concern is to develop the best systems for the suitable delivery of such drugs at the sites and in the whole body, with a dose that won't compromise with the basic treatment (Yadav et al. 2011). In the future, herbal nanoparticle concepts for infectious disease and cancer drug delivery may also entice some potential research groups and create attention-grabbing results. Therefore, using "herbal/plant extract" enclosed in nanocarriers will elevate its potential for the treatment of several chronic diseases and health benefits. Plant extracts are also potent antioxidants and constituents that can be made useful in the food industry (Sethiya et al. 2010). This type of integrative research between the traditional "herbal/plant extract" and modern drug delivery system, i.e., "nanotechnology," has established attraction to the pharmaceutical industry which may be taken advantage of in the near future that will promote peoples' health.

12 Conclusions

Research in this area is still at the exploratory stage. Many problems in the research, production, and application need to be solved. In addition, more attention should be paid to the research on the carrier materials in order to develop more suitable carriers which can reduce the toxicity of drugs, enhance their activity, and improve the overall quality of the agents. Herbal drugs have enormous therapeutic potential, which should be explored through some value-added drug delivery systems. Lipid solubility and molecular size are the major limiting factors for drug molecules to pass the biological membrane to be and phytomolecules, despite having excellent bioactivity in vitro demonstrate less or no in vivo actions due to their poor lipid solubility or improper molecular size or both, resulting to poor absorption and poor bioavailability. Standardized plant extracts or mainly polar phytoconstituents like flavonoids, terpenoids, tannins, and xanthones when administered through novel drug delivery system show much better absorption profile which enables them to cross the biological membrane, resulting in enhanced bioavailability. Hence more amount of active constituent becomes present at the site of action (liver, brain, heart, kidney, etc.) at similar or less dose as compared to the conventional plant extract or phytomolecule. Hence, pharmaceutical nanotechnology is the most ideal and suitable carrier systems for the improvement of pharmacokinetics and bioavailability of plant actives and extracts.

References

Adam F, Abert-Vian M, Peltier G et al (2012) "Solvent-free" ultrasound-assisted extraction of lipids from fresh microalgae cells: a green, clean and scalable process. Bioresour Technol 114:457–465

Adil I, Cetin H, Yener M et al (2007) Subcritical (carbon dioxide + ethanol). extraction of polyphenols from apple and peach pomaces, and determination of the antioxidant activities of the extracts. J Supercrit Fluid 43(1):55–63

Agrawal R, Kaur IP (2010) Inhibitory effect of encapsulated curcumin on ultraviolet-induced photoaging in mice. Rejuvenation Res 13(4):397–410. http://doi.org/10.1089/rej.2009.0906

Ahmad HA (2013) Special issue on polymer and hybrid particles for biomedical applications. J Colloid Sci Biotechnol 2:153–154

Ajila CM, Brar SK, Verma M et al (2011) Extraction and analysis of polyphenols: recent trends. Crit Rev Biotechnol 31(3):227–249

Akay S, Alpak I, Yesil-Celiktas O (2011) Effects of process parameters on supercritical CO_2 extraction of total phenols from strawberry (*Arbutus unedo* L.) fruits: an optimization study. J Sep Sci 34(15):1925–1931

Ali J, Akhtar N, Sultana Y et al (2008) Antipsoriatic microemulsion gel formulations for topical delivery of babchi oil (*Psoralea corylifolia*). Methods Find Exp Clin Pharmacol 30:1–9

Alonso-Salces RM, Korta E, Barranco A et al (2001) Pressurized liquid extraction for the determination of polyphenols in apple. J Chromatogr A 933(1):37–43

Alupului A, Călinescu L, Lavric V (2012) Microwave extraction of active principles from medicinal plants. U.P.B. Sci Bull, Ser B 74(2):129–142

Ameer K, Shahbaz Shahbaz AM, Kwon JH (2017) Green extraction methods for polyphenols from plant matrices and their byproducts: a review. Compre Rev Food Sci Food Saf 16:295–315

Andrade CAS, Correia MTS, Coelho LCBB et al (2004) Antitumor activity of *Cratylia mollis* lectin encapsulated into liposomes. Int J Pharm 278(2):435–445

Andrade KS, Gonalvez RT, Maraschin M et al (2012) Supercritical fluid extraction from spent coffee grounds and coffee husks: antioxidant activity and effect of operational variables on extract composition. Talanta 88:544–552

Andrade KS, Poncelet D, Ferreira S (2017) Sustainable extraction and encapsulation of pink pepper oil. J Food Eng 204:38–45

Anekpankul T, Goto M, Sasaki M et al (2007) Extraction of anti-cancer damnacanthal from roots of Morinda citrifolia by subcritical water. Sep Purif Technol 55(3):343–349

Annamalai A, Christina VLP, Sudha D et al (2013) Green synthesis, characterization and antimicrobial activity of Au NPs using *Euphorbia hirta* L. leaf extract. Colloids Surf B: Biointerfaces 108:60–65

Arapitsas P, Turner C (2008) Pressurized solvent extraction and monolithic column-HPLC/DAD analysis of anthocyanins in red cabbage. Talanta 74(5):1218–1223

Armendáriz-Barragán B, Zafar N, Badri W et al (2016) Plant extracts: from encapsulation to application. Expert Opin Drug Deliv 13:1165–1175. http://doi.org/10.1080/17425247.2016.1182487

Arokiyaraj S, Saravanan M, Udaya Prakash NK et al (2013) Enhanced antibacterial activity of iron oxide magnetic nanoparticles treated with *Argemone mexicana* L. leaf extract: an *in vitro* study. Mater Res Bull 48:3323–3327

Aroonsri P, Jintanaporn W, Saengrawee S et al (2008) Applications of novel drug delivery system for herbal formulations. Nanomed Nanotechnol Biol Med 4:70–78

Arunachalam KD, Annamalai SK, Arunachalam AM et al (2013) Green synthesis of crystalline silver nanoparticles using *Indigofera aspalathoides*-medicinal plant extract for wound healing applications. Asian J Chem 25:S311–S314

Asbahani AE, Miladi K, Badri W et al (2015) Essential oils: from extraction to encapsulation. Int J Pharm 483:220–243

Asghari J, Ondruschka B, Mazaheritehrani M (2011) Extraction of bioactive chemical compounds from the medicinal Asian plants by microwave irradiation. J Med Plants Res 5(4):495–506

Aziz AH, Khiang Peh K, Fung Tan YT (2011) Herbal delivery system for treatment of obesity administration of encapsulated khat-extracts on body weight of rats. Obes Res Clin Prac 5:e305–e312

Badri W, Miladi K, Nazari QA et al (2016) Encapsulation of NSAIDs for inflammation management: overview, progress, challenges and prospects. Int J Pharm 515:757–773

Badri W, Miladi K, Nazari QA et al (2017) Effect of process and formulation parameters on polycaprolactone nanoparticles prepared by solvent displacement. Colloids Surf Physicochem Eng Asp 516:238–244

Bai XL, Yue TL, Yuan YH (2010) Optimization of microwave-assisted extraction of polyphenols from apple pomace using response surface methodology and HPLC analysis. J Sep Sci 33(23):3751–3758

Bairwa NK, Sethiya NK, Mishra SH (2010) Protective effect of stem bark of *Ceiba pentandra* linn against paracetamol-induced hepatotoxicity in rats. Pharmacognosy Res 2:26–30

Ballard TS, Mallikarjunan P, Zhou K et al (2010) Microwave-assisted extraction of phenolic antioxidant compounds from peanut skins. Food Chem 120(4):1185–1192

Barakat N, Khil M, Omran A (2009) Extraction of pure natural hydroxyapatite from the bovine bones bio waste by three different methods. J Mater Process Tech 209(7):3408–3415

Basarkar A, Devineni D, Palaniappan R et al (2007) Preparation, characterization, cytotoxicity and transfection efficiency of poly (DL-lactide-co-glycolide) and poly (DL-lactic acid) cationic nanoparticles for controlled delivery of plasmid DNA. Int J Pharm 343:247–254

Bernhoft A (2010) A brief review on bioactive compounds in plants. In: Proceedings from a symposium held at the Norwegian Academy of Science and Letters, Oslo

Bhatia A, Shard P, Chopra D et al (2011) Chitosan nanoparticles as carrier of immunorestoratory plant extract: synthesis, characterization and immunorestoratory efficacy. Int J Drug Del 3:381–385

Bhattacharjee P, Singhal RS, Tiwari SR (2006) Supercritical carbon dioxide extraction of cottonseed oil. J Food Eng 79(3):892–989

Bhattacharya S (2009) Phytosomes: emerging strategy in delivery of herbal drugs and nutraceuticals. Pharm Times 41:9–12

Basniwal RK, Buttar HS, Jain VK et al (2011) Curcumin nanoparticles: preparation, characterization, and antimicrobial study. J Agric Food Chem 59(5):2056–2061

Bilati U, Allémann E, Doelker E (2005) Development of a nanoprecipitation method intended for the entrapment of hydrophilic drugs into nanoparticles. Eur J Pharm Sci 24:67–75

Billon A, Chabaud L, Gouyette A et al (2005) Vancomycin biodegradable poly(lactide-co-glycolide) microparticles for bone implantation. Influence of the formulation parameters on the size, morphology, drug loading and *in vitro* release. J Microencapsul 22:841–852
Bisht S, Feldmann G, Soni S et al (2007) Polymeric nanoparticle-encapsulated curcumin ("nanocurcumin"): a novel strategy for human cancer therapy. J Nanobiotechnol 5(3):1–18
Bitar A, Zafar N, Valour JP et al (2015) Elaboration of sponge-like particles for textile functionalization and skin penetration. Colloid Polym Sci 293:2967–2977
Braga ME, Santos RM, Seabra IJ et al (2008) Fractioned SFE of antioxidants from maritime pine bark. J Supercrit Fluid 47(1):37–48
Briones-Labarca V, Plaza-Morales M, Giovagnoli-Vicuña C et al (2015) High hydrostatic pressure and ultrasound extractions of antioxidant compounds, sulforaphane and fatty acids from Chilean papaya (*Vasconcellea pubescens*) seeds: effects of extraction conditions and methods. LWT-Food Sci Technol 60(1):525–534
Buckton G (2000) Interfacial phenomena in drug delivery and targeting. CRC Press, New York
Budrat P, Shotipruk A (2009) Enhanced recovery of phenolic compounds from bitter melon (*Momordica charantia*) by subcritical water extraction. Sep Purif Technol 66(1):125–129
Burris JN, Lenaghan SC, Zhang M et al (2012) Nanoparticle biofabrication using English ivy (*Hedera helix*). J Nanobiotechnol 10:41–47
Butler MS, Buss AD (2006) Natural products-the future scaffolds for novel antibiotics. Biochem Pharmacol 71:919–929
Calvo TRA, Busch VM, Santagapit PR (2017) Stability and release of an encapsulated solvent-free lycopene extract in alginate-based beads. LWT-Food Sci Tech 77:406–412
Campos E, Melo N, Paula E et al (2013) Screening of conditions for the preparation of poly(ε-caprolactone). Nanocapsules containing the local anesthetic articaine. J Colloid Sci Biotec 2(2):106–111
Carrera C, Ruiz-Rodríguez A, Palma M et al (2012) Ultrasound-assisted extraction of phenolic compounds from grapes. Anal Chim Acta 732:100–104
Carvalho R (2005) Supercritical fluid extraction from rosemary (*Rosmarinus officinalis):* kinetic data, extract's global yield, composition, and antioxidant activity. J Supercrit Fluid 35(3):197–204
Carvalho AGS, Silva VM, Hubinger MD (2014) Microencapsulation by spray drying of emulsified green coffee oil with two-layered membranes. Food ResInt 61:236–245
Cavero S, Garćıa-Risco M, Mańn F, Jaime L et al (2006) Supercritical fluid extraction of antioxidant compounds from oregano: chemical and functional characterization via LC-MS and *in vitro* assays. J Supercrit Fluid 38(1):62–69
Cenni E, Granchi D, Avnet S et al (2008) Biocompatibility of poly (D,L-lactide-co-glycolide) nanoparticles conjugated with alendronate. Biomaterials 29:1400–1411
Chakraborty B, Nath A, Saikia H et al (2014) Bactericidal activity of selected medicinal plants against multidrug resistant bacterial strains from clinical isolates. Asian Pac J Trop Med 7S1:435–441
Chan K, Ismail M (2009) Supercritical carbon dioxide fluid extraction of *Hibiscus cannabinus* L. seed oil: a potential solvent-free and high antioxidative edible oil. Food Chem 114(3):970–975
Chandini SK, Rao LJ, Gowthaman MK et al (2011) Enzymatic treatment to improve the quality of black tea extracts. Food Chem 127(3):1039–1045
Chao P, Deshmukh M, Kutscher HL et al (2010) Pulmonary targeting microparticulate camptothecin delivery system: anticancer evaluation in a rat orthotopic lung cancer model. Anti-Cancer Drugs 21:247–238
Chaturvedi M, Kumar M, Sinhal A (2011) Recent development in novel drug delivery systems of herbal drugs. Int J Green Pharm 5:87–94
Cheigh CI, Chung EY, Chung MS (2012) Enhanced extraction of flavanones hesperidin and narirutin from *Citrus unshiu* peel using subcritical water. J Food Eng 110(3):472–477
Chen H, Chang X, Weng T et al (2004) A study of microemulsion systems for transdermal delivery of triptolide. J Control Release 98:427–436

Chen H, Chang X, Du D et al (2006) Podophyllotoxin-loaded solid lipid nanoparticles for epidermal targeting. J Control Release 110:296–306
Chen F, Sun Y, Zhao G et al (2007) Optimization of ultrasound-assisted extraction of anthocyanins in red raspberries and identification of anthocyanins in extract using high-performance liquid chromatography-mass spectrometry. Ultrason Sonochem 14(6):767–778
Chen S, Xing XH, Huang JJ et al (2010a) Enzyme-assisted extraction of flavonoids from Ginkgo biloba leaves: improvement effect of flavonol transglycosylation catalyzed by *Penicillium decumbens* cellulase. Enzyme Microb Technol 48(1):100–105
Chen ZP, Sun J, Chen HX et al (2010b) Comparative pharmacokinetics and bioavailability studies of quercetin, kaempferol and isorhamnetin after oral administration of Ginkgo biloba extracts, Ginkgo biloba extract phospholipid complexes and *Ginkgo biloba* extract solid dispersions in rats. Fitoterapia 81:1045–1052
Chen Y, Luo H, Gao A et al (2011) Ultrasound-assisted extraction of polysaccharides from litchi (*Litchi chinensis* Sonn). seed by response surface methodology and their structural characteristics. Innovat Food Sci Emerg Tech 12(3):305–309
Chen X, Deng Y, Xue Y et al (2012) Screening of bioactive compounds in Radix Salviae *miltiorrhizae* with liposomes and cell membranes using HPLC. J Pharm Biomed Anal 70:194–201
Chiara S, Alessandro DL, Francesco L et al (2005) Preparation and characterization of *Artemisia arborescens* liposomes. Eur J Pharm Biopharm 59:161–168
Chiremba C, Rooney LW, Trust BJ (2012) Microwave-assisted extraction of bound phenolic acids in bran and flour fractions from sorghum and maize cultivars varying in hardness. J Chromato A 1012(2):119–128
Cintra e Silva DO, Estevanato LLC, Simioni AR et al (2012) Successful strategy for targeting the central nervous system using magnetic albumin nanospheres. J Biomed Nanotechnol 8:182–189
Clark AM (1996) Natural products as a resource for new drugs. Pharm Res 13:1133–1144
Croteau R, Kutchan TM, Lewis NG (2000) Natural products (secondary metabolites). In: Buchanan B, Gruissem W, Jones R (eds) Biochemistry and molecular biology of plants. American Society of Plant Physiologists, Rockville, pp 1250–1318
Das J, Das S, Samadder A (2012) Poly (lactide-co-glycolide). encapsulated extract of *Phytolacca decandra* demonstrates better intervention against induced lung adenocarcinoma in mice and on A549 cells. Eur J Pharm Sci 47(2):313–324
Detoni CB, Oliveira DM, Santo IE et al (2012) Evaluation of thermal-oxidative stability and antiglioma activity of *Zanthoxylum tingoassuiba* essential oil entrapped into multi and unilamellar liposomes. J Liposome Res 22:1–7
Devi VK, Jain N, Valli KS (2010) Importance of novel drug delivery systems in herbal medicines. Pharmacogn Rev 4:27–31
Dey S, Rathod VK (2013) Ultrasound-assisted extraction of β-carotene from *Spirulina platensis*. Ultrason Sonochem 20(1):271–276
Dhanaraju MD, Jayakumar R, Vamsadhara C (2004) Influence of manufacturing parameters on development of contraceptive steroid loaded injectable microspheres. Chem Pharm Bull 52:976–979
Dhobi M, Mandal V, Hemalatha S (2009) Optimization of microwave assisted extraction of bioactive flavolignan–silybinin. J Chem Metro 3(1):13–23
Dolatabadi JEN, Hamishehkar H, Eskandani M (2014) Formulation, characterization and cytotoxicity studies of alendronate sodium-loaded solid lipid nanoparticles. Colloids Surf B: Biointerfaces 17:21–28
Dominguez H, Ntiiiez MJ, Lema JM (1995) Enzyme-assisted hexane extraction of soybean oil. Food Chem 54(2):223–231
Drulis-Kawa Z, Dorotkiewicz-Jach A (2010) Liposomes as delivery systems for antibiotics. Int J Pharm 387:187–198
Du FY, Xiao XH, Luo XJ et al (2009) Application of ionic liquids in the microwave-assisted extraction of poly phenolic compounds from medicinal plants. Talanta 78(3):1177–1184

Duan Y, Zhang B, Chu L et al (2016) Evaluation *in vitro* and *in vivo* of curcumin-loaded mPEG-PLA/TPGS mixed micelles for oral administration. Colloids Surf B Biointerfaces 141:345–354

Durante M, Lenucci S, Marresa PP et al (2016) α-Cyclodextrin encapsulation of supercritical CO2 extracted oleoresins from different plant matrices: A stability study. Food Chem 199:684–693

Egydio J, Moraes M, Rosa P (2010) Supercritical fluid extraction of lycopene from tomato juice and characterization of its antioxidation activity. J Supercrit Fluid 54(2):159–164

Einbond LS, Mighty J, Redenti S et al (2013) Actein induces calcium release in human breast cancer cells. Fitoterapia 91:28–38

Eley JG, Mathew P (2007) Preparation and release characteristics of insulin and insulin-like growth factor-one from polymer nanoparticles. J Microencapsul 24:225–234

El-Mowafy M, Viitala T, Ibrahim HM et al (2013) Silymarin loaded liposomes for hepatic targeting: *in vitro* evaluation and HepG2 drug uptake. Eur J Pharm Sci 50:161–171

El-Samaligy MS, Afifi NN, Mahmoud EA (2006a) Evaluation of hybrid liposome-encapsulated silymarin regarding physical stability and *in vivo* performance. Int J Pharm 319:121–129

El-Samaligy MS, Afifi NN, Mahmoud EA (2006b) Increasing bioavailability of silymarin using a buccal liposomal delivery system: preparation and experimental design investigation. Int J Pharm 308:140–148

Environmental Protection Agency, USA. http://www.epa.gov/greenchemistry/pubs/about_gc.html. Accessed 22 Sept 2017

Erdogan S, Ates B, Durmaz G et al (2011) Pressurized liquid extraction of phenolic compounds from *Anatolia propolis* and their radical scavenging capacities. Food Chem Toxicol 49(7):1592–1597

Fabiana TMCV, Thaís RMS, Jose ODC et al (2008) Quercetin in w/o microemulsion: *in vitro* and *in vivo* skin penetration and efficacy against UVB-induced skin damages evaluated *in vivo*. Eur J Pharm Biopharm 69:948–957

Fabiano A, Mattii L, Braca A et al (2016) Nanoparticles based on quaternary ammonium-chitosan conjugate: a vehicle for oral administration of antioxidants contained in red grapes. J Drug Deli Sci Tech 32:291–297

Fabricant DS, Farnsworth NR (2001) The value of plants used in traditional medicine for drug discovery. Environ Health Perspect 109(1S):69–75

Fang JY, Hung CF, Hwang TL et al (2005) Physicochemical characteristics and *in vivo* deposition of liposome-encapsulated tea catechins by topical and intratumor administrations. J Drug Target 13:19–27

Fang JY, Hwang TL, Huang YL et al (2006) Enhancement of the transdermal delivery of catechins by liposomes incorporating anionic surfactants and ethanol. Int J Pharm 310:131–138

Fang JY, Fang CL, Liu CH (2008) Lipid nanoparticles as vehicles for topical psoralen delivery: solid lipid nanoparticles (SLN) versus nanostructured lipid carriers (NLC). Eur J Pharm Biopharm 70:633–640

Farías-Campomanes AM, Rostagno MA, Meireles MA (2013) Production of polyphenol extracts from grape bagasse using supercritical fluids: yield, extract composition and economic evaluation. J Supercrit Fluid 77:70–78

Feng-Lin Y, Tzu-Hui W, Liang-Tzung L et al (2008) Preparation and characterization of Cuscuta chinensis nanoparticles. Food Chem Toxicol 46:1771–1777

Feng-Lin Y, Tzu-Hui W, Liang-Tzung L et al (2009) Preparation of naringenin nanoparticles. Pharm Res 26(4):893–902

Fessi H, Puisieux F, Devissaguet JP et al (1989) Nanocapsule formation by interfacial polymer deposition following solvent displacement. Int J Pharm 55:R1–R4

Fu RQ, He FC, Meng DS et al (2006) Taxol PLA nanoparticles. Acta Acad Med Mils Tertiae 28:1573–1574

Fu YJ, Liu W, Zu YG et al (2008) Enzyme-assisted extraction of luteolin and apigenin from pigeonpea [*Cajanus cajan* (L.) Millsp.] leaves. Food Chem 111(2):508–512

Gagliardi M, Bardi G, Bifone A (2012) Polymeric nanocarriers for controlled and enhanced delivery of therapeutic agents to the CNS. Ther Deliv 3:875–887

Galindo-Rodriguez SA, Allemann E, Fessi E et al (2005) Polymeric nanoparticles for oral delivery of drugs and vaccines: a critical evaluation of *in vivo* studies. Crit Rev Ther Drug Carrier Syst 22:419–464

Ganesan P, Arulselvan P, Choi DK (2017) Phytobioactive compound-based nanodelivery systems for the treatment of type 2 diabetes mellitus – current status. Int J Nanomedicine 12:1097–1111

García-Marino M, Rivas-Gonzalo J, Ibáñez E et al (2006) Recovery of catechins and proanthocyanidins from winery by-products using subcritical water extraction. Anal Chim Acta 563(1–2):44–50

Garti N (1997) Double emulsions: scope, limitations and new achievements. Colloids Surf Physicochem Eng Asp 123–124:233–246

Garti N, Aserin A (1996) Double emulsions stabilized by macromolecular surfactants. Adv Colloid Interface Sci 65:37–69

Garti N, Bisperink CGJ (1998) Double emulsions: progress and applications. Curr Opin Colloid Interface Sci 3:657–667

Gavini E, Alamanni MC, Cossu M et al (2005) Tabletted microspheres containing *Cynara scolymus* (var. Spinoso sardo) extract for the preparation of controlled release nutraceutical matrices. J Microencapsul 22:487–499

Gelmez N, Kincal N, Yener M (2009) Optimization of supercritical carbon dioxide extraction of antioxidants from roasted wheat germ based on yield, total phenolic and tocopherol contents, and antioxidant activities of the extracts. J Supercrit Fluid 48(3):217–224

Ghafoor K, Choi YH, Jeon JY et al (2009) Optimization of ultrasound-assisted extraction of phenolic compounds, antioxidants and anthocyanins from grape (*Vitis vinifera*) seeds. J Agricul Food Chem 57(11):4988–4994

Ghafoor K, Hui T, Choi YH (2011) Optimization of ultrasound-assisted extraction of total anthocyanins from grape peel. J Food Biochem 35:735–746

Ghayempour SA, Montazer M, Rad MM (2015) Tragacanth gum as a natural polymeric wall for producing antimicrobial nanocapsules loaded with plant extract. Int J Bio Macromol 81:514–520

Ghayempour S, Montazer M, Mahmoudi Rad M (2016) Encapsulation of Aloe Vera extract into natural Tragacanth Gum as a novel green wound healing product. Int J Bio Macromol 93(A):344–349

Ghoreishi S, Shahrestani R (2009) Subcritical water extraction of mannitol from olive leaves. J Food Eng 93(4):474–481

Giannuzzo AN, Boggetti HJ, Nazareno MA (2003) Supercritical fluid extraction of naringin from the peel of Citrus *paradise*. Phytochem Anal 14(4):221–223

Giri TK, Choudhary C, Alexander A et al (2013) Prospects of pharmaceuticals and biopharmaceuticals loaded microparticles prepared by double emulsion technique for controlled delivery. Saudi Pharm J 21:125–141

Glenn GM, Klamczynski AP, Woods DF et al (2010) Encapsulation of plant oils in porous starch microspheres. J Agric Food Chem 58:4180–4184

Godin B, Touitou E (2004) Mechanism of bacitracin permeation enhancement through the skin and cellular membrane from an ethosomal carrier. J Control Release 94:365–379

Gómez-García R, Martínez-Ávila GCG, Aguilar CN (2012) Enzyme-assisted extraction of antioxidative phenolics from grape (*Vitis vinifera* L.) residues. 3 Biotech 2(4):297–300

González-Paredes A, Clarés-Naveros B, Ruiz-Martínez MA et al (2011) Delivery systems for natural antioxidant compounds: archaeosomes and archaeosomal hydrogels characterization and release study. Int J Pharm 421:321–331

Gortzi O, Lalas S, Chinou I et al (2008) Reevaluation of bioactivity and antioxidant activity of *Myrtus communis* extract before and after encapsulation in liposomes. Eur Food Res Technol 226:583–590

Gulati M, Grover M, Singh S et al (1998) Lipophilic drug derivatives in liposomes. Int J Pharm 165:129–168

Guo C, Yang C, Li Q et al (2012) Development of a quercetin-loaded nanostructured lipid carrier formulation for topical delivery. Int J Pharm 430(1–2):292–298

Habbu P, Madagundi S, Kulkarni R et al (2013) Preparation and evaluation of Bacopa–phospholipid complex for antiamnesic activity in rodents. Drug Invent Today 5:13–21
Han GT, Yang ZM, Peng Z et al (2013) Preparation and properties analysis of slow-release microcapsules containing patchouli oil. Adv Mate Res 641–642:935–938
Hao S, Wang B, Wang Y et al (2013) Preparation of Eudragit L 100-55 enteric nanoparticles by a novel emulsion diffusion method. Colloids Surf B Biointerfaces 108:127–133
Hassas-Roudsari M, Chang P, Pegg R et al (2009) Antioxidant capacity of bioactives extracted from canola meal by subcritical water, ethanolic and hot water extraction. Food Chem 114(2):717–726
Hayat K, Hussain S, Abbas S et al (2009) Optimized microwave-assisted extraction of phenolic acids from citrus mandarin peels and evaluation of antioxidant activity *in vitro*. Sep Purif Technol 70(1):63–70
He ZF, Liu DY, Zeng S et al (2008) Study on preparation of *Ampelopsin liposomes*. J Chine Mat Med 33:27–30
Heneweer C, Gendy SE, Peñate-Medina O (2012) Liposomes and inorganic nanoparticles for drug delivery and cancer imaging. Ther Deliv 3:645–656
Herrera MC, Luque de Castro MD (2004) Ultrasound-assisted extraction for the analysis of phenolic compounds in strawberries. Analy Bioana Chem 379(7–8):1106–1112
Herrera MC, Luque de Castro MD (2005) Ultrasound-assisted extraction of phenolic compounds from strawberries prior to liquid chromatographic separation and photodiode array ultraviolet detection. J Chroma 1100(1):1–7
Herrero M, Cifuentes A, Iba ez E (2006) Sub and supercritical fluid extraction of functional ingredients from different natural sources: plants, food-by-products, algae and microalgae: a review. Food Chem 98(1):136–148
Herrero EP, Alonso MJ, Csaba N (2012) Polymer-based oral peptide nanomedicines. Ther Deliv 3:657–668
Hino T, Kawashima Y, Shimabayashi S (2000) Basic study for stabilization of w/o/w emulsion and its application to transcatheter arterial embolization therapy. Adv Drug Deliv Rev 45:27–45
Hong W, Chen DW, Zhao XL et al (2008) Preparation and study in vitro of longcirculating nanoliposomes of curcumin. China J Chine Mat Med 33:988–992
Hou J, Zhou SW (2008) Formulation and preparation of glycyrrhizic acid solid lipid nanoparticles. ACTA Academiae Medicinae Militaris Tertiae 30:1043–1045
Howard L, Pandjaitan N (2008) Pressurized liquid extraction of flavonoids from spinach. J Food Sci 73(3):C151–C157
Hu Q, Pan B, Xu J et al (2007) Effects of supercritical carbon dioxide extraction conditions on yields and antioxidant activity of *Chlorella pyrenoidosa* extracts. J Food Eng 80(4):997–1001
Hu LD, Xing Q, Meng J, Shang C (2010) Preparation and enhanced bioavailability of cryptotanshinone-loaded solid lipid nanoparticles. AAPS Pharm Sci Tech 11(2):582–587
Huaneng XU, Zhang Y, Chaohong H (2007) Ultrasonically assisted extraction of isoflavones from stem of *Pueraria lobata* (Willd.) Ohwi and its mathematical model. Chinese J Chem Eng 15(6):861–867
Hung CF, Chen JK, Liao MH et al (2006) Development and evaluation of emulsion-liposome blends for resveratrol delivery. J Nanosci Nanotechnol 6:2950–2958
Ibraheem D, Fessi H, Elaissari A (2013) DNA encapsulation via double emulsion like process. J Colloid Sci Biotechnol 2:328–333
Iqbal M, Zafar N, Fessi H et al (2015) Double emulsion solvent evaporation techniques used for drug encapsulation. Int J Pharm 496:173–190
Jahanshahi M (2008) Google search. https://www.google.com/search?q=Mohsen +Jahanshahi%2C+2008andie=utf-8andoe=utf-8andclient=firefox-b-ab. Accessed 23 Sept 2017.
Ju ZY, Howard LR (2003) Effects of solvent and temperature on pressurized liquid extraction of anthocyanins and total phenolics from dried red grape skin. J Agric Food Chem 51(18):5207–5213
Juqun X, Rong G (2007) Interactions between flavonoids and hemoglobin in lecithin liposomes. Int J Biol Macromol 40:305–311

Kaity S, Isaac J, Ghosh A (2013) Interpenetrating polymer network of locust bean gumpoly (vinyl alcohol) for controlled release drug delivery. Carbohydr Polym 94:456–467

Kakkar V, Singh S, Singla D et al (2011) Exploring solid lipid nanoparticles to enhance the oral bioavailability of curcumin. Mol Nutr Food Res 55(3):495–503

Katara R, Majumdar DK (2013) Eudragit RL 100-based nanoparticulate system of aceclofenac for ocular delivery. Colloids Surf B Biointerfaces 103:455–462

Kaufmann B, Christen P (2002) Recent extraction techniques for natural products: microwave-assisted extraction and pressurized solvent extraction. Phytochem Anal 13(2):105–113

Ke X, Xu Y, Yan F et al (2007) The liposomes of wogonin and rats *in vivo* pharmacokinetics. J China Pharm Univ 38:502–506

Ke Z, Zhang Z, Wu H et al (2017) Optimization and evaluation of Oridonin-loaded Soluplus-Pluronic P105 mixed micelles for oral administration. Int J Pharm 518:193–202

Khan MK, Abert-Vian M, Fabiano-Tixier AS et al (2010) Ultrasound-assisted extraction of polyphenols (flavanone glycosides) from orange (*Citrus sinensis* L.) peel. Food Chem 119(2):851–858

Khandare V, Walia S, Singh M et al (2010) Black carrot (*Daucus carota* ssp. sativus) juice: processing effects on antioxidant composition and color. Food Bioprod Process 89(4):482–486

Khoee S, Yaghoobian M (2009) An investigation into the role of surfactants in controlling particle size of polymeric nanocapsules containing penicillin-G in double emulsion. Eur J Med Chem 44:2392–2399

Khoee S, Sattari A, Atyabi F (2012) Physico-chemical properties investigation of cisplatin loaded polybutyladipate (PBA) nanoparticles prepared by w/o/w. Mater Sci Eng C 32:1078–1086

Khorassani MA, Taylor LT (2004) Sequential fractionation of grape seeds into oils, polyphenols, and procyanidins via a single system employing CO2-based fluids. J Agri Food Chem 52(9):2440–2444

Khuda-Bukhsh AR, Bhattacharyya SS, Paul S et al (2010) Polymeric nanoparticle encapsulation of a naturally occurring plant scopoletin and its effects on human melanoma cell A375. Zhong Xi Yi Jie He Xue Bao 8(9):853–862

Kim J, Nagaoka T, Ishida Y et al (2009a) Subcritical water extraction of nutraceutical compounds from citrus pomaces. Separ Sci Technol 44(11):2598–2608

Kim W, Kim J, Veriansyah B et al (2009b) Extraction of bioactive components from *Centella asiatica* using subcritical water. J Supercrit Fluid 48(3):211–216

Kim W, Veriansyah B, Lee Y et al (2010) Extraction of mangiferin from mahkota dewa (*Phaleria macrocarpa*). using subcritical water. J Ind Eng Chem 16(3):425–430

Klejdus B, Lojková L, Plaza M et al (2010) Hyphenated technique for the extraction and determination of isoflavones in algae: ultrasound-assisted supercritical fluid extraction followed by fast chromatography with tandem mass spectrometry. J Chromatogr 1217(51):7956–7965

Ko M, Cheigh C, Cho S et al (2011) Subcritical water extraction of flavonol quercetin from onion skin. J Food Eng 102(4):327–333

Kong Y, Fu YJ, Zu YG et al (2009) Ethanol modified supercritical fluid extraction and antioxidant activity of cajaninstilbene acid and pinostrobin from pigeonpea [*Cajanus cajan* (L.) Millsp.] leaves. Food Chem 117(1):152–159

Kong Y, Zu YG, Fu YJ et al (2010) Optimization of microwave-assisted extraction of cajaninstilbene acid and pinostrobin from pigeonpea leaves followed by RP-HPLC-DAD determination. J Food Comps Anal 23(4):382–388

Kotakadi VS, Rao YS, Gaddam SA et al (2013) Simple and rapid biosynthesis of stable silver nanoparticles using dried leaves of *Catharanthus roseus*. Linn. G. Donn and its antimicrobial activity. Colloids Surf B Biointerfaces 105:194–198

Kristl J, Teskač K, Caddeo C et al (2009) Improvements of cellular stress response on resveratrol in liposomes. Eur J Pharm Biopharm 73:253–259

Kumari A, Yadav SK, Yadav SC (2010) Biodegradable polymeric nanoparticles based drug delivery systems. Colloids Surf B Biointerfaces 75:1–18

Kurmudle NN, Bankar SB, Bajaj IB et al (2010) Enzyme-assisted three phase partitioning: a novel approach for extraction of turmeric oleoresin. Process Biochem 46(1):423–426

Labouebe ZM, Lange N, Gurny R et al (2006) Hypericin-loaded nanoparticles for the photodynamic treatment of ovarian cancer. Int J Pharm 326:174–181

Lacatusu I, Badea N, Stan R et al (2012) Novel bio-active lipid nanocarriers for the stabilization and sustained release of sitosterol. Nanotechnology 23:455702 (13pp)

Landbo AK, Meyer A (2001) Enzyme-assisted extraction of antioxidative phenols from black currant juice press residues (*Ribes nigrum*). J Agri Food Chem 49(7):3169–3177

Laouini A, Jaafar-Maalej C, Limayem-Blouza I et al (2012) Preparation, characterization and applications of liposomes: state of the art. J Colloid Sci Biotech 1:147–168

Laroze L, Soto C, Zúñiga ME (2010) Phenolic antioxidants extraction from raspberry wastes assisted by-enzymes. Elec J Biotech 13(6):1–11

Latif S, Anwar F (2009) Physicochemical studies of hemp (Cannabis sativa) seed oil using enzyme-assisted cold-pressing. Eur J Lipid Sci Technol 111(10):1042–1048

Leal P, Kfouri M, Alexandre F et al (2010) Brazilian ginseng extraction via LPSE and SFE: global yields, extraction kinetics, chemical composition and antioxidant activity. J Supercrit Fluid 54(1):38–45

Lee JJ, Nama S, Park JH et al (2016) Nanocomposites based on Soluplus and Angelica gigas Nakai extract fabricated by an electrohydrodynamic method for oral administration. J Colloid Interf Sci 484:146–154

Lertsutthiwong P, Rojsitthisak P, Nimmannit U (2008) Preparation of turmeric oil-loaded chitosan-alginate biopolymeric nanocapsules. Mater Sci Engn C 29(3):856–860

Levchenko TS, Hartner WC, Torchilin VP (2012) Liposomes for cardiovascular targeting. Ther Deliv 3:501–514

Li H, Chen B, Yao S (2005) Application of ultrasonic technique for extracting chlorogenic acid from *Eucommia ulmoides* Oliv. Ultraso Sonochem 12(4):295–300

Li BB, Smith B, Hossain MM (2006a) Separation and purification in the food industry extraction of phenolics from citrus peels: II. Enzyme-assisted extraction method. Sepa Puri Tech 48(2):189–196

Li Y, Dong L, Jia A et al (2006b) Preparation and characterization of solid lipid nanoparticles loaded traditional Chinese medicine. Int J Biol Macromol 38:296–299

Li L, Wang DK, Li LS et al (2007a) Preparation of docetaxel submicron emulsion for intravenous administration. J Shenyang Pharm Univ 12:736–739

Li YC, Dong L, Jia K et al (2007b) Preparation and anti-fibrotic effects of solid lipid nanoparticles loaded with silibinin. Jiaotong Univ (Med Sci) 28:517–520

Li H, Zhao X, Ma Y et al (2009) Enhancement of gastrointestinal absorption of quercetin by solid lipid nanoparticles. J Control Release 133(3):238–244

Li H, Wu J, Rempel CB et al (2010) Effect of operating parameters on oil and phenolic extraction using supercritical CO_2. J Am Oil Chem Soc 87(9):1081–1089

Li H, Deng Z, Wu T et al (2011a) Microwave-assisted extraction of phenolics with maximal antioxidant activities in tomatoes. Food Chem 130(4):928–936

Li Y, Skouroumounis GK, Elsey GM et al (2011b) Microwave-assistance provides very rapid and efficient extraction of grape seed polyphenols. Food Chem 129(2):570–576

Li N, Taylor LS, Mauer LJ (2011c) Degradation kinetics of catechins in green tea powder: effects of temperature and relative humidity. J Agric Food Chem 59(11):6082–6090

Li GY, Zhong M, Zhou ZD et al (2011d) Formulation optimization of chelerythrine loaded O-carboxymethylchitosan microspheres using response surface methodology. Int J Biol Macromol 49:970–978

Li W, Yalcin M, Lin Q et al (2017) Self-assembly of green tea catechin derivatives in nanoparticles for oral lycopene delivery. J Control Release 248:117–124

Liau B, Shen C, Liang F et al (2010) Supercritical fluid extraction and anti-solvent purification of carotenoids from microalgae and associated bioactivity. J Supercrit Fluid 55(1):169–175

Lin AH, Li HY, Liu YM et al (2007) Characterisation of berberine nanoparticles. China Pharm 18:755–757

Lince F, Marchisio DL, Barresi AA (2008) Strategies to control the particle size distribution of poly-epsilon-caprolactone nanoparticles for pharmaceutical applications. J Colloid Interface Sci 322:505–515

Lira MCB, Ferraz MS, da Silva DGVC et al (2009) Inclusion complex of usnic acid with β-cyclodextrin: characterization and nanoencapsulation into liposomes. J Incl Phenom Macrocycl Chem 64:215–224
Liu M, Dong J, Yang Y et al (2005) Anti-inflammatory effects of triptolide loaded poly (d.l-lactic acid) nanoparticles on an adjuvant-induced arthritis in rats. J Ethnopharmacol 97:219–225
Liu M, Li H, Luo G et al (2008) Pharmacokinetics and biodistribution of surface modification polymeric nanoparticles. Arch Pharm Res 31(4):547–554
Liu D, Hu H, Lin Z et al (2013) Quercetin deformable liposome: preparation and efficacy against ultraviolet B induced skin damages *in vitro* and *in vivo*. J Photochem Photobiol B 127C:8–17
Londóno-Londóno J, de Lima VR, Lara O et al (2010) Clean recovery of antioxidant flavonoids from citrus peel: optimizing an aqueous ultrasound-assisted extraction method. Food Chem 119(1):81–87
Loureiro JA, Andrade S, Duarte A et al (2017) Resveratrol and grape extract-loaded solid lipid nanoparticles for the treatment of Alzheimer's disease. Molecules 22(2):1–11
Lun Su Y, Leung LK, Huang Y et al (2003) Stability of tea theaflavins and catechins. Food Chem 83(2):189–195
Luthria DL (2008) Influence of experimental conditions on the extraction of phenolic compounds from parsley (*Petroselinum crispum*) flakes using a pressurized liquid extractor. Food Chem 107(2):745–752
Ma Y, Ye X, Hao Y et al (2008) Ultrasound-assisted extraction of hesperidin from Penggan (*Citrus reticulata*) peel. Ultrason Sonochem 15(3):227–232
Machida Y, Onishi H, Kurita A et al (2000) Pharmacokinetics of prolonged-release CPT-11-loaded microspheres in rats. J Control Release 66:159–175
Maćias-Śanchez M, Mantell C, Rodŕiguez M et al (2007) Supercritical fluid extraction of carotenoids and chlorophyll a from Synechococcus sp. J Supercrit Fluid 39(3):323–329
Maćias-Śanchez M, Serrano C, Rodŕiguez M et al (2009) Kinetics of the supercritical fluid extraction of carotenoids from microalgae with CO_2 and ethanol as cosolvent. Chem Eng J 150(1):104–113
Maćias-Śanchez M, Fernandez-Sevilla J, Acíen Fernández F et al (2010) Supercritical fluid extraction of carotenoids from *Scenedesmus almeriensis*. Food Chem 123(1):928–935
Maier T, Göppert A, Kammerer DR et al (2008) Optimization of a process for enzyme-assisted pigment extraction from grape (*Vitis vinifera* L.) pomace. Eur Food Res Technol 227(1):267–275
Maiti K, Mukherjee K, Gantait A et al (2005) Enhanced therapeutic benefit of quercetin phospholipid complex in carbon tetrachloride-induced acute liver injury in rats: a comparative study. Iran J Pharmacol Ther 4:84–90
Maiti K, Mukherjee K, Gantait A et al (2006) Enhanced therapeutic potential of naringenin-phospholipid complex in rats. J Pharm Pharmacol 58:1227–1233
Maiti K, Mukherjee K, Gantait A (2007) Curcumin–phospholipid complex: preparation, therapeutic evaluation and pharmacokinetic study in rats. Int J Pharm 330:155–163
Makeshwar KB, Wasankar SR (2013) Niosome: a novel drug delivery system. Asian J Pharm Res 3(1):16–20
Manach C, Williamson G, Morand C et al (2005) Bioavailability and bioefficacy of polyphenols in humans. I. Review of 97 bioavailability studies. Am J Clin Nutri 81(Suppl 1):230–242
Mandal AK, Das N (2005) Sugar coated liposomal flavonoid: a unique formulation in combating carbontetrachloride induced hepatic oxidative damage. J Drug Target 13:305–315
Marianecci C, Rinaldi F, Mastriota M et al (2012) Anti-inflammatory activity of novel ammonium glycyrrhizinate/niosomes delivery system: human and murine models. J Control Release 164:17–25
Marsanasco M, Márquez AL, Wagner JR et al (2011) Liposomes as vehicles for vitamins E and C: an alternative to fortify orange juice and offer vitamin C protection after heat treatment. Food Res Int 44:3039–3046
Martinez-Correa H, Magalhaes P, Queiroga C et al (2010) Extracts from pitanga (*Eugenia uniflora* L.) leaves: influence of extraction process on antioxidant properties and yield of phenolic compounds. J Supercrit Fluid 55(3):998–1006

Mei Z, Chen H, Weng T et al (2003) Solid lipid nanoparticle and microemulsion for topical delivery of triptolide. Eur J Pharm Biopharm 56:189–196

Meyer AS, Jepsen SM, Sorensen NS (1998) Enzymatic Release of Antioxidants for Human Low-Density Lipoprotein from Grape Pomace. J Agric Food Chem 46(7):2439–2446

Memvanga PB, Tona GL, Mesia GK et al (2015) Antimalarial activity of medicinal plants from the Democratic Republic of Congo: a review. J Ethnopharmacol 169:76–98

Mezadri H (2010) Development of nanoemulsions containing extracts of fruits of *Syagrus romanzoffiana* (Cham.) Glassman and phytochemical study of these extracts. Faculdade de Ciências Farmacêuticas UFOP, Portuguese

Mignet N, Seguin J, Ramos Romano M et al (2012) Development of a liposomal formulation of the natural flavonoid fisetin. Int J Pharm 423:69–76

Miladi K, Sfar S, Fessi H et al (2013) Drug carriers in osteoporosis: preparation, drug encapsulation and applications. Int J Pharm 445:181–195

Miladi K, Ibraheem D, Iqbal M et al (2014) Particles from preformed polymers as carriers for drug delivery. EXCLI J 13:28–57

Miladi K, Sfar S, Fessi H et al (2016) Nanoprecipitation process: from particle preparation to *in vivo* applications. In: Vauthier C, Ponchel G (eds) Polym nanoparticles nanomedicines. Springer, Cham, pp 17–53

Min KH, Park K, Kim YS et al (2008) Hydrophobically modified glycol chitosan nanoparticles-encapsulated camptothecin enhance the drug stability and tumor targeting in cancer therapy. J Control Release 127:208–218

Moghassemi S, Hadjizadeh A (2014) Nano-niosomes as nanoscale drug delivery systems: an illustrated review. J Control Release 185:22–36

Mohammadi A, Jafari SM, Esfanjani F et al (2016) Application of nano-encapsulated olive leaf extract in controlling the oxidative stability of soybean oil. Food Chem 190:513–519

Moore J, Cheng Z, Su L et al (2006) Effects of solid-state enzymatic treatments on the antioxidant properties of wheat bran. J Agric Food Chem 54(24):9032–9045

Mora-Huertas CE, Fessi H, Elaissari A (2010) Polymer-based nanocapsules for drug delivery. PubMed NCBI. http://www.ncbi.nlm.nih.gov/pubmed/19825408. Accessed 28 June 2016.

Moreno LCGI, Puerta E, Suarez-Santiago JE et al (2017) Effect of the oral administration of nanoencapsulated quercetin on a mouse model of Alzheimer's disease. Int J Pharm 517:50–57

Moulari B, Lboutounne H, Pellequer Y et al (2005) Vectorization *of Harungana madagascariensis* Lam. Ex Poir. (Hypericaceae) ethanolic leaf extract by using PLG-nanoparticles: antibacterial activity assessment. Drug Dev Res 65(1):26–33

Moulari B, Lboutounne H, Chaumont J et al (2006) Potentiation of the bacterial activity of *Harungana madagascariensis* Lam. Ex Poir. (Hypericaceae) leaf extract against oral bacteria using poly (D,L-lactide-co-glycolide) nanoparticles: in vitro study. Acta Odentol Scand 64(3):153–158

Mukerjee A, Vishwanatha JK (2009) Formulation, characterization and evaluation of curcumin-loaded PLGA nanospheres for cancer therapy. Anticancer Res 29(10):3867–3875

Müller RH, Radtke M, Wissing SA (2002) Solid lipid nanoparticles (SLN) and nanostructured lipid carriers (NLC) in cosmetic and dermatological preparations. Adv Drug Deliv Rev 54:S131–S155

Munoz O, Sepulveda M, Schwartz M (2004) Effects of enzymatic treatment on anthocyanic pigments from grape skins from Chilean wine. Food Chem 87(4):487–490

Nacke C, Schrader J (2011) Liposome based solubilisation of carotenoid substrates for enzymatic conversion in aqueous media. J Mol Catal B Enzym 71:133–138

Nagaraj B, Ci obanu CS, Malaka B et al (2012) Synthesis of plant mediated gold nanoparticles using flower extracts of Carthamus *tinctorius* l. (safflower) and evaluation of their biological activities. Dig J Nanomater Biostruct 7:1289–1296

Naik SR, Panda VS (2008) Hepatoprotective effect of Ginkgoselect Phytosome® in rifampicin induced liver injurym in rats: evidence of antioxidant activity. Fitoterapia 79:439–445

Nalawade PB, Gajjar AK (2016) Microencapsulation of lutein extracted from marigold flowers (*Tagetes erecta* L.) using full factorial design. J Drug Deliv Sci Tech 33:75–87

Namdaria M, Eatemadib A, Soleimaninejade M et al (2017) A brief review on the application of nanoparticle enclosed herbal medicine for the treatment of infective endocarditis. Biomed Pharmacother 87:321–331

Naseer N, Fatima H, Asghar A et al (2014) Magnetically responsive hybrid polymer colloids for ultrasensitive molecular imaging. J Colloid Sci Biotech 3:19–29

Nedovic V, Kalusevic A, Manojlovic V et al (2011) An overview of encapsulation technologies for food applications. Procedia Food Sci 1:1806–1815

Nesterenko A, Alric I, Silvestre F et al (2012) Influence of soy protein's structural modifications on their microencapsulation properties: α-tocopherol microparticle preparation. Food Res Int 48:387–396

Nieto A, Borrull F, Pocurull E et al (2010) Pressurized liquid extraction: a useful technique to extract pharmaceuticals and personal-care products from sewage sludge. TrAC Trends Anal Chem 29(7):752–764

Njimoh DL, Assob JCN et al (2015) Antimicrobial activities of a Plethora of medicinal plant extracts and hydrolates against human pathogens and their potential to reverse antibiotic resistance. Int J Microbiol. Online data. Available from: http://www.wilsoncenter.org/nano. Accessed 1 Sept 2015

Oszmianski J, Wojdylo A, Kolniak J (2011) Effect of pectinase treatment on extraction of antioxidant phenols from pomace, for the production of puree-enriched cloudy apple juices. Food Chem 127(1):100–105

Ota A, Istenič K, Skrt M et al (2018) Encapsulation of pantothenic acid into liposomes and into alginate or alginate–pectin microparticles loaded with liposomes. J Food Eng 229:21–31

Pallerla SM, Prabhakar B (2013) A review on solid lipid nanoparticles. Int J Pharm Sci Rev Res 20:196–206

Pan X, Niu G, Liu H (2003) Microwave-assisted extraction of tea polyphenols and tea caffeine from green tea leaves. Chem Eng Process 42(2):129–133

Pan G, Yu G, Zhu C et al (2011) Optimization of ultrasound-assisted extraction (UAE) of flavonoids compounds (FC) from hawthorn seed (HS). Ultrason Sonochem 19(3):486–490

Panda VS, Naik SR (2008) Cardioprotective activity of *Ginkgo biloba* phytosomes in isoproterenol-induced myocardial necrosis in rats: a biochemical and histoarchitectural evaluation. Exp Toxicol Pathol 60:397–404

Paré JJR, Bélanger JMR, Stafford SS (1994) Microwave-assisted process (MAP™): a new tool for the analytical laboratory. TrAC Trends Anal Chem 13(4):176–184

Parrado J, Miramontes E, Jover M et al (2006) Preparation of a rice bran enzymatic extract with potential use as functional food. Food Chem 98(4):742–748

Pasquet V, Chérouvrier JR, Farhat F et al (2011) Study on the microalgal pigments extraction process: performance of microwave assisted extraction. Process Biochem 46(1):59–67

Passerini N, Perissutti B, Albertini B et al (2012) A new approach to enhance oral bioavailability of *Silybum Marianum* dry extract: association of mechanochemical activation and spray congealing. Phytomedicine 19:160–168

Passos C, Silva R, Da Silva F et al (2010) Supercritical fluid extraction of grape seed (*Vitis vinifera* L.) oil. Effect of the operating conditions upon oil composition and antioxidant capacity. Chem Eng J 160(2):634–640

Patten GS, Abeywardena MY, Bennett LE (2016) Inhibition of angiotensin converting enzyme, angiotensin II receptor blocking, and blood pressure lowering bioactivity across plant families. Crit Rev Food Sci Nutr 56:181–214

Patwardhan B, Vaidya ADB, Chorghade M (2004) Ayurveda and natural products drug discovery. Curr Sci 86:789–799

Paul S, Bhattacharyya SS, Boujedaini N et al (2011) Anticancer potentials of root extract of polygala senega and its PLGA nanoparticles-encapsulated form. Evid Based Complement Alternat Med 2011:1–13

Péan JM, Venier-Julienne MC, Boury F et al (1998) NGF release from poly (d,l-lactide-co-glycolide) microspheres. Effect of some formulation parameters on encapsulated NGF stability. J Control Release 56:175–187

Périno-Issartier S, Abert-Vian M, Chemat F (2011) Solvent free microwave-assisted extraction of antioxidants from sea buckthorn (*Hippophae rhamnoides*) food by-products. Food Bioprocess Technol 4(6):1020–1028
Pinto JF (2010) Site specific drug delivery systems within gastro intestinal tract: from the mouth to the colon. Int J Pharm 395:44–52
Pitsiree P, Chomto P, Phaechamud T (2013) Silymarin solid lipid nanoparticle containing shellac wax fabricated with hot melt emulsification research. J Pharm Biol Chem Sci 4:784–801
Plaza M, Amigo-Benavent M, del Castillo M et al (2010a) Facts about the formation of new antioxidants in natural samples after subcritical water extraction. Food Res Int 43(10):2341–2348
Plaza M, Amigo-Benavent M, Del Castillo M et al (2010b) Neoformation of antioxidants in glycation model systems treated under subcritical water extraction conditions. Food Res Int 43(4):1123–1129
Plaza M, Santoyo S, Jaime L et al (2010c) Screening for bioactive compounds from algae. J Pharm Biomed Anal 51(2):450–545
Poletto FS, Fiel LA, Lopes MV et al (2012) fluorescent-labeled poly(ε-caprolactone) lipid-core nanocapsules: synthesis, physicochemical properties and macrophage uptake. J Colloid Sci Biotechnol 1:89–98
Pourali O, Asghari F, Yoshida H (2010) Production of phenolic compounds from rice bran biomass under subcritical water conditions. Chem Eng J 160(1):259–266
Prado JM, Dalmolin I, Carareto ND et al (2012) Supercritical fluid extraction of grape seed: process scale-up, extract chemical composition and economic evaluation. J Food Eng 109(2):249–257
Proestos C, Komaitis M (2008) Application of microwave-assisted extraction to the fast extraction of plant phenolic compounds. LWT-Food Sci Technol 41(4):652–659
Pulicharla R, Marques C, Das RK et al (2016) Encapsulation and release studies of strawberry polyphenols in biodegradable chitosan nanoformulation. Int J Biol Macromol 88:171–178
Pustovidko AV, Rokitskaya TI, Severina II et al (2012) Derivatives of the cationic plant alkaloids berberine and palmatine amplify protonophorous activity of fatty acids in model membranes and mitochondria. Mitochondrion 13(5):520–525
Quintanar-Guerrero D, Fessi H, Allémann E et al (1996) Influence of stabilizing agents and preparative variables on the formation of poly (D,l-lactic acid) nanoparticles by an emulsification-diffusion technique. Int J Pharm 143:133–141
Rajendran R, Radhai R, Kotresh TM et al (2013) Development of antimicrobial cotton fabrics using herb loaded nanoparticles. Carbohydr Polym 91(2):613–617
Ram DT, Debnath S, Babu MN (2012) A review on solid lipid nanoparticles. Res J Pharm Technol 5:1359–1368
Rane S, Prabhakar B (2009) Influence of liposome composition on paclitaxel entrapment and pH sensitivity of liposomes. Int J Pharm Technol Res 1:914–917
Rangsriwong P, Rangkadilok N, Satayavivad J et al (2009) Subcritical water extraction of polyphenolic compounds from *Terminalia chebula* Retz. fruits. Sep Purif Technol 66(1):51–56
Ratcharin PWN, Indranupakorn R (2012) Preparation of *Zingiber officinale* extract loaded solid lipid nanoparticles. Adv Materials Res 506:389–392
Rieux des A, Fievez V, Garinot M (2006) Nanoparticles as potential oral delivery systems of proteins and vaccines: a mechanistic approach. J Control Release 116:1–27
Rijo P, Pedro L, Falé ML et al (2014) Optimization of medicinal plant extraction methods and their encapsulation through extrusion technology. Measurement 58:249–255
Rivas M, Tarhini M, Badri W et al (2017) Nanoprecipitation process: from encapsulation to drug delivery. Int J Pharm 532(1):66–81
Rodriguez-Jasso RM, Mussatto SI, Pastrana L (2011) Microwave-assisted extraction of sulfated polysaccharides (fucoidan) from brown seaweed. Carbohydrate Polymers 86(3):1137–1144
Rodríguez-Meizoso I, Marin F, Herrero M et al (2006) Subcritical water extraction of nutraceuticals with antioxidant activity from oregano. Chemical and functional characterization. J Pharm Biomed Anal 41(5):1560–1565

Rodríguez-Meizoso I, Jaime L, Santoyo S et al (2010) Subcritical water extraction and characterization of bioactive compounds from *Haematococcus pluvialis* microalga. J Pharm Biomed Anal 51(2):456–463
Rodríguez-Rojo S, Visentin A, Maestri D et al (2012) Assisted extraction of rosemary antioxidants with green solvents. J Food Eng 109(1):98–103
Rosenthal A, Pyle DL, Niranjan K (1996) Aqueous and enzymatic processes for edible oil extraction. Enz Microbial Tech 19(6):402–420
Rostagno MA, Palma M, Barroso CG (2003) Ultrasound-assisted extraction of soy isoflavones. J Chroma A 1012(2):119–128
Rostagno MA, Palma M, Barroso CG (2007) Microwave-assisted extraction of soy isoflavones. Anal Chim Acta 588(2):274–282
Rout P, Naik S, Rao Y (2008) Subcritical CO_2 extraction of floral fragrance from *Quisqualis indica*. J Supercrit Fluid 45(2):200–205
Routray W, Orsat V (2014) MAE of phenolic compounds from blueberry leaves and comparison with other extraction methods. Ind Crops Prod 58:36–45
Rubió L, Motilva MJ, Romero MP (2013) Recent advances in biologically active compounds in herbs and spices: a review of the most effective antioxidant and anti-inflammatory active principles. Crit Rev Food Sci Nutr 53:943–953
Rutkowska J, Stolyhwo A (2009) Application of carbon dioxide in subcritical state (LCO_2) for extraction/fractionation of carotenoids from red paprika. Food Chem 115(2):745–752
Sáhin S, Sámli R (2013) Optimization of olive leaf extract obtained by ultrasound-assisted extraction with response surface methodology. Ultrason Sonochem 20(1):595–602
Sahoo SK, Parveen S, Panda JJ (2007) The present and future of nanotechnology in human health care. Nanomed Nanotechnol Biol Med 3:20–31
Sahu A, Bora U, Kasoju N et al (2008) Synthesis of novel biodegradable and self-assembling methoxy poly (ethylene glycol)-palmitate nanocarrier for curcumin delivery to cancer cells. Acta Biomaterialia 4:1752–1761
Saldaña MDA, Mohamed RS, Baer MG et al (1999) Extraction of purine alkaloids from maté (*Ilex paraguariensis*) using supercritical CO_2. J Agr Food Chem 47(9):3804–3808
Sandhya S, Chandra SJ, Vinod KR et al (2012) Preclinical studies of a novel polyherbal phyto–complex hair growth promoting cream. Asian Pac J Trop Biomed 2:S296–S304
Santos DT, Veggi PC, Meireles MA (2012) Optimization and economic evaluation of pressurized liquid extraction of phenolic compounds from jabuticaba skins. J Food Eng 108(3):444–452
Santos FK, Oyafuso MH, Kiill CP et al (2013) Nanotechnology-based drug delivery systems for treatment of hyperproliferative skin diseases – a review. Curr Nanosci 9(1):159–167
Schuch A, Deiters P, Henne J et al (2013) Production of W/O/W (water-in-oil-in-water) multiple emulsions: droplet breakup and release of water. J Colloid Interface Sci 402:157–164
Semalty A, Semalty M, Singh D et al (2012) Phyto-phospholipid complex of catechin in value added herbal drug delivery. J Incl Phenom Macrocycl Chem 73:377–386
Seremeta KP, Chiappetta DA, Sosnik A (2013) Poly(ε-caprolactone), Eudragit® RS 100 and poly(ε-caprolactone)/Eudragit® RS 100 blend submicron particles for the sustained release of the antiretroviral efavirenz. Colloids Surf B Biointerfaces 102:441–449
Serra A, Seabra I, Braga M et al (2010) Processing cherries (*Prunus avium*) using supercritical fluid technology. Part 1: recovery of extract fractions rich in bioactive compounds. J Supercrit Fluid 55(1):184–191
Sethiya NK, Trivedi A, Patel MB et al (2010) Comparative pharmacognostical investigation on four ethanobotanicals traditionally used as Shankhpushpi in India. J Adv Pharm Technol Res 1:388
Sharma AT, Mitkare SS, Moon RS (2011) Multicomponent herbal therapy: a review. Int J Pharm Sci Rev Res 6:185–187
Shimada S (2008) Composition comprising nanoparticle *Ginkgo biloba* extract with the effect of brain function activation IPC8 Class-AA61K914FI, USPC Class-424489
Shu YY, Ko MY, Chang YS (2003) Microwave-assisted extraction of ginsenosides from ginseng root. Microchem J 74(2):131–139

Shyam KR, Mruthunjaya K, Kumar GM (2012) Hepatoprotective effect of gallic acid and gallic acid phytosome against carbon tetrachloride induced damage in albino rats. Res J Pharm Technol 5:677–681

Sihvonen M, Järvenpää E, Hietaniemi V et al (1999) Advances in supercritical carbon dioxide technologies. Trends Food Sci Tech 10(6–7):217–222

Silva PI, Stringheta PC, Teófilo RF et al (2013) Parameter optimization for spray-drying microencapsulation of jaboticaba (*Myrciaria jaboticaba*) peel extracts using simultaneous analysis of responses. J Food Eng 117:538–544

Singh D, Rawat MS, Semalty A et al (2012) Rutin-phospholipid complex: an innovative technique in novel drug delivery system- NDDS. Curr Drug Deliv 9:305–314

Sinico C, De Logu A, Lai F et al (2005) Liposomal incorporation of *Artemisia arborescens* L. essential oil and *in vitro* antiviral activity. Eur J Pharm Biopharm 59(1):161–168

Siqueira-Moura MP, Primo FL, Espreafico EM et al (2013) Development, characterization, and photocytotoxicity assessment on human melanoma of chloroaluminum phthalocyanine nanocapsules. Mater Sci Eng C 33:1744–1752

Smith RM (2003) Before the injection—modern methods of sample preparation for separation techniques. J Chroma A 1000(1–2):3–27

Soares ASP (2013) A special issue on applications of microencapsulation. J Colloid Sci Biotech 2:77–77

Song YM, Ping QN, Wu ZH (2005) Preparation of silybin nanoemulsion and its pharmacokinetics in rabbits. J China Pharm Univ 5:427–431

Sou K, Inenaga S, Takeoka S et al (2008) Loading of curcumin into macrophages using lipid-based nanoparticles. Int J Pharm 352:287–293

Souguir H, Salaün F, Douillet P et al (2013) Nanoencapsulation of curcumin in polyurethane and polyurea shells by an emulsion diffusion method. Chem Eng J 221:133–145

Su YL, Fu ZY, Zhang JY et al (2008) Preparation of radix salvia nanoparticles. Powder Technol 184:114–121

Sun HW, Ouyang WQ (2007) Preparation, quality and safety evaluation of berberine nanoemulsion for oral application. J Shanghai Jiaotong Univ (Agric Sci) 1:60–65

Sun P, Den SH, Yu WP (2009) Evaluation of garlicin liposomes. J Shan Univ TCM 31:37–39

Suresh RN, Vandana SP (2008) Hepatoprotective effect of ginkgo select Phytosome® in rifampicin induced liver injurym in rats: evidence of antioxidant activity. Fitoterapia 79:439–445

Surya S, Salam AD, Tomy DV et al (2014) Diabetes mellitus and medicinal plants-a review. Asian Pac J Trop Dis 4:337–347

Svarc-Gajić J, Stojanović Z, Carretero AS (2013) Development of a microwave-assisted extraction for the analysis of phenolic compounds from *Rosmarinus officinalis*. J Food Eng 119(3):525–532

Szydlowska-Czerniak A, Karlovits G, Hellner G et al (2010) Effect of enzymatic and hydrothermal treatments of rapeseeds on quality of the pressed rapeseed oils: art I: antioxidant capacity and antioxidant content. Process Biochem 45(1):7–17

Tan SN, Yong JWH, Teo CC et al (2011) Determination of metabolites in *Uncaria sinensis* by HPLC and GCMS after green solvent microwave-assisted extraction. Talanta 83(3):891–898

Tan SP, Kha TC, Parks SE et al (2015) Effects of the spray-drying temperatures on the physiochemical properties of an encapsulated bitter melon aqueous extract powder. Powder Tech 281:65–75

Tao Y, Wang P, Wang JD et al (2017) Combining various wall materials for encapsulation of blueberry anthocyanin extracts: optimization by artificial neural network and genetic algorithm and a comprehensive analysis of anthocyanin powder properties. Powder Tech 311:77–87

Tiyaboonchai W, Tungpradit W, Plianbangchang P (2007) Formulation and characterization of curcuminoids loaded solid lipid nanoparticles. Int J Pharm 337:299–306

Tong-Un T, Wannanon P, Wattanathorn J et al (2010) Quercetin liposomes via nasal administration reduce anxiety and depression-like behaviors and enhance cognitive performances in rats. Am J Pharmacol Toxicol 5:80–88

Tsai WC, Li WC, Yin HY et al (2012) Constructing liposomal nanovesicles of ginseng extract against hydrogen peroxide-induced oxidative damage to L929 cells. Food Chem 132:744–751

Tse MT, Blatchford C, Oya Alpar H (2009) Evaluation of different buffers on plasmid DNA encapsulation into PLGA microparticles. Int J Pharm 370:33–40

Tzu-Hui W, Feng-Lin Y, Liang-Tzung L et al (2008) Preparation, physicochemical, characterization and antioxidant effects of quercetin nanoparticles. Int J Pharm 346(1–2):160–168

Uner M, Yener G (2007) Importance of solid lipid nanoparticles (SLN) in various administration routes and future perspectives. Int J Nanomed 2:289–300

Upadhyay R, Ramalakshmi K, Jagan Mohan Rao L (2012) Microwave-assisted extraction of chlorogenic acids from green coffee beans. Food Chem 130(1):184–188

Van de Ven H, Vermeersch M, Matheeussen A et al (2011) PLGA nanoparticles loaded with the antileishmanial saponin β-aescin: factor influence study and *in vitro* efficacy evaluation. Int J Pharm 420:122–132

Vandana SP, Suresh RN (2008) Cardioprotective activity of *Ginkgo biloba* phytosomes in isoproterenol-induced myocardial necrosis in rats: a biochemical and histopathological approach. Exp Toxicol Pathol 60:397–404

Verma A, Hartonen K, Riekkola ML (2008) Optimisation of supercritical fluid extraction of indole alkaloids from *Catharanthus roseus* using experimental design methodology comparison with other extraction techniques. Phytochem Anal 19(1):52–63

Vijayaraghavan K, Nalini SPK, Prakash NU et al (2012) One step green synthesis of silver nano/microparticles using extracts of *Trachyspermum ammi* and *Papaver somniferum*. Colloids Surf B Biointerfaces 94:114–117

Voinea M, Simionescu M (2002) Designing of "intelligent" liposomes for efficient delivery of drugs. J Cell Mol Med 6:465–474

Walde P, Ichikawa S (2001) Enzymes inside lipid vesicles: preparation, reactivity and applications. Biomol Engn 18:143–177

Wang L, Weller CL (2006) Recent advances in extraction of nutraceuticals from plants. Trends Food Sci Tech 17(6):300–312

Wang X, Jiang Y, Wang YW et al (2008a) Enhancing anti-inflammation activity of curcumin through O/W nanoemulsions. Food Chem 108:419–424

Wang JX, Xiao XH, Li GK (2008b) Study of vacuum microwave-assisted extraction of polyphenolic compounds and pigment from Chinese herbs. J Chroma A 1198:45–53

Wang J, Sun B, Cao Y et al (2008c) Optimisation of ultrasound-assisted extraction of phenolic compounds from wheat bran. Food Chem 106(2):804–810

Wang T, Jonsdottir R, Kristinsson HG (2010) Enzyme-enhanced extraction of antioxidant ingredients from red algae *Palmaria palmata*. LWT-Food Sci Technol 43(9):1387–1393

Wang Y, Xu H, Fu Q et al (2011) Protective effect of resveratrol derived from *Polygonum cuspidatum* and its liposomal formon nigral cells in Parkinsonian rats. J Neurol Sci 304:29–34

Wang F, Li J, Wang C (2012) Hydrophilic and fluorescent colloidal nanorods of mwnts as effective targeted drug carrier. J Colloid Sci Biotech 1:192–200

Watanabe M, Kawano K, Toma K et al (2008) *In vivo* antitumor activity of camptothecin incorporated in liposomes formulated with an artificial lipid and human serum albumin. J Control Release 127(3):321–328

WHO (2015) World Health Organization. http://www.who.int/medicines/publica%20tions/traditional/trm_strategy14_23/en/. Accessed 22 Sept 2017

Wink M (2003) Evolution of secondary metabolites from an ecological and molecular phylogenetic perspective. Phytochemistry 64(1):3–19

Wissing SA, Kayser O, Müller RH (2004) Solid lipid nanoparticles for parenteral drug delivery. Adv Drug Deliv Rev 56:1257–1272

Wu TZ, Yen FL, Lin LT et al (2008) Preparation, physicochemical characterization, and antioxidant effects of quercetin nanoparticles. Int J Pharm 346(1–2):160–168

Xia EQ, Ai XX, Zang SY et al (2011) Ultrasound-assisted extraction of phillyrin from *Forsythia suspensa*. Ultrason Sonochem 18(2):549–552

Xiao L, Zhang YH, Xu JC et al (2008a) Rutin chitosan – alginate microcapsules floating research. Chin Tradit Herb Drugs 39:209–2012
Xiao W, Han L, Shi B (2008b) Optimization of microwave-assisted extraction of flavonoid from *radix astragali* using response surface methodology. Separ Sci Technol 43(3):671–681
Xiao B, Si X, Zhang M et al (2015) Oral administration of pH-sensitive curcumin-loaded microparticles for ulcerative colitis therapy. Colloids Surf B Biointerf 135:379–385
Xiaoyan A, Jun Y, Min W et al (2008) Preparation of chitosan gelatin scaffold containing tetrandrine-loaded nano-aggregates and its controlled release behavior. Int J Pharm 350(1–2):257–264
Xie Y, Peng J, Fan G et al (2008) Chemical composition and antioxidant activity of volatiles from *Patrinia villosa* Juss obtained by optimized supercritical fluid extraction. J Pharm Biomed Anal 48(3):796–801
Xue M, Jiang ZZ, Wu T et al (2011) Protective effects of tripterygium glycoside-loaded solid lipid nanoparticles on male reproductive toxicity in rats. Arzneimittelforschung 61:571–576
Xue M, Zhao Y, Li X et al (2012) Comparison of toxicokinetic and tissue distribution of triptolide-loaded solid lipid nanoparticles vs free triptolide in rats. Eur J Pharm Sci 47:713–717
Yadav D, Suri S, Choudhary AA et al (2011) Novel approach: herbal remedies and natural products in pharmaceutical science as nano drug delivery systems. Int J Pharm Technol 3:3092–3116
Yang Y, Zhang F (2008) Ultrasound-assisted extraction of rutin and quercetin from *Euonymus alatus* (Thunb.) Sieb. Ultraso Sonochem 15(4):308–313
Yang Y, Kayan B, Bozer N et al (2007) Terpene degradation and extraction from basil and oregano leaves using subcritical water. J Chromatogr 1152(1–2):262–267
Yang L, Wang H, Yuan-gang Z et al (2011) Ultrasound-assisted extraction of the three terpenoid indole alkaloids vindoline, catharanthine and vinblastine from *Cathara nthus roseus* using ionic liquid aqueous solutions. Chem Eng J 172(2–3):705–712
Yang HG, Kim HJ, Kim HS et al (2013) Ethosome formulation for enhanced transdermal delivery of Artemisia *princeps* Pampanini extracts. Appl Chem Eng 24:190–195
Yanyu X, Yunmei S, Zhipeng C et al (2006) The preparation of silybin–phospholipid complex and the study on its pharmacokinetics in rats. Int J Pharm 307:77–78
Yen FL, Wu TH, Lin LT et al (2008) Nanoparticles formulation of *Cuscuta chinensis* prevents acetaminophen-induced hepatotoxicity in rats. Food Chem Toxicol 46:1771–1777
Yen FY, Wu TH, Lin LT et al (2009) Naringenin-loaded nanoparticles improve the physicochemical properties and the hepatoprotective effects of naringenin in orally-administered rats with CCl_4-induced acute liver failure. Pharm Res 26(4):893–902
Yepez B, Espinosa M, Ĺopez S et al (2002) Producing antioxidant fractions from herbaceous matrices by supercritical fluid extraction. Fluid Phase Equilib 194:879–884
Yesil-Celiktas O, Cetin-Uyanikgil EO (2012) *In vitro* release kinetics of polycaprolactone encapsulated plant extract fabricated by supercritical anti-solvent process and solvent evaporation method. J Supercrit Fluids 62:219–225
Yi C, Shi J, Xue S et al (2009) Effects of supercritical fluid extraction parameters on lycopene yield and antioxidant activity. Food Chem 113(4):1088–1094
Yilmaz EE, Vural H, Őzvural BE (2010) Extraction and identification of proanthocyanidins from grape seed (*Vitis vinifera*) by using supercritical carbon dioxide. J Supercrit Fluid 55(3):924–928
Yin J, Becker EM, Andersen ML et al (2012) Green tea extract as food antioxidant. Synergism and antagonism with alpha-tocopherol in vegetable oils and their colloidal systems. Food Chem 135:2195–2202
Yoshida PA, Yokota D, Foglio MA et al (2010) Liposomes incorporating essential oil of Brazilian cherry (*Eugenia uniflora* L.): characterization of aqueous dispersions and lyophilized formulations. J Microencapsul 27:416–425
You J, Cui FD, Han X et al (2006) Study of the preparation of sustained-release microspheres containing zedoary turmeric oil by the emulsion–solvent-diffusion method and evaluation of the self-emulsification and bioavailability of the oil. Colloids Surf B Biointerfaces 48:35–41
Youfang C, Xianfu L, Hyunjin P et al (2009) Artemisinin nanoparticles. Nanomed Nanotechnol Biol Med 5:316–322

Yourdkhani M, Leme-Kraus AA, Aydin B et al (2017) Encapsulation of grape seed extract in polylactide microcapsules for sustained bioactivity and time-dependent release in dental material applications. Dent Mater 33:630–636
Yuan J, Liu J, Hu Y et al (2012) The immunological activity of propolis flavonoids liposome on the immune response against ND vaccine. Int J Biol Macromol 51:400–405
Zafar N, Fessi H, Elaissari A (2014) Cyclodextrin containing biodegradable particles: From preparation to drug delivery applications. Int J Pharm 461:351–366
Zakeri-Milani P, Loveymi BD, Jelvehgari M et al (2013) The characteristics and improved intestinal permeability of vancomycin PLGA-nanoparticles as colloidal drug delivery system. Colloids Surf B Biointerfaces 103:174–181
Zambaux MF, Bonneaux F, Gref R et al (1998) Influence of experimental parameters on the characteristics of poly(lactic acid) nanoparticles prepared by a double emulsion method. J Control Release 50:31–40
Zarena A, Udaya Sankar K (2009) Supercritical carbon dioxide extraction of xanthones with antioxidant activity from *Garcinia mangostana*: characterization by HPLC/LC-ESI-MS. J Supercrit Fluid 49(3):330–337
Zeng H, Wang Y, Nie C et al (2012) Preparation of magnetic molecularly imprinted polymers for separating rutin from Chinese medicinal plants. Analyst 137:2503–2512
Zhai Y, Zhai G (2014) Advances in lipid-based colloid systems as drug carrier for topic delivery. J Control Release 193:90–99
Zhang L, Kosaraju SL (2007) Biopolymeric delivery system for controlled release of polyphenolic antioxidants. Eur Poly J 4:32956–32966
Zhang G, He L, Hu M (2011a) Optimized ultrasonic-assisted extraction of flavonoids from *Prunella vulgaris* L. and evaluation of antioxidant activities *in vitro*. Innovat Food Sci Emerg Tech 12(1):18–25
Zhang HF, Yang XH, Wang Y (2011b) Microwave assisted extraction of secondary metabolites from plants: current status and future directions. Trends Food Sci Technol 22(12):672–688
Zhang Y, Cheng X, Zhang Y et al (2013a) Biosynthesis of silver nanoparticles at room temperature using aqueous aloe leaf extract and antibacterial properties. Colloids Surf A Physicochem Eng Asp 423:63–68
Zhang J, Tang Q, Xu X et al (2013b) Development and evaluation of a novel phytosome-loaded chitosan microsphere system for curcumin delivery. Int J Pharm 448:168–174
Zhao H, Gagnon J, Häfeli UO (2007) Process and formulation variables in the preparation of injectable and biodegradable magnetic microspheres. Biomagn Res Technol 5:2(11pp), 2
Zhao Y, Wanga C, Albert HL et al (2010) Self-nanoemulsifying drug delivery system (SNEDDS) for oral delivery of Zedoary essential oil: formulation and bioavailability studies. Int J Pharm 383(1–2):170–177
Zhao L, Temelli F, Chen L (2017) Encapsulation of anthocyanin in liposomes using supercritical carbon dioxide: effects of anthocyanin and sterol concentrations. J Func Foods 34:159–167
Zheng L, Song J (2009) Curcumin multi-wall carbon nanotubes modified glassy carbon electrode and its electrocatalytic activity towards oxidation of hydrazine. Sens Actuators B Chem 135:650–655
Zheng X, Kan B, Gou M et al (2010) Preparation of MPEG-PLA nanoparticle for honokiol delivery *in vitro*. Int J Pharm 386(1–2):262–267
Zheng FY, Chen LH, Li SX et al (2013) Effect of edible plants combination on mineral bioaccessibility and bioavailability, using *in vitro* digestion and liposome affinity extraction. Food Res Int 53:174–179
Zheng B, Teng L, Xing G et al (2015) Proliposomes containing a bile salt for oral delivery of Ginkgo biloba extract: formulation optimization, characterization, oral bioavailability and tissue distribution in rats. Eur J Pharma Sci 77:254–264
Zhinan M, Huabing C, Ting W et al (2003) Solid lipid nanoparticle and microemulsion for topical delivery of triptolide. Eur J Pharm Biopharm 56:189–196
Zhinan M, Xiaokuan L, Qunrong W et al (2005) Research on the anti-inflammatory activity and hepatotoxicity of triptolide loaded solid lipid nanoparticle. Pharmacol Res 51:345–351

Zhong H, Deng Y, Wang X et al (2005) Multivesicular liposome formulation for the sustained delivery of breviscapine. Int J Pharm 301:15–24

Zhou T, Xiao X, Li G et al (2011) Study of polyethylene glycol as a green solvent in the microwave-assisted extraction of flavone and coumarin compounds from medicinal plants. J Chromatogr A 1218(23):3608–3615

Zill-e-Huma, Abert-Vian MA, Fabiano-Tixier AS et al (2011) A remarkable influence of microwave extraction: enhancement of antioxidant activity of extracted onion varieties. Food Chem 127(4):1472–1480

Zougagh M, Valcárcel M, Ríos A (2004) Supercritical fluid extraction: a critical review of its analytical usefulness. Trends Anal Chem 23(5):399–405

Zu G, Zhang R, Yang L et al (2012) Ultrasound-assisted extraction of carnosic acid and rosmarinic acid using ionic liquid solution from *Rosmarinus officinalis*. Int J Mol Sci 13(9):11027–11043

Zuckerman GB, Bielory L (2002) Complementary and alternative medicine herbal therapies for atopic disorders. Am J Med 113(Suppl):47S–51S

Zuo YB, Zeng AW, Yuan XG et al (2008) Extraction of soybean isoflavones from soybean meal with aqueous methanol modified supercritical carbon dioxide. J Food Eng 89(4):384–389

NMR Identification of Biologically Active Natural Products: Strategies and Challenges

Gloria Ivonne Hernández-Bolio and Luis Manuel Peña-Rodríguez

1 Introduction

Anytime a pure metabolite is obtained at the end of a phytochemical study or a bioassay-guided isolation, it is necessary to establish its chemical structure before proceeding further, either to study its biological activity, to be used as a chemotaxonomical marker, or as a substrate for the preparation of novel semisynthetic derivatives. Up to a few decades ago, structure elucidation was a challenging task requiring chemical degradation and/or the total synthesis of the isolated product to establish or confirm the proposed structure (Bross-Walch et al. 2005). Currently, novel spectroscopic techniques, particularly nuclear magnetic resonance (NMR), offer reproducibility, faster times of analysis, minimum sample requirements, and software for the simulation of spectra or the prediction of the structure (Emwas et al. 2015). With this, the identification of a natural product, which could take years and require the collaboration of several research groups in the past, can now be carried out in 24 h using less than 1 mg of sample (Reynolds and Mazzola 2015).

NMR is a nondestructive spectroscopic technique based on the absorption of energy by important atom isotopes such as ^{1}H, ^{13}C, ^{15}N, etc., when placed in an intense and homogenous magnetic field (Joseph-Nathan 1982). The absorption of energy, followed by energy release and relaxation, causes the nucleus to produce a "resonance" signal (Hornak 2017). While each signal corresponds to an atom in the molecule, the level of energy required for each atom to "resonate" is related to its position or chemical shift in the spectrum. Often NMR resonances display fine structures characterized by split peaks (e.g., doublets, triplets, or multiplets) resulting from magnetic coupling between adjacent or covalently bonded atoms or nuclei. This peak-splitting phenomenon is called scalar, spin-spin, or *J*-coupling and can be

G. I. Hernández-Bolio · L. M. Peña-Rodríguez (✉)
Unidad de Biotecnología, Centro de Investigación Científica de Yucatán, Mérida, México
e-mail: lmanuel@cicy.mx

V. Cechinel Filho (ed.), *Natural Products as Source of Molecules with Therapeutic Potential*, https://doi.org/10.1007/978-3-030-00545-0_12

used to obtain detailed information about the chemical structure of a given molecule (Wishart 2013).

Presently, NMR spectroscopy has contributed significantly to the discovery of new natural products (Halabalaki et al. 2014) and is currently considered the most important spectroscopic technique in chemistry; the continuous development of NMR pulse sequences has made it possible to analyze specific features of the active nuclei (e.g., heteroatom bonding, spatial orientation, long-range coupling, etc.), providing information about both the measured nuclei and the surrounding atoms to give a complete outlook of the entire molecule of interest.

In this chapter, we briefly discuss the main techniques used in the NMR analyses of natural products and subsequently present some of the strategies applied in the structure elucidation of challenging molecules isolated during our investigations on bioactive metabolites from medicinal plants of the Yucatán Peninsula.

2 Identification of New and Structurally Common Natural Products

Once the NMR spectra of a recently purified metabolite are obtained, it is important to revise and analyze each and every one of the signals taking into account their chemical shift and splitting pattern; often, it is possible to recognize some structural features that can lead to the proposal of an initial structure. In these cases, it is necessary to compare the experimental data with those already reported in the literature for the same or a similar class of metabolites. In some cases, most of the signals in ^{1}H-NMR are similar, with the exception of one or two. These differences are frequently dismissed, even though they represent an interesting opportunity to discover a new or unusual natural product. One such case is *epi-flemistrictin B*, a chalcone isolated from *Lonchocarpus xuul* (Leguminosae). Previous phytochemical studies of this native plant of the Yucatán Peninsula led to the isolation of a number of flavonoids with antiprotozoal and cytotoxic activity (Borges-Argaez et al. 2007). The chromatographic separation of the n-hexane fraction of the roots of *L. xuul* yielded a metabolite with a $C_{20}H_{20}O_4$ molecular formula. The signals in its ^{1}H and ^{13}C-NMR spectra were very similar to those reported for flemistrictin B, a known derivative of isocordoin isolated from the leaves of *Flemingia stricta* (Yam-Puc and Pena-Rodriguez 2009). However, a detailed analysis of the ^{1}H-NMR spectrum of flemistrictin B showed the H-2″ signal as a doublet of doublets (J = 8.4, 9.6 Hz) at δ 4.78, while the same proton appeared as a triplet centered at δ 3.89 in the new metabolite; this finding suggested that the two metabolites were epimeric in their C-2″ stereochemistry (Fig. 1). To define the stereochemistry at the C-2″ position of the epimeric chalcones, a theoretical calculation of the H-2″ chemical shift was carried out using density functional theory (DFT), a quantum chemistry program which allows the calculation of many NMR parameters with remarkable precision and yields deeper insight into their origin (Günther 2013). The results obtained confirmed the significant differences in the chemical shift value for H-2″ in each of the epimeric

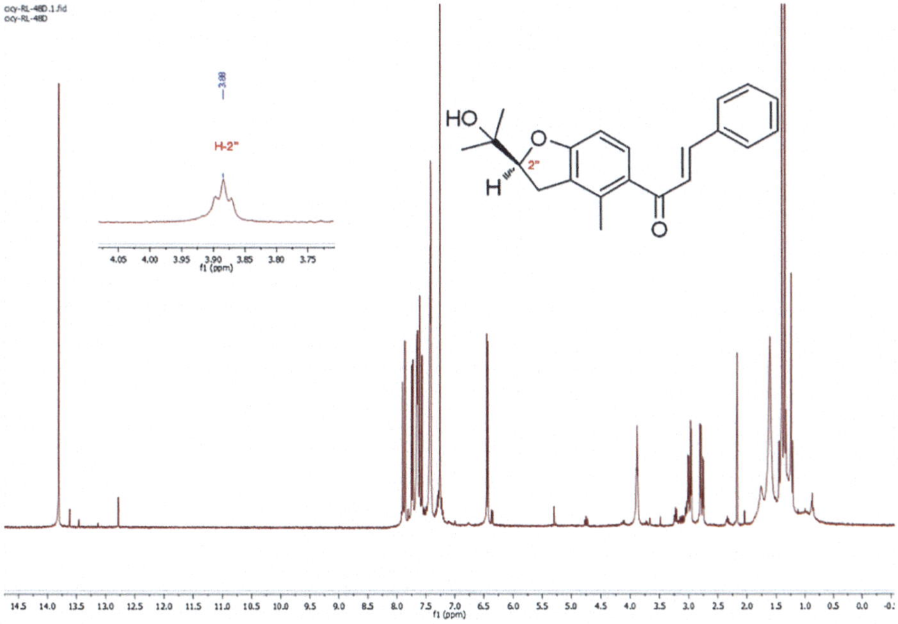

Fig. 1 ^{1}H-NMR ($CDCl_3$, 400 MHz) spectrum of epi-flemistrictin B with a close-up of the H-2″ proton signal

structures and showed that the H-2″ in the alpha orientation appears at a significantly higher field (δ 3.41) than that of the H-2″ in the beta orientation (δ 4.72). On this basis, the new metabolite was named as *epi-flemistrictin B* (Escalante-Erosa et al. 2012).

Another example of little variations in the recorded NMR spectra can be found in the case of *3-O-acetyl ceanothic acid*, a molecule with trypanocidal activity isolated from the medium polarity fraction of the root extract of *Colubrina greggii* var. *yucatanensis* (Rhamnaceae). The spectroscopic data of this metabolite proved to be very similar to those reported for ceanothic acid, a ceanothane-type triterpene also known as emmolic acid. However, the presence of a low-field carbinol proton (δ 5.07) and of an acetyl methyl singlet (δ 2.03) in the ^{1}H-NMR spectrum of the isolated metabolite (Fig. 2) suggested it being the 3-*O*-acetyl derivative of ceanothic acid. This was confirmed by the strong HMBC correlations observed between both the carbinol and acetyl methyl protons with the ester carbonyl carbon at δ 172.5. The presence of this metabolite in the original root extract of *C. greggii* ruled out its being an artifact of the isolation procedure (Dominguez-Carmona et al. 2011).

A different and interesting case was found with ***dinimbidiol ether***, a secondary metabolite with strong antioxidant activity isolated from *Cnidoscolus souzae* (Euphorbiaceae), a plant commonly known as "chaya" and used for its analgesic and anti-inflammatory properties (Zapata-Estrella et al. 2014). A number of ^{1}H and ^{13}C-NMR signals of the new metabolite coincided with those reported for the previously

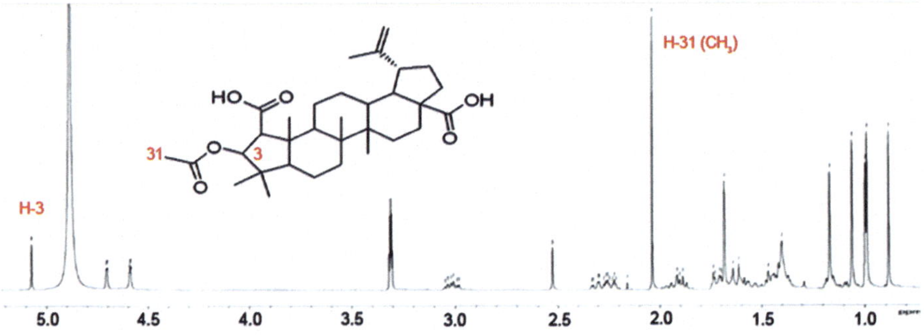

Fig. 2 ^{1}H-NMR (CD_3OD, 400 MHz) spectrum of 3-*O*-acetyl ceanothic acid

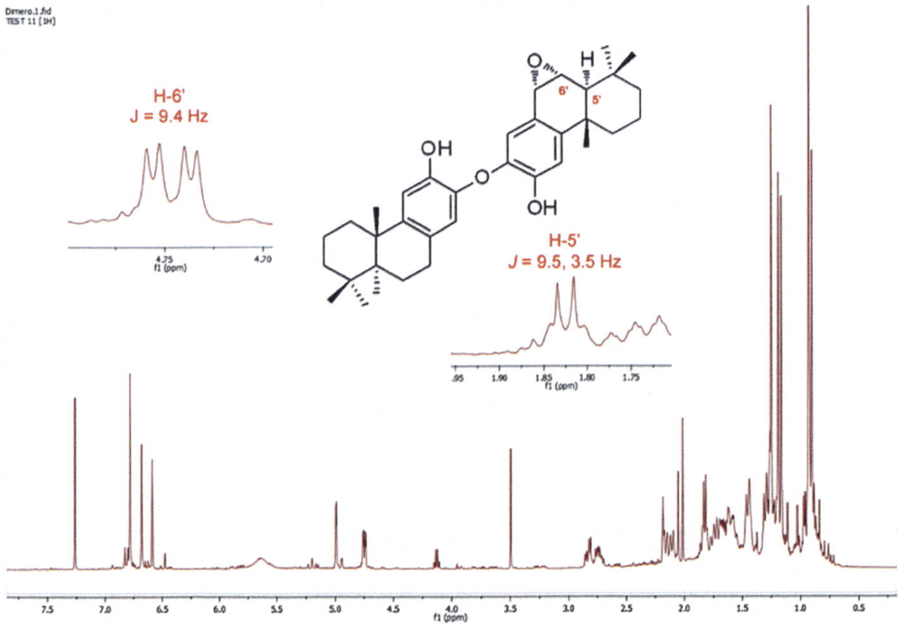

Fig. 3 ^{1}H-NMR (CDCl3, 400 MHz) spectrum of dinimbidiol ether with a close-up of the H-5′ and H-6′ proton signals

reported 7-deoxynimbidiol, while a second set of signals could be assigned to an epoxidated nimbidiol derivative. The dimeric nature of the new metabolite was confirmed by an LC-MS analysis which showed a single component at t_R 12.68 min, with a pseudomolecular ion peak at *m/z* 515 [M^+-H], corresponding to a molecular formula of $C_{34}H_{44}O_4$. The orientation of the epoxide group was suggested by the characteristic *trans*-diaxial coupling constant value of 9.4 Hz between the H-5′ and H-6′ protons at δ 1.84 and δ 4.63, respectively (Fig. 3). The unusual phenoxy-ether

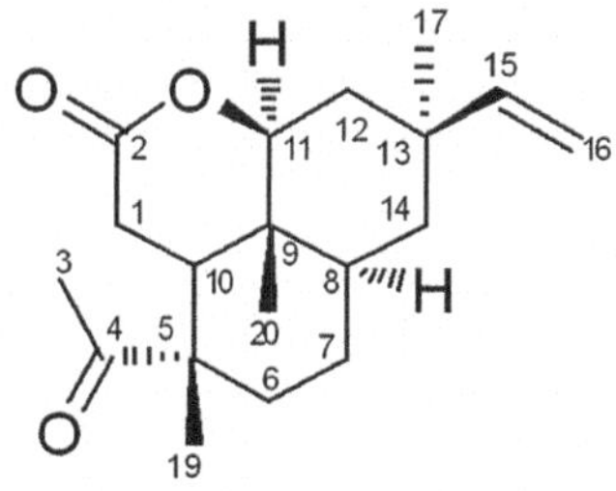

a

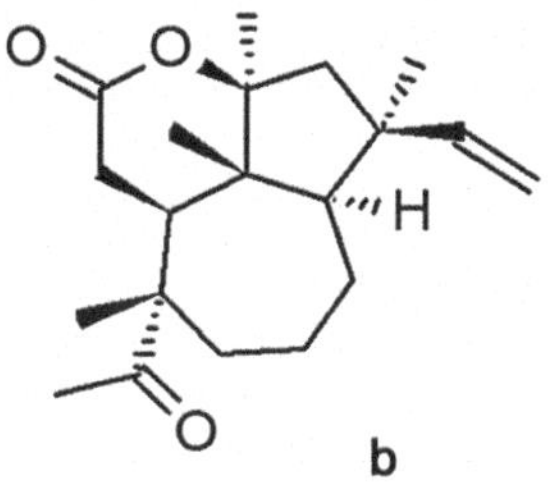

b

H Position	HMBC	
	^2J	^3J
1 ax	C2, C10	C9, C5
1eq		
2		
3	C4	
4		
5		
6eq		C8, C10
6ax		
7		
8		
9		
10	C1, C5, C9	C11, C19, C20
11		C10, C20
12ax	C13	C14, C15, C17
12eq		
13		
14eq		
14ax		
15	C13	C14
16	C15	C13
17	C13	C14, C15
19	C5	C4, C6, C10
20	C9	C3, C8, C10, C11

Fig. 4 Proposed structures (**a** and **b**) of merilactone and correlations observed in the HMBC experiment

linkage between the two monomers was confirmed when methylation of the natural product yielded a dimethylated derivative (Garcia-Sosa et al. 2017).

Sometimes, the analysis of the spectra available can result in the proposal of more than one possible structure. In this case, it is important to select those NMR methods that can be most helpful in solving a specific problem. During the search of bioactive metabolites from *Chiococca alba* (Rubiaceae), a plant used in the Yucatecan traditional medicine to cure dysentery and other ailments such as asthma, headaches, and diarrhea (Kan et al. 1944), the chromatographic purification of the root extract yielded a pure metabolite with a molecular formula of $C_{19}H_{28}O_3$. The detailed analysis of the ^1H and ^{13}C NMR spectra, together with the results from the HC-COBI and HMBC experiments, appeared to be consistent with two possible structures (**a** and **b**; Fig. 4). However, structure **a** proved to be the correct one on the basis of the following reasons: First, it was noted that all observable ^1H-^1H couplings corresponded to normal six-membered ring conformations; second, the HMBC experiment showed a ^3J correlation between one of the H-12 methylene protons (1.73 ppm) and C-14 (37.5 ppm) and between H-15 (5.80 ppm) and C-14 (37.5 ppm). The relative stereochemistry of the isolated metabolite was revealed by the NOESY experiment (Fig. 5), which provides information about proton-proton connectivity through space. On this basis, the new and unusual nor-*seco*-pimarane was identified as 4-acetyl-4,8,9b-trimethyl-8-vinyldecahidrobenzo[*de*]chromen-2-one and designated with the trivial name ***merilactone*** (Borges-Argaez et al. 2001).

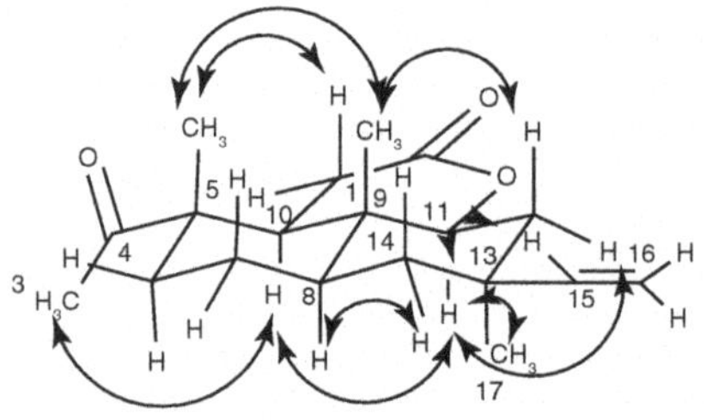

H Position	NOESY correlations	H Position	NOESY correlations
1ax	Me-19	11	H-10, H-8, H-12eq, Me-17
1eq		12ax	Me-20
2		12eq	H-11
3	H-19	13	
4		14eq	H-15
5		14ax	
6eq		15	H-14eq
6ax		16	
7		17	H-11
8	H-11	19	H-1ax, Me-20
9		20	H-12ax, Me-19
10	H-11, Me-3		

Fig. 5 Correlations observed in the NOESY experiment of merilactone

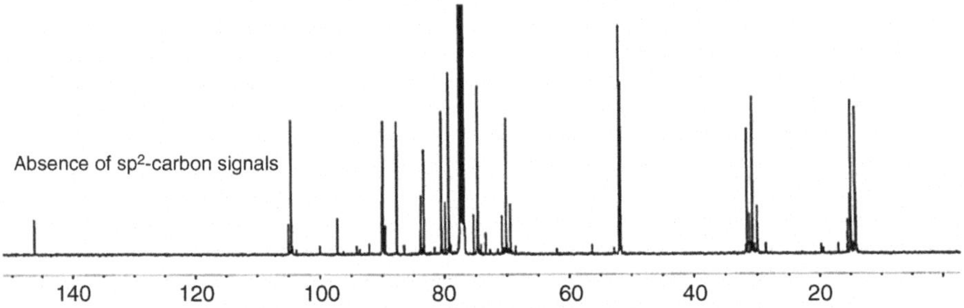

Fig. 6 ^{13}C-NMR ($CDCl_3$, 100 MHz) spectrum of urechitol A

3 Identification of Novel Metabolites: The Case of Urechitol A

Frequently, after a metabolite is purified and identified, it is found that the structure has already been reported from a different source. Even though literature data can be of great assistance, there are cases in which the identification of a particular structure requires a detailed analysis of all the spectroscopic evidence. In addition to ^{1}H and ^{13}C-NMR spectra, it is necessary to take into account the results from 2D experiments to build the initial fragments which can be put together to produce the final structure (Breton and Reynolds 2013). This approach has been used to identify two structurally related tri-nor-sesquiterpenes from the root extract of *Pentalinon andrieuxii* (Apocynaceae), a plant used in Yucatecan traditional medicine to treat leishmaniasis (Chan-Bacab et al. 2003). The HREIMS of the first product suggested a molecular formula of $C_{14}H_{22}O_7$, which implied the presence of four unsaturation sites in the structure. The absence of sp^2 carbon signals in the ^{13}C NMR spectrum of the isolated metabolite (Fig. 6) indicated that the four unsaturation sites implied by the molecular formula corresponded to a tetracyclic structure (Yam-Puc et al. 2009). While the initial analysis of the ^{1}H-NMR spectrum of the isolated metabolite allowed the construction of the first structural subunits A and B (Fig. 7), the remaining subunits were built after analyzing the results from 2D experiments such as COSY (^{1}H-^{1}H Correlation Spectroscopy) and HSQC (Heteronuclear Single Quantum Correlation).

COSY is a 2D experiment showing crosspeaks whenever two protons are directly coupled via two or three bonds (i.e., geminal or vicinal couplings). In this case, a detailed analysis of the COSY experiment allowed the construction of the structural subunits C, D, and E (Fig. 8), where the methine proton signal at δ 5.28 in subunit

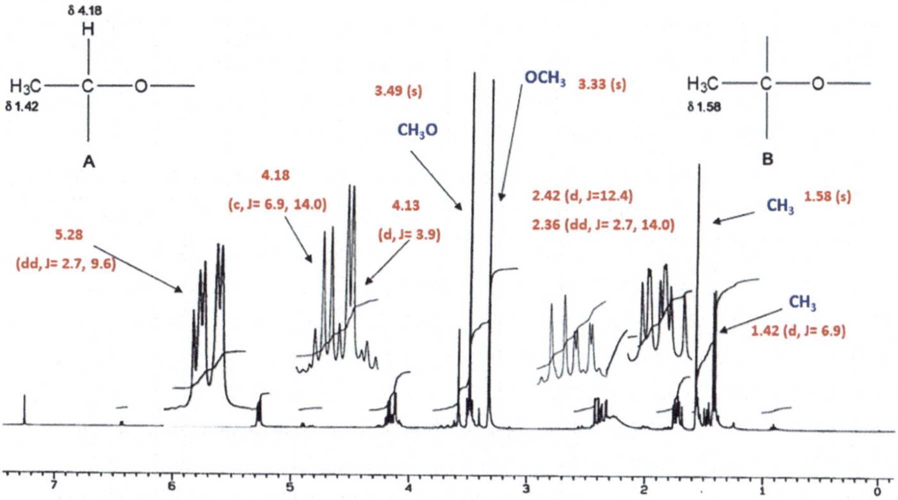

Fig. 7 ^{1}H-NMR ($CDCl_3$, 400 MHz) spectrum of urechitol A showing substructures **A** and **B**

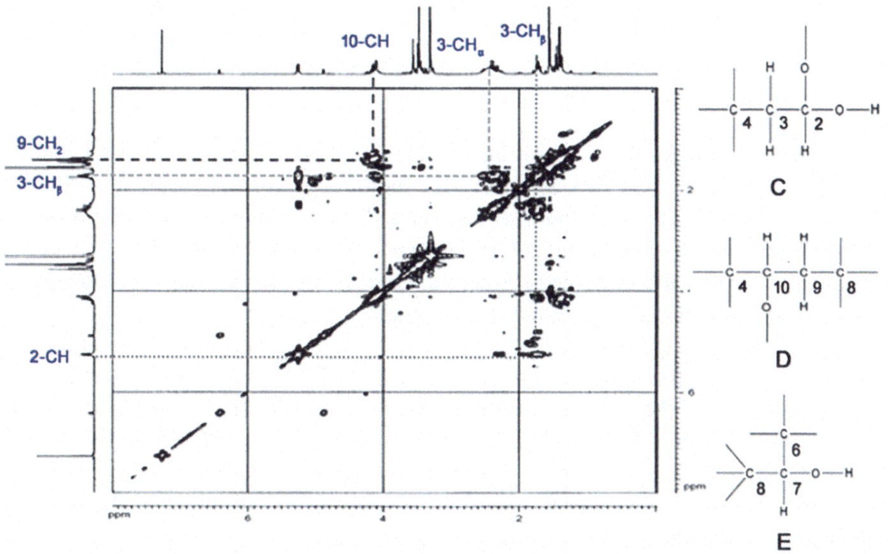

Fig. 8 Correlations observed in the COSY experiment of urechitol A and substructures **C**, **D**, and **E**

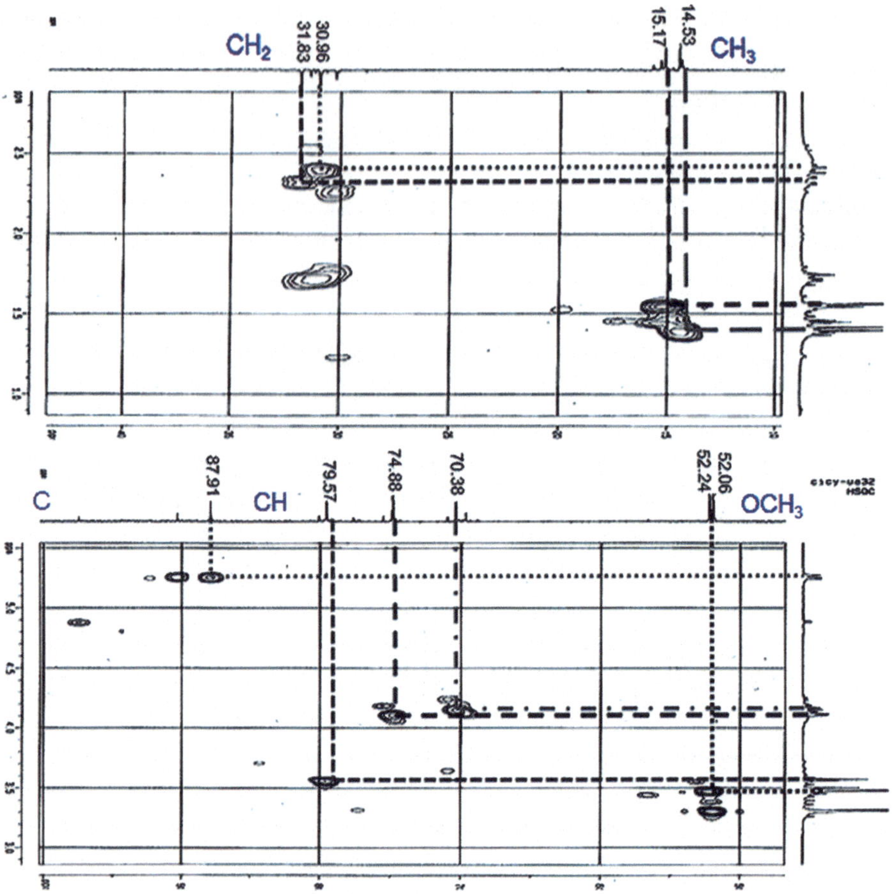

Fig. 9 Correlations observed in the HSQC experiment of urechitol A

C showed a coupling to the two protons of an isolated methylene group at δ 1.72 (m) and δ 2.36 (dd, $J = 2.7$, 14 Hz). Similarly, the methine proton appearing as a doublet ($J = 3.9$ Hz) at δ 4.13 in the subunit D showed a clear correlation to the protons of a second isolated methylene group at δ 179 (d, $J = 12.8$ Hz) and δ 2.42 (d, $J = 12.4$ Hz). Finally, the fact that the methine proton of subunit E appeared as one-proton singlet at δ 3.59 indicated its being bonded to two quaternary carbons.

The results of the HSQC experiment, which identifies protons bonded to carbon by correlating the signals of protons in the ^{1}H-NMR spectrum with those of carbons in the ^{13}C-NMR, were combined with those from DEPT-135 and DEPT-90 experiments, which allow to distinguish between C, CH, CH_2, and CH_3 signals in the ^{13}C-NMR spectrum, to identify four methyl groups, two methylenes, four methines, and four quaternary carbons (Fig. 9) in the structure of the isolated metabolite. Additionally, the chemical shift values of all quaternary carbons (δ

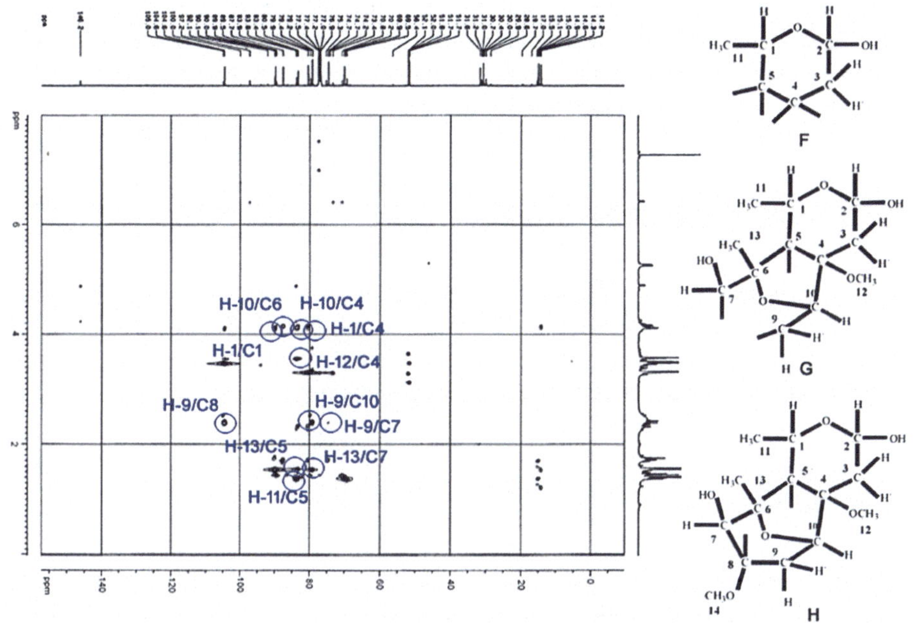

Fig. 10 Correlations observed in the HMBC experiment of urechitol A and substructures **F**, **G**, and **H**

80.5, δ 83.7, δ 89.9, and δ 104.6) and the carbons of all methine groups (δ 70.4, δ 74.9, δ 79.6, and δ 87.9) clearly indicated that they were all bonded to oxygen. Similarly, the chemical shift values of two of the four methyl group signals (δ 51.9 and δ 52.2) also indicated their being bonded to oxygen.

The HMBC (Heteronuclear Multiple Bond Correlation) experiment allows the correlation between protons and carbons separated by two (^2J) and three (^3J) bonds and is commonly used to connect structural subunits. In this case, a ^3J correlation between H-1 and C-2/C-4 indicated that subunits A and C could be connected through an oxygen atom (Fig. 10). Similarly, a ^3J correlation between H-11 and an unassigned quaternary carbon (C-5) suggested that C-1 and C-5 could connect to complete the six membered ring of structural subunit F. Linking of structural subunits F and D was confirmed by a ^2J correlation between H-10 and C-4. Furthermore, the ^3J correlation observed between the protons of one of the methoxyl groups (H-12) and C-4 allowed placement of the ether group at C-4. The ^3J correlation between H-10 and C-6 indicated that structural subunits B and D are connected through an oxygen atom, while the ^3J correlation between H-13 and C-7/C-5 allowed the connection of the structural subunits B and E to produce the five-membered ring subunit G. Finally, a ^2J correlation between the C-9 methylene protons and C-8/C-10, together with a ^3J correlation between the three protons of the second methoxyl group and C-8 and a ^3J correlation between H-9 and C-7, resulted in the construction of tricyclic structural subunit H. Both the structure and relative stereochemistry of this product were confirmed unambiguously by a single-crystal X-ray diffraction experiment and reported as a new natural product, with a novel skeleton, designated with the common name of ***urechitol A*** (Yam-Puc et al. 2009).

4 Identification of Terpenoids

Even though terpenoids represent the largest and more structurally diverse group of secondary metabolites (Dewick 2002), there are no simple rules or strategies for their identification. However, it is always possible to recognize structural features, which can lead to the identification of a given metabolite.

4.1 Ent-kaurenes

Ent-kaurenes are diterpenoids with an exocyclic double bond which can be readily detected by ^{1}H and ^{13}C-NMR; the prefix *ent* is used to indicate that these metabolites are enantiomeric to those of the kaurane series. The phytochemical study of the roots of *Chiococca alba* (Rubiaceae) led to the isolation of a metabolite whose ^{13}C-NMR spectrum (Fig. 11) showed the characteristic signals for the two sp^{2}

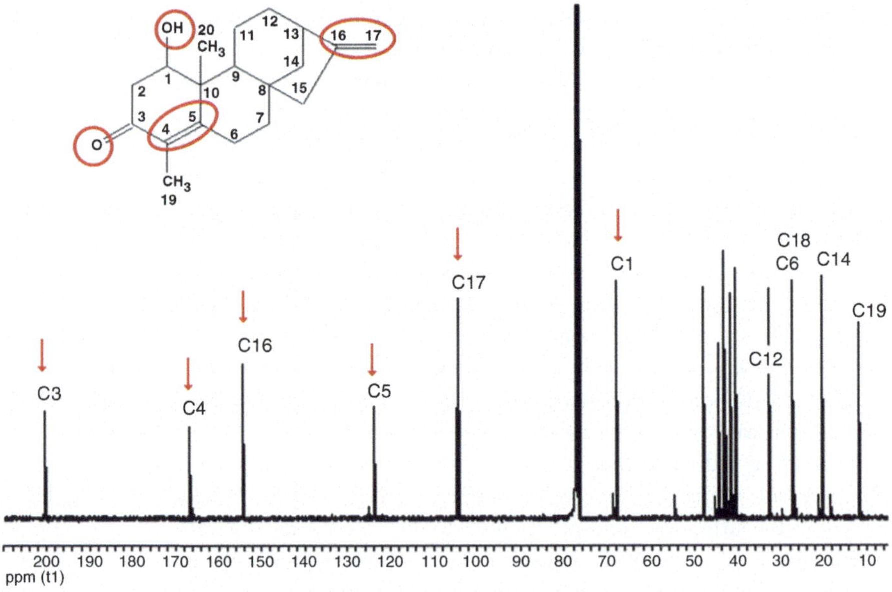

Fig. 11 ^{13}C-NMR (CDCl$_3$, 100 MHz) spectrum of 1-hydroxy-18-nor-kaur-4,16-dien-3-one

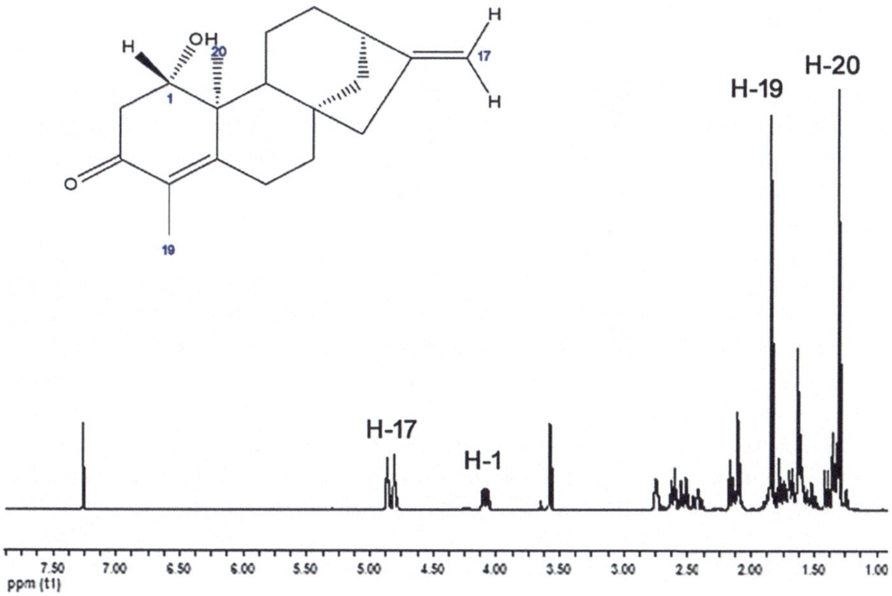

Fig. 12 ^{1}H-NMR ($CDCl_3$, 400 MHz) spectrum of 1-hydroxy-18-nor-kaur-4,16-dien-3-one

carbons of an *exo*-methylene at δ 154.4 (s) and δ 104.6 (t), together with signals for two additional olefinic carbons at δ 166.8 (s) and δ 124.0 (s), a carbonyl carbon at δ 200.5 (s), and an oxygen-bearing methine carbon at δ 68.2 (d).

Similarly, the ^{1}H-NMR spectrum of the isolated metabolite (Fig. 12) showed two broad singlets at δ 4.81 and δ 4.86 corresponding to the two vinylic protons of the exocyclic double bond, together with an axial-oriented carbinol proton at δ 4.09 (ddd, $J = 13.8, 6.1, 2.0$), two methylene protons at δ 2.53 (dd, $J = 13.8, 6.1$ Hz) and δ 1.34 (d, $J = 13.9$ Hz), and a vinylic methyl group at δ 1.85. This spectroscopic evidence strongly suggested an *ent*-kaurene structure for the isolated metabolite (Nagashima et al. 2003). A detailed analysis of the HMBC experiment allowed the construction of the full structure when ^2J and ^3J correlations could be observed between the protons of the methylene group at δ 2.53/1.34 and the carbonyl carbon at δ 200.5 and the sp^2 carbon at δ 166.8, respectively, together with ^3J and ^2J correlations observed between the two carbons at δ 200.5/166.8 and the vinylic methyl group at δ 1.85. Alternatively, the ^3J correlation observed between the carbinol proton at δ 4.09 and the methyl group at δ 27.4 indicated that they were both located in the same ring. Similar correlations allowed the construction of the remaining tetracyclic structure, which could not be found reported in the literature. The new nor-diterpene, with an *ent*-kaurene skeleton, was designated as ***1-hydroxy-18-nor-kaur-4,16-dien-3-one*** (Dzib-Reyes et al. 2012).

4.2 Pregnanes

Pregnanes are C21 steroidal metabolites having the cyclopentane perhydro phenanthrene ring system, with β-oriented angular methyl groups at C-10 and C-13 and a two-carbon atom side chain at C-17. Plant pregnanes, like many other naturally occurring steroids, usually show the biogenetically favored *β*-hydroxyl group at the C-3 position (Deepak et al. 1989).

While searching for novel leishmanicidal metabolites, two pregnanes were isolated from the root extract of *Pentalinon andrieuxii*. The ^{1}H-NMR spectrum (Fig. 13) of the first product revealed the presence of two oxygen-bearing methine groups at δ 3.49 (m) and δ 4.39 (d, *J* = 6.6 Hz), where the low-field chemical shift of the second proton suggested its being part of an ester or lactone functionality. Additional signals in the ^{1}H-NMR spectrum included a vinylic proton signal at δ 5.37 (dd, *J* = 2.8 Hz), together with a three-proton singlet at δ 1.03 and a three-proton doublet (*J* = 6.6 Hz) at δ 1.33, indicating the presence of a trisubstituted double bond and two methyl groups, one bonded to a quaternary carbon and another to a methine, respectively. The spectroscopic data of this product proved to be identical to that of ***3β,14β,20-trihydroxypregn-5-ene-18-oic-(18–20) lactone*** (Yam-Puc et al. 2012), the aglycone of amaloside C previously reported from *Amalocalyx yunnanensis* (Apocynaceae) (Xiao-Ling et al. 1993).

The ^{1}H-NMR spectrum of the second product (Fig. 14) showed no vinylic proton signals and the presence of a single carbinol proton at δ 4.11 (dd, *J* = 2.6 Hz), in addition to two three-proton singlets at δ 1.58 and δ 0.94 corresponding to two

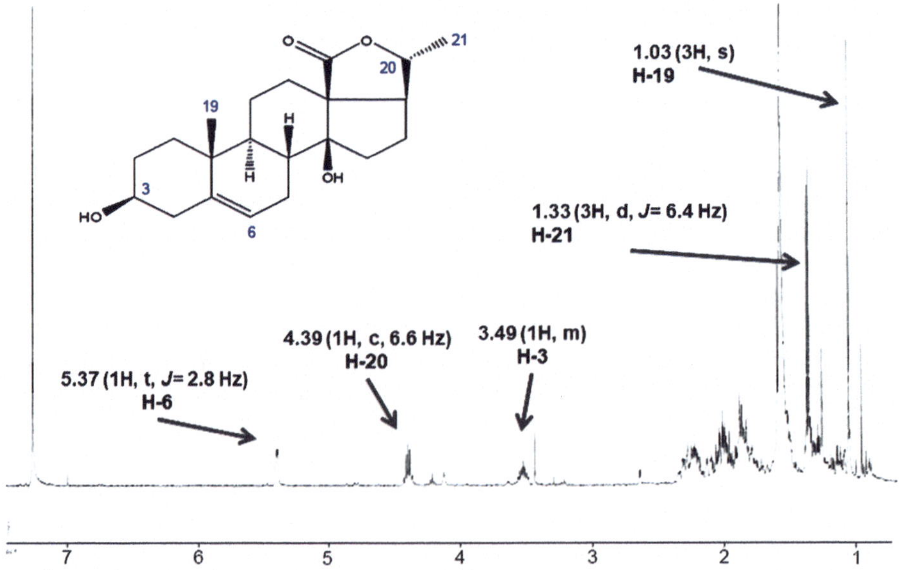

Fig. 13 ^{1}H-NMR ($CDCl_3$, 400 MHz) spectrum of 3*β*,14*β*,20-trihydroxypregn-5-ene-18-oic-(18–20) lactone

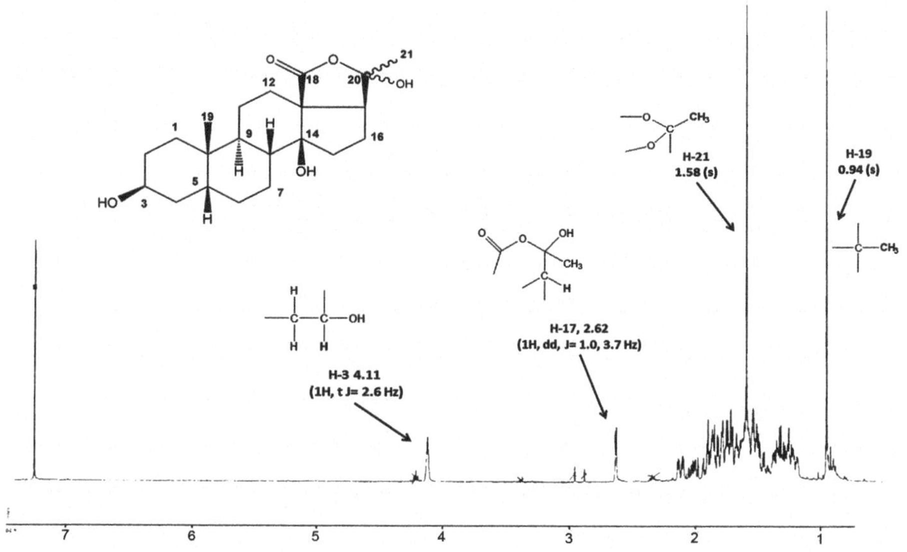

Fig. 14 ^{1}H-NMR (CDCl$_3$, 400 MHz) spectrum of 3*β*,14*β*,20, 20-tetrahydroxypregn-18-oic-(18–20) lactone

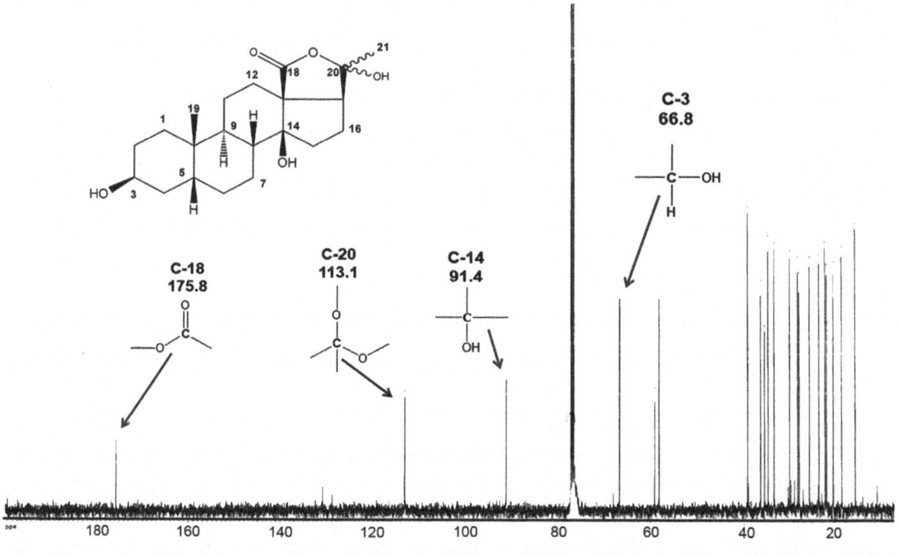

Fig. 15 ^{13}C-NMR (CDCl$_3$, 100 MHz) spectrum of 3*β*,14*β*,20, 20-tetrahydroxypregn-18-oic-(18–20) lactone

methyl groups bonded to quaternary carbons. The ^{13}C-NMR spectrum of the second pregnane (Fig. 15) showed the expected methine carbon at δ 66.8, in addition to an oxygen-bearing quaternary carbon at δ 91.4, an ester carbonyl carbon at δ 175.8, and a hemiketal carbon at δ 113.1. The presence of the hemiketal group in the structure of the pregnane was confirmed when treatment of the natural product with $NaBH_4$ yielded the expected reduction product, which showed a new carbinol proton at δ 4.38 and a three-proton doublet at δ 1.35. The assignment of the *cis* configuration at the A- and B-ring junction was based on the chemical shift observed for the C-19 methyl group carbon (δ 23.6) in the ^{13}C-NMR spectrum; it is known that the signal for the ring junction methyl group carbon in the *cis* isomer occurs δ 11–δ 12 downfield when compared to that of the same methyl group in the *trans* isomer (Wehrli and Nishida 1979). On the basis of these results, the second metabolite was identified as ***3β,14β,20, 20-tetra-hydroxy-5β-pregn-18-oic-(18–20)-lactone***, a new natural product (Yam-Puc et al. 2012).

4.3 Cycloartanes

Cycloartanes can be readily detected by ^{1}H-NMR because of the characteristic high-field (δ 0.60–δ 0.30) signals corresponding to the C-19 methylene protons of the cyclopropane ring. These resonances are usually displayed as an AX system with a germinal coupling constant of 4 Hz (Pistelli 2002).

An interesting exercise on the identification of cycloartanes originates from the investigation of the epicuticular wax of *Cocos nucifera*. The TLC analysis of the hexane extract from the pines of *C. nucifera* showed the presence of two main components, while the most polar component was identified as ***lupeol methyl ether*** by direct comparison with an authentic sample, the less polar component proved to be a mixture of two metabolites that could only be separated by using AgNO3-impregnated silica gel plates. Successive $AgNO_3$-impregnated column chromatography and preparative TLC yielded both components in pure form. The ^{1}H-NMR spectrum of the less polar product (Fig. 16) showed two signals at δ 0.32 (d, $J = 4.5$ Hz) and δ 0.56 (d, $J = 4.0$ Hz), suggesting a cycloartane skeleton. Additionally, the presence of a sharp singlet at δ 3.36 clearly indicated the presence of a methoxyl group in the structure, while the signals corresponding to four methyl groups [δ 0.90 (d, $J = 6.5$ Hz), δ 0.95 (s), δ 0.97 (s), and δ 1.05 (t, $J = 7.5$ Hz)] and two vinylic protons [δ 4.77 (bd, $J = 1$ Hz) and δ 4.79 (bs)] were in agreement with an eight-carbon side chain having a gem-dimethyl group and a 1,1 disubstituted double bond as substituents. This data proved to be identical to those previously reported for ***skimmiwallin*** [3β-methoxy-25-ethyl-9,19-cyclolanost- 24(24^1)-ene], a cycloartane isolated from the petrol ether extract of *Skimmia wallichii* (Kostova et al. 1996).

The ^{1}H and ^{13}C-NMR spectra of the second component (Fig. 17) appeared to be very similar to those of ***skimmiwallin***, suggesting an isomeric structure. Differences could only be detected after a detailed analysis of the results from HMBC experiments; while the HMBC experiment on skimmiwallin showed a clear J^3 correlation between the C-27-methyl signal (δ 1.05) and the sp^3-quaternary carbon at δ 39.51 (C-25), the same experiment (Fig. 18) on the new metabolite showed a J^3 correlation between the C-27-methyl signal (δ 1.05) and the sp^2-quaternary carbon (C-25, δ 157.92) of the 1,1 disubstituted double bond. This data allowed the identification of

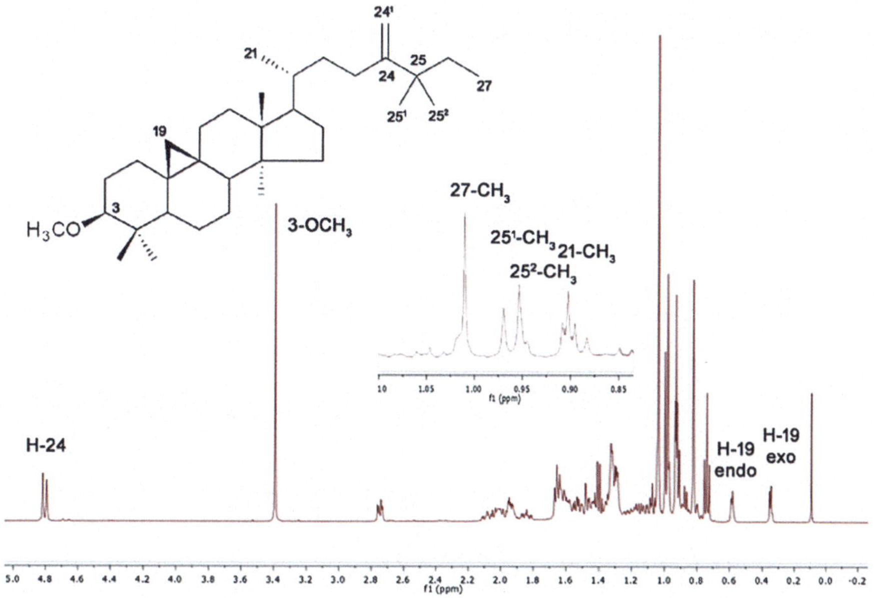

Fig. 16 ^{1}H-NMR ($CDCl_3$, 500 MHz) spectrum of skimmiwallin with a close-up of 21-CH_3, 25[1]-CH_3, 25[2]-CH_3 and 27-CH_3 signals

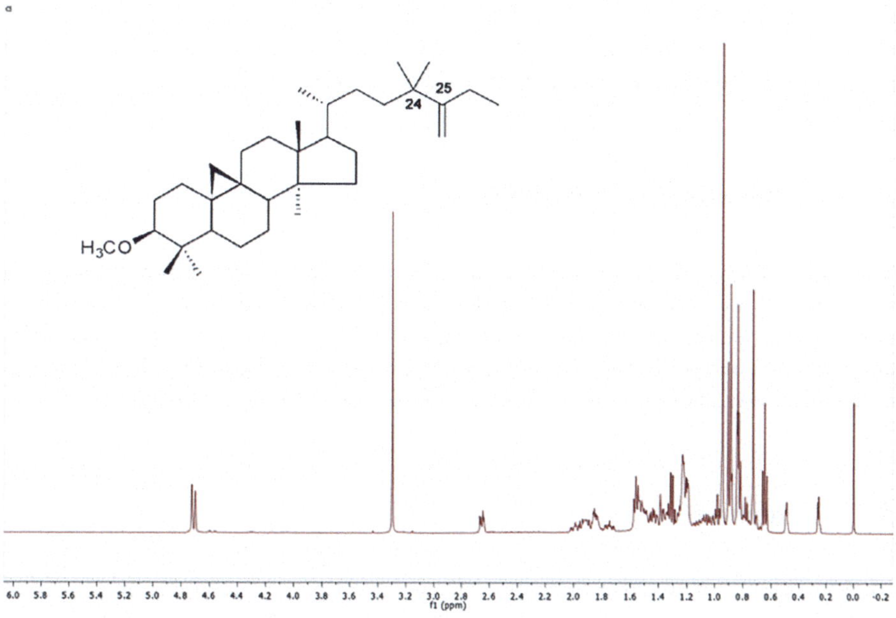

Fig. 17 ^{1}H-NMR ($CDCl_3$, 500 MHz) spectrum of isoskimmiwallin

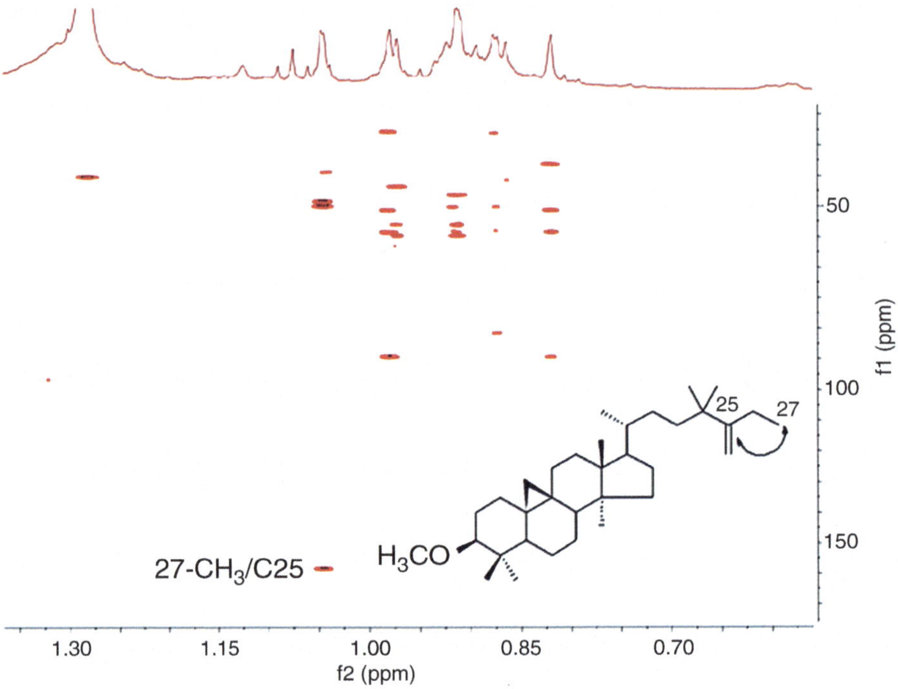

Fig. 18 Key correlations observed in the HMBC experiment of isoskimmiwallin

the new isomeric metabolite, designated as ***isoskimmiwallin*** because of its isomeric relation to ***skimmiwallin*** (Escalante Erosa et al. 2002).

5 Identification of Flavonoids

The native flora of Yucatán is rich in species of the Leguminosae family; some of these species are endemic or quasi-endemic and represent an interesting and abundant source of both known and unknown bioactive flavonoids. Being that NMR analyses are commonly used to identify and elucidate the structure of novel flavonoids, there are a number of key features that facilitate their identification.

5.1 Chalcones

Chalcones are considered as precursors of a great variety of flavonoids; their general structure includes two aromatic rings connected through a three-carbon bridge bearing an α,β-unsaturated ketone; chalcones are frequently found in plants, and they

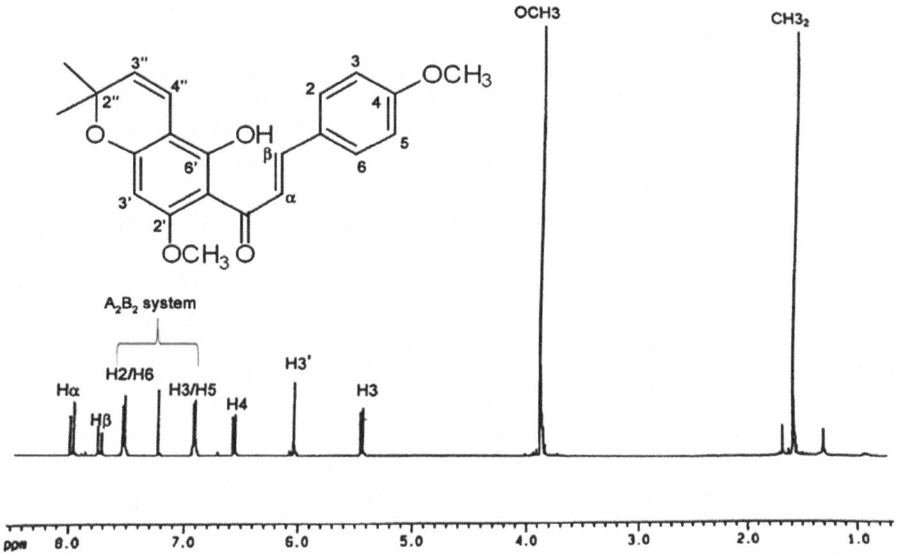

Fig. 19 ^{1}H-NMR (CDCl$_3$, 500 MHz) spectrum of 2′,4-dimethoxy-6′-hydroxylonchocarpin

have been reported to exhibit antibacterial, antiproliferative, and anti-inflammatory properties (Hwang et al. 2011).

The chromatographic purification of the hexane-soluble fraction of the leaf extract of both *Lonchocarpus xuul* (Leguminosae) and *L. yucatanensis* led to the isolation of a metabolite whose ^{1}H-NMR spectrum (Fig. 19) showed two *trans*-olefinic protons at δ 7.79 and δ 8.04 (ea $J = 15.6$ Hz). These signals, characteristic of the *α* ανδ *β* protons of an *α,β*-unsaturated ketone, suggested a chalcone structure for the isolated metabolite. Additional signals in the ^{1}H-NMR spectrum included two six-proton singlets at δ 3.86 and δ 1.56, indicating the presence of two methoxyl and two methyl groups, respectively, as well as an isolated aromatic proton at δ 6.07 and four aromatic protons (δ 7.57 and δ 6.95; 2H ea.; d, $J = 8.5$ ea) showing the A_2B_2 pattern typical of a 4-substituted B-ring. The presence of two *cis*-coupled olefinic protons, together with two methyl group resonances and a quaternary carbon at δ 78.0 in the ^{13}C-NMR spectrum, indicated a 2,2-dimethyl-dehydro-pyran system. The arrangement of the substituents and the placement of the pyran system were established through an HMBC experiment, which showed a correlation between H-4″ and C-6′ (δ 155.9), and between H-3′ and C-5′ (δ 103.3), while H-3′ showed the expected correlations to C-2′ (δ 161.3) and C-4 (δ 167.7). On the basis of this data, the chalcone was identified as ***2′,4-dimethoxy-6′-hydroxylonchocarpin*** and reported as new metabolite (Borges-Argáez et al. 2002).

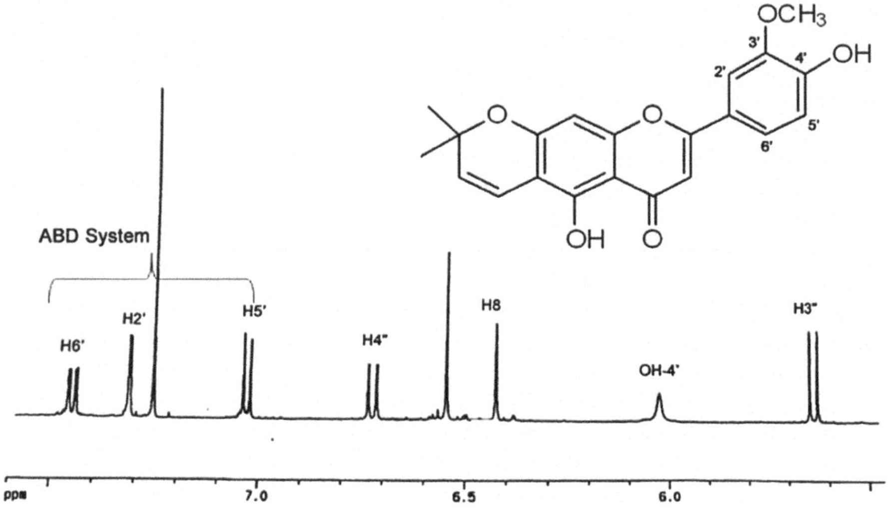

Fig. 20 ^{1}H-NMR ($CDCl_3$, 500 MHz) spectrum of 5,4′-dihidroxy-3′-methoxy-(7,6:2″,3″)-6″, 6″-dimethyl-pyranoflavone

5.2 *Flavones*

Flavones have the general backbone of 2-phenylchromen-4-one (2-phenyl-1-benzopyran-4-one); therefore their ^{1}H-NMR spectrum commonly shows aromatic protons with different spin systems, which depend on the substitution pattern of the aromatic rings. A metabolite isolated from the leaves of both *L. xuul* and *L. yucatanensis* revealed resonances from a methoxyl group (δ 4.01, s), a 2,2- dimethyl-dehydropyran ring (δ 1.48, 6H, s; δ 5.62 and δ 6.73, 1H ea., d $J = 10$ Hz ea), one vinylic (δ 6.55), and one isolated aromatic proton (δ 6.43), as well as an ABD system corresponding to three protons (δ 7.03, d, $J = 8.5$ Hz; δ 7.33 d $J = 2.0$ Hz; δ 7.47 dd $J = 8.5$, 2.0 Hz) in a trisubstituted aromatic ring (Fig. 20). The HMBC correlations observed between the isolated aromatic proton H-8 and the oxygen-bonded aromatic carbon C-7 at δ 159.4, and between the hydrogen-bonded hydroxyl proton at δ 13.09 and the C-6 (δ 105.4) and C-7 carbons, indicated that the 2,2-dimethyl-dihydro-pyran group was connected to the A-ring in a linear manner. The substitution pattern in the B-ring was established from the results observed in the NOE (nuclear Overhauser effect) experiment, which is based on the irradiation of a particular proton at its resonance frequency and the observation of enhancement in the signals located in the same "face" or proximity of the irradiated proton (Silverstein et al. 2014). In this case, irradiation of the methoxyl group resonance caused a significant enhancement of the H-2′ signal, indicating that the methoxyl group was located in the 3′-position of the B-ring. On the basis of these results, this metabolite

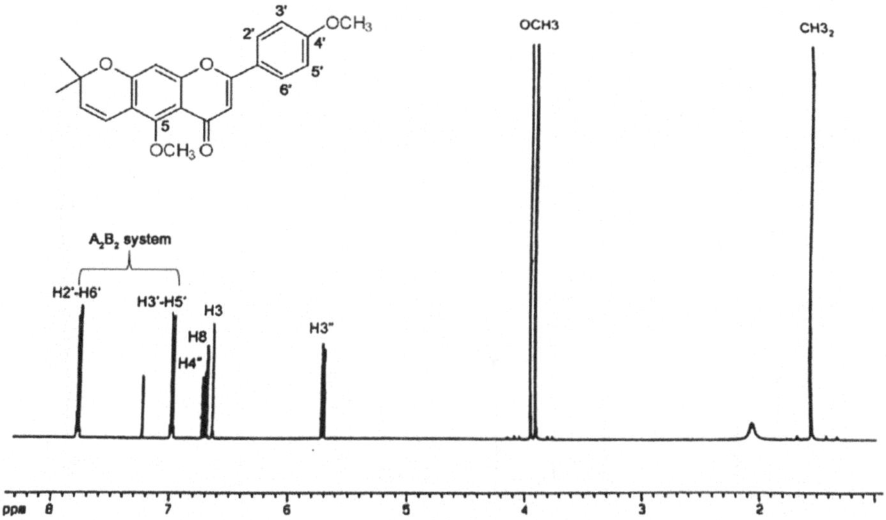

Fig. 21 ^{1}H-NMR ($CDCl_3$, 500 MHz) spectrum of 5,4′-dimethoxy-(7,6:2″,3″)-6″,6″-dimethylpyranoflavone

was unambiguously identified as the novel ***5,4′-dihidroxy-3′-methoxy-(6:7)-2′,2-dimethyl-pyranoflavone*** (Borges-Argáez et al. 2002).

A second flavone isolated from the same extracts revealed the presence of two methoxyl groups (δ 3.89 and δ 3.93) in its ^{1}H-NMR spectrum (Fig. 21); however, the same spectrum did not show the expected low-field hydrogen-bonded hydroxyl proton, suggesting that one of the methoxyl groups was located at the C-5 position of the structure. Additionally, the signals for four aromatic protons (δ 7.01 and δ 7.83, 2H ea., d $J = 8.6$ Hz ea), appearing as an A_2B_2 system (δ 7.01 and δ 7.83, 2H ea., d $J = 8.6$ Hz ea) in the ^{1}H-NMR spectrum of the second flavone, suggested a *para*-disubstituted B-ring. This was confirmed by the results of the NOE experiment which showed that irradiation of the methoxyl signal at δ 3.89 enhanced the two-proton signal at δ 7.01. These arguments allowed the identification of the second flavone as ***5,4′-dimethoxy-(6:7)-2′,2-dimethyl-pyranoflavone*** (Borges-Argáez et al. 2002).

5.3 *Flavans*

Even though flavans are also benzopyran derivatives, theirs is a 2-phenyl-3,4-dihydro-2H-chromene skeleton. Since hydroxyl groups frequently occur at the C-3 and C-4 positions, carbinol protons and/or methoxyl group signals are commonly observed in the ^{1}H-NMR spectra of these metabolites.

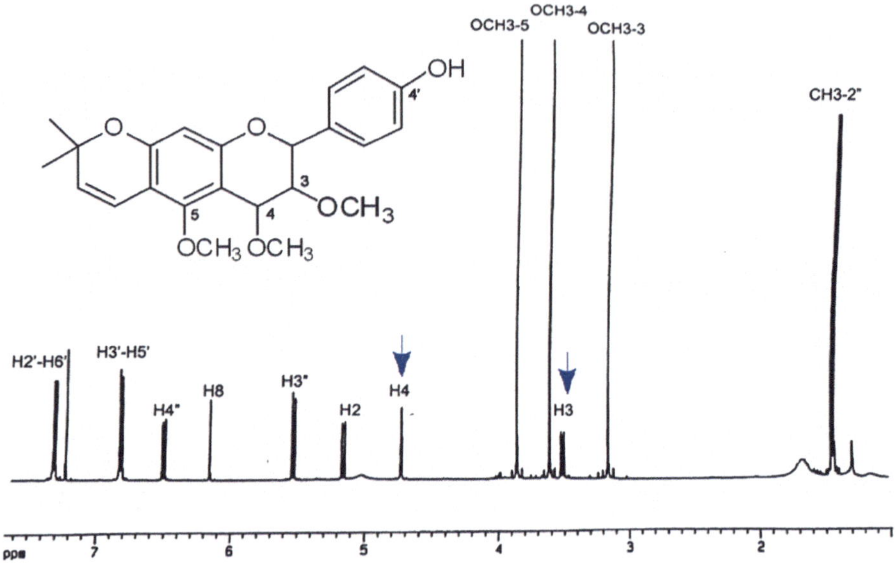

Fig. 22 ^{1}H-NMR ($CDCl_3$, 500 MHz) spectrum of 3β,4β,5-trimethoxy-4′-hydroxy-(7,6:2″,3″)-6″, 6″-dimethylpyranoflavan

The ^{1}H-NMR spectrum (Fig. 22) of a novel flavan isolated from the hexane-soluble fraction of the stem bark extract of *L. xuul* displayed an A_2B_2 system (δ 6.86 and δ 7.35, 2H ea., d J = 8.5 Hz ea) corresponding to the four protons of a *para*-disubstituted B-ring, as well as a single aromatic proton (δ 6.18, s) indicating a pentasubstituted A-ring, three methoxyl group singlets (δ 3.16, δ 3.62, δ 3.86), and three oxymethine protons at δ 5.17 (d, J = 10.3 Hz), δ 3.51 (dd, J = 10.3, 2.9 Hz), and δ 4.74 (d, J = 2.9 Hz). The location of the methoxyl groups at positions C-3, C-4, and C-5 was established from the results of an HMBC experiment, while the 3,4-dioxygenated flavan skeleton was confirmed by the three oxygenated carbon signals (δ 75.4, δ 82.0, and δ 68.6) observed in the ^{13}C-NMR spectrum of the flavan. The large coupling constant (10.3 Hz) observed between the proton signals at δ 5.17 and δ 3.51 indicated their having a trans-diaxial arrangement, while the small coupling constant (2.9 Hz) between the proton signals at δ 3.51 and δ 4.74 is characteristic for an axial-equatorial arrangement. This allowed the identification of ***3β,4β,5-trimethoxy-4′-hydroxy-(6:7)-2″,2″-dimethylpyranoflavan*** as a new flavan from *Lonchocarpus* spp. (Borges-Argáez et al. 2000, 2002).

6 Assignment of Protons in Sugar Moieties

The assignment of protons belonging to a sugar moiety in the ^{1}H NMR spectra of glycosylated metabolites is always a challenge. Often, with the exception of the anomeric proton, carbinol protons in sugars have similar chemical shifts (between δ

3.0 and δ 4.2), and the multiplicity of the signals frequently results in overlapping, making it difficult to identify the structure. There are however strategies and NMR experiments which can be used to successfully assign the sugar protons in glycosylated metabolites.

The first step when analyzing the ^{1}H NMR spectrum of a glycosylated metabolite is to identify the anomeric proton or protons in the structure; these are the most deshielded protons in a sugar moiety and usually appear between δ 4.4 and δ 5.5. The number of anomeric protons and carbons, which appear as signals around 100 ppm in the corresponding ^{13}C-NMR, indicates the number of sugar residues in the structure (Duus et al. 2000). The chemical shift and coupling constant of the anomeric proton signal in the ^{1}H NMR spectrum can also suggest the relative configuration of the anomeric carbon; the proton of the α-anomer is usually more deshielded and shows a smaller coupling constant (2–3 Hz) than that of the β-anomer (6–7 Hz) (Bubb 2003). It is advisable to search for the number of protons of the sugar moiety so as to identify the type of sugar (i.e., hexose, pentose) or to identify key alkyl groups in the molecule (e.g., C6-CH_3 in rhamnose, C6-CH_2OH in glucose). When there are two or more sugar residues attached to the molecule of interest, it is important to establish the link between them using the results from an HMBC experiment (Duus et al. 2000).

A 2D-NMR experiment which can help determine the coupling constant values of overlapped proton signals is *J*-resolved; this experiment yields a 2D map correlating *J*-coupling with conventional chemical shifts (Parella 2010), making it possible to establish the relative configuration of each proton and facilitating the identification and quantification of a glycosylated metabolite. This experiment was particularly useful when assigning the sugar protons in the natural product *arbutin*, a glycosylated hydroquinone derivative with anthelmintic activity, isolated from the leaf extract of *Lysiloma latisiliquum* (Hernández-Bolio et al. 2017). The ^{1}H-NMR spectrum of arbutin showed the anomeric proton at δ 4.74 and the hydroxylated methylene proton signals at δ 3.88 and δ 3.70; however, the remaining carbinol proton signals overlapped in the region between δ 3.50 and δ 3.35 (Fig. 23). A detailed analysis of the *J*-resolved experiment of arbutin (Fig. 24) allowed the assignment of the chemical shift and multiplicity of the different protons as follows: δ 3.47 (t, J = 8.4 Hz, H-2′), 3.40 (d, J = 8.6 Hz, H-3′), 3.44 (d, J = 8.5 Hz, H-4′), and 3.42 (t, J = 8.0 Hz, H-5′) confirming glucose as the sugar moiety of the hydroxyquinone.

The problem of assigning individual proton signals to their corresponding sugar residue represents an additional challenge, especially when more than one sugar units are linked to the aglycone. This difficulty can be overcome by using the 2D TOCSY (Total Correlation Spectroscopy) experiment, which can identify ^{1}H spin systems that can be associated with individual sugar units located in the molecule (Gheysen et al. 2008). In the TOCSY experiment, a series of 180° pulses eliminate the effect of the external magnetic field (chemical shifts), without affecting the scalar coupling. Short time of magnetization (20–50 ms) yields primarily crosspeaks of strongly coupled protons, while longer spin lock times (100–300 ms) allow magnetization transfer to remote protons of the spin system (Günther 2013). A TOCSY

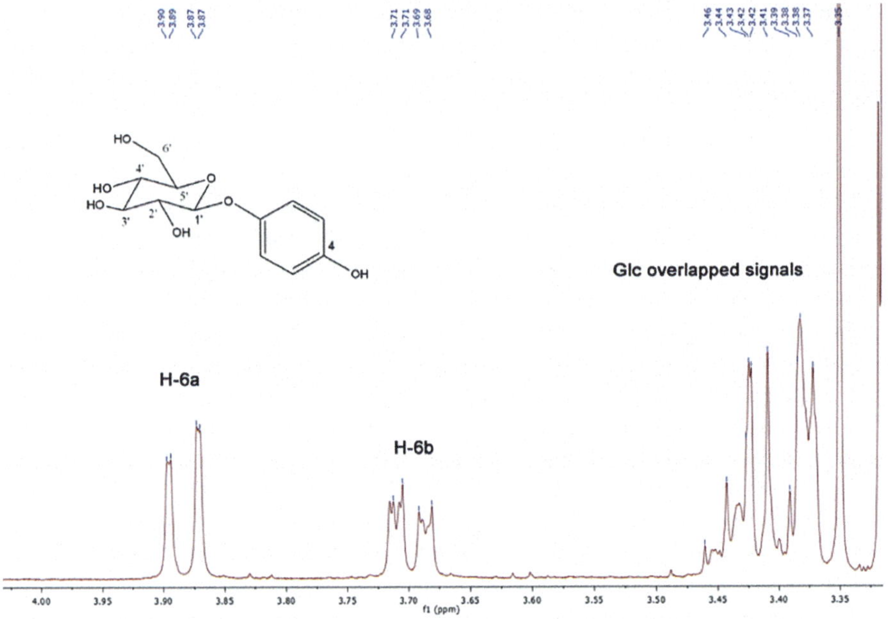

Fig. 23 Close-up of the ^{1}H-NMR (CD_3OD, 500 MHz) spectrum of arbutin showing the region between δ 3.30 and δ 4.00

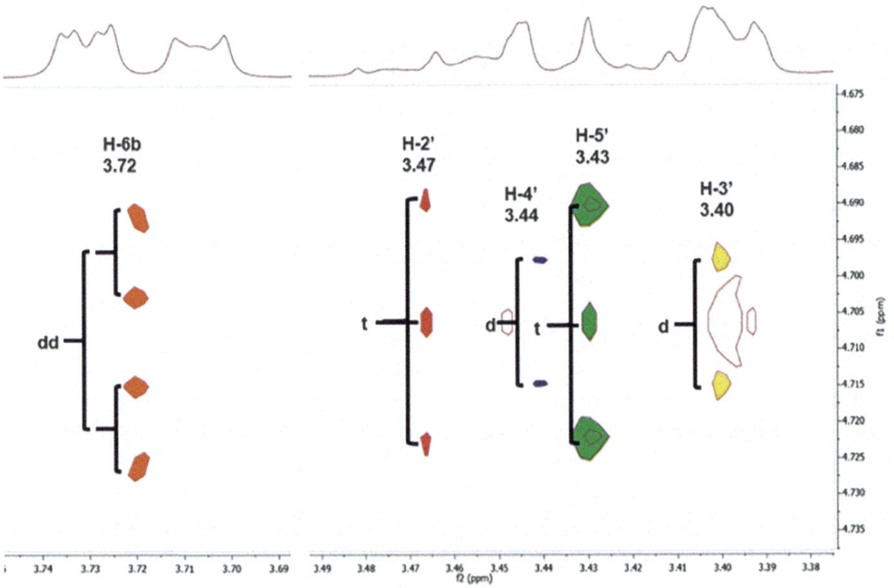

Fig. 24 Coupling patterns determined from *J*-resolved experiment of arbutin

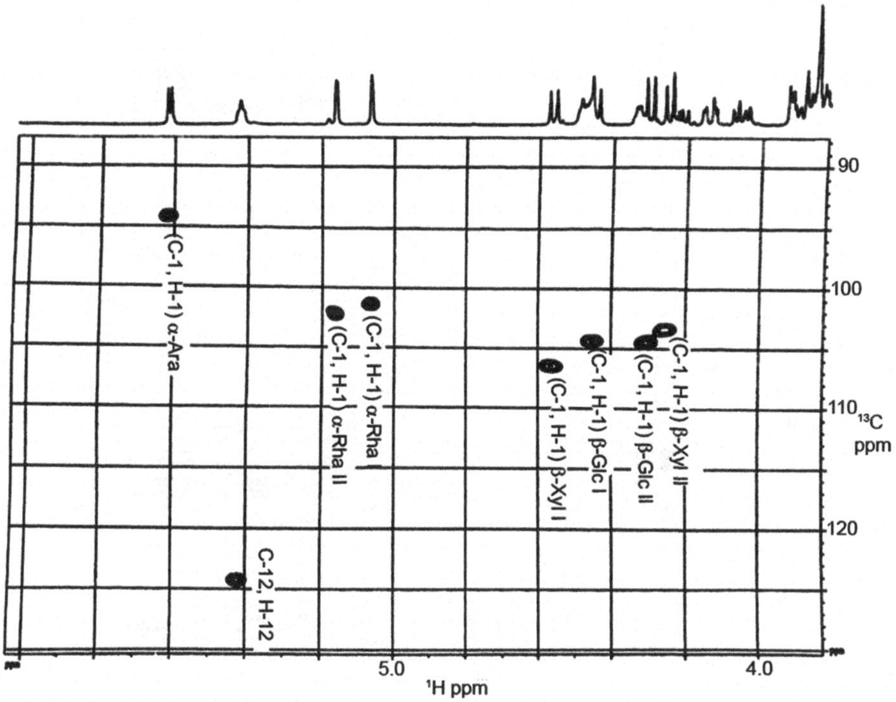

Fig. 25 Correlations observed in the HSQC experiment of 3-*O*-(β-D-glucopyranosyl-(1 → 6)- β-D-glucopyranosyl)-28-*O*-(α-L-rhamnopyranosyl-(1 → 3)[β -D-xylopyranosyl-(1 → 4)]- β-D-xylopyranosyl-(1 → 4)α-L-rhamnopyranosyl-(1 → 2)-α-L-arabinopyranosyl)-16α-hydroxy-protobassic acid

experiment, carried out at two different mixing times (60 ms and 100 ms), was used to assign the protons in the sugar moieties of the triterpenoid saponins isolated from *Sideroxylon foetidissimum* subsp. *gaumeri* (Sanchez-Medina et al. 2009). The shorter mixing time allowed the analysis of the correlations between H-1, H-2, and H-3 of the sugars, whereas with the longer mixing time, the correlation between H-1 and H-5 (Ara, Xyl) and C6-CH_2 (Glc) and C6-CH (Rha) could be observed. The sugar units were identified as α-Ara*p*, α-Rha*p* (two units), β-Xyl*p*, and β-Glc*p* (two units). These results were confirmed by the HSQC experiment of the isolated saponin, which showed seven anomeric carbon signals (Fig. 25). Finally, the results from the HMBC experiment (Fig. 26) indicated a bisdesmosidic structure, with a β-D-Glc*p*-(1 → 6)- β-D-Glc*p* *O*-linked at C-3. Similarly, it was established that the remaining five sugars were linked as a pentasaccharide at C-28. On the basis of these results, the isolated saponin was identified as ***3-O-(β-D-glucopyranosyl-(1 → 6)-β-D-glucopyranosyl)-28-O-(α-L-rhamnopyranosyl-(1 → 3)[β -D-xylopyranosyl-(1 → 4)]- β-D-xylopyranosyl-(1 → 4)α-L-rhamnopyranosyl-(1 → 2)-α-L-arabinopyranosyl)-16α-hydroxyprotobassic acid***.

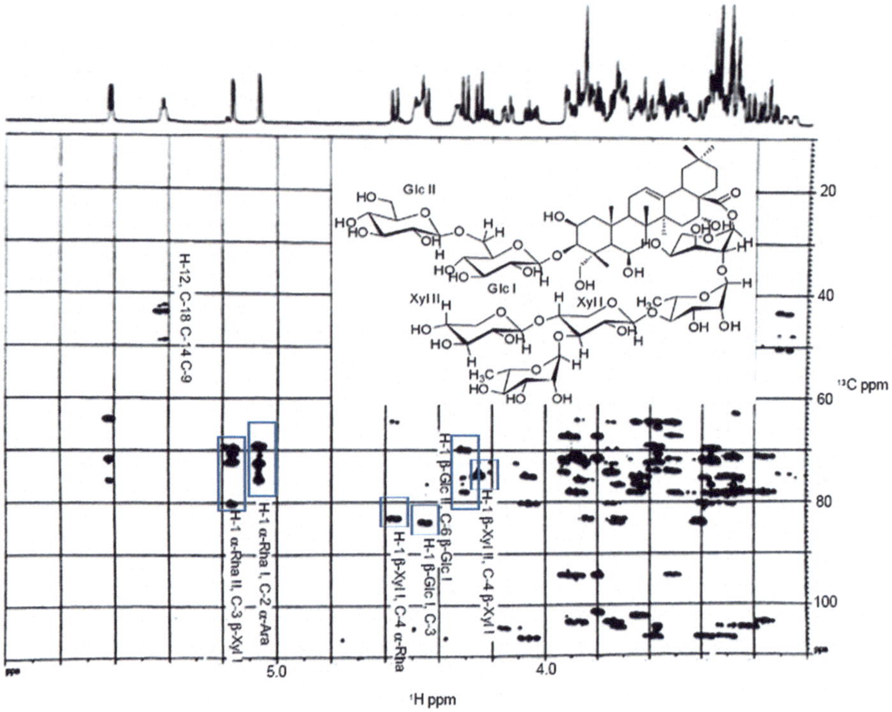

Fig. 26 Correlations observed in the HMBC experiment of 3-*O*-(β-D-glucopyranosyl-(1 → 6)-β-D-glucopyranosyl)-28-*O*-(α-L-rhamnopyranosyl-(1 → 3)[β -D-xylopyranosyl-(1 → 4)]-β-D-xylopyranosyl-(1 → 4)α-L-rhamnopyranosyl-(1 → 2)-α-L-arabinopyranosyl)-16α-hydroxy-protobassic acid

7 NMR Spectroscopy as a Tool to Establish the Biosynthetic Origin of Natural Products

Recently, ^{13}C-NMR has become a key spectroscopic technique for the elucidation of biosynthetic pathways and the study of metabolite fluxes via quantitative assessment of multiple isotopologues (Eisenreich and Bacher 2007). In these experiments, a specific isotopologue (molecules that differ only in its isotopic composition) or isotopologue mixture is enriched through the incorporation of an adequate ^{13}C-labeled precursor chosen according to the metabolic pathway being investigated. Afterward, it is possible to establish the biosynthetic origin of a given molecule by analyzing the coupling patterns of the ^{13}C atoms in its structure.

A good example of this kind of analysis is given by the elucidation of the biosynthetic origin of lupeol-3-(3′-R-hydroxy)-stearate (aka procrim b), obtained from *P. andrieuxii*. For this investigation, potted plants were exposed to $^{13}CO_2$ for 5 h, and after a chase period of 6 days, the triterpene ester was isolated from the stem extract. After confirming the structure of the molecule by 1D- and 2D-NMR analysis (Yam-

Puc et al. 2013), the positional assignments of the $^{13}C_2$-pairs were established by a careful analysis of the coupling constants observed for each of the enriched carbons and confirmed by the correlations observed in the 2D experiments INADEQUATE (Fig. 27) and ADEQUATE (Fig. 28).

The *INADEQUATE* (Incredible Natural Abundance Double Quantum Transfer Experiment) was originally developed to measure ^{13}C-^{13}C coupling constants more easily, since these parameters are difficult to determine because of the well-known low natural abundance of ^{13}C. More recently, the introduction of 2D-INADEQUATE made it possible to analyze the carbon skeleton of a molecular structure via their ^1J (^{13}C, ^{13}C) couplings (Günther 2013). The INADEQUATE of procrim b (Fig. 27) displayed horizontal signal pairs of [$^{13}C_2$]-isotopologues, confirming that [$^{13}C_2$]-labeled precursor units had been specifically incorporated into the molecule during the biosynthetic process. This assignment was further substantiated by the ADEQUATE (Adequate Double Quantum Transfer Experiment) (Fig. 28) which yielded a few but highly significant correlations based on the magnetization transfer to a proton attached to one of the carbons in the [$^{13}C_2$]-pair. In summary, the combined analysis of the ^{13}C-NMR spectrum and the INADEQUATE and ADEQUATE of procrim b made it possible to assign 11 pairs of $^{13}C_2$-isotopologues in the lupeol skeleton and 5 [$^{13}C_2$]-pairs in the 3′-hydroxystearate moiety of the molecule. These results confirmed [$^{13}C_2$]acetyl-CoA being the precursor of the fatty acid chain and the formation of the triterpene skeleton via the mevalonate pathway (Pena-Rodriguez et al. 2014).

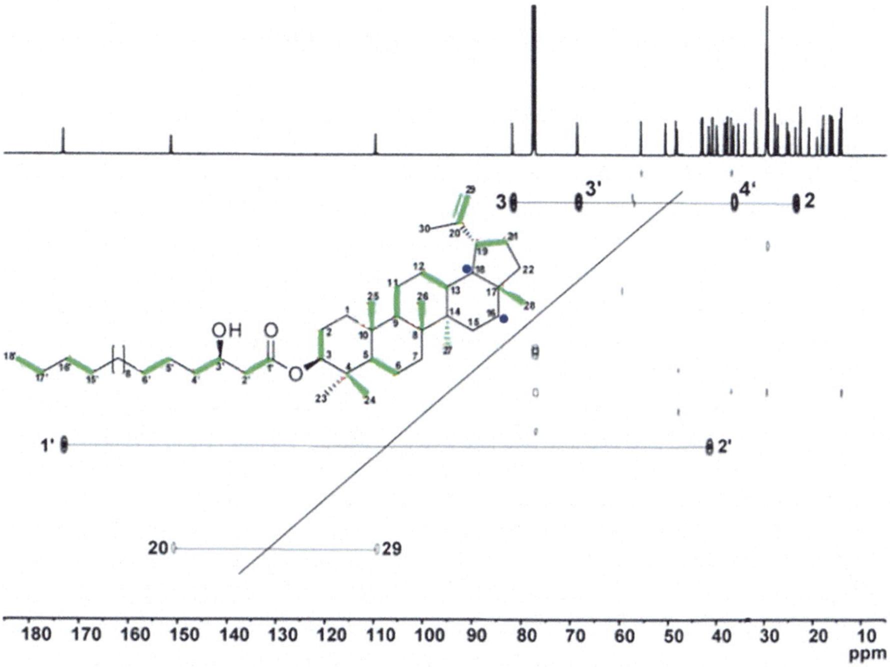

Fig. 27 Correlations observed in the INADEQUATE of procrim b (Peña-Rodríguez et al. 2014)

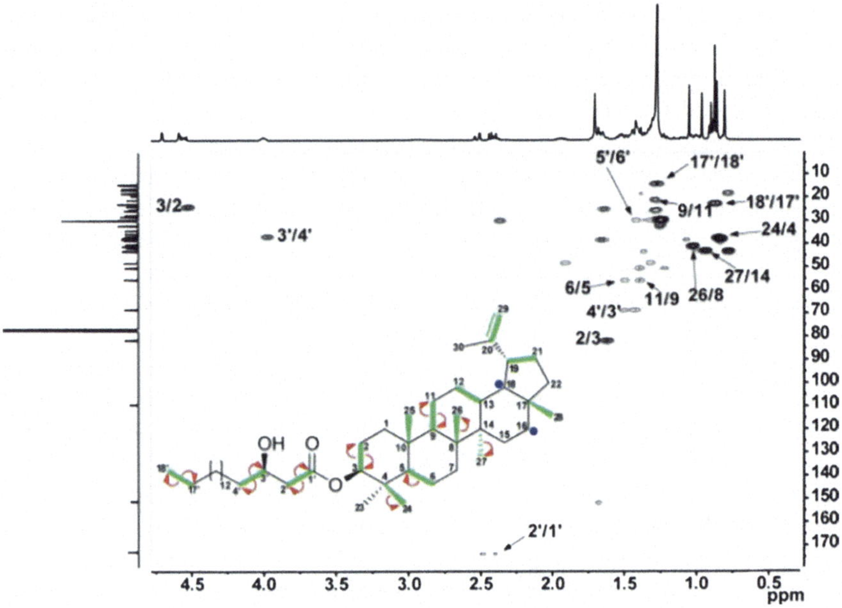

Fig. 28 Correlations observed in the ADEQUATE of procrim b (Peña-Rodríguez et al. 2014)

8 Identification of Epimers: The Case of Quinic Acid Esters

Quinic acid esters, formed by esterification with one or more units of cinnamic acid, occur widely in the plant kingdom and often exhibit various degrees of diasterotopism (i.e., protons belonging to diastereomers) in the C-2 methylene protons. The chirality of the hydroxylated ciclohexanoic acid skeleton induces small but important chemical differences in the methylene protons of the structure (Simmler et al. 2014) which can be detected by ^{1}H NMR and used to identify the correct chemical structure.

In the search for natural analgesics from the Yucatecan flora, a number of dicaffeoylquinic acid esters were isolated from the root extract of *Calea urticifolia* (Asteraceae). The fact that three of the metabolites isolated from the medium polarity fraction showed the same parent ion peak at *m/z* 530 ($C_{26}H_{26}O_{12}$) in their mass spectra suggested their having isomeric structures. While the first metabolite was identified as ***3,4-O-dicaffeoylquinic acid methyl ester*** by comparing its spectroscopic data with those reported in the literature (Zhang et al. 2000), the ^{1}H and ^{13}C-NMR spectra of the second metabolite showed a number of key differences in the chemical shift values of C-1 (δ 74.19 vs δ 75.56) and the ester carbonyl carbon (δ 173.66 vs 176.12) in its ^{13}C-NMR spectrum; these differences coincided with those reported for the same carbons in 3,5-*O*-dicaffeoylquinic acid and its C-1 epimer (Wang et al. 2009). Additionally, the small downfield chemical shift observed for

Fig. 29 Newman projection showing the γ-steric effect in quinic acid derivatives

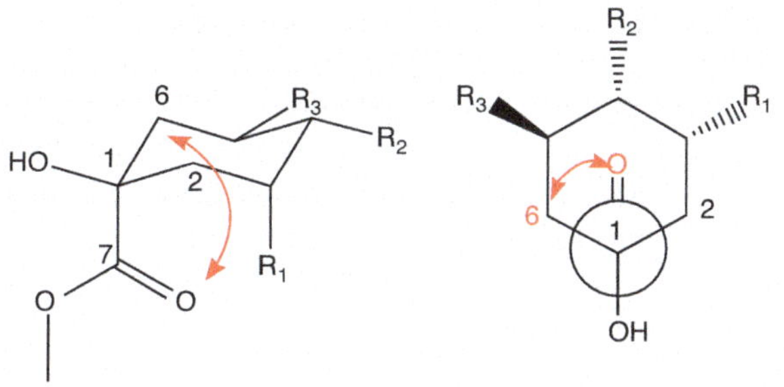

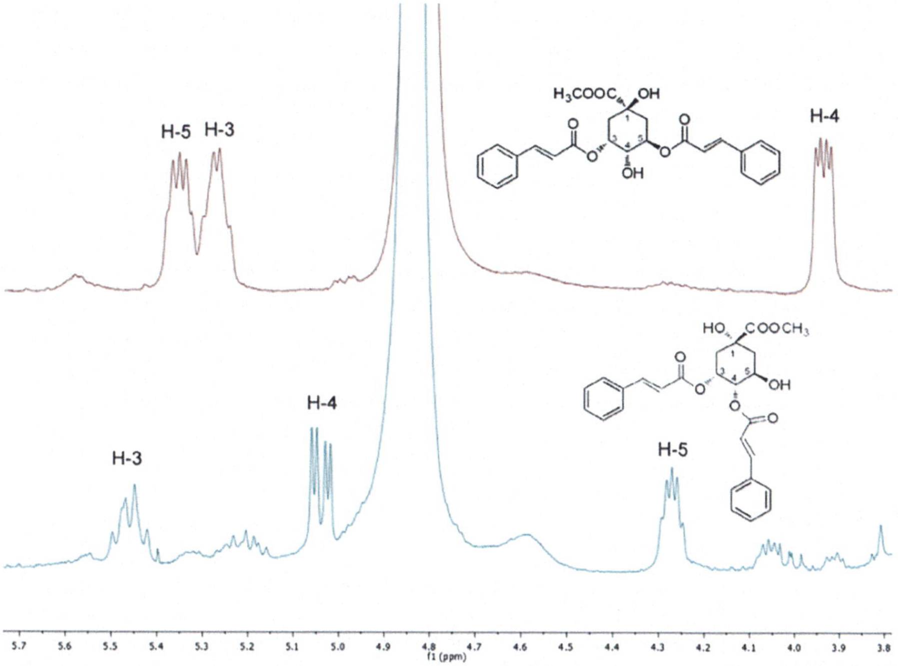

Fig. 30 Close-up of the ^{1}H-NMR (CD_3OD, 500 MHz) spectrum of ***3,4-O-dicaffeoylquinic acid methyl ester*** (bottom) and ***3,5-O-dicaffeoyl-epi-quinic acid methyl ester*** (top), showing the region between δ 3.80 and δ 5.70

C-3 in the ^{13}C-NMR of the second metabolite (δ 67.00 vs δ 69.82) is in agreement with that resulting from a γ-steric effect that occurs when a heteroatom and the carbon in the gamma position are in a gauche conformation (Fig. 29). This evidence allowed the identification of the second metabolite as the novel quinic acid derivative ***3,4-O-dicaffeoyl-epi-quinic acid methyl ester*** (Mijangos-Ramos et al. 2018).

A careful analysis of the ^{1}H-NMR spectrum of the third metabolite showed that while the H-3, H-4, and H-5 protons of the quinic acid moiety appeared at δ 5.45, δ 5.03, and δ 4.26 in the first metabolite, the same protons appeared at δ 5.23, δ 3.90, and δ 5.31 in the spectrum of the third metabolite (Fig. 30). The downfield shift observed for the H-5 proton, together with the upfield shift observed for the H-4 proton, in the ^{1}H-NMR spectrum of the third metabolite suggested that the hydroxyl groups at these positions were esterified. An HMBC experiment confirmed C-3 and C-5 as the esterified positions and allowed the identification of the third metabolite as ***3,5-O-dicaffeoyl-epi-quinic acid methyl ester*** (Zhang et al. 2000; Wang et al. 2009).

9 Perspectives

The structural elucidation of biologically active natural products is an important but challenging task, requiring a level of expertise that can only be achieved through practice and persistency. The knowledge of a researcher, which derives from information collected during years of investigating different areas, may support and strengthen his or her skills and assist him or her when facing the many challenges presented by a continuously growing diversity of chemical structures.

In the area of NMR, the new pulse sequence developments and spectral edition currently allow the efficient resolution of complex spin systems, including the characterization of mixtures (Castañar 2017; Dal Poggetto et al. 2016). Other techniques as nonuniform sampling (NUS) reduce significantly experimental times and costs (Castañar and Parella 2015). All of this experience and these advances should lead us to a better understanding of secondary metabolites chemistry and, most important, to a comprehensive knowledge about biologically active natural products.

References

Borges-Argáez R, Peña-Rodríguez LM, Waterman P (2000) Flavonoids from the stem bark of Lonchocarpus xuul. Phytochemistry (54):611–614

Borges-Argaez R, Medina-Baizabal L, May-Pat F et al (2001) Merilactone, an unusual C-19 metabolite from the root extract of *Chiococca alba*. J Nat Prod 64(2):228–231

Borges-Argáez R, Peña-Rodrıguez LM, Waterman PG (2002) Flavonoids from two *Lonchocarpus* species of the Yucatan peninsula. Phytochemistry 60(5):533–540

Borges-Argaez R, Balnbury L, Flowers A et al (2007) Cytotoxic and antiprotozoal activity of flavonoids from *Lonchocarpus* spp. Phytomedicine 14(7–8):530–533

Breton RC, Reynolds WF (2013) Using NMR to identify and characterize natural products. Nat Prod Rep 30(4):501–524

Bross-Walch N, Kuhn T, Moskau D et al (2005) Strategies and tools for structure determination of natural products using modern methods of NMR spectroscopy. Chem Biodivers 2(2):147–177

Bubb WA (2003) NMR spectroscopy in the study of carbohydrates: characterizing the structural complexity. Concepts Magn Reson A 19a(1):1–19

Castañar L (2017) Pure shift 1H NMR: what is next? Magn Reson Chem 55(1):47–53

Castañar L, Parella T (2015) Broadband 1H homodecoupled NMR experiments: recent developments, methods and applications. Magn Reson Chem 53(6):399–426

Chan-Bacab MJ, Balanza E, Deharo E et al (2003) Variation of leishmanicidal activity in four populations of Urechites andrieuxii. J Ethnopharmacol 86(2–3):243–247

Dal Poggetto G, Castañar L, Morris G et al (2016) A new tool for NMR analysis of complex systems: selective pure shift TOCSY. RSC Adv 6(102):100063–100066

Deepak D, Khare A, Khare MP (1989) Plant pregnanes. Phytochemistry 28(12):3255–3263

Dewick PM (2002) Medicinal natural products: a biosynthetic approach. Wiley, Chichester

Dominguez-Carmona D, Escalante-Erosa F, Garcia-Sosa K et al (2011) Metabolites from roots of *Colubrina greggii* var. yucatanensis and evaluation of their antiprotozoan, cytotoxic and antiproliferative activities. J Braz Chem Soc 22(7):1279–1285

Duus J, Gotfredsen C, Bock K (2000) Carbohydrate structural determination by NMR spectroscopy: modern methods and limitations. Chem Rev 100(12):4589–4614

Dzib-Reyes EV, García-Sosa K, Simá-Polanco P et al (2012) Diterpenoids from the root extract of *Chiococca alba*. Revista latinoamericana de química 40(3):123–129

Eisenreich W, Bacher A (2007) Advances of high-resolution NMR techniques in the structural and metabolic analysis of plant biochemistry. Phytochemistry 68(22–24):2799–2815

Emwas A-H, Luchinat C, Turano P et al (2015) Standardizing the experimental conditions for using urine in NMR-based metabolomic studies with a particular focus on diagnostic studies: a review. Metabolomics 11(4):872–894

Escalante Erosa F, Gamboa-León MR, Lecher JG et al (2002) Major components from the epicuticular wax of Cocos nucifera. Rev Soc Quim Mex 46(3):247–250

Escalante-Erosa F, González-Morales B, Quijano-Quiñones RF et al (2012) Dihydrospinochalcone-A and epi-flemistrictin-B, natural isocordoin derivatives from the root extract of Lonchocarpus xuul. Nat Prod Commun 7(12):1589–1590

Garcia-Sosa K, Aldana-Perez R, Moo V et al (2017) Dinimbidiol ether, a novel bioactive dimeric diterpene from the root extract of *Cnidoscolus souzae*. Nat Prod Commun 12(9):1391–1392

Gheysen K, Mihai C, Conrath K et al (2008) Rapid identification of common hexapyranose monosaccharide units by a simple TOCSY matching approach. Chem Eur J 14(29):8869–8878

Günther H (2013) NMR spectroscopy: basic principles, concepts and applications in chemistry. Wiley, Weinheim an der Bergstrasse

Halabalaki M, Vougogiannopoulou K, Mikros E et al (2014) Recent advances and new strategies in the NMR-based identification of natural products. Curr Opin Biotechnol 25:1–7

Hernández-Bolio GI, Kutzner E, Eisenreich W et al (2017) The use of 1H-NMR metabolomics to optimise the extraction and preliminary identification of anthelmintic products from the leaves of *Lysiloma latisiliquum*. Phytochem Anal 29(4):413–420

Hornak JP (2017). The basics of NMR. http://wwwcisritedu/htbooks/nmr/bnmrhtm. Accessed 25 Jan 2018

Hwang D, Hyun J, Jo G et al (2011) Synthesis and complete assignment of NMR data of 20 chalcones. Magn Reson Chem 49(1):41–45

Joseph-Nathan P (1982) Resonancia magnetica nuclear de hidrogeno-1 y de carbono-13. OEA Ser G CP Organization of American States, Washington, DC

Kan FL, Villanueva LHH, Trujillo CAE, Comisión Reeditora de la Enciclopedia Y (1944) Enciclopedia yucatanense. Gobierno de Yucatán

Kostova I, Simeonov M, Iossifova T et al (1996) Three 9,19-cyclotetracyclic triterpenes from *Skimmia wallichii*. Phytochemistry 43:643–648

Mijangos-Ramos IF, Zapata-Estrella HE, Ruiz-Vargas JA, Escalante-Erosa F, Gómez-Ojeda N, García-Sosa K, Cechinel-Filho V, Meira-Quintão NL, Peña-Rodríguez LM (2018) Bioactive dicaffeoylquinic acid derivatives from the root extract of *Calea urticifolia*. Rev Bras Farm 28(3):339–343

Nagashima F, Kasai W, Kondoh M et al (2003) New ent-kaurene-type diterpenoids possessing cytotoxicity from the New Zealand liverwort *Jungermannia* species. Chem Pharm Bull 51(10):1189–1192

Parella T (2010) Pulse program catalog: I. 1D & 2D NMR experiments, vol 1. Bruker Biospin GmbH. Barcelona, Spain

Pena-Rodriguez L, Yam-Puc A, Knispel N et al (2014) Isotopologue profiling of triterpene formation under physiological conditions. Biosynthesis of Lupeol-3-(3′-R-hydroxy)-stearate in *Pentalinon andrieuxii*. JOC 79(7):2864–2873

Pistelli L (2002) Secondary metabolites of genus Astragalus: structure and biological activity. Stud Nat Prod Chem 27:443–545

Reynolds WF, Mazzola EP (2015) Nuclear magnetic resonance in the structural elucidation of natural products. In: Progress in the chemistry of organic natural products 100. Springer,Vienna, Austria, pp 223–309

Sanchez-Medina A, Stevenson P, Habtemariam S et al (2009) Triterpenoid saponins from a cytotoxic root extract of *Sideroxylon foetidissimum* subsp gaumeri. Phytochemistry 70(6):765–772

Silverstein RM, Webster FX, Kiemle DJ et al (2014) Spectrometric identification of organic compounds. Wiley, Hoboken

Simmler C, Napolitano JG, McAlpine JB et al (2014) Universal quantitative NMR analysis of complex natural samples. Curr Opin Biotechnol 25:51–59

Wang FN, Ma ZQ, Liu Y et al (2009) New phenylethanoid glycosides from the fruits of *Forsythia suspense* (Thunb.) vahl. Molecules 14(3):1324–1331

Wehrli F, Nishida T (1979) The use of carbon-13 nuclear magnetic resonance spectroscopy in natural products chemistry. In: Fortschritte der Chemie Organischer Naturstoffe [Progress in the chemistry of organic natural products]. Springer, Vienna, pp 1–229

Wishart DS (2013) Characterization of biopharmaceuticals by NMR spectroscopy. TrAC Trends Anal Chem 48:96–111

Xiao-Ling S, Ying-Jie H, Yin-Ling A et al (1993) Steroids from Amalocalyx yunnanensis. Phytochemistry 33(3):687–689

Yam-Puc A, Pena-Rodriguez L (2009) Isocordoin derivatives from the root extract of *Lonchocarpus* xuul. J Mex Chem Soc 53(1):12–14

Yam-Puc A, Escalante-Erosa F, Pech-Lopez M et al (2009) Trinorsesquiterpenoids from the root extract of *Pentalinon andrieuxii*. J Nat Prod 72(4):745–748

Yam-Puc A, Chee-González L, Escalante-Erosa F et al (2012) Steroids from the root extract of *Pentalinon andrieuxii*. Phytochem Lett 5(1):45–48

Yam-Puc AF, Escalante-Erosa K, García-Sosa FG et al (2013) A case of mistaken identity. Lupeol-3-(3′*R*)-hydroxy-stearate can be mistakenly identified as lupeol acetate when only analyzed by GC-MS. Phytochem Lett 6:649–652

Zapata-Estrella H, Sanchez-Pardenilla A, Garcia-Sosa K et al (2014) Bioactive metabolites from *Cnidoscolus souzae* and *Acmella pilosa*. Nat Prod Commun 9(9):1319–1321

Zhang WD, Bang Tam HT, Chen WS et al (2000) Two new caffeoyl conjugation from Erigeron breviscapus. J Asian Nat Prod Res 2(4):283–288

Herbal Medicine and Public Healthcare: Current and Future Challenges

Dâmaris Silveira, Jose M. Prieto, Marcela M. Freitas, and Andre L. D. A. Mazzari

Abbreviations

ADME	Absorption, distribution, metabolism and excretion
ADR	Adverse drug reactions
ANVISA	National Sanitary Surveillance Agency
BCM	Brazilian Central of Medicines
CAM	Complementary and alternative medicines
CHP	Conventional healthcare practitioners
GC	Gas chromatography
HDI	Herbal-drug interaction
HPLC	High-performance liquid chromatography
MHRA	Medicines and Healthcare Products Regulatory Agency
NCCAM	National Center for Complementary and Alternative Medicine
NDMC	National Drug Monitoring Centre
NHPD	Natural Health Products Directorate
NHPs	Natural health products
NHS	National Health Systems
NNHPD	Natural and Non-prescription Health Products Directorate
PNPIC	National Policy on Complementary and Integrative Practices
PNPMF	National Policy on Medicinal Plants and Herbal Medicines
QC	Quality control
RENAME	Brazilian list of essential medicines

D. Silveira (✉) · M. M. Freitas
Faculty of Health Sciences, University of Brasilia, Brasilia, Brazil
e-mail: damaris@unb.br

J. M. Prieto
Faculty of Health Sciences, University of Brasilia, Brasilia, Brazil

School of Pharmacy, University College London, London, UK

A. L. D. A. Mazzari
School of Pharmacy, University College London, London, UK

V. Cechinel Filho (ed.), *Natural Products as Source of Molecules with Therapeutic Potential*, https://doi.org/10.1007/978-3-030-00545-0_13

RENISUS	Plant species considered significant to SUS
SUS	Brazilian public health system
TAIM	Traditional Arabic and Islamic medicine
TCM	Traditional Chinese medicine
THMP	Traditional herbal medicinal product
TKM	Traditional Korean medicine
TLC	Thin-layer chromatography
TMP	Traditional medicine practitioners
UK	United Kingdom
USA	United States of America
VIGIPÓS/Notivisa	National Notification System to the Sanitary Surveillance
WHO	World Health Organization

1 Introduction

Traditional treatments (including herbal medicines) and traditional practitioners are the primary source of healthcare for many millions of people – and sometimes the only one (WHO 2013). Since the Alma-Ata Declaration, embraced by the World Health Organization (WHO) (WHO 1978), initiatives have been done to incorporate traditional medicine in public health services. Indeed, Ayurveda, Kampo, traditional Chinese medicine (TCM), traditional Korean medicine (TKM) and Unani should be all considered as a valuable repository of human knowledge where the use of medicinal plants is a central and common practice. They have blossomed into conventional medicine systems in their home countries (Yuan et al. 2016), and the number of WHO Member States with national policies for traditional medicine grew from 25 in 1999 to 65 in 2012 (WHO 2013).

Traditional medicine (TM) concerns mostly indigenous health traditions. Complementary and alternative medicines (CAM) are a more complex term and include more modern practices such as osteopathy. The National Center for Complementary and Alternative Medicine (NCCAM, USA) considers complementary medicine as those treatments used together with conventional medicine, whilst alternative medicine refers to those practices that are used in place of conventional medicine (NCCIH 2016; Nishimura et al. 2009).

Elements of both TM and CAM (T&CAM thereafter) have been included, even though slowly, in the agenda of public health policies and medical researchers during the past 30 years. The integration of such practices in their healthcare systems not only contributes to improve the health and well-being of patients but to the respect of their traditional lifestyles and beliefs. Although the adopted T&CAM therapies differ from country to country, and even from regions to regions, the most recognised medicines usually are TCM, Ayurveda, acupuncture and mind-body and chiropractic medicine (Peregoy et al. 2014).

The increasing popularity and acceptance of T&CAM reflect the rise of the number of elderly population and, consequently, the prevalence of chronic diseases, the concerns about adverse effects and cost of conventional medicine and the easy access to health information (Hussain and Malik 2013). The public, however, ignores or accepts that for most of T&CAM practices, information about efficacy, safety and quality is poor, and the control on the proper utilisation of these practices is scarce or inexistent, so the evaluation of the results of treatment (efficacy) is quite tricky. Moreover, the established standards, certification requirements, competency testing (quality) or legal requirements for a business licence (regulation) are often absent in a range of T&CAM 'clinics' (e.g. yoga, meditation instruction). As a result, most individuals offer these practices without proper authorisation (Eisenberg et al. 2016). Therefore, integrating T&CAM into conventional medicine still poses huge challenges.

After all the above considerations, it is clear that the successful integration of traditional medicine in public health services depends on a coordinated approach by the following stakeholders:

- **Governments** need to pass national laws (or accept international regulations) to regulate the standards and scope of the use of traditional medicines that are to be integrated into the National Health Services. Importantly, they also need to allocate budgets to implement this decision.
- **Providers of traditional medicine** need to achieve the quality, safety and efficacy standards that are required by the above-mentioned laws. Ideally, it must become as 'comparable' to 'modern' medicine as possible.
- **Healthcare professionals and/or traditional practitioners** need now to achieve a satisfactory level of understanding, training and competence to use -or allow the use- of Traditional medicine within National Health Services.
- Finally, **the patients and the general public** need to be educated and informed about the possibilities (and limitations) that traditional medicine offers to them.

A clear and sustained commitment from governmental instances is the key stone, as illustrated in Fig. 1. All efforts put in education, information and quality of the traditional medicine are of no use if there is not a legal and budgetary framework to support it. In other words, the integration of TM into the National Health Systems (NHS) is just a political decision which needs money to be effectively implemented.

In the following pages, we provide the reader with an overall picture of the challenges that each stakeholder faces and what are the current approaches to solve them. We deliberately will restrict the discussion to herbal medicine or Phytotherapy, one of the many types of traditional practices. As other T&CAM, the insertion of Phytotherapy into public healthcare faces challenges in regard to regulation, recognition, health workers' capacitation, insurance of safety and efficacy, quality, sustainability, environmental obligation and value addition (Hussain and Malik 2013). However, herbal medicines have the potential to integrate faster into the National Health Systems by placing themselves in the market as 'medicinal products' following similar regulations to those in use for the registration of drugs (or licenced medicines).

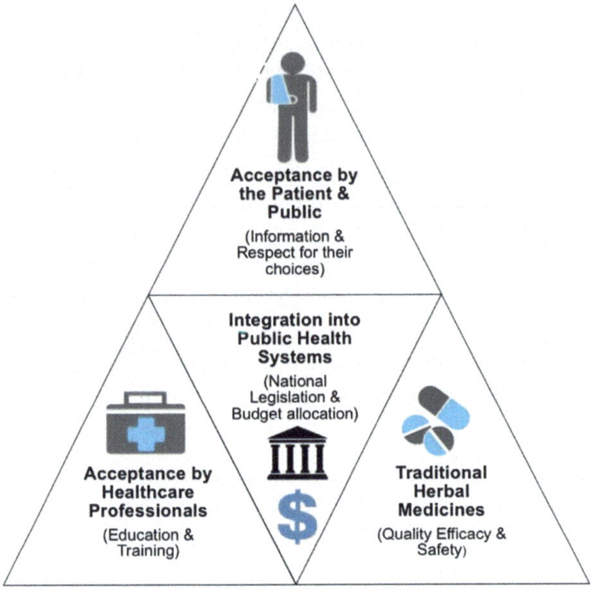

Fig. 1 Overview of the challenges for traditional herbal medicines to integrate into public health

2 Challenge 1: The Herbal Medicines

One the most affordable, popular and frequent form of T&CAM is herbal medicine or Phytotherapy. It is used worldwide and is part of the heritage of almost all human communities, cultures and traditions. The use of plants to treat diseases has been followed by humans since ancient times. Aromatic and medicinal species have been found during an archaeological excavation of graveyards dated 15,000–11,000 years ago (Nadel et al. 2013), as well as in a Middle Palaeolithic grave dated 50,000 years ago (Lietava 1992; Sommer 1999). Phytotherapy shares the same origin with conventional medicine and seems to be more 'acceptable' by physicians, nurses and other prescribers, as well as by the patients that perceive herbal medicinal products (HMP) as a more drug-like modality than others like T&CAM. Indeed, the status of Phytotherapy has shifted from the empirical basis to evidence-based research into efficacy, safety, interactions with drugs and quality control (Ben-Arye et al. 2011).

Moreover, the effects of the treatment based on Phytotherapy can be monitored and evaluated, because HMP can easier be submitted to safety, quality and efficacy evaluation than most of other T&CAM.

HMP include plant drugs, herbal materials, herbal preparations and finished herbal products. Finished herbal products may contain excipients in addition to the active ingredients, but cannot be added with synthetic compounds or isolated phyto-constituents (WHO 2017). These therapeutic tools have been playing a huge role in the primary healthcare of individuals and public health systems, not only among poor and unassisted communities but also in several developed countries. However, herbal medicine products, mainly those with an origin outside Europe, are still far to be recognised so valuable as chemically well-defined pharmaceutical drugs are.

2.1 *Efficacy*

The evaluation of the efficacy and safety of any medicine is a long and expensive stepwise process in which the active ingredient/s move from preclinical to clinical evaluation before registration. If licenced, a larger population is exposed to the drug that continues to be evaluated by means of pharmacovigilance. It is not unusual that after being licenced and marketed, a medicine is retired due to either lack of 'self-efficacy' or safety concerns. Therefore, only after decades and billions of investment may a drug be fully characterised for its efficacy and safety.

Herbal medicines attract much less investment because they are less profitable, and therefore very few manufacturers will go into this demanding process. Most of the herbal products are therefore not marketed as medicines but as unlicenced medicines and/or food supplements. In some cases, certain countries or regions have established a special licence for herbal medicines. These schemes are usually less stringent in terms of efficacy and safety evidence but require pharmaceutical quality all along the manufacturing process from the herbal drug to the final herbal medicinal product. This is the case in Europe and Brazil, where there is in place a regulatory framework for 'traditional' herbal medicinal products (THMP) in the aims to increase the safety and the quality of these products.

The efficacy of a medicinal product is a very complex concept, and healthcare professionals and public alike tend to be biased about its interpretation. In the 1970s, the introduction of modern clinical trials was regarded as the gold standard to measure efficacy, but soon it was evident that many clinical trials on the same product were yielding different results. In the 1990s there was a major shift towards abandoning the absolute term of 'efficacy' as measured by clinical trials in favour of a more flexible concept of 'levels of efficacy' where the meta-analysis of all available clinical evidence for a medicine superseded clinical trials as the ultimate proof for efficacy (Fig. 2).

It is unfair to say that all traditional herbal medicines lack efficacy or evidence. It exists for most of them, although mostly at lower levels of evidence. Moreover, many medicinal plants (in the form of proprietary extracts) have been subjected to clinical trials (of variable power and design). The fact that they are not medicines in a given country does not mean that in others it has been licenced. For example, the apolar extracts of *Serenoa repens* are sold all over Europe as full licenced medicines (Permixon® in France, Spain, Germany) backed up by clinical trials but just as a THMP (Prostasan®) in the United Kingdom. This is due to marketing considerations only, as the manufacturers do not regard the British market as profitable as continental Europe.

Fig. 2 Current hierarchy of levels of evidence-based efficacy

2.2 *Quality*

The quality of herbal medicines still lays on a grey zone and can only be assured if the product is registered under a robust regulatory framework that ensures the production is done according to the Good Manufacturing Practices and marketed within an efficient sanitary surveillance system enforcing broad pharmacovigilance systems.

The active ingredient/s in herbal medicine is/are herbal drug/s. Herbal drugs are by definition 'complex chemical entities'. In contrast, most conventional medicines have as active ingredients synthetic drugs which are 'single chemical entities'.

The first step in the quality control (QC) of herbal medicines is to ascertain the identity of the herbal drug, namely, the plant species and the part of the plant. This is accomplished by macroscopy and microscopy assays. Although recently DNA barcoding is being introduced as an identity test, it cannot distinguish if the plant material comes from the right part of the plant. The second step is to run a series of tests for foreign matter, ashes, dry matter, pesticides and others. They ensure the plant material is pure and well processed. The third step is to analyse if the plant material contains the right chemistry. This is accomplished by different techniques depending on the herbal drug (colorimetric reactions, TLC, HPLC, GC).

This chemical analysis has to be reduced to one or a few compounds among the thousands present in the plant. After all, we know the exact active compound/s for a few medicinal plants only. This is the case of senna (sennosides, chiefly sennosides A and B, ipecac alkaloids, chiefly emetine). For all the other medicinal plants, the active compound/s is/are partially known or not known at all. For example, *Hypericum perforatum* may be quantified on hypericins or hyperforins as both compounds have been related to its anxiolytic activity, whilst in herbs with non-identified active principles such as *Arctium lappa,* the quantification is performed on an 'arbitrarily' chosen compound ('phytomarker'). This phytomarker do not generally account for any clear pharmacological activity and ideally has to be as exclusive of this plant species as possible.

Due to all the variables involved in the raw material production, the batch-to-batch consistency is one of the biggest challenges in the development of an active herbal ingredient.

The pharmacological value of a herbal formulation depends on the chemical composition of the active ingredient. Development and validation of robust analytical methods, comprising qualitative and quantitative evaluation of phytochemical profile and markers, remain as one of the most significant challenges. However, without a proper quality assurance, to expect a consistent therapeutic effect or the assurance of safety is not possible.

2.3 Safety

In addition to the dangers related to the quality of the herbal medicines, an important safety aspect should be considered regarding the consumption of this type of product: the interaction with synthetic drugs. Adverse drug reactions (ADR) caused by herbal-drug interaction (HDI) are possible to happen, and they can cause serious health problems as a result of bad therapeutic practices. Medicinal plants and herbal medicines, when consumed, can cause pharmacodynamic disorders, producing increase or decrease of the pharmacological effects and/or pharmacokinetic effects, leading to problems related to the absorption, distribution, metabolism and excretion (ADME) of conventional drugs, if they are co-administered, due to inductive and/or inhibitory effects in metabolising enzymes and transporters (Skalli and Soulaymani Bencheikh 2012).

To prevent these undesirable and harmful effects to the life of patients, HDI should be evaluated carefully by means of preclinical and clinical pharmacology methods and, mainly, by the pharmacovigilance system of the pertinent country (Skalli and Soulaymani Bencheikh 2012). The need of developing a programme to monitor the adverse reactions of medicines closely has become a priority for the WHO after the tragedy with thalidomide in 1961. Spontaneous adverse notification monitoring systems were implemented in countries such as Australia, Canada, the United States, Ireland, Japan, New Zealand, Germany and the United Kingdom (Rozenfeld and Rangel 1988). In Brazil, with the implementation of the National Sanitary Surveillance Agency (Anvisa) in 1999, a most robust pharmacovigilance system was created. Also, the National Drug Monitoring Centre (NDMC) was founded, and after some years of improvements, the National Notification System to the Sanitary Surveillance (VIGIPÓS/Notivisa) was established (Mazzari and Prieto 2014).

Even with a pharmacovigilance system in compliance with the establishments of the WHO, Brazil still has difficulty in obtaining and reporting HDI, which triggers, therefore, concerns about the risks associated with the use of medicinal plants and herbal medicines in conjunction with synthetic drugs.

Spontaneous reporting and experimental studies are the main sources of HDI data, and both of them have their pros and cons. Spontaneous reporting is a relatively inexpensive way to collect and interpret voluntary reports, which will serve as a warning mechanism by the pharmacovigilance systems. In many European countries, companies that produce and market herbal medicines are obligated to conduct pharmacovigilance for the products and report any case of ADR to the health authorities. The issues about spontaneous reporting mostly involve the health professional who prescribes the treatment to the patients. Sometimes the interaction reports made by doctors and pharmacists are quite variable, and thus a conclusion is practically impossible to be drawn. Spontaneous reporting of preclinical and clinical studies is a very good source of HDI data. Well-conducted clinical studies (especially the ones that are conducted based on preclinical data) are generally more valuable than case reports. However, the high costs of those trials (especially the clinical ones) are a drawback, and therefore, in many cases, spontaneous reporting collected by the pharmacovigilance systems is the only way to discover cases of HDI.

After the publication of the International Drug Monitoring Programme by WHO, national pharmacovigilance centres of the organisation's member countries were responsible for collecting, processing and analysing suspected cases of HDI. An article published in 2012 has revealed, through official data collected in the VigiSearch (tool used by the WHO to store the ADR cases received by the pharmacovigilance centres of member countries), that 811 cases of HDI have been received by the system, which represents 4.6% of all herbal reports (17,754) and 0.012% of the total number of reports (7,017,658) in the official WHO database (Skalli and Soulaymani Bencheikh 2012).

Countries like the United States, Germany, Australia, Canada, Switzerland and the United Kingdom are responsible for the highest number of HDI reports. In total, 27 countries reported cases of HDI to the WHO. However, Brazil was not found to be part of that list, but that does not mean that HDI does not occur in the country.

For example, about 40% of the plant species of RENISUS (a list of plant species considered significant to Brazilian public health system – SUS) could cause PK interactions that interfere directly with the metabolic enzymes responsible for the phase I and phase II metabolisms and the P-gp expression/activity (Mazzari and Prieto 2014). For herbal medicines available on both Anvisa's simplified registry list (IN 02/14) and the Brazilian list of essential medicines (RENAME), such as *Allium sativum*, *Glycine max*, *Mentha* x *piperita* and *Zingiber officinale*, available data indicated that they can cause PK disturbances in various metabolic enzymes, as well as P-gp, and, consequently, cause HDI (Ajith et al. 2007; Le Bon et al. 2003; Shon and Nam 2004; Unger and Frank 2004).

Underreported HDI could also be a result of self-medication. Due to the belief of the 'harmless nature' of medicinal plants and herbal remedies, many patients do not report its use to doctors and other health professionals, which contribute, therefore, to the lack of such data (Balbino and Dias 2010).

3 Challenge 2: The Healthcare Workforce

3.1 Traditional Practitioners

Due to the urgency of the use of medicinal plants and herbal medicines in the world, the WHO began a formulation and implementation of strategies for these products to be used correctly and safely by the population. Among the measures produced, we can highlight the use of herbalists or professionals specialised in traditional medicine (traditional medicine practitioners or TMP) (Amole 2012). This practice has been used in African countries that depend almost exclusively on medicinal plants as a primary source of medical treatment for diseases, integrating the TMP in local health systems conjointly with local health systems to work in cooperation with the conventional healthcare practitioners (CHP). For the implementation of this measure, the WHO published in 1995 a guide for training of TMP professionals for their inclusion in local health systems. The TMP has its competence recognised by local communities to diagnose and help with the physical, mental and social well-being of members of their respective communities (WHO 1995). The TMP is also trained to detect and treat diseases considered as prevalent in their respective regions, in addition to identifying adverse reactions that may occur during the treatment.

In 2001, the WHO also issued the document 'Promoting the Role of Traditional Medicine in Health Systems: A Strategy for the African Region' promoting the effective collaboration between the TMP and CHP initially for the control of AIDS in the African continent (Busia and Kasilo 2010). Countries like Mali, Senegal, Zambia, Uganda, Botswana, Malawi, Mozambique, South Africa and Central African Republic have observed benefits from this cooperation. For example, in Mali, the collaboration between TMP and CHP resulted in a reduction in mortality caused by malaria from 5% in 1997 to 2% in 1998, in addition to the reduction in mortality for the severe form of the disease, which fell from 38% to 10% in 2008 (King and Homsy 1997).

The trend in Ethiopia, like several developing countries, is to integrate and harmonise the beneficial indigenous medical traditions and the conventional healthcare system. In Ethiopia, the government recognises the herbal medicine, and in 1986, 6000 traditional medicine practitioners were registered with the Ethiopian Ministry of Health. Currently, the Directorate of Traditional and Modern Medicine Research, in partnership with the Ethiopian Public Health Institute and traditional healers, is carrying on research about medicinal plants used for diseases considered priority or with public health importance (Abay 2009; Asmelashe Gelayee et al. 2017).

In South Africa, traditional healers are part of the official occupational classification and are included in counts in this report where data are available (WHO 2006).

In Europe, there is also a movement for the regulation of traditional medicine health professionals. The MHRA (Medicines and Healthcare products Regulatory Agency), which is the UK regulatory agency, initiated the formulation of proposals for the official registration of these professionals, with the intent of contributing to the safety of the use of medicinal plants and herbal remedies in the country. With this change, it is expected that the medicinal plants and herbal medicines will be used in a better effective and rational manner, also avoiding its co-administration with conventional drugs and, therefore, helping to prevent possible cases of HDI (MHRA 2016).

Countries such as Brazil have been under a multicultural influence, starting by the use of indigenous traditional herbal medicines to which European plant species were added through the colonisation and Jesuitical missions. Then African traditional medicine was incorporated through centuries of slavery practices, a significant immigration from Japan brought Kampo practices, and recently, due to the 'globalisation', it is also possible to observe the influence of Ayurveda and Chinese traditional medicine in the use of herbal medicines.

Despite the ancient and widely use of medicinal plants, only in 1981, through the Ordinance 212, the Brazilian Ministry of Health defined the medicinal plant research as a public health priority (Brazil 1981), leading to the launch of CEME (Central of Medicines) Medicinal Plant Program (Amaral et al. 2006). However, the insertion of the herbal medicine in the Brazilian Health System (SUS) only was effective after 2006, with the publication of two national policies: the National Policy on Complementary and Integrative Practices (PNPIC) (Brazil 2006b) and National Policy on Medicinal Plants and Herbal Medicines (PNPMF) (Brazil 2006b). The inclusion of TMP is highlighted in the PNPIC and PNPMF. The PNPMF guideline five of the policy mentions the valorisation of the practitioners of traditional knowledge and their participation in the expansion of understanding of the traditional use of medicinal plants. The PNPIC still points the need for the inclusion of TMP in SUS, which is congruent with the need of permanent education of the health professionals on the use of medicinal plants and herbal remedies, which is also emphasised in the PNPMF guideline document 3 (Brazil 2006a, b).

3.2 Conventional Healthcare Practitioners

When patients search herbal medicine-based treatment, they may encounter profound differences in the professional performance and background among the health workers. According to WHO survey, 58 Member States described the lack of education and training for T&CAM providers as one of the challenges faced, and only 30% of surveyed countries offer high-level (bachelor's, master's and doctoral degrees) education programmes in T&CAM (WHO 2013). Little attention has been paid to the formal education in health science areas (medicine, pharmacy, nursery, dentistry and others), to integrate Phytotherapy and herbal medicines into the curriculum.

In Brazil, as for in Japan (Nishimura et al. 2009), there is no separate licence for physicians, nurses, pharmacists, dentists and other health workers of traditional medicine (including Phytotherapy) or conventional medicine. All of them have permission to practise Phytotherapy and prescribe herbal medicines. However, among more than 2600 formal health curricula (in universities and faculties), only a few offers mandatory content or course on Phytotherapy. For example, in Brazil, there are more than 150 faculties of Medicine. However, students of only one of them have access to a mandatory course on Phytotherapy. The same situation occurs with faculties of nursery, nutrition and dentistry (Barreto and Silveira 2014).

In Europe almost all School of Pharmacy have a module on Pharmacognosy and Phytotherapy, but all other health professions completely lack any formal education on these aspects.

In Ethiopia, although over 80% of the inhabitants use herbal medicines, public health policies emphasise the need to develop and carry on research on traditional medicine, the level of knowledge and skill on Phytotherapy by formal educators is not beyond research and herbalists were not trained in formal programmes (Abay 2009). For example, the curriculum for undergraduate pharmacy programme comprises few-hour course on alternative and complementary medicine, which is not sufficient to prepare pharmacists as experts in herbal medicines (Asmelashe Gelayee et al. 2017).

Kampo medicine originated from traditional Chinese medicine (TCM) and has been used for more than 1500 years. Around 80% of Japanese physicians use Kampo in daily practice, even in university hospitals (Nishimura et al. 2009; WHO 2013). The prescriptions of Kampo medicines are regulated, and more than 140 formulas are under the Japanese insurance programme. Moreover, 243 Kampo plant species for decoction are available and covered by the Japanese National Insurance System (Sahashi 2005).

Due to the insertion of the traditional medicine (gSo-ba Rig-pa) in the Bhutanese public health system, health professionals were necessary, and, in 1971, formal training for traditional doctors (Drungtshos) and traditional compounders (sMenpas) was initiated (Wangchuk and Tobgay 2016; Wangchuk et al. 2007).

It is essential to ensure that the knowledge, qualification and training of health professionals, even those well trained in conventional medicine, are adequate to grant the right attention to the patients that ask for Phytotherapy as the therapeutical treatment. They should be able to deal with issues that may arise from herbal medicine uses, such as intoxication, herbal-drug interaction and others.

4 Challenge 3: The Regulators

4.1 Regulation

In a scenario of low health spending in some countries and the high cost of health in others (WHO 2014), the adoption of herbal medicine as a part or as a replacement of conventional treatments seems to be an attractive alternative. Considering the WHO data for 2012, approximately 70 Member States had launched public policies about T&CAM, and herbal medicine regulation is available in more than 110.

However, it is a work in progress (WHO 2013). Although cooperation among regulatory agencies is increasing, a lack of harmonisation of the herbal medicine regulatory chain still can be observed, due to the specificity of countries' laws, as well as the several categories of herbal products around the world. Countries with well-established sanitary legislation tend to have a regulatory framework including herbal medicines in the minor or high degree of control. Whatever the strength of the rules, if Good Manufacturing Practices are mandatory, the quality and safety of herbal medicines can be assured.

In countries with a well-recognised traditional medicine, e.g. China, herbal medicines have been widely used, and the integration with public health services is quite accomplished. In other countries, herbal medicine has been, partially or entirely, integrated into the healthcare system. In the same way PNPIC and PNPMF contributed to insertion of herbal medicine into the Brazilian public health system, these policies have been promoting the research and use of medicinal plants and herbal medicines according to quality, safety and efficacy statements (Carvalho et al. 2011), stimulating the pharmaceutical industry and, consequently, improving the Brazilian regulation of herbal medicine framework (Carvalho et al. 2018). Before these regulations, herbal products were traded on an unregulated market. Now Anvisa recognises herbal medicines manufactured under the new regulatory framework (Carvalho et al. 2011) produced under GMP, by licenced pharmaceutical laboratories. All other products (plant drug derivatives or pharmaceutical dosage forms) manufactured or sold without Anvisa licence are considered irregular products. The exception is for those produced by Farmácias Vivas (Living Pharmacies), a programme launched in 2009 by Brazilian Ministry of Health, to stimulate the production of medicinal plants and herbal products, which are distributed in the public health services (Batista and Gondim 2012). Similarly to Brazil, Kampo medicines are made in Japan by pharmaceutical companies under the Pharmaceutical Affairs Law (Nishimura et al. 2009).

The traditional medicine system in Bhutan is gSo-ba Rig-pa (that means literally 'the science of nourishment') has been also included in the National Health System since 1967. gSo-ba Rig-pa contains more than 1000 formulas and recipes, of which 98 were selected to be included in Bhutanese list of essential medicines (Wangchuk et al. 2007). Currently, there are 58 hospitals and medical units offering Bhutanese traditional medicines, following the mainstream medicine regulations in conformity with the integrated policy of quality assurance system (Wangchuk and Tobgay 2016).

In Ghana, the Ministry of Health created the Department of Traditional and Alternative Medicine in 1992, and a 'Traditional Medicine Practice Act' was promulgated later in 2000. In 2005, the Policy Guideline on Traditional Medicine Development was published, with the recommendation of integration of herbal medicine into conventional practice (GHANA 2000, 2005). Since 2011, this traditional herbal medicine has been used at 17 conventional hospitals, although with restrictions regarding government support and national health insurance coverage (Boateng et al. 2016).

In 2000, the Ministry of Ayurveda, Yoga and Naturopathy, Unani, Siddha and Homeopathy established the National Medicinal Plants Board as well the State Medicinal Plants Board in India. This Board aims to support policies and programmes related to all aspects concerning medicinal plants, from good agricultural practices to international trade market (India 2017; Kala and Sajwan 2007). Medicinal plants play a major role in India, not only concerning public health but also for the domestic and foreign markets. India exports more than 40,000 tons/year of medicinal plants, to 95 countries for a value of US$ 62 million/year (Dar et al. 2017).

Arabian medicine played a significant role in the development of modern medicine. Arabian medicine successfully merged ancient practices of Mesopotamia, Greece, Rome, Persia, India and China and had considerable influence upon European medicine practice throughout the Middle Ages (Azaizeh et al. 2010; Pan et al. 2014). It reached its peak during the Arabian Empire period (632–1258), when more than 1400 different herbal medicines were used by Islamic physicians (Pan et al. 2014). A new term 'traditional Arabic and Islamic medicine (TAIM)' was proposed by Azaizeh et al. (2010), recognising both traditional Arabic and Islamic medicine as one system, containing similar historical roots, with religious influences of Islam comprising several practices, from medicinal herbs to spiritual healing, and connecting Islamic medical systems, religion and regional healing practices (Al-Rawi and Fetters 2012; Azaizeh et al. 2010). Currently, it is estimated that 250 plant species are used in TAIM, for treatment of several diseases (Lev and Amar 2002; Said et al. 2002). However, it is important to note that each Middle East nation presents a different level of acceptance of TAIM. As an example, in Egypt, the National Centre for Medicinal Plants, linked to health ministry, was established in 1995, and T&CAM is part of the national drug policy (WHO 2005).

The European Union established in 2004 a unique regulatory framework which includes herbal medicines as a 'traditional herbal medicinal product' (THMP) with efficacy based on 'traditional use only' (EU 2004). This is a de facto recognition of 'traditional use' as a level of evidence (see Fig. 3), an innovation which is still the matter of heated debates within the scientific and medical community (Colquhoun 2011). This registration is granted to any medicinal product claiming the relief, cure or prevention of minor, self-limiting conditions in which active ingredients are exclusively herbal drugs. It, therefore, excludes the use of isolated plant chemicals as active or functional ingredients. The product is to be administered topically or orally only (European 2004). The registration and authorisation process depends on the available evidence of traditional or well-established use of the herbal ingredient(s).

Under this scheme, if the medicinal herb has been in use within the European for at least 30 years (or at least 15 years in the European and 15 years outside the European), then it is eligible for a 'traditional herbal medicinal product' (THMP) registration. This requires a dossier reviewing the scientific literature to justify the 30/15 years of medicinal use, its putative efficacy and safety and the manufacturing quality standards. For herbal drugs already listed in the European Pharmacopoeia, such application could be *c.a.* $30,000 (which includes both registration and

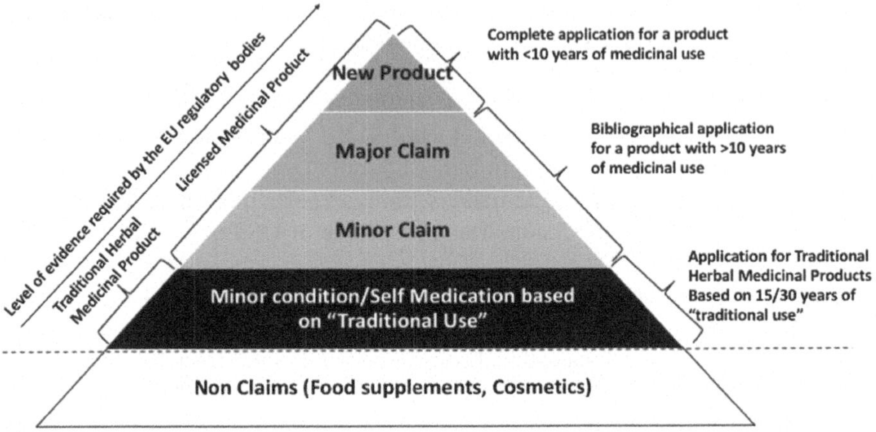

Fig. 3 Different routes and levels of evidence and documentation demanded by European Medicine Agencies

consultancy fees) according to the National Health Federation (2011) if quality controls can be done by applying established pharmacopoeial standards and methods, so only stability testing is required to be developed for the product (NHF 2011). In the case the herbal ingredient is not listed in any of the European monographs, then the development of quality control methods must be done (many times from scratch), which will double or triple the cost of the application dossier. Many small companies without the economic means of affording this registration and/or investing in the capacity of pharmaceutical quality controls for their herbal ingredients and final products cannot access the THMP status. Moreover, the need of 30/15 years of traditional use within European resulted in a 'closed list' of possible active ingredient which excludes many herbs from non-European traditions.

Canada is following this example by promoting a new pathway for licencing 'natural health products used as traditional medicines'. Accordingly, the Canadian Natural Health Products Directorate (NHPD) changed its name in 2012 to the Natural and Non-prescription Health Products Directorate (NNHPD) 'to include the oversight of non-prescription and disinfectant drugs in addition to natural health products (NHPs)' (Canada 2012).

The United States continues to make a clear division between food and drugs, and the introduction of 'botanical drugs' is not to be confused with herbal medicines. It is true that in this case, a door was opened to (highly standardised) herbal extracts, only if they are accompanied by clinical trials. Medicinal herbs remain considered food (supplements), and therefore pharmaceutical quality is not enforced, and medicinal claims are not allowed. However, the creation in 1998 of the National Center for Complementary and Integrative Health (NCCIH) as the federal government's lead agency for scientific research on the medical and healthcare systems, practices and products that are not generally considered part of conventional medicine is a bold statement in favour of their future integration in public health (NCCIH 2018).

The Chinese government approved Laws in 2016 for 'carrying forward traditional Chinese medicine, guaranteeing and promoting the development of the Traditional Chinese Medicine undertaking, and protecting the health of the people' (PRC 2016). The government embarks on building a TCM service system and making it available to all its citizens. To this end, the Chinese state encourages social forces to invest in TCM, establish a TCM education system and promote scientific research and technical development. Importantly this law envisages the protection of intellectual property of TCM therapies towards its 'exportation' to the international market.

4.2 Public Funding and Cost of Herbal Medicines

Public funding is allocated for cost-effective therapeutic interventions only. Therefore, herbal medicines will have to prove efficacy (which is difficult due to their intrinsic chemical variability, as previously discussed) and at the same time beat the cost of synthetic drugs (which is many times lower). Despite these difficulties, a meta-analysis by Kennedy, Hart and Seely (Kennedy et al. 2009) concluded that the cost-effectiveness of some herbal medicines is encouraging in certain areas (such as post-operative surgery) but still needs confirmation from further research. A controversial report by Smallwood (Smallwood 2005) making a case for the cost-effectiveness of T&CAM towards integration in the British National Health System gained heavy criticisms from academics mostly due to the lack of proof of efficacy for most of the alternative therapeutic interventions (Ernst 2006).

The use of THMPs in Europe is not funded by any European public system. Although may be prescribed by healthcare professionals, the patient must pay in full for them. It is noteworthy that THMP have an elevated price tag when compared with conventional drugs. In the United Kingdom, a pack of 12 tablets of devil's claw (indicated for the treatment of rheumatic pain, backache or lumbago) costs around $10, whereas 12 tablets of ibuprofen (an analgesic with similar indications) would be available at $0.30 only. This is due to the cost of maintaining a traceable quality along the very long agricultural and industrial processes which are characteristic of the herbal medicines vs. the low cost of old, well-established generic synthetic drugs. In the United States, many health insurers offer at least one form of alternative healthcare coverage. Most often this is for chiropractic care (87%) or acupuncture (47%) with herbal medicine only covered by a few insurers, even if its cost is lower: an initial 'herbal' consultation fee may range from $30 to $60, the follow-up consultation costing around $30 and a month's supply of herbs between $30 and $50 (Hafner 2018).

After 10 years that PNPIC and PNPMF were launched, several T&CAM practices had been included in the Brazilian Health System (SUS). Twelve herbal medicines were incorporated into RENAME (Brazil 2017). These products are available for free to the patients in the health centres and pharmacies.

5 Challenge 4: The Public

The famous catchphrase '80% of the global population relies on Traditional Medicine', coined in the 1980s, is not valid anymore. A number of public using T&CAM as the first point of contact are much less compelling according to the latest studies (Oyebode et al. 2016).

The perception of the public regarding the use of herbal medicines is influenced by a number of factors, of which cultural background, economic status and literacy are perhaps the most defining ones. However, even within a culturally and economically homogeneous social group, the expectations towards the efficacy of herbal medicines will vary with the health status. This is exemplified by the fact that chronic patients dissatisfied with the lack of complete response from 'modern' medicine are more likely to accept T&CAMs in their quest for both physical and spiritual wellness (van den Brink-Muinen and Rijken 2006).

However, the trend in developed countries is a continuous increase of people deciding to take charge of T&CAM in order to enhance their health, despite the low expenditure scenario after the 2008 economic crisis as shown in many reports and reviews (Ekor 2013; Ventola 2010; WHO 2003), thus indicating a widespread positive perception in wealthy and literate societies. In developing countries, there is an active academic work surveying this aspect. In Malaysia, all recent studies confirm that the use of plants for healing is still very important in Malay culture, even in urban environments (Adnan and Othman 2012). However, the use of traditional medicine is decreasing as literacy increases among the population (Nasir et al. 2012), and new generations do not transmit traditional culture onto the newer ones. In Nigeria a survey among diabetic patients has shown that half of them took herbs alongside their current orthodox medications. Strikingly, half of those did not know the contents of these herbal preparations. In most cases, the herbal preparations were obtained from herbalists in the community or through a relative in the village. Others purchased the plant materials from the market and made the preparation at home. More common antidiabetic herbal plants were grown by these patients and used as part of their diet. The different reasons given by patients for the use of these herbal preparations alongside their orthodox medications seem to be constant across both developed and under development countries: (a) general perception that herbs are safe and (b) better feeling of 'wellness' when herbs are taken alongside the orthodox medications than when either of them is taken alone. Interestingly, the presence of diabetic complications was not identified as a predictor for herb use among patients with type 2 diabetes (Ezuruike and Prieto 2016). In Africa, not only adult patients are highly reliant on herbal medicines but also children: in Kenya, a study showed that herbal medicine use among under 5-year-old children is high (89.4%) and raised concerns about how the concomitant use of herbs with conventional medicine is disseminated among the general public (Nzuki 2016).

6 Future Trends and Challenges

WHO is committed to supporting the harmonisation of T&CAM use at national, regional, interregional and international levels in the years to come, by promoting the protection and scientifically driven integration of traditional medicines into public health. The two biggest economies of the planet, the United States and China, are unequivocally working towards these objectives, by means of the creation of public institutes to foster the evidence-based use of such therapeutic approaches. The consolidation of the Traditional Herbal Medicines Directive in Europe is an example of how the concept 'traditional' may come to terms with strict pharmaceutical requirements. Countries such Canada and Brazil are following closely with similar approaches. All these initiatives are however under continuous attacks by some of the 'nontraditional' medical profession. It is true that many T&CAMs lack evidence, but even the most 'provocative evidence-based hardliners' cannot ignore the effectiveness of many herbal medicines (Singh and Ernst 2009).

The public seems to be increasingly less reliant on T&CAM than previously thought as a result of the enormous progress in the worldwide availability of both information and synthetic medicines. However, the demand for high-quality T&CAM to increase wellness is strong in rich countries where an increasingly health-conscious society has been established. In mild-income countries, newer generations are less prone to T&CAM, but this is still a common option for chronic patients.

The current scenario of decreased global use of T&CAM (Oyebode et al. 2016) and increased regulatory, professional and public scrutiny of the quality and efficacy of herbal medicines may lead to the creation/integration of a 'new traditional herbal medicine' into public health. For example, the traditional use of ginkgo leaves in Kampo medicine was inexistent, the seeds being used for respiratory diseases only. European companies are investing considerable amounts of money on clinical trials using proprietary extracts of ginkgo leaves such as EGb 761. Initial attempts were not conclusive due to lack of power and short duration (Birks and Grimley Evans 2009), but recent trials involving thousands of people during decades just demonstrated that these extracts decrease the risk of dying before dementia and provide longer lifetime without dementia than patients taking other drugs for the same indication (Dartigues et al. 2017).

A world where herbal medicines are wholly accepted in mainstream medicine may create a new challenge: the increased demand for certain botanical drugs may not be sustainable. The readers are directed to the concept of 'ecopharmacognosy', a term coined by Professor G. Cordell in 2013 as the 'study of sustainable, biologically active natural resources' (Cordell 2014). Right now, whilst the cultivation of some botanical drugs such as echinacea is being scaled up almost at will, others such as golden root (*Rhodiola rosea*) are being depleted in their natural environment. Research on alternative resourcing must become a priority for all natural product research, for example, by providing with evidence that these rare plants can be grown under controlled conditions to obtain consistent, high-quality extracts and generate authentic germplasm (Peschel et al. 2013).

7 Conclusions

Our conclusion is that all challenges towards integration of Herbal Medicine in public health are being met by all stakeholders. The outcome may be, however, that of a 'new traditional use' based on modern evidence and, certainly, that the need for sustainable sources of herbal medicines will create a new 'ecopharmacy' model benefiting all sectors of a more health-discerning global population.

References

Abay SM (2009) Ethiopian herbal medicine practice and the recognition with modern medicine. Pharmacogn Rev 3:37–40

Adnan N, Othman N (2012) The relationship between plants and the Malay culture. Procedia-Social and Behavioral Sciences 42:231–41.

Ajith TA, Hema U, Aswathy MS (2007) *Zingiber officinale* Roscoe prevents acetaminophen-induced acute hepatotoxicity by enhancing hepatic antioxidant status. Food Chem Toxicol 45:2267–2272

Al-Rawi S, Fetters MD (2012) Traditional Arabic & Islamic medicine: a conceptual model for clinicians and researchers. Global J Health Sci 4:164–169

Amaral ACF, Rodrigues AG, Ribeiro JEG et al (2006) A fitoterapia no SUS e o Programa de Pesquisa de Plantas Medicinais da Central de Medicamentos. Série B. Textos Básicos de Saúde. Secretaria de Ciência, Tecnologia e Insumos Estratégicos, Departamento de Assistência Farmacêutica, Brasilia

Amole O (2012) The role of traditional medicine in primary health care. Journal of Immunodeficiency & Disorders 1(1):2

Asmelashe Gelayee D, Binega Mekonnen G, Asrade Atnafe S et al (2017) Herbal medicines: personal use, knowledge, attitude, dispensing practice, and the barriers among community pharmacists in Gondar, Northwest Ethiopia. J Evid Based Complementary Altern Med 2017:1–6

Azaizeh H, Saad B, Cooper E et al (2010) Traditional Arabic and Islamic medicine, a re-emerging health aid. J Evid Based Complementary Altern Med 7:419–424

Balbino EE, Dias MF (2010) Farmacovigilância: um passo em direção ao uso racional de plantas medicinais e fitoterápicos. Rev Bras Farmacogn 20:992–1000

Barreto BB, Silveira D (2014) Inclusion of courses on phytotherapy in undergraduate curriculum of health-related courses. J Med Plants Res 8:1374–1386

Batista LM, Gondim VAM (2012) Phytotherapy in primary care in SUS: realities and perspectives. Pesquisa Brasileira em Odontopediatria e Clinica Integrada 12:293–296

Ben-Arye E, Lev E, Keshet Y, Schiff E (2011) Integration of herbal medicine in primary care in Israel: a Jewish-Arab cross-cultural perspective. J Evid Based Complement Altern Med 2011:1–8

Birks J, Grimley Evans J (2009) *Ginkgo biloba* for cognitive impairment and dementia. Cochrane Database Syst Rev 1:CD003120

Boateng MA, Danso-Appiah A, Turkson BK et al (2016) Integrating biomedical and herbal medicine in Ghana – experiences from the Kumasi South Hospital: a qualitative study. BMC Complement Altern Med 16:189

BRAZIL (1981) Medicinal plant study is one of the priorities for health public research. D.O.U., Brasilia

BRAZIL (2006a) National policy on complementary and integrative practices in the Unified Health System. Ministry of Health, Brasilia

BRAZIL (2006b) National policy on medicinal plants and herbal medicines. Presidence Chamber, Brasília

BRAZIL (2017) RENAME. National list of essential medicines. Ministry of Health, Brasilia

Busia K, Kasilo O (2010) Collaboration between traditional health practitioners and conventional health practitioners: some country experiences. 2010. Disponível em http://www.aho.afro.who.int/en/ahm/issue/13/reports/collaboration-between-traditional-health-practitioners-and-conventional-health. Acesso em 04 de agosto de 2014

CANADA (2012) Pathway for licensing natural health products used as traditional medicines, v1.0. Health Canada. Accessed 9 Feb 2018

Carvalho ACB, Perfeito JPS, Costa e Silva LV et al (2011) Regulation of herbal medicines in Brazil: advances and perspectives. Braz J Pharm Sci 47:467–473

Carvalho ACB, Lana TN, Perfeito JPS et al (2018) The Brazilian market of herbal medicinal products and the impacts of the new legislation on traditional medicines. J Ethnopharmacol 212:29–35

Colquhoun D (2011) Why does the MHRA refuse to label herbal products honestly? Kent Woods and Richard Woodfield tell me. DC's improbable science. http://www.dcscience.net/2011/04/01/why-does-the-mhra-refuse-to-label-herbal-products-honestly-kent-woods-and-richard-woodfield-tell-me/. Accessed 9 Feb 2018

Cordell GA (2014) Ecopharmacognosy: exploring the chemical and biological potential of nature for human health. Biol Med Nat Prod Chem 3:1–14

Dar AK, ul Hassan W, Lone AH et al (2017) Study to assess high demand and high commercial value medicinal plants of Jammu and Kashmir India-with special focus on routes of procurement and identification. IJRDPL 6:2576–2585

Dartigues JF, Grasset L, Helmer C et al (2017) *Ginkgo biloba* extract consumption and long-term occurrence of death and dementia. J Prev Alzheimers Dis 4:16–20

Eisenberg DM, Kaptchuk TJ, Post DE et al (2016) Establishing an integrative medicine program within an Academic Health Center: essential considerations. Acad Med 91:1223–1230

Ekor M (2013) The growing use of herbal medicines: issues relating to adverse reactions and challenges in monitoring safety. Front Pharmacol 4:177

Ernst E (2006) The 'Smallwood report': method or madness? Br J Gen Pract 56:64–65

EU (2004) Directive 2004/24/EC of the European Parliament and of the Council of 31 March 2004 amending, as regards traditional herbal medicinal products, Directive 2001/83/EC on the Community code relating to medicinal products for human use. European Union Parliament, Official Journal of the European Union

Ezuruike U, Prieto JM (2016) Assessment of potential herb-drug interactions among Nigerian adults with type-2 diabetes. Front Pharmacol 7(248):8

GHANA (2000) Traditional Medicine Practice Act Ministry of Health, Acra

GHANA (2005) Policy guidelines on traditional medicine development. Ministry of Health, Acra

Hafner C (2018) How can I pay for TCM treatments? University of Minnesota's Earl E. Bakken Center for Spirituality & Healing. https://www.takingcharge.csh.umn.edu/explore-healing-practices/what-traditional-chinese-medicine/how-can-i-pay-tcm-treatments

Hussain S, Malik F (2013) Integration of complementary and traditional medicines in public health care systems: challenges and methodology. J Med Plants Res 7:2952–2959

INDIA (2017) National Medicinal Plants Board. http://nmpb.nic.in/. Minister of AYUSH

Kala CP, Sajwan BS (2007) Revitalizing Indian systems of herbal medicine by the National Medicinal Plants Board through institutional networking and capacity building. Curr Sci 93:797–806

Kennedy DA, Hart J, Seely D (2009) Cost effectiveness of natural health products: a systematic review of randomized clinical trials. Evid Based Complement Alternat Med 6:7

King R, Homsy J (1997) Involving traditional healers in AIDS education and counselling in sub-Saharan Africa: a review. AIDS (London, England) 11(Suppl A):S217–S225

Le Bon AM, Vernevaut MF, Guenot L et al (2003) Effects of garlic powders with varying alliin contents on hepatic drug metabolizing enzymes in rats. J Agric Food Chem 51:7617–7623

Lev E, Amar Z (2002) Ethnopharmacological survey of traditional drugs sold in the Kingdom of Jordan. J Ethnopharmacol 82:131–145
Lietava J (1992) Medicinal plants in a Middle Paleolithic grave Shanidar IV? J Ethnopharmacol 35:263–266
Mazzari ALDA, Prieto JM (2014) Herbal medicines in Brazil: Pharmacokinetic profile and potential herb-drug interactions. Front Pharmacol 5(162):12
MHRA (2016) Herbal medicines regulation: unlicensed herbal medicines supplied by a practitioner following a one-to-one consultation. Disponível em: http://www.mhra.gov.uk/Howweregulate/Medicines/Herbalmedicinesregulation/Unlicensedherbalmedicinessuppliedbyapractitionerfollowingaonetooneconsultation/index.htm. Accessed Acesso em 15 de julho de 2014
Nadel D, Danin A, Power RC et al (2013) Earliest floral grave lining from 13,700–11,700-y-old Natufian burials at Raqefet Cave, Mt. Carmel, Israel. Proc Natl Acad Sci U S A 110(29): 11774–11778
Nasir MA, Hashmi RI, Ahmad NS (2012) Drug utilization patterns in Rawalpindi and Islamabad, Pakistan. Journal of Pakistan Medical Association 62(5):426–29
NCCIH (2016) Complementary, alternative, or integrative health: what's in a name? https://nccih.nih.gov/health/integrative-health#cvsa. National Center for Complementary and Alternative Medicine. 2017
NCCIH (2018) NCCIH facts-at-a-glance and mission. https://nccih.nih.gov/about/ataglance
NHF (2011) Herbs trampled by goose-stepping bureaucrats The EU Herbal Medicinal Regime. Health Freedom News. https://www.thenhf.com/hfn-magazine/health-freedom-articles/2011-articles/88herbs-trampled-by-goose-stepping-bureaucrats-the-eu-herbal-medicinal-regime. Accessed 9 Feb 2018
Nishimura K, Plotnikoff GA, Watanabe K (2009) Kampo medicine as an integrative medicine in Japan. Japan Med Assoc J 52:147–149
Nzuki DM (2016) Utilization of herbal medicine among children under 5 years of age in Tharaka Nithi county. Kennyatta University, Kenya
Oyebode O, Kandala NB, Chilton PJ et al (2016) Use of traditional medicine in middle-income countries: a WHO-SAGE study. Health Policy Plan 31:984–991
Pan SY, Litscher G, Gao SH et al (2014) Historical perspective of traditional indigenous medical practices: the current renaissance and conservation of herbal resources. J Evid Based Complement Altern Med 2014:20
Peregoy JA, Clarke TC, Jones LI, Stussman BJ, Nahin RL (2014) Regional variation in use of complementary health approaches by U.S. adults, 146th edn. U.S. Department of Health and Human Services, Centers for Disease Control and Prevention, National Center for Health Statistics, Los Angeles
Peschel W, Prieto JM, Karkour C, Williamson EM (2013) Effect of provenance, plant part and processing on extract profiles from cultivated European *Rhodiola rosea* L. for medicinal use. Phytochemistry 86:92–102
PRC (2016) Law of the people's republic of China on traditional Chinese medicine. Standing Committee of National People's Congress. People's Republic of China, order number 59 of the President. Data Issued December 12, 2016
Rozenfeld S, Rangel ITM (1988) A farmacovigilância. Cad de Saúde Pública 4:336–341
Sahashi Y (2005) Herbs covered by health insurance in Japan. KAIM 1:70–84
Said O, Khalil K, Fulder S et al (2002) Ethnopharmacological survey of medicinal herbs in Israel, the Golan Heights and the West Bank region. J Ethnopharmacol 83:251–265
Shon YH, Nam KS (2004) Inhibition of cytochrome P450 isozymes and ornithine decarboxylase activities by polysaccharides from soybeans fermented with *Phellinus igniarius* or *Agrocybe cylindracea*. Biotechnol Lett 26:159–163
Singh S, Ernst E (2009) Trick or treatment?: Alternative medicine on trial. New York: Corgi Books. 410 p
Skalli S, Soulaymani Bencheikh R (2012) Safety monitoring of herb-drug interactions: a component of pharmacovigilance. Drug Saf 35(10):785–791

Smallwood C (2005). The role of complementary and alternative medicine in the NHS: an investigation into the potential contribution of main-stream alternative therapies to healthcare in the UK

Sommer JD (1999) The Shanidar IV 'flower burial': a re-evaluation of Neanderthal burial ritual. Camb Archaeol J 9:127–129

Unger M, Frank A (2004) Simultaneous determination of the inhibitory potency of herbal extracts on the activity of six major cytochrome P450 enzymes using liquid chromatography/mass spectrometry and automated online extraction. Rapid Commun Mass Spectrom 18:2273–2281

Van den Brink-Muinen A, Rijken PM (2006) Does trust in health care influence the use of complementary and alternative medicine by chronically ill people? BMC Public Health 6:188–188

Ventola CL (2010) Current issues regarding complementary and alternative medicine (CAM) in the United States: part 1: the widespread use of CAM and the need for better-informed health care professionals to provide patient counseling. PT 35:461–468

Wangchuk P, Tobgay T (2016) Quality assurance of the university medical education, hospital services and traditional pharmaceutical products of the Bhutanese So-wa-rig-pa health care system. BMC Complement Altern Med 16:283

Wangchuk P, Wangchuk D, Aagaard-Hansen J (2007) Traditional Bhutanese medicine (gSo-BA Rig-PA): an integrated part of the formal health care services. Southeast Asian J Trop Med Public Health 38:161–167

WHO (1978) Declaration of Alma Ata. International conference on primary health care, Alma-Ata, USSR, 6–12 September. www.who.int/hpr/NPH/docs/declaration_almaata.pdf. World Health Organization, Geneva

WHO (1995) Traditional practitioners as primary health care workers: a study of effectiveness of four training projects in Ghana, Mexico and Bangladesh. Division of strengthening of health services and Traditional medicine programme, Geneva

WHO (2003) Traditional medicine-Document A56/18

WHO (2005) National policy on traditional medicine and regulation of herbal medicines: report of a WHO global survey. World Health Organization, Geneva

WHO (2006) The world health report 2006: working together for health. World health report. World Health Organization, Geneva

WHO (2013) Traditional medicine strategy: 2014–2023. World Health Organization, Hong Kong

WHO (2014) Twelfth general programme of work 2014–2019. Not merely the absence of disease. General Programme of Work World Health Organization, Geneva

WHO (2017) Traditional medicine: definitions. World Health Organization. http://who.int/medicines/areas/traditional/definitions/en/. Accessed 20 Sept 2017

Yuan H, Ma Q, Ye L et al (2016) The traditional medicine and modern medicine from natural products. Molecules 21:559

Index

V. Cechinel Filho (ed.), *Natural Products as Source of Molecules with Therapeutic Potential*, https://doi.org/10.1007/978-3-030-00545-0

N

O

P

Q

R

S